Handbook of Biological Control and Horticultural Crops

Handbook of Biological Control and Horticultural Crops

Vol. 1

Dr. J.S. Bohra

AGROTECH PRESS

JAIPUR ▪ NEW DELHI

AGROTECH PRESS

Regd. Off. "Bony Residency", Gate No. 2
Opp. Tilak Public School, Vishweswariya Nagar,
Gopalpura Road, Jaipur - 302018, India
Ph: 0141 - 2761280, Telefax: 2761381
E-mail : agrotechpress@gmail.com

Delhi Office: 102, 1st Floor, Satyam House,
4327/3, Ansari Road, Darya Ganj,
New Delhi - 110002, India
Ph: 011-45652440
visit us: www.agrotechpress.com

Distributed By
KEYA INTERNATIONAL PRESS LLP
E-mail: keyainternationalpressllp@gmail.com

Published in India, 2025

ISBN : 9789384568559 (3 Vols. Set)

Typeset by Thomson Press

Printed and bound in India by Thomson Press Ltd.,
New Delhi

Contents

Preface

Pests are considered as important limiting factors productivity of horticultural crops, resulting to lower productivity of horticultural crops. Pests like coconut mice, spiralling whitefly on fruit crops, coffee berry borer, diamond back moth on crucifers, serpentine leaf miners on several crops, fruit flies in fruits and cucurbits are some of the examples. The annual crop losses due to pests all over the world are in billions. The excessive dependence on chemical pesticides leads to the development of pest (insects, mites, nematodes and diseases) resistance, outbreak of secondary pests, toxicity towards natural enemies and pesticide residues in the food chain. Considering these limitations, there has been a growing awareness to develop biological control that effectively reduce pest populations and ensure sustainability of production, cost effectiveness and healthy environment.

Biomanagement or Biological control is the beneficial action of predators, parasites, pathogens, and competitors in controlling pests and their damage. Biomanagement provided by these living organisms is especially important for reducing the numbers of pest insects. Biological control is a component of an integrated pest management strategy. It is defined as the reduction of pest populations by natural enemies and typically involves an active human role. Plant pathogens, nematodes, and vertebrates also have many natural enemies, but this biological control is often harder to recognize, less-well understood, or more difficult to manage. Conservation, augmentation, and classical biological control (also called importation) are tactics for harnessing the effects of natural enemies. The use of biological control for the management of pest insects pre-dates the modern pesticide era. The first major successes in biological control occurred with exotic pests controlled by natural enemy species collected from the country or area of origin of the pest (classical control). Augmentative control has been successfully applied against a range of open-field and greenhouse pests, and conservation biological control schemes have been developed with indigenous predators and parasitoids.

The commercial scale production of biological control agents is intense. Several practices were internationally developed for pest control using commercially available biological control agents, specifically microorganisms. The introduction of rational approaches for the environmental risk assessment of non-native control agents is an essential step in the wider application of biological control, but future success is strongly dependent on a greater level of investment in research and development by governments and related organizations that are committed to a reduced reliance on chemical control.

At the start of the twenty-first century, the opportunities and need for effective biological control are greater than ever: pest resistance continues to be a problem, pesticides are being withdrawn on environmental grounds without suitable replacements, and in some areas of the world, including Europe, there is uncertainty over the durability and public acceptance of genetically modified (GM) pest-resistant crops. Against this background, this book defines the terms used to describe different types of biomanagement of pest and reviews landmark events in the development of this approach. The advantages and limitations of using natural enemy species in pest management are evaluated, as well as the common features of the most successful agents. The contribution of biological control to sustainable agriculture is reviewed on a global scale, together with recent information on the possible effects of GM crops on natural enemies of pests, and developments in the regulatory framework and risk assessment for the import and release of non-native species. It offers a panoramic perspective of how biological control can contribute to sustainable agriculture and increase biodiversity.

Though it is evident that biological control programmes have been successfully implemented in a wide range of crop environments in all parts of the world, and the potential to increase the role of biological control is great, it remains the case that the adoption and implementation of this method of control is slow. Owing to technical and economic, but mainly attitudinal barriers, ecologically sound pest management has not been used for a wide variety of pests and diseases. Although the funding for research on biological control is limited, it is certain that the main constraints are related to the attitudes of advisory personnel and farmers, combined with the disinterest in anything other than chemical control among the pesticide industries, and the hollow endorsement of alternative methods made by policy makers in which words are yet to be matched by a realistic investment in research and development. However, it is certain that future pest management will depend strongly on biological control because it is the most sustainable, cheapest and environmentally safest system of pest management, with additional benefits for growers and consumers.

Dr. J.S. Bohra

1

Introduction

A pest is "a plant or animal detrimental to humans or human concerns (as agriculture or livestock production)"; alternative meanings include organisms that cause nuisance and epidemic disease associated with high mortality (specifically: plague). In its broadest sense, a pest is a competitor of humanity. In the past, the term might have been used for detrimental animals only, thus for example, causing confusion where the generic term 'pesticide' meant 'insecticide' to some people. It is any living organism which is invasive or prolific, detrimental, troublesome, noxious, destructive, a nuisance to either plants or animals, human or human concerns, livestock, human structures, wild ecosystems etc. It is a loosely defined term, often overlapping with the related terms vermin, weed, plant and animal parasites and pathogens. It is possible for an organism to be a pest in one setting but beneficial, domesticated or acceptable in another.

Usage of the Term

Often animals are derided as pests as they cause damage to agriculture by feeding on crops or parasitising livestock, such as codling moth on apples, or boll weevil on cotton. An animal could also be a pest when it causes damage to a wild ecosystem or carries germs within human habitats. Examples of these include those organisms which vector human disease, such as rats and fleas which carry the plague disease, mosquitoes which vector malaria, and ticks which carry Lyme disease.

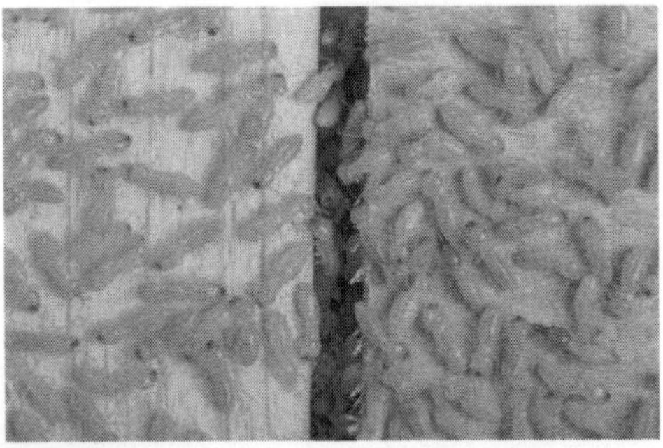

Figure: *Pests often occur in high densities, making the damage they do even more detrimental*

The term pest may be used to refer specifically to harmful animals but is also often taken to mean all harmful organisms including weeds, plant pathogenic fungi and viruses. Pesticides are chemicals and other agents (*e.g.* beneficial micro-organisms) that are used to control or protect other organisms from pests. The related term vermin has much overlap with pest, but generally only includes those creatures that are seen to be vectors of diseases.

It is possible for an animal to be a pest in one setting but beneficial or domesticated in another (for example, European rabbits introduced to Australia caused ecological damage beyond the scale they inflicted in their natural habitat). Many weeds (plant pests) are also seen as useful under certain conditions, for instance Patterson's curse is often valued as food for honeybees and as a wildflower, even though it can poison livestock.

The term "plant pest" has a very specific definition in terms of the International Plant Protection Convention and phytosanitary measures worldwide. A pest is any species, strain or biotype of plant, animal, or pathogenic agent injurious to plants or plant products.

Plants may be considered pests if an invasive species. Any prolific animal or plant may be considered pests.

The greatest importance as pests (in the order of economic importance) are insects, mites, nematodes and gastropods.

Invertebrate Pests

Insects

- Ants, cockroaches, flies and wasps are household pests, as they typically consume human food
- Aphids, larvae, grasshoppers and crickets cause damage to crop plants

- Lice, fleas and bed bugs can all cause skin irritation
- Mosquitoes, tsetse flies and kissing bugs cause irritation and carry disease
- Termites, woodworm and wood ants cause structural damage
- Bookworms, silverfish, carpet beetles and clothes moths cause non-structural damage
- Gypsy moths attack hardwood trees

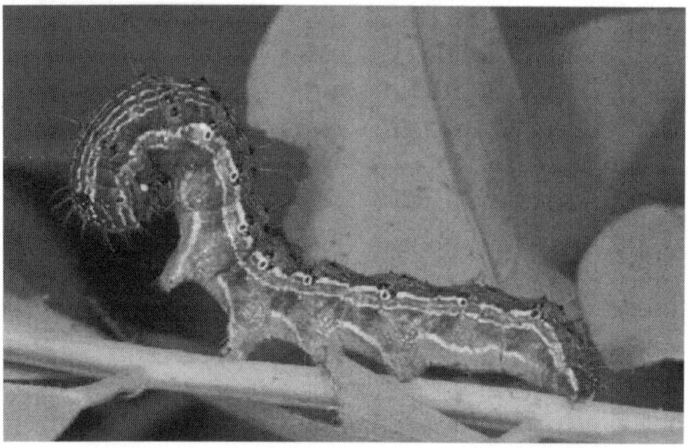

Figure: *Caterpillars cause crop damage*

Figure: *Termites cause structural damage*

Nematodes

- Root-knot nematode
- Soybean cyst nematode
- Potato cyst nematode

Parasites

- Chiggers cause skin irritation
- *Sarcoptes scabiei* causes scabies
- Ticks and mites cause irritation and can spread disease

Gastropods

Some slugs are pests in both agriculture and gardens. Their significance is increasing drastically. *Deroceras reticulatum* is a worldwide distributed slug pest. Local importance slug pests include: *Deroceras* spp., *Milax* spp., *Tandonia* sp., *Limax* spp., *Arion* spp. and some species of Veronicellidae: *Veronicella sloanei.*

Land snail pests include:

- *Helix aspersa* damages citrus fruits in California,
- *Cernuella virgata, Theba pisana* and *Cochlicella* spp. decrease quality of grains when harvested with the product in South Australia.
- *Achatina fulica* damages vegetables and ornamental plants in the Pacific region.
- *Succinea costaricana* damages ornamental plants in Costa Rica.
- *Ovachlamys fulgens* damages ornamental plants and orchids in Costa Rica.
- Other species considered to be pests include: *Amphibulima patula dominicensis*, *Zachrysia provisoria* and *Bradybaena similaris.*

Freshwater snail pests include:

- *Pomacea canaliculata* damages rice in Southeast Asia.
- *Bulinus* sp., *Biomphalaria* spp. and *Oncomelania* are intermediate hosts of schistosomes causing schistosomiasis.
- Various species in Lymnaeidae are intermediate hosts of fasciolosis.

Weeds

A weed is a plant considered undesirable. Weeds are commonly unwanted plants in human-controlled settings, such as farm fields, gardens, lawns, and parks. Weeds have no botanical classification value, since a plant that is a weed in one context is *not* a weed when growing where it is wanted. A number of plants that many consider weeds are often intentionally grown in gardens and other cultivated settings. It is applied to any plant that grows or reproduces aggressively, or is outside its native habitat. The term is occasionally used to broadly describe species outside the plant kingdom that can live in diverse environments and reproduce quickly, and has even been applied to humans.

Ecological Role

Weeds generally share similar adaptations that give them advantages and allow them to proliferate in disturbed environments where soil or natural vegetative cover has been damaged. Different types of habitat and disturbances will result in colonization by different communities of weed species.

***Figure:** A dandelion is a common plant all over the world, especially in Europe, Asia, and the Americas. It is a well-known example of a plant that is considered a weed in some contexts (such as lawns) but not a weed in others (such as when it is used as a leaf vegetable or herbal medicine).*

A number of plant species are adapted to naturally occurring disturbed environments include dunes and other windswept areas with shifting soils, alluvial flood plains, river banks and deltas, and areas that are often burned, and plant scientists call then ruderal species. Since human agricultural practices often mimic these natural environments where weedy species have evolved, some weeds have adapted to grow and proliferate in human-disturbed areas such as agricultural fields, lawns, roadsides, and construction sites. The weedy nature of these species often gives them an advantage over more desirable crop species because they often grow quickly and reproduce quickly, have seeds that persist in the soil seed bank for many years, or have short lifespans with multiple generations in the same growing season. Perennial weeds often have underground stems that spread out under the soil surface or, like ground ivy (*Glechoma hederacea*), have creeping stems that root and spread out over the ground.

Some plants become dominant when introduced into new environments because the animals which specialize in feeding on them are absent; in what is sometimes called the "natural enemies hypothesis", plants freed from these specialist consumers may increase their competitive ability. In locations where predation and mutual competitive relationships no longer exist, some plants are able to increase allocation of resources into growth or reproduction. The weediness of some species that are introduced into new environments may be caused by their production of allelopathic chemicals which indigenous plants are not yet adapted to, a scenario sometimes called the "novel weapons hypothesis". These chemicals may limit the growth of established plants or the germination and growth of seeds and seedlings.

Dispersal

Many weed species have moved out of their natural geographic ranges and spread around the world in tandem with human migrations and commerce. Weed seeds are often collected and transported with crops after the harvesting of grains, so humans are a vector of transport as well as a producer of the disturbed environments to which weed species are well adapted, resulting in many weeds having a close association with human activities.

Some weed species have been classified as noxious weeds by government authorities because, if left unchecked, they often compete with native or crop plants or cause harm to livestock. They are often foreign species accidentally or imprudently imported into a region where there are few natural controls to limit their population and spread.

Competition with Cultivated and Endemic Plants

A number of native or non-native plants are unwanted in a specific location for a number of reasons. An important one is that they interfere with food and fibre production in agriculture, wherein they must be controlled in order to prevent lost or diminished crop yields. Other important reasons are that they interfere with other cosmetic, decorative, or recreational goals, such as in lawns, landscape architecture, playing fields, and golf courses. Similarly, they can be of concern for environmental reasons whereby introduced species out-compete for resources or space with desired endemic plants. For all these reasons; horticulture, both functional and cosmetic, and environmental, - weeds interfere by:

- competing with the desired plants for the resources that a plant typically needs, namely, direct sunlight, soil nutrients, water, and (to a lesser extent) space for growth;
- providing hosts and vectors for plant pathogens, giving them greater opportunity to infect and degrade the quality of the desired plants;
- providing food or shelter for animal pests such as seed-eating birds and Tephritid fruit flies that otherwise could hardly survive seasonal shortages;

- offering irritation to the skin or digestive tracts of people or animals, either physical irritation via thorns, prickles, or burs, or chemical irritation via natural poisons or irritants in the weed (for example, the poisons found in *Nerium* species);
- causing root damage to engineering works such as drains, road surfaces, and foundations, blocking streams and rivulets.

In weed ecology some authorities speak of the relationship between "the three Ps": plant, place, perception. These have been very variously defined, but the weed traits listed by H.G. Baker are widely cited.

Weeds have long been a concern, perhaps as long as humans have cultivated plants. They are mentioned in various historic texts, such as a Shakespearean sonnet:

> *"To thy fair flower add the rank smell of weeds: / But why thy odour matcheth not thy show, / The soil is this, that thou dost common grow."*

and the Bible:

> *"Cursed is the ground because of you; through painful toil you will eat of it all the days of your life. It will produce thorns and thistles for you, and you will eat the plants of the field. By the sweat of your brow you will eat your food until you return to the ground,"*

While the term "weed" generally has a negative connotation, many plants known as weeds can have beneficial properties. A number of weeds, such as the dandelion (*Taraxacum*), are edible, and their leaves or roots may be used for food or herbal medicine. Burdock is common over much of the world, and is sometimes used to make soup and medicine in East Asia. Some weeds attract beneficial insects, which in turn can protect crops from harmful pests. Weeds can also prevent pest insects from finding a crop, because their presence disrupts the incidence of positive cues which pests use to locate their food.

Weeds may also act as a "living mulch", providing ground cover that reduces moisture loss and prevents erosion. Weeds may also improve soil fertility; dandelions, for example, bring up nutrients like calcium and nitrogen from deep in the soil with their tap root, and clover hosts nitrogen-fixing bacteria in its roots, fertilizing the soil directly. The dandelion is also one of several species which break up hardpan in overly cultivated fields, helping crops grow deeper root systems. Some garden flowers originated as weeds in cultivated fields and have been selectively bred for their garden-worthy flowers or foliage. An example of a crop weed that is grown in gardens is the corncockle, (*Agrostemma githago*), which was a common weed in European wheat fields, but is now sometimes grown as a garden plant.

Some people have appreciated weeds for their tenacity, their wildness and even the work and connection to nature they provide. As Christopher Lloyd wrote in *The Well-Tempered Garden*

"Many gardeners will agree that hand-weeding is not the terrible drudgery that it is often made out to be. Some people find in it a kind of soothing monotony. It leaves their minds free to develop the plot for their next novel or to perfect the brilliant repartee with which they should have encountered a relative's latest example of unreasonableness."

Weeds as Adaptable Species

An alternate definition often used by biologists is any species, not just plants, that can quickly adapt to any environment. Some traits of weedy species are the ability to reproduce quickly, disperse widely, live in a variety of habitats, establish a population in strange places, succeed in disturbed ecosystems and resist eradication once established.

Such species often do well in human-dominated environments as other species are not able to adapt. Common examples include the common pigeon, brown rat and the raccoon. Other weedy species have been able to expand their range without actually living in human environments, as human activity has damaged the ecosystems of other species. These include the coyote, the white-tailed deer and the brown headed cowbird.

In response to the idea that humans may face extinction due to environmental degradation, paleontologist David Jablonsky counters by arguing that humans are a weed species. Like other weedy species, humans are widely dispersed in a wide variety of environments, and are highly unlikely to go extinct no matter how much damage the environment faces.

Role in Mass Extinctions

A mass extinction is generally caused by some abrupt disruption to the entire planet's environment. This results in major changes in habitat worldwide, and most endemic species, specially adapted to a single habitat, cannot survive in the new habitats. Thus only weedy species survive, and they dominate the planet in the immediate aftermath. Cockroaches, for example, have survived several mass extinctions. The current Holocene extinction event, then, could lead to a planet inhabited entirely by what are known today as weeds. The fossil record indicates that after mass extinctions, a weed-dominated planet persists for five to ten million years before life re-diversifies.

Biological Pest Control

Biological control is a bioeffector-method of controlling pests (including insects, mites, weeds and plant diseases) using other living organisms. It relies on predation, parasitism, herbivory, or other natural mechanisms, but typically also involves an active human management role. It can be an important component of integrated pest management (IPM) programmes. There are three basic types of biological pest control strategies: importation (sometimes called classical biological control), augmentation and conservation.

Natural enemies of insect pests, also known as biological control agents, include predators, parasitoids, and pathogens. Biological control agents of plant diseases are most often referred to as antagonists. Biological control agents of weeds include seed predators, herbivores and plant pathogens.

Types of Biological Pest Control

There are three basic types of biological pest control strategies: importation (sometimes called classical biological control), augmentation and conservation.

Importation

Importation (or "classical biological control") involves the introduction of a pest's natural enemies to a new locale where they do not occur naturally. This is usually done by government authorities. In many instances the complex of natural enemies associated with a pest may be inadequate, a situation that can occur when a pest is accidentally introduced into a new geographic area, without its associated natural enemies. These introduced pests are referred to as exotic pests and comprise about 40% of the insect pests in the United States.

The process of importation involves determining the origin of the introduced pest and then collecting appropriate natural enemies associated with the pest or closely related species. Selected natural enemies are then passed through a rigorous assessment, testing and quarantine process, to ensure that they will work and that no unwanted organisms (such as hyperparasitoids) are introduced. If these procedures are passed, the selected natural enemies are mass-produced and then released. Follow-up studies are conducted to determine if the natural enemy becomes successfully established at the site of release, and to assess the long-term benefit of its presence.

To be most effective at controlling a pest, a biological control agent requires a colonizing ability which will allow it to keep pace with the spatial and temporal disruption of the habitat. Its control of the pest will also be greatest if it has temporal persistence, so that it can maintain its population even in the temporary absence of the target species, and if it is an opportunistic forager, enabling it to rapidly exploit a pest population. However an agent with such attributes is likely to be non-host specific, which is not ideal when considering its overall ecological impact, as it may have unintended effects on non-target organisms.

There are many examples of successful importation programmes, including:

- Joseph Needham noted a Chinese text dating from 304AD, *Records of the Plants and Trees of the Southern Regions*, by Hsi Han, which describes mandarin oranges protected by biological pest control techniques that are still in use today.
- One of the earliest successes in the west was in controlling *Icerya purchasi*, the cottony cushion scale, a pest that was devastating the California citrus industry in the late 19th century. A predatory insect *Rodolia cardinalis* (the Vedalia Beetle), and a parasitoid fly were introduced from Australia by Charles Valentine Riley. Within a few years the cottony cushion scale was completely controlled by these introduced natural enemies.
- Damage from *Hypera postica* Gyllenhal, the alfalfa weevil, a serious introduced pest of forage, was substantially reduced by the introduction of several natural enemies. 20 years after their introduction the population of weevils in the alfalfa area treated for alfalfa weevil in the Northeastern United States was reduced by 75 percent.
- A small wasp, *Trichogramma ostriniae*, was introduced from China to help control the European corn borer (*Ostrinia nubilalis*), one of the most destructive insects in North America, making it a recent example of a long history of classical biological control efforts for this major pest.
- The population of *Levuana iridescens* (the Levuana moth), a serious coconut pest in Fiji, was brought under control by a classical biological control programme in the 1920s.

Classical biological control is long lasting and inexpensive. Other than the initial costs of collection, importation, and rearing, little expense is incurred. When a natural enemy is successfully established it rarely requires additional input and it continues to kill the pest with no direct help from humans and at no cost. However importation does not always work. It is usually most effective against exotic pests and less so against native insect pests. The reasons for failure are not often known but may include the release of too few individuals, poor adaptation of the natural enemy to environmental conditions at the release location, and lack of synchrony between the life cycle of the natural enemy and host pest.

Augmentation

Augmentation involves the supplemental release of natural enemies, boosting the naturally occurring population. Relatively few natural enemies may be released at a critical time of the season (inoculative release) or millions may be released (inundative release). An example of inoculative release occurs in greenhouse production of several crops. Periodic releases of the parasitoid, *Encarsia formosa*, are used to control greenhouse whitefly, and the predatory mite *Phytoseiulus persimilis* is used for control of the two-spotted spider mite. Lady beetles, lacewings, or parasitoids such as those from the genus *Trichogramma* are frequently released in large numbers (inundative release). Recommended release rates for Trichogramma in vegetable or field crops range from 5,000 to 200,000 per acre (1 to 50 per square metre) per week depending on level of pest infestation. Similarly, entomopathogenic nematodes are released at rates of millions and even billions per acre for control of certain soil-dwelling insect pests.

The spraying of octopamine analogues (such as 3-FMC) has been suggested as a way to boost the effectiveness of augmentation. Octopamine, regarded as the invertebrate counterpart of dopamine plays a role in activating the insects' flight-or-fight response. The idea behind using octopamine analogues to augment biological control is that natural enemies will be more effective in their eradication of the pest, since the pest will be behaving in an unnatural way because its flight-or-fight mechanism has been activated. Octopamine analogues are purported to have two desirable characteristics for this type of application: (1) they affect insects at very low dosages (2) they do not have a physiological effect in humans (or other vertebrates).

Conservation

The conservation of existing natural enemies in an environment is the third method of biological pest control. Natural enemies are already adapted to the habitat and to the target pest, and their conservation can be simple and cost-effective. Lacewings, lady beetles, hover fly larvae, and parasitized aphid mummies are almost always present in aphid colonies.

Cropping systems can be modified to favour the natural enemies, a practice sometimes referred to as habitat manipulation. Providing a suitable habitat, such as a shelterbelt, hedgerow, or beetle bank where beneficial insects can live and reproduce, can help ensure the survival of populations of natural enemies. Things as simple as leaving a layer of fallen down leaves or mulch in place provides a suitable food source for worms and provides a shelter for small insects, in turn also providing a food source for hedgehogs and shrew mice. Compost pile(s) and containers for making leaf compost also provide shelter, as long as they are accessible by the animals (not fully closed). A stack of wood may provide a shelter for voles, hedgehogs, shrew mice, some species of butterflies, ... Long grass and ponds provide shelters for frogs and toads (which themselves eat snails).

Not cutting any annual or other non-hardy plants before winter (but instead in spring) allows many insects to make use of their hollow stems during winter. In California prune trees are sometimes planted in grape vineyards to provide an improved overwintering habitat or refuge for a key grape pest parasitoid. The prune trees harbour an alternate host for the parasitoid, which could previously overwinter only at great distances from most vineyards. The provisioning of artificial shelters in the form of wooden caskets, boxes or flowerpots is also sometimes undertaken, particularly in gardens, to make a cropped area more attractive to natural enemies. For example, the stimulation of the natural predator *Dermaptera* is done in gardens by hanging upside-down flowerpots filled with straw or wood wool. Green lacewings are given housing by using plastic bottles with an open bottom and a roll of cardboard inside of it. Birdhouses provide housing for birds, some of whom eat certain pests. Attracting the most useful birds can be done by using a correct diametre opening in the birdhouse (just large enough for the specific species of bird that needs to be attracted to fit through, but not other species of birds).

Besides the provisioning of natural or artificial housing, the providing of nectar-rich plants is also beneficial. Often, many species of plants are used so as to provide food for many natural predators, and this for a long period of time (this is done by using different types of plants as each species only blooms for a short period). It should be mentioned that many natural predators are nectivorous during the adult stage, but parasitic or predatory as larvae. A good example of this is the soldier beetle which is frequently found on flowers as an adult, but whose larvae eat aphids, caterpillars, grasshopper eggs, and other beetles. Letting certain plants (as Helianthus spp, Rudbeckia spp,

Dipsacus spp, Echinacea spp) come into seed is also advised, to supply food for birds. Having some trees or shrubs in place that carry berries is also practised and provide a source of food for birds. Often, trees/ shrubs are used that do not produce berries fit for human consumption, avoiding food competition. Examples are Sorbus spp, Amelanchier spp, Crataegus spp, Sambucus nigra, Ilex aquifolium, Rhamnus frangula. Obviously for this to work, these trees can not be pruned/trimmed until after the birds and other animals have eaten all of the berries.

Also, the providing of host plants (plants on which organisms can lay their eggs) may also be necessairy. These organisms for which host plants can be foreseen can be certain natural predators, caterpillars, and even a limited amount of host plants for pests can be tolerated. The latter ensures that natural predators remain in the vicinity and tolerating a certain amount of loss to pests would be needed anyhow since no chemical pesticides can be used (organic pesticides can be used but often can, on itself, not eliminate all pests during an infestation). This, as natural predators are susceptible to the same pesticides used to target pests. Plants for caterpillars are optional and only ensure that sufficient amounts of moths are produced which form a source of food to bats. Bats may be wanted as they also consume large amounts of mosquitoes, which despite not targeting any plants, can still be a nuisance to people in areas where there is much standing water nearby (i.e. pond, creek, ...).

Conservation strategies such as mixed plantings and the provision of flowering borders can be more difficult to accommodate in large-scale crop production. There may also be some conflict with pest control for the large producer, because of the difficulty of targeting the pest species, also refuges may be utilised by the pest insects as well as by natural enemies. Some plants that are attractive to natural enemies may also be hosts for certain plant diseases, especially plant viruses that could be vectored by insect pests to the crop.

Biological Control Agents

Predators

Predators are mainly free-living species that directly consume a large number of prey during their whole lifetime. Ladybugs, and in particular their larvae which are active between May and July in the northern hemisphere, are voracious predators of aphids, and will also consume mites, scale insects and small caterpillars. The larvae of many hoverfly species principally feed upon greenfly, one larva devouring up to fifty a day, or 1000 in its lifetime. They also eat fruit tree spider mites and small caterpillars. Adults feed on nectar and pollen, which they require for egg production.

***Figure:** Lacewings are available from biocontrol dealers.*

***Figure:**Predatory Polistes wsp looking for bollworms or other caterpillars on a cotton plant*

Dragonflies are important predators of mosquitoes, both in the water, where the dragonfly naiads eat mosquito larvae, and in the air, where adult dragonflies capture and eat adult mosquitoes. Community-wide mosquito control programmes that spray adult mosquitoes also kill dragonflies, thus reducing an important biocontrol agent.

Several species of entomopathogenic nematode are important predators of insect pests. *Phasmarhabditis hermaphrodita* is a microscopic nematode that kills slugs, thereafter feeding and reproducing inside. The nematode is applied by watering onto moist soil, and gives protection for up to six weeks in optimum conditions.

Other useful garden predators include lacewings, pirate bugs, rove and ground beetles, aphid midge, centipedes, spiders, predatory mites, as well as larger fauna such as frogs, toads, lizards, hedgehogs, slow-worms and birds. Cats and rat terriers kill field mice, rats, June bugs, and birds. Dachshunds are bred specifically to fit inside tunnels underground to kill badgers.

Parasitoid Insects

Parasitoids lay their eggs on or in the body of an insect host, which is then used as a food for developing larvae. The host is ultimately killed. Most insect parasitoids are wasps or flies, and usually have a very narrow host range.

Four of the most important groups are:

- Ichneumonid wasps: (5–10 mm). Prey mainly on caterpillars of butterflies and moths.
- Braconid wasps: Tiny wasps (up to 5 mm) attack caterpillars and a wide range of other insects including greenfly. A common parasite of the cabbage white caterpillar- seen as clusters of sulphur yellow cocoons bursting from collapsed caterpillar skin.
- Chalcid wasps: Among the smallest of insects (<3 mm). Parasitize eggs/larvae of greenfly, whitefly, cabbage caterpillars, scale insects and Strawberry Tortrix Moth (*Acleris comariana*).
- Tachinid flies: Parasitize a wide range of insects including caterpillars, adult and larval beetles, true bugs, and others.

Examples of parasitoids:

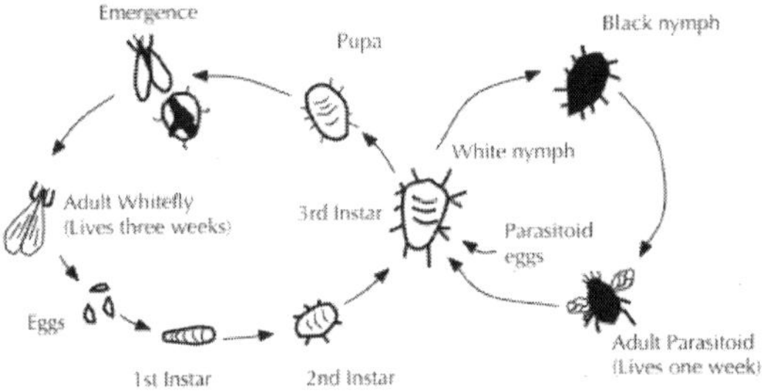

Figure: *Diagram illustrating the life cycles of Greenhouse whitefly and its parasitoid wasp Encarsia formosa*

- *Encarsia formosa* A small predatory chalcid wasp which is a parasitoid of whitefly, a sap-feeding insect which can cause wilting and black sooty moulds. It is most effective when dealing with low level infestations, giving protection over a long period of time. The wasp lays its eggs in young whitefly 'scales', turning them black as the parasite larvae pupates.

- *Eretmocerus* spp. (against white flies)
- *Aphidius colemani* (against aphids)
- *Gonatocerus ashmeadi* (Hymenoptera: Mymaridae) has been introduced to control the glassy-winged sharpshooter *Homalodisca vitripennis* (Hemipterae: Cicadellidae) in French Polynesia and has successfully controlled ~95% of the pest density.

Parasitoids are one of the most widely used biological control agents. Commercially there are two types of rearing systems: short-term daily output with high production of parasitoids per day, and long-term low daily output with a range in production of 4-1000million female parasitoids per week. Larger production facilities produce on a yearlong basis, whereas some facilities will produce only seasonally.

Rearing facilities are usually a significant distance from where the agents will be used in the field, and transporting the parasitoids from the point of production to the point of use can pose problems. Shipping conditions can be too hot, and even vibrations from planes or trucks can disrupt the parasitoids.

Micro-organisms

Pathogenic micro-organisms include bacteria, fungi, and viruses. They kill or debilitate their host and are relatively host-specific. Various microbial insect diseases occur naturally, but may also be used as biological pesticides. When naturally occurring, these outbreaks are density-dependent in that they generally only occur as insect populations become denser.

Bacteria

Bacteria used for biological control infect insects via their digestive tracts, so insects with sucking mouth parts like aphids and scale insects are difficult to control with bacterial biological control. *Bacillus thuringiensis* is the most widely applied species of bacteria used for biological control, with at least four sub-species used to control Lepidopteran (moth, butterfly), Coleopteran (beetle) and Dipteran (true flies) insect pests. The bacteria is available in sachets of dried spores which are mixed with water and sprayed onto vulnerable plants such as brassicas and fruit trees. Bacillus thuringiensis has also been incorporated into crops, making them resistant to these pests and thus reducing the use of pesticides.

Fungi

Fungi that cause disease in insects are known as entomopathogenic fungi, including at least fourteen species that attack aphids. *Beauveria bassiana* is used to manage a wide variety of insect pests including: whiteflies, thrips, aphids and weevils. A remarkable additional feature of some fungi is their effect on plant fitness. *Trichoderma* species may enhance biomass production promoting root development, dissolving insoluble phosphate containing minerals.

Examples of entomopathogenic fungi:

- *Beauveria bassiana* (against white flies, thrips, aphids and weevils)
- *Paecilomyces fumosoroseus* (against white flies, thrips and aphids)
- *Metarhizium* spp. (against beetles, locusts and grasshoppers, Hemiptera, spider mites and other pests)
- *Lecanicillium* spp. (against white flies, thrips and aphids)
- *Cordyceps* species (includes teleomorphs of the above genera: that infect a wide spectrum of arthropods)
- *Trichoderma* species are used to manage certain plant pathogens. *Trichoderma viride* has been used against Dutch Elm disease, and to treat the spread of fungal and bacterial growth on tree wounds. It may also have potential as a means of combating silver leaf disease.

Several members of Chytridiomycota and Blastocladiomycota have been explored as agents of biological control. From Chytridiomycota, *Synchytrium solstitiale* is being considered as a control agent of the yellow star thistle (*Centaurea solstitialis*) in the United States. *Synchytrium minutum* occasionally parasitizes kudzu and was considered as a control agent against this weed outside of its native range, but *S. minutum* parasitizes agricultural crop plants more frequently than it parasitizes kudzu. Batrachochytrium dendrobatidis was briefly considered and soundly rejected as a means of controlling invasive frog populations in Hawaii. From Blastocladiomycota, certain members of Coelomomyces were explored as possible agents of biological control of mosquitoes.

Viruses

- A viral biological control that can be introduced in order to control the overpopulation of European rabbit in Australia is the rabbit haemorrhagic disease virus that causes the rabbit haemorrhagic disease.

Combined Use of Parasitoids and Pathogens

In cases of massive and severe infection of invasive pests, techniques of pest control are often used in combination. An example being, that of the emerald ash borer (*Agrilus planipennis* Fairmaire, family

Buprestidae), an invasive beetle from China, which has destroyed tens of millions of ash trees in its introduced range in North America. As part of the campaign against the emerald ash borer (EAB), American scientists in conjunction with the Chinese Academy of Forestry searched since 2003 for its natural enemies in the wild leading to the discovery of several parasitoid wasps, namely *Tetrastichus planipennisi*, a gregarious larval endoparasitoid, *Oobius agrili*, a solitary, parthenogenic egg parasitoid, and *Spathius agrili*, a gregarious larval ectoparasitoid. These have been introduced and released into the United States of America as a possible biological control of the emerald ash borer. Initial results have shown promise with *Tetrastichus planipennisi* and it is now being released along with *Beauveria bassiana*, a fungal pathogen with known insecticidal properties.

Plants

The legume vine *Mucuna pruriens* is used in the countries of Benin and Vietnam as a biological control for problematic *Imperata cylindrica* grass. *Mucuna pruriens* is said not to be invasive outside its cultivated area. *Desmodium uncinatum* can be used in push-pull farming to stop the parasitic plant, *Striga*.

Indirect Control

Pests may be controlled by biological control agents that do not prey directly upon them. For example the Australian bush fly, *Musca vetustissima*, is a major nuisance pest in Australia, but native decomposers found in Australia are not adapted to feeding on cow dung, which is where bush flies breed. Therefore the Australian Dung Beetle Project (1965-1985,) led by Dr. George Bornemissza of the Commonwealth Scientific and Industrial Research Organisation, released forty-nine species of dung beetle, with the aim of reducing the amount of dung and therefore also breeding sites of the fly.

Effects of Biological Control

Effects on native biodiversity:

Figure: *The cane toad,* Bufo marinus

Biological control can potentially have positive and negative effects on biodiversity. The most common problems with biological control occur via predation, parasitism, pathogenicity, competition, or other attacks on non-target species. Often a biological control agent is imported into an area to reduce the competitive advantage of an exotic species that has previously invaded or been introduced there, the aim being to thereby protect the existing native species and ecology. However the introduced control does not always target only the intended species; it can also target native species. In Hawaii during the 1940s parasitic wasps were introduced to control a lepidopteran pest and the wasps are still found there today. This may have a negative impact on the native ecosystem, however, host range and impacts need to be studied before declaring their impact on the environment.

Over the past 15 years with the rise in biological control interest there has become a greater focus on the non-target impacts that could occur. In the past many biological control releases were not thoroughly examined and agents of biological control were released without any consideration. When introducing a biological control agent to a new area, a primary concern is its host-specificity. Generalist feeders (control agents that are not restricted to preying on a single species or a small range of species) often make poor biological control agents, and may become invasive species themselves. For this reason potential biological control agents should be subject to extensive testing and quarantine before release into any new environment. If a species is introduced and attacks a native species, the biodiversity in that area can change dramatically. When one native species is removed from an area, it may have filled an essential ecological niche. When this niche is absent it may directly affect the entire ecosystem.

Vertebrate animals tend to be generalist feeders, and seldom make good biological control agents; many of the classic cases of "biocontrol gone awry" involve vertebrates. For example the cane toad, *Bufo marinus*, was intentionally introduced to Australia to control the introduced French's Cane Beetle and the Greyback Cane Beetle, pests of sugar cane. 102 toads were obtained from Hawaii and bred in captivity to increase their numbers until they were released into the sugar cane fields of the tropic north in 1935. It was later discovered that the toads could not jump very high and so they could not eat the cane beetles which stayed up on the upper stalks of the cane plants. However the toad thrived by feeding on other insects and it soon spread very rapidly; it took over native amphibian habitat and brought foreign disease to native toads and frogs, dramatically reducing their populations. Also when it is threatened or handled, the cane toad releases poison from parotid glands on its shoulders; native Australian species such as goannas, tiger snakes, dingos and northern quolls that attempted to eat the toad were harmed or killed. This example shows how small mis-introduced organisms can alter the native biodiversity in large ecosystems. If native species are reduced or eradicated, a domino effect can take place until a new equilibrium is reached.

Other examples of biological control agents that subsequently crossed over to native species are:

- *Rhinocyllus conicus*, a seed-feeding weevil, was introduced to North America to control exotic thistles (Musk and Canadian). However the weevil does not target only the exotic thistles; it also targets native thistles that are essential to various native insects which rely solely on native thistles and do not adapt to other plant species.
- The mongoose was introduced to Hawaii in order to control the rat population. However it preyed on the endemic birds of Hawaii, especially their eggs, more often than it ate the rats. (Note, however, that the introduction of the mongoose was not undertaken based on scientific—or perhaps any—understanding of the consequences of such an action. The introduction of a generalist mammal for biocontrol of anything would be unthinkable by any reasonable standards today.)
- 5 cats brought to the subantarctic Prince Edward Islands to catch mice in 1949 multiplied to 3,400 in about two decades and started to threaten local extinction of birds. They had to be infected with feline distemper virus. The rest were shot and completely eliminated by the 1990s.
- The sturdy and prolific mosquito fish was introduced from around the Gulf of Mexico to around the world in the 1930s and 40s to combat malaria; however, it was found to cause the decline of local fish and frogs through competition of other food source as well as eating their eggs.

Living organisms, through the process of evolution, may achieve increased resistance to biological, chemical, and physical methods of control over time. In the event the target pest population is not completely exterminated or is still capable of reproduction (were the pest control means a form of sterilization), the surviving population could acquire a tolerance to the applied pressures - this can result in an evolutionary arms race with the control method. Successful biological control reduces the population density of the target species over several years, thus providing the potential for native species to re-establish. In addition, regeneration and reestablishment programmes can aid the recovery of native species. To develop or find a biological control that exerts control only on the targeted species is a very lengthy process of research and experiments.

Effects on Invasive Species

Biological control programmes aim to reduce or eliminate populations of ecologically and agriculturally harmful invasive species. Examples where this has been achieved include:

- The alligator weed was introduced to the United States from South America. This aquatic weed spreads rapidly and causes many problems in lakes and rivers. The weed takes root in shallow water causing major problems for navigation, irrigation, and flood control. The alligator weed flea beetle and two other biological controls were released in Florida. Because of their success, Florida banned the use of herbicides to control alligator weed three years after the controls were introduced.
- *Galerucella calmariensis*, a leaf beetle, has been introduced in North America as a control agent for purple loosestrife (*Lythrum salicaria*).
- In the late 19th century cottony cushion scale was discovered in the Californian citrus industry, and it was feared that severe economic losses would result. However the vedalia beetle and, subsequently, *Cryptochaetum iceryae*, a parasitoid fly, were introduced to control the pest. Within a few years the cottony cushion scale was controlled and the citrus industry suffered little financial loss.
- *Salvinia molesta* is a major aquatic weed. It covers many waterways causing damage to water flow and the ecosystem. This weed was incorrectly identified at first. Scientists found weevils eating this weed. They named this species of weevils *C. salviniae*. The weevils have become a great biological control success for all countries.
- Prickly pear cacti, Opuntia spp., were introduced into Queensland, Australia as an ornamental plant. It quickly spread to cover over 25 million hectares of Australia. Two control agents were used to help control the spread of the plant. These were *C. cactorum*, a Lepidopteran species, and Dactylopius spp., a Hemipteran species.

Grower Education

A potential obstacle to the adoption of biological pest control measures is growers sticking to the familiar use of pesticides, particularly since it has been claimed that many of the pests that are controlled today using pesticides, actually became pests because pesticide use reduced or eliminated natural predators.

A method of increasing grower adoption of biocontrol involves letting growers learn by doing, for example showing them simple field experiments, having observations of live predation of pests, or collections of parasitised pests. In the Philippines, early season sprays against leaf folder caterpillars were common practice, but growers were asked to follow a 'rule of thumb' of not spraying against leaf folders for the first 30 days after transplanting; participation in this resulted in a reduction of insecticide use by 1/3 and a change in grower perception of insecticide use.

Beneficial Insects

Beneficial insects (sometimes called beneficial bugs) are any of a number of species of insects that perform valued services like pollination and pest control. The concept of *beneficial* is subjective and only arises in light of desired outcomes from a human perspective. In farming and agriculture, where the goal is to raise selected crops, insects that hinder the production process are classified as pests, while insects that assist production are considered beneficial. In horticulture and gardening; pest control, habitat integration, and 'natural vitality' aesthetics are the desired outcome with beneficial insects.

Encouraging beneficial insects, by providing suitable living conditions, is a pest control strategy, often used in organic farming, organic gardening or Integrated Pest Management. Companies specializing in biological pest control sell many types of beneficial insects, particularly for use in enclosed areas, like greenhouses.

Types of Beneficial Insects

Some species of bee are beneficial as pollinators, although generally only efficient at pollinating plants from the same area of origin, facilitating propagation and fruit production for many plants. Also, some bees are predatory or parasitic, killing pest insects. This group includes not only honeybees, but also many other kinds that are more efficient at pollinating. Bees can be attracted by many companion plants, especially bee balm and pineapple sage for honeybees, or apiaceae like queen anne's lace and parsley, for predatory bees.

Ladybirds are generally thought of as beneficial because they eat large quantities of aphids, mites and other arthropods that feed on various plants.

Bee

Bees are flying insects closely related to wasps and ants, and are known for their role in pollination and for producing honey and beeswax. Bees are a monophyletic lineage within the superfamily Apoidea, presently considered as a clade Anthophila. There are nearly 20,000 known species of bees in seven to nine recognized families, though many are undescribed and the actual number is probably higher. They are found on every continent except Antarctica, in every habitat on the planet that contains insect-pollinated flowering plants.

Bees are adapted for feeding on nectar and pollen, the former primarily as an energy source and the latter primarily for protein and other nutrients. Most pollen is used as food for larvae.

Bees have a long proboscis (a complex "tongue") that enables them to obtain the nectar from flowers. They have antennae almost universally made up of 13 segments in males and 12 in females, as is typical for the superfamily. Bees all have two pairs of wings, the hind pair being the smaller of the two; in a very few species, one sex or caste has relatively short wings that make flight difficult or impossible, but none are wingless.

The smallest bee is *Trigona minima*, a stingless bee whose workers are about 2.1 mm (5/64") long. The largest bee in the world is *Megachile pluto*, a leafcutter bee whose females can attain a length of 39 mm (1.5"). Members of the family Halictidae, or sweat bees, are the most common type of bee in the Northern Hemisphere, though they are small and often mistaken for wasps or flies.

The best-known bee species is the European honey bee, which, as its name suggests, produces honey, as do a few other types of bee. Human management of this species is known as beekeeping or apiculture.

Bees are the favourite meal of *Merops apiaster*, the bee-eater bird. Other common predators are kingbirds, mockingbirds, beewolves, and dragonflies.

Bees play an important role in pollinating flowering plants, and are the major type of pollinator in ecosystems that contain flowering plants. Bees either focus on gathering nectar or on gathering pollen depending on demand, especially in social species. Bees gathering nectar may accomplish pollination, but bees that are deliberately gathering pollen are more efficient pollinators. It is estimated that one third of the human food supply depends on insect pollination, most of which is accomplished by bees, especially the domesticated European honey bee. Contract pollination has overtaken the role of honey production for beekeepers in many countries. Monoculture and the massive decline of many bee species (both wild and domesticated) have increasingly caused honey bee keepers to become migratory so that bees can be concentrated in seasonally varying high-demand areas of pollination.

Most bees are fuzzy and carry an electrostatic charge, which aids in the adherence of pollen. Female bees periodically stop foraging and groom themselves to pack the pollen into the scopa, which is on the legs in most bees, and on the ventral abdomen on others, and modified into specialized pollen baskets on the legs of honey bees and their relatives. Many bees are opportunistic foragers, and will gather pollen from a variety of plants, while others are oligolectic, gathering pollen from only one or a few types of plant. A small number of plants produce nutritious floral oils rather than pollen, which are gathered and used by oligolectic bees. One small subgroup of stingless bees, called "vulture bees," is specialized to feed on carrion, and these are the only bees that do not use plant products as food. Pollen and nectar are usually combined to form a "provision mass", which is often soupy, but can be firm. It is formed into various shapes (typically spheroid), and stored in a small chamber (a "cell"), with the egg deposited on the mass. The cell is typically sealed after the egg is laid, and the adult and larva never interact directly (a system called "mass provisioning").

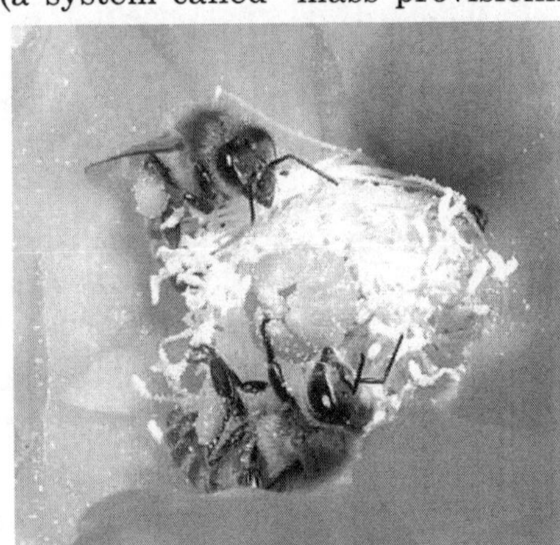

***Figure:** Honey Bees immersed in Cactus flower pollen*

In New Zealand scientists discovered that three genera of native bees have evolved to open flower buds of the native mistletoe *Peraxilla tetrapetala*. The buds cannot open themselves but are visited by birds such as the tui and bellbird which twist the top of the ripe bud. That action releases a mechanism which causes the petals to suddenly spring open, giving access to the nectar and pollen. However, when observing the native bees in the Canterbury province in the South Island, the scientists were astonished to see the bees biting the top off the buds, then pushing with their legs, occasionally popping open the buds to allow the bees to harvest the nectar and pollen, and therefore aid in the pollination of the mistletoe which is in decline in New Zealand. Nowhere else in the world have bees demonstrated ability to open explosive bird-adapted flowers.

Visiting flowers can be a dangerous occupation. Many assassin bugs and crab spiders hide in flowers to capture unwary bees. Other bees are lost to birds in flight. Insecticides used on blooming plants kill many bees, both by direct poisoning and by contamination of their food supply. A honey bee queen may lay 2000 eggs per day during spring buildup, but she also must lay 1000 to 1500 eggs per day during the foraging season, mostly to replace daily casualties, most of which are workers dying of old age. Among solitary and primitively social bees, however, lifetime reproduction is among the lowest of all insects, as it is common for females of such species to produce fewer than 25 offspring.

The population value of bees depends partly on the individual efficiency of the bees, but also on the population itself. Thus while bumblebees have been found to be about ten times more efficient pollinators on cucurbits, the total efficiency of a colony of honey bees is much greater due to greater numbers. Likewise during early spring orchard blossoms, bumblebee populations are limited to only a few queens, and thus are not significant pollinators of early fruit.

Coccinellidae

The Coccinellidae are a family of small beetles, ranging from 0.8 to 18 mm (0.0315 to 0.708 inches). They are commonly yellow, orange, or scarlet with small black spots on their wing covers, with black legs, heads and antennae. Such colour patterns vary greatly, however; for example, a minority of species, such as *Vibidia duodecimguttata*, a twelve-spotted species, have whitish spots on a brown background. Coccinellids are found worldwide, with over 5,000 species described, more than 450 native to North America alone.

Coccinellidae are known colloquially as ladybirds (in Britain, Ireland, the Commonwealth, and the southern United States), ladybugs (originating in North America) or lady cows, among other names. When they need to use a common name, entomologists widely prefer the names ladybird beetles or lady beetles as these insects are not true bugs.

The Coccinellidae are generally considered useful insects, because many species feed on aphids or scale insects, which are pests in gardens, agricultural fields, orchards, and similar places. Within the colonies of such plant-eating pests, they will lay hundreds of eggs, and when these hatch the larvae will commence feeding immediately. However, some species do have unwelcome effects. Among these, the most prominent are the subfamily Epilachninae, which are plant eaters. Usually, Epilachninae are only mild agricultural pests, eating the leaves of grain, potatoes, beans, and various other crops, but their numbers can increase explosively in years when their natural enemies are few, such as parasitoid wasps that attack their eggs. When that happens, they can do major crop damage. They occur in practically all the major crop-producing regions of temperate and tropical countries.

Other insects commonly identified as beneficial include:

- Minute pirate bug
- Earwig
- Big-eyed bug
- Beneficial nematodes
- Assassin bug
- Damsel bug
- Mealybug destroyer
- Soldier beetle
- Green lacewing
- Syrphid fly
- Tachinid fly
- Ichneumon wasp
- Trichogramma wasp

Attractive Plants

Plants in the Apiaceae and Asteraceae families are general valuable companions.

2

Biological Pesticide

Biopesticides, a contraction of 'biological pesticides', include several types of pest management intervention: through predatory, parasitic, or chemical relationships. The term has been associated historically with biological control - and by implication - the manipulation of living organisms. Regulatory positions can be influenced by public perceptions, thus:

- in the EU, biopesticides have been defined as "a form of pesticide based on micro-organisms or natural products".
- the US EPA states that they "include naturally occurring substances that control pests (biochemical pesticides), microorganisms that control pests (microbial pesticides), and pesticidal substances produced by plants containing added genetic material (plant-incorporated protectants) or PIPs".

They are typically created by growing and concentrating naturally occurring organisms and/or their metabolites including bacteria and other microbes, fungi, nematodes, proteins, *etc.* They are often considered to be important components of integrated pest management (IPM) programmes, and have received much practical attention as substitutes to synthetic chemical plant protection products (PPPs).

Biopesticides fall into three major classes:

- Microbial pesticides which consist of bacteria, entomopathogenic fungi or viruses (and sometimes includes the metabolites that bacteria or fungi produce). Entomopathogenic nematodes are also often classed as microbial pesticides, even though they are multi-cellular.

- Biochemical pesticides or herbal pesticides are naturally occurring substances that control (or monitor in the case of pheromones) pests and microbial diseases.
- Plant-incorporated protectants (PIPs) have genetic material from other species incorporated into their genetic material (*i.e.* GM crops). Their use is controversial, especially in many European countries.

Biopesticides have usually no known function in photosynthesis, growth or other basic aspects of plant physiology; however, their biological activity against insect pests, nematodes, fungi and other organisms is well documented. Every plant species has developed a built-in unique chemical complex structure that protects it from pests. The plant kingdom offers a diverse array of complex chemical structures and almost every imaginable biological activity. These biodegradable, economical and renewable alternatives are used especially under organic farming systems.

Examples: *Bacillus thuringiensis*, a bacterial disease of Lepidoptera, Coleoptera and Diptera, is a well-known insecticide example. The toxin from *B. thuringiensis* (Bt toxin) has been incorporated directly into plants through the use of genetic engineering. The use of Bt Toxin is particularly controversial. Its manufacturers claim it has little effect on other organisms, and is more environmentally friendly than synthetic pesticides. However, at least one scientific study has suggested that it may lead to slight histopathological changes on the liver and kidneys of mammals with Bt toxin in their diet.

Other microbial control agents include products based on:

- entomopathogenic fungi (*e.g. Beauveria bassiana, Lecanicillium* spp., *Metarhizium* spp.),
- plant disease control agents: include *Trichoderma* spp. and *Ampelomyces quisqualis* (a hyper-parasite of grape powdery mildew); *Bacillus subtilis* is also used to control plant pathogens.
- beneficial nematodes attacking insect (*e.g. Steinernema feltiae*) or slug (*e.g. Phasmarhabditis hermaphrodita*) pests
- entomopathogenic viruses (*e.g.. Cydia pomonella* granulovirus).
- weeds and rodents have also been controlled with microbial agents.

Various naturally occurring materials, including fungal and plant extracts, have been described as biopesticides. Products in this category include:

- Insect pheromones and other semiochemicals
- Fermentation products such as Spinosad (a macro-cyclic lactone)
- Chitosan: a plant in the presence of this product will naturally induce systemic resistance (ISR) to allow the plant to defend itself against disease, pathogens and pests.
- Biopesticides may include natural plant-derived products, which include alkaloids, terpenoids, phenolics and other secondary chemicals. Certain vegetable oils such as canola oil are known to have pesticidal properties. Products based on extracts of plants such as garlic have now been registered in the EU and elsewhere.
- Naturally occurring minerals such as baking soda may also have pesticidal applications.

Applications

Biopesticides are typically microbial biological pest control agents that are applied in a manner similar to chemical pesticides. In order to implement these environmentally friendly pest control agents effectively, it can be important to pay attention to the way they are formulated and applied.

Biopesticides for use against crop diseases have already established themselves on a variety of crops. For example, biopesticides already play an important role in controlling downy mildew diseases. Their benefits include: a 0-Day Pre-Harvest Interval, the ability to use under moderate to severe disease pressure, and the ability to use as a tank mix or in a rotational programme with other registered fungicides. Because some market studies estimate that as much as 20% of global fungicide sales are directed at downy mildew diseases, the integration of biofungicides into grape production has substantial benefits in terms of extending the useful life of other fungicides, especially those in the reduced-risk category.

A major growth area for biopesticides is in the area of seed treatments and soil amendments. Fungicidal and biofungicidal seed treatments are used to control soil borne fungal pathogens that cause seed rots, damping-off, root rot and seedling blights. They can also be used to control internal seed–borne fungal pathogens as well as fungal pathogens that are on the surface of the seed. Many biofungicidal products also show capacities to stimulate plant host defences and other physiological processes that can make treated crops more resistant to a variety of biotic and abiotic stresses.

Advantages

- Harmful residues not detected
- Can be cheaper than chemical pesticides when locally produced.
- Can be more effective than chemical pesticides in the long-term (as demonstrated, for example, by the LUBILOSA Programme)
- Biodegradable

Disadvantages

- High specificity: which may require an exact identification of the pest/pathogen and the use of multiple products to be used; although this can also be an advantage in that the biopesticide is less likely to harm species other than the target
- Often slow speed of action (thus making them unsuitable if a pest outbreak is an immediate threat to a crop)
- Often variable efficacy due to the influences of various biotic and abiotic factors (since biopesticides are usually living organisms, which bring about pest/pathogen control by multiplying within the target insect pest/pathogen)
- Living organisms evolve and increase their resistance to biological, chemical, physical or any other form of control. If the target population is not exterminated or rendered incapable of reproduction, the surviving population can acquire a tolerance of whatever pressures are brought to bear, resulting in an evolutionary arms race.

Bioherbicide

A bioherbicide is a biologically based control agent for weeds. Among the three major types of pesticides (agricultural pest-control agents) herbicides are used to control weeds, or undesirable plants. (The other major classes of pesticides are insecticides, which control insects; and fungicides, which control fungi. Normally, any pesticide can only be expected to control, but not eradicate, agricultural pests.)

Almost every agricultural pest has at least one naturally occurring enemy that will reduce its population. Bioherbicides are weed-control agents that utilize such naturally occurring enemies, rather than depending on man-made chemicals. This can be important because agents of biological control ordinarily have many fewer, and much milder, effects on the environment than do synthetic chemicals. What is more, they tend not to lead to the public health problems that chemicals are associated with. These two advantages of biological control agents, including bioherbicides, do not however mean that they need not also be subjected to careful tests for environmental and public health safety.

Bioherbicides may be compounds derived from microbes such as fungi, bacteria or protozoa; or phytotoxic plant residues, extracts or single compounds derived from other plant species.

A bioherbicide based on a fungus is called a mycoherbicide.

In the industry, bioherbicides and other biopesticides are often referred to as "naturals".

LUBILOSA

LUBILOSA was the name of a research programme that aimed at developing a biological alternative to the chemical control of locusts. This name is an acronym of the French title of the programme: Lutte Biologique contre les Locustes et les Sauteriaux (biological control of locusts and grasshoppers). During its 13 year life (November 1989 to December 2002), the programme identified an isolate of *Metarhizium* fungus virulent to locusts and went through all the necessary steps to develop the commercial biopesticide product 'Green Muscle' based on its spores.

Collaborators

The programme was conceived by Chris Prior and David Greathead of the International Institute of Biological Control (IIBC), a former research institute of CAB International that was based at Silwood Park in the UK at that time. IIBC requested and obtained the collaboration of the former Biological Control Centre for Africa in Cotonou, Benin, that was owned and managed by the International Institute of Tropical Agriculture (IITA). When IIBC asked for financial support from the Dutch Directoraat Generaal voor Internationale Samenwerking (DGIS: Directorate General of International Cooperation), DGIS contributed by paying for a Dutch locust expert based at the Département de Formation en Protection des Végétaux (DFPV: Department of Crop Protection Training) of the AGRHYMET Regional Centre (ARC) in Niamey, Niger, which is part of the Comité permanent Inter-états de Lutte contre la Sécheresse dans le Sahel (CILSS: Permanent Interstate Committee for Drought Control in the Sahel). Since DFPV was given its own budget within LUBILOSA, it became a full partner within the programme. For a number of years, LUBILOSA collaborated with GTZ's locust programme. Permanent collaboration was established with the crop protection agencies of Niger, Benin, Burkina Faso, Chad, Mali, Senegal and The Gambia.

Donors

The LUBILOSA programme was supported financially by the following donors:

- CIDA: Canadian International Development Agency
- DFID: Department for International Development, UK
- DGIS: Dutch Directorate General of International Cooperation
- SDC: Swiss Agency for Development and Cooperation
- USAID: United States Agency for International Development

Over the life of the programme, these donors contributed UK£10,200,000 (approximately US$17,000,000 at the time).

Four Phases

The LUBILOSA programme has been funded in phases of three or four years each:

- First phase 1990–1992
- Second phase 1993–1995
- Third phase 1996–1998
- Fourth phase 1999–2002

First Phase

The project was started late in 1989, with Chris Prior and David Greathead, obtaining funding and forming a team to develop a biological means of controlling locusts and grasshoppers. While examining the various options for biological control, it soon became apparent that oil formulations of the spores of certain fungi belonging to the form phylum Deuteromycota (Anamorphic fungi) offered the most promising option. Such fungi will grow on artificial substrates and so can be mass-produced relatively quickly in large quantities. Their spores are lipophilic and therefore suspend much more readily in oils than in water. As most locust control is carried out using oil formulations at ultra-low volume (ULV) rates of application, developing an oil formulation of fungal spores enabled operators to use the same equipment for applying a chemical pesticide and maximise efficacy.

Usually, fungi are most active under moist conditions and in early attempts to use them as control agents, they were therefore applied in water formulations. However, the results in practice were often disappointing and the approach was largely abandoned during most of the 20th century. When interest in microbial control was revived, water formulations of fungi started to be used successfully in greenhouses, but the results in open fields continued to disappoint. In the 1980s, Prior had previously discovered that some anamorphic fungi are more effective when applied in oil.

During the first phase of LUBILOSA, the technical feasibility of using such oil formulations against locusts was demonstrated in the laboratory, then field cage and "arena" trials. An extensive survey was launched in West Africa and the Arabian Peninsula to look for isolates virulent to locusts and grasshoppers, because few such isolates were available in public collections. The survey made use of a network of collaborators and came up with some 180 isolates, many of which belonged to *Metarhizium acridum*, some to other *Metarhizium* species and to *Beauveria bassiana* with a few records of *Sygliocladium acridiorum* (syn. *Sorosporella*). Laboratory screening of these isolates showed that *M. acridum* was easily the most virulent species under warm conditions (30°C), and confirmed the selection of isolate IMI 330189 for further development. Subsequent trials in cages and arenas confirmed that oil formulations were infective even under very dry conditions. IMI 330189 and similar isolates were assigned to *Metarhizium flavoviride* in early papers, then *Metarhizium anisopliae* var. *acridum*, but are now described as *Metarhizium acridum* (Driver & Milner) J.F. Bisch., Rehner & Humber.

Second Phase

During phase 2, the oil formulation was tested in the field and shown to be effective. Field trials proved to be difficult to carry out on such highly mobile insects as grasshoppers and locusts. With locusts, fixed plots, unless several square kilometres in size, cannot be used, so it is necessary to follow hopper bands. Although there were many locust infestations during this phase, of both desert locusts (*Schistocerca gregaria*) and brown locusts (*Locustana pardalina*), the chemical treatment teams were also very active and little headway was therefore made in developing the techniques necessary to measure the effect of *Metarhizium* on unrestrained locusts. However, it was possible to scale up treatments against the variegated grasshopper (*Zonocerus variegatus*) in the forest zone and against Sahelian grasshoppers, especially the Senegalese grasshopper (*Oedaleus senegalensis*).

Contacts were made with several internationally renowned biopesticide producers to ascertain whether they would be able and willing to reliably meet LUBILOSA's spore requirements during phase 3, but none were able to make that commitment. Accordingly, part of the IITA station in Cotonou was converted into a spore production unit and the human and technical capacity enhanced to meet the expected spore requirements. The facility proved an excellent research unit and made it possible to refine realistic technical specifications and to test production, contamination control, spore separation, drying and packaging.

Third Phase

Field Trials: The LUBILOSA team continued to look for opportunities to test *Metarhizium* against locusts and grasshoppers. Field trials were conducted against Senegalese grasshoppers, African rice grasshoppers (*Hieroglyphus daganensis*), variegated grasshoppers, Sahelian tree locusts (*Anacridium melanorhodon*), brown locusts and desert locusts. These trials demonstrated that *Metarhizium* at a dosage of 50 g/ha could reduce grasshopper populations and those of tree locusts by 80-90% within two to three weeks. Population level control was more difficult to achieve with other locusts, but a significant effect was demonstrated on hopper bands notwithstanding the great difficulties in tracking individual bands.

Some trials against Senegalese grasshoppers included the chemical insecticide fenitrothion for comparison. Because of its rather short persistence, this chemical turned out to be much less effective than commonly believed. It was highly effective in producing an immediate sharp reduction in population densities, but continued hatching and remigration into the treated plots caused these densities to start increasing again after less than a week, until they reached or exceeded the original levels after about two to three weeks. *Metarhizium*, on the other hand, reduced population levels at a slower rate, but maintained low levels for at least one and a half months. The half-life of its spores on the vegetation was estimated at more than 7 days under rainy season conditions in the Sahel and, of course, some conidia must survive between seasons.

Grasshopper Behaviour

In the course of these field trials, it became clear that locusts and grasshoppers are able to detect that they have become infected and alter their behaviour accordingly. In particular, they spend more time basking in the sun, even in the middle of the day. Uninfected grasshoppers also bask in the sun when their body temperature is below the preferred 38-40°C. However, detailed investigations showed that infected grasshoppers increase their body temperature to up to 4° higher. This effect has been dubbed "behavioural fever" and is thought to slow down the infection just like a fever in warm-blooded animals does. This basking behaviour poses a potential problem for the fungus.

The latter's preferred temperature is 28-30°C and it stops growing above 35°. Measurements of body temperatures in the field have shown that the basking behaviour reduces the time available for growth by many hours per day. Under certain conditions, this enables the insects to lay at least one egg pod before dying. Interestingly, healthy locusts and grasshoppers, especially females, take some time after fledging to reach maturity and lay eggs, during which time they build up sufficient fat to produce up to three egg pods. However, infected ones usually dispense with this fattening period and go on to produce eggs immediately after fledging. This generally wears them out so much that they die after laying one pod. So even if *Metarhizium* does not kill before or shortly after fledging, it usually allows the formation of only a single egg pod and thereby reduces the fecundity of infected females.

(Eco)toxicology

During phase 3, the programme started investigating the environmental impact of the chosen *Metarhizium* isolate (IMI 330189). Standard toxicological tests carried out by certified laboratories showed that the strain had practically no effect on mammals, birds and fish (e.g. LD_{50} for rats >2000 mg active ingredient per kg of body weight). Other toxicological and ecotoxicological tests were performed by the LUBILOSA team. An important goal was to show that its isolate only infected locusts and grasshoppers. Infection tests were therefore performed on a large number of invertebrates. The great majority could not be infected even by high doses. However, some species, like termites, honey bees, certain parasitoid wasps and silk worms, did get infected under laboratory conditions. Further experiments were carried out under semi-field and full field conditions, which showed that only silk worms suffered significant infection under those conditions. The conclusion from all these tests was that the LUBILOSA strain of *Metarhizium* was incapable of infecting vertebrates and infected only a small number of insect species under artificial conditions. The available studies indicate that under natural conditions, the isolate only infects species belonging to the Acrididea and the domesticated silk worm *Bombyx mori*.

Mass Production

The spore production unit in Cotonou made it possible to optimise the mass production of spores of the LUBILOSA strain at an appropriate level of technology. The production process chosen was one with a liquid phase and a solid phase. During the liquid phase, the fungus biomass is bulked up in a solution of sugar and yeast and may produce submerged conidia. Subsequently, the resulting broth is used to inoculate partially boiled and sterilised rice. The fungal mycelium invades the rice and then starts to sporulate. At this point, rice and fungus are slowly dried before separating the spores from the rice. A novel machine was invented to achieve this in an efficient way, which was subsequently further developed into the 'MycoHarvester'. The study and optimisation of the mass production process led to the adoption of quality standards for the produced spores, which could be imposed on those who wanted to produce the spores commercially under licence. Various aspects of quality control are important, including: levels of contamination (especially the absence of human pathogens), virulence to target pests, particle size spectrum and, not least, viable spore count. Extensive research was carried out on optimising storage of spores, which should be dry (<5% moisture content) and ideally be maintained under cool conditions.

Commercialisation

Two commercial companies agreed to look into the feasibility and economic viability of producing *Metarhizium* for the locust control market. These were Biological Control Products (BCP) near Durban, South Africa, and Natural Plant Protection (NPP) of Noguères, France (part of Calliope S.A.). The former registered the product under the name Green Muscle (deposited by CAB International), first in South Africa and subsequently in other southern and eastern African countries. BCP is the leading producer of biopesticides in Africa. Its main market is South Africa but its products are increasingly being sold in other countries. The company continues its production and promotion of Green Muscle. It was taken over by Becker Underwood in 2010. NPP, however, only obtained a temporary sales permit for the CILSS countries in West Africa, but was unable to produce Green Muscle using its own production process.

Fourth Phase

The last phase of LUBILOSA was characterised by promotion activities and stewardship of the commercial producer. Launching the product on the market and informing potential clients was not sufficient. The market for locust control products was (and still is) peculiar and difficult to penetrate. Most users are government or intergovernmental agencies involved in locust and/or grasshopper control. Especially in the beginning, it proved difficult to convince the professionals in these agencies of the need for an alternative product and to convince them that the product was as effective as those they had been using so far. Moreover, the large pesticide companies offered all kinds of incentives to those in charge of procurement in order that they buy their products, a practice that was difficult to follow for a small company like BCP. Until the end of the programme and even beyond, there was a constant need to do more field trials or demonstrations. Towards the end, the Locust Group of the Food and Agriculture Organization of the United Nations (FAO) finally became interested and involved in some of the trials: especially against desert locusts.

Apart from the Sahelian zone, grasshoppers are generally not considered serious pests in Africa and are therefore rarely controlled with pesticides. For that reason, it was only in the Sahel that serious efforts were undertaken to promote Green Muscle for grasshopper control. In a number of countries, and especially Niger, cereal growers are organised in village brigades that are charged with controlling pests that surpass the capacity of individual growers. Green Muscle was given to some of these after training in its correct application. After the initial disappointment with its slow action, the villagers usually became satisfied with the final results. However, the price proved to be a serious obstacle. At that time, the cost of producing one kg of Green Muscle was about $200. At a dosage of 50 g/ha, that translated into a price of $10 per hectare. The price of conventional products was then approximately $5/ha. Though growers were prepared to pay a premium of up to $2 for the fact that Green Muscle was harmless to their health and that of their livestock, they were clearly not going to pay $10. Another problem was that no producer had yet been identified for the West African market and the availability of the product could therefore not be guaranteed.

At the end of the programme in December 2002, the LUBILOSA collaborators were faced with a situation where commercial uptake of Green Muscle was still non-existent, all quantities purchased so far exclusively for trial purposes. BCP was left alone to promote the product with little support from CABI and IITA, who had run out of funds for serious activities. Fortunately, the FAO had become interested, which made it possible to carry out a few more field trials.

Post-LUBILOSA

After the termination of the LUBILOSA programme, activities to promote the use of Green Muscle continued, including field trials against red locusts and desert locusts with support from the FAO. A new project in West Africa, PRéLISS (Regional Project for Integrated Grasshopper Control in the Sahel), aimed at developing an integrated pest management (IPM) strategy that included the use of Green Muscle. This project conducted field trials with reduced dosages of Green Muscle and proved that 25 g/ha was as effective as the registered dosage of 50 g/ha. It also tested mixtures with lambda-cyhalothrin (a pyrethroid insecticide) to try to solve the problem of the slow action of Green Muscle. It was demonstrated that application of both products together, each at a quarter of its registered dose, caused mortality almost as quick as lambda-cyhalothrin alone and maintained low population levels for weeks, and all this at a much lower cost.

After it had become clear that NPP could not produce Green Muscle, the search for a second producer continued. A number of candidate producers were approached, but none was ready to take up the challenge. This situation only changed when desert locusts started invading West African countries again in late 2003.

The president of Senegal, Abdoulaye Wade, and his wife Viviane were very concerned about the prospect of huge quantities of chemical insecticides being sprayed to contain the upsurge. The first lady instructed her technical advisor, Sébastien Couasnet, to find out whether an alternative existed. Apparently, the information about Green Muscle had remained stuck at the professional level of the crop protection agency of Senegal (and most likely other countries as well), notwithstanding all the efforts of LUBILOSA and PRéLISS to make the existence of the new product widely known.

Fortunately, Couasnet stumbled upon the LUBILOSA website, which was still hosted by IITA and maintained by the then project leader of PRéLISS, Christiaan Kooyman. The two met in Cotonou and immediately set about to design a production plant for Green Muscle. Couasnet was able to convince the first lady to organise the financing and construction of the plant started in 2005. It took a while to start up the production and get rid of all sources of contamination, but this was eventually achieved with the backstopping of IITA, CAB International and Roy Bateman of IPARC, who installed the latest version of the MycoHarvester v.3. The production licence was granted in late 2007. For the following four years, production and marketing were taken care of by the Fondation Agir pour l'Education et la Santé, of which Mme. Wade was president. Unfortunately, FAES did not last but went under because of internal disagreements, especially concerning the extent to which private capital should be brought in to perpetuate the production of Green Muscle and similar products. After the defeat of President Wade in the presidential elections of 2012, FAES was closed down. Production of fungi had already stopped 8 months earlier. Currently, the only remaining producer of Green Muscle is Becker Underwood South Africa, which was acquired by BASF SE in 2012.

Chitosan (Natural Biocontrol for Agricultural & Horticultural use)

Chitosan

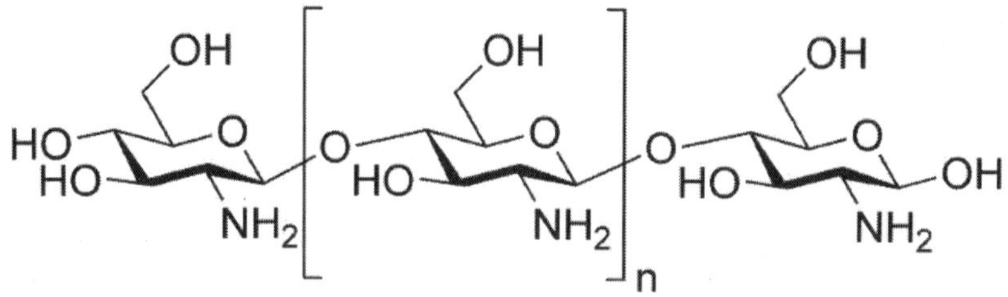

Figure: *Chitosan is a linear polysaccharide composed of randomly distributed β-(1-4)-linked D-glucosamine (deacetylated unit) and N-acetyl-D-glucosamine (acetylated unit). It is made by treating shrimp and other crustacean shells with the alkali sodium hydroxide.*

Chitosan has a number of commercial and possible biomedical uses. It can be used in agriculture as a seed treatment and biopesticide, helping plants to fight off fungal infections. In winemaking it can be used as a fining agent, also helping to prevent spoilage. In industry, it can be used in a self-healing polyurethane paint coating. In medicine, it may be useful in bandages to reduce bleeding and as an antibacterial agent; it can also be used to help deliver drugs through the skin.

More controversially, chitosan has been asserted to have use in limiting fat absorption, which would make it useful for dieting, but there is evidence against this.

Other uses of chitosan that have been researched include use as a soluble dietary fibre.

Manufacture and Properties

Chitosan is produced commercially by deacetylation of chitin, which is the structural element in the exoskeleton of crustaceans (such as crabs and shrimp) and cell walls of fungi. The degree of deacetylation (%DD) can be determined by NMR spectroscopy, and the %DD in commercial chitosans ranges from 60 to 100%. On average, the molecular weight of commercially produced chitosan is between 3800 and 20,000 Daltons. A common method for the synthesis of chitosan is the deacetylation of chitin using sodium hydroxide in excess as a reagent and water as a solvent. This reaction pathway, when allowed to go to completion (complete deacetylation) yields up to 98% product.

The amino group in chitosan has a pKa value of ~6.5, which leads to a protonation in acidic to neutral solution with a charge density dependent on pH and the %DA-value. This makes chitosan water soluble and a bioadhesive which readily binds to negatively charged surfaces such as mucosal membranes. Chitosan enhances the transport of polar drugs across epithelial surfaces, and is biocompatible and biodegradable. It is not approved by FDA for drug delivery though. Purified quantities of chitosans are available for biomedical applications.

Chitosan and its derivatives, such as trimethylchitosan (where the amino group has been trimethylated), have been used in nonviral gene delivery. Trimethylchitosan, or quaternised chitosan, has been shown to transfect breast cancer cells, with increased degree of trimethylation increasing the cytotoxicity; at approximately 50% trimethylation, the derivative is the most efficient at gene delivery. Oligomeric derivatives (3-6 kDa) are relatively nontoxic and have good gene delivery properties.

Usage

Agricultural and Horticultural Use

The agricultural and horticultural uses for chitosan, primarily for plant defence and yield increase, are based on how this glucosamine polymer influences the biochemistry and molecular biology of the plant cell. The cellular targets are the plasma membrane and nuclear chromatin. Subsequent changes occur in cell membranes, chromatin, DNA, calcium, MAP Kinase, oxidative burst, reactive oxygen species, callose pathogenesis-related (PR) genes and phytoalexins.

Natural Biocontrol and Elicitor

In agriculture, chitosan is used primarily as a natural seed treatment and plant growth enhancer, and as an ecologically friendly biopesticide substance that boosts the innate ability of plants to defend themselves against fungal infections. The natural biocontrol active ingredients, chitin/chitosan, are found in the shells of crustaceans, such as lobsters, crabs, and shrimp, and many other organisms, including insects and fungi. It is one of the most abundant biodegradable materials in the world.

Degraded molecules of chitin/chitosan exist in soil and water. Chitosan applications for plants and crops are regulated by the EPA, and the USDA National Organic Programme regulates its use on organic certified farms and crops. EPA-approved, biodegradable chitosan products are allowed for use outdoors and indoors on plants and crops grown commercially and by consumers.

The natural biocontrol ability of chitosan should not be confused with the effects of fertilizers or pesticides upon plants or the environment. Chitosan active biopesticides represent a new tier of cost-effective biological control of crops for agriculture and horticulture. The biocontrol mode of action of chitosan elicits natural innate defence responses within plant to resist insects, pathogens, and soil-borne diseases when applied to foliage or the soil. Chitosan increases photosynthesis, promotes and enhances plant growth, stimulates nutrient uptake, increases germination and sprouting, and boosts plant vigour. When used as seed treatment or seed coating on cotton, corn, seed potatoes, soybeans, sugar beets, tomatoes, wheat and many other seeds, it elicits an innate immunity response in developing roots which destroys parasitic cyst nematodes without harming beneficial nematodes and organisms.

Agricultural applications of chitosan can reduce environmental stress due to drought and soil deficiencies, strengthen seed vitality, improve stand quality, increase yields, and reduce fruit decay of vegetables, fruits and citrus crops. Horticultural application of chitosan increases blooms and extends the life of cut flowers and Christmas trees. The US Forest Service has conducted research on chitosan to control pathogens in pine trees and increase resin pitch outflow which resists pine beetle infestation.

Chitosan has a rich history of being researched for applications in agriculture and horticulture dating back to the 1980s. By 1989, chitosan salt solutions were applied to crops for improved freeze protection or to crop seed for seed priming. Shortly thereafter, chitosan salt received the first ever biopesticide label from the EPA, then followed by other intellectual property applications.

Chitosan has been used to protect plants in space, as well, exemplified by NASA's experiment to protect adzuki beans grown aboard the space shuttle and Mir space station in 1997. NASA results revealed chitosan induces increased growth (biomass) and pathogen resistance due to elevated levels of beta 1-3 glucanase enzymes within plant cells. NASA confirmed chitosan elicits the same effect in plants on earth.

Nontoxic, low molecular weight chitosan polymer solutions appear to be safe enough for broad-spectrum agricultural and horticultural uses. In 2008, the EPA approved natural broad-spectrum elicitor status for an ultralow molecular active ingredient of 0.25% chitosan.

A natural chitosan elicitor solution for agriculture and horticultural uses was granted an amended label for foliar and irrigation applications by the EPA in 2009. Given its low potential for toxicity and abundance in the natural environment, chitosan does not harm people, pets, wildlife, or the environment when used according to label directions. The US Forest Service tested chitosan as an ecofriendly biopesticide to prearm pine trees to defend themselves against mountain pine beetles.

Companion Planting

Companion planting in gardening and agriculture is the planting of different crops in proximity for pest control, pollination, providing habitat for beneficial creatures, maximizing use of space, and to otherwise increase crop productivity. Companion planting is a form of polyculture.

Companion planting is used by farmers and gardeners in both industrialized and developing countries for many reasons. Many of the modern principles of companion planting were present many centuries ago in cottage gardens in England and home gardens in Asia, and thousands of years ago in Mesoamerica.

In China, mosquito ferns (*Azolla* spp.) have been used for at least one thousand years as companion plants for rice crops. They host a cyanobacterium that fixes nitrogen from the atmosphere, and they block light from plants that would compete with the rice.

Companion planting was practised in various forms by the indigenous peoples of the Americas prior to the arrival of Europeans. These peoples domesticated squash 8,000 to 10,000 years ago, then maize, then common beans, forming the Three Sisters agricultural technique. The cornstalk served as a trellis for the beans to climb, and the beans fixed nitrogen, benefitting the maize.

Companion planting was widely promoted in the 1970s as part of the organic gardening movement. It was encouraged for pragmatic reasons, such as natural trellising, but mainly with the idea that different species of plant may thrive more when close together. It is also a technique frequently used in permaculture, together with mulching, polyculture, and changing of crops.

Examples of Companion Plants

Nasturtium (*Tropaeolum majus*) is a food plant of some caterpillars which feed primarily on members of the cabbage family (brassicas), and some gardeners claim that planting them around brassicas protects the food crops from damage, as eggs of the pests are preferentially laid on the nasturtium. This practice is called trap cropping.

The smell of the foliage of marigolds is claimed to deter aphids from feeding on neighbouring crops. Marigolds with simple flowers also attract nectar-feeding adult hoverflies, the larvae of which are predators of aphids.

Versions

There are a number of systems and ideas using companion planting.

Square foot gardening attempts to protect plants from many normal gardening problems by packing them as closely together as possible, which is facilitated by using companion plants, which can be closer together than normal.

Another system using companion planting is the forest garden, where companion plants are intermingled to create an actual ecosystem, emulating the interaction of up to seven levels of plants in a forest or woodland.

Organic gardening may make use of companion planting, since many synthetic means of fertilizing, weed reduction and pest control are forbidden.

Host-finding Disruption

Recent studies on host-plant finding have shown that flying pests are far less successful if their host-plants are surrounded by any other plant or even "decoy-plants" made of green plastic, cardboard, or any other green material.

The host-plant finding process occurs in phases:

- The first phase is stimulation by odours characteristic to the host-plant. This induces the insect to try to land on the plant it seeks. But insects avoid landing on brown (bare) soil. So if only the host-plant is present, the insects will quasi-systematically find it by simply landing on the only green thing around. This is called (from the point of view of the insect) "appropriate landing". When it does an "inappropriate landing", it flies off to any other nearby patch of green. It eventually leaves the area if there are too many 'inappropriate' landings.

- The second phase of host-plant finding is for the insect to make short flights from leaf to leaf to assess the plant's overall suitability. The number of leaf-to-leaf flights varies according to the insect species and to the host-plant stimulus received from each leaf. The insect must accumulate sufficient stimuli from the host-plant to lay eggs; so it must make a certain number of consecutive 'appropriate' landings. Hence if it makes an 'inappropriate landing', the assessment of that plant is negative, and the insect must start the process anew.

Thus it was shown that clover used as a ground cover had the same disruptive effect on eight pest species from four different insect orders. An experiment showed that 36% of cabbage root flies laid eggs beside cabbages growing in bare soil (which resulted in no crop), compared to only 7% beside cabbages growing in clover (which allowed a good crop). Simple decoys made of green cardboard also disrupted appropriate landings just as well as did the live ground cover.

Companion Plant Categories

The use of companion planting can be of benefit to the grower in a number of different ways, including:

- Hedged investment – the growing of different crops in the same space increases the odds of some yield being given, even if one crop fails.
- Increased level interaction – when crops are grown on different levels in the same space, such as providing ground cover or one crop working as a trellis for another, the overall yield of a plot may be increased.
- Protective shelter is when one type of plant may serve as a wind break or provide shade for another.
- Pest suppression – some companion plants may help prevent pest insects or pathogenic fungi from damaging the crop, through chemical means.
- Predator recruitment and positive hosting – The use of companion plants that produce copious nectar or pollen in a vegetable garden (insectary plants) may help encourage higher populations of beneficial insects that control pests, as some beneficial predatory insects only consume pests in their larval form and are nectar or pollen feeders in their adult form.
- Trap cropping – some companion plants are claimed to attract pests away from others.

- Pattern disruption – in a monoculture pests spread easily from one crop plant to the next, whereas such easy progress may be disrupted by surrounding companion plants of a different type.

Insectary Plants

Insectary plants are those that attract insects. As such, beneficial insectary plants are intentionally introduced into an ecosystem to increase pollen resources and nectar resources required by the natural enemies of harmful or unwanted insect pests. Beyond an effective natural control of pests, the *friendly insects* also assist in pollination.

The "friendly insects" include ladybeetles, bees, ground beetles, hoverflies, and parasitic wasps. Other animals that are frequently considered beneficial include lizards, spiders, toads, and humming birds. Beneficial insects are as much as ten times more abundant in the insectary plantings area. Mortality of scale insects (caused by natural enemies) can be double with insectary plantings. In addition, a diversity of insectary plants can increase the population of beneficial insects such that these levels can be sustained even when the insectary plants are removed or die off.

For maximum benefit in the garden, insectary plants can be grown alongside desired garden plants that do not have this benefit. The insects attracted to the insectary plants will also help the other nearby garden plants.

Many members of the Apiaceae (formerly known as Umbelliferae) family are excellent insectary plants. Fennel, angelica, coriander (cilantro), dill, and wild carrot all provide in great number the tiny flowers required by parasitic wasps. Various clovers, yarrow, and rue also attract parasitic and predatory insects. Low-growing plants, such as thyme, rosemary, or mint, provide shelter for ground beetles and other beneficial insects. Composite flowers (daisy and chamomile) and mints (spearmint, peppermint, or catnip) will attract predatory wasps, hoverflies, and robber flies. The wasps will catch caterpillars and grubs to feed their young, while the predatory and parasitic flies attack many kinds of insects, including leafhoppers and caterpillars.

Other insectary plants include: mustard (Brassica juncea), phacelia (Phacelia tanacetifolia), buckwheat (Fagopyrum esceulentum), marigold (Tagetes patula), elderberry (sambucus mexicana), and Korean licorice mint (Agastache rugosa), Rubus fruticosa, Convolvulus tricolour, Crataegus monogyna, Anthriscus sylvestris, Chrysanthemum segetum, Scrophularia, Rosa canina, Hedera helix, Centaurea cyanus, Phacelia tanacetifolia, Escholzia californica, Prunus spinosa, Lobularia maritima

Inundative Application

Inundative biological control refers to the release of overwhelming numbers of a mass-produced biological control agent in the expectation of achieving a rapid reduction of a pest population without necessarily achieving continuing impact.

Japanese Beetle

The beetle species *Popillia japonica* is commonly known as the Japanese beetle. It is about 15 millimetres (0.6 in) long and 10 millimetres (0.4 in) wide, with iridescent copper-coloured elytra and green thorax and head. It is not very destructive in Japan, where it is controlled by natural predators, but in America it is a serious pest of about 200 species of plants, including rose bushes, grapes, hops, canna, crape myrtles, birch trees, linden trees and others.

It is a clumsy flier, dropping several centimetres when it hits a wall. Japanese beetle traps therefore consist of a pair of crossed walls with a bag or plastic container underneath, and are baited with floral scent, pheromone, or both. However, studies conducted at the University of Kentucky and Eastern Illinois University suggest beetles attracted to traps frequently do not end up in the traps, but alight on plants in the vicinity, thus causing more damage along the flight path of the beetles and near the trap than may have occurred if the trap were not present.

These insects damage plants by skeletonizing the foliage, that is, consuming only the leaf material between the veins, and may also feed on fruit on the plants if present.

Control

During the larval stage, the Japanese beetle lives in lawns and other grasslands, where it eats the roots of grasses. During that stage, it is susceptible to a fatal disease called milky spore disease, caused by a bacterium called milky spore, *Paenibacillus* (formerly *Bacillus*) *popilliae*. The USDA developed this biological control and it is commercially available in powder form for application to lawn areas. Standard applications (low density across a broad area) take from one to five years to establish maximal protection against larval survival (depending on climate), expanding through the soil through repeated rounds of infection.

On field crops such as squash, floating row covers can be used to exclude the beetles, however this may necessitate hand pollination of flowers. Kaolin sprays can also be used as barriers.

Research performed by many US extension service branches has shown pheromone traps attract more beetles than they catch. Traps are most effective when spread out over an entire community, and downwind and at the borders (i.e., as far away as possible, particularly upwind), of managed property containing plants being protected. Natural repellents include catnip, chives, garlic, and tansy, as well as the remains of dead beetles, but these methods have limited effectiveness. Additionally, when present in small numbers, the beetles may be manually controlled using a soap-water spray mixture, shaking a plant in the morning hours and disposing of the fallen beetles, or simply picking them off attractions such as rose flowers, since the presence of beetles attracts more beetles to that plant.

Natural predators of the beetle are the anchor bug and the blue-winged wasp.

Mating Disruption

Mating disruption (MD) is a pest management technique designed to control certain insect infestations. Specifically, mating disruption involves the use of synthesized sex pheromones to disrupt the reproductive cycle of insects.

In many insect species of interest to agriculture, such as those in the order Lepidoptera, females emit an airborne trail of a specific chemical blend constituting that species' sex pheromone. This aerial trail is referred to as a pheromone plume.

Males of that species use the information contained in the pheromone plume to locate the emitting female (known as a "calling" female). Mating disruption exploits the male insects' natural response to follow the plume by introducing a synthetic pheromone into the insects' habitat. The synthetic pheromone is a volatile organic chemical designed to mimic the species-specific sex pheromone produced by the female insect. The general effect of mating disruption is to confuse the male insects by masking the natural pheromone plumes, causing the males to follow "false pheromone trails" at the expense of finding mates, and affecting the males' ability to respond to "calling" females.

Consequently, the male population experiences a reduced probability of successfully locating and mating with females, which leads to the eventual cessation of breeding and collapse of the insect infestation. The California Department of Pesticide Regulation, the California Department of Food and Agriculture, and the United States Environmental Protection Agency consider mating disruption to be among the most environmentally friendly treatments used to eradicate pest infestations. Mating disruption works best if large areas are treated with pheromones. Ten acres is a good minimum size for a successful MD programme, but larger areas are preferable

Advantages of Mating Disruption

Pheromone programmes are most effective when controlling low to moderate pest population densities. MD has also been identified as a pest control method in which the insect does not become resistant. The scientific community, together with governmental agencies throughout the world, understands the benefits of mating disruption using species-specific sex pheromones, and consider sex-pheromone-based insect control programmes among the most environmentally friendly treatments to be used to manage and control insect pest populations. Insect pheromone has been successfully used as an effective tool to slow the spread and to eradicate pests from very large areas in the US; for example to control the Gypsy moth (Lymantria dispar), a devastating forestry pest, and to eradicate the boll weevil and pink bollworm, two of the most damaging pest of cotton. Conventional pesticide based control methods, kill insects directly, whereas mating disruption confuses male insects from accurately locating a mating partner, leading to the eventual collapse of the mating cycle. Mating disruption, due to the specificity of the sex pheromone of the insect species, has the benefit of only affecting the males of that species, while leaving other non target species unaffected. This allows for very targeted pest management, promoting the suppression of a single pest species, leaving the populations of beneficial insects (pollinators and natural enemies) intact. Mating disruption, like most pest management strategies, is a useful technique, but should not be considered a stand-alone treatment programme for it targets only a single species in plant production systems that usually have several pests of concern. Mating disruption is a valuable tool that should be used in Integrated Pest Management(IPM) programmes.

Pheromone programmes have been used for several decades around the globe and to date (2009) there is no documented public health evidence to suggest that agricultural use of synthetic pheromones is harmful to humans or to any other non-target species. However, continuing research is being conducted.

Disadvantages of Mating Disruption

Over the decades that pheromone pest programmes have been used several disadvantages have been argued when compared to the use of conventional pesticides. Most pheromones target a single species, so a specific mating disruption formulation controls only the species that uses that pheromone blend; whereas pesticides usually kill indiscriminately a plethora of species, including multiple species with a single application. Some synthetic pheromones have high developmental and production costs, causing the mating disruption technique to be too costly to be adopted by conventional commercial growers. Furthermore most commercial pheromone mating disruption formulations must be applied by hand, which can be an expensive and time consuming. Novel pheromone formulations recently developed to be mechanically applied provide long lasting mating disruption effects (e.g., depending on the target pest a single application of SPLAT controls the target pest for a complete reproductive cycle, or for the entire season

Methods of Dispersal

Microencapsulated Pheromones

Microencapsulated pheromones (MECs) are small droplets of pheromone enclosed within polymer capsules. The capsules control the release rate of the pheromone into the surrounding environment. The capsules are small enough to be applied in the same method as used to spray insecticides. The effective field longevity of the microencapsulated pheromone formulations ranges from a few days to slightly more than a week, depending on climatic conditions, capsule size and chemical properties. Microcapsules in the pheromone formulations are usually kept above a prescribed diametre to avoid the risk of inhalation by humans.

Hand Applied Dispensers

- Hollow tube dispensers are plastic twist-tie type dispensers, plastic hollow fibres or plastic hollow microfibres fibres, filled with synthetic sex pheromone and placed throughout the area to be protected.
- Pheromone Baits and Stations are a stationary, attract and kill type of dispensers. Some are relatively large platform, containing a pheromone lure inside a glue board that ensnares the attracted insect. Other pheromone bait stations contain a pheromone lure in conjunction with a surface containing a dose of insecticide that reduces the attracted insect's fitness, thus reducing its ability to mate and reproduce.
- High-emission dispensers There are several very high dose pheromone dispensers, some do it passively, like pheromone sachets and large dollops of SPLAT pheromone formulations,

others do it be actively releasing bursts of sex pheromone at timed intervals.

Monolithic Flowable Dispensers

A new, effective and economical concept in pheromone delivery using a flowable formulation to create long lasting monolithic pheromone dispensers has been brought to the market in the past decade. These novel SPLAT pheromone mating disruption formulations can provide effective season long suppression effect (e.g., depending on the target pest a single application of SPLAT controls the target pest for a complete reproductive cycle, or for the entire season) and can be manually or mechanically applied. Although mechanical dispersal techniques require specialized off-the-shelf application technology and/or equipment, once the application system is made to work it allows protection of extensive areas using pheromones, one of the most benign and effective pest management techniques available today. A benefit of SPLAT is that the dollop anchors where it lands, avoiding unwanted drift of the formulation once applied in the field, and, depending on the mode of application, the cured dollops are retrievable.

Aerial Dispersal

In November 2007, a controversial aerial approach was used to spray microencapsulated LBAM pheromone in urban and rural areas of the counties of Santa Cruz and Monterey California to combat the invasive light brown apple moth. Usually the effect of disruption of orientation of the male moths to females (or monitoring pheromone traps) can be detected by the reduction in moth capture in monitoring pheromone traps.

The government campaign using areawide aerial microencapsulated pheromone applications failed to show any sign of mating disruption on the light brown apple moth populations in the treated area. It was found that the first aerial campaign was performed using an incomplete (the wrong) pheromone blend of the light brown apple moth (the wrong blend decreased tremendously the likelihood of success of the mating disruption programme), and the LBAM microencapsulated formulation was untested, and finally, microencapsule formulations are notoriously known for their short field life, weak and erratic performance. Furthermore it is possible that the LBAM microencapsulated formulation used in the government campaign was unfit for aerial delivery in urban areas; although pheromone is safe, the formulation used had microcapsules of very small diametre which made it into a possible inhalation hazard that seems to be linked to an increase in allergenic reactions of the population in the target area. This set of LBAM mating disruption aerial applications done by the government has created tremendous dissent of the public in general as well as of several sectors of the scientific community. Now, several years later, the affected communities as well as the nascent US pheromone industry (which provides safer, yet very effective, alternatives to the use of conventional pesticides) are still suffering the ripple effects of these disastrous Bay Area LBAM eradication campaigns.

But there are numerous, successful pest suppression programmes that rely on aerial dispersal of pheromone mating disruptants. One of the largest pheromone mating disruption programmes in the globe is the Gypsy Moth Slow the Spread. Gypsy Moth Slow the Spread has been implemented across the 1,200-mile (1,900 km) gypsy moth frontier from Wisconsin to North Carolina.

The programme area is located ahead of the advancing front of the gypsy moth population. The STS programme focuses on early detection and suppression of the low–level populations along this advancing front, disrupting the natural progress of population buildup and spread. Every year hundreds of thousands of acres are aerially sprayed with two pheromone Gypsy moth pheromone mating disruption formulations, Flakes and SPLAT. A single mating disruption formulation application promotes season-long suppression of gypsy moth in the treated areas. With a crew of 8 people it was possible to aerially treat with SPLAT GM over 20,000 acres (81 km^2) of forest in a single day. The consortium of Federal and State participants have been able to do the following:

- decrease the new territory invaded by the gypsy moth each year from 15,600 square miles (40,000 km^2) to 6,000 square miles (16,000 km^2);
- protect forests, forest–based industries, urban and rural parks, and private property; and
- avoid at least $22 million per year in damage and management costs.

It seems that the tremendous success of the Gypsy Moth Slow the Spread programme is related to extremely well planned campaigns, which involves communication, transparency and clarity of objectives: in advance to an application STS holds meetings that include the area population in general, concerned citizens, public officials, scientists and technical personnel to discuss strategies of management of Gypsy moth in the areas of concern. There is a movement requesting that new government invasive species eradication campaigns model their pest suppression actions on the existing successful suppression programmes like GM STS, and embrace a more effective policy of communication, transparency and clarity of objectives.

With the involvement and education of the public, areawide eradication campaigns will be better planned and more able to deliver decisive end effective pest eradication actions.

Nematophagous Fungus

Nematophagous fungi are carnivorous fungi specialized in trapping and digesting nematodes. Around 160 species are known. There exist both species that live inside the nematodes from the beginning and others that catch them mostly with glue traps or in rings, some of which constrict on contact.

Some species possess both types of traps. Another technique is to stun the nematodes using toxins, which is a method employed by *Coprinus comatus*, *Stropharia rugosoannulata*, and the family Pleurotaceae.

Nematophagous fungus can be useful in controlling nematodes. *Paecilomyces* for example can be used as a bio-nematicide.

Nematicide

A nematicide is a type of chemical pesticide used to kill plant-parasitic nematodes. Nematicides have tended to be broad-spectrum toxicants possessing high volatility or other properties promoting migration through the soil. Temik (Aldicarb), a carbamate insecticide marketed by Bayer CropScience, is an example of a commonly used commercial nematicide. It is important in potato production, where it has been used for control of soil-borne nematodes.

Aldicarb is a cholinesterase inhibitor, which prevents the breakdown of acetylcholine in the synapse. In case of severe poisoning, the victim dies of respiratory failure. It is no longer authorised for use in the EU and, in August, 2010, Bayer CropScience announced that it plans to discontinue Aldicarb by 2014. Human health safety and environmental concerns have resulted in the widespread deregistration of several other agronomically important nematicides.

Prior to 1985, the persistent halocarbon DBCP was a widely used nematicide and soil fumigant.

However, it was banned from use after being linked to sterility among male workers; the Dow Chemical company was subsequently found liable for more than $600 million in damages.

Natural Nematicides

Several natural nematicides are known. An environmentally benign garlic-derived polysulfide product is approved for use in the European Union (under Annex 1 of 91/414) and the UK as a nematicide. Another common natural nematicide is obtained from neem cake, the residue obtained after cold-pressing the fruit and kernels of the neem tree. Known by several names in the world, the tree was first cultivated in India in ancient times and is now widely distributed throughout the world. The root exudate of marigold (*Tagetes*) is also found to have nematicidal action. Nematophagous fungi, a type of carnivorous fungi, can be useful in controlling nematodes, *Paecilomyces* being one example.

Besides chemicals, soil steaming can be used in order to kill nematodes. Superheated steam is induced into the soil, which causes almost all organic material to deteriorate.

3

Organic Horticulture

Organic horticulture is the science and art of growing fruits, vegetables, flowers, or ornamental plants by following the essential principles of organic agriculture in soil building and conservation, pest management, and heirloom variety preservation.

The Latin words *hortus* (garden plant) and *cultura* (culture) together form *horticulture*, classically defined as the culture or growing of garden plants. *Horticulture* is also sometimes defined simply as "agriculture minus the plough." Instead of the plough, horticulture makes use of human labour and gardener's hand tools, although some small machine tools like rotary tillers are commonly employed now.

Mulches, cover crops, compost, manures, vermicompost, and mineral supplements are soil-building mainstays that distinguish this type of farming from its commercial counterpart. Through attention to good healthy soil condition, it is expected that insect, fungal, or other problems that sometimes plague plants can be minimized. However, pheromone traps, insecticidal soap sprays, and other pest-control methods available to organic farmers are also utilized by organic horticulturists.

Horticulture involves five areas of study. These areas are floriculture (includes production and marketing of floral crops), landscape horticulture (includes production, marketing and maintenance of landscape plants), olericulture (includes production and marketing of vegetables), pomology (includes production and marketing of fruits), and postharvest physiology (involves maintaining quality and preventing spoilage of horticultural crops). All of these can be, and sometimes are, pursued according to the principles of organic cultivation.

Organic horticulture (or organic gardening) is based on knowledge and techniques gathered over thousands of years. In general terms, organic horticulture involves natural processes, often taking place over extended periods of time, and a sustainable, holistic approach - while chemical-based horticulture focuses on immediate, isolated effects and reductionist strategies.

Organic Gardening Systems

There are a number of formal organic gardening and farming systems that prescribe specific techniques. They tend to be more specific than, and fit within, general organic standards. Forest gardening, a fully organic food production system which dates from prehistoric times, is thought to be the world's oldest and most resilient agroecosystem.

Biodynamic farming is an approach based on the esoteric teachings of Rudolf Steiner. The Japanese farmer and writer Masanobu Fukuoka invented a no-till system for small-scale grain production that he called Natural Farming. French intensive and biointensive methods and SPIN Farming (Small Plot INtensive) are all small scale gardening techniques. These techniques were brought to the United States by Alan Chadwick in the 1930s. This method has since been promoted by John Jeavons, Director of Ecology Action. A garden is more than just a means of providing food, it is a model of what is possible in a community - everyone could have a garden of some kind (container, growing box, raised bed) and produce healthy, nutritious organic food, a farmers market, a place to pass on gardening experience, and a sharing of bounty, promoting a more sustainable way of living that would encourage their local economy. A simple 4' x 8' (32 square feet) raised bed garden based on the principles of bio-intensive planting and square foot gardening uses fewer nutrients and less water, and could keep a family, or community, supplied with an abundance of healthy, nutritious organic greens, while promoting a more sustainable way of living.

Organic gardening is designed to work with the ecological systems and minimally disturb the Earth's natural balance. Because of this organic farmers have been interested in reduced-tillage methods. Conventional agriculture uses mechanical tillage, which is plowing or sowing, which is harmful to the environment. The impact of tilling in organic farming is much less of an issue. Ploughing speeds up erosion because the soil remains uncovered for a long period of time and if it has a low content of organic matter the structural stability of the soil decreases. Organic farmers use techniques such as mulching, planting cover crops, and intercropping, to maintain a soil cover throughout most of the year. The use of compost, manure mulch and other organic fertilizers yields a higher organic content of soils on organic farms and helps limit soil degradation and erosion.

Other methods can also be used to supplement an existing garden. Methods such as composting, or vermicomposting. These practices are ways of recycling organic matter into some of the best organic fertilizers and soil conditioner. Vermicompost is especially easy. The byproduct is also an excellent source of nutrients for an organic garden.

Pest Control Approaches

Differing approaches to pest control are equally notable. In chemical horticulture, a specific insecticide may be applied to quickly kill off a particular insect pest. Chemical controls can dramatically reduce pest populations in the short term, yet by unavoidably killing (or starving) natural control insects and animals, cause an increase in the pest population in the long term, thereby creating an ever increasing problem. Repeated use of insecticides and herbicides also encourages rapid natural selection of resistant insects, plants and other organisms, necessitating increased use, or requiring new, more powerful controls.

In contrast, organic horticulture tends to tolerate some pest populations while taking the long view. Organic pest control requires a thorough understanding of pest life cycles and interactions, and involves the cumulative effect of many techniques, including:

- Allowing for an acceptable level of pest damage
- Encouraging predatory beneficial insects to flourish and eat pests
- Encouraging beneficial microorganisms
- Careful plant selection, choosing disease-resistant varieties
- Planting companion crops that discourage or divert pests
- Using row covers to protect crop plants during pest migration periods
- Rotating crops to different locations from year to year to interrupt pest reproduction cycles
- Using insect traps to monitor and control insect populations

Each of these techniques also provides other benefits, such as soil protection and improvement, fertilization, pollination, water conservation and season extension. These benefits are both complementary and cumulative in overall effect on site health. Organic pest control and biological pest control can be used as part of integrated pest management (IPM). However, IPM can include the use of chemical pesticides that are not part of organic or biological techniques.

Ecoagriculture

Eco friendly agriculture describes landscapes that support both agricultural production and biodiversity conservation, working in harmony together to improve the livelihoods of rural communities.

While many rural communities have independently practised ecoagriculture for thousands of years, over the past century many of these landscapes have given way to segregated land use patterns, with some areas employing intensive farming practices without regard to biodiversity impacts, and other areas fenced off completely for habitat or watershed protection. A new ecoagriculture movement is now gaining momentum to unite land managers and other stakeholders from diverse environments to find compatible ways to conserve biodiversity while also enhancing agricultural production.

Approach and Practitioners

Ecoagriculture is both a conservation strategy and a rural development strategy. Ecoagriculture recognizes agricultural producers and communities as key stewards of ecosystems and biodiversity and enables them to play those roles effectively.

Ecoagriculture applies an integrated ecosystem approach to agricultural landscapes to address all three pillars—conserving biodiversity, enhacing agricultural production, and improving livelihoods—drawing on diverse elements of production and conservation management systems. Meeting the goals of ecoagriculture usually requires collaboration or coordination between diverse stakeholders who are collectively responsible for managing key components of a landscape.

Landscape Scale

Ecoagriculture uses the landscape as a unit of management. A landscape is a cluster of local ecosystems with a particular configuration of topography, vegetation, land use, and settlement. The goals of ecoagriculture -- to maintain biodiversity and ecosystem services, manage agricultural production sustainably, and contribute to improved livelihoods among rural people -- cannot be achieved at just a farm or plot level, but are linked at the landscape scale. Therefore, to make impact, all of the elements of a landscape as a whole must be considered.

Defining a landscape depends on the local context. Landscapes may be defined or delimited by natural, historical, and/or cultural processes, activities or values. Landscapes can incorporate many different features, but all of the various features have some influence or effect on each other.

Landscapes can vary greatly in size, from the Congo Basin in west-central Africa where landscapes are often huge because there are vast stretches of apparently undifferentiated land, to western Europe where landscapes tend to be much smaller because of the wide diversity of topographies and land use activities occurring close to each other.

Importance of Agriculture Areas for Biodiversity Conservation

Agriculture is the most dominant human influence on earth. Nearly one-third of the world's land area is heavily influenced by cropland or planted pastures. An even greater area is being fallowed as part of an agricultural cycle or is in tree crops, livestock grazing systems, or production forestry.

In addition, most of the world's 100,000+ protected areas contain significant amounts of agricultural land. And over half of the most species-rich areas in the world contain large human populations whose livelihoods depend on farming, forestry, herding, or fisheries.

Agriculture as it is often practised today threatens wild plant and animal species and the natural ecosystem services upon which both humans and wildlife depend. Over 70% of the fresh water withdrawn by humans goes to irrigation for crops, causing a profound impact on the hydrological cycles of ecological systems. Moreover, fertilizers, pesticides, and agricultural waste threaten habitats and protected areas downstream. Landclearing for agriculture also disrupts sources of food and shelter for wild biodiversity, and unsustainable fishing practices deplete freshwater and coastal fisheries.

Additionally, an increase in the planting and marketing of monoculture crops across the globe has decreased diversity in agricultural products, to the extent that many local varieties of fruits, vegetables, and grains have now become extinct.

Given that demands on global agricultural production are increasing, it is imperative that the management of agricultural landscapes be improved to both increase productivity and enhance biodiversity conservation. Wild biodiversity increasingly depends on agricultural producers to find ways to better protect habitats, and agriculture critically needs healthy and diverse ecosystems to sustain productivity.

Bridging Conservation and Agriculture

Traditionally there has existed a divide between conservationists, who want to set land aside for the protection of wild biodiversity, and agriculturalists, who want to use land for production. Because more than half of all plant and animal species exist principally outside protected areas -- mostly in agricultural landscapes -- there is a great need to close the gap between conservation efforts and agricultural production. For example, conservation of wetlands within agricultural landscapes is critical for wild bird populations. Such species require initiatives by and with farmers. Ecoagriculture provides a bridge for these two communities to come together.

Farmers as Stewards in Ecoagriculture

Farming communities play a vital role as stewards of their ecosystems and biodiversity. Their dependence on their land and natural resources necessitates a conservation ethic. Agricultural productivity critically depends upon a range of ecosystem services. Wild species often also play an important role in providing livestock fodder, fuel, veterinary medicines, soil nutrient supplements and construction materials to farmers, as well constituting an essential element of cultural, religious, and spiritual practices. The dominance of agriculture in global land use requires that ecoagriculture approaches be fostered by rural producers and their communities on a globally significant scale. To do this, farmers need to be able to conserve biodiversity more consistently in ways that benefit their livelihoods. Experiences from around the world suggest that there are a number of incentives to encourage and enable farmers and their communities to preserve or transition towards ecoagriculture landscapes:

- Many management practices that improve ecosystem health also benefit farmers by reducing production costs, raising or stabilizing yields, or improving product quality. Intensive rotation grazing systems practised in Europe, the United States, and Zimbabwe have been shown to reduce dairy production costs compared to stall-fed systems, while also reducing risks of land degradation and improving wildlife habitat.
- Farming communities are especially motivated to conserve biodiversity and ecosystem services critical to their own livelihoods and cultural, spiritual, or aesthetic values. To protect their access to local water sources and medicinal plants, for example, farmers in western Kenya have mobilized to protect threatened forests in and near their communities. And in some

agricultural landscapes in West Africa, 'sacred groves' are the principal remaining areas of native forest.

- Farmers are seeking new income opportunities from product markets that value supplies from biodiversity-friendly production systems. More than 80 eco-certification programmes now provide opportunities for farmers to receive higher prices for products produced with environmentally friendly practices.
- Farmers can gain new income opportunities from payment for ecosystem services provided by non-farm beneficiaries of their ecological stewardship. These opportunities include carbon emission offset payments for carbon sequestration in soils and trees and water quality protection, among others.
- Farmers are seeking ways to comply with the goals of environmental regulation, in ways that also maintain or improve their agricultural livelihoods. In the US, farmers in the Chesapeake Bay watershed are incorporating perennial vegetative buffer strips around stream banks, which provide habitat niches for birds and wildlife, to both help meet water quality regulations and to diversify their output.

Ecoagriculture Land Management Practices

Agricultural landscapes that aim to achieve the objectives of ecoagriculture -- enhanced biodiversity conservation, increased food production, and improved rural livelihoods -- should be managed in ways that protect and expand natural areas and improve wildlife habitats and ecosystem functions, in collaboration with local communities to insure their benefit. Specific land management practices that may be incorporated include:

- Plan and manage protected areas together with local farming, pastoralist, and forest communities in their landscapes. Community-conserved areas on lands owned by farmers and pastoralists are important for ecosystem-wide management of biodiversity. The more ownership/engagement these communities have in the management of protected areas, the more successful the landscape will be overall in contributing to the three goals of ecoagriculture.
- Link unfarmed areas, forest fragments, and wetlands within agricultural landscapes to develop habitat networks and corridors that support and expand the range of wild species. This approach is particularly useful to migratory species, which can include pollinators and natural enemies of agricultural pests.

- Reduce or reverse conversion of natural areas to agricultural areas by improving the productivity of currently utilized agricultural, forestry, grazing lands and fisheries.
- Modify farming systems so they mimic natural vegetation and ecological processes. Integrating trees, shrubs, and grasses into agricultural production systems can improve ecosystem services across the whole landscape.
- Manage agricultural wastes to protect the surrounding ecosystem by encouraging shifts from input-intensive to 'knowledge-intensive' agricultural practices. These may include integration of crop, livestock, and nutrient systems; more precise application of organic and non-organic fertilizers; and crop rotations to improve soil fertility.
- Encourage soil, water, and vegetation management strategies that limit negative impacts on surrounding ecosystems. These practices include conservation tillage, improved fallow systems, on-farm crop or fertilizer trees, inter-cropping, and livestock diversification.

Role of Traditional and Local Knowledge

Many indigenous peoples and rural communities have developed, maintained, and adapted different types of ecoagriculture systems for centuries. Local farmers, pastoralists, fishers, forest users, and other community members are the foundation of rural land stewardship. Their knowledge, traditions, land use practices, and resource-management institutions are essential to the development of viable ecoagriculture systems for their landscapes.

The mainstreaming of ecoagriculture approaches will be crucially dependent upon mobilizing local communities to become leaders in ecoagriculture, as teachers and as advocates for political and institutional change. Communities facing similar challenges can share questions, ideas, and solutions with each other. Local communities also need effective processes for sharing their expertise with national policymakers and the international community and thus play a more central role in setting ecoagriculture objectives in policy and programme development.

Contribution of Ecoagriculture to the Millennium Development Goals

The Millennium Development Goals (MDGs), eight ambitious targets which range from halving extreme poverty to halting the spread of HIV/AIDS and providing universal primary education, were put forth by the United Nations in 2000, to be achieved by 2015. Ecoagriculture strategies will be essential to achieving the MDGs, particularly for hunger and poverty, water and sanitation, and environmental sustainability.

The MDGs will not be reached without securing the ability of the rural poor to feed their families and gain income, while at the same time protecting the biodiversity and ecosystem services that sustain their livelihoods. Of the estimated 800 million people who do not have access to sufficient food, half are smallholder farmers, one-fifth are rural landless, and one-tenth are principally dependent on rangelands, forests and fisheries.

For most of them, reducing poverty and hunger will depend centrally on their ability to sustain and increase crop, livestock, forest, and fishery production.

A key opportunity for enhancing progress towards the MDGs is investment in locally-driven land management approaches -- such as ecoagriculture strategies -- that build upon synergies between rural livelihoods, environmental sustainability, and food security.

Related Fields

The values and/or principles of ecoagriculture have much in common with existing concepts, such as sustainable agriculture, permaculture, agroecology, integrated natural resource management, organic agriculture, agroforestry, conservation agriculture, protected area management, and others. In fact, 'ecoagriculture' landscapes often feature many of these approaches. Ecoagriculture draws heavily on these and many other innovations in rural land use planning and management. The landscape management framework defined by ecoagriculture has four particularly important characteristics:

- Large scale: Ecoagriculture moves beyond the management of individual farms and/or protected areas to help detect and plan for interactions among different land uses at the landscape scale. In addition, important attributes such as wildlife population dynamics and watershed functions can be meaningfully understood only at the landscape scale. Also, in recognition of the fact that short-term tradeoffs may lead to long-term synergies, ecoagriculture advocates conducting analyses over longer temporal scales than is commonly done.
- Emphasis on synergies: Ecoagriculture emphasizes both the need and the opportunity to foster synergies among conservation, agricultural production, and rural livelihoods. The ecoagriculture research and monitoring agenda seeks, in part, to identify and document these synergies.

- Emphasis on stakeholder collaboration: Ecoagriculture can not be achieved by individual land managers. The management of ecoagriculture landscapes requires processes that support a variety of land managers (within the landscape) with diverse environmental and socio-economic goals to collaboratively develop coordinated conservation and production management approaches that collectively achieve conservation, production, and livelihood goals at a landscape scale.
- Importance of both conservation and agricultural production: Building on the Millennium Ecosystem Assessment, ecoagriculture brings conservation fully into the agricultural and rural development discourse by highlighting the importance of ecosystem services in supporting continued agricultural production.

Ecoagriculture also identifies the conservation of native biodiversity and ecosystems as an equally important goal in its own right. It also supports conservationists to more effectively conserve nature within and outside protected areas by working with the agricultural community and developing conservation-friendly livelihood strategies for rural land users.

Pest Control

Pest control refers to the regulation or management of a species defined as a pest, usually because it is perceived to be detrimental to a person's health, the ecology or the economy.

Pest control is at least as old as agriculture, as there has always been a need to keep crops free from pests. In order to maximize food production, it is advantageous to protect crops from competing species of plants, as well as from herbivores competing with humans.

The conventional approach was probably the first to be employed, since it is comparatively easy to destroy weeds by burning them or plowing them under, and to kill larger competing herbivores, such as crows and other birds eating seeds. Techniques such as crop rotation, companion planting (also known as intercropping or mixed cropping), and the selective breeding of pest-resistant cultivars have a long history.

In the UK, following concern about animal welfare, humane pest control and deterrence is gaining ground through the use of animal psychology rather than destruction. For instance, with the urban Red Fox which territorial behaviour is used against the animal, usually in conjunction with non-injurious chemical repellents. In rural areas of Britain, the use of firearms for pest control is quite common. Airguns are particularly popular for control of small pests such as rats, rabbits and grey squirrels, because of their lower power they can be used in more restrictive spaces such as gardens, where using a firearm would be unsafe.

Chemical pesticides date back 4,500 years, when the Sumerians used sulfur compounds as insecticides. The Rig Veda, which is about 4,000 years old, also mentions the use of poisonous plants for pest control. It was only with the industrialization and mechanization of agriculture in the 18th and 19th century, and the introduction of the insecticides pyrethrum and derris that chemical pest control became widespread. In the 20th century, the discovery of several synthetic insecticides, such as DDT, and herbicides boosted this development. Chemical pest control is still the predominant type of pest control today, although its long-term effects led to a renewed interest in traditional and biological pest control towards the end of the 20th century.

Causes

Many pests have only become a problem because of the direct actions of humans. Modifying these actions can often substantially reduce the pest problem. In the United States, raccoons caused a nuisance by tearing open refuse sacks. Many householders introduced bins with locking lids, which deterred the raccoons from visiting. House flies tend to accumulate wherever there is human activity and is virtually a global phenomenon, especially where food or food waste is exposed. Similarly, seagulls have become pests at many seaside resorts. Tourists would often feed the birds with scraps of fish and chips, and before long, the birds would become dependent on this food source and act aggressively towards humans.

Living organisms evolve and increase their resistance to biological, chemical, physical or any other form of control. Unless the target population is completely exterminated or is rendered incapable of reproduction, the surviving population will inevitably acquire a tolerance of whatever pressures are brought to bear - this results in an evolutionary arms race.

Types of Pest Control

Biological pest control: Biological pest control is the control of one through the control and management of natural predators and parasites. For example: mosquitoes are often controlled by putting *Bt Bacillus thuringiensis* ssp. *israelensis*, a bacterium that infects and kills mosquito larvae, in local water sources. The treatment has no known negative consequences on the remaining ecology and is safe for humans to drink. The point of biological pest control, or any natural pest control, is to eliminate a pest with minimal harm to the ecological balance of the environment in its present form.

Mechanical Pest Control

Mechanical pest control is the use of hands-on techniques as well as simple equipment, devices, and natural ingredients that provide a protective barrier between plants and insects. For example: weeds can be controlled by being physically removed from the ground. This is referred to as tillage and is one of the oldest methods of weed control.

Physical Pest Control

Physical pest control is a method of getting rid of insects and small rodents by removing, attacking, or setting up barriers that will prevent further destruction of one's plants.

Elimination of Breeding Grounds

Proper waste management and drainage of still water, eliminates the breeding ground of many pests.

Garbage provides food and shelter for many unwanted organisms, as well as an area where still water might collect and be used as a breeding ground by mosquitoes. Communities that have proper garbage collection and disposal, have far less of a problem with rats, cockroaches, mosquitoes, flies and other pests than those that don't.

Open air sewers are ample breeding ground for various pests as well. By building and maintaining a proper sewer system, this problem is eliminated.

Certain spectrums of LED light can "disrupt insects' breeding.

Poisoned Bait

Poisoned bait is a common method for controlling rat populations, however is not as effective when there are other food sources around, such as garbage. Poisoned meats have been used for centuries for killing off wolves, birds that were seen to threaten crops, and against other creatures.

This can be a problem, since a carcass which has been poisoned will kill not only the targeted animal, but also every other animal which feeds on the carcass. Humans have also been killed by coming in contact with poisoned meat, or by eating an animal which had fed on a poisoned carcass.

This tool is also used to manage several caterpillars e.g. Spodoptera litura, fruit flies, snails and slugs, crabs etc.

Field Burning

Traditionally, after a sugar cane harvest, the fields are all burned, to kill off any insects or eggs that might be in the fields. Examples of these pests are Rodents.

Hunting

Historically, in some European countries, when stray dogs and cats became too numerous, local populations gathered together to round up all animals that did not appear to have an owner and kill them. In some nations, teams of rat catchers work at chasing rats from the field, and killing them with dogs and simple hand tools. Some communities have in the past employed a bounty system, where a town clerk will pay a set fee for every rat head brought in as proof of a rat killing.

Traps

With the many traps available on the market today you can easily remove mice and rats from homes. You must first know what rodent needs to be removed, you can then decide what type of trap is the best suited to your needs. The snap trap is the most widely used, it utilizes a trigger (sometimes shaped like cheese) to hold bait, and kills the rodent by striking it behind the head with a wire rod or jaw. In some instances you may wish to use glue traps also called glue boards. This type of trap requires the mouse or rat to attempt to cross the trap so the glue can hold the rodent. After a catch is made you can euthanize the rodent and dispose of it trap and all, or some glue boards will release the catch when you pour vegetable oil on them, as the oil reacts with the glue to lose its grip. The last type of trap are live catch traps, this type of trap is typically a repeating style so more than one animal can be caught at a time, they can also be released from this trap in a new location if desired.

Pesticides

Spraying pesticides by planes, handheld units, or trucks that carry the spraying equipment, is a common method of pest control. Crop dusters commonly fly over farmland and spray pesticides to kill off pests that would threaten the crops. However, some pesticides may cause cancer and other health problems, as well as harming wildlife.

Space Fumigation

A project that involves a structure be covered or sealed airtight followed by the introduction of a penetrating, deadly gas at a killing concentration a long period of time (24-72hrs.). Although expensive, space fumigation targets all life stages of pests.

Space Treatment

A long term project involving fogging or misting type applicators. Liquid insecticide is dispersed in the atmosphere within a structure. Treatments do not require the evacuation or airtight sealing of a building, allowing most work within the building to continue but at the cost of the penetrating effects. Contact insecticides are generally used, minimizing the long lasting residual effects.

Laboratory studies conducted with U-5897 (3-chloro-1,2-propanediol) where attempted in the early 1970s although these proved unsuccessful. Research into sterilization bait is ongoing.

Another effective method of soil sterilization is soil steaming. Pest is killed through hot steam which is induced into the soil.

Destruction of Infected Plants

Forest services sometimes destroy all the trees in an area where some are infected with insects, if seen as necessary to prevent the insect species from spreading. Farms infested with certain insects, have been burned entirely, to prevent the pest from spreading elsewhere.

Natural Rodent Control

Several wildlife rehabilitation organizations encourage natural form of rodent control through exclusion and predator support and preventing secondary poisoning altogether.

The United States Environmental Protection Agency agrees, noting in its Proposed Risk Mitigation Decision for Nine Rodenticides that "without habitat modification to make areas less attractive to commensal rodents, even eradication will not prevent new populations from recolonizing the habitat."

Repellents

- Balsam fir oil from the tree *Abies balsamea* is an EPA approved non-toxic rodent repellent.
- *Acacia polyacantha* subsp. *campylacantha* root emits chemical compounds that repel animals including crocodiles, snakes and rats.

Electronic Pest Control

Electronic pest control is the name given to the use of any of the several types of electrically powered devices designed to repel or eliminate pests, usually rodents or insects. Since these devices are not regulated under the Federal Insecticide, Fungicide and Rodenticide Act (FIFRA) in the United States, the US EPA does not require the same kind of efficacy testing that it does for chemical pesticides.

Types of Devices

There are two types of electronic pest control devices widely available, these are Ultrasonic and Electromagnetic.

Ultrasonic devices operate through emitting short wavelength, high frequency sound waves that are too high in pitch to be heard by the human ear (all frequencies greater than 20,000 Hz). Humans are unable to hear sounds higher than 20 kHz due to physiological limitations of the cochlea. Some animals such as dogs, bats, and rodents can hear well into the ultrasonic range. Some insects, such as grasshoppers and locusts can detect frequencies from 50,000 Hz to 100,000 Hz, and moths and lacewings can detect ultrasound as high as 240,000 Hz produced by insect-hunting bats. Contrary to popular belief, birds cannot hear ultrasonic sound.

Insects detect sound by special hairs or sensilla located on the antennae (mosquitoes) or genitalia (cockroaches), or by more complicated tympanal organs (grasshoppers, locusts, moths and butterflies). Electromagnetic pest repelling devices claim to affect the nervous system of spiders, ants, mice, and other rodents. There have been similar studies on effects of EM radiation emitted by cellphones on humans.

Radio Wave Pest Control

The concept of radio wave (RW) or radio frequency (RF) to control the behaviour of living organisms has shown promise. According to Drs. Juming Tang and Shaojin Wang at Washington State University (WSU) with colleagues at the University of California-Davis and USDA's Agricultural Research Service in Parlier, California, RF energy generates heat through agitation of bound water molecules, This process generates heat through ionic conduction and agitation of free water molecules in insects. As a result, more thermal energy is converted in insects.

RF treatments control insect pests without negatively affecting food stuffs and storage locations. RF treatments may serve as a non-chemical alternative to chemical fumigants for post-harvest pest control in commodities (such as almonds, pecans, pistachios, lentils, peas, and soybeans), reducing the long-term impact on the environment, human health and competitiveness of agricultural industries.

Effects on Pests

"Ultrasound and Arthropod Pest Control" an extensive Kansas State University study confirmed that ultrasonic sound devices do have both a repellent effect as well as a reduction in mating and reproduction of insects. However, the results were mixed and ultrasonic sound had little or no effect on some pests. Ultrasonic devices were highly effective on crickets while the same devices had little repellent effect on cockroaches. Additionally the results were mixed with some devices being effective while others having no effect depending on the test subject. The study also concluded there was no effect on ants or spiders in any of the tests. They concluded, based on the mixed results, that more research is needed to improve these devices.

A 2002 study by Genesis Laboratories Inc. does lend some credence to the ability of electronic repellent devices to repel certain pests in controlled environments. "Preliminary study of white-footed mice behaviour in the test apparatus demonstrated a significant preference for the non-activated chamber among both sexes." This study was sponsored by Global Instruments, the maker of the "Pest-a-cator" series of electronic repellent devices.

In 2003, the Federal Trade Commission required Global Instruments, the maker of the Pest-A-Cator/Riddex series of electromagnetic pest control devices, to discontinue any claims for their efficacy until they are backed by credible scientific evidence. This ban continues to be in effect.

Victor Pest obtained positive results from independent researchers which resulted in two ultrasonic devices being granted registration by the Canadian EPA(PMRA). The results from the tests were the device "successfully repelled the rodents from the protected area in 13 of the 17 sites. This represents a 81.3% success rate...the average number of days before rodent activity was stopped was six days."

Cockroaches respond to electronic pest control devices by moving about a bit more than usual, but don't appear eager to escape from the sound waves. This includes devices that emit uniform frequency as well as changing frequencies of ultrasound. Rodents adjust to the ultrasound (or any new sound) and eventually ignore it. However, researchers were able to use the increased cockroach activity to good effect by increasing the rate they caught the roaches in sticky traps. At best, ultrasonic waves have only a partial or temporary effect on rodents. Numerous studies have rejected ultrasonic sound as a practical means of rodent control.

Tests of commercial ultrasonic devices have indicated that rodents may be repelled from the immediate area of the ultrasound device for a few minutes to a few days, but they will nearly always return and resume normal activities. Other tests have shown that the degree of repellence depends on the frequency, intensity, and the pre-existing condition of the rodent infestation. The intensity of such sounds must be so great that damage to humans or domestic animals would also be likely. Commercial ultrasonic pest control devices do not produce sounds of such intensity.

Effects on Mosquitoes

- Bart Knols, an entomologist who chairs the advisory board of the Dutch Malaria Foundation and edits the website Malaria World, claims that there is "no scientific evidence whatsoever" that ultrasound repels mosquitoes.
- A 2010 review article examined 10 field studies, in which ultrasonic repellent devices had been put to the test, and concluded that they "have no effect on preventing mosquito bites" and "should not be recommended or used".
- It goes on: "Given these findings from 10 carefully conducted studies, it would not be worthwhile to conduct further research on EMRs [electronic mosquito repellents] in preventing mosquitoes biting or in trying to prevent the acquisition of malaria."
- In 2005, the British consumer magazine Holiday Which tested a range of mosquito deterrents. The four appliances that used a buzzer were described by the magazine's editor Lorna Cowan as "a shocking waste of money" which "should be removed from sale". One, the Lovebug, a ladybird-shaped gadget designed to be clipped on to a baby's cot or child's pushchair - was singled out as a particular cause for concern, because of the likelihood that parents would trust it to keep mosquitoes away, and their children would be hurt as a result. The Lovebug is still readily available in Europe, though it was withdrawn from the US market after manufacturer Prince Lionheart was reprimanded by the Federal Trade Commission.

Safety

Professor Tim Leighton from the Institute of Sound and Vibration Research, University of Southampton, United Kingdom produced an 83 page paper entitled "What is Ultrasound" which was published in 2007. In the paper Leighton expressed concern about the growth in commercial products which exploit the discomforting effects of in-air ultrasound (to pests for whom it is within their audible frequency range, or to humans for whom it is not, but who can experience unpleasant subjective effects and, potentially, shifts in the hearing threshold). Leighton claims that commercial products are often advertised with cited levels which cannot be critically accepted due to lack of accepted measurement standards for ultrasound in air, and little understanding of the mechanism by which they may represent a hazard.

The UK's independent Advisory Group on Non-ionising Radiation (AGNIR) produced a 180 page report on the health effects of human exposure to ultrasound and infrasound in 2010. The UK Health Protection Agency (HPA) published their report, which recommended an exposure limit for the general public to airborne ultrasound sound pressure levels (SPL) of 70 dB (at 20 kHz), and 100 dB (at 25 kHz and above).

Permaculture Zone 5

Permaculture is a branch of ecological design, ecological engineering, environmental design, construction and Integrated Water Resources Management that develops sustainable architecture, regenerative and self-maintained habitat and agricultural systems modelled from natural ecosystems. The term permaculture (as a systematic method) was first coined by Australians Bill Mollison and David Holmgren in 1978. The word *permaculture* originally referred to "permanent agriculture" but was expanded to stand also for "permanent culture," as it was seen that social aspects were integral to a truly sustainable system as inspired by Masanobu Fukuoka's natural farming philosophy.

Core Tenets

The core tenets of permaculture are:

- *Care for the earth*: Provision for all life systems to continue and multiply. This is the first principle, because without a healthy earth, humans cannot flourish.
- *Care for the people*: Provision for people to access those resources necessary for their existence.
- *Return of surplus*: Reinvesting surpluses back into the system to provide for the first two ethics. This includes returning waste back into the system to recycle into usefulness.

Permaculture design emphasizes patterns of landscape, function, and species assemblies. It determines where these elements should be placed so they can provide maximum benefit to the local environment. The central concept of permaculture is maximizing useful connections between components and synergy of the final design. The focus of permaculture, therefore, is not on each separate element, but rather on the relationships created among elements by the way they are placed together; the whole becoming greater than the sum of its parts.

Permaculture design therefore seeks to minimize waste, human labour, and energy input by building systems with maximal benefits between design elements to achieve a high level of synergy. Permaculture designs evolve over time by taking into account these relationships and elements and can become extremely complex systems that produce a high density of food and materials with minimal input.

The design principles which are the conceptual foundation of permaculture were derived from the science of systems ecology and study of pre-industrial examples of sustainable land use. Permaculture draws from several disciplines including organic farming, agroforestry, integrated farming, sustainable development, and applied ecology. Permaculture has been applied most commonly to the design of housing and landscaping, integrating techniques such as agroforestry, natural building, and rainwater harvesting within the context of permaculture design principles and theory.

Theory

Twelve Design Principles: Twelve Permaculture design principles articulated by David Holmgren in his *Permaculture: Principles and Pathways Beyond Sustainability*:

1. *Observe and interact*: By taking time to engage with nature we can design solutions that suit our particular situation.
2. *Catch and store energy*: By developing systems that collect resources at peak abundance, we can use them in times of need.
3. *Obtain a yield*: Ensure that you are getting truly useful rewards as part of the work that you are doing.
4. *Apply self-regulation and accept feedback*: We need to discourage inappropriate activity to ensure that systems can continue to function well.
5. *Use and value renewable resources and services*: Make the best use of nature's abundance to reduce our consumptive behaviour and dependence on non-renewable resources.
6. *Produce no waste*: By valuing and making use of all the resources that are available to us, nothing goes to waste.
7. *Design from patterns to details*: By stepping back, we can observe patterns in nature and society. These can form the backbone of our designs, with the details filled in as we go.
8. *Integrate rather than segregate*: By putting the right things in the right place, relationships develop between those things and they work together to support each other.

9. *Use small and slow solutions*: Small and slow systems are easier to maintain than big ones, making better use of local resources and producing more sustainable outcomes.
10. *Use and value diversity*: Diversity reduces vulnerability to a variety of threats and takes advantage of the unique nature of the environment in which it resides.
11. *Use edges and value the marginal*: The interface between things is where the most interesting events take place. These are often the most valuable, diverse and productive elements in the system.
12. *Creatively use and respond to change*: We can have a positive impact on inevitable change by carefully observing, and then intervening at the right time.

Patterns

Permaculture is a design philosophy which seeks to exploit and imitate naturally occurring patterns. Patterns occur in nature in many forms: From studying Fractals, patterns emerge as mathematical principles from apparently random things such as the shape of plants.

Permaculture designers are encouraged to develop a sensitivity to the patterns that exist in nature, to determine their complex functions and the various interrelations. Since humans have complex intentions and a wide variety of goals, permaculture design draws upon a variety patterns, building a pattern language as defined by Christopher Alexander. These natural patterns can often be utilized—either exploited or imitated or both exploited and imitated—to satisfy specific design goals. For example, "the application of pattern on a design site involves the designer recognizing the shape and potential to fit these patterns or combinations of patterns comfortably onto the landscape". Patterns such as spiral, branching, wave, net and honeycomb are structural patterns that are repeated throughout nature. "A lot of the structural patterns combine strength and beauty with efficiency of space through large surface area or extensive edges. Looking at the benefits of these characteristics provides us with attitudes that we can emulate in our design work."

Layers

Layers are one of the tools used to design functional ecosystems that are both sustainable and of direct benefit to humans. A mature ecosystem has a huge number of relationships between its component parts: trees, understory, ground cover, soil, fungi, insects, and animals. Because plants grow to different heights, a diverse community of life is able to grow in a relatively small space, as each layer is stacked one on top of another. There are generally seven recognized layers in a food forest, although some practitioners also include fungi as an eighth layer.

1. The canopy: the tallest trees in the system. Large trees dominate but typically do not saturate the area, i.e. there exist patches barren of trees.
2. Understory layer: trees that revel in the dappled light under the canopy.
3. Shrubs: a diverse layer of woody perennials of limited height. Includes most berry bushes.
4. Herbaceous: Plants in this layer die back to the ground every winter (if winters are cold enough, that is). They do not produce woody stems as the Shrub layer does. Many cullinary and medicinal herbs are in this layer. A large variety of beneficial plants fall into this layer. May be annuals, biennials or perennials
5. Soil surface/Groundcover: There is some overlap with the Herbaceous layer and the Groundcover layer; however plants in this layer grow much closer to the ground, grow densely to fill bare patches of soil, and often can tolerate some foot traffic. Cover crops retain soil and lessen erosion, along with green manures that add nutrients and organic matter to the soil, especially nitrogen
6. Rhizosphere: Root layers within the soil. The major components of this layer are the soil and the organisms that live within it such as plant roots (including root crops such as potatoes and other edible tubers), fungi, insects, nematodes, worms, etc.
7. Vertical layer: climbers or vines, such as runner beans and lima beans (vine varieties)

Guilds

A guild is any group of species where each provides a unique set of diverse functions that work in conjunction, or harmony. Guilds are groups of plants, animals, insects, etc. that work well together. Some plants may be grown for food production, some have tap roots that draw nutrients up from deep in the soil, some are nitrogen-fixing legumes, some attract beneficial insects, and others repel harmful insects. When grouped together in a mutually beneficial arrangement, these plants form a guild.

Edge Effect

The edge effect in ecology is the effect of the juxtaposition or placing side by side of contrasting environments on an ecosystem.

Permaculturists argue that, where vastly differing systems meet, there is an intense area of productivity and useful connections. An example of this is the coast; where the land and the sea meet there is a particularly rich area that meets a disproportionate percentage of human and animal needs. So this idea is played out in permacultural designs by using spirals in the herb garden or creating ponds that have wavy undulating shorelines rather than a simple circle or oval (thereby increasing the amount of edge for a given area).

Zones

Zones are a way of intelligently organizing design elements in a human environment on the basis of the frequency of human use and plant or animal needs. Frequently manipulated or harvested elements of the design are located close to the house in zones 1 and 2. Less frequently used or manipulated elements, and elements that benefit from isolation (such as wild species) are farther away. Zones are about positioning things appropriately. Zones are numbered from 0 to 5.

Zone 0

The house, or home centre. Here permaculture principles would be applied in terms of aiming to reduce energy and water needs, harnessing natural resources such as sunlight, and generally creating a harmonious, sustainable environment in which to live and work. Zone 0 is an informal designation, which is not specifically defined in Bill Mollison's book.

Zone 1

The zone nearest to the house, the location for those elements in the system that require frequent attention, or that need to be visited often, such as salad crops, herb plants, soft fruit like strawberries or raspberries, greenhouse and cold frames, propagation area, worm compost bin for kitchen waste, etc. Raised beds are often used in zone 1 in urban areas.

Zone 2

This area is used for siting perennial plants that require less frequent maintenance, such as occasional weed control or pruning, including currant bushes and orchards, pumpkins, sweet potato, etc. This would also be a good place for beehives, larger scale composting bins, and so on.

Zone 3

The area where main-crops are grown, both for domestic use and for trade purposes. After establishment, care and maintenance required are fairly minimal (provided mulches and similar things are used), such as watering or weed control maybe once a week.

Zone 4

A semi-wild area. This zone is mainly used for forage and collecting wild food as well as production of timber for construction or firewood.

Zone 5

A wilderness area. There is no human intervention in zone 5 apart from the observation of natural ecosystems and cycles. Through this zone we build up a natural reserve of bacteria, moulds and insects that can aid the zones above it.

People and Permaculture

Permaculture uses observation of nature to create regenerative systems, and the place where this has been most visible has been on the landscape. There has been a growing awareness though that firstly, there is the need to pay more attention to the peoplecare ethic, as it is often the dynamics of people that can interfere with projects, and secondly that the principles of permaculture can be used as effectively to create vibrant, healthy and productive people and communities as they have been in landscapes.

Domesticated Animals

Domesticated animals are often incorporated into site design.

Common Practices

Agroforestry: Agroforestry is an integrated approach of using the interactive benefits from combining trees and shrubs with crops and/ or livestock. It combines agricultural and forestry technologies to create more diverse, productive, profitable, healthy and sustainable land-use systems. In agroforestry systems, trees or shrubs are intentionally used within agricultural systems, or non-timber forest products are cultured in forest settings.

Forest gardening is a term permaculturalists use to describe systems designed to mimic natural forests. Forest gardens, like other permaculture designs, incorporate processes and relationships that the designers understand to be valuable in natural ecosystems. The terms forest garden and food forest are used interchangeably in the permaculture literature. Numerous permaculturists are proponents of forest gardens, such as Graham Bell, Patrick Whitefield, Dave Jacke, Eric Toensmeier and Geoff Lawton. Bell started building his forest garden in 1991 and wrote the book *The Permaculture Garden* in 1995,

Whitefield wrote the book *How to Make a Forest Garden* in 2002, Jacke and Toensmeier co-authored the two volume book set *Edible Forest Gardening* in 2005, and Lawton presented the film *Establishing a Food Forest* in 2008.

Tree Gardens, such as Kandyan tree gardens, in South and Southeast Asia, are often hundreds of years old. Whether they derived initially from experiences of cultivation and forestry, as is the case in agroforestry, or whether they derived from an understanding of forest ecosystems, as is the case for permaculture systems, is not self-evident. Many studies of these systems, especially those that predate the term permaculture, consider these systems to be forms of agroforestry. Permaculturalists who include existing and ancient systems of polycropping with woody species as examples of food forests may obscure the distinction between permaculture and agroforestry.

Food forests and agroforestry are parallel approaches that sometimes lead to similar designs.

Hügelkultur

Hügelkultur is the practice of burying large volumes of wood to increase soil water retention. The porous structure of wood acts as a sponge when decomposing underground. During the rainy season, masses of buried wood can absorb enough water to sustain crops through the dry season. This technique has been used by permaculturalists Sepp Holzer, Toby Hemenway, and Masanobu Fukuoka.

Natural Building

A natural building involves a range of building systems and materials that place major emphasis on sustainability. Ways of achieving sustainability through natural building focus on durability and the use of minimally processed, plentiful or renewable resources, as well as those that, while recycled or salvaged, produce healthy living environments and maintain indoor air quality.

The basis of natural building is the need to lessen the environmental impact of buildings and other supporting systems, without sacrificing comfort, health or aesthetics. To be more sustainable, natural building uses primarily abundantly available, renewable, reused or recycled materials. In addition to relying on natural building materials, the emphasis on the architectural design is heightened. The orientation of a building, the utilization of local climate and site conditions, the emphasis on natural ventilation through design, fundamentally lessen operational costs and positively impact the environment. Building compactly and minimizing the ecological footprint is common, as are on-site handling of energy acquisition, on-site water capture, alternate sewage treatment and water reuse.

Rainwater Harvesting

Rainwater harvesting is the accumulating and storing of rainwater for reuse before it reaches the aquifer. It has been used to provide drinking water, water for livestock, water for irrigation, as well as other typical uses. Rainwater collected from the roofs of houses and local institutions can make an important contribution to the availability of drinking water. It can supplement the subsoil water level and increase urban greenery. Water collected from the ground, sometimes from areas which are especially prepared for this purpose, is called stormwater harvesting.

Greywater is wastewater generated from domestic activities such as laundry, dishwashing, and bathing, which can be recycled on-site for uses such as landscape irrigation and constructed wetlands. Greywater is largely sterile, but not potable (drinkable). Greywater differs from water from the toilets which is designated sewage or blackwater, to indicate it contains human waste. Blackwater is septic or otherwise toxic and cannot be reused.

Sheet Mulching

In agriculture and gardening, mulch is a protective cover placed over the soil. Any material or combination can be used as mulch, stones, leaves, cardboard, wood chips, gravel, etc., though in permaculture mulches of organic material are the most common because they perform more functions. These include: absorbing rainfall, reducing evaporation, providing nutrients, increasing organic matter in the soil, feeding and creating habitat for soil organisms, suppressing weed growth and seed germination, moderating diurnal temperature swings, protecting against frost, and reducing erosion. Sheet mulching is an agricultural no-dig gardening technique that attempts to mimic natural processes occurring within forests. Sheet mulching mimics the leaf cover that is found on forest floors. When deployed properly and in combination with other Permacultural principles, it can generate healthy, productive and low maintenance ecosystems.

Sheet mulch serves as a "nutrient bank," storing the nutrients contained in organic matter and slowly making these nutrients available to plants as the organic matter slowly and naturally breaks down. It also improves the soil by attracting and feeding earthworms, slaters and many other soil micro-organisms, as well as adding humus. Earthworms "till" the soil, and their worm castings are among the best fertilizers and soil conditioners. Sheet mulching can be used to reduce or eliminate undesirable plants by starving them of light, and can be more advantageous than using herbicide or other methods of control.

Managed Intensive Rotational Grazing

Grazing has long been blamed for much of the destruction we see in the environment. However, it has been shown that when grazing is modelled after nature, the opposite effect can be seen. Also known as cell grazing, managed intensive rotational grazing (MIRG) is a system of grazing in which ruminant and non-ruminant herds and/or flocks are regularly and systematically moved to fresh pasture, range, or forest with the intent to maximize the quality and quantity of forage growth. This disturbance is then followed by a period of rest which allows new growth. MIRG can be used with cattle, sheep, goats, pigs, chickens, rabbits, geese, turkeys, ducks and other animals depending on the natural ecological community that is being mimicked. Sepp Holzer and Joel Salatin have shown how the disturbance caused by the animals can be the spark needed to start ecological succession or prepare ground for planting. Allan Savory's holistic management technique has been likened to "a permaculture approach to rangeland management". One variation on MIRG that is gaining rapid popularity is called eco-grazing. Often used to either control invasives or re-establish native species, in eco-grazing the primary purpose of the animals is to benefit the environment and the animals can be, but are not necessarily, used for meat, milk or fibre.

Keyline Design

Keyline design is a technique for maximizing beneficial use of water resources of a piece of land developed in Australia by farmer and engineer P. A. Yeomans. The *Keyline* refers to a specific topographic feature linked to water flow which is used in designing the drainage system of the site.

Fruit Tree Management

Masanobu Fukuoka, as part of early experiments on his family farm in Japan, experimented with no-pruning methods, noting that he ended up killing many fruit trees by simply letting them go, which made them become convoluted and tangled, and thus unhealthy. Then he realised this is the difference between natural-form fruit trees and the process of change of tree form that results from abandoning previously-pruned unnatural fruit trees. He concluded that the trees should be raised all their lives without pruning, so they form healthy and efficient branch patterns that follow their natural inclination. This is part of his implementation of the Tao-philosophy of Wú wéi translated in part as no-action (against nature), and he described it as no unnecessary pruning, nature farming or "do-nothing" farming, of fruit trees, distinct from non-intervention or literal no-pruning. He ultimately achieved yields comparable to or exceeding standard/intensive practices of using pruning and chemical fertilisation.

Mollison and Holmgren

In the mid-1970s, Bill Mollison and David Holmgren started developing ideas about stable agricultural systems on the southern Australian island state of Tasmania. This was a result of the danger of the rapidly growing use of industrial-agricultural methods. In their view, highly dependent on non renewable resources, these methods were additionally poisoning land and water, reducing biodiversity, and removing billions of tons of topsoil from previously fertile landscapes. A design approach called *permaculture* was their response and was first made public with the publication of their book *Permaculture One* in 1978.

By the early 1980s, the concept had broadened from agricultural systems design towards sustainable human habitats. After *Permaculture One*, Mollison further refined and developed the ideas by designing hundreds of permaculture sites and writing more detailed books, notably *Permaculture: A Designers Manual*. Mollison lectured in over 80 countries and taught his two-week Permaculture Design Course (PDC) to many hundreds of students.

In 1991, a four-part television documentary by ABC productions called "The Global Gardener" showed permaculture applied to a range of worldwide situations, bringing the concept to a much broader public. In 2012, the UMass Permaculture Initiative won the White House "Champions of Change" sustainability contest, which declared that "they demonstrate how permaculture can feed a growing population in an environmentally sustainable and socially responsible manner".

In 1997, Holmgren explained that the primary agenda of the permaculture movement is to assist people to become more self-reliant through the design and development of productive and sustainable gardens and farms.

Trademark and Copyright Issues

There has been contention over who if anyone controls the legal rights to the word *permaculture*, meaning is it trademarked or copyrighted, and if so, who holds the legal rights to the use of the word. For a long time Bill Mollison claimed to have copyrighted the word, and his books said on the copyright page, "The contents of this book and the word PERMACULTURE are copyright." These statements were largely accepted at face-value within the permaculture community. However, copyright law does not protect names, ideas, concepts, systems, or methods of doing something; it only protects the expression or the description of an idea, not the idea itself. Eventually Mollison acknowledged that he was mistaken and that no copyright protection existed for the word *permaculture.*

In 2000 Mollison's US based Permaculture Institute sought a service mark (a form of trademark) for the word permaculture when used in educational services such as conducting classes, seminars, or workshops. The service mark would have allowed Mollison and his two Permaculture Institutes (one in the US and one in Australia) to set enforceable guidelines as to how permaculture could be taught and who could teach it, particularly with relation to the PDC, despite the fact that he had instituted a system of certification of teachers to teach the PDC in 1993.This certification was granted to teachers like April Sampson-kelly and others in 1993. The service mark failed and was abandoned in 2001. Also in 2001 Mollison applied for trademarks in Australia for the terms "Permaculture Design Course" and "Permaculture Design". These applications were both withdrawn in 2003. In 2009 he sought a trademark for "Permaculture: A Designers' Manual" and "Introduction to Permaculture", the names of two of his books. These applications were withdrawn in 2011. There has never been a trademark for the word *permaculture* in Australia.

Criticisms

General Criticisms: In 2011, Owen Hablutzel argued that "permaculture has yet to gain a large amount of specific mainstream scientific acceptance," and that "the sensitiveness to being perceived and accepted on scientific terms is motivated in part by a desire for Permaculture to expand and become increasingly relevant." Bec-Hellouin permaculture farm engaged in a research programme in partnership with INRA and AgroParisTech to collect scientific data.

In his books *Sustainable Freshwater Aquaculture* and *Farming in Ponds and Dams*, Nick Romanowski expresses the view that the presentation of aquaculture in Bill Mollison's books is unrealistic and misleading. However *Sustainable Freshwater Aquaculture* has likewise received criticism as being "too basic", "for primary school children" and "negative" in its approach.

Linda Chalker-Scott alleges that Toby Hemenway's views regarding invasive species in the permaculture book *Gaia's Garden* are pseudoscience.

Agroforestry

Greg Williams argues that forests cannot be more productive than farmland because the net productivity of forests decline as they mature due to ecological succession. Proponents of permaculture respond that this is true only if one compares data from between woodland forest and climax vegetation, but not when comparing farmland vegetation with woodland forest. For example, ecological succession generally results in a forest's productivity rising after its establishment only until it reaches the *woodland state* (67% tree cover), before declining until *full maturity*.

Sterile Insect Technique

The sterile insect technique is a method of biological control, whereby overwhelming numbers of sterile insects are released. The released insects are normally male as it is the female that causes the damage, usually by laying eggs in the crop, or, in the case of mosquitoes, taking blood from humans. The sterile males compete with the wild males for female insects. If a female mates with a sterile male then it will have no offspring, thus reducing the next generation's population. Repeated release of insects can diminish small populations, though it could be impossible to eradicate it and is not efficient against dense insect populations.

The technique has successfully been used to eradicate the Screw-worm fly (*Cochliomyia hominivorax*) in areas of North America. There have also been many successes in controlling species of fruit flies, most particularly the Medfly (*Ceratitis capitata*), and the Mexican fruit fly (*Anastrepha ludens*).

Insects are mostly sterilized with radiation, which might weaken the newly sterilized insects if doses are not correctly applied, making them less able to compete with wild males. However, other sterilization techniques are under development which would not affect the insects' ability to compete for a mate.

Development of the Sterile Insect Technique

Raymond Bushland and Edward Knipling first developed the technique to eliminate screwworms preying on warm-blooded animals, especially cattle herds. With larvae that invade open wounds and eat into animal flesh, the flies were capable of killing cattle within 10 days of infection. In the 1950s, screwworms caused annual losses to American meat and dairy supplies that were projected at above $200 million. Screwworm maggots are also known to parasitize human flesh. Since a female screwworm mates only once in her lifetime, this physiological phenomenon has been exploited by biologists in breaking its life cycle. After mating with a sterile male, the screwworm female will not mate again or lay any eggs.

The quest of Bushland and Knipling to find an alternative to chemical pesticides in controlling the devastation wrought by these insects began in the late 1930s when both scientists were working at the United States Department of Agriculture Laboratory in Menard, Texas. At that time, the screwworm was decimating livestock herds across the American South. Red meat and dairy supplies were also affected across Mexico, Central America, and South America.

While Bushland initially researched chemical treatment of screwworm-infested wounds in cattle, Knipling developed the theory of autocidal control – breaking the life cycle of the pest itself. Bushland's enthusiasm for Knipling's theory sparked both men to intensify the search for a way to rear large numbers of flies in a "factory" setting, and most importantly, to find an effective way to sterilize flies.

Their work in this area was interrupted by World War II, but Drs. Bushland and Knipling resumed their efforts in the early 1950s with their successful tests on the screwworm population of Sanibel Island, Florida. The sterile insect technique worked; near eradication was achieved using X-ray sterilized flies.

In 1954, the technique was used to completely eradicate screwworms from the 176-square-mile (460 km^2) island of Curaçao, off the coast of Venezuela. Screwworms were eliminated in a span of only seven weeks, saving the domestic goat herds that were a source of meat and milk for the island people.

During the 1960s and 1970s, SIT was used to control the screwworm population in the United States. The 1980s saw Mexico and Belize eliminate their screwworm problems through the use of SIT, and eradication programmes have progressed through all of Central America, with a biological barrier having been established in Panama to prevent reinfestation from the south. In 1991, Knipling and Bushland's technique halted a serious outbreak in northern Africa. Similar programmes against the Mediterranean fruit fly in Mexico and California use the same principles. In addition, the technique was used to eradicate the melon fly from Okinawa and has been used in the fight against the tsetse fly in Africa.

The technique has been able to suppress insects threatening livestock, fruit, vegetable, and fibre crops. The technique has also been lauded for its many environmentally sound attributes: it uses no chemicals, leaves no residues, and has no negative effect on non-target species.

Proven effective in controlling outbreaks of a wide range of insect pests throughout the world, the technique has been a boon in protecting the agricultural products to feed the world's human population. Both Bushland and Knipling received worldwide recognition for their leadership and scientific achievements, including the World Food Prize. Their research and the resulting Sterile Insect Technique were hailed by former U.S. Secretary of Agriculture Orville Freeman as "the greatest entomological achievement of (the 20th) century."

Sterile Fly for African Trypanosomiasis

Sleeping sickness or the African trypanosomiasis is a parasitic disease in humans. Caused by protozoa of genus *Trypanosoma* and transmitted by the Tsetse fly, the disease is endemic in certain regions of Sub-Saharan Africa, covering about 36 countries and 60 million people. It is estimated that 50,000 - 70,000 people are infected, and about 40,000 die every year. Three major epidemics have occurred in the past hundred years, in 1896 - 1906, 1920, and 1970.

Studies of the tsetse fly show that females generally only mate once in their lifetimes and very rarely mate a second time. Once a female fly has mated, she can then produce continual offspring throughout her short life.

The sterile fly is an innovative solution to the problem of the African trypanosomiasis. Specially bred male Tsetse flies are sterilized through irradiation process. These sterilized male flies are then released into areas where sleeping sickness is prevalent, and then mate with the females. Because the male is sterile, and the females mate only once, the population of Tsetse flies in the affected area will drop. Studies have shown that this process has been very effective in preventing sleeping sickness in people who live in the area.

Since sleeping sickness is fatal without treatment and infected people can be without symptoms for months, the release of sterile flies into affected areas leads to greater levels of health and economic activity.

Success Stories

- Screwworm fly (*Cochliomyia hominivorax*) eradicated from the United States, Mexico, and Libya.
- Mexican fruit fly (*Anastrepha ludens*, Loew) eradicated from most of northern Mexico.
- Tsetse fly eradicated from Zanzibar.

- Medfly (*Ceratitis capitata*, Wiedemann) eradicated from northern part of Chile and southern part of Peru and southern part of Mexico.
- Melon fly (*Bactrocera cucurbitae*, Coquillett) eradicated from Okinawa, Japan.

Current Targets

- *Anopheles* mosquito - Malaria vector, example *Anopheles arabiensis.*
- Tsetse fly (*Glossina spp*) - sleeping sickness vector.
- Painted Apple Moth (Lepidoptera: Lymantriidae) in Auckland, New Zealand.
- Codling moth (*Cydia pomonella*) (Lepidoptera: Tortricidae) in British Columbia, Canada.
- *Aedes* mosquitoes, vectors for filariasis, dengue and yellow fever.
- Continuing use across the world against various fruit fly species, including Medfly, Caribbean and Mexican fruit fly (Nth, Sth and Central America), Queensland fruit fly in Australia (*Bactrocera tryoni*) and several other *Bactrocera* sp. across Australia, Asia and Oceania. The IAEA lists 41 different SIT facilities (DIR-SIT) across the globe, aiding the retrieval of information on mass production of sterile pest insect and also provides information on the doses of radiation (IDIDAS) used in the control of pest insects and mites of agricultural, commercial or quarantine significance. It includes data on both the radiation dose required for the disinfestation of generic commodity groups, fresh and durable, and also the radiation dose used to induce sterility for pest control through the sterile insect technique.

Drawbacks

- As with insecticide treatment, repeated treatment is sometimes required to suppress the population before the use of sterile insects.
- Sex separation could be difficult for some species, though this can be easily performed on Medfly and screwworm, for example.
- Radiation treatment in some cases affects the health of males, so sterilized insects in such cases are at a disadvantage when competing for females.
- The technique is species specific. For instance, there are 22 species of Tsetse fly in Africa, and the technique must be implemented separately for each.

- Standard operating procedures of mass rearing and irradiation do not leave room for mistakes. Since the fifties, when SIT was first used as a means for pest control, several failures have occurred in different places around the world where non-sterilized artificial produced insects were released before the problem was spotted.
- Application to large areas should be long lasting, otherwise migration of wild insects from outside the control area could repopulate.
- The major drawback to this technique is that the cost of producing such a large number of sterile insects is often prohibitive in poorer countries.

Genetic Modification

A method using recombinant DNA technology to create genetically modified insects called RIDL ("release of insects carrying a dominant lethal") is under development by a company called *Ox*ford *I*nsect *Tec*hnologies (Oxitec), UK. The method works by introducing a repressible "dominant lethal" gene into the insects. This gene kills the insects but it can be repressed by an external additive (tetracycline), which allows the insects to be reared in manufacturing facilities. This external additive is commonly administered orally, and so can be an additive to the insect food. The insects can also be given genetic markers, such as DsRED fluorescence, that make monitoring the progress of eradication easier, preferably under the field conditions.

There are potentially several types of RIDL, but the more advanced forms have a female-specific dominant lethal gene. This avoids the need for a separate sex separation step, as the repressor can be withdrawn from the final stage of rearing, leaving only males.

These males are then released in large numbers into the affected region. The released males are not sterile, but any female offspring their mates produce will have the dominant lethal gene expressed and so will have a high probability to die; but about 5% of the descendants survive. Using RIDL means that the males will not have to be sterilized by radiation before release, making the males more healthy when they need to compete with the wild males for mates. Progress towards applying this technique to mosquitoes has been made by researchers at Imperial College London who created the world's first transgenic malaria mosquito.

A similar technique is the daughterless carp, a genetically modified organism produced in Australia by the CSIRO in the hope of eradicating the introduced carp from the Murray River system. As of 2005, it was undergoing tests to assess the risks of releasing it into the wild .

Biotechnological approaches based on genetically modified organism (transgenic organisms) are still under development. However, since no legal framework exists to authorize the release of such organisms in the nature, sterilization by irradiation remains the most used technique.

Genetically Modified Insect

A genetically modified insect is an insect that has been genetically modified for various reasons such as agricultural production, oil production and pest control.

In biological research, transgenic fruit flies (*Drosophila melanogaster*) are model organisms used to study the effects of genetic changes on development. Fruit flies are often preferred over other animals due to their short life cycle, low maintenance requirements, and relatively simple genome compared to many vertebrates.

The first release of modified mosquitoes were performed in the 1970s. It was mosquitoes of the species that transmit the dengue virus that where sterilised by irradiation.

The British company Oxitec use a technique called RIDL that produce fertile adults but induce a high mortality of the descendants. The adults generated with this technique and released in the environment are not sterile but their descendants have a survival rate of only about 5% (or much higher in presence of tetracycline).

4

Sustainable Farming

Sustainable agriculture is the act of farming using principles of ecology, the study of relationships between organisms and their environment. The phrase was reportedly coined by Australian agricultural scientist Gordon McClymont. It has been defined as "an integrated system of plant and animal production practices having a site-specific application that will last over the long term" For Example:

- Satisfy human food and fibre needs
- Enhance environmental quality and the natural resource base upon which the agricultural economy depends
- Make the most efficient use of non-renewable resources and on-farm resources and integrate, where appropriate, natural biological cycles and controls
- Sustain the economic viability of farm operations
- Enhance the quality of life for farmers and society as a whole

Farming and Natural Resources

Sustainable agriculture can be understood as an ecosystem approach to agriculture. Practices that can cause long-term damage to soil include excessive tilling of the soil(leading to erosion) and irrigation without adequate drainage (leading to salinization). Long-term experiments have provided some of the best data on how various practices affect soil properties essential to sustainability. In the United States a federal agency, USDA-Natural Resources Conservation Service, specializes in providing technical and financial assistance for those interested in pursuing natural resource conservation and production agriculture as compatible goals.

The most important factors for an individual site are sun, air, soil, nutrients, and water. Of the five, water and soil quality and quantity are most amenable to human intervention through time and labour.

Although air and sunlight are available everywhere on Earth, crops also depend on soil nutrients and the availability of water. When farmers grow and harvest crops, they remove some of these nutrients from the soil. Without replenishment, land suffers from nutrient depletion and becomes either unusable or suffers from reduced yields. Sustainable agriculture depends on replenishing the soil while minimizing the use or need of non-renewable resources, such as natural gas (used in converting atmospheric nitrogen into synthetic fertilizer), or mineral ores (e.g., phosphate). Possible sources of nitrogen that would, in principle, be available indefinitely, include:

1. recycling crop waste and livestock or treated human manure
2. growing legume crops and forages such as peanuts or alfalfa that form symbioses with nitrogen-fixing bacteria called rhizobia
3. industrial production of nitrogen by the Haber Process uses hydrogen, which is currently derived from natural gas, (but this hydrogen could instead be made by electrolysis of water using electricity (perhaps from solar cells or windmills)) or
4. genetically engineering (non-legume) crops to form nitrogen-fixing symbioses or fix nitrogen without microbial symbionts.

The last option was proposed in the 1970s, but is only recently becoming feasible. Sustainable options for replacing other nutrient inputs (phosphorus, potassium, etc.) are more limited. More realistic, and often overlooked, options include long-term crop rotations, returning to natural cycles that annually flood cultivated lands (returning lost nutrients indefinitely) such as the Flooding of the Nile, the long-term use of biochar, and use of crop and livestock landraces that are adapted to less than ideal conditions such as pests, drought, or lack of nutrients.

Crops that require high levels of soil nutrients can be cultivated in a more sustainable manner if certain fertilizer management practices are adhered to.

Water

In some areas sufficient rainfall is available for crop growth, but many other areas require irrigation. For irrigation systems to be sustainable, they require proper management (to avoid salinization) and must not use more water from their source than is naturally replenishable. Otherwise, the water source effectively becomes a non-renewable resource. Improvements in water well drilling technology and submersible pumps, combined with the development of drip irrigation and low pressure pivots, have made it possible to regularly achieve high crop yields in areas where reliance on rainfall alone had previously made successful agriculture unpredictable. However, this progress has come at a price. In many areas, such as the Ogallala Aquifer, the water is being used faster than it can be recharged.

Several steps must be taken to develop drought-resistant farming systems even in "normal" years with average rainfall. These measures include both policy and management actions: 1) improving water conservation and storage measures, 2) providing incentives for selection of drought-tolerant crop species, 3) using reduced-volume irrigation systems, 4) managing crops to reduce water loss, or 5) not planting crops at all.

Indicators for Sustainable Water Resource Development

Internal Renewable Water Resources: This is the average annual flow of rivers and groundwater generated from endogenous precipitation, after ensuring that there is no double counting. It represents the maximum amount of water resource produced within the boundaries of a country. This value, which is expressed as an average on a yearly basis, is invariant in time (except in the case of proved climate change). The indicator can be expressed in three different units: in absolute terms (km3/yr), in mm/yr (it is a measure of the humidity of the country), and as a function of population (m3/person per yr).

Global Renewable Water Resources: This is the sum of internal renewable water resources and incoming flow originating outside the country. Unlike internal resources, this value can vary with time if upstream development reduces water availability at the border. Treaties ensuring a specific flow to be reserved from upstream to downstream countries may be taken into account in the computation of global water resources in both countries.

Dependency Ratio: This is the proportion of the global renewable water resources originating outside the country, expressed in percentage. It is an expression of the level to which the water resources of a country depend on neighbouring countries.

Water Withdrawal: In view of the limitations described above, only gross water withdrawal can be computed systematically on a country basis as a measure of water use. Absolute or per-person value of yearly water withdrawal gives a measure of the importance of water in the country's economy. When expressed in percentage of water resources, it shows the degree of pressure on water resources. A rough estimate shows that if water withdrawal exceeds a quarter of global renewable water resources of a country, water can be considered a limiting factor to development and, reciprocally, the pressure on water resources can have a direct impact on all sectors, from agriculture to environment and fisheries.

Soil

Figure: *Walls built to avoid water run-off*

Soil erosion is fast becoming one of the worlds greatest problems. It is estimated that "more than a thousand million tonnes of southern Africa's soil are eroded every year. Experts predict that crop yields will be halved within thirty to fifty years if erosion continues at present rates." Soil erosion is not unique to Africa but is occurring worldwide. The phenomenon is being called *Peak Soil* as present large scale factory farming techniques are jeopardizing humanity's ability to grow food in the present and in the future. Without efforts to improve soil management practices, the availability of arable soil will become increasingly problematic.

Some Soil Management techniques

1. No-till farming
2. Keyline design
3. Growing wind breaks to hold the soil
4. Incorporating organic matter back into fields
5. Stop using chemical fertilizers (which contain salt)
6. Protecting soil from water run off(soil erosion)

Phosphate

Phosphate is a primary component in the chemical fertilizer which is applied in modern agricultural production. However, scientists estimate that rock phosphate reserves will be depleted in 50–100 years and that Peak phosphorus will occur in about 2030. The phenomenon of Peak phosphorus is expected to increase food prices as fertilizer costs increase as rock phosphate reserves become more difficult to extract. In the long term, phosphate will therefore have to be recovered and recycled from human and animal waste in order to maintain food production.

Land

As the global population increases and demand for food increases, there is pressure on land resources. Land can also be considered a finite resource on Earth. Expansion of agricultural land has an impact on biodiversity and contributes to deforestation. The Food and Agriculture Organisation of the United Nations estimates that in coming decades, cropland will continue to be lost to industrial and urban development, along with reclamation of wetlands, and conversion of forest to cultivation, resulting in the loss of biodiversity and increased soil erosion.

Energy for Agriculture

Energy is used all the way down the food chain from farm to fork. In industrial agriculture, energy is used in on-farm mechanisation, food processing, storage, and transportation processes. It has therefore been found that energy prices are closely linked to food prices. Oil is also used as an input in agricultural chemicals. Higher prices of non-renewable energy resources are projected by the International Energy Agency. Increased energy prices as a result of fossil fuel resources being depleted may therefore impact negatively on the global food security unless action is taken to 'decouple' fossil fuel energy from food production, with a move towards 'Energy-Smart' agricultural systems.

Economics

Socioeconomic aspects of sustainability are also partly understood. Regarding less concentrated farming, the best known analysis is Netting's study on smallholder systems through history. The Oxford Sustainable Group defines sustainability in this context in a much broader form, considering effect on all stakeholders in a 360 degree approach

Given the finite supply of natural resources at any specific cost and location, agriculture that is inefficient or damaging to needed resources may eventually exhaust the available resources or the ability to afford and acquire them. It may also generate negative externality, such as pollution as well as financial and production costs.

The way that crops are sold must be accounted for in the sustainability equation. Food sold locally does not require additional energy for transportation (including consumers). Food sold at a remote location, whether at a farmers' market or the supermarket, incurs a different set of energy cost for materials, labour, and transport.

Methods

What grows where and how it is grown are a matter of choice. Two of the many possible practices of sustainable agriculture are crop rotation and soil amendment, both designed to ensure that crops being cultivated can obtain the necessary nutrients for healthy growth. Soil amendments would include using locally available compost from community recycling centres. These community recycling centres help produce the compost needed by the local organic farms.

Many scientists, farmers, and businesses have debated how to make agriculture sustainable. Using community recycling from yard and kitchen waste utilizes a local area's commonly available resources. These resources in the past were thrown away into large waste disposal sites, are now used to produce low cost organic compost for organic farming. Other practices includes growing a diverse number of perennial crops in a single field, each of which would grow in separate season so as not to compete with each other for natural resources.

This system would result in increased resistance to diseases and decreased effects of erosion and loss of nutrients in soil. Nitrogen fixation from legumes, for example, used in conjunction with plants that rely on nitrate from soil for growth, helps to allow the land to be reused annually. Legumes will grow for a season and replenish the soil with ammonium and nitrate, and the next season other plants can be seeded and grown in the field in preparation for harvest.

Monoculture, a method of growing only one crop at a time in a given field, is a very widespread practice, but there are questions about its sustainability, especially if the same crop is grown every year. Today it is realized to get around this problem local cities and farms can work together to produce the needed compost for the farmers around them. This combined with growing a mixture of crops (polyculture) sometimes reduces disease or pest problems but polyculture has rarely, if ever, been compared to the more widespread practice of growing different crops in successive years (crop rotation) with the same overall crop diversity. Cropping systems that include a variety of crops (polyculture and/or rotation) may also replenish nitrogen (if legumes are included) and may also use resources such as sunlight, water, or nutrients more efficiently.

Replacing a natural ecosystem with a few specifically chosen plant varieties reduces the genetic diversity found in wildlife and makes the organisms susceptible to widespread disease. The Great Irish Famine (1845–1849) is a well-known example of the dangers of monoculture. In practice, there is no single approach to sustainable agriculture, as the precise goals and methods must be adapted to each individual case. There may be some techniques of farming that are inherently in conflict with the concept of sustainability, but there is widespread misunderstanding on impacts of some practices.

Today the growth of local farmers' markets offer small farms the ability to sell the products that they have grown back to the cities that they got the recycled compost from. By using local recycling this will help move people away from the slash-and-burn techniques that are the characteristic feature of shifting cultivators are often cited as inherently destructive, yet slash-and-burn cultivation has been practised in the Amazon for at least 6000 years; serious deforestation did not begin until the 1970s, largely as the result of Brazilian government programmes and policies. To note that it may not have been slash-and-burn so much as slash-and-char, which with the addition of organic matter produces terra preta, one of the richest soils on Earth and the only one that regenerates itself.

There are also many ways to practice sustainable animal husbandry. Some of the key tools to grazing management include fencing off the grazing area into smaller areas called paddocks, lowering stock density, and moving the stock between paddocks frequently.

Several attempts have been made to produce an artificial meat, using isolated tissues to produce it in vitro; Jason Matheny's work on this topic, which in the New Harvest project, is one of the most commented.

Soil Treatment

Soil steaming can be used as an ecological alternative to chemicals for soil sterilization. Different methods are available to induce steam into the soil in order to kill pests and increase soil health. Community and farm composting of kitchen, yard, and farm organic waste can provide most if not all the required needs of local farms. This composting could potentially be a reliable source of energy.

Off-farm Impacts

A farm that is able to "produce perpetually", yet has negative effects on environmental quality elsewhere is not sustainable agriculture. An example of a case in which a global view may be warranted is over-application of synthetic fertilizer or animal manures, which can improve productivity of a farm but can pollute nearby rivers and coastal waters (eutrophication). The other extreme can also be undesirable, as the problem of low crop yields due to exhaustion of nutrients in the soil has been related to rainforest destruction, as in the case of slash and burn farming for livestock feed.In Asia, specific land for sustainable farming is about 12.5 acres which includes land for animal fodder, cereals productions lands for some cash crops and even recycling of related food crops.In some cases even a small unit of aquaculture is also included in this number (AARI-1996)

Sustainability affects overall production, which must increase to meet the increasing food and fibre requirements as the world's human population expands to a projected 9.3 billion people by 2050. Increased production may come from creating new farmland, which may ameliorate carbon dioxide emissions if done through reclamation of desert as in Israel and Palestine, or may worsen emissions if done through slash and burn farming, as in Brazil.

International Policy

Sustainable agriculture has become a topic of interest in the international policy arena, especially with regards to its potential to reduce the risks associated with a changing climate and growing human population.

The Commission on Sustainable Agriculture and Climate Change, as part of its recommendations for policy makers on achieving food security in the face of climate change, urged that sustainable agriculture must be integrated into national and international policy. The Commission stressed that increasing weather variability and climate shocks will negatively affect agricultural yields, necessitating early action to drive change in agricultural production systems towards increasing resilience. It also called for dramatically increased investments in sustainable agriculture in the next decade, including in national research and development budgets, land rehabilitation, economic incentives, and infrastructure improvement.

Urban Planning

There has been considerable debate about which form of human residential habitat may be a better social form for sustainable agriculture.

Many environmentalists advocate urban developments with high population density as a way of preserving agricultural land and maximizing energy efficiency. However, others have theorized that sustainable ecocities, or ecovillages which combine habitation and farming with close proximity between producers and consumers, may provide greater sustainability. The use of available city space (e.g., rooftop gardens, community gardens, garden sharing, and other forms of urban agriculture) for cooperative food production is another way to achieve greater sustainability.

One of the latest ideas in achieving sustainable agriculture involves shifting the production of food plants from major factory farming operations to large, urban, technical facilities called vertical farms. The advantages of vertical farming include year-round production, isolation from pests and diseases, controllable resource recycling, and on-site production that reduces transportation costs. While a vertical farm has yet to become a reality, the idea is gaining momentum among those who believe that current sustainable farming methods will be insufficient to provide for a growing global population.

Sustainable Gardening

Sustainable gardening (which is taken here to include sustainable landscapes, sustainable landscape design, sustainable landscape architecture and sustainable sites) comprises a disparate group of horticultural interests that share, to a greater or lesser extent, the aims and objectives associated with the international post-1980s sustainable development and sustainability programmes developed to address the fact that humans are now using natural biophysical resources faster than they can be replenished by nature. Included within this compass are those home gardeners, and members of the landscape and nursery industries and municipal authorities, that integrate environmental, social and economic factors in an attempt to create a more sustainable future.

Organic gardening and the use of native plants are integral to sustainable gardening.

After the establishment of sustainable agriculture in the early 1980s it was some time before the emergence of Sustainable Horticulture (as sustainable *production* horticulture) at the International Society of Horticultural Science's First International Symposium on Sustainability in Horticulture held at the International Horticultural Congress in Toronto in 2002. This symposium produced "conclusions ... on Sustainability in Horticulture and a Declaration for the 21st Century".

The principles and objectives outlined at this conference were discussed in more practical terms at the following conference at Seoul in 2006.

Many of the eco-friendly principles and ideas espoused by sustainable gardens, landscapes and sites perpetuate sustainable practices established as a reaction to resource-intensive industrial agriculture. These practices were established as movements for self-sufficiency and small-scale farming based on a holistic systems approach and ecological principles. Included here would be: biodynamic agriculture, no-till farming, agroecology, Fukuoka farming, forest gardening, organic gardening and others. On a larger scale there is the more recent "whole farm planning" which was established in 1995, and ecoagriculture established in 2000, and other variants of sustainable agricultural systems. Perhaps the most influential of these approaches is permaculture, established by Australians Bill Mollison and David Holmgren as both a design system and a loosely defined philosophy or lifestyle ethic. Permaculture shares many principles and practices of the above but not the broad philosophical base as indicated by the title of the 2002 publication *Permaculture, principles and pathways beyond sustainability*. The application of sustainability principles to the horticultural sphere has now becoming broadly accepted in commerce and academia.

Definition

The American Sustainable Sites Initiative is an interdisciplinary approach used by the American Society of Landscape Architects, the Lady Bird Johnson Wildflower Centre and the United States Botanic Garden to create voluntary national guidelines and performance benchmarks for sustainable land design, construction and maintenance practices: it was founded in 2005.

Principles and Concepts

Managing global biophysical cycles and ecosystem services for the benefit of humans, other organisms and future generations has now become a global human responsibility. The method of applying sustainability to gardens, landscapes and sites is still under development and varies somewhat according to the context under consideration. However, there are a number of basic and common underlying biological and operational principles and practices in the sustainable sites literature.

Biological Principles

Sustainable management of man-made landscapes emulates the natural processes that sustain the biosphere and its ecosystems. First and foremost is the harnessing the energy of the Sun and the cycling of materials thereby minimising waste and energy use. Running within, and dependent on, the natural economy there is the production and consumption of goods and services in the "human economy" which has now significantly altered, in a detrimental way, natural biogeochemical cycles (notable here are the water cycle, carbon cycle and nitrogen cycle so sustainable practices maximise support for ecosystem services.

Native Plants

The use of native plants in a garden or landscape can both preserve and protect natural ecosystems, and reduce the amount of care and energy required to maintain a healthy garden or landscape. Native plants are adapted to the local climate and geology, and often require less maintenance than exotic species. Native plants also support populations of native birds, insects, and other animals that they coevolved with, thus promoting a healthy community of organisms.

Plants in a garden or maintained landscape often form a source population from which plants can colonize new areas. Avoiding the use of invasive species helps to prevent such plants from establishing new populations. Similarly, the use of native species can provide a valuable source to help these plants colonise new areas.

Some non-native species can form an ecological trap in which native species are lured into an environment that appears attractive but is poorly suited to them. However, in Britain research by the University of Sheffield as part of the BUGS project (Biodiversity in Urban Gardens in Sheffield) has revealed that for many invertebrates - the majority of wild animals in most gardens - it is not just native plants which can sustain them.

The findings were published in popular form in Ken Thompson's book 'No Nettles Required: The truth about wildlife gardening'. He confirms the approach which Chris Baines had promoted in 'How to Make a Wildlife Garden'.

Operational Principles

Enhancement of ecosystem services is encouraged throughout the lifecycle of any site by providing clear design, construction, (operations), and management criteria. To be sustainable over the long term requires environmental, social and economic demands are integrated to provide intergenerational equity by providing regenerative sustainable systems. Operational guidelines will link to and supplement existing guidelines for the built environment (supplementing existing green building and landscape guidelines), the wider environment, and they will include metrics (benchmarks, audits, criteria, indexes etc.) that give some measure of sustainability (a rating system) by clarifying what is sustainable or not sustainable or, more likely, what is more or less sustainable.

Scale

Impacts of a site can be assessed and measured over any spatio-temporal scale or context.

Direct and Indirect Environmental Impact

Impacts of a site may be *direct* by having direct measurable impacts on biodiversity and ecology at the site itself or *indirect* when impacts occur away from the site.

Site Principles

The following are some site principles for sustainable gardening:

- Do no harm
- Use the Precautionary principle
- Design with nature and culture
- Use a decision-making hierarchy of preservation, conservation, and regeneration
- Provide regenerative systems as intergenerational equity
- Support a living process
- Use a system thinking approach
- Use a collaborative and ethical approach
- Maintain integrity in leadership and research
- Foster environmental stewardship

Measuring Site Sustainability

One major feature distinguishing the approach of sustainable gardens, landscapes and sites from other similar enterprises is the quantification of site sustainability by establishing performance benchmarks. Because sustainability is such a broad and inclusive concept the environmental impacts of sites can be categorised in numerous ways depending on the purpose for which the figures are required. The process can include minimising negative environmental impacts and maximising positive impacts.

As currently applied the environment is usually given priority over social and economic factors which may be added in or regarded as an inevitable and integral part of the management process. A home gardener is likely to use simpler metrics than a professional landscaper or ecologist.

Three methodologies for measuring site sustainability include BREEAM developed by the BRE organisation in the UK, Leed, developed in America and the Oxford 360 degree sustainability Index used in Oxford Park and developed by the Oxford Sustainable Group in Scandinavia.

The *Sustainable Sites Initiative* is producing recommendations for the American Landscape Industry. The standards and guidelines finally adopted will lead to a uniform national standard, which does not currently exist. Sustainable Sites is currently in the pilot programme stage, and will formally introduce its first rating system by 2013.

The U.S. Green Building Council supports the project and plans to adopt the Sustainable Sites metrics into future versions of its Leadership in Energy and Environmental Design Green Building Rating System. Sites are rated according to their impact on ecosystem services: The following ecosystem services have been identified by the study group:

- Local climate regulation
- Air and water cleansing
- Water supply and regulation
- Erosion and sediment control
- Hazard mitigation
- Pollination
- Habitat functions
- Waste decomposition and treatment
- Global climate regulation
- Human health and well-being benefits
- Food and renewable non-food products
- Cultural benefits

Inputs

- Fossil fuels
- Embodied energy and water
- Compost
- Mulch
- Ecology & biodiversity
- Fertilizer
- Hard landscape materials

- Equipment
- Products

Outputs

- Energy & water
- Food
- Green waste
- Ecology & biodiversity
- Chemicals
- Old hard landscape materials
- Old equipment
- Old products

Constraints

Any kind of auditing or benchmarking will depend on the selection and weighting of the metrics chosen; the depth and detail of analysis required; the purpose for which the figures are required; and the environmental circumstances of the particular site.

Pests in Various Parts of Plants

Bulbs, Corms and Tubers

In botany, a bulb is a short stem with fleshy leaves or leaf bases that function as food storage organs during dormancy. (In gardening, other kinds of storage organ are also called "bulbs" or "ornamental bulbs".)

A bulb's leaf bases, also known as scales, generally do not support leaves, but contain food reserves to enable the plant to survive adverse conditions. At the centre of the bulb is a vegetative growing point or an unexpanded flowering shoot. The base is formed by a stem, and plant growth occurs from this basal plate. Roots emerge from the underside of the base, and new stems and leaves from the upper side. Tunicate bulbs have dry, membranous outer scales that protect the continuous lamina of fleshy scales. Species in the genera *Allium*, *Hippeastrum*, *Narcissus*, and *Tulipa* all have tunicate bulbs. Non-tunicate bulbs, such as *Lilium* and *Fritillaria* species, lack the protective tunic and have looser scales.

Other types of storage organs (such as corms, rhizomes, and tubers) are sometimes referred to as bulbs, although as the term is used in botany, they are not. The technical term for plants that form underground storage organs, including bulbs as well as tubers and corms, is geophyte.

Some epiphytic orchids (family Orchidaceae) form above-ground storage organs called pseudobulbs, that superficially resemble bulbs.

Nearly all plants that form true bulbs are monocotyledons, and include:

- *Amaryllis*, *Hippeastrum*, *Narcissus*, and several other members of the amaryllis family Amaryllidaceae. This includes onion, garlic, and other alliums, members of the Amaryllid subfamily Allioideae.
- Lily, tulip, and many other members of the lily family Liliaceae.
- Two groups of *Iris* species, family Iridaceae: subgenus *Xiphium* (the "Dutch" irises) and subgenus *Hermodactyloides* (the miniature "rock garden" irises).

Oxalis, in the family Oxalidaceae, is the only dicotyledon genus that produces true bulbs.

Bulbous plant species cycle through vegetative and reproductive growth stages; the bulb grows to flowering size during the vegetative stage and the plant flowers during the reproductive stage. Certain environmental conditions are needed to trigger the transition from one stage to the next, such as the shift from a cold winter to spring. Once the flowering period is over, the plant enters a foliage period of about six weeks during which time the plant absorbs nutrients from the soil and energy from the sun for setting flowers for the next year. Bulbs dug up before the foliage period is completed will not bloom the following year but then should flower normally in subsequent years.

After the foliage period is completed, bulbs may be dug up for replanting elsewhere. Any surface moisture should be dried, then the bulbs may be stored up to about 4 months for a fall planting. Storing them much longer than that may cause the bulbs to dry out inside and become nonviable.

Bulbil

A bulbil is a small bulb, and may also be called a bulblet, bulbet, or bulbel.

Small bulbs can develop that replace or propagate a large bulb. If one or several moderate-sized bulbs form to replace the original bulb, they are called renewal bulbs. Increase bulbs are small bulbs that develop either on each of the leaves inside a bulb, or else on the end of small underground stems connected to the original bulb.

Some lilies form small bulbs, called bulbils, in their leaf axils. Several members of the onion family, Alliaceae, including *Allium sativum* (garlic), form bulbils in their flower heads, sometimes as the flowers fade, or even instead of the flowers. The so-called tree onion (*Allium cepa* var. *proliferum*) forms small onions which are large enough for pickling.

Some ferns, such as Hen and Chicken Fern produce new plants at the tips of the fronds' pinnae, which are sometimes referred to a bulbils.

Bulb Pests and Diseases

As with all garden plants bulbs are susceptible to attack by certain garden pests. If left unchecked these pests can cause huge amounts of disease and damage. Resolva Bug Killer can be applied all year round for the control of aphids, whitefly, thrips and mealy bugs It also contains a foliar feed that provides vital support to your plants while the pest problem is being solved. Resolva Bug Killer kills on contact meaning troublesome pests are dealt with quick and easily. For larger areas it is also available in concentrate formulation.

Slugs can also cause damage to bulbs, particularly tulips, gladioli, dahlias and lilies. For slugs choose to use Eraza Slug and Snail Killer at the first sign of trouble and supplement with the barrier Earth Matters Slug Blocker Granules. Eraza Slug and Snail Killer offer a complete and effective control for slugs and snails. The pellets are rain proof and pet repellent so that if applied at the recommended rate they have longer staying power without posing a risk to pets and wildlife.

Roots

The root is the organ of a plant that typically lies below the surface of the soil. However, roots can also be aerial or aerating (growing up above the ground or especially above water). Furthermore, a stem normally occurring below ground is not exceptional either.Therefore, the root is best defined as the non-leaf, non-nodes bearing parts of the plant's body. However, important internal structural differences between stems and roots exist.

The first root that comes from a plant is called the radicle. The four major functions of roots are 1) absorption of water and inorganic nutrients, 2) anchoring of the plant body to the ground, and supporting it, 3) storage of food and nutrients, 4) vegetative reproduction. In response to the concentration of nutrients, roots also synthesise cytokinin, which acts as a signal as to how fast the shoots can grow. Roots often function in storage of food and nutrients. The roots of most vascular plant species enter into symbiosis with certain fungi to form mycorrhizae, and a large range of other organisms including bacteria also closely associate with roots.

Root Growth

Early root growth is one of the functions of the apical meristem located near the tip of the root. The meristem cells more or less continuously divide, producing more meristem, root cap cells (these are sacrificed to protect the meristem), and undifferentiated root cells. The latter become the primary tissues of the root, first undergoing elongation, a process that pushes the root tip forward in the growing medium. Gradually these cells differentiate and mature into specialized cells of the root tissues.

There is correlation of roots using the process of plant perception to sense their physical environment to grow, including the sensory of light, and physical barriers. Plant roots will generally grow in any direction where the correct environment of air, mineral nutrients and water exists to meet the plant's needs. Roots will shy or shrink away from dry, or other poor soil conditions.

Over time, given the right conditions, roots can crack foundations, snap water lines, and lift sidewalks. At germination, roots grow downward due to gravitropism, the growth mechanism of plants that also causes the shoot to grow upward. In some plants (such as ivy), the "root" actually clings to walls and structures.

Growth from apical meristems is known as primary growth, which encompasses all elongation. Secondary growth encompasses all growth in diametre, a major component of woody plant tissues and many nonwoody plants. For example, storage roots of sweet potato have secondary growth but are not woody. Secondary growth occurs at the lateral meristems, namely the vascular cambium and cork cambium. The former forms secondary xylem and secondary phloem, while the latter forms the periderm.

In plants with secondary growth, the vascular cambium, originating between the xylem and the phloem, forms a cylinder of tissue along the stem and root. The vascular cambium forms new cells on both the inside and outside of the cambium cylinder, with those on the inside forming secondary xylem cells, and those on the outside forming secondary phloem cells. As secondary xylem accumulates, the "girth" (lateral dimensions) of the stem and root increases. As a result, tissues beyond the secondary phloem (including the epidermis and cortex, in many cases) tend to be pushed outward and are eventually "sloughed off" (shed).

At this point, the cork cambium begins to form the periderm, consisting of protective cork cells containing suberin. In roots, the cork cambium originates in the pericycle, a component of the vascular cylinder. The vascular cambium produces new layers of secondary xylem annually. The xylem vessels are dead at maturity but are responsible for most water transport through the vascular tissue in stems and roots.

Identifying Vegetable Root Maggots and Root Maggot Control

Vegetable root maggots are the larva of a type of fly called a root maggot fly. There are several kinds with different preferred host plants. The eggs of these root eating insects are laid in the soil and hatch into larva. The larva are the little worms you see on the roots of your plant. The larva will come to the surface to pupate and then they are adults who will start the process all over again. Eggs can survive the winter in the soil. If a plant is unexplainably stunted or if it starts to wilt for no reason, there may be vegetable root maggots in the soil. Root maggots are more likely to attack in cool weather.

The best way to tell is to gently lift the plant from the soil and examine their roots. If vegetable root maggots are the culprit, the roots will be eaten away or tunneled through in the case of larger rooted plants like turnips. And, of course, root maggot larva will be present.

Root maggots typically attack either legume plants (beans and peas) or cruciferous plants (cabbage, broccoli, turnips, radishes, etc.) but they are not exclusive to those plants and can be found on almost any type of vegetable.

Root Maggot Control

These root eating insects will stay in your garden beds and attack other plants unless you take steps to get rid of them. There are several things you can do for root maggot control.

The first thing to do is to get rid of infested plants. Dying plants will attract the root maggot fly and should either be disposed of in the trash or burned. Do not compost them. Once a plant is infested, it cannot be saved, but you can do several things to keep the next plants from becoming infected.

Organic root maggot control can be:

- Dusting the plants with diatomaceous earth
- Adding beneficial nematodes to the soil
- Releasing predatory rove beetles into your garden
- Covering plants with floating row covers
- Solarizing infected beds

If you wish to use chemicals for root maggot control, apply a liquid pesticide to your garden bed at the beginning of the growing season. Make sure that you soak the soil. This will kill off the vegetable root maggots. Keep in mind that anything else in treated soil, such as worms, will also be killed.

Root Weevil

Root weevils are a plant pest both indoors and outdoors. These destructive little insects will invade the root system of a healthy plant and then proceed to eat the plant from the roots up. Identifying and controlling root weevil in your garden and houseplants can keep your plants from suffering unnecessary damage.

Root weevils can be one of several kinds. The most common in the garden is the black vine root weevil or the strawberry root weevil. The black vine weevil attacks shrubs and strawberry weevils attack strawberries. While these are the most common, they are far from the only kind. All plants in your home or garden are susceptible to weevil infestation.

Larval root weevils will look like white grubs or worms and will be found in the soil. Adult weevils are beetle like insects that can be black, brown or grey. If root weevils are present in your garden or houseplants, there will be damage to both the roots and the leaves. The leaves of the plant will be irregular, as though someone has been taking bites out of the edges. This damage will appear in the night, as root weevils come out to feed at night.

Controlling root weevil is possible. Organic root weevil control methods include purchasing parasitic nematodes or predatory beetles can be bought to hunt down the weevils. You can also hand pick the adults off the plant at night while they are eating. Weevils are also attracted to moisture, so a shallow pan of water can be set out at night and the weevils will climb into it and drown.

Non-organic root weevil control methods are to spray the leaves of the plant with a pesticide and to thoroughly soak the soil with a liquid pesticide. But, remember, when you do this, you may also be killing off beneficial insects and small animals as well.

Finding these insects in the roots and leaves of your plants is unpleasant, but it can be fixed. As always, the best root weevil control is to make sure you never get any in the first place. Be sure to practice good garden hygiene and clean up dead plants and do not over mulch.

Stems and Shoots

A stem is one of two main structural axes of a vascular plant, the other being the root. The stem is normally divided into nodes and internodes. The nodes hold buds which grow into one or more leaves, conifer cones, roots, other stems, or flowers (inflorescences); the internodes distance one node from another. The term "shoots" is often confused with "stems"; "shoots" generally refers to new fresh plant growth including both stems and other structures like leaves or flowers. In most plants stems are located above the soil surface but some plants have underground stems. A stem develops buds and shoots and usually grows above the ground. Inside the stem, materials move up and down the tissues of the transport system.Stems are often green.Stems have thin pipes to carry water around the plant.Stem consists of xylem and phloem tissue. Stems and roots hold the plant upright.

Buds

In botany, shoots consist of stems including their appendages, the leaves and lateral buds, flowering stems and flower buds. The new growth from seed germination that grows upward is a shoot where leaves will develop. In the spring, perennial plant shoots are the new growth that grows from the ground in herbaceous plants or the new stem and/or flower growth that grows on woody plants.

Figure: *Edible shoots of Sachaline*

In everyday speech, shoots are often synonymous with stems. Stems, which are an integral component of shoots, provide an axis for buds, fruits, and leaves.

Young shoots are often eaten by animals because the fibres in the new growth have not yet completed secondary cell wall development, making the young shoots softer and easier to chew and digest. As shoots grow and age, the cells develop secondary cell walls that have a hard and tough structure. Some plants (e.g. bracken) produce toxins that make their shoots inedible or less palatable.

Leaf

A leaf is an organ of a vascular plant, and is the principal appendage of the vascular plant stem. The leaves and stem together form the shoot. Foliage is a mass noun that refers to leaves collectively.

Typically a leaf is a thin, dorsiventrally flattened organ, borne above ground and specialized for photosynthesis. Most leaves have distinctive upper (adaxial) and lower (abaxial) surfaces that differ in colour, hairiness, the number of stomata and other features. The palisade mesophyll almost always occurs on the upper side of the blade or lamina of the leaf but in some species, including the mature foliage of *Eucalyptus* palisade occurs on both sides and the leaves are said to be isobilateral. Many types of leaves are adapted in ways almost unrecognisable in those terms: some are not flat (for example many succulent leaves and the needles of conifers), some are not above ground (such as bulb scales), and some are without major photosynthetic function (consider for example cataphylls, and spines). Furthermore, several kinds of leaf-like structures found in vascular plants are not totally homologous with them. Examples include phyllodes, cladodes, and phylloclades that differ from leaves in their structure and origin. Conversely, many structures of non-vascular plants, and even of some lichens, which are not plants at all (in the sense of being members of the kingdom Plantae), do look and function much like leaves.

Leafminers

Leafminers are pests which tunnel through the mesophyll of leaves creating distinctive and unsightly squiggly patterns. They are, collectively, mostly comprised of members of two insect orders: Lepidoptera (moths and butterflies) and Diptera (flies). The latter will be the focus of this page as caterpillar pests are covered in the previous page. There are about 98,000 members of the order Diptera, and many are pests in some way, shape or form, but only a handful are active as leafminers. In the United States they are not as common as pests like aphids and caterpillars, but do deserve attention. Two decades ago, and we don't know why, they were a more serious problem than they are today, but are still a concern for growers and, thus, mentioned on this website.

How They Become a Problem

Leafminers can attack outdoor and indoor crops of many kinds, but the damage they cause is not usually life-threatening. Aesthetically speaking, though, these flies can destroy the value of a crop. Larger plants, shrubs and trees can be infested by leafminers every year without noticeable growth setbacks unless the infestation is serious-in which case photosynthesis can be hindered. And cosmetic issues are not always a concern unless the plants, shrubs or tree are to be sold outright. Growers' concerns usually involve leafminer infestations on smaller specimens which will be marketed as plants for their appearance and not for their fruits, etc. Chrysanthemums are one example of such a crop. On mums and other such flowering plants, leafminers are a more serious concern.

A Serious Pest

The infestation process begins as the female flies lands on your plants and begin to feed. They do this by puncturing the leaf with their sucking mouth parts. This activity creates a tiny white spot which, by itself, is not of great concern to growers-though a lot of feeding activity can make the plant's leaves unsightly. The real damage begins when the female fly lays an egg in the puncture wound. The egg hatches and the resulting larva or maggot begins to tunnel through the leaf, eating the cells of which the mesophyll is composed, creating a void, tunnel or mine as they go. The outer leaf surfaces remain intact, but with no cells between the upper and lower surfaces the feeding becomes very noticeable and ugly. As the maggot continues to feed, it tunnels randomly. Moreover, the maggot increases in size so the mines get larger and more noticeable as progress is made.

Additionally, the maggot's feces is forced out of the mine's entrance creating yet another unsightly mess (and this is exacerbated when a maggot has been parasitized). The growth process continues until the maggot is ready, through complete metamorphosis, to pupate. At this point they usually drop to the ground. On the ground, control of these pests can obtained by employing one or more soil-pest controls. When the entire process is complete, the pupae give forth a new generation of adult flies ready to mate and lay eggs in leaves so that the cycle may continue.

For the Scout

You will probably not be bothered with large numbers of leafminers and, unless they are a known problem-pest in your area, we wouldn't recommend special treatments or scouting measures be taken. Become visually familiar with them-which can be difficult as they are unremarkable in appearance and look like, well, little two to three millimetre flies with a yellow spot on each side of their thorax, in essence, stay alert and do nothing else. Utilizing yellow sticky traps during a regular scouting regimen should be the only step you'd take in an effort to scout for them, aside, that is, from looking for those telltale mines during your normal plant inspections. Which, by the way, if you do find mined leaves, remove and destroy them.

Some Solutions

There are few biological pest control agents available to specifically address leafminer pests from the order Diptera. The parasitoids which are available have been known to do a very good job, though. If possible, use them preventively or, at least, at the very first sign of these flies. Other biocontrol agents, those used for soil pests, for example, can also help you with leafminers while being used primarily to thwart other pests like fungus gnats. Chemical controls can also be effective, but expect difficulty when trying to effectively treat the leaf-protected larvae and soil-protected pupae.

Flowers

A flower, sometimes known as a bloom or blossom, is the reproductive structure found in flowering plants (plants of the division Magnoliophyta, also called angiosperms). The biological function of a flower is to effect reproduction, usually by providing a mechanism for the union of sperm with eggs. Flowers may facilitate outcrossing (fusion of sperm and eggs from different individuals in a population) or allow selfing (fusion of sperm and egg from the same flower). Some flowers produce diaspores without fertilization.

Common Pests: The most common pests of foliage and flowering plants are spider mites, mealy bugs, fungus gnats, whitefly and aphids. The use of plant material for outdoor as well as indoor decoration exposes them to all manner of garden pests, such as caterpillars, slugs, snails and thrips, many of which thrive once the plants are moved back indoors. For this reason, plants which are being moved from decks, patios, etc. to the home or office should be washed thoroughly with a spray of clear water.

Pest Control: Despite the best control programmes employed by growers and retailers, it sometimes happens that a new plant harbors pests whose eggs have survived to hatch in you home conditions. A period of isolation (10-14 days) may be the answer. Yet some pests may not show up for a longer time, being present in small numbers until the indoor climate becomes more favourable; for example, spider mites thrive in dry air, so they may not become evident until the humid season is past.

Regular and frequent checking for abnormal growth and insects can be combined with your checks for watering needs. Most pests lurk under leaves and close to the growing tip where tissue is softest. Most of these pests can be washed off with a spray of water. Several washes over a period of two or three weeks, plus isolation for the infested plant to prevent the pest from moving to other plants, generally eliminates the problem. Chemical sprays are not pleasant to use, and are generally recommended for outdoor application.. Some pesticides are hazardous, not only to the general environment, but to the plant tissue. Ferns are extremely sensitive to all forms of chemical spray, and infested fronds are best removed and destroyed.

Diseases: Diseases are rare on foliage and flowering plants. The only ones likely to be encountered are Botrytis, the gray mold fungus that invades dead and dying tissues, and soft leaf spot or crown rot fungi which thrive in moist atmospheres. The best cure is to remove affected parts, and avoid splashing the plant when watering; give more space for air circulation, or move infected plants to a drier place.

Fruit

In botany, a fruit is a part of a flowering plant that derives from specific tissues of the flower, one or more ovaries, and in some cases accessory tissues. Fruits are the means by which these plants disseminate seeds. Many of them that bear edible fruits, in particular, have propagated with the movements of humans and animals in a symbiotic relationship as a means for seed dispersal and nutrition, respectively; in fact, humans and many animals have become dependent on fruits as a source of food.

Fruits account for a substantial fraction of the world's agricultural output, and some (such as the apple and the pomegranate) have acquired extensive cultural and symbolic meanings.

In common language usage, "fruit" normally means the fleshy seed-associated structures of a plant that are sweet or sour and edible in the raw state, such as apples, oranges, grapes, strawberries, bananas, and lemons. On the other hand, the botanical sense of "fruit" includes many structures that are not commonly called "fruits", such as bean pods, corn kernels, wheat grains, and tomatoes.

The section of a fungus that produces spores is also called a fruiting body.

Natural predators such as lacewings, leatherwing beetles, ladybird beetles, ground beetles, wasps, praying mantis and pirate bugs will control or contain most pest populations at an acceptable level, especially if trees are kept vigorous, orchard areas are kept clean of trash and weeds, and trees are well pruned to facilitate good air movement. Home orchardists have little need to completely eradicate pests.

Most insect and mite pests of fruit trees are controlled by many beneficial species of insects and mites found in the orchard. Do not spray pests unless you are certain they are present in damaging numbers or this publication suggests you do so. Unnecessary sprays reduce control provided by beneficial species and may result in added damage from pests freed from their natural controls.

Several of the most common fruit tree diseases may be controlled by using the proper fixed copper spray during the dormant season. These diseases include: bacterial canker, brown rot, coryneum blight and peach leaf curl.

How Pests, Diseases, and Hosts Fit Together

The tables, below, are brief formats designed only as guides to be used with Integrated Pest Management (IPM) and other research-based references. If one or more of these pest and disease problems are not present, do not apply preventatives. Many do not become problems until trees begin to bear.

How Pests, Diseases, and Hosts Fit Together

Pest/Disease	*Hosts(s)*
Codling Moth	Apple, pear, sometimes walnut, plum
Peach Twig Borer (PTB)	Peach & nectarine, almond, apricot, plum & prune
Brown Rot Fungus	Peach & nectarine, cherry, plum & prune, almond
Fireblight	Pear, apple, quince
Scab	Apple, pear
Bacterial Canker	Peach & nectarine, cherry, plum & prune, pear (blast)
Peach Leaf Curl	Peach & nectarine

Goals for reducing insect and mite damage in home orchards are to:

- Protect trees and crops from substantial damage
- Make it simple and easy
- Make it safe
- Use methods that enhance natural control factors

Dormant season treatments are most important since they will not kill beneficial insects. One spray controls many orchard pests. The following tables of Orchard Pests by Season list several common home orchard pests and include information on damage, description, life cycle, and control options. More importantly, it provides information on the seasons when these pests should be addressed if they are problems in the home orchard. When damage becomes obvious, it is often too late to treat with successful results for that year.

Seeds

A seed is an embryonic plant enclosed in a protective outer covering called the seed coat, usually with some stored food. It is a characteristic of spermatophytes (gymnosperm and angiosperm plants) and the product of the ripened ovule which occurs after fertilization and some growth within the mother plant. The formation of the seed completes the process of reproduction in seed plants (started with the development of flowers and pollination), with the embryo developed from the zygote and the seed coat from the integuments of the ovule.

Seeds have been an important development in the reproduction and spread of gymnosperm and angiosperm plants, relative to more primitive plants such as ferns, mosses and liverworts, which do not have seeds and use other means to propagate themselves. This can be seen by the success of seed plants (both gymnosperms and angiosperms) in dominating biological niches on land, from forests to grasslands both in hot and cold climates.

The term "seed" also has a general meaning that antedates the above—anything that can be sown, e.g. "seed" potatoes, "seeds" of corn or sunflower "seeds". In the case of sunflower and corn "seeds", what is sown is the seed enclosed in a shell or husk, whereas the potato is a tuber.

Many structures commonly referred to as "seeds" are actually dry fruits. Plants producing berries are called baccate. Sunflower seeds are sometimes sold commercially while still enclosed within the hard wall of the fruit, which must be split open to reach the seed. Different groups of plants have other modifications, the so-called stone fruits (such as the peach) have a hardened fruit layer (the endocarp) fused to and surrounding the actual seed. Nuts are the one-seeded, hard-shelled fruit of some plants with an indehiscent seed, such as an acorn or hazelnut.

Seed Treatment

In agriculture and horticulture, a seed treatment or seed dressing is a chemical, typically antimicrobial or fungidal, with which seeds are treated (or "dressed") prior to planting. Less frequently insecticides are added. Seed treatments can be an environmentally more friendly way of using pesticides as the amounts used can be very small. It is usual to add colour to make treated seed less attractive to birds if spilt and easier to see and clean up in the case of an accidental spillage.

One seed treatment, imidacloprid, from the neonicotinoid family of insecticides, is controversial and was banned in France for use on maize, due to that government's belief that the chemical was implicated in recent dramatic drops in bee counts, and possibly in Colony Collapse Disorder. Dust from treated seed is known to have caused at least some problems particularly from crops such as maize drilled during the main honey flows. Improvements to pneumatic drills to reduce dust release, and improvements to seed treatment compounds to prevent the compound breaking up into dust have been introduced in Europe led by Germany and Holland from 2009 to 2012. Information on seed treatments including the information above can be seen on the registration authority databases.

Seed coating is a thicker form of covering of seed and may contain fertiliser, growth promoters and or seed treatment as well as an inert carrier and a polymer outer shell. Seed dressing is also used to refer to the process of removing chaff, weed seeds and straw from a seed stock. Care is needed not to confuse the two.

Insect Pests of Malvaceous Crops

The principle economic use of Malvaceous plants is as a source of natural fibres, the family providing perhaps the world's 3 most important fibre crops. Plants of the family are also used for food, beverages, timber, in traditional medicine, and in horticulture.

Fibre

The most important crop is cotton, which is obtained from the seed hairs of 4 species of *Gossypium*, the Old World diploids *Gossypium arboreum* and *Gossypium herbaceum*, and the New World tetraploids *Gossypium hirsutum* and *Gossypium barbadense*. Cotton is the world's preeminent natural fibre, of which in excess of 20 million tonnes are produced annually. This dwarfs the production of animal fibres (wool, about 1½ million tonnes annually) and is comparable to the total production of synthetic fibres. The largest producers are China, the United States of America, India and Pakistan. Production is also extensive in Central Asia, the Middle East, Australia, Brasil, and Mediterranean Europe.

Jute, obtained from the phloem fibres of *Corchorus capsularis* and *Corchorus olitorius*, is also an important fibre crop; with a production in excess of 2 million tonnes annually it is the second largest vegetable fibre. India and Bangladesh are by far the largest producers. Kenaf, obtained from the phloem fibres of *Hibiscus cannabinus*, has an annual production of about 1 million tonnes. Thailand, Burma, China and Bangladesh are the major producers.

Several other species are exploited for their phloem fibres, including devil's cotton, *Abroma angusta*, China jute, *Abutilon theophrastii*, roselle, *Hibiscus sabdariffa* and bast, *Tilia* sp., and many others are or were used at the subsistence level.

Kapok, the seed hairs of a number of trees belonging to Bombacoideae, particularly *Bombax ceiba* (250,000 tonnes annually), is not spinnable and therefore can't be used for textiles, but is waterproof. It is used primarily for insulation.

Food

The most important food crops are durian, the fruit of *Durio zibethinus*, which is an important plantation crop in South East Asia, okra, the fruit of *Abelmoschus esculentus*, which is used as a vegetable, and cacao, the fruit of *Theobroma cacao*, which is the source of chocolate and cacao butter. The fruits of a number of other species are also eaten. Several species, including *Abelmoschus manihot* and species of *Corchorus* and *Malva*, are grown as leaf vegetables. Marshmallows were originally made from the candied roots of *Althaea officinalis*.

Vegetable Oils

Vegetable oil is produced as a byproduct (cottonseed oil, kenaf oil) of the cultivation of cottons and kenaf for fibre. A vegetable oil is also produced from the baobab (*Adansonia*).

Beverages

Kola nuts, the fruits of *Cola nitida* and other species is used commercially as a flavouring ingredient in drinks (colas or kolas). They contain stimulant alkaloids - caffeine and theobromine. Cacao is used in the production of beverages, in addition to chocolate.

Timber

Many arborescent species of Malvaceae are used for timber, some of them commercially. Perhaps the most important are *Triplochiton*, *Mansonia* and *Heritiera*, but *Tilia*, *Talipariti* and many other genera are used. Balsa, *Ochroma pyramidale*, is noted for the low density of its wood.

Medicinal

Many species of Malvoideae have been exploited for the demulcent nature of their mucilaginous sap. The best known source is marsh-mallow, *Althaea officinalis*.

Horticulture

Many species of Malvaceae have showy flowers, and have a place in horticulture, the number of species used in at least a small way exceeding one hundred. In temperate regions the predominant malvaceous horticultural subjects are hollyhocks, *Alcea rosea*, musk mallows, *Malva moschata*, shrub Lavateras, *Lavatera ×clementii*, annual Lavateras, *Lavatera trimestris*, hardy hibiscus, *Hibiscus moscheutos* and allies, and Rose of Sharon, *Hibiscus syriacus*.

Biological Control

Biological control represents one alternative to the use of insecticides. Biological control is the conscious use of living beneficial organisms, called natural enemies, for the control of pests. Virtually all pests have natural enemies and appropriate management of natural enemies can effectively control many pests. The natural enemies of insects are a diverse group of organisms that includes predators, parasitic insects, nematodes, and various mircoorganisms. Although biological control will not control all pests all of the time, it is a key component of integrated pest management. The intent of biological control is not to eradicate pests, but to keep them at tolerable levels at which they cause no appreciable harm.

In the upper Midwest, literally hundreds of types of natural enemies occur naturally in the environments of agricultural crops, managed forests, and managed urban and suburban landscapes. These have a major impact on pest populations, and often result in pests being kept below damaging levels. When there is no human activity involved inmanaging these natural enemies, we refer to the results as "natural control" of the pests. Natural control is of major importance in pest management, but the benefits frequently go unrecognized because the natural enemies are often very small and their activity is easily overlooked.

To get the greatest benefit from both natural control and biological control, it is necessary to recognize natural enemies, understand their biological characteristics (such as life cycle, preferred host insects, and efficiency at controlling pests). Some natural enemies, such as lady beetles, are readily recognized in the adult stage by most people but may not be recognized in the egg, larval, or pupal stages. Other natural enemies, including most parasites, are much more difficult to recognize.

Although birds, mammals, frogs, and other higher animals can be important as natural enemies, they can rarely be effectively managed for biological control. These animals lie outside the scope of this discussion, which deals primarily with the predatory or parasitic insects of pest insects and mites.

Insect Pests of Cole Crops

Cole crops are a common sight in the home garden, especially in cooler weather. But some gardeners may not know what cole crops are. Whether you know what cole crop plants are or not, chances are you enjoy them on a regular basis.

Cole crops, at a basic level, are plants that belong to the mustard (Brassica) family and are all descendants of wild cabbage. As a group, these plants grow better in cool weather. This leads many people to think that the word "cole" is a variation of the word "cold" and they may even refer to these plants as cold crops. Actually, the word "cole" is a variation of a Latin word that means stem.

The following is a list of the most common of these plants:

- Brussels sprout
- Cabbage

- Cauliflower
- Collards
- Kale
- Kohlrabi
- Mustard
- Broccoli
- Turnip
- Watercress

When to Plant Cole Crops

The specific time when to plant cole crops will be different depending on which one you are growing. For example, most cabbage varieties can be planted out much earlier than broccoli or cauliflower because cabbage plants can tolerate a much lower temperature. In general, these crops grow best when daytime temperatures are below 80 degrees F. and nighttime temperatures are below 60 degrees F. at night. Temperatures higher than this can lead to buttoning, bolting or poor head formation.

Growing Cole Crop Plants

For best results, cole crops should be grown in full sun, but due to their need for cooler temperatures, if you have a partially shaded garden, the vegetables in this family will do okay here as well. Also, if you live in an area that has a short cool season, planting them in part shade can help alleviate daytime temperatures by keeping direct sun from falling on the plants.

Cole crop plants usually need significant amounts of nutrients, particularly micro-nutrients that may not be found in standard fertilizers. Therefore, it is important to work organic material into the beds you plan on growing cole crops in prior to planting them.

Since many of these crops are susceptible to the same types of diseases and pests, rotating plants at least every few years is a good idea. This will help to cut down on diseases and pests that overwinter in the soil and attack the plants.

Insect Pests of Solanaceous Vegetables

The Solanaceae, or nightshades, is an economically important family of flowering plants. The family ranges from annual and perennial herbs to vines, lianas, epiphytes, shrubs and trees, and includes a number of important agricultural crops, medicinal plants, spices, weeds, and ornamentals. Many members of the family contain potent alkaloids, and some are highly toxic, but many cultures eat nightshades, in some cases as staple foods. The family belongs to the order Solanales, in the asterid group dicotyledons (Magnoliopsida). The Solanaceae consists of approximately 98 genera and some 2,700 species, with a great diversity of habitats, morphology and ecology.

The name Solanaceae derives from the genus *Solanum*, "the nightshade plant". The etymology of the Latin word is unclear. The name may come from a perceived resemblance of certain solanaceous flowers to the sun and its rays. At least one species of *Solanum* is known as the "sunberry". Alternatively, the name could originate from the Latin verb *solari*, meaning "to soothe", presumably referring to the soothing pharmacological properties of some of the psychoactive species of the family.

The family has a worldwide distribution, being present on all continents except Antarctica. The greatest diversity in species is found in South America and Central America.

The Solanaceae include a number of commonly collected or cultivated species. The most economically important genus of the family is *Solanum*, which contains the potato (*Solanum tuberosum*, in fact, another common name of the family is the "potato family"), the tomato (*Solanum lycopersicum*), and the eggplant or aubergine (*Solanum melongena*). Another important genus, *Capsicum*, produces both chili peppers and bell peppers.

The genus *Physalis* produces the so-called groundcherries, as well as the tomatillo (*Physalis philadelphica*), the Cape gooseberry and the Chinese lantern. The genus *Lycium* contains the boxthorns and the wolfberry *Lycium barbarum*. *Nicotiana* contains, among other species, the plant that produces tobacco. Some other important members of Solanaceae include a number of ornamental plants such as *Petunia*, *Browallia* and *Lycianthes*, the source of psychoactive alkaloids, *Datura*, *Mandragora* (mandrake), and *Atropa belladonna* (deadly nightshade). Certain species are universally known for their medicinal uses, their psychotropic effects, or for being poisonous.

With the exception of tobacco (*Nicotiana tabacum*, Nicotianoideae) and petunia (*Petunia x hybrida*, Petunioideae), most of the economically important genera are contained in the subfamily Solanoideae. Finally, but not less importantly, the Solanaceae include many model organisms which are important in the investigation of fundamental biological questions at cellular, molecular, and genetic levels, such as tobacco and the petunia.

Description

Plants in the Solanaceae can take the form of herbs, shrubs, trees, vines and lianas and sometimes epiphytes. They can be annuals, biennials or perennials, upright or decumbent. Some have subterranean tubers. They do not have laticifers, nor latex, nor coloured saps. They can have a basal or terminal group of leaves or neither of these types. The leaves are generally alternate or alternate to opposed (that is, alternate at the base of the plant and opposed towards the inflorescence). The leaves can be herbaceous, leathery or transformed into spines.

The leaves are generally petiolate or subsessile, rarely sessile. They are frequently inodorous, but on occasions they are aromatic or fetid. The foliar lamina can be either simple or compound, the latter can be either pinnatifid or ternate. The leaves have reticulated venation and lack a basal meristem. The laminae are generally dorsiventral and lack secretory cavities. The stomata are generally confined to one of a leaf's two sides, they are rarely found on both sides.

The flowers are generally hermaphrodite, although some are monoecious, andromonoecious, or dioecious species (such as some *Solanum* or *Symonanthus*). Pollination is entomophilous. The flowers can be solitary or grouped into terminal, cymose, or axillary inflorescences. The flowers are medium-sized, fragrant (*Nicotiana*), fetid (*Anthocercis*) or inodorous. The flowers are usually actinomorphic, slightly zygomorphic, or markedly zygomorphic (for example, in flowers with a bilabial corolla in *Schizanthus* species).

The irregularities in symmetry can be due to the androecium, to the perianth, or both at the same time. In the great majority of species, the flowers have a differentiated perianth with a calyx and corolla (with five sepals and five petals, respectively) an androecium with five stamens and two carpels forming a gynoecium with a superior ovary (they are therefore referred to as pentamers and tetracyclic).

The stamens are epipetalous and are typically present in multiples of four or five, most commonly four or eight. They usually have a hypogynous disk. The calyx is gamosepalous (as the sepals are joined together forming a tube), with the (4)5(6) segments equal, it has five lobes, with the lobes shorter than the tube, it is persistent and often accrescent.

The corolla usually has five petals that are also joined together forming a tube. Flower shapes are typically rotate (wheel-shaped, spreading in one plane, with a short tube) or tubular (elongated cylindrical tube), campanulated or funnel-shaped.

The androecium has (2)(4)5(6) free stamens within it, oppositsepals (that is, they alternate with the petals), they are usually fertile or, in some cases (for example in Salpiglossideae) they have staminodes. In the latter case, there is usually either one staminode (*Salpiglossis*) or three (*Schizanthus*).

The anthers touch on their upper end forming a ring, or they are completely free, dorsifixed or basifixed with poricide dehiscence or through small longitudinal cracks. The stamen's filament can be filliform or flat. The stamens can be inserted inside the coralline tube or exserted. The plants demonstrate simultaneous microsporogenesis, the microspores are tetrad, tetrahedral or isobilateral. The pollen grains are bicellular at the moment of dehiscence, usually open and angular.

The gynoecium is bicarpelar (rarely three- or five-locular) with a superior ovary and two locules, which may be secondarily divided by false septa, as is the case for Nicandreae and Datureae. The gynoecium is located in an oblique position relative to the flower's median plane. They have one style and one stigma; the latter is simple or bilobate. Each locule has one to 50 ovules that are anatropous or hemianatropous with axillar placentation.

The development of the embryo sack can be the same as for *Polygonum* or *Allium* species. The embryo sack's nuclear poles become fused before fertilization. The three antipodes are usually ephemeral or persistent as in the case of *Atropa*.

The fruit can be a berry as in the case of the tomato or wolfberry, a dehiscent capsule as in *Datura*, or a drupe. The fruit has axial placentation. The capsules are normally septicidal or rarely loculicidal or valvate.

The seeds are usually endospermic, oily (rarely starchy), and without obvious hairs. The seeds of most Solanaceae are round and flat, about 2–4 mm (0.079–0.157 in) in diametre. The embryo can be straight or curved and has two cotyledons. Most species in the Solanaceae have 2n=24 chromosomes, but the number may be a higher multiple of 12 due to polyploidy.

Wild potatoes, of which there are about 200, are predominantly diploid (2 × 12 = 24 chromosomes), but triploid (3 × 12 = 36 chromosomes), tetraploid (4 × 12 = 48 chromosomes), pentaploid (5 × 12 = 60) and even hexaploid (6 × 12 = 72 chromosome) species or populations exist. The cultivated species *Solanum tuberosum* has 4 × 12 = 48 chromosomes. Some *Capsicum* species have 2 × 12 = 24 chromosomes, while others have 26 chromosomes.

The Diversity of Some Characteristics

Despite the previous description, the Solanaceae exhibit a large morphological variability, even in their reproductive characteristics. Examples of this diversity include:

The number of carpels that form the gynoecium: In general, the Solanaceae have a gynoecium (the female part of the flower) formed of two carpels. However, *Melananthus* has a monocarpelar gynoecium, there are three or four carpels in *Capsicum*, three to five in *Nicandra*, some species of *Jaborosa* and *Trianaea* and four carpels in *Iochroma umbellatum*.

The number of locules in the ovary: The number of locules in the ovary is usually the same as the number of carpels. However, some species occur in which the numbers are not the same due to the existence of false septa (internal walls that subdivide each locule), such as in *Datura* and some members of the Lycieae (the genera *Grabowskia* and *Vassobia*).

Type of ovules and their number: The ovules are generally inverted, folded sharply backwards (anatropous), but some genera have ovules that are rotated at right angles to their stalk (campilotropous) as in *Phrodus*, *Grabowskia* or *Vassobia*), or are partially inverted (hemitropous as in *Cestrum*, *Capsicum*, *Schizanthus* and *Lycium*). The number of ovules per locule also varies from a few (two pairs in each locule in *Grabowskia*, one pair in each locule in *Lycium*) and very occasionally only one ovule is in each locule as for example in *Melananthus*.

The type of fruit: The fruits of the great majority of the Solanaceae are berries or capsules (including pyxidia) and less often drupes. Berries are common in the subfamilies Cestroideae, Solanoideae (with the exception of *Datura*, *Oryctus*, *Grabowskia* and the tribe Hyoscyameae) and the tribe Juanulloideae (with the exception of *Markea*). Capsules are characteristic of the subfamilies Cestroideae (with the exception of *Cestrum*) and Schizanthoideae, the tribes Salpiglossoideae and Anthocercidoideae, and the genus *Datura*. The tribe Hyoscyameae has pyxidia. Drupes are typical of the Lycieae tribe and in Iochrominae.

Economic Importance

The solanaceas include such important food species as the potato (*Solanum tuberosum*), the tomato (*Solanum lycopersicum*), the pepper (*Capsicum annuum*) and the aubergine or egg plant (*Solanum melongena*). *Nicotiana tabacum,* originally from South America, is now cultivated throughout the world to produce tobacco. Many solanaceas are important weeds in various parts of the world. Their importance lies in the fact that they can host pathogens or diseases of the cultivated plants, therefore their presence increases the loss of yield or the quality of the harvested product.

An example of this can be seen with *Acnistus arborescens* and *Browalia americana* that host thrips, which cause damage to associated cultivated plants, and certain species of *Datura* that play host to various types of virus that are later transmitted to cultivated solanaceas. Some species of weeds such as, for example *Solanum mauritianum* in South Africa represent such serious ecological and economic problems that studies are being carried out with the objective of developing a biological control through the use of insects.

Various solanaceas species are grown as ornamental trees or shrubs. Examples include *Brugmansia x candida* ("Angel's Trumpet") grown for its large pendulous trumpet-shaped flowers, or *Brunfelsia latifolia*, whose flowers are very fragrant and change colour from violet to white over a period of 3 days. Other shrub species that are grown for their attractive flowers are *Lycianthes rantonnetii* (Blue Potato Bush or Paraguay Nightshade) with violet-blue flowers and *Nicotiana glauca* ("Tree Tobacco") Other solanacea species and genera that are grown as ornamentals are the petunia *(Petunia* × *hybrida), Lycium, Solanum, Cestrum, Calibrachoa* × *hybrida* and *Solandra*. There is even a hybrid between *Petunia* and *Calibrachoa* (which constitutes a new nothogenus called ×*Petchoa* G. Boker & J. Shaw) that is being sold as an ornamental. Many other species, in particular those that produce alkaloids, are used in pharmacology and medicine *(Nicotiana, Hyoscyamus*, and *Datura*).

Solanaceas and the Genome

Many of the species belonging to this family, among them tobacco and the tomato, are model organisms that are used for research into fundamental biological questions. One of the aspects of the solanaceas' genomics is an international project that is trying to understand how the same collection of genes and proteins can give rise to a group of organisms that are so morphologically and ecologically different. The first objective of this project was to sequence the genome of the tomato. In order to achieve this each of the 12 chromosomes of the tomato's haploid genome was assigned to different sequencing centres in different countries. So chromosomes 1 and 10 were sequenced in the United States, 3 and 11 in China, 2 in Korea, 4 in Britain, 5 in India, 7 in France, 8 in Japan, 9 in Spain and 12 in Italy. The sequencing of the mitochondrial genome was carried out in Argentina and the chloroplast genome was sequenced in the European Union.

Key Pests of Solanaceous Vegetables

Crop	Insect pests	Diseases
Tomato	Thrips; Aphids; Mites; Whitefly; Beetles; Fruit fly; Cluster caterpillars; Looper caterpillars; Potato moth; Heliothis (Helicoverpa).	Tomato spotted wilt virus (TSWV); Tobacco mosaic virus (TMV); Bacterial canker; Bacterial speck; Bacterial spot; Damping-off; Powdery mildew; Tomato yellow leaf curl virus; Tomato leaf curl virus; Early blight; Bacterial wilt; Grey mould; Sclerotinia rot; Anthracnose; Fusarium wilt; Nematodes.
Eggplant	Aphids; Thrips; Leafhoppers; Two-spotted mite; Beetles; Whitefly; Heliothis (Helicoverpa); Ceropid (spittle) bug; Leafminers; Potato moth.	Damping-off; Tomato spotted wilt virus; Root rots; Tobacco mosaic virus; Tobamoviruses; Bacterial wilt; Nematodes.
Capsicum and chilli	Aphids; Thrips; Beetles; Queensland fruit fly; Whitefly; Mites.	Bacterial spot; Bacterial wilt; Anthracnose; Cercospora spot; Powdery mildew; Stem rot; Sudden wilt; Tomato spotted wilt virus (TSWV); Tomato yellow leaf curl virus; Capsicum chlorosis virus (CaCV); Cucumber mosaic virus (CMV); Pepper mild mottle virus (PMMV); Potato mosaic virus (OVY); Tobacco mosaic virus (TMV); Nematodes.
Potato	Thrips (Melon thrips, Western Flower thrips, Onion thrips); Aphids; Potato moth; Whitefly; Beetles; Looper caterpillars; Leafhoppers; Bugs; Potato moth.	Armillaria root rot; Bacterial wilt; Black leg; Soft rot; Potato cyst nematode; Root-knot nematode; Late blight; Tomato spotted wilt virus; Potato leaf roll virus; Potato virus Y; Common scab of potatoes; Rhizoctonia; Powdery mildew; Fusarium wilt; Anthracnose.

Pest Management

- Integrated Pest Management (IPM) – IPM methods discourage the use of broad-spectrum chemical pesticides and instead recommend growers employ a combination of plant protection strategies, such as best-farm hygiene, biological controls, and a targeted application of selective chemicals.
- Key control method considerations for viral diseases include – the use of healthy seedlings; destruction of harvested capsicum and tomato crops before planting new blocks nearby; maintenance of high crop and farm hygiene standards; and use of virus resistant varieties in disease management strategies, where possible. For crops such as capsicum that can be infected with Tobacco mosaic virus (TMV), it is important than anyone who has been smoking wash their hands before coming into contact with the crop, particularly before picking.
- Chemical control – Pests can be controlled with an appropriately registered chemical. The APVMA maintains a database of all chemicals registered for the control of pests in Australia. However,

growers need to ensure that they meet the relevant Maximum Residue Limits (MRLs) for the chemicals in the end market, be it domestic or export.

- Weed management – Effective weed management strategies are limited for capsicum and chilli producers. Weed management practices include the use of plastic mulch, selective grass herbicides, hand weeding, or tillage. It is important to control weeds that may be host to a virus or insect pests.
- Control of insect pests, mites and diseases can be achieved using biological, cultural and/or chemical methods. Cost-effective, safe strategies such as irrigation at strategic times and maintaining good crop hygiene should be used. It is important to examine the history of a paddock and determine what insect pests, mites and diseases were common in previous years. Preventative measures should be taken for the pests of most concern.

Insect Pests of Mango

Hopper (*Idioscopus clypealis, Idioscopus nitidulus* and *Amritodus atkinson*) : Of all the mango pests, hoppers are considered as the most serious and widespread pest throughout the country. Large number of nymphs and adult insects puncture and suck the sap of tender parts, thereby reducing the vigour of the plants. Heavy puncturing and continuous draining of the sap causes curling and drying of the infested tissue. They also damage the crop by secreting a sweet sticky substance which facilitate the development of the fungus *Maliola managiferae,* commonly known as sooty mould affecting adversely the photosynthetic activities of the leaves. Hopper population shoots up in February-April and June-August.

Control: Three sprays of 0.15% Carbaryl or 0.04% Monocrotophos or 0.05% Phosphamidon or 0.05% Methyl Parathion are effective in controlling the hoppers. First spray should be given at the early stage of panicle formation. The second spray at full-length stage of panicles but before full bloom and the third spray after the fruits set at pea size stage are recommended. Biological control agents such as the predators *Mallada boninensis* and *Chrysopa lacciperda,* the egg parasite *Polynema* sp. and a preparation of the fungus *Beauveria bassiana* are important and useful in controlling this pest.

Mealy Bugs (Drosicha mangiferae): Nymphs and adults suck the plant sap and reduce the vigour of the plant. Excessive and continuous draining of plant sap causes wilting and finally drying of infested tissue. They also secrete honey dew, a sticky substance, which facilitate the development of the fungus *Maliola mangiferae* (sooty mould). The female adult crawls down the tree in the month of April-May and enter in the cracks in the soil for laying eggs. Just after hatching, the minute newly hatched pink to brown coloured nymphs crawl up the tree. After climbing up the tree, they start sucking the sap of tender plant parts. They are considered more important because they infest the crop during the flowering season.

Control: Flooding of orchards with water in the month of October kills the eggs. Ploughing the orchards in the month of November exposes the eggs to suns heat. Polythene bands of 400 gauge and 25 cm width fastened around the tree trunk have been found effective barrier to stop the ascent of nymphs to the trees. The band should be fastened well in advance before the hatching of eggs, i.e., during November- December. Application of 250 g per tree of Methyl Parathion 2% dust in soil around the trunk kills the newly hatched nymphs.Spraying of 0.05% Monocrotophos or 0.2% Carbaryl or 0.05% Methyl Parathion is useful in controlling young nymphs of the mealy bug. The entomogenous fungus Beauveria bassiana is found to be an effective bioagent in controlling the nymphs of this pest.

Inflorescence Midge (Erosomyia indica): The midge infests and damages the crop in three different stages. The first attack is at the floral bud burst stage. The eggs are laid on newly emerging inflorescence; the larvae tunnel the axis and thus destroy the inflorescence completely. The mature larvae make small exit holes in the axis of the inflorescence and slip down into the soil for pupation. When the tender fruits are attacked they slowly turn yellow and finally drop. The third attack is on tender new leaves encircling the inflorescence. The most damaging one is the first attack in which the entire inflorescence is destroyed even before flowering and fruiting. The inflorescence shows stunted growth and its axis bends at the entrance point of the larvae.

Control: As the larvae pupate in the soil, ploughing of the orchard expose pupating larvae to sun helps in killing them. Soil application of Heptachlor or Methyl Parathion @ 25 to 30 kg/ha kills pupating larvae in the soil. Spraying of 0.05% Fenetrothion or 0.045% Dimethoate at the bud burst stage of the inflorescence is effective.

Fruitfly (Daccus dorsalis;, D. zonatus and D. correctus): The oriental fruitfly is one of the most serious pests of mango in the country, which has created problem in the export of fresh fruits. The female punctures the outer wall of the mature fruits with the help of its pointed ovipositor and insert eggs in small clusters inside the mesocarp of mature fruits. After hatching, the larva feeds on the pulp of fruit which appears normal from outside, but drops down finally. The mature maggots fall down into the soil for pupation. The emergence of fruitfly starts from April onwards and the maximum population is recorded during May- July, which coincides with fruit maturity. The population declines slowly from August to September after that it is non-existent up to March.

Control: Bait sprays of Carbaryl (0.2%)+protein hydrolysate (0.1%) or molasses starting at first week of April and repeated once after 21 days or hanging traps containing 100 ml water emulsion of Methyl Euginol (0.1%) + Malathion (0.1%) during fruiting is effective. As per the experiments conducted in IIHR, Bangalore against fruit fly, hot water treatment at 480C for one hour gave total control in Alphonso and Totapuri.

Scale Insects: (*Pulvinanapolygonata, Aspidiotus destructor, Ceroplastis* sp. And *Rastococcus* sp): The nymphs and adult scales suck the sap of the leaves and other tender parts and reduce the vigour of the plants. They also secrete honeydew, which helps in the development of sooty mould on leaves and other tender parts of the tree. In case of severe scale infestation, growth and fruit bearing capacity of the tree is affected adversely.

Control: Pruning of the heavily infested plant parts and their immediate destruction followed by two sprays of Monocrotophos (0.04%) or Dimethoate (0.06%) at an interval of 20 days is very effective in controlling the scale population.

Shoot Borer (Chlumetia transversa): This pest is found all over the country. Larvae of this moth bore into the young shoot resulting in dropping of leaves and wilting of shoots. Larvae also bore into the inflorescence stalk. Female moths lay egg on tender leaves. After hatching, young larvae enter the midrib of leaves and then enter into young shoots through the growing points by tunneling downwards.

Control: The attacked shoots may be clipped off and destroyed. Spraying of Carbaryl (0.2%) or Quinalphos (0.5%) or Monocrotophos (0.04%) at fortnightly intervals from the commencement of new flush gives effective control of the pest. A total of 2 to 3 sprays may be given depending on the intensity of infestation.

Bark Eating Caterpillar (Indarbella quadrinotata): The old, shady and neglected orchards are more prone to attack by this pest. Larvae of this moth feed on the bark and weaken the tree. A single female lays about 300-400 eggs in batches on the bark. The caterpillar spins brown silken web on the tree, which consists of their excreta and wood particles. Larvae also make shelter tunnels inside the stem in which they rest. Larvae generally feed from April to December.

Control: Removal of the webs from tree trunks and injecting emulsion of Monocrotophos (0.05%) or DDVP (0.05%) in each hole and plugging them with mud can control the pest.

Stem Borer *(Batocera rufomaculata):* The grub of this beetle causes damage as it feeds inside the stems, boring upward resulting in drying of branches and in severe cases attacked stem is killed. Eggs are laid either in the slits of tree trunk or in the cavities in main branches and stems covered with a viscous fluid. Pupation takes place within the stem. Beetle emerges in July-August

Control: Cleaning the tunnel with a hard wire, pouring kerosene oil, creosote, petrol, crude oil or formalin and subsequently closing entrance of the tunnel with mud or plugging it with cotton wool soaked in any of the above substances kills the grub.

Shoot Gall Psylla (Apsylla cistellata): It is a very serious pest of mango in many parts of India, particularly in Tarai region of U.P., north Bihar and West Bengal. This pest creates green conical galls in leaf axis. The activity of the pest starts from August. The galls dry out after emergence of adults in March. The adult females lay eggs in the midribs as well as in lateral axis of new leaves. Nymphs emerge from eggs during August-September and crawl to the adjacent buds to suck cell sap. As a result of feeding, the buds develop into hard conical green galls. The galls are usually seen during September-October. Consequently, there is no fruit set.

Control: The galls with nymphs inside should be collected and destroyed to prevent carryover of the pest. The pest can effectively be controlled by spraying of Monocrotophos (0.05%), Parathion (0.04%), Metasystox (0.1%) or Quinalphos (0.05%) at 2 week intervals starting from the middle of August. Repeated use of same chemical for every spray should be avoided.

Leaf Webber (Ortbaga euadrusalis): The infestation starts from the month of April and continues up to December. Eggs are laid singly or in clusters within silken webbings on leaves. Upon hatching, the caterpillars feed on leaf surface by scrapping. Later, they make web on tender shoots and leaves together and feed within. Old orchards with lesser space between tree canopy have more infestation than open orchards.

Control: Pruning of infested shoots and their burning in the month of April to July is found effective. Raking of the soil around the base of the trees in January, after the last generation has pupated, helps in checking the pest population. Three sprays starting from the last week of July at 15 days interval with Carbaryl (0.2%) or Monocrotophos (0.05%) or Quinalphos (0.05%) effectively controls the pest.

Stone Weevil (Sternochetus mangiferae): It is a common pest of mango in southern India. Varieties with high TSS and sugar such as Alphonso, Bangalora, Neelum, etc. are more prone to attack by this pest. Female lay eggs under the rind of ripening fruits. Newly emerged grubs bore through the pulp, feed on seed coat and later cause damage to cotyledons. Pupation takes place inside the seed. Pulp is discoloured around the affected portion.

Control: The pest population can be kept under check by destroying the affected fruits and exposing the hibernating weevils by digging the soil. Spraying the trees with Fenthion (0.01%) is found effective. In Alphonso and Banganapalli, a single spray of Monocrotophos 36 EC 1.5ml/litre of water at marble stage gave 100% and 97.5% control of stone weevil respectively. In Totapuri, Carbaryl 50 % WDP @ 4 g/ litre of water is effective in controlling stone weevil.

5

Entomopathogenic Bacteria and Viruses

The main insect pests of cucurbits are: Cucurbit Fruit Fly, Red Pumpkin Beetle and Epilachna Beetle. Their damage symptoms and control measures are described as follows.

Cucurbit Fruit Fly

Fruit flies are most damaging for cucurbitaceous crops.

Nature of damage:

- Adult female flies select soft and young fruits for oviposition by puncturing the rind with the ovipositor. Such damaged fruits show signs of raised and brown resinous encrustation at the sites of ovipositional punctures due to discharge and drain out fruit juice through punctures
- Infested fruits are either rotten or deformed in shape due to the microbial infection. Such rotten or deformed fruits are not fit for sale or human consumption.
- Infested fruits with apparent firmness and growth show internal decay and foul smell when cut open.

Control measures:

- Growing of resistant or early maturing varieties and infested fruits should be collected and destroyed.
- Use of sex pheromone like "Cuelure".

- Fruit flies can be controlled by using poison bait (100g smashed ripened pumpkin + 0.5g Dipterex + 100ml water).
- Apply carbaryl 10% DP in pit before sowing of seeds to destroy the pupae.

Red Pumpkin Beetle

The red pumpkin beetle is a serious pest of cucurbitaceous plants, except bitter gourd.

Nature of damage:

- Adult beetles feed voraciously on the leaf lamina by making irregular holes.
- The maximum damage is done when the crop is in the cotyledon stage. The first generation is therefore more injurious than the subsequent generations.
- The adult insect also feed on the leaves of grown up plant by scrapping off their chlorophyll and make the leaves net like appearance.
- The attacked plants may wither away and re-sowing of the crop may become necessary in certain cases.
- The larvae cause damage in various ways by boring into the roots and the underground stem portion and by feeding on the leaves and fruits line in contact with the soil.
- The damaged roots and the underground roots and the underground stems portion may rot due to infection by the saprophytic fungi.
- The young and smaller fruits of the infested creepers may dry up, whereas the bigger and mature fruits become unfit for human consumption.

Control measures:

- Spraying of contact insecticides like carbaryl 10% DP has been effective in controlling beetles, but repeated application at an interval of 7 days are necessary.
- Using neem oil cake in the soil has been reported to kill the pest larvae.
- Sown early that the plant passes the cotyledon stage by the time the beetles become active.

Epilachna beetle: Epilachna beetle are major insect pest of teasle gourd, cucumber, bitter gourd, watermelon, muskmelon and other cucurbitaceous crops.

Nature of damage:

- Many grubs and adults can be seen on the ventral surface of the leaves.
- Both the adult and grub stages feed on the leaf surfaces and skeletonize the leaves which present a lace like appearance.
- The attacked leaves turn brown, dry up and fall off.
- In case of severe infestation the crop presents a very unhealthy look.
- The vigour of the plant and thereby its yield is adversely affected.

Control measures:

- The pest population can be suppressed effectively and its infestation can be squeeze in the bud, by regular killing and picking of eggs, grubs and adults, if the cropped area is small.
- Also the larvae and adults can be shaken down in container of kerosinized water early in the morning.
- The pest can be killed by spraying Malathion/ Fyfanon/ Zythiol 50 EC @2 ml of water. The treatment should be applied as soon as the pest appears and repeated at 15 days interval.

Entomopathogenic Viruses

Baculoviruses are a large group of double-stranded DNA viruses (almost 1000 species have been described); the majority have been isolated from a few insect orders: Lepidoptera, Diptera, Hymenoptera and Coleoptera. Viral genome ranges in size from 80 to 200 kb. Individual baculoviruses usually have a narrow host range limited to a few closely related species. The most widely studied baculovirus is the Autographa californica nuclepolyhedrovirus (AcMNPV). Baculoviruses are arthropod viruses well known due to their potential as agents of biological control of pests in agriculture and forestry. They are also widely used as expression vectors in biotechnology. The family Baculoviridae contains diverse members and in the past the classification was based on virus morphology. It was divided into two genera: the Nucleopolyhedrovirus (NPVs) and the Granulovirus (GVs). Recently, this division was challenged because the comparison of 29 fully sequenced baculoviral genomes indicated that virus phylogeny followed more closely the classification of the hosts than the virion morphological traits, but the traditional division into two genera is still widely used.Baculoviruses infect arthropods and they do not replicate in vertebrates, plants and microorganisms. Although they do not replicate, they may, under special conditions, enter animal cells. This unexpected property made them a valuable tool in the last few years for studies of transient expression of foreign genes under vertebrate promoters introduced into baculovirus genome. The circular DNA genome of AcMNPV is surrounded by a small basic protein which neutralizes the negative charge of the DNA. This structure is protected by proteins forming a nucleocapsid. Virions consist of one or more nucleocapsids embedded in a membranous envelope. Two morphologically distinct, but genetically identical, viral forms are produced at different times post-infection:

- Budded virus particles (BV) which serve for the transmission of the virus to other tissues of the caterpillar body.
- Occlusion bodies (OB) which are responsible for the survival of the virus in the environment and the spread of the virus from insect to insect.

Baculovirus system of expression of foreign genes has many advantages over other systems because high level of foreign gene expression is usually achieved compared to other eukaryotic expression systems. Baculovirus genome can accommodate large pieces (up to 50 kbp) of foreign DNA, so it is possible to express more than one foreign gene. Additionally, the insertion of specific signal sequences in front of a foreign gene leads very often to the export of the gene product outside of the infected cell.

Recombinant baculoviruses are usually constructed in two steps. Initially, a heterologous gene is introduced into a baculovirus transfer vector. It consists of a bacterial replicon of a multicopy plasmid, a selection marker gene, promoter and terminator regions along with flanking baculovirus sequences from a non-essential locus, and a multiple cloning site (or a single unique restriction site) downstream from a viral promoter. Most often the promoters and the flanking DNA originate from one of the late genes: polyhedrin or p10 gene. The latter is another viral gene coding for a protein which is produced in large quantities late in the infection. It is the main component of the fibrillar structures which accumulate in the nucleus and in the cytoplasm of infected cells.

Life Cycle

Viruses are unable to reproduce without a host - they are *obligate parasites*. Baculoviruses are no exception. The cells of the host's body are taken over by the genetic message carried within each virion, and forced to produce more virus particles until the cell, and ultimately the insect, dies. Most baculoviruses cause the host insect to die in a way that will maximize the chance that other insects will come in contact with the virus and become infected in turn. Infection by baculovirus begins when an insect eats virus particles on a plant - perhaps from a sprayed treatment. The infected insect dies and "melts" or falls apart on foliage, releasing more virus. This additional infective material can infect more insects, continuing the cycle.

Prognoses for Future Use of Baculovirus Pesticides

Large-scale application of AgMNPV in Brazil has proven that the baculovirus protection can be done at relatively low cost. It is very likely that the growing awareness of the benefits of the environment-friendly pesticides will result in the reevaluation of the prospects for biological protection with baculovirus preparations. However, until today, baculovirus insecticides have not met their full potential to control pest insects worldwide. The development of recombinant baculovirus was efficiently completed by researchers in several countries, but the in vitro commercial technology still lags behind today, due to technical problems. Future development of baculovirus pesticides will probably depend on the attitude towards genetically modified organisms. In countries where use of genetically modified organisms is restricted, only naturally occurring baculoviruses will be used for protection of crops. In this case the improvements will be at the level of diagnostics of infection, development of the in vitro cultures and changes in the formulations of the biopesticide.

In countries which favour the introduction of genetically modified organisms, the improvements will be achieved by introduction of exogenous genes into baculovirus genome, thus greatly enhancing the killing activity of bioinsecticide formulations. The activity of baculoviruses against their natural hosts may be enhanced by introduction of insect-specific toxins or by interference with insect physiology. Baculovirus genome modifications by introduction of exogenous toxin genes were extensively studied in many laboratories. Most of the research was devoted to the studies of arthropod toxin genes isolated from the scorpion or spiders. The most potent insect-specific toxin gene used for construction of baculovirus recombinants was the gene coding for a toxin from scorpion Androctonus australis. The feeding damage caused by larvae infected with this modified baculovirus was reduced by about 60% in comparison to a wild type baculovirus. Baculovirus recombinants that produced occlusion bodies incorporating Bacillus thuringiensis toxin were constructed by making a fusion protein consisting of polyhedrin and Bt toxin. The pathogenicity of the recombinant was remarkably increased compared to wild-type virus. These studies proved that it is possible to construct a biopesticide which combines the advantages of the virus and the bacterial toxin.

Entomopathogenic viruses can infect and kill pest insects. In nature there are entomopathogenic fungi, nematodes, bacteria and viruses everywhere. They contribute to the natural regulation of many populations of pest insects.

In vitro Production

The entomopathogenic virus LdMNPV for example is highly effective against gypsy moth, *Lymantria dispar.* This pest causes losses in forestry in different climatic zones.

So far a LdMNPV production is possible in vivo using the pest insect as host to meet the market requirements. BIOCOMES will focus on the development of the in vitro production of this LdMNPV virus to allow a significant further up-scaling of the production. The in vitro production of the LdMNPV agent will be developed in cell cultures to replace in vivo production depending on insect rearing.

Entomopathogenic Fungi

An entomopathogenic fungus is a fungus that can act as a parasite of insects and kills or seriously disables them.

Typical Life Cycle

These fungi usually attach to the external body surface of insects in the form of microscopic spores (usually asexual, mitosporic spores also called conidia). Under permissive conditions of temperature and (usually high) moisture, these spores germinate, grow as hyphae and colonize the insect's cuticle; eventually they bore through it and reach the insects' body cavity (hemocoel). Then, the fungal cells proliferate in the host body cavity, usually as walled hyphae or in the form of wall-less protoplasts (depending on the fungus involved). After some time the insect is usually killed (sometimes by fungal toxins) and new propagules (spores) are formed in/on the insect if environmental conditions are again permissive; usually high humidity is required for sporulation.

Groups

The entomopathogenic fungi include taxa from several of the main fungal groups and do not form a monophyletic group. Many common and/or important entomopathogenic fungi are in the order Hypocreales of the Ascomycota: the asexual (anamorph) phases *Beauveria, Metarhizium, Nomuraea, Paecilomyces* = *Isaria, Hirsutella* and the sexual (teleomorph) state *Cordyceps*; others (*Entomophthora, Zoophthora, Pandora, Entomophaga*) belong in the order Entomophthorales of the Zygomycota.

Pest Control

Since they are considered natural mortality agents and environmentally safe, there is worldwide interest in the use and manipulation of entomopathogenic fungi for biological control of insects and other arthropod pests. In particular, the asexual phases of Ascomycota (*Beauveria* spp., *Lecanicillium* spp., *Metarhizium* spp., *Paecilomyces* spp. and others) are under intense scrutiny due to the traits favouring their use as biological insecticides.

Production. Most entomopathogenic fungi can be grown on artificial media. However, some require extremely complex media; others, like *Beauveria bassiana* and exploitable species in the genus *Metarhizium*, can be grown on starch-rich substrates like cereal grains (rice, wheat).

Virulence. The Entomophthorales are often reported as causing high levels of mortality (epizootics) in nature. These fungi are highly virulent. The anamorphic Ascomycota (*Metarhizium*, *Beauveria* etc.) are reported as causing epizootics less frequently in nature.

Also important are their properties regarding specificity (host range), storage, formulation, and application.

Predation

In ecology, predation is a biological interaction where a predator (an organism that is hunting) feeds on its prey (the organism that is attacked). Predators may or may not kill their prey prior to feeding on them, but the act of predation often results in the death of its prey and the eventual absorption of the prey's tissue through consumption. Other categories of consumption are herbivory (eating parts of plants), mycophagy (eating parts of fungi) and detritivory, the consumption of dead organic material (detritus).

All these consumption categories fall under the rubric of consumer-resource systems. It can often be difficult to separate various types of feeding behaviours. For example, some parasitic species prey on a host organism and then lay their eggs on it for their offspring to feed on it while it continues to live or on its decaying corpse after it has died. The key characteristic of predation however is the predator's direct impact on the prey population. On the other hand, detritivores simply eat dead organic material arising from the decay of dead individuals and have no direct impact on the "donor" organism(s).

Selective pressures imposed on one another often leads to an evolutionary arms race between prey and predator, resulting in various antipredator adaptations. Ways of classifying predation surveyed here include grouping by trophic level or diet, by specialization, and by the nature of the predator's interaction with prey.

Predators may be put to use in conservation efforts to control introduced species. Although the aim in this situation is to remove the introduced species entirely, keeping its abundance down is often the only possibility. Predators from its natural range may be introduced to control populations, though in some cases this has little effect, and may even cause unforeseen problems. Besides their use in conservation biology, predators are also important for controlling pests in agriculture. Natural predators are an environmentally friendly and sustainable way of reducing damage to crops, and are one alternative to the use of chemical agents such as pesticides.

Recent Research Trends in the Use of Predators for Biological Control

The arthropod predators of insects and mites include beetles, true bugs, lacewings, flies, midges, spiders, wasps, and predatory mites. Insect predators can be found throughout plants, including the parts below ground, as well as in nearby shrubs and trees. Some predators are specialized in their choice of prey, others are generalists. Some are extremely useful natural enemies of insect pests. Unfortunately, some prey on other beneficial insects as well as pests.

Insect predators can be found in almost all agricultural and natural habitats. Each group may have a different life cycle and habits. Although the life history of some common predators is well studied, information on the biology and relative importance of many predatory species is lacking. In this document, we have included the more common and better understood beneficial predators.

Major characteristics of arthropod predators:

- adults and immatures are often generalists rather than specialists
- they generally are larger than their prey
- they kill or consume many prey
- males, females, immatures, and adults may be predatory
- they attack immature and adult prey

Relative Effectiveness

Most beneficial predators will consume many pest insects during their development, but some predators are more effective at controlling pests than others. Some species may play an important role in the suppression of some pests. Others may provide good late season control, but appear too late to suppress the early season pest population. Many beneficial species may have only a minor impact by themselves but contribute to overall pest mortality. Often too, the role of the beneficial predators has not been adequately studied.

Surveys of agricultural systems give an indication of the potential number and diversity of predators in a crop. For example, over 600 species of predators in 45 families of insects and 23 families of spiders and mites have been recorded in Arkansas cotton. Eighteen species of predatory insects (not including spiders and mites) have been found in potatoes in the northeastern United States. There may be thousands of predators per acre, in addition to many parasitoids. Although the impact of any one species of natural enemy may be minor, the combined impact of predators, parasitoids, and insect pathogens can be considerable.

Parasitoid

Figure: *SEM image of endoparasitoid ciliates of the genus* Collinia, *which can cause mass mortality in affected krill populations*

A parasitoid is an organism that spends a significant portion of its life history attached to or within a single host organism in a relationship that is in essence parasitic; unlike a true parasite, however, it ultimately sterilises or kills, and sometimes consumes, the host. Thus parasitoids are similar to typical parasites except in the more dire prognosis for the host.

The term *parasitoid* was coined in 1913 by the German writer O. M. Reuter (and adopted in English by his reviewer, William Morton Wheeler) to describe the strategy in which, during its development, the parasite lives in or on the body of a single host individual, eventually killing that host, the adult parasitoid being free-living. Since that time however, the concept has been variously generalised and widely applied. In practice it is not always necessary to distinguish parasitoidy from parasitism, nor is it always even possible to do so cleanly. However, when it is appropriate to do so, a typically parasitic relationship is one in which parasite and host interact without lethal harm to the host, and without dramatically reducing the host's reproductive success. In most such relationships, the parasite arrogates enough nutrients or other resources to thrive without preventing the host from reproducing. In contrast, in a *parasitoidal* relationship the exploiting organism kills or sterilises the host, typically before it can produce offspring. A non-lethal parasite sometimes is termed a *biotroph*. In contrast, when a parasitoidal relationship is regarded as a form of parasitism, the parasitoid may be called a *necrotroph*.

When an organism sterilises its host without directly killing it, then whether to term it a parasitoid or a parasite, is a matter of context and preference. Often for a parasite to prevent reproduction of the host is incidental, but various forms of systematic parasitic castration occur among parasitoids, and many such biological strategies are highly sophisticated. Crustacean parasites or parasitoids include several impressive examples.

Protelean is a term that various authors use to denote organisms that live as parasites only during the early, growing, phases of their lives; typically they then begin by behaving as internal parasites; also typically they end that phase of their lives parasitoidally by killing or consuming the host. Finally they emerge as free-living adults, with or without an intervening phase of diapause or metamorphosis.

Protelean organisms are widely regarded as a special class of parasites, or more usually parasitoids. The most typical examples of proteleans are the parasitoidal Hymenoptera, Diptera, Strepsiptera, and some other insects. Some that do not necessarily kill the host, such as the Strepsiptera, may nonetheless be counted as parasitoids because they generally functionally sterilize it. Usually such insects are holometabolous. It is reasonable to regard holometaboly as preadaptation for the protelean life history because it implies that their larval stage of life differs drastically from the adult stage, both functionally and morphologically.

Idiobiont parasitoids are those that prevent further development of the host after initially immobilizing it, and, almost without exception, develop outside the host. Koinobiont parasitoids allow the host to continue its development while feeding upon it, and may parasitize any host life stage. In turn, koinobionts can be subdivided further into endoparasitoids, which develop inside body of the host, and ectoparasitoids, which develop outside the host body, though the parasitoids frequently are attached or embedded in the host's tissues.

It is fairly common for a parasitoid itself to serve as the host for another parasitoid's offspring. The latter is commonly termed a hyperparasite, but in most cases this term is slightly misleading, as both the host *and* the primary parasitoid are killed. A better term might be secondary parasitoid, or hyperparasitoid.

Most known specialist hyperparasite and hyperparasitoid species are in the insect order Hymenoptera, but a fair amount of incidental hyperparasitoidy results when a single host or a single food stash happens to house multiple guests and rations run short. Some members of the flesh fly family, Sarcophagidae, subfamily Miltogramminae, for example members of the genus *Craticulina*, are kleptoparasites of wasps in the subfamilies Bembicinae and Philanthinae (both currently placed in the family Crabronidae). Both those subfamilies tend to build nests by digging tunnels in sand, which they then stock with prey such as flies or bees, depending on the species. Kleptoparasitic flies such as *Craticulina* are much smaller than the host wasp and lay their eggs on the prey as the wasp returns to the nest on a victualing flight. The fly larvae are small, though faster-growing than the wasp larva, and if there is only one, the wasp is likely to complete its metamorphosis successfully, but when there are several it might suffer from malnutrition or even get eaten itself, which amounts to incidental kleptoparasitoidy.

In contrast though, as described in the following section, some insects, such as some members of the Trigonalidae, not only are specialist hyperparasitoids, but have advanced behavioural adaptations to support their speciality.

Note once again that there is no clear separation between the concepts of parasitism and parasitoidy. Many species of true parasites can cause the death of their host if for example they are present in overwhelming numbers or the host is in poor condition, or other compromising circumstances develop, such as secondary infections. For example, blood-sucking mites sometimes overwhelm nestlings of birds such as swallows to the point that the young birds cannot fledge successfully,.

Infestations of other mites cause various kinds of mange in mammals. Mange mites are generally in the families Demodicidae that cause Demodicosis or demodectic mange, Sarcoptidae that cause scabies or sarcoptic mange, and Psoroptidae that cause scab in sheep and rabbits. Severe mange can debilitate animals to the point that they cannot feed themselves adequately, so that in unfavourable circumstances they may die.

Again, various species of paralytic ticks sometimes kill dogs if the owners are insufficiently alert, and soft ticks can fatally poison a host such as a horse that might rest in an infested shady spot because it does not know the local hazards. Conversely, some parasitoids do somewhat shorten the lives of their hosts or constrain their reproduction, but without necessarily killing them as a part of their interaction. Almost any microbial disease could be defined as a parasitic condition, and some could be argued to amount to clear examples of parasitoidy. One converse argument is that when the death of the host is neither a logical nor necessarily even a desirable consequence from the point of view of the parasite, the relationship should be regarded as parasitic rather than parasitoidal. This certainly would apply to examples such as mange, and diseases in which the living victim acts as a natural reservoir or even a vector.

In their extreme forms the categories of parasitism and parasitoidy patently are distinct; one is in no doubt whether the larva of a Tarantula hawk wasp behaves more like a parasitoid, or even a predator, than a parasite; and similarly the biting midges that suck blood from large insects plainly are simply ectoparasites. However, there is a continuum of intermediate and contingent conditions that bridge those categories in practically every respect. This should not be taken too seriously as a material problem in terminology; the terms are useful in particular contexts and should not be abused by inappropriate application in contexts in which they create confusion rather than clarity. Many examples of species that are technically parasitoidal, at least facultatively, are not generally referred to as parasitoidal. Many microbial diseases and the aforementioned soft ticks constitute instructive examples.

Nor do those complete the list of examples that constitute the requirement for fuzzy distinctions in such matters; at the opposite extreme from parasitism, parasitoidy in turn grades into predation. Differences between various kinds of hunting wasps provide convenient illustrations. Predatory social wasps hunt flies, caterpillars and the like, grab them, butcher them, carry them home and feed them to their young. That is patent predation. Some solitary wasps, such as bee pirates, sting prey, sometimes fatally, before saving it, usually entire, in a nest or burrow for the young to feed on. That too is predation, fairly clearly.

In contrast, the best-known protelean solitary hunting wasps sting prey to paralyse it before storing it for the young in the nest. The larvae then proceed to eat the stored prey alive, sometimes according to very sophisticated schedules that delay killing the victim sooner than necessary, thereby avoiding having their rations rot before they could be consumed. Some authorities regard such larval behaviour as having a strong element of parasitoidy. That view is based largely on the view that the young larvae begin with small exactions like any parasite, then proceed to the point where they eat at such a rate that they might as well be predators.

Other wasps paralyse prey in the plant or other environment in which it feeds, before laying eggs nearby. The emerging young attack and feed on the paralysed prey organism in its own home. Some solitary parasitoids among insects lay their eggs on or in their live prey and any of a wide range of consumption schedules might follow. Some parasitoids even lay their eggs where the larvae must locate the prey for themselves when they hatch from the eggs. Examples include flies in the families Tachinidae and Bombyliidae. The physiological and strategic sophistication of such relationships, whether parasitoidal or parasitic, often are impressive.

Patently there is no point to trying to draw arbitrary lines of distinction between such vague, and often variable, life histories. In each ecological or ethological study the terms applied should reflect the facts in the contexts relevant to the matter in question. Such studies need not in all cases use the identical terminology, and there is no reason they should. All that is necessary is that the terminology in each study should be clear, useful and relevant.

Types of Parasitoids

The parasitoidal type of relationship seems to occur largely in organisms that have fast reproduction rates, such as insects or (perhaps more rarely) mites or nematodes. Workers in this field have pointed out that parasitoids often are closely coevolved with their hosts, which is inarguably true. To maintain a sound perspective of the matter though, one must remember that coevolution might reasonably be expected to develop to even higher degrees of sophistication in the more intimate classes of parasitic relationships. In fact advanced degrees of coevolution occur in the complex interplay between simultaneously extant predator-prey relationships as well.

In using the term *parasitoid* it is common to think in terms of parasitoidal insects such as Tachinidae and Pompilidae. Some writers recognise or discuss no other classes of parasitoidy. More realistically however, the life histories of several other groups of organisms are equally parasitoidal. In general there is no logical basis for excluding wider use of the term. For example, among the so-called worms, many Nematoda are important parasitoids of insects, snails and similar commercially important pest organisms. Under favourable circumstances they commonly multiply in the host until the carcase is a shell overflowing with a pullulating mass of worms.

Other organisms that might merit the term include certain Pteromalid gall wasps that abort host plant inflorescences, seed weevils, certain plants largely regarded as parasitic, and certain bacteria and viruses (e.g., bacteriophages), in relationships where the beneficiaries obligately destroy their hosts.

Not all such organisms regularly behave quite so parasitoidally of course; for example some bacteriophages establish complex life cycles in which phage particles do get released catastrophically, but only at intervals of many generations of the host, whereas other bacterial viruses emerge intermittently but fairly harmlessly in small numbers at a time.

The clearest cases of fully functional parasitoidy are the likes of many parasitoidal wasps and flies that consume their hosts as completely as any spider or hawk that summarily eats its prey. However, there also are many species of parasitoid that frequently or even routinely kill their "host" or "prey" without consuming much of it. This apparently wasteful strategy sometimes might have the effect of reducing the risk that the prey could escape or offer resistance. In other cases the residue of the victim simply might be difficult to eat or not very nutritious. For example various Phorid flies such as Apocephalus species, are parasitoids of particular species of ants. Various species attack ant genera such as "big-headed ants", Fire ants or *Solenopsis*, *Paraponera*, and leaf-cutter ants. However, the larvae of most such Phoridae eat mainly the contents of the ants' "head capsules" abandoning the rest of the carcases when pupating. In laying her eggs, the parent fly selects the largest ant workers, which have just the size of head to produce an adequate adult phorid. Presumably the large head-capsule contains the most concentrated nutritious muscle and brain tissue. One also could think of the Rabies virus in similar terms, or the aforementioned soft ticks. Both commonly or invariably cause the death of the host, after consuming at most a trivial fraction of the host's resources.

Influence on Host Behaviour

In another strategy, some parasitoids influence the host's behaviour in ways that favour the propagation of the parasitoid, often at the cost of the host's own life. A spectacular example is the endoparasitoid Dicrocoelium dendriticum, the Lancet Liver Fluke that causes host ants to die clinging to grass stalks where grazers or birds may be expected to eat them and complete the parasitoidal fluke's life cycle in its definitive host. Similarly, as strepsipteran parasitoids of ants mature, they cause the hosts to dawdle high on grass stalks, positions that are risky, but favour the emergence of the Strepsipterans. Other species of endoparasitoids cause infected woodlice and land Amphipoda such as *Talitroides* to run about in the open by day, where predators such as birds can catch them and continue the cycle.

Figure: *Female Apocephalus borealis ovispositing into the abdomen of a worker honey bee.*

Returning to the case of the rabies virus and the disease, one could rationalise the death of the host similarly. The virus affects the host's central nervous system with eventually fatal effects. That could be seen as a consequence of the strategy for dissemination of the virus by affecting the host behaviour. Similar principles might apply to, for example, *Vibrio cholerae*, the cholera bacterium and other, often fatal, enteric pathogens that induce diarrhoea and spread by contagion.

For the soft ticks the benefit of the paralysis they inflict might be seen as influencing behaviour in that it prevents the host from wandering away while they feed, which they do very quickly and in large numbers, some species emerging from their hiding places at night. Other species hide in sandy patches in the shade of trees in semi-desert such as the Kalahari, and emerge to feed as soon as any large animal settles down in the shade during the heat of the day.

In parasitic birds such as cuckoos, the young often are adapted to act as "super solicitors", with loud, persistent voices and with large, vividly coloured gapes and behaviour that stimulate the feeding instincts of the foster parents to the utmost. Consequently the legitimate chicks, even if they are not evicted, often starve because they are less well-equipped for soliciting for food.

Mosquitoes carrying Dengue bite their victims in daylight hours as against twilight and night hours of uninfected mosquitoes.

Parasitoidal Microbial Diseases

As mentioned, some microbial parasitoids waste most of the host's resources when it dies, but there are other parasitoidal strategies among microbes as well. One more conceptually economical form of parasitoidy is exemplified by microbial pathogens of various invertebrates such as many insects. The most notorious might well be Microsporidiosis in the form of nosema in silkworms.

This infection is highly virulent and the tissues of the victims contain huge numbers of infectious spores. In effect the pathogen in its role of parasitoid has used up most of the resources of the host to propagate and spread its offspring. Similarly, many viruses, bacterial and other, continue to propagate inside a host cell until it physically ruptures. In doing so they too consume effectively the whole of the host's resources.

Parasitoidal fungi such as Entomophthora species carry this principle as far as is possible. Having infected and killed an insect, they continue to grow on the carcase and release spores for as long as any resources remain. In this such microbes resemble the aforementioned propagation of some Nematoda in snails and insects.

Parasitoidal Plants

There are parasitoidal plants as well. Various species of dodder indiscriminately parasitise wide ranges of host plants, and debilitate or kill the branches that they infect, and commonly the whole host plant as well. Mistletoes in families such as Santalaceae and Loranthaceae commonly accumulate on host trees till they stunt and eventually kill them, sometimes after many decades. Occasionally a freak condition can arise where the (strictly speaking "hemiparasitic") plant can supply sufficient photosynthetic power to support the root system of a small host tree for several years after the live host shoots have effectively disappeared.

A related example is where the parasitoid plant is not strictly a parasite in the normal sense, but nonetheless exploits the host's resources of space, support and light. The best-known are the so-called "strangler figs". Some of them will grow on and round the trunk of the host tree and squeeze it or starve it of light until, after perhaps decades, it dies. The strangler eventually replaces the host utterly as the original trunk rots from within the stems of the strangler, leaving a hollow framework.

Parasitoidal Crustaceans

The subphylum Crustacea includes a surprising range of parasitoidal species and strategies. As with many other parasitoids, the killing of the host often is incidental. For example, in the family Ergasilidae, the "gill lice", most adult females live as parasites in the gills of fish. The harm they do the host is incidental to the parasitism, but it often is fatal or at least debilitates the fish so badly as to prevent breeding.

A particularly startling genus of Cirripedia, or barnacles is Sacculina. It literally injects itself into a crab of a suitable species and by complex processes converts itself into an egg-laying bag. In the process it disrupts the reproductive system of the host, an act of parasitic castration that qualifies it for classification as a parasitoid rather than just a parasite.

Those examples are just a few of many among the Crustacea.

Parasitoidal Insects

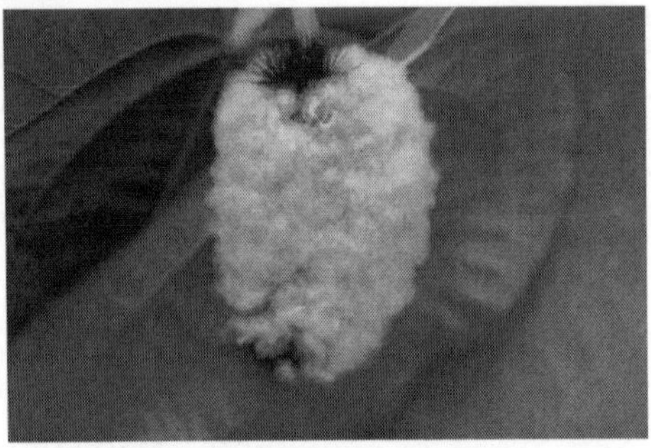

Figure: *Braconidae wasp cocoons on giant leopard larva*

About 10% of described insect species are entomophagous parasitoids. There are four insect orders that are particularly renowned for this type of life history. By far the majority are in the order Hymenoptera.

The largest and best-known group comprises the so-called "Parasitica" within the Hymenopteran suborder Apocrita: the largest subgroups of these are the chalcidoid wasps (superfamily Chalcidoidea) and the ichneumon wasps (superfamily Ichneumonoidea), followed by the Proctotrupoidea and Platygastroidea. Outside of the Parasitica, many other Hymenopteran lineages that include parasitoids, such as most of the Chrysidoidea and Vespoidea, and the rare Symphytan family Orussidae.

The flies (order Diptera) include several families of parasitoids, the largest of which is the family Tachinidae, and also smaller families such as Pipunculidae, Conopidae, and others. Other families of flies that are not primarily parasitoids or parasites, or at least not primarily protelean, do nonetheless include protelean species. For example Phoridae have already been mentioned as parasitoidal on ants, and at least some flesh fly species, such as Emblemasoma auditrix, are parasitoidal on cicadas, and have raised great interest because they locate their hosts by sound. The kleptoparasitic flesh fly genus *Craticulina* has already been mentioned and logically qualifies as a protelean fly genus.

Two other orders with parasitoidal members are the "twisted-wing parasites" (order Strepsiptera), which is a small group consisting entirely of parasitoids (though they generally sterilize the host rather than killing it), and the beetles (order Coleoptera), which includes at least two families, Ripiphoridae and Rhipiceridae, that are largely parasitoids, and rove beetles (family Staphylinidae) of the genus *Aleochara.* Occasional members of other orders can be parasitoids; one of the more remarkable is the moth family Epipyropidae, which are ectoparasitoids of planthoppers and Cicadas. The genus Cyclotorna has even more elaborate habits, beginning its growth period parasitising plant bugs, and concluding by feeding on ant larvae in their colonies.

Hymenopteran parasitoids often have unique life cycles. In one family, the Trigonalidae, the female wasps deposit eggs into small pockets they cut into the edge of leaves with their ovipositor. A caterpillar chewing these leaves may unknowingly swallow some of the eggs, and when they get into the caterpillar's gut, they hatch and burrow through the gut wall and into the body cavity. Later they search the caterpillar's body cavity for other parasitoid larvae, and it is these they attack and feed on. Some trigonalids, once in a caterpillar or sawfly larva, need their vehicle to fall prey to a social wasp. The wasp carries the caterpillar back to its nest, and there it is butchered and fed to the wasp's young; they will serve as the host for the trigonalid, the eggs of which are in the butchered caterpillar. Another example is the Ichneumon wasp, which parasitizes the butterfly *Phengaris rebeli* butterfly by directly seeking out Myrmica ant nests that the butterfly larvae previously parasitized through smell and ovipositing into the *P. rebeli* larvae. Once the wasps' eggs hatch, they feast on the carcass of the dead caterpillar.

Parasitoidal and Parasitic Vertebrates

Perhaps because they are less specialised and their relationships with their hosts are less intimate than is the case with many invertebrates, it often is more difficult to distinguish parasitism from parasitoidy in vertebrates. In fact many of their relationships of such types do not immediately suggest parasitism to most people at all. However, the very concept is so open to interpretation that it emerges frequently in vertebrate biology.

Kleptoparasitism for example is ubiquitous, and is a major constraint on reproduction or even survival among vertebrate predators, especially in times of famine. Male lions in a pride for example, largely leave hunting for non-threatening prey to females. However, prides that specialise in very large prey such as giraffes, elephants, or buffalo, may behave differently.

Other predators such as cheetah, leopard, and even lions sometimes may be chased from their kills by hyaenas. Hyaena may sometimes follow such predators so routinely in the hope of confiscating their kills, that the hunters spend more effort on avoiding hyaena than on hunting.

Ethologists could multiply examples of kleptoparasitism many-fold; it may be intraspecific or interspecific; it ranges from the smallest foragers and predators to the largest, and may combine with predation, where the robber is happy to eat both hunter and prey. Curiously though, interspecific robbers often show at least some constraint as though they were robbing conspecifics, and do not necessarily attack the host as directly as they would have done had there not been a "robbery" situation. Interpretation and speculation about the nature of such behaviour is beyond the scope of this article however.

It is not easy to classify such relationships, because many of them involve degrees of payment in terms of protection and other benefits; for example the male lions who preempt the females' kills do at least offer protection from hyaenas and rival males.

Kleptoparasitism occurs in many other forms among vertebrates, but for it to lead to the death of the host is not so common, and this would seem to disqualify it from the category of parasitoidy. Still, when the hosts are hard pressed in hard circumstances, the resulting injury and famine could cause reduced reproduction and even death.

Lampreys present both parasitic and parasitoidal examples. Most species are not parasitic, but among the North American species for example, there are several species ectoparasitic on freshwater fishes. They rasp away the skin of the host and suck the blood, but most do only superficial damage. In contrast, the most notorious species is the sea lamprey, *Petromyzon marinus*. Its rasping wounds can extend deep into the host's flesh, and the muscle damage and loss of blood commonly weaken the host severely, affecting its reproduction unfavourably. Often the harm is severe enough to kill the host.

Hagfish, are distant relatives of lampreys. They are largely carrion feeders and predators of large worms and similar small creatures, but various species also attack weakened fishes much as some lampreys do, and accordingly rank as opportunistic parasitoids under at least some conditions.

The sabre-toothed blenny presents a curiously difficult example of parasitism to classify. It parasitises the relationship between some cleaner fish and their client fishes, more than it parasitises either party to the relationship; it attacks the client fish, approaching it in the guise of cleaner wrasse and snatches a mouthful of scales or other convenient tissue. Clients often react violently, and thereafter trust neither wrasse nor the wrasse-mimicking blenny. In its violence and the pernicious effect on a valuable relationship, it suggests parasitoidy as well as parasitism.

Another form of parasitism that can approach parasitoidy occurs in the Perissodini, Cichlids from Lake Tanganyika. Seven species in the genus Perissodus are specialised in eating scales from other fish. Their teeth are variously suited to being able to grab bits of skin with the scales attached, and such bits of skin and scale formed major components of the stomach contents. At least some of the species also have adaptations in their behaviour to enable them to approach potential hosts They also have an adaptation of the jaw that enables them to lash out sideways in passing a victim; the jaw is asymmetrical, and there is continuous selection for the asymmetry that currently is less frequent in the population, because host fishes are more alert to defend themselves on the side on which they have been attacked in the past.

Such a lifestyle is reminiscent of sharks of the genus *Isistius*, which is known as the cookiecutter shark because of the circular wounds it leaves in the skins of whales and large fish that it has bitten in passing. Isistius species have been referred to as partly ectoparasitic, but they sometimes overwhelm their hosts and kill them, which by definition amounts to parasitoidy.

Candiru and related fishes in the Family Trichomycteridae, subfamilies Vandelliinae and Stegophilinae, present unusual examples of vertebrate parasitism, and occasionally parasitoidy. Most popular accounts are obsessed with the idea of candiru entering the human urethra and other orifices, but they are very varied in their habits. Some burrow partway into the skin of larger fish, apparently largely for purposes of protection and transport rather than food. Several at least are haematophagous, commonly entering the gill cavities of larger fishes and feeding on blood drawn from the gill filaments. At least when large fishes are tethered by fishermen where large numbers of the parasites occur, the hosts may die. Possibly this effect is analogous to the effect of soft ticks on hosts that do not avoid the sand patches where they assemble.

Among birds the best-known forms of parasitism are brood parasitism by various species of cuckoos, honey-guides, cowbirds, and several more. They qualify as parasitoids because many of them will cause the starvation of the host's chicks by competing with them for food, and many others either will remove host eggs when laying eggs in host nests, (sometimes eating the eggs removed), or the chick will eject or kill the eggs or chicks of the host when they hatch. Some hatchlings actually have hooked beaks adapted to attacking the host chicks and eggs, hooks that vanish before fledging.

Wasp Parasitoidal Oviposition

Oviposition is a process which is complex, as it not only depends on finding the specific host and certain environmental conditions, but also on evading the different types of host defence mechanisms. There are parasitoidal plants, crustaceans, insects, and vertebrates. Among parasitoidal insects, parasitoidal wasp oviposition is a topic that has been studied thoroughly.

Parasitoidal wasps are classified under the order Hymenoptera, suborder Apocrita and infraorder Parasitica. Parasitoidal wasps can be classified as either endoparasitic or ectoparasitic according to the locations where they lay their eggs. Endoparasitic wasps insert their eggs inside their host, while ectoparasitic wasps lay theirs outside the host body.

Parasitoid wasps have a different range of obstacles that have to be overcome in order to successfully oviposit inside or on their hosts, according to their endoparasitic or ectoparasitic nature. These barriers include behavioural, morphological, physiological or immunological defences. An example of immunological defence towards endoparasitic wasps would be the encapsulation of their eggs. To thwart this immunological protection system, wasps inundate their host with their eggs so as to overload the encapsulation response. Another way of hindering a host's defence is to introduce a virus which interferes with the development of defences.

Parasitoid wasps are able to locate hosts by detecting certain plant indirect defence mechanisms against insect herbivores. When attacked by insect herbivores, some plants release chemicals which attract parasitoid wasps but do not cause any harm to these predators. Some wasps insert the ovipositor organ into a host, while other wasps place their eggs onto a host egg. The ovipositor is a relatively long tubelike organ used to inject eggs into hosts. This organ consists of the genital structures that are made from segments eight and nine of the wasp's body.

Once the host has been located, the wasp either uses the horizontal and vertical alighting approach from behind. In the horizontal method, the wasp maintains its body horizontally until it grabs on the ant's metasoma (upper abdomen) with its tarsi (lower legs). After that it places the rest of its legs on the ant's abdomen, folds its wings, and proceeds to inject its victim with the ovipositor. The vertical approach is more complicated, and requires more steps. The wasp attaches its front legs onto the ant's abdomen, then rotates 180 degrees so that it becomes upside down. After that it rotates once more so that its head and legs switch places, before inserting its ovipositor.

Ants are often aware of the wasps' presence, as they violently turn around, moving their mandibles and legs to hit their attacker. This example of behavioural defence displayed by ants is not unusual, and sometimes prevents the wasps from depositing their eggs successfully. To try to beat such behavioural diversion, wasps close in very fast to inject their hosts. To prevent missing the target when making contact with the host, the wasp may wait for the ant to stop moving, and then attack suddenly.

Parasitoid Wasp

The term parasitoid wasp refers to a large evolutionary grade of hymenopteran superfamilies, mainly in the Apocrita. The parasitic or parasitoidal Apocrita are divided into some dozens of families. They are parasitoids of various animals, mainly other arthropods. Many of them are considered beneficial to humans because they control populations of agricultural pests. Others are unwelcome because they are hyperparasitoids, attacking beneficial parasitoids.

Parasitoidal wasps range from some of the smallest species of insects, to wasps about an inch long. Some are parasitoids that complete their metamorphosis in a single small egg of a small insect, and such a wasp is necessarily less than 1 mm long. Most females have a 'spine-like' ovipositor at the tip of the abdomen (Drees and Jackman). The egg and larval stage are usually not observed unless dissected from the host in which the adult female parasitized, except in species that practically fill the skin of the host with parasitoid larvae.

Distinguishing Features: Ichneumonidae and Braconidae

Braconidae: Most are brownish or black, not brightly coloured (Borror and White). The family is cosmopolitan. They also have a wide range of hosts that they parasitize. Some species of Braconids pupate in silken cocoons on the outside of their dead host (Miller and Ball). Wing venation: Only one recurrent vein, or no recurrent veins.

Ichneumonidae

Ichneumonids vary greatly in size and their colour varies from brightly coloured yellow to uniform black. The family is cosmopolitan. Most are very specific in the species of the host they parasitize. Wing venation: Two recurrent veins.

Hosts

Caterpillars provide major examples of larval Lepidoptera as a class of host, but various species parasitoid wasps in various Hymenopteran families parasitize their own favoured life stages [egg, larva, adult] of particular species in many other orders of insects including Coleoptera, Diptera, Hemiptera and other Hymenoptera. Some attack other Arthropoda, such as spiders.

Adult female wasps parasitize insect species, most species ovipositing into their hosts' bodies or eggs. The females of some parasitoid species also insert secretory products (combinations that may include polydnaviruses, ovarian proteins, and venom) that protect the egg from the immune system of the host. Once a host of a parasitoid that expresses polydnavirus particles has been parasitised, the virus that accompanied the egg during oviposition, infects the cells of the host in ways that benefit the parasitoid.

Life Cycle

Inside the host the egg hatches into a larva or larvae. The larva feed inside the host until ready to pupate, which by then is generally either dead or moribund. Depending on its species, the parasitoid then may eat its way out of the host or remain in the more or less empty skin. In either case it then generally spins a cocoon and pupates. As adults, parasitoid wasps feed primarily on nectar from flowers. If they happen to be species that rely on polydnavirus then all adults include the DNA for their associated species of the virus in their genomes.

Polydnavirus (PDV)

Description

Polydnaviruses are a unique group of insect viruses that have a mutualistic relationship with some parasitic wasps. Wasps benefit from this mutualism because the virus provides certain protection of the parasitic larva inside the host by weakening the host's immune system and by altering the cells of the host to be more beneficial to the parasite. The polydnavirus, like all viruses, needs a host to replicate and in this case is the oviducts of the adult female. The relationship between these viruses and the wasp is obligatory in the sense that all individuals are 'infected' with the viruses; the virus has been incorporated in the wasp's genome (Miller and Ball). These relationships between virus and parasitoid have been and are currently studied as model systems to study parasitoid-host immune interactions (Beckage).

Mechanisms

Some species have direct aggressive mechanisms for parasitizing a host by actively targeting the host immune system and then suppressing it; polydnaviruses are responsible for this approach (Miller and Ball). Others have more passive approaches to parasitism by ovipositing into a part of the host that has little exposure to the host's immune system or by avoiding activation of the host's immune system (Miller and Ball); ovarian proteins are more involved in this approach and PDV less so.

Evolution

There are two especially recognized genera of polydnaviruses: Ichnoviruses (IV) and Bracoviruses (BV). The ichnoviruses occur in ichneumonid wasp species and bracoviruses in braconid wasps. The genome of the virus is composed of multiple segments of double-stranded, super-helical DNA packaged in capsid proteins and a double layer (IV) or single layer (BV) envelope. The large genome of polydnaviruses is what distinguishes PDV from other viruses (Feming and Summers). While both have segmented DNA genomes, little or no sequence homology exists between BV and IV, suggesting that the two genera evolved independently. Specifically, BV and IV radically differ in morphology, methods of nucleocapsid release from cells, and possible packaging of multi-genomic DNAs; it is also unknown whether the two follow the same strategy of replication.

Pest Control

Typically, parasitoid wasps are not considered agricultural pests and are considered a beneficial insect as they control the population of host insects. They are also increasingly being released directly into regions specifically for the use of agricultural pest control.

"A number of parasitic wasp species are commercially available from insectaries and are purchased and released in augmentative biological control programmes. Other species have been imported from other countries from which pests have been accidentally introduced without their natural enemies and released to reintroduce the natural enemy with its host, a practice called importation, or "classical" biological control and which occasionally results in sustained suppression."

Host Defences

The hosts of parasitoids have developed several levels of defence. Many hosts try to hide from the parasitoids in inaccessible habitats. They may also get rid off their frass (body wastes) and avoid plants that they have chewed on as both can attract parasitoids. The egg shells and cuticles of the prey are thickened to prevent the parasitoid from penetrating them. When they encounter the egg laying female, prey use defences like dropping off the plant they are on, twisting and thrashing so as to dislodge or kill the female and even regurgitating onto the wasp to entangle it. The wriggling can sometimes help by causing the wasp to "miss" laying the egg on the host and instead place it nearby. Wriggling of pupae can cause the wasp to lose its grip on the smooth hard pupa or get trapped in the silk strands. Some caterpillars even bite the female wasps that approach it. Some insects secrete poisonous compounds that kill or drive away the parasitoid. Ants that are in a symbiotic relationship with caterpillars, aphids or scale insects may protect them from attack by wasps.

Even parasitoid wasps are vulnerable to hyperparasitoid wasps. Some parasitoid wasps change the behaviour of the infected host to build a silk web around the pupa of the wasps after they emerge from its body to protect them from hyperparasitoids.

In endoparasitoids, host immune cells can encapsulate the eggs and larvae of parasitoid wasps. In aphids, the presence of a secondary endosymbiont, Buchnera aphidicola that carries a particular latent phage makes the aphid relatively immune to their parasitoid wasps by killing many of the eggs. However, wasps counter this by laying more eggs in aphids that have the endosymbiont so that at least one of them can hatch and parasitize the aphid.

Certain caterpillars eat plants that are toxic to both themselves and the parasite to cure themselves. Drosophila melanogaster larvae also *self-medicate* with ethanol to treat parasitism. D. melanogaster females lay their eggs in food containing toxic amounts of alcohol if they detect parasitoid wasps nearby. Despite the alcohol retarding the growth of the flies, it protects them from the wasps.

Tachinidae

Tachinidae is a large and rather variable family of true flies within the insect order Diptera, with more than 8,200 known species and many more to be discovered. There are over 1300 species in North America. Insects in this family are commonly called tachina flies or simply tachinids. As far as is known, they all are Protelean parasitoids, or occasionally parasites, of Arthropoda. The family is cosmopolitan. Species occur in many habitats in many regions, including Neotropical, Nearctic, Afrotropical, Palaearctic, Oriental, Australasian and Oceanic.

Life Cycle

Reproductive strategies vary greatly between Tachinid species, largely, but not always clearly, according to their respective life cycles. This means that they tend to be generalists rather than specialists. Comparatively few are restricted to a single host species, so there is little tendency towards the close co-evolution one finds in the adaptations of many specialist species to their hosts, such as are typical of protelean parasitoids among the Hymenoptera.

Larvae (maggots) of most members of this family are parasitoids (developing inside a living host, ultimately killing it). In contrast a few are parasitic (not generally killing the host). Tachinid larvae feed on the host tissues, either after having been injected into the host by the parent, or penetrating the host from outside. Various species have different modes of oviposition and of host invasion. Typically, Tachinid larvae are endoparasites (internal parasites) of caterpillars of butterflies and moths, or the eruciform larvae of sawflies, but some species attack adult beetles and some attack beetle larvae. Others attack various types of true bugs, and others attack grasshoppers; a few even attack centipedes.

Oviposition and Ovoviviparity

Probably the majority of female Tachinids lay white, ovoid eggs with flat undersides onto the skin of the host insect. Imms mentions the genera *Gymnosoma*, *Thrixion*, *Winthemia*, and *Eutachina* as examples. In a closely related strategy some genera are effectively ovoviviparous (some authorities prefer the term *ovolarviparous*) and deposit a hatching larva onto the host. The free larvae immediately bore into the host's body. illustrative genera include: *Exorista*, *Voria*, and *Plagia*. Many Tachinid eggs hatch quickly, having partly developed inside the mother's uterus, which is long and often coiled for retaining developing eggs. However, it is suggested that the primitive state probably is to stick unembryonated eggs to the surface of the host.

Many other species inject eggs into the host's body, using the extensible, penetrating part of their ovipositor, sometimes called the oviscapt, which literally means something like "egg digger". Species in the genera *Ocyptera*, *Alophora*, and *Compsilura* are examples. Yet another strategy of oviposition among some Tachinidae is to lay large numbers of small, darkly coloured eggs on the food plants of the host species. *Sturmia*, *Zenillia*, and *Gonia* are such genera. Usually only one egg is laid on or in any individual host, and accordingly such an egg tends to be large, as is typical for eggs laid in small numbers. They are large enough to be clearly visible if stuck onto the outside of the host, and they generally are so firmly stuck that eggs cannot be removed from the skin of the host without killing them.

Many Tachinids are important natural enemies of major insect pests, and some species actually are used in biological pest control; for example, many species of Tachinid flies have been introduced into North America from their native lands as biocontrols to suppress populations of alien pests. Conversely, certain tachinid flies that prey on useful insects are themselves considered as pests; they can present troublesome problems in the sericulture industry by attacking silkworm larvae. One particularly notorious silkworm pest is the Uzi fly (*Exorista bombycis*).

Another reproductive strategy is to leave the eggs in the host's environment, for example the female might lay on leaves, where the host is likely to ingest them. Some tachinids that are parasitoids of stem-boring caterpillars deposit eggs outside the host's burrow, letting the first instar larvae do the work of finding the host for themselves. In other species, the maggots use an ambush technique, waiting for the host to pass and then attacking it and burrowing into its body.

Adult Tachinids are not parasitic, but either do not feed at all or visit flowers, decaying matter, or similar sources of energy to sustain themselves until they have concluded their procreative activities. Their non-parasitic behaviour after eclosion from the pupa is what justifies the application of the term "protelean".

Appearance

Figure: *Most tachinids are dull coloured, resembling house flies*

Tachinid flies are extremely varied in appearance. Some adult flies may be brilliantly coloured and then resemble blow-flies (family Calliphoridae). Most however are rather drab, some resembling house flies. However, Tachinid flies commonly are more bristly and more robust. Also, they usually have a characteristic appearance. They have three-segmented antennae, a diagnostically prominent postscutellum bulging beneath the scutellum (a segment of the mesonotum). They are aristate flies, and the arista usually is bare, though sometimes plumose. The calypters (small flaps above the halteres) are usually very large. Their fourth long vein bends away sharply.

Adult flies feed on flowers and nectar from aphids and scale insects. As many species typically feed on pollen, they can be important pollinators of some plants, especially at higher elevations in mountains where bees are relatively few.

The taxonomy of this family presents many difficulties. It is largely based on morphological characters of the adult flies, but also on reproductive habits and on the immature stage.

6

Biological Control Agents for Lepidoptera and Coleopteran Pests

Lepidoptera is a large order of insects that includes moths and butterflies (both called lepidopterans). It is one of the most widespread and widely recognizable insect orders in the world, encompassing moths and the three superfamilies of butterflies, skipper butterflies, and moth-butterflies. The term was coined by Linnaeus in 1735 and is derived from Ancient Greek scale and wing. Comprising an estimated 174,250 species, in 126 families and 46 superfamilies, the Lepidoptera show many variations of the basic body structure that have evolved to gain advantages in lifestyle and distribution. Recent estimates suggest that the order may have more species than earlier thought, and is among the four most speciose orders, along with the Hymenoptera, Diptera, and the Coleoptera.

Lepidopteran species are characterized by more than three derived features, some of the most apparent being the scales covering their bodies and wings, and a proboscis. The scales are modified, flattened "hairs", and give butterflies and moths their extraordinary variety of colours and patterns. Almost all species have some form of membranous wings, except for a few that have reduced wings or are wingless. Like most other insects, butterflies and moths are holometabolous, meaning they undergo complete metamorphosis. Mating and the laying of eggs are carried out by adults, normally near or on host plants for the larvae. The larvae are commonly called caterpillars, and are completely different from their adult moth or butterfly form, having a cylindrical body with a well-developed head, mandible mouth parts, and from 0 to 11 (usually 8) pairs of prolegs. As they grow, these larvae will change in appearance, going through a series of stages called instars.

Once fully matured, the larva develops into a pupa, referred to as a chrysalis in the case of butterflies and a cocoon in the case of moths. A few butterflies and many moth species spin a silk case or cocoon prior to pupating, while others do not, instead going underground.

The Lepidoptera have, over millions of years, evolved a wide range of wing patterns and colouration ranging from drab moths akin to the related order Trichoptera, to the brightly coloured and complex-patterned butterflies. Accordingly, this is the most recognized and popular of insect orders with many people involved in the observation, study, collection, rearing of and commerce in these insects. A person who collects or studies this order is referred to as a lepidopterist.

Butterflies and moths play an important role in the natural ecosystem as pollinators and as food in the food chain; conversely, their larva are considered very problematic to vegetation in agriculture, as their main source of food is often live plant matter. In many species, the female may produce anywhere from 200 to 600 eggs, while in others the number may go as high as 30,000 eggs in one day. The caterpillars hatching from these eggs can cause damage to large quantities of crops. Many moth and butterfly species are of economic interest by virtue of their role as pollinators, the silk they produce or as pest species.

The word Lepidoptera comes from the Latin word for "scaly wing", from the Ancient Greek *lepis* meaning scale and *pteron* meaning wing. Sometimes the term Rhopalocera is used to group the species that are butterflies, derived from the Ancient Greek *rhopalon* and *kœras* meaning "club" and "horn", respectively; coming from the shape of the antennae of butterflies.

The origins of the common names "butterfly"and "moth" are varied and often obscure. The English word butterfly is from Old English *buttorfleoge*, with many variations in spelling. Other than that, the origin is unknown, although it could be derived from the pale yellow colour of many species' wings suggesting the colour of butter. The species of Heterocera are commonly called moths. The origins of the English word moth are more clear, deriving from the Old English *moððe*" from Common Germanic (compare Old Norse *motti*, Dutch *mot* and German *Motte* all meaning "moth"). Perhaps its origins are related to Old English *maða* meaning "maggot" or from the root of "midge", which until the 16th century was used mostly to indicate the larva, usually in reference to devouring clothes.

The etymological origins of the word "caterpillar", the larval form of butterflies and moths, are from the early 16th century, from Middle English *catirpel, catirpeller*, probably an alteration of Old North French *catepelose*: *cate*, cat (from Latin *cattus*) + *pelose*, hairy (from Latin *pilôsus*).

Distribution and Diversity

Lepidoptera are among the most successful groups of insects. They are found on all continents, except Antarctica. Lepidoptera inhabit all terrestrial habitats ranging from desert to rainforest, from lowland grasslands to montane plateaus but almost always associated with higher plants, especially angiosperms (flowering plants). Among the most northern dwelling species of butterflies and moths is the Arctic Apollo (*Parnassius arcticus*), which is found in the Arctic Circle in northeastern Yakutia, at an altitude of 1500 metres above sea level. In the Himalayas, various Apollo species such as *Parnassius epaphus*, besides others, have been recorded to occur up to an altitude of 6,000 metres above sea level.

Some lepidopteran species exhibit symbiotic, phoretic, or parasitic life-styles, inhabiting the bodies of organisms rather than the environment. Coprophagous pyralid moth species, called sloth moths, such as *Bradipodicola hahneli* and *Cryptoses choloepi*, are unusual in that they are exclusively found inhabiting the fur of sloths, mammals found in Central and South America. Two species of *Tinea* moths have been recorded as feeding on horny tissue and have been bred from the horns of cattle. The larva of *Zenodochium coccivorella* is an internal parasite of the coccid *Kermes* species. Many species have been recorded as breeding in natural materials or refuse such as owl pellets, bat caves, honey-combs or diseased fruit.

Of the approximately 174,250 lepidopteran species described until 2007, butterflies and skippers are estimated to comprise approximately 17,950, with moths making up the rest. The vast majority of Lepidoptera are to be found in the tropics, but substantial diversity exists on most continents. North America has over 700 species of butterflies and over 11,000 species of moths, while there are about 400 species of butterflies and 14,000 species of moths reported from Australia. The diversity of Lepidoptera in each faunal region has been estimated by John Heppner in 1991 based partly on actual counts from the literature, partly on the card indices in the Natural History Museum (London) and the National Museum of Natural History (Washington), and partly on estimates:

Diversity of Lepidoptera in each faunal region					
	Palearctic	***Nearctic***	***Neotropic***	***Afrotropic***	***Indo-Australian (comprising Indomalayan and Australian regions)***
Estimated number of species	22,465	11,532	44,791	20,491	47,286

External Morphology

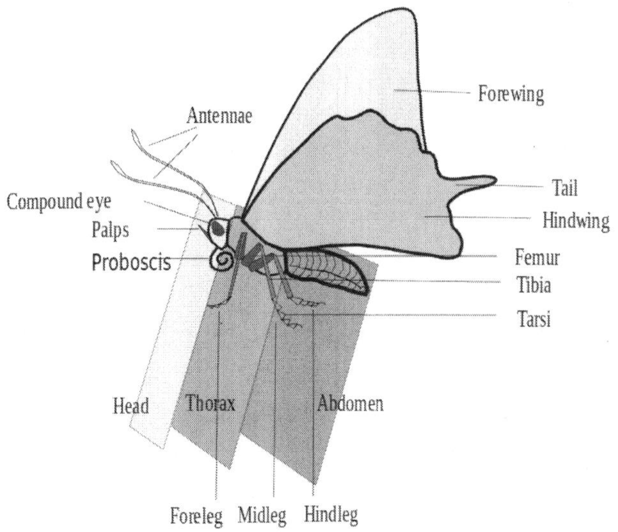

Figure: *Parts of an adult butterfly*

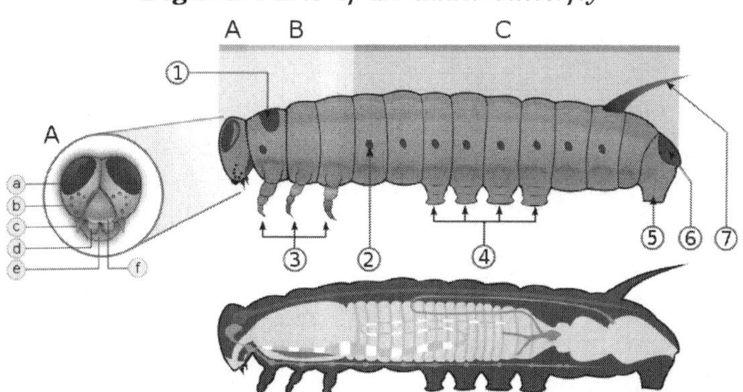

Figure: *A – head, B – thorax, C – abdomen, 1 – prothoracic shield, 2 – spiracle, 3 – true legs, 4 – midabdominal prolegs, 5 – anal proleg, 6 – anal plate, 7 – tentacle, a – frontal triangle, b – stemmata (ocelli), c – antenna, d – mandible, e – labrum.*

Lepidoptera are morphologically distinguished from other orders principally by the presence of scales on the external parts of the body and appendages, especially the wings. Butterflies and moths vary in size from microlepidoptera only a few millimetres long, to conspicuous animals with a wingspan of many inches, such as the Monarch butterfly and Atlas moth. Lepidopterans undergo a four-stage life cycle: egg; larva or caterpillar; pupa or chrysalis; and imago (plural: *imagines*) / adult and show many variations of the basic body structure, which have evolved to gain advantages in lifestyle and distribution.

Head

The head is where many sensing organs and the mouth parts are found. Like the adult, the larva also have a toughened, or sclerotized head capsule. Here, there are two compound eyes, and *chaetosema*, raised spots or clusters of sensory bristles unique to Lepidoptera, even though many taxa have lost one or both of these spots. The antennae have a wide variation in form among species and even between different sexes. The antennae of butterflies are usually filiform and shaped like clubs, those of the skippers are hooked while those of moths have flagellar segments variously enlarged or branched. Some moths have antennae that are enlarged, or tapered and hooked at the ends.

The maxillary galeae are modified and form an elongated proboscis. The proboscis consists of one to five segments, usually kept coiled up under the head by small muscles when it is not being used to suck up nectar from flowers or other liquids. Some basal moths still have mandibles, or separate moving jaws, like their ancestors and these form the family Micropterigidae.

The larvae, called caterpillars, have a toughened head capsule. Caterpillars lack the proboscis and have separate chewing mouthparts. These mouthparts, called mandibles, are used to chew up the plant matter that the larvae eat. The lower jaw, or labium, is weak but may carry a spinneret, an organ used to create silk. The head is made of large lateral lobes, each having an ellipse of up to six simple eyes.

Thorax

The thorax is made of three fused segments, the prothorax, mesothorax, and metathorax, each with a pair of legs. The first segment contains the first pair of legs. In some males of the butterfly family Nymphalidae, the fore-legs are greatly reduced and are not used for walking or perching. The three pairs of legs are covered with scales. Lepidoptera also have olfactory organs on their feet, which aid the butterfly in "tasting" or "smelling" out its food. In the larval form there are 3 pairs of true legs, with 0–11 pairs of abdominal legs (usually 8) and hooklets, called apical crochets.

The two pairs of wings are found on the middle and third segment, or mesothorax and metathorax respectively. In the more recent genera, the wings of the second segment are much more pronounced, although some more primitive forms have similarly sized wings of both segments. The wings are covered in scales arranged like shingles, which form an extraordinary variety of colours and patterns. The mesothorax is designed to have more powerful muscles to propel the moth or butterfly through the air, with the wing of this segment (forewing) having a stronger vein structure. The largest superfamily, Noctuidae, have the wings modified to act as Tympanal or hearing organs.

The caterpillar has an elongated soft body that may have hair-like or other projections, 3 pairs of true legs, with 0–11 pairs of abdominal legs (usually 8) and hooklets, called apical crochets. The thorax will usually have a pair of legs on each segment. The thorax is also lined with many spiracles on both the mesothorax and metathorax, except for a few aquatic species, who instead have a form of gill.

Abdomen

The abdomen, which is less sclerotized than the thorax, consists of 10 segments with membranes in between allowing for articulated movement. The sternum, on the first segment, is small in some families and is completely absent in others. The last 2 or 3 segments form the external parts of the species' sex organs. The genitalia of Lepidoptera are highly varied and are often the only means of differentiating between species. Male genitals include a valva, which is usually large, as it is used to grasp the female during mating. Female genitalia include three distinct sections.

In the females of basal moths, there is only one sex organ, which is used for copulation and as an ovipositor, or egg laying organ. 98% of moth species have a separate organ for mating, and an external duct that carries the sperm from the male.

The abdomen of the caterpillar has 4 pairs of prolegs, normally located on the third to sixth segments of the abdomen, and a separate pair of prolegs by the anus, which have a pair of tiny hooks called crotchets. These aid in gripping and walking, especially in species that lack many prolegs (e. g. larvae of Geometridae). In some basal moths, these prolegs may be on every segment of the body, while prolegs may be lost completely in other groups, which are more adapted to boring and living in sand (e. g., Prodoxidae and Nepticulidae respectively).

Scales

The wings, head, parts of the thorax and abdomen of Lepidoptera are covered with minute scales, a feature from which the order 'Lepidoptera' derives its name. Most scales are lamellar, or blade-like and attached with a pedicel, while other forms may be hair-like or specialized as secondary sexual characteristics.

The lumen or surface of the lamella, has a complex structure. It gives colour either by coloured pigments that it contains, or through structural colouration with mechanisms that include photonic crystals and diffraction gratings. Scales provide functions including insulation, thermoregulation, producing pheromones (in males only), and aiding gliding flight, but the most important is the large diversity of vivid or indistinct patterns they provide, which help the organism protect itself by camouflage or mimicry, and which act as signals to other animals including rivals and potential mates.

Internal Morphology

In the reproductive system of butterflies and moths, the male genitalia are complex and unclear. In females there are three types of genitalia based on the relating taxa: *monotrysian*, *exoporian*, and *ditrysian*. In the *monotrysian* type there is an opening on the fused segments of the sterna 9 and 10, which act as insemination and oviposition.

In the *exoporian* type (in Hepaloidae and Mnesarchaeoidea) there are two separate places for insemination and oviposition, both occurring on the same sterna as the *monotrysian* type, i.e. 9 and 10. The *ditrysian* groups have an internal duct that carry sperm, with separate openings for copulation and egg-laying In most species the genitalia are flanked by two soft lobes, although they may be specialized and sclerotized in some species for ovipositing in area such as crevices and inside plant tissue.

Hormones and the glands that produce them run the development of butterflies and moths as they go through their life cycle, called the endocrine system. The first insect hormone PTTH (Prothoracicotropic hormone) operates the species life cycle and diapause. This hormone is produced by corpora allata and corpora cardiaca, where it is also stored. Some glands are specialized to perform certain task such as producing silk or producing saliva in the palpi. While the corpora cardiaca produce PTTH, the corpora allata also produces jeuvanile hormones, and the prothorocic glands produce moulting hormones.

In the digestive system, the anterior region of the foregut has been modified to form a pharyngeal sucking pump as they need it for the food they eat, which are for the most part liquids. An esophagus follows and leads to the posterior of the pharynx and in some species forms a form of crop. The midgut is short and straight, with the hindgut being longer and coiled. Ancestors of lepidopteran species, stemming from Hymenoptera, had midgut ceca, although this is lost in current butterflies and moths. Instead, all the digestive enzymes other than initial digestion, are immobilized at the surface of the midgut cells. In larvae, long-necked and stalked goblet cells are found in the anterior and posterior midgut regions, respectively. In insects, the goblet cells excrete positive potassium ions, which are absorbed from leaves ingested by the larvae. Most butterflies and moths display the usual digestive cycle, however species that have a different diet require adaptations to meet these new demands.

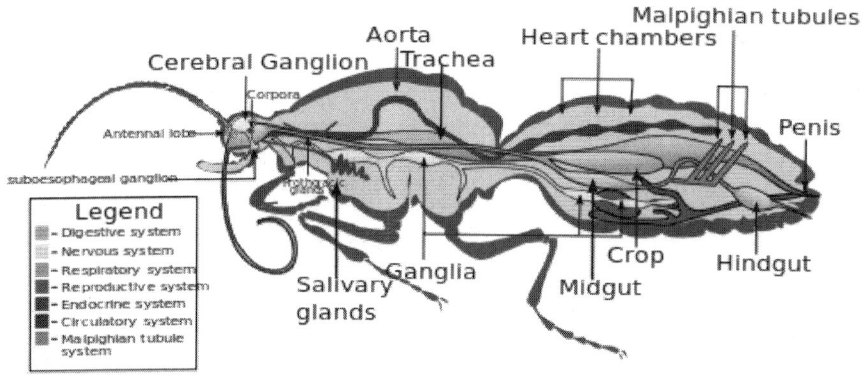

***Figure:** Internal morphology of adult male in the family Nymphalidae, showing most of the major organ systems, with characteristic reduced forelegs of that family. The corpora include the corpus allatum and the corpus cardiaca.*

In the circulatory system, hemolymph, or insect blood, is used to circulate heat in a form of thermoregulation, where muscles contraction produces heat, which is transferred to the rest of the body when conditions are unfavourable. In lepidopteran species, hemolymph is circulated through the veins in the wings by some form of pulsating organ, either by the heart or by the intake of air into the trachea. Air is taken in through spiracles along the sides of the abdomen and thorax supplying the trachea with oxygen as it goes through the lepidopteran's respiratory system. There are three different tracheae supplying and diffusing oxygen throughout the species body: The dorsal, ventral, and visceral. The dorsal tracheae supply oxygen to the dorsal musculature and vessels, while the ventral tracheae supply the ventral musculature and nerve cord, and the visceral tracheae supply the guts, fat bodies, and gonads.

Polymorphism

Figure: *The* Heliconius *butterflies from the tropics of the Western Hemisphere are the classical model for Müllerian mimicry.*

Polymorphism is the appearance of forms or "morphs", which differ in colour and number of attributes within a single species. In Lepidoptera, polymorphism can be seen not only between individuals in a population, but also between the sexes as *sexual dimorphism*, between geographically separated populations in *geographical polymorphism* and also between generations flying at different seasons of the year (*seasonal polymorphism* or polyphenism). In some species, the polymorphism is limited to one sex, typically the female. This often includes the phenomenon of mimicry when mimetic morphs fly alongside non-mimetic morphs in a population of a particular species. Polymorphism occurs both at specific level with heritable variation in the overall morphological design of individuals as well as in certain specific morphological or physiological traits within a species.

Environmental polymorphism, in which traits are not inherited, is often termed as polyphenism. Polyphenism in Lepidoptera is commonly seen in the form of seasonal morphs especially in the butterfly families of Nymphalidae and Pieridae. An Old World pierid butterfly, the Common Grass Yellow (*Eurema hecabe*) has a darker summer adult morph, triggered by a long day exceeding 13 hours in duration, while the shorter diurnal period of 12 hours or less induces a paler morph in the post-monsoon period. Polyphenism also occurs in caterpillars, an example being the Peppered Moth, *Biston betularia*.

Geographical polymorphism is where geographical isolation causes a divergence of a species into different morphs. A good example is the Indian White Admiral *Limenitis procris*, which has five forms, each geographically separated from the other by large mountain ranges. An even more dramatic showcase of geographical polymorphism is the

Apollo butterfly (*Parnassius apollo*). Because the Apollos live in small local populations, thus having no contact with each other, coupled with their strong stenotopic nature and weak migration ability, interbreeding between populations of one species practically does not occur; by this, they form over 600 different morphs, with the size of spots on the wings of which varies greatly.

Seasonal Diphenism in the Common Grass Yellow, Eurema Hecabe

Sexual dimorphism is the occurrence of differences between males and females in a species. In Lepidoptera, sexual dimorphism is widespread and almost completely set by genetic determination. Sexual dimorphism is present in all families of the Papilionoidea and more prominent in the Lycaenidae, Pieridae and certain taxa of the Nymphalidae. Apart from colour variation, which may differ from slight to completely different colour-pattern combinations, secondary sexual characteristics may also be present. Different genotypes maintained by natural selection may also be expressed at the same time. Polymorphic and/or mimetic females occur in the case of some taxa in the Papilionidae primarily to obtain a level of protection not available to the male of their species. The most distinct case of sexual dimorphism is that of adult females of many Psychidae species who have only vestigial wings, legs, and mouthparts as compared to the adult males who are strong fliers with well-developed wings and feathery antennae.

Reproduction and Development

Species of Lepidoptera undergo holometabolism or “complete metamorphosis”. Their life cycle normally consists of an egg, larva, pupa, and an imago or adult. The larvae are commonly called caterpillars, and the pupae of moths that are encapsulated in silk are called cocoons while the uncovered pupae of butterflies are called chrysalides.

Mating

Males usually get a head start, and start eclosion or emergence, earlier than females and peak in numbers before females. Both of the sexes are sexually mature by the time of eclosion. Butterflies and moths normally don’t associate with each other, except for migrating species, staying relatively asocial. Mating begins with an adult (female or male) attracting a mate, normally using visual stimuli, especially in diurnal species like most butterflies. However, the females of most nocturnal species, including almost all moth species, use pheromones to attract males, sometimes from long distances. Some species engage in a form of acoustic courtship, or attract mates using sound or vibration such as the polka-dot wasp moth, *Syntomeida epilais*.

Adaptations include undergoing one seasonal generation, two or even more, called voltinism (Univoltism, bivoltism and multivism respectively). Most lepidoptera in temperate climates are univoltine, while in tropical climates most have two seasonal broods. Some others may take advantage of any opportunity they can get, and mate continuously throughout the year.

These seasonal adaptations are controlled by hormones, and these delays in reproduction are called diapause. Many lepidopteran species, after mating and laying their eggs, die shortly afterwards, having only lived for a few days after eclosion. Others may still be active for several weeks and then overwinter and become sexually active again when the weather becomes more favourable, or diapause. The sperm of the male that mated most recently with the female is most likely to have fertilized the eggs but the sperm from a prior mating may still prevail.

Life Cycle

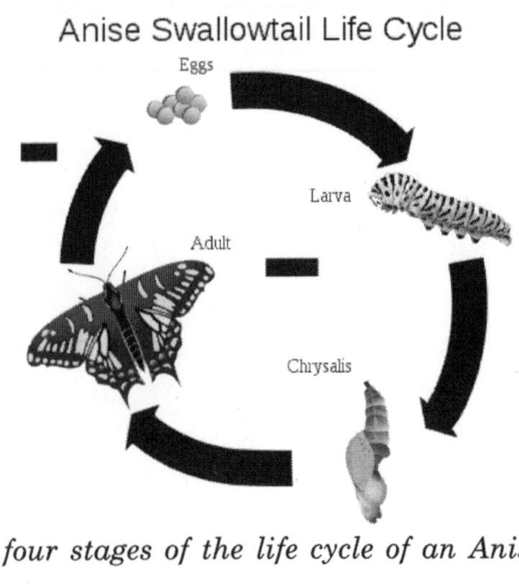

Figure: *The four stages of the life cycle of an Anise Swallowtail*

Eggs

Lepidoptera usually reproduce sexually and are oviparous (egg-laying), though some species exhibit live birth in a process called ovoviviparity. There are a variety of differences in egg-laying and the number of eggs laid. Some species simply drop their eggs in flight (these species normally have polyphagous larvae, meaning they eat a variety of plants e. g., hepialids and some nymphalids) while most Lepidoptera will lay their eggs near or on the host plant that the larvae feed on. The number of eggs laid may vary from only a few to several thousand. The females of both butterflies and moths select the host plant instinctively, and primarily, by chemical cues.

The egg is covered by a hard-ridged protective outer layer of shell, called the chorion. It is lined with a thin coating of wax, which prevents the egg from drying out before the larva has had time to fully develop. Each egg contains a number of micropyles, or tiny funnel-shaped openings at one end, the purpose of which is to allow sperm to enter and fertilize the egg. Butterfly and moth eggs vary greatly in size between species, but they are all either spherical or ovate.

The egg stage lasts a few weeks in most butterflies but eggs laid close to winter, especially in temperate regions, go through a diapause, and hatching may be delayed until spring. Other butterflies may lay their eggs in the spring and have them hatch in the summer. These butterflies are usually northern species (e. g. *Nymphalis antiopa*).

Larvae

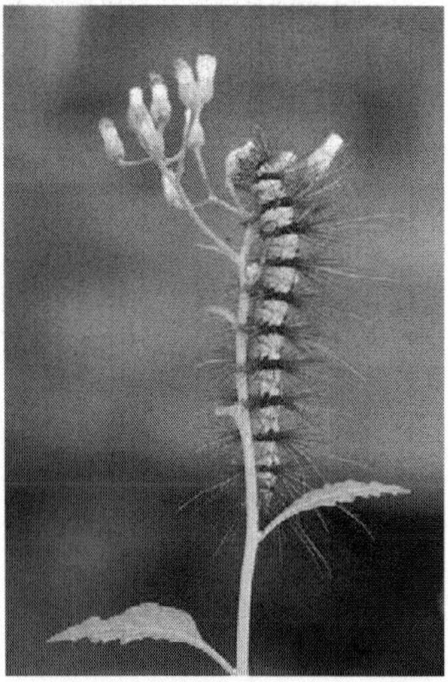

Figure: *The larval form typically lives and feeds on plants.*

The larvae or caterpillars are the first stage in the life cycle after hatching. Caterpillars, are "characteristic polypod larvae with cylindrical bodies, short thoracic legs and abdominal prolegs (pseudopods)". They have a toughened (sclerotised) head capsule with an adfrontal suture formed by medial fusion of the sclerites, mandibles (mouthparts) for chewing, and a soft tubular, segmented body, that may have hair-like or other projections, three pairs of true legs, and additional prolegs (up to five pairs). The body consists of thirteen segments, of which three are thoracic and ten are abdominal. Most larvae are herbivores, but a few are carnivores (some eat ants or other caterpillars) and detritivores.

Different herbivore species have adapted to feed on every part of the plant and are normally considered pests to their host plant; some species have been found to lay their eggs on the fruit and other species lay their eggs on clothing or fur (e. g., *Tineola bisselliella*, the common clothes moth). Some species are carnivorous and others are even parasitic. Some lycaenid species such as *Maculinea rebeli* are social parasites of *Myrmica* ants nests. A species of Geometridae from Hawaii has carnivorous larvae that catch and eat flies. Some pyralid caterpillars are aquatic.

The larvae develop rapidly with several generations in a year; however, some species may take up to 3 years to develop and exceptional examples like *Gynaephora groenlandica* take as long as seven years. The larval stage is where the feeding and growing stages occur, and the larvae periodically undergo hormone-induced ecdysis, developing further with each instar, until they undergo the final larval-pupal molt. Lepidoptera pupa, known as chrysalis in the case of butterflies, have functional mandibles and with appendages fused or glued to the body in most species, while the pupal mandibles are not functional in others.

The larvae of both butterflies and moths exhibit mimicry to deter potential predators. Some caterpillars have the ability to inflate parts of their head to appear snake-like. Many have false eye-spots to enhance this effect. Some caterpillars have special structures called osmetreia (Papilionidae family), which are exposed to produce smelly chemicals. These are used in defence.

Host plants often have toxic substances in them and caterpillars are able to sequester these substances and retain them into the adult stage. This helps making them unpalatable to birds and other predators. Such unpalatability is advertised using bright red, orange, black or white warning colours. The toxic chemicals in plants are often evolved specifically to prevent them from being eaten by insects. Insects in turn develop countermeasures or make use of these toxins for their own survival. This "arms race" has led to the coevolution of insects and their host plants.

Wing Development

No form of wing is externally visible on the larva, however when larvae are dissected, developing wings can be seen as disks, which can be found on the second and third thoracic segments, in place of the spiracles that are apparent on abdominal segments. Wing disks develop in association with a trachea that runs along the base of the wing, and are surrounded by a thin peripodial membrane, which is linked to the outer epidermis of the larva by a tiny duct. Wing disks are very small until the last larval instar, when they increase dramatically in size, are invaded by branching tracheae from the wing base that precede the formation of the wing veins, and begin to develop patterns associated with several landmarks of the wing.

Near pupation, the wings are forced outside the epidermis under pressure from the hemolymph, and although they are initially quite flexible and fragile, by the time the pupa breaks free of the larval cuticle they have adhered tightly to the outer cuticle of the pupa (in obtect pupae). Within hours, the wings form a cuticle so hard and well-joined to the body that pupae can be picked up and handled without damage to the wings.

Pupa

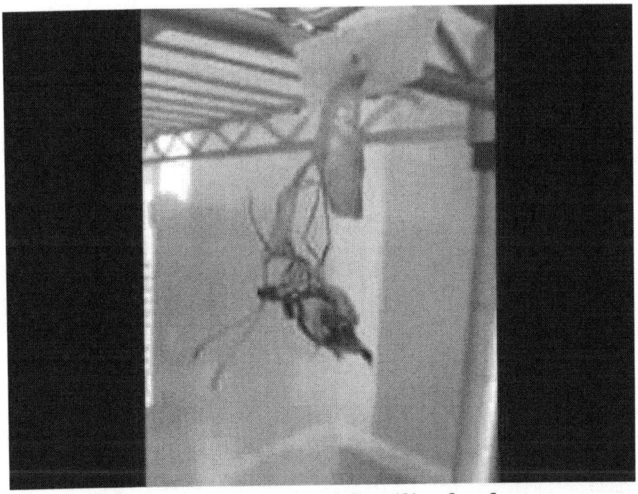

***Figure:** Eclosion of* Papilio dardanus

After about 5 to 7 instars, or molts, certain hormones, like prothoracicotropic hormone, stimulate the production of ecdysone, which initiates insect molting. Then, the larva puparium, a sclerotized or hardened cuticle of the last larval instar, develops into the pupa. Depending on the species, the pupa may be covered in silk and attached to many different types of debris or may not be covered at all. The pupa stays attached to the leaf by silk spun by the caterpillar before it spins the silk for the full pupa. Features of the imago are externally recognizable in the pupa. All the appendages on the adult head and thorax are found cased inside the cuticle (antennae, mouthparts, etc.), with the wings wrapped around, adjacent to the antennae.

While encased, some of the lower segments are not fused, and are able to move using small muscles found in between the membrane. Moving may help the pupa, for example, escape the sun, which would otherwise kill it. The pupa of the Mexican jumping bean moth (*Cydia deshaisiana*) does this. The larvae cut a trapdoor in the bean (species of Sebastiania) and use the bean as a shelter. When there is a sudden rise in temperature, the pupa inside twitches and jerks, pulling on the threads inside. Wiggling may also help to deter parasitoid wasps from laying eggs on the pupa. Other species of moth are able to make clicks to deter predators.

The length of time before the pupa ecloses (emerges) varies greatly. The monarch butterfly may stay in its chrysalis for two weeks, while other species may need to stay for more than 10 months in diapause. The adult will emerge from the pupa either by using abdominal hooks or from projections located on the head. The mandibles found in the most primitive moth families are used to escape from their cocoon (e. g., Micropterigoidea).

Adult

Most lepidopteran species do not live long after eclosion, only needing a few days to find a mate and then lay their eggs. Others may remain active for a longer period (from one to several weeks), or go through diapause and overwintering as monarch butterflies do, or waiting out environmental stress. Some adult species of Microlepidoptera go through a stage where there is no reproductive-related activity lasting through summer and winter, followed by mating and oviposition, or egg laying, in the early spring.

While most butterflies and moths are terrestrial, many species of Pyralidae are truly aquatic with all stages except the adult occurring in water. Many species from other families such as Arctiidae, Nepticulidae, Cosmopterygidae, Tortricidae, Olethreutidae, Noctuidae, Cossidae and Sphingidae are aquatic or semi-aquatic.

Behaviour

Flight: Flight is an important aspect of the lives of butterflies and moths and is used for evading predators, searching for food and finding mates in a timely manner as lepidopteran species do not live long after eclosion. It is the main form of locomotion in most species. In lepidoptera, the forewings and hindwings are mechanically coupled and flap in synchrony. Flight is anteromotoric, or being driven primarily by action of the forewings. Although it has been reported that lepidopteran species can still fly when their hindwings are cut off, it reduces their linear flight and turning capabilities.

Lepidopteran species have to be warm, about 77 to 79 °F (25 to 26 °C) in order to fly. They depend on their body temperature being sufficiently high and since they can't regulate it themselves, this is dependent on their environment. Butterflies living in cooler climates may use their wings to warm their bodies. They will bask in the sun, spreading out their wings so that they get maximum exposure to the sunlight. In hotter climates butterflies can easily overheat, so they are usually active only during the cooler parts of the day, early morning, late afternoon or early evening. During the heat of the day they rest in the shade. Some larger thick-bodied moths (e. g. Sphingidae) can generate their own heat to a limited degree by vibrating their wings. The heat generated by the flight muscles warms the thorax while the temperature of the abdomen is unimportant for flight. To avoid overheating some moths rely on hairy scales, internal air sacs, and other structures to separate the thorax and abdomen and keep the abdomen cooler.

Some species of butterfly can reach fast speeds, such as the Southern Dart, which can go as fast as 48.4 km/h. Sphingids are some of the fastest flying insects, some are capable of flying at over 50 km/h (30 miles per hour), having a wingspan of 35–150 mm. In some species, there is sometimes a gliding component to their flight. Flight occurs either as hovering, or as forward or backward motion. In butterfly and in moth species, like hawk moths, hovering is important as they need to maintain a certain stability over flowers when feeding on the nectar.

Navigation

Navigation is important to lepidoptera species, especially for those that migrate. Butterflies, which have more species that migrate, have been shown to navigate using time compensated sun compasses. They can see polarized light and therefore can orient even in cloudy conditions. The polarized light in the region close to the ultraviolet spectrum is suggested to be particularly important. It is suggested that most migratory butterflies are those that live in semi-arid areas where breeding seasons are short. The life-histories of their host plants also influence the strategies of the butterflies. Other theories include the use of landscapes. Lepidoptera may use coastal lines, mountains and even roads to orient themselves. Above sea it has been observed that the flight direction is much more accurate if the coast is still visible.

Many studies have also shown that moths navigate. One study showed that many moths may use the Earth's magnetic field to navigate, as a study of the moth Heart and Dart suggests. Another study, this time of the migratory behaviour of the Silver Y, showed that even at high altitudes the species can correct its course with changing winds, and prefers flying with favourable winds, suggesting a great sense of direction. *Aphrissa statira* in Panama loses its navigational capacity when exposed to a magnetic field, suggesting it uses the Earth's magnetic field.

Moths exhibit a tendency to circle artificial lights repeatedly. This suggests that they use a technique of celestial navigation called transverse orientation.

By maintaining a constant angular relationship to a bright celestial light, such as the Moon, they can fly in a straight line. Celestial objects are so far away, that even after travelling great distances, the change in angle between the moth and the light source is negligible; further, the moon will always be in the upper part of the visual field or on the horizon.

When a moth encounters a much closer artificial light and uses it for navigation, the angle changes noticeably after only a short distance, in addition to being often below the horizon. The moth instinctively attempts to correct by turning towards the light, causing airborne moths to come plummeting downwards, and—at close range—which results in a spiral flight path that gets closer and closer to the light source. Other explanations have been suggested, such as the idea that moths may be impaired with a visual distortion called a Mach band by Henry Hsiao in 1972. He stated that they fly towards the darkest part of the sky in pursuit of safety and are thus inclined to circle ambient objects in the Mach band region.

Migration

Lepidopteran migration is usually seasonal, the insects moving to escape dry seasons or other disadvantageous conditions. Most lepidopterans that migrate are butterflies, the distance travelled varying from short to very long journeys. Some butterflies that migrate include the Mourning Cloak, Painted Lady, American Lady, Red Admiral, and the Common Buckeye. Particularly famous migrations are those of the Monarch butterfly from Mexico to northern United States and southern Canada, a distance of about 4,000–4,800 km (2,500–3,000 mi). Other well-known migratory species include the Painted Lady and several of the danaine butterflies. Spectacular and large-scale migrations associated with the Monsoons are seen in peninsular India. Migrations have been studied in more recent times using wing tags and also using stable hydrogen isotopes.

Moths also undertake migrations, an example being the uraniids. *Urania fulgens* undergoes population explosions and massive migrations that may be not surpassed by any other insect in the Neotropics. In Costa Rica and Panama, the first population movements may begin in July and early August and, depending on the year, may be very massive, continuing unabated for as long as five months.

Communication

Pheromones are commonly involved in mating rituals among species, especially moths, but they are also an important aspect of other forms of communication. Usually the pheromones are produced by either the male or the female and detected by members of the opposite sex with their antennae. In many species, a gland between the eighth and ninth segment under the abdomen in the female produces the pheromones. Communication can also occur through stridulation, or producing sounds by rubbing various parts of the body together.

Moths are known to engage in acoustic forms of communication; most often as courtship, attracting mates using sound or vibration. Like most other insects, moths pick up these sounds using tympanic membranes in the abdomen. An example is that of the polka-dot wasp moth (*Syntomeida epilais*), which produce sounds with a frequency above that normally detectable by humans (~20 kHz). These sounds also function as tactile communication, or communication through touch, as they stridulate, or vibrate a substrate like leaves and stems.

Most moths lack bright colours as many species use colouration as camouflage but butterflies engage in visual communication. Female cabbage butterflies, for example, use ultraviolet light to communicate, with scales coloured in this range on the dorsal wing surface. When they fly, each down stroke of the wing creates a brief flash of ultraviolet light that the males apparently recognize as the flight signature of a potential mate. These flashes from the wings may attract several males who engage in aerial courtship displays.

Ecology

Moths and Butterflies are important in the natural ecosystem. They are integral participants in the food chain, having co-evolved with flowering plants and predators, lepidopteran species have formed a network of trophic relationships between autotrophs and heterotrophs, which are included in the stages of Lepidoptera larvae, pupae and adults. Larvae and pupae are links in the diet of birds and parasitic entomophagous insects. The adults are included in food webs in a much broader range of consumers (including birds, small mammals, reptiles, etc.).

Defence and Predation

Lepidopteran species are soft bodied, fragile and almost defenceless while the immature stages move slowly or are immobile, hence all stages are exposed to predation. Adult butterflies and moths are predated upon by birds, lizards, amphibians, dragonflies and spiders, besides others. Caterpillars and pupa fall prey, not only to birds but invertebrate predators, small mammals, as well as fungi and bacteria. Parasitoid and parasitic wasps and flies may lay eggs in the caterpillar, which eventually kill it as they hatch inside its body and eat its tissues. Insect-eating birds are probably the worst predators. Lepidoptera, especially the immature stages, are an ecologically important food to many insectivorous birds, such as the Great Tit in Europe.

An "evolutionary arms race" can be seen between predator and prey species. Lepidoptera have developed a number of strategies for defence and protection including evolution of morphological characters and changes in ecological life-style and in behaviour. These include aposematism, mimicry, camouflage, development of threat patterns and displays and so on. Only a few birds, such as the nightjars, hunt nocturnal lepidoptera. Their main predators are bats. Again, an "evolutionary race" exists, which has led to numerous evolutionary adaptations of moths to escape from their main predators, such as the ability to hear ultrasonic sounds, or even to emit sounds in some cases. Lepidoptera eggs are also predated upon. Some caterpillars, such as the zebra swallowtail butterfly larvae, are cannibalistic and may eat other larvae of the same species. Lepidopteran species rely on a variety of strategies.

Some species of lepidoptera are poisonous to predators, such as the Monarch butterfly in the Americas, Atrophaneura species (roses, windmills etc.) in Asia, as well as Papilio antimachus and the birdwings, the largest butterflies in Africa and Asia respectively. They obtain their toxicity by sequestering the chemicals from the plants they eat into their own tissues. Some Lepidoptera manufacture their own toxins. Predators that eat poisonous butterflies and moths may become sick and vomit violently, learning not to eat those types of species. A predator who has previously eaten a poisonous lepidopteran may avoid other species with similar markings in the future, thus saving many other species as well. Toxic butterflies and larvae tend to develop bright colours, striking patterns as an indicator to predators about their toxicity. This phenomenon is known as aposematism. Other caterpillars emit bad smells to ward off predators. Some caterpillars, especially members of Papilionidae, contain an osmeterium, a Y-shaped protrusible gland found in the prothoracic segment of the larvae. When threatened, the caterpillar emits unpleasant smells from the organ to ward off the predators.

Camouflage is also an important defence strategy, which involves the use of colouration or shape to blend into the surrounding environment.

Some lepidopteran species blend with its surroundings, making them difficult to be spotted by predators. Caterpillars can exhibit shades of green that match its host plant. Others look like inedible objects, such as twigs or leaves. For instance, the Mourning Cloak will fade into the backdrop of trees when it folds its wings back. The larvae of some species, such as the Common Mormon (Papilio polytes) and the Western Tiger Swallowtail look like bird droppings. For example, adult Sesiidae species (also known as *clearwing moths*) have a general appearance that is sufficiently similar to a wasp or hornet to make it likely that the moths gain a reduction in predation by Batesian mimicry. Eyespots are a type of automimicry used by some butterflies and moths. In butterflies, the spots are composed of concentric rings of scales in different colours. The proposed role of the eyespots is to deflect attention to predators. Their resemblance to eyes provokes the predator's instinct to attack these wing patterns.

Batesian and Müllerian mimicry complexes are commonly found in Lepidoptera. Genetic polymorphism and natural selection give rise to otherwise edible species (the mimic) gaining a survival advantage by resembling inedible species (the model). Such a mimicry complex is referred to as *Batesian* and is most commonly known in the example between the limenitidine Viceroy butterfly in relation to the inedible danaine Monarch. Later research has discovered that the Viceroy is, in fact more toxic than the Monarch and this resemblance should be considered as a case of Müllerian mimicry. In Müllerian mimicry, inedible species, usually within a taxonomic order, find it advantageous to resemble each other so as to reduce the sampling rate by predators who need to learn about the insects' inedibility. Taxa from the toxic genus *Heliconius* form one of the most well known Müllerian complexes. The adults of the various species now resemble each other so well that the species cannot be distinguished without close morphological observation and, in some cases, dissection or genetic analysis.

There is evidence moths are able to hear the range emitted by bats, which in effect causes flying moths to make evasive maneuvers because bats are a main predator of moths. Ultrasonic frequencies trigger a reflex action in the noctuid moth that cause it to drop a few inches in its flight to evade attack. Tiger moths in a defence emit clicks within the same range of the bats, which interfere with the bats, and foil their attempts to echolocate it.

Pollination

Most species of Lepidoptera engage in some form of entomophily (more specifically psychophily and phalaenophily for butterflies and moths respectively), or the pollination of flowers. Most adult butterflies and moths feed on the nectar inside flowers, using their proboscis to reach the nectar hidden at the base of the petals. In the process, the adult brushes against the flower's stamen, on which the flower's reproductive pollen is made and stored. The pollen is transferred on appendages on the adult, who flies to the next flower to feed and unwittingly deposits the pollen on the stigma of the next flower, where the pollen germinates and fertilizes the seeds.

Flowers pollinated by butterflies tend to be large and flamboyant, being pink or lavender in colour, frequently having a landing area, and are usually scented, as butterflies are typically day-flying. Since butterflies do not digest pollen (except for Heliconid species), more nectar is offered than pollen. The flowers have simple nectar guides with the nectaries usually hidden in narrow tubes or spurs, reached by the long "tongue" of the butterflies. Butterflies like the *Thymelicus flavus* have been observed to engage in flower constancy, which means that they are more likely to transfer pollen to other conspecific plants. This can be beneficial for the plants being pollinated, as flower constancy prevents the loss of pollen during different flights and the pollinators from clogging stigmas with pollen of other flower species.

Among the more important moth pollinator groups are the hawk moths of the family Sphingidae. Their behaviour is similar to hummingbirds, i.e., using rapid wing beats to hover in front of flowers. Most hawk moths are nocturnal or crepuscular, so moth-pollinated flowers (e.g., *Silene latifolia*) tend to be white, night-opening, large and showy with tubular corollas and a strong, sweet scent produced in the evening, night or early morning. A lot of nectar is produced to fuel the high metabolic rates needed to power their flight. Other moths (e.g., noctuids, geometrids, pyralids) fly slowly and settle on the flower. They do not require as much nectar as the fast-flying hawk moths, and the flowers tend to be small (though they may be aggregated in heads).

Mutualism

Mutualism is a form of biological interaction where each individual involves benefits in some shape or form. An example of a mutualistic relationship would be the relationship shared by yucca moths (Tegeculidae) and their host, yucca flowers (Liliaceae). Female yucca moths enter the host flowers, collect the pollen into a ball using specialized maxillary palps, then move to the apex of the pistil where pollen is deposited on the stigma, and lay eggs into the base of the pistil where seeds will develop. The larvae develop in the fruit pod and feed on a portion of the seeds. Thus, both insect and plant benefit, forming a highly mutualistic relationship. Another form of mutualism occurs between some larvae of butterflies and certain species of ants (e. g. Lycaenidae). The larvae communicate with the ants using vibrations that are transmitted through a substrate, such as the wood of a tree or stems, as well as using chemical signals. The ants provide some degree of protection to these larvae and they in turn gather honeydew secretions.

Parasitism

There are only 41 known species of parasitoid lepidoptera (1-Pyralidae; 40-Epipyropidae). The larvae of the Greater and Lesser wax moths feed on the honeycomb inside bee nests and may become pests; they are also found in bumblebee and wasp nests, albeit to a lesser extent. In northern Europe the wax moth is regarded as the most serious parasitoid of the bumblebee, and is found only in bumblebee nests. In some areas in southern England as many as eighty percent of nests can be destroyed. Other parasitic larvae are known to prey upon cicadas and leaf hoppers.

In reverse, moths and butterflies may be subject to parasitic wasps and flies, which may lay eggs on the caterpillars, which hatch and feed inside its body resulting in death. Although, in a form of parasitism called idiobiont, the adult paralyzes the host, so as not to kill it but for it to live as long as possible, in order for the parasitic larvae to benefit the most. Another form of parasitism, is koinobiont, where the species live off their host while inside or endoparasitic. These parasites live inside the host caterpillar throughout its life cycle, or may affect it later on as an adult. In other orders, koinobionts include flies, a majority of coleopteran, and many hymenopteran parasitoids. Some species may be subject to a variety of parasites, such as the Gypsy moth (*Lymantaria dispar*), which is attacked by a series of 13 species, in 6 different taxa throughout its life cycle.

In response to a parsitoid egg or larvae in the caterpillar's body, the plasmatocytes, or simply the host's cells can form a multilayered capsule that eventually cause the endoparasite to asphyxiate, and die. The process is called encapsulation, and is one of the caterpillar's only means of defence against parasitoids.

Other Biological Interactions

A few species of Lepidoptera are secondary consumers, or predators. These species typically prey upon the eggs of other insects, aphids, scale insects, or ant larvae. Some caterpillars are cannibals, and others prey on caterpillars of other species (e. g. Hawaiian *Eupithecia*). Those of the 15 species in Eupithecia that mirror inchworms, are the only known species of butterflies and moths that are ambush predators. There are 4 known species that eat snails. For example, the Hawaiian caterpillar, (*H. molluscivora*), uses silk traps, in a manner similar to that of spiders to capture certain species of snails (typically Tornatellides).

Larvae of some species of moths in the Tineidae, Gelechioidae and Noctuidae (family/superfamily/family respectively), besides others, feed on detritus, or organic material that is not living, such as fallen leaves and fruit, fungi, and animal products and turn it into humus. Well known species include the cloth moths (*Tineola bisselliella*, *T. pellionella*, and *T. tapetzella*), which feed on detritus containing keratin, including hair, feathers, cobwebs, bird nests (particularly of Domestic Pigeons, *Columba livia domestica*) and fruits or vegetables. These species are important to ecosystems as they remove substances that would otherwise take a long time to decompose.

Pests

The larvae of many Lepidopteran species are major pests in agriculture. Some of the major pests include Tortricidae, Noctuidae, and Pyralidae. The larvae of the Noctuidae genus *Spodoptera* (armyworms), *Helicoverpa* (corn earworm), or *Pieris brassicae* can cause extensive damage to certain crops. *Helicoverpa zea* larvae (cotton bollworms or tomato fruitworms) are polyphagous, meaning they eat a variety of crops, including tomatoes and cotton.

Butterflies and moths are one of the largest taxa to solely feed and be dependent on living plants, in terms of the number of species, and they are in many ecosystems, making up the largest biomass to do so. In many species, the female may produce anywhere from 200 to 600 eggs, while in some others it may go as high as 30,000 eggs in one day. This can create many problems for agriculture, where many caterpillars can affect acres of vegetation. Some reports estimate that there have been over 80,000 caterpillars of several different taxa feeding on a single oak tree. In some cases, phytophagous larvae can lead to the destruction of entire trees in relatively short periods of time.

Ecological ways of removing pest lepidoptera species are becoming more economically viable, as research has shown ways like introducing parasitic wasp and flies. For example *Sarcophaga aldrichi*, a fly which deposited larvae feed upon the pupae of the Forest Tent Caterpillar Moth. Pesticides can affect other species other than the species they are targeted to eliminate, damaging the natural ecosystem. Another good biological pest control method is the use of pheromone traps. A pheromone trap is a type of insect trap that uses pheromones to lure insects. Sex pheromones and aggregating pheromones are the most common types used. A pheromone-impregnated lure is encased in a conventional trap such as a Delta trap, water-pan trap, or funnel trap.

Species of moths that are detritivores would naturally eat detritus containing keratin, such as hairs or feathers. Well known species are cloth moths (*T. bisselliella, T. pellionella,* and *T. tapetzella*), feeding on foodstuffs that people find economically important, such as cotton, linen, silk and wool fabrics as well as furs; furthermore they have been found on shed feathers and hair, bran, semolina and flour (possibly preferring wheat flour), biscuits, casein, and insect specimens in museums.

Beneficial Insects

Even though most butterflies and moths affect the economy negatively, some species are a valuable economic resource. The most prominent example is that of the Domesticated silkworm moth (*Bombyx mori*), the larvae of which make their cocoons out of silk, which can be spun into cloth. Silk is and has been an important economic resource throughout history. The species *Bombyx mori* has been domesticated to the point where it is completely dependent on mankind for survival. A number of wild moths such as *Bombyx mandarina*, and *Antheraea* species, besides others, provide commercially important silks.

The preference of the larvae of most Lepidopteran species to feed on a single species or limited range of plants is used as a mechanism for biological control of weeds in place of herbicides. The pyralid cactus moth was introduced from Argentina to Australia, where it successfully suppressed millions of acres of Prickly pear cactus. Another species of the Pyralidae, called the alligator weed stem borer (*Arcola malloi*), was used to control the aquatic plant known as alligator weed (*Alternanthera philoxeroides*) in conjunction with the alligator weed flea beetle; in this case, the two insects work in synergy and the weed rarely recovers.

Breeding butterflies and moths, or butterfly gardening/rearing, has become an ecologically viable process of introducing species into the ecosystem to benefit it. Butterfly ranching in Papua New Guinea permits nationals of that country to 'farm' economically valuable insect species for the collectors market in an ecologically sustainable manner.

Food

Lepidoptera feature prominently in entomophagy as food items on almost every continent. While in most cases, adults, larvae or pupae are eaten as staples by indigenous people, beondegi or silkworm pupae are eaten as a snack in "Korean cuisine while Maguey worm is considered a delicacy in Mexico. In the Carnia region of Italy, children catch and eat ingluvies of the toxic *Zygaena* moths in early summer.

The ingluvies, despite having a very low cyanogenic content, serves as a convenient, supplementary source of sugar to the children who can include this resource as a seasonal delicacy at minimum risk.

Bio-control

aterpillars are major pests in forests, stored grains, and fibre and food crops. Pest resistance to insecticide is an increasing problem and moths are among the most feared invasive species.

The SIT is increasingly being developed as a component of AW-IPM programmes for a number of moth pests. It is being applied to control the codling moth (*Cydia pomonella*) in apples and pears in Canada and the false codling moth (*Thaumatotibia leucotreta*) in citrus in South Africa, the pink bollworm (*Pectinophora gossypiella*) in cotton in California, and more recently to eradicate this pest from the south-western USA and north-western Mexico, and to eradicate the invasive Australian painted apple moth (*Teia anartoides*) in New Zealand.

It is also being developed to control the invasive cactus moth (*Cactoblastis cactorun*) in Alabama, USA and Yucatan, Mexico, the carob moth (*Ectomyelois caratonia*) in dates in North Africa, and the diamondback moth (*Plutella xylostella*) in crucifer vegetables.

For moth pests there is considerable potential for expanded implementation of inherited sterility, involving the release of substerilized moths, which has shown a significant synergistic interaction with biological control.

Attract-and-Kill Strategies for Control of Lepidopteran Pests

Traditional control methods for lepidopteran pests and hops often involve foliar treatments with broad-spectrum organophosphates or pyrethroids. However, as many of these high-risk chemicals face increased regulation, growers have shifted towards using softer insecticide programmes including reduced risk insecticides and biopesticides, which are typically more effective against early instar larvae than on later instar larvae.

Attract-and-kill technology was explored as a reduced-risk alternative control strategy against lepidopteran pests of grapes and hops. A feeding attractant (1:1 acetic acid: 3- methyl-1-butanol (AAMB)) derived from fermented molasses attracts a large number of noctuid species.

Current Control Methods for Diamondback Moth and other Brassica Insect Pests

The diamondback moth (DBM), Plutella xylostella (L.), remains a major pest of brassica crops worldwide. DBM has been estimated globally to cost US$ 1 billion in direct losses and control costs. Chemical control of this pest remains difficult due to the rapid development of resistance to insecticides and to their effect on natural enemies. These problems are especially severe in South Asia and Africa where lack of knowledge, limited access to newer and safer insecticides, and a favourable climate result in DBM

remaining a serious year-round pest which substantially increases the cost and uncertainty of crop production. Despite these problems, application of synthetic insecticides remains overwhelmingly the most common control strategy. Biologically- based efforts to control DBM in Africa and Asia have focused strongly on parasitoid introductions. However, despite the identification and deployment of promising parasitoids in many regions, these efforts have had limited impact, often because farmers continue early season spraying of broad-spectrum insecticides that are lethal to parasitoids and thus exacerbate DBM outbreaks. A significant driver for this pattern of insecticide use is the presence of aphids and other pests whose appearance initiates inappropriate spraying. Despite often extensive training of producers in farmer field schools, many growers seem loath to discard calendar or prophylactic spraying of insecticides.

The introduction of an IPM technology that could replace the use of broad-spectrum insecticides for DBM and other key Lepidoptera is crucial if the benefits of parasitoid introduction are to be fully realised. The deployment of DBM-resistant brassicas expressing proteins from Bacillus thuringiensis could help to break this cycle of insecticide misuse and crop loss, but their deployment should be part of an integrated pest management (IPM) package, which recognises the constraints of farmers while addressing the requirement to control other Lepidoptera, aphids and other secondary pests.

Biological Control of Coleopteran Pests

The Coleoptera order of insects is commonly called beetles. The word "coleoptera" is from the Greek *koleos*, meaning "sheath"; and ðôåñüí, *pteron*, meaning "wing", thus "sheathed wing", because most beetles have two pairs of wings, the front pair, the "elytra", being hardened and thickened into a sheath-like, or shell-like, protection for the rear pair, and for the rear part of the beetle's body. The superficial consistency of most beetles' morphology, in particular their possession of elytra, has long suggested that the Coleoptera are monophyletic, but growing evidence indicates this is unjustified, there being arguments, for example, in favour of allocating the current suborder Adephaga their own order, or very likely even more than one.

The Coleoptera include more species than any other order, constituting almost 25% of all known types of animal life-forms. About 40% of all described insect species are beetles (about 400,000 species), and new species are discovered frequently. Some estimates put the total number of species, described and undescribed, at as high as 100 million, but a figure of one million is more widely accepted. The largest taxonomic family is commonly thought to be the Curculionidae (the weevils or snout beetles), but recently the Staphylinidae (the rove beetles) have claimed this title.

The diversity of beetles is very wide. They are found in all major habitats, except marine and the polar regions. They have many classes of ecological effects; particular species are adapted to practically every kind of diet. Some are nonspecialist detritus feeders, breaking down animal and plant debris; some feed on particular kinds of carrion such as flesh or hide; some feed on wastes such as dung; some feed on fungi, some on particular species of plants, others on a wide range of plants. Some are generalist pollen, flower and fruit eaters. Some are predatory, usually on other invertebrates; some are parasites or parasitoids. Many of the predatory species are important controls of agricultural pests. For example, beetles in the family Coccinellidae ("ladybirds" or "ladybugs") consume aphids, scale insects, thrips, and other plant-sucking insects that damage crops.

Conversely, beetles are prey of various invertebrates and vertebrates, including insects, fish, reptiles, birds, and mammals. The Coleoptera are not generally serious pests, but they include agricultural and industrial pests, such as the Colorado potato beetle *Leptinotarsa decemlineata*, the boll weevil *Anthonomus grandis*, the red flour beetle *Tribolium castaneum*, and the mungbean or cowpea beetle *Callosobruchus maculatus*. Also included is the death-watch beetle, the larvae of which can cause serious structural damage to buildings by boring into the timbers.

Species in the Coleoptera have a hard exoskeleton, particularly on their forewings (elytra, singular elytron). These elytra distinguish beetles from most other insect species, except for the Dermaptera. The hemelytra of Heteroptera have a slight resemblance, but are not the same and their function is largely different.

Like all armoured insects, beetles' exoskeletons comprise numerous plates called sclerites, some fused, and some separated by thin sutures. This combines armored defences with maintaining flexibility. The general anatomy of a beetle is superficially uniform, but specific organs and appendages may vary greatly in appearance and function between the many families in the order, and even more so between the suborders (such as Adephaga) that currently seem increasingly to be separate orders in their own right. All insects' bodies are divided into three sections: the head, the thorax, and the abdomen, and the Coleoptera are no exception. Their internal morphology and physiology also resemble those of other insects.

Beetles are endopterygotes; they undergo complete metamorphosis, a biological process by which an animal physically develops after a birth or hatching, undergoing a series of conspicuous and relatively abrupt changes in its body structure. Males may fight for females in various ways, and such species tend to display marked sexual dimorphism.

Distribution and Diversity

Figure: *Coccinellidaes in County Wexford, Ireland.*

Beetles are by far the largest order of insects, with 350,000–400,000 species in four suborders (Adephaga, Archostemata, Myxophaga, and Polyphaga), making up about 40% of all insect species described, and about 30% of all animals. Though classification at the family level is a bit unstable, about 500 families and subfamilies are recognized. One of the first proposed estimates of the total number of beetle species on the planet is based on field data rather than on catalogue numbers. The technique used for this original estimate, possibly as many as 12 million species, was criticized, and was later revised, with estimates of 850,000–4,000,000 species proposed. Some 70–95% of all beetle species, depending on the estimate, remain undescribed. The beetle fauna is not equally well known in all parts of the world. For example, the known beetle diversity of Australia is estimated at 23,000 species in 3265 genera and 121 families. This is slightly lower than reported for North

America, a land mass of similar size with 25,160 species in 3526 genera and 129 families. While other predictions show there could be as many as 28,000 species in North America, including those currently undescribed, a realistic estimate of the little-studied Australian beetle fauna's true diversity could vary from 80,000 to 100,000.

Coleoptera are found in nearly all natural habitats, including freshwater and marine habitats, everywhere vegetative foliage is found, from trees and their bark to flowers, leaves, and underground near roots- even inside plants in galls, in every plant tissue, including dead or decaying ones.

Lifecycle

Egg

A single female may lay from several dozen to several thousand eggs during her lifetime. Eggs are usually laid according to the substrate on which the larvae will feed upon hatching. Among others, they can be laid loose in the substrate (for example, flour beetle), laid in clumps on leaves (for example, Colorado potato beetle), individually attached (for example, mungbean beetle and other seed borers), or buried in the medium (for example, carrot weevil).

Parental care varies between species, ranging from the simple laying of eggs under a leaf to certain scarab beetles, which construct underground structures complete with a supply of dung to house and feed their young. Other beetles are leaf rollers, biting sections of leaves to cause them to curl inwards, then laying their eggs, thus protected, inside.

Larva

The larva is usually the principal feeding stage of the beetle lifecycle. Larvae tend to feed voraciously once they emerge from their eggs. Some feed externally on plants, such as those of certain leaf beetles, while others feed within their food sources. Examples of internal feeders are most Buprestidae and longhorn beetles. The larvae of many beetle families are predatory like the adults (ground beetles, ladybirds, rove beetles). The larval period varies between species, but can be as long as several years. The larvae are highly varied amongst species, with well-developed and sclerotized heads, and have distinguishable thoracic and abdominal segments (usually the tenth, though sometimes the eighth or ninth).

Beetle larvae can be differentiated from other insect larvae by their hardened, often darkened heads, the presence of chewing mouthparts, and spiracles along the sides of their bodies. Like adult beetles, the larvae are varied in appearance, particularly between beetle families. Beetles whose larvae are somewhat flattened and are highly mobile are the ground beetles, some rove beetles, and others; their larvae are described as campodeiform.

Some beetle larvae resemble hardened worms with dark head capsules and minute legs. These are elateriform larvae, and are found in the click beetle (Elateridae) and darkling beetle (Tenebrionidae) families. Some elateriform larvae of click beetles are known as wireworms. Beetles in the families of the Scarabaeoidea have short, thick larvae described as scarabaeiform, but more commonly known as grubs.

All beetle larvae go through several instars, which are the developmental stages between each moult. In many species, the larvae simply increase in size with each successive instar as more food is consumed. In some cases, however, more dramatic changes occur. Among certain beetle families or genera, particularly those that exhibit parasitic lifestyles, the first instar (the planidium) is highly mobile to search out a host, while the following instars are more sedentary and remain on or within their host. This is known as hypermetamorphosis; examples include the blister beetles (family Meloidae) and some rove beetles, particularly those of the genus *Aleochara*.

Pupa

As with all endopterygotes, beetle larvae pupate, and from these pupae emerge fully formed, sexually mature adult beetles, or imagos. Adults have extremely variable lifespans, from weeks to years, depending on the species. In some species, the pupa may go through all four forms during its development, called hypermetamorphosis (for example, Meloidae). Pupae always have no mandibles (are adecticous). In most, the appendages are not attached to the pupae, or they are exarate; with most being obtect in form.

Behaviour

Locomotion: Aquatic beetles use several techniques for retaining air beneath the water's surface. Beetles of the family Dytiscidae hold air between the abdomen and the elytra when diving. Hydrophilidae have hairs on their under surface that retain a layer of air against their bodies. Adult crawling water beetles use both their elytra and their hind coxae (the basal segment of the back legs) in air retention, while whirligig beetles simply carry an air bubble down with them whenever they dive.

The elytra allow beetles and weevils to both fly and move through confined spaces, doing so by folding the delicate wings under the elytra while not flying, and folding their wings out just before take off. The unfolding and folding of the wings is operated by muscles attached to the wing base; as long as the tension on the radial and cubital veins remains, the wings remain straight. In day-flying species (for example, Buprestidae, Scarabaeidae), flight does not include large amounts of lifting of the elytra, having the metathorac wings extended under the lateral elytra margins.

Communication

Beetles have a variety of ways to communicate, some of which include a sophisticated chemical language through the use of pheromones. From the host tree, mountain pine beetles have many forms of communication. They can emit both an aggregative pheromone and an anti-aggregative pheromone. The aggregative pheromone attracts other beetles to the tree, and the anti-aggregative pheromone neutralizes the aggregative pheromone. This helps to avoid the harmful effects of having too many beetles on one tree competing for resources. The mountain pine beetle can also stridulate to communicate, or rub body parts together to create sound, having a "scraper" on their abdomens that they rub against a grooved surface on the underside of their left wing cover to create a sound that is not audible to humans. Once the female beetles have arrived on a suitable pine tree host, they begin to stridulate and produce aggregative pheromones to attract other unmated males and females. New females arrive and do the same as they land and bore into the tree. As the males arrive, they enter the galleries that the females have tunneled, and begin to stridulate to let the females know they have arrived, and to also warn others that the female in that gallery is taken. At this point, the female stops producing aggregative pheromones and starts producing anti-aggregative pheromone to deter more beetles from coming.

Since species of Coleoptera use environmental stimuli to communicate, they are affected by the climate. Microclimates, such as wind or temperature, can disturb the use of pheromones; wind would blow the pheromones while they travel through the air. Stridulating can be interrupted when the stimulus is vibrated by something else.

Parental care

Among insects, parental care is very uncommon, only found in a few species. Some beetles also display this unique social behaviour. One theory states parental care is necessary for the survival of the larvae, protecting them from adverse environmental conditions and predators. One species, a rover beetle (*Bledius spectabilis*) displays both causes for parental care: physical and biotic environmental factors. Said species lives in salt marshes, so the eggs and/or larvae are endangered by the rising tide. The maternal beetle will patrol the eggs and larvae and apply the appropriate burrowing behaviour to keep them from flooding and from asphyxiating. Another advantage is that the mother protects the eggs and larvae from the predatory carabid beetle *Dicheirotrichus gustavi* and from the parasitoid wasp *Barycnemis blediator*. Up to 15% of larvae are killed by this parasitoid wasp, being only protected by maternal beetles in their dens.

Some species of dung beetle also display a form of parental care. Dung beetles collect animal feces, or "dung", from which their name is derived, and roll it into a ball, sometimes being up to 50 times their own weight; albeit sometimes it is also used to store food. Usually it is the male that rolls the ball, with the female hitch-hiking or simply following behind. In some cases the male and the female roll together. When a spot with soft soil is found, they stop and bury the dung ball. They will then mate underground. After the mating, one or both of them will prepare the brooding ball. When the ball is finished, the female lays eggs inside it, a form of mass provisioning. Some species do not leave after this stage, but remain to safeguard their offspring.

Feeding

Besides being abundant and varied, beetles are able to exploit the wide diversity of food sources available in their many habitats. Some are omnivores, eating both plants and animals. Other beetles are highly specialized in their diet. Many species of leaf beetles, longhorn beetles, and weevils are very host-specific, feeding on only a single species of plant. Ground beetles and rove beetles (family Staphylinidae), among others, are primarily carnivorous and will catch and consume many other arthropods and small prey, such as earthworms and snails. While most predatory beetles are generalists, a few species have more specific prey requirements or preferences.

Decaying organic matter is a primary diet for many species. This can range from dung, which is consumed by coprophagous species (such as certain scarab beetles of the family Scarabaeidae), to dead animals, which are eaten by necrophagous species (such as the carrion beetles of the family Silphidae). Some of the beetles found within dung and carrion are in fact predatory. These include the clown beetles, preying on the larvae of coprophagous and necrophagous insects.

Ecology

Defence and Predation

Beetles and their larvae have a variety of strategies to avoid being attacked by predators or parasitoids. These include camouflage, mimicry, toxicity, and active defence. Camouflage involves the use of colouration or shape to blend into the surrounding environment. This sort of protective colouration is common and widespread among beetle families, especially those that feed on wood or vegetation, such as many of the leaf beetles (family Chrysomelidae) or weevils. In some of these species, sculpturing or various coloured scales or hairs cause the beetle to resemble bird dung or other inedible objects. Many of those that live in sandy environments blend in with the colouration of the substrate. The giant African longhorn beetle (*Petrognatha gigas*) resembles the moss and bark of the tree it feeds on. Another defence that often uses colour or shape to deceive potential enemies is mimicry. A number of longhorn beetles (family Cerambycidae) bear a striking resemblance to wasps, which helps them avoid predation even though the beetles are in fact harmless. This defence is an example of Batesian mimicry and, together with other forms of mimicry and camouflage occurs widely in other beetle families, such as the Scarabaeidae. Beetles may combine their colour mimicry with behavioural mimicry, acting like the wasps they already closely resemble. Many beetle species, including ladybirds, blister beetles, and lycid beetles can secrete distasteful or toxic substances to make them unpalatable or even poisonous. These same species often exhibit aposematism, where bright or contrasting colour patterns warn away potential predators, and there are, not surprisingly, a great many beetles and other insects that mimic these chemically protected species.

Chemical defence is another important defence found amongst species of Coleoptera, usually being advertised by bright colours. Others may utilize behaviours that would be done when releasing noxious chemicals (for example, Tenebrionidae). Chemical defence may serve purposes other than just protection from vertebrates, such as protection from a wide range of microbes, and repellents. Some species release chemicals in the form of a spray with surprising accuracy, such as ground beetles (Carabidae), may spray chemicals from their abdomen to repel predators. Some species take advantage of the plants from which they feed, and sequester the chemicals from the plant that would protect it and incorporate into their own defence. African carabid beetles (for example, *Anthia* and *Thermophilum*) employ the same chemicals used by ants, while Bombardier beetles have a their own unique separate gland, spraying potential predators from far distances.

Large ground beetles and longhorn beetles may defend themselves using strong mandibles and/or spines or horns to forcibly persuade a predator to seek out easier prey. Many species have large protrusions from their thorax and head such as the Rhinoceros beetle, which can be used to defended themselves from predators. Many species of weevil that feed out in the open on leaves of plants react to attack by employing a "drop-off reflex". Even further, some will combine it with thanatosis, which they will close up their legs, antennae, mandibles, etc. and use their cryptic colouration to blend in with the background. Species with varied colouration do not do this as they can not camaflouge.

Parasitism

Over 1000 species of beetles are known to be either parasitic, predatory, or commensals in the nests of ants. Technically, many species of beetles and their larvae could be considered to be ectoparasites, because they feed on plants and some live inside the bark or wood of trees, but such relationships are generally regarded as herbivory (plant eating) rather than parasitism.

A few species of beetles are actually ectoparasitic on mammals. One such species, *Platypsyllus castoris*, parasitises beavers (*Castor* spp.). This beetle lives as a parasite both as a larva and as an adult, feeding on epidermal tissue and possibly on skin secretions and wound exudates. They are strikingly flattened dorsoventrally, no doubt as an adaptation for slipping between the beavers' hairs. They also are wingless and eyeless, as are many other ectoparasites.

Other parasitic beetles include those that are kleptoparasites of other invertebrates, such as the small hive beetle (*Aethina tumida*) that infests honey bee hives. The larvae tunnel through comb towards stored honey or pollen, damaging or destroying cappings and comb in the process. Larvae defecate in honey and the honey becomes discoloured from the feces, which causes fermentation and a frothiness in the honey; the honey develops a characteristic odor of decaying oranges. Damage and fermentation cause honey to run out of combs, destroying large amounts of it both in hives and sometimes also in honey extracting rooms. Heavy infestations cause bees to abscond; though the beetle is only a minor pest in Africa, beekeepers in other regions have reported the rapid collapse of even strong colonies.

Pollination

Beetle-pollinated flowers are usually large, greenish or off-white in colour, and heavily scented. Scents may be spicy, fruity, or similar to decaying organic material. Most beetle-pollinated flowers are flattened or dish-shaped, with pollen easily accessible, although they may include traps to keep the beetle longer. The plants' ovaries are usually well protected from the biting mouthparts of their pollinators. Beetles may be particularly important in some parts of the world such as semiarid areas of southern Africa and southern California and the montane grasslands of KwaZulu-Natal in South Africa.

Mutualism

Amongst most orders of insects, mutualism is not common, but some examples occur in species of Coleoptera, such as the ambrosia beetle, the ambrosia fungus, and probably bacteria. The beetles excavate tunnels in dead trees in which they cultivate fungal gardens, their sole source of nutrition. After landing on a suitable tree, an ambrosia beetle excavates a tunnel in which it releases spores of its fungal symbiont. The fungus penetrates the plant's xylem tissue, digests it, and concentrates the nutrients on and near the surface of the beetle gallery, so the weevils and the fungus both benefit. The beetles cannot eat the wood due to toxins, and uses its relationship with fungi to help overcome its host tree defences and to provide nutrition for their larvae. Chemically mediated by a bacterially produced polyunsaturated peroxide, this mutualistic relationship between the beetle and the fungus is coevolved.

Commensalism

Pseudoscorpions are small arachnids with flat, pear-shaped bodies and pincers that resemble those of scorpions (only distant relatives), usually ranging from 2 to 8 millimetres (0.08 to 0.31 in) in length. Their small size allows them to hitch rides under the elytra of giant harlequin beetles to be dispersed over wide areas while simultaneously being protected from predators. They may also find mating partners as other individuals join them on the beetle. This would be a form of parasitism if the beetle were harmed in the process, but the beetle is, presumably, unaffected by the presence of the hitchhikers.

Beetles as Pests

About 75% of beetle species are phytophagous in both the larval and adult stages, living in or on plants, wood, fungi, and a variety of stored products, including cereals, tobacco, and dried fruits. Because many of these plants are important for agriculture, forestry, and the household, beetles can be considered pests. Some of these species cause significant damage, such as the boll weevil, which feeds on cotton buds and flowers. The boll weevil crossed the Rio Grande near Brownsville, Texas, to enter the United States from Mexico around 1892, and had reached southeastern Alabama by 1915.

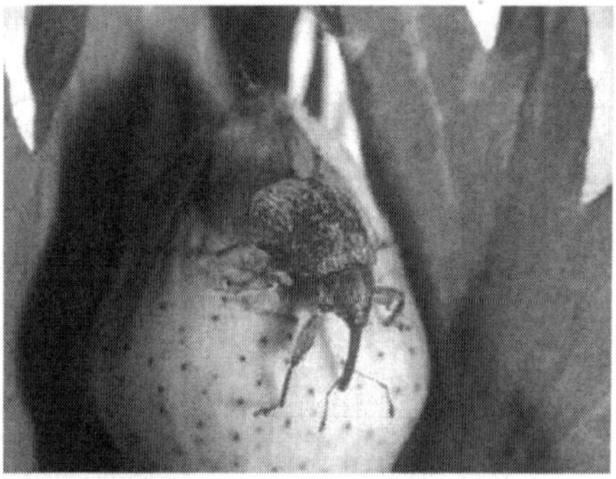

Figure: *Cotton boll weevil*

By the mid-1920s, it had entered all cotton-growing regions in the US, travelling 40 to 160 miles (60–260 km) per year. It remains the most destructive cotton pest in North America. Mississippi State University has estimated, since the boll weevil entered the United States, it has cost cotton producers about $13 billion, and in recent times about $300 million per year. Many other species also have done extensive damage to plant populations, such as the bark beetle and elm leaf beetle. The bark beetle and elm leaf beetle, among other species, have been known to nest in elm trees. Bark beetles in particular carry Dutch elm disease as they move from infected breeding sites to feed on healthy elm trees. The spread of Dutch elm disease by the beetle has led to the devastation of elm trees in many parts of the Northern Hemisphere, notably in Europe and North America.

Figure: *Larvae of the Colorado potato beetle,* Leptinotarsa decemlineata

Situations in which a species has developed immunity to pesticides are worse, as in the case of the Colorado potato beetle, *Leptinotarsa decemlineata*, which is a notorious pest of potato plants. Crops are destroyed and the beetle can only be treated by employing expensive pesticides, to many of which it has begun to develop resistance. Suitable hosts can include a number of plants from the potato family (Solanaceae), such as nightshade, tomato, eggplant and capsicum, as well as potatoes. The Colorado potato beetle has developed resistance to all major insecticide classes, although not every population is resistant to every chemical.

Pests do not only affect agriculture, but can also even affect houses, such as the death watch beetle. The death watch beetle, *Xestobium rufovillosum* (family Anobiidae), is of considerable importance as a pest of older wooden buildings in Great Britain. It attacks hardwoods such as oak and chestnut, always where some fungal decay has taken or is taking place. The actual introduction of the pest into buildings is thought to take place at the time of construction.

Other pest include the coconut hispine beetle, *Brontispa longissima*, which feeds on young leaves and damages seedlings and mature coconut palms. On September 27, 2007, Philippines' Metro Manila and 26 provinces were quarantined due to having been infested with this pest (to save the $800-million Philippine coconut industry). The mountain pine beetle normally attacks mature or weakened lodgepole pine. It can be the most destructive insect pest of mature pine forests. The current infestation in British Columbia is the largest Canada has ever seen.

Beetles as Beneficial Resources

Figure: Coccinella septempunctata, *a beneficial beetle*

Beetles are not only pests, but can also be beneficial, usually by controlling the populations of pests. One of the best, and widely known, examples are the ladybugs or ladybirds (family Coccinellidae). Both the larvae and adults are found feeding on aphid colonies. Other ladybugs feed on scale insects and mealybugs. If normal food sources are scarce, they may feed on small caterpillars, young plant bugs, or honeydew and nectar. Ground beetles (family Carabidae) are common predators of many different insects and other arthropods, including fly eggs, caterpillars, wireworms, and others.

Dung beetles (Scarabidae) have been successfully used to reduce the populations of pestilent flies and parasitic worms that breed in cattle dung. The beetles make the dung unavailable to breeding pests by quickly rolling and burying it in the soil, with the added effect of improving soil fertility, tilth, and nutrient cycling. The Australian Dung Beetle Project (1965–1985), led by Dr. George Bornemissza of the Commonwealth Scientific and Industrial Research Organization, introduced species of dung beetle to Australia from South Africa and Europe, and effectively reduced the bush fly (*Musca vetustissima*) population by 90%.

Dung beetles play a remarkable role in agriculture. By burying and consuming dung, they improve nutrient recycling and soil structure. They also protect livestock, such as cattle, by removing dung, which, if left, could provide habitat for pests such as flies. Therefore, many countries have introduced the creatures for the benefit of animal husbandry. In developing countries, the beetle is especially important as an adjunct for improving standards of hygiene. The American Institute of Biological Sciences reports that dung beetles save the United States cattle industry an estimated US$380 million annually through burying above-ground livestock feces.

Some beetles help in a professional setting, doing things that people cannot; those of the family Dermestidae are often used in taxidermy and preparation of scientific specimens to clean bones of remaining soft tissue. The beetle larvae are used to clean skulls because they do a thorough job of cleaning, and do not leave the tool marks that taxidermists' tools do. Another benefit is, with no traces of meat remaining and no emulsified fats in the bones, the trophy will not develop the unpleasant dead odor. Using the beetle larvae means that all cartilage is removed along with the flesh, leaving the bones spotless.

Biotechnology and Derived Products

The requirement of fruits and vegetables is increasing proportionally with the increasing population in the country. How do we keep horticultural production on par with the burgeoning population? Although conventional plant breeding techniques have made considerable progress in the development of improved varieties, they have not been able to keep pace with the increasing demand for vegetables and fruits in the developing countries. Therefore an immediate need is felt to integrate biotechnology to speed up the crop improvement programmes. Biotechnological tools have revolutionized the entire crop improvement programmes by providing new strains of plants, supply of planting material, more efficient and selective pesticides and improved fertilizers. Many genetically modified fruits and vegetables are already in the market in developed countries. Modern biotechnology encompasses broad areas of biology from utilization of living organisms or substances from those organisms to make or to modify a product, to improve plant or animal or to develop micro-organisms for specific use. It is a new aspect of biological and agricultural science which provides new tools and strategies in the struggle against world's food production problem. The major areas of biotechnology which can be adopted for improvement of horticultural crops are –

Tissue Culture

One of the widest applications of biotechnology has been in the area of tissue culture and micro propagation in particular. It is one of the most widely used techniques for rapid asexual in vitro propagation. This technique is economical in time and space affords greater output and provides disease free and elite propagules. It also facilitates safer and quarantined movements of germplasm across nations. When the traditional methods are unable to meet the demand for propagation material this technique can produce millions of uniformly flowering and yielding plants. Micropropagation of almost all the fruit crops and vegetables is possible now. Production of virus free planting material using meristem culture has been made possible in many horticultural crops. Embryo rescue is another area where plant breeders are able to rescue their crosses which would otherwise abort. Culture of excised embryos of suitable stages of development can circumvent problems encountered in post zygotic incompatibility. This technique is highly significant in intractable and long duration horticultural species. Many of the dry land legume species have been successfully regenerated from cotyledons, hypocotyls, leaf, ovary, protoplast, petiole root, anthers, etc., Haploid generation through anther/pollen culture is recognized as another important area in crop improvement. It is useful in being rapid and economically feasible. Complete homozygosity of the offspring helps in phenotype selection for quantitative characters and particularly for qualitatively inherited characters making breeding much easier successful isolation, culture and fusion of plant protoplasts has been very useful in transferring cytoplasmic male sterility for obtaining hybrid vigour through mitochondrial recombination and for genetic transformation in plants.

In vitro germplasm conservation is of great significance in providing solutions and alternative approaches to overcoming constrains in management of genetic resources. In crops which are propagated vegetatively and which produce recalcitrant seeds and perennial crops which are highly heterozygous seed storage is not suitable. In such crops especially, *in vitro* storage is of great practical importance. These techniques have successfully been demonstrated in a number of horticultural crops and there are now various germplasm collection centres. In vitro germplasm also assures the exchange of pest and disease free material and helps in better quarantine.

Plant breeders are continually searching for new genetic variability that is potentially useful in cultivar improvement. A portion of plants regenerated by tissue culture often exhibits phenotypic variation atypical of the original phenotype. Such variation, termed somaclonal variation may be heritable i.e. genetically stable and passed on to the next generation. Alternatively, the variation may be epigenetic and disappear following sexual reproduction. These heritable variation are potentially useful to plant breeders.

Genetic Engineering of Plants

Genetic Engineering involves three major steps:

i) Identification and isolation of suitable genes for transfer

ii) Delivery system to insert desired gene into recipient cells.

iii) Expression of new genetic information in recipient cells.

Using techniques of genetic engineering many useful genes have been introduced into plants and many transgenic plants have been developed in which the foreign DNA has been stably integrated and resulted in the synthesis of appropriate gene product. Transgenic plants have covered about 52.6 m hectares in the Industrial and developing countries upto 2001. Genes for the following traits have been introduced to the crop plants.

Herbicide tolerance: Transgenic plants are developed that are resistant to herbicides allowing farmers to spray crops so as to kill only weeds but not their crops. Many herbicide tolerant plants have been developed in tomato, tobacco, potato, soybean, cotton, corn oilseed rape, petunia, etc. Glyphosate is one of the most potent broad spectrum environment friendly herbicide known, it is marketed under the trade name Round up. Glyphosate kills plants by blocking the action of an enzyme (5-enolpyruvyl shikimate-3-phosphate synthase) (EPSPS) an essential enzyme in the biosynthesis of aromatic amino acids, tyrosine, phenylalanine and tryptophan. Amino acids are building blocks of protein. Transgenic plants resistant to Glyphosate have been developed by transferring gene of EPSPS that over prodoce this enzyme thus inhibiting the effect of Glyphosate. A number of detoxifying enzymes have been identified in plants as well as in microbes. Some of these include glutahthione-s-transferase or GST in maize and other plants which detoxifies the herbicide bromoxynil and phosphinothricin acetyl transferase (PAT) which detoxifies the herbiside PPT (L-phosphinothricine). Transgenic plants using *bxn* gene from Klebsiella and *bar* gene from Strepotomyces have been obtained in potato, oilseed, sugarbeet, soybean, cotton and corn and are found to be herbicide resistance. These transgenic plants reduce the use of weeding labour, farmers cost and increase yield.

Engineering pathogen resistance: Viruses are the major pests of crop plants which cause considerable yield losses. Many strategies have been applied to control virus infection using coat protein and satellite RNA. Viruses are submicroscopic pockets of nucleic acid (DNA or RNA) enclosed in a protein coat and can multiply within a host cell. Use of viral coat protein as a transgene for producing virus resistant plants is one of the most spectacular successes achieved in plant biotechnology. Coat protein gene from tobacco mosaic virus (TMV) classified as a positive strand RNA virus has been transferred to tobacco, making it nearly resistant against TMV. Using gene for nucelocapsid protein resistance has been introduced in crops like tomato, tobacco, lettuce, groundnut, pepper and in ornaments like *Impatiens*, *Ageratum* and *Crysnathemum* against tomato spotted wilt virus. Use of satellite RNA (SATRNA) makes many transgenic plants resistant to Cucumber Mosaic Virus (CMV). Transgenic resistant plants have also been developed against alfalfa mosaic virus, potato virus X, Rice tungro virus, tobacco rattle virus and Papaya ring spot virus.

During the last decade many resistance genes whose products are involved in recognizing the invading pathogens have been identified and cloned. A number of signaling pathways which follow the pathogen infection have been dissected. Many of the antifungal compounds synthesized by plants which combat fungal infections have been identified. The major strategies for developing fungal resistance have been production of transgenic plants with antifungal molecules like proteins and toxins, and generation of hypersensitive response through R genes or by manipulating genes of SAR pathway. A chitinase gene from bean plants in tobacco and *Brassica napus* showed enhanced resistance to *Rhizoctonia solani*. In another case chitinase gene obtained from *Serratia marcescens* (soil bacterium) is introduced in tobacco making it resistant to *Alternaria longipes* which causes brown spot diseases. Acetyl transferase gene is introduced in tobacco making it resistant to *Pseudomonas syringea*, a causal agent of wild fire disease.

Stress resistance : A number of genes responsible for providing resistance against stresses such as to water stress heat, cold, salt, heavy metals and phytohormones have been identified. Studies are also being conducted on metabolites like proteins and betains that have been implicated in stress tolerance. Resistance against chilling was introduced into tobacco plants by introducing gene for glycerol-1-phosphate acyl-transferase enzyme from *Arabidopsis.* Many plants respond to drought stress by synthesizing a group of sugar derivatives called polyols (Mannitol, Sorbitol and Sion). Plants that have more polyols are more resistant to stress. Using a bacterial gene capable of synthesizing mannitols it is possible to raise the level of mannitol very high making plants resistant to drought.

Fruit Quality: Tomatoes which ripen slowly are helpful in transportation process. Transgenic tomato with reduced pectin methyl esterase activity and increased level of soluble solids and higher pH increases processing quality. Tomatoes exhibiting delayed ripening have been produced either by using antisense RNA against enzymes involved in ethylene production (Eg ACC synthase) or by using gene for deaminase which degraded l-aminocyclopropane-l-carboxylic acid (ACC) an immediate precursor of ethylene. This increases the shelf life of tomatoes. These tomatoes can also stay on the plant long giving more time for accumulation of sugars and acids for improving flavour. It is produced at commercial level in European and American countries. Tomatoes with elevated sucrose and reduced starch could also be produced using sucrose phosphate synthase gene. Starch content in potatoes has been increased by 20-40% by using a bacterial ADP glucose pyrophosphorylase gene.

Pest resistance: The insecticidal beta endotoxin gene (bt gene) has been isolated from *Bacillus thuringiensis* the commonly occurring soil bacteria and transferred to number of plants like cotton, tobacco, tomato, soybean, potato, etc. to make them resistant to attack by insects. These genes produce insecticidal crystal proteins which affect a range of lepidopteran, coleopteran, dipteran insects. These crystals upon ingestion by the insect larva are solubilised in the highly alkaline midgut into individual protoxins which vary from 133 to 136 kDa in molecular weight. Insecticidal crystal protein produced during vegetative growth of the cells (VIP)are also found to be highly effective against insect control. Bt resistant plants are already in the market.

Male sterility and Fertility restoration: This is helpful in hybrid seed production. Transgenic plants with male sterility and fertility restoration genes have become available in *Brassica napus.* It facilitates production of hybrid seed without manual emasculation and controlled pollination as often done in maize. In 1990, Mariani and others from

Belgium have successfully used a gene construct having another specific promoter from TA29 gene of tobacco and bacterial coding sequence for a ribonuclease gene from Bacillus Sp. (barnase gene) for production of transgenic plants in *Brassica napus.* Here the translated gene prevented normal pollen development leading to male sterilily.

Molecular Diagnostics

Nucleic acid probes:- It is now possible to detect the plant diseases even before onset of symptoms by using cDNA probes. Probes are nucleic acid sequences of pathogen causing organisms labelled with certain markers. cDNA probes corresponding to specific regions of the pathogens can be generated using standard recombinant DNA technique.

Monoclonal antibodies (McAb): Immunochemical techniques are extremely useful for the rapid and accurate routine detection of plant pathogens and ultimately the diagnosis of plant disease and their relatedness, The introduction of hybridoma technology has provided methods for the production of homologous and biochemically defined immunological reagents of identical specificity which are produced by a single cell line and are directed against a unique epitope of the immunizing antigen. The great potential of McAbs in phytopathological diagnostics is essential because of homogeneous antibody preparations with defined activity and specificity can be produced in large quantities over long periods. Even though hybridoma technology is a laborious and expensive enterprise compared to standard immunization procedures it is going to be widely used for large scale diagnosis.

Molecular Markers

The possibilities of using gene tags of molecular makers for selecting agronomic traits has made the job of breeder easier. It has been possible to score the plants for different traits or disease resistance at the seedling stage itself. The use of RFLP (Restriction Fragment Length polymorphism), RAPD (Random Amplified Polymorphic DNA) , AFLP (Amplified Fragment Length Polymorphism) and isozyme markers in plant breeding are numerous. RFLPs are advantageous over morphological and isozyme markers primarily because their number is limited only by genome size and they are not environmentally or developmentally influenced. Molecular maps now exist for a number of crop plants including corn, tomato, potato, rice, lettuce, wheat, Brassica species and barley. RFLPs have wide ranging applications including cultivar finger printing, identification of quantitative trait loci, analysis of genome organization, germplasm introgression and map-based cloning. AFLP is becoming the tool of choice for fingerprinting because of its reproducibility compared to RAPD. Microsatellile or simple sequence repeats (SSRS) markers have also become the choice for a wide range of applications in genotyping, genome mapping and genome analysis.

Development of Microbial Inoculan

Indiscriminate and injudicious use of chemical fertilizers and pesticides for the crop production and control of insect-pests has resulted in pollution of the environment deterioration of soil health and development of resistance by many insects and residue problems. Hence there is a great concern world wide to use safer biofertilisers and biopesticdies in the integrated nutrient management and pest management systems.

Biofertilizers are micro-organisms which fix atmospheric nitrogen or solubilise fixed phosphorus in the soil and make more nutrients available to the plant. Some of the organisms providing major inputs are the biological nitrogen fixing organisms like *Rhizobium, Azotobacter, Azospirillum* and phosphate solubilising organisms like *Bacillus polymyxa, B. magaterium, Pseudomonas striata* and certain fungal species of Aspergillus and *Penicillium.*

The benefits of using micro-organisms as fertilizers are many fold. They are less expensive, nontoxic to plants, do not pollute the ground water nor render the soil acidic and unfit for growth of plants. *Rhizobium* forms nodules on the roots of leguminous plants and help in fixing nitrogen from the atmosphere to ammonium irons which get converted to amino acids in the plant system. Inoculation with this bacteria helps in reducing addition of nitrogenous fertilizers to the soil. *Azospirillum* is also found colonizing inter cellular spaces inside the root system. These bacteria also contribute substantially to the nitrogen requirement of the plant.

Phosphate solubilising bacteria are another group of micro-organisms which solubilise the insoluble phosphorus in the soil and make them readily available to the crop.

Mycorrhiza is the symbiotic association of the roots of crop plants with non-pathogenic fungus. They provide nutrients absorbed from deeper layers of soil to the plants. They help the plants in better plant establishment and growth when inoculated. Many fruit crops like papaya, mango, banana, citrus, pomegranate are found to be dependent on this association and are greatly benefited by its inoculation in procuring higher phosphate and other nutrient from the soil. These mycorrhizal associations help the plants in overcoming pathogen attack also. They improve soil characters too.

Genetic modification of microbes: By using DNA recombination technique it has been possible to genetically manipulate different strains of these bacteria suitable to different environmental conditions and to develop strains with traits with capacity for better competitiveness and nodulation.

Biopesticides are biological organisms which can be formulated as that of the pesticides for the control of pests. Biopesticides are gaining importance in agriculture, horticulture and in public heatlh programmes for the control of pests. The advantages of using biopesticides are many. They are specific to target pests and do not harm the non target organisms such as bees, butterflies and are safe to humans and live stocks, they do not disturb the food-chain nor leave behind toxic residues.

Some of the microbial pesticides used to control insect pests are *Bacillus thuringiensis* species to control various insect pests. Insecticidal property of these bacteria are due to crystals of insecticidal proteins produced during sporulation. These proteins are stomach poisons and are highly insect specific. Bt toxins could kill plant parasitic nematode too. Number of baculoviruses (BV) nuclear polyhedrosis virus (NPV) is being developed as microbial pesticides both nationally and internationally, A few examples of these are *Heliothis, Spodoptera, Plusia, Agrotis, Trichoplusia*, etc.

Biocontrol agents: These are other microbes which are antagonistic to several pathogenic fungus and are good substitutes to fungicides or insecticide. These are Bacillus sps. *Pseudomonas fluorescens, Trichoderma, Verticillium sp., Streptromyces* sps. etc. These organisms are commercially available.

The extent of commercial application of plant biotechnology is the important mark for measuring the vitality of this newly emerging technology. Small and marginal farmers can adopt less expensive technologies like the use of biofertilizers and biopesticides while capital intensive technologies can be adopted by rich farmers

Improvement of Biological Control Agents through Molecular Strategies

Molecular biology is the branch of biology that deals with the molecular basis of biological activity. This field overlaps with other areas of biology and chemistry, particularly genetics and biochemistry. Molecular biology chiefly concerns itself with understanding the interactions between the various systems of a cell, including the interactions between the different types of DNA, RNA and protein biosynthesis as well as learning how these interactions are regulated.

Relationship to Other Biological Sciences

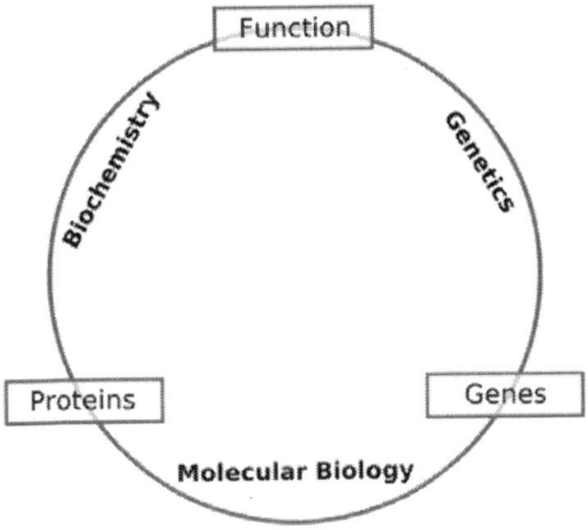

Figure: *Schematic relationship between biochemistry, genetics and molecular biology.*

Researchers in molecular biology use specific techniques native to molecular biology but increasingly combine these with techniques and ideas from genetics and biochemistry. There is not a defined line between these disciplines.

The figure above is a schematic that depicts one possible view of the relationship between the fields:

- *Biochemistry* is the study of the chemical substances and vital processes occurring in living organisms. Biochemists focus heavily on the role, function, and structure of biomolecules. The study of the chemistry behind biological processes and the synthesis of biologically active molecules are examples of biochemistry.
- *Genetics* is the study of the effect of genetic differences on organisms. This can often be inferred by the absence of a normal component (e.g. one gene). The study of "mutants" – organisms which lack one or more functional components with respect to the so-called "wild type" or normal phenotype. Genetic interactions (epistasis) can often confound simple interpretations of such "knockout" studies.
- *Molecular biology* is the study of molecular underpinnings of the processes of replication, transcription, translation, and cell function. The central dogma of molecular biology where genetic material is transcribed into RNA and then translated into protein, despite being an oversimplified picture of molecular biology, still provides a good starting point for understanding the field. This picture, however, is undergoing revision in light of emerging novel roles for RNA.

Much of the work in molecular biology is quantitative, and recently much work has been done at the interface of molecular biology and computer science in bioinformatics and computational biology. As of the early 2000s, the study of gene structure and function, molecular genetics, has been among the most prominent sub-field of molecular biology.

Increasingly many other loops of biology focus on molecules, either directly studying their interactions in their own right such as in cell biology and developmental biology, or indirectly, where the techniques of molecular biology are used to infer historical attributes of populations or species, as in fields in evolutionary biology such as population genetics and phylogenetics. There is also a long tradition of studying biomolecules "from the ground up" in biophysics.

Techniques of Molecular Biology

Since the late 1950s and early 1960s, molecular biologists have learned to characterize, isolate, and manipulate the molecular components of cells and organisms. These components include DNA, the repository of genetic information; RNA, a close relative of DNA whose functions range from serving as a temporary working copy of DNA to actual structural and enzymatic functions as well as a functional and structural part of the translational apparatus; and proteins, the major structural and enzymatic type of molecule in cells.

Expression Cloning

One of the most basic techniques of molecular biology to study protein function is expression cloning. In this technique, DNA coding for a protein of interest is cloned (using PCR and/or restriction enzymes) into a plasmid (known as an expression vector). A vector has 3 distinctive features: an origin of replication, a multiple cloning site (MCS), and a selective marker (usually antibiotic resistance). The origin of replication will have promoter regions upstream from the replication/transcription start site.

This plasmid can be inserted into either bacterial or animal cells. Introducing DNA into bacterial cells can be done by transformation (via uptake of naked DNA), conjugation (via cell-cell contact) or by transduction (via viral vector). Introducing DNA into eukaryotic cells, such as animal cells, by physical or chemical means is called transfection. Several different transfection techniques are available, such as calcium phosphate transfection, electroporation, microinjection and liposome transfection. DNA can also be introduced into eukaryotic cells using viruses or bacteria as carriers, the latter is sometimes called bactofection and in particular uses *Agrobacterium tumefaciens*. The plasmid may be integrated into the genome, resulting in a stable transfection, or may remain independent of the genome, called transient transfection.

In either case, DNA coding for a protein of interest is now inside a cell, and the protein can now be expressed. A variety of systems, such as inducible promoters and specific cell-signaling factors, are available to help express the protein of interest at high levels. Large quantities of a protein can then be extracted from the bacterial or eukaryotic cell. The protein can be tested for enzymatic activity under a variety of situations, the protein may be crystallized so its tertiary structure can be studied, or, in the pharmaceutical industry, the activity of new drugs against the protein can be studied.

Polymerase Chain Reaction (PCR)

The polymerase chain reaction is an extremely versatile technique for copying DNA. In brief, PCR allows a single DNA sequence to be copied (millions of times), or altered in predetermined ways. For example, PCR can be used to introduce restriction enzyme sites, or to mutate (change) particular bases of DNA, the latter is a method referred to as "Quick change". PCR can also be used to determine whether a particular DNA fragment is found in a cDNA library. PCR has many variations, like reverse transcription PCR (RT-PCR) for amplification of RNA, and, more recently, quantitative PCR which allow for quantitative measurement of DNA or RNA molecules.

Gel electrophoresis

Gel electrophoresis is one of the principal tools of molecular biology. The basic principle is that DNA, RNA, and proteins can all be separated by means of an electric field and size. In agarose gel electrophoresis, DNA and RNA can be separated on the basis of size by running the DNA through an agarose gel. Proteins can be separated on the basis of size by using an SDS-PAGE gel, or on the basis of size and their electric charge by using what is known as a 2D gel electrophoresis.

Macromolecule Blotting and Probing

The terms *northern*, *western* and *eastern* blotting are derived from what initially was a molecular biology joke that played on the term *Southern blotting*, after the technique described by Edwin Southern for the hybridisation of blotted DNA. Patricia Thomas, developer of the RNA blot which then became known as the *northern blot*, actually didn't use the term. Further combinations of these techniques produced such terms as *southwesterns* (protein-DNA hybridizations), *northwesterns* (to detect protein-RNA interactions) and *farwesterns* (protein-protein interactions), all of which are presently found in the literature.

Southern Blotting

Named after its inventor, biologist Edwin Southern, the Southern blot is a method for probing for the presence of a specific DNA sequence within a DNA sample. DNA samples before or after restriction enzyme (restriction endonuclease) digestion are separated by gel electrophoresis and then transferred to a membrane by blotting via capillary action. The membrane is then exposed to a labelled DNA probe that has a complement base sequence to the sequence on the DNA of interest. Most original protocols used radioactive labels, however non-radioactive alternatives are now available. Southern blotting is less commonly used in laboratory science due to the capacity of other techniques, such as PCR, to detect specific DNA sequences from DNA samples. These blots are still used for some applications, however, such as measuring transgene copy number in transgenic mice, or in the engineering of gene knockout embryonic stem cell lines.

Northern Blotting

The northern blot is used to study the expression patterns of a specific type of RNA molecule as relative comparison among a set of different samples of RNA. It is essentially a combination of denaturing RNA gel electrophoresis, and a blot. In this process RNA is separated based on size and is then transferred to a membrane that is then probed with a labelled complement of a sequence of interest. The results may be visualized through a variety of ways depending on the label used; however, most result in the revelation of bands representing the sizes of the RNA detected in sample. The intensity of these bands is related to the amount of the target RNA in the samples analyzed. The procedure is commonly used to study when and how much gene expression is occurring by measuring how much of that RNA is present in different samples. It is one of the most basic tools for determining at what time, and under what conditions, certain genes are expressed in living tissues.

Western Blotting

Antibodies to most proteins can be created by injecting small amounts of the protein into an animal such as a mouse, rabbit, sheep, or donkey (polyclonal antibodies) or produced in cell culture (monoclonal antibodies). These antibodies can be used for a variety of analytical and preparative techniques.

In western blotting, proteins are first separated by size, in a thin gel sandwiched between two glass plates in a technique known as SDS-PAGE (sodium dodecyl sulfate polyacrylamide gel electrophoresis). The proteins in the gel are then transferred to a PVDF, nitrocellulose, nylon or other support membrane. This membrane can then be probed with solutions of antibodies. Antibodies that specifically bind to the protein of interest can then be visualized by a variety of techniques, including coloured products, chemiluminescence, or autoradiography. Often, the antibodies are labelled with enzymes. When a chemiluminescent substrate is exposed to the enzyme it allows detection. Using western blotting techniques allows not only detection but also quantitative analysis.

Analogous methods to western blotting can be used to directly stain specific proteins in live cells or tissue sections. However, these *immunostaining* methods, such as FISH, are used more often in cell biology research.

Eastern Blotting

Eastern blotting technique is to detect post-translational modification of proteins. Proteins blotted on to the PVDF or nitrocellulose membrane are probed for modifications using specific substrates.

Arrays

A DNA array is a collection of spots attached to a solid support such as a microscope slide where each spot contains one or more single-stranded DNA oligonucleotide fragment. Arrays make it possible to put down large quantities of very small (100 micrometre diametre) spots on a single slide. Each spot has a DNA fragment molecule that is complementary to a single DNA sequence (similar to Southern blotting). A variation of this technique allows the gene expression of an organism at a particular stage in development to be qualified (expression profiling). In this technique the RNA in a tissue is isolated and converted to labelled cDNA. This cDNA is then hybridized to the fragments on the array and visualization of the hybridization can be done. Since multiple arrays can be made with exactly the same position of fragments they are particularly useful for comparing the gene expression of two different tissues, such as a healthy and cancerous tissue. Also, one can measure what genes are expressed and how that expression changes with time or with other factors. For instance, the common baker's yeast, *Saccharomyces cerevisiae*, contains about 7000 genes; with a microarray, one can measure qualitatively how each gene is expressed, and how that expression changes, for example, with a change in temperature. There are many different ways to fabricate microarrays; the most common are silicon chips, microscope slides with spots of ~ 100 micrometre diametre, custom arrays, and arrays with larger spots on porous membranes (macroarrays). There can be anywhere from 100 spots to more than 10,000 on a given array.

Arrays can also be made with molecules other than DNA. For example, an antibody array can be used to determine what proteins or bacteria are present in a blood sample.

Allele-specific Oligonucleotide

Allele-specific oligonucleotide (ASO) is a technique that allows detection of single base mutations without the need for PCR or gel electrophoresis. Short (20-25 nucleotides in length), labelled probes are exposed to the non-fragmented target DNA. Hybridization occurs with high specificity due to the short length of the probes and even a single base change will hinder hybridization. The target DNA is then washed and the labelled probes that didn't hybridize are removed. The target DNA is then analyzed for the presence of the probe via radioactivity or fluorescence. In this experiment, as in most molecular biology techniques, a control must be used to ensure successful experimentation. The Illumina Methylation Assay is an example of a method that takes advantage of the ASO technique to measure one base pair differences in sequence.

Antiquated Technologies

In molecular biology, procedures and technologies are continually being developed and older technologies abandoned. For example, before the advent of DNA gel electrophoresis (agarose or polyacrylamide), the size of DNA molecules was typically determined by rate sedimentation in sucrose gradients, a slow and labour-intensive technique requiring expensive instrumentation; prior to sucrose gradients, viscometry was used.

Aside from their historical interest, it is often worth knowing about older technology, as it is occasionally useful to solve another new problem for which the newer technique is inappropriate.

Biotechnological Approaches for the Control of Insect Pests in Crop Plants

Agricultural pests include plant pathogens (e.g. fungi, oomycetes, bacteria, viruses, nematodes), weeds, arthropods (primarily insects and mites), molluscs (slugs and snails) and a small number of vertebrates. They reduce the yield and quality of produce by feeding on crops, by transmitting diseases, or by competition with crop plants for space and other resources (weeds, for example).

There are estimated to be about 67,000 different pest species worldwide. They are a significant constraint on agricultural production, responsible for around 40% loss of potential global crop yields. These losses occur despite the considerable efforts made at pest control, and they suggest that improvements in pest management are significant way forward for improving yields and access to food.

Pest problems are an almost inevitable part of agriculture. They occur largely because agricultural systems ('agro-ecosystems') are simplified, less stable modifications of natural ecosystems. The creation and management of agricultural land disrupts the ecological forces that regulate potential pest species in natural ecosystems: these include physico- chemical conditions, food availability, predation, and competition. Thus, growing crops in monoculture provides a concentrated food resource that allows pest populations to achieve far higher densities than they would in natural environments. New food resources for pests are provided when a crop is introduced into a country. Cultivation can make the physico-chemical environment more favourable for pest activity, for example though irrigation or the warm conditions found in glasshouses. Finally, using broad spectrum pesticides will destroy natural predators that help keep pests under control.

Some of the most important problems occur when pests develop resistance to chemical pesticides. For example, over 500 arthropod species now show resistance to one or more types of chemical. Other serious problems are caused by alien (i.e. non-native) species that are accidentally introduced to a new country or continent and which escape their co- evolved natural predators More than 11,000 alien species have been documented. These cause highly significant damage to crops, biodiversity and landscape valued at billions of dollars per annum. There are also threats from emerging pests, such as new strains of plant pathogens that evolve to overcome varietal resistance in crops.

Farmers and growers are under immense pressure to reduce the use of chemical pesticides without sacrificing yields or crop quality, but at the same time the control of pests is becoming increasingly difficult due to pesticide resistance and the decreasing availability of products. Alternative control methods are needed urgently. These need to be used as part of Integrated Pest Management. There is a particular need to preserve the effectiveness of chemical pesticides, which will remain a key component of pest management for the foreseeable future.

Biological Control

Biological control is the use by man of a living organism to help manage the population density of a pest organism. Natural enemies are organisms that kill or debilitate another organism. Biological control programmes operate throughout the world in agriculture and forestry.

Biological control agents (BCAs) include the following: (1) predatory insects and mites, which eat their prey; (2) parasitoids, which are insects with a free living adult stage and a larval stage that is parasitic on another insect; (3) parasites and microbial pathogens, such as nematodes, fungi, bacteria, viruses and protozoa, which cause lethal infections. In Europe, typical examples of where these agents are being used include: (1) the application of parasitoids to control whiteflies in glasshouses; (2) parasitic nematodes against slugs; (3) mycoparasitic fungi to control plant diseases of horticultural crops; (4) use of viruses to control codling moth in apple orchards; and (5) building habitats on farms to increase natural populations of predators and other beneficial organisms.

Natural enemies represent a large component of the world's biodiversity, for example parasitoids account for about 10% of the world's species, and each agricultural pest species is fed upon by up to hundreds of species of natural enemies. However, only a small proportion of the available species have been investigated for crop protection. Thus, while c. 750 species of insect pathogenic fungi are known, less than 20 have received serious attention as control agents of insect pests. The natural enemies inhabiting an agro- ecosystem play a key role in preventing pests reaching damaging levels.

The ways in which biological control agents are used vary according to the type of pest (plant, microorganism, vertebrate or invertebrate), the biological characteristics of the control agent, as well as the agricultural setting. There are three broad biocontrol strategies: Introduction (release of an alien control agent to control an alien pest), augmentation (application of natural enemies that already live in the area of use), and conservation (manipulating agricultural practices or the environment to enhance natural control). People who use biocontrol are bound by the FAO code of conduct on the import and release of biological control agents as well as national legislation on releases into the environment. The use of microbial agents is governed by Plant Protection Product regulations, in which authorisations proceed along similar lines to those for chemical pesticides although there are also some specific schemes.

BCAs have a range of attractive properties that include host specificity, lack of toxic residue, no phytotoxic effects, human safety, and the potential for pest management to be self sustaining. Many are able to actively locate their prey. BCAs can also be produced locally which can be important in terms of choosing and matching natural enemies to small scale needs.

Successful use requires fundamental knowledge of the ecology of both the natural enemy and the pest. When this condition is satisfied, and the agent is used firmly within IPM, then biological control can sometimes be more cost effective than purely chemical control. However, under the present market system, many biological control products have not competed well with less expensive and more effective synthetic pesticides.

The down sides of BCAs are that most are niche products, pest control is not immediate, there can be lack of environmental persistence, and efficacy can be low and unpredictable particularly in outdoor environments. The approvals process used for microbial agents results in significantly greater authorisation costs than for macro- agents.

Entomopathogens

Many farmers and growers are now familiar with the use of predators and parasitoids for biological control of arthropod (insect and mite) pests, but it is also possible to use specific micro-organisms that kill arthropods. These include entomopathogenic fungi, nematodes, bacteria and viruses. These are all widespread in the natural environment and cause infections in many pest species.

Entomopathogens contribute to the natural regulation of many populations of arthropods.

Much of the research in this area concerns the causal agents of insect diseases and their exploitation for biological pest control. Many entomopathogens can be mass produced, formulated, and applied to pest populations in a manner analogous to chemical pesticides, i.e. as nonpersistent remedial treatments that are released inundatively. Entomopathogens have also been used as classical biological control agents of alien insect pests, and natural pest control by entomopathogens has been enhanced by habitat manipulation.

Helicoverpa Species

Helicoverpa species are polyphagous pests of at least 181 plant species from 49 families including cotton, corn, soybeans, tobacco and chick-pea. They are one of the most serious pests in cotton-producing countries like Australia, India and China, causing enormous economic problems.

The lepidopteran species, *H. armigera*, progresses though four stages of development; egg, six larval instars, pupal and adult. The time frame of each of these stages varies with environmental conditions. Over the warmer months, the life cycle can be completed in 30-40 days and each female moth can lay from 500-3000 eggs.

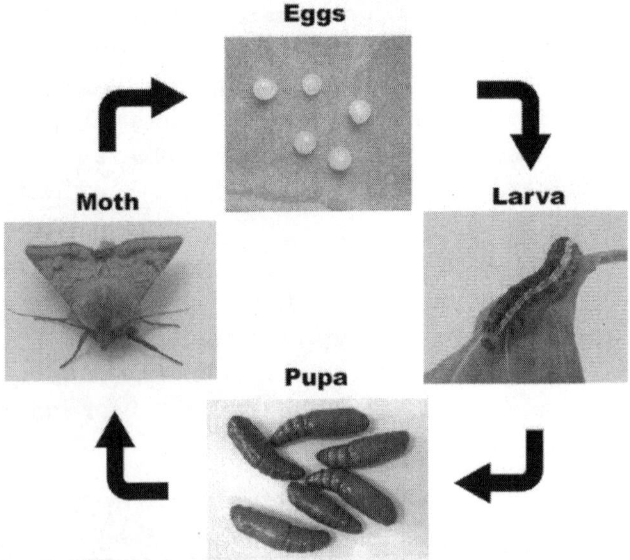

Helicoverpa Armigera Life Cycle

One of the reasons these pests are so damaging is the larva's feeding preference for plant structures that are high in nitrogen, principally reproductive structures and growing points such as cotton buds and bolls, corn ears, tobacco buds, and sorghum heads. Damage to these structures has a direct influence on yield. *H. armigera* larvae are foliar feeders at the early instar stage and shift to developing seeds or bolls at later stages. *H. armigera* is a major problem in Australia because it has developed resistance to many of the chemical insecticides that have been used for its control. Unlike other lepidopteran species, *H. armigera* larvae don't migrate far from their original host plant, consequently their populations in agricultural areas are exposed to consistent selection pressure, leading to greater resistance to insecticides.

In the 1995/96 growing season, transgenic cotton known as Ingard that expressed the *Cry1Ac* gene became commercially available in Australia. To preserve the susceptibility of lepidopterans to Bt toxins, a conservative resistance management plan was imposed, where planting of Ingard cotton was restricted to 30% of the cotton production area per farm. The average amount of insecticide used per hectare was 44% lower on Ingard cotton compared to conventional cotton. In the 2004/05 growing season, Ingard cotton was replaced by Bollgard II, which expressed both the *Cry1Ac* and *Cry2Ab* genes. Restrictions were not placed on this new variety and Bollgard II cotton comprised around 80% of the total cotton area planted in Australia during the 2004/05 and 2005/06 seasons and 95% of the total cotton area in the 2010/2011 season. This reduced the average amount of insecticide used per hectare by 85% compared to conventional cotton. So far, there have been no reported field failures of Bollgard II due to resistance. However, while alleles that confer resistance to Cry1Ac in *H. armigera* are rare in the field, alleles that confer resistance to Cry2Ab are more common.

The Use of Genetically Modified Plants for Control of Lepidopteran Insects

As mentioned previously, insects are responsible for major crop losses worldwide. In addition to direct impacts on yield, insects also reduce yields by making crops more susceptible to disease causing pathogens. Last decade, most control measures focused on the use of chemical pesticides, a curative pest control strategy that was useful for rapid control of certain pest outbreaks. However, excessive and indiscriminate large-scale use of pesticides has led to development of pesticide-resistant insects. Additionally, the long-term and extensive use of synthetic chemicals has led to concerns regarding their impact on food safety, associated human health and the environment. As the use of pesticides for prevention of insect-associated losses cannot be overlooked in agriculture, there is a greater need to develop alternative or additional technologies which would allow a more selective use of pesticides and provide sustainable crop protection. To achieve this objective, it is necessary to enhance the resistance of plants to pests and pathogens through integrated pest management (IPM) programmes. They will need to consist of a combination of control strategies including (A) the use of natural biocontrol factors such as pathogens, predators or parasites; (B) various preventive pest control strategies including crop rotation, intercropping, and cultivation of pest-resistant varieties of plants and (C) genetic control via the release of sterile insects and also the use of natural insecticides. The latter includes secondary metabolites, viruses and transgenes.

As the products of most transgenes are ingested by the insect pest and therefore act through the gut, most of the focus has been on transgene encoded proteins that target the insect midgut and/or the peritrophic membrane to disrupt digestion or nutrition. Generally, the detrimental effects on larval and insect growth result from limited assimilation of nutrients. Furthermore, any severe delay in growth and development, in a natural setting, lengthens the period in which the larvae are vulnerable to natural predators such as mice, spiders and predaceous insects. The use of transgenic plants that express insecticidal agents thus reduces the population of insect pests and reduces the usage of chemical insecticides. This extends the useful life of the insecticides and also reduces the ecological damage they may cause. As with any new method of insect control, the impact of transgenic plants on non-target and beneficial insects, particularly pollinators such as honey bees, needs to be assessed. Since the discovery that dsRNA can silence genes, RNA interference (RNAi) has been developed as an effective tool for regulating gene expression in plants and animals. RNA interference or gene silencing has been used to inhibit virus replication and spread in transgenic plants and has potential to be developed commercially for disease control. The use of RNAi for insect control is less well developed. Insect genes can be down-regulated by injection of dsRNA or by oral administration of high concentrations of exogenously supplied dsRNA as part of an artificial diet, but a much more efficient method of delivering dsRNA is needed before RNAi technology can be used to control pests in the field. To date, the most successful transgenes for insect control have been the genes encoding insecticidal toxins from the soil bacterium *Bacillus thuringiensis.*

The Bacillus Thuringiensis Endotoxins

The use of genes encoding endotoxins from *Bacillus thuringiensis* is now a well-established technology for producing transgenic plants with enhanced resistance to the larvae of lepidopteran insect pests. Bt cotton was first released for commercial production in the USA in 1996 and subsequently grown in several countries including Argentina, Australia, China, Colombia, Indonesia, Mexico, South Africa, and India. Since then other transgenic crop species producing Bt toxins have been commercialized including maize, tomato and potato (http://cera-gmc.org). The adoption of Bt crop varieties by farmers has been rapid reflecting the benefits of these crops such as reduced insecticide use, lower production costs and higher yields. Only two Bt crops are grown in Australia. In the most recent season (2011/2012) approximately 80% of the cotton grown in Australia was Bollgard II.

B. thuringiensis, a Gram-positive soil bacterium, produces a proteinaceous parasporal crystalline inclusion during sporulation. There are two main categories of Bt toxins: Cry and Cyt. These two groups are classified further by a detailed nomenclature system that describes groups Cry1 to Cry55 and Cyt1 to Cyt2. The Cry toxins are divided into three larger families that are not related phylogenetically. The largest Cry family is the three domain family, and genes from this family are present in the majority of commercialised Bt crops.

The larvae of insect orders primarily affected by Bt toxins are Lepidoptera (butterflies and moths), Diptera (mosquitoes) and Coleoptera (larval and adult beetles). However, Bt toxins are not toxic to people, wildlife, or most beneficial insects and therefore the opportunities for biological control are great. The effect of Bt toxins on a range of lepidopteran insects has been studied including: *Bombyx mori*, *Helicoverpa armigera*, *Heliothis virescens*, *Manduca sexta*, *Ostrinia nubilalis*, *Plutella xylostella*, *Sesamia nonagrioides*, *Spodoptera exigua*, *Spodoptera frugiperda* and *Spodoptera littoralis*. The Cry toxins produced in Bt crops generally target lepidopteran pests, although some also target coleopteran pests. The first commercialised Bt crops contained only one Cry toxin, but second generation Bt crops have between two to six different toxins.

Table: *This table lists the transgenic crops in Australia producing Bt proteins.*

Trade name	***Crop***	***Bt protein***	***Company***	***Year released***
Ingard ®	cotton	Cry1Ac	Monsanto	1996
Bollgard II ®	cotton	Cry1Ac, Cry2Ab	Monsanto	2003

Mechanism of Action

The Bt toxin mechanism of action is described by two models: The pore formation model and the signal transduction model. The initial steps of both models are the same. Upon ingestion by insects the crystalline inclusion is solubilised in the midgut. Most target insects have a high gut pH that is crucial for the efficacy of Bt toxins since most Bt-protoxins are only soluble above pH 9.5. The 130 kDa protoxins are activated by insect gut proteases, which typically cleave from both the C- and N-termini resulting in a 43-65 kDa protease-resistant active core.

The pore formation model has been the accepted mode of action for 20 years and is supported by numerous publications. In this model the activated toxins bind to the primary receptors in the brush border membrane of the midgut epithelium columnar cells. The major receptors for Cry toxins in lepidopterans are cadherin-like proteins. The binding site of Cry toxins varies depending on the structure of the Cry toxin. Binding to cadherin facilitates further proteolytic cleavage of the toxin and promotes the formation of oligomers.

The toxins then interact with secondary receptors in the midgut larval membrane. These secondary receptors are GPI-anchored proteins; either aminopeptidases or alkaline phosphatases. Following secondary receptor binding, the toxin inserts into the membrane and creates pores. These pores lead to the disruption of membrane integrity and cause an electrolyte imbalance that ultimately leads to death by starvation or septicaemia. It is likely that there are more receptors involved in Bt toxicity since insects lacking the cadherin receptor are still killed by modified Bt toxins.

An alternative model for the Bt toxin mechanism of action proposes that Cry toxins trigger a signalling cascade pathway. This model differs from the pore formation model in that it does not involve toxin oligomerisation, secondary receptors or the formation of pores in the membrane. Instead, in this model, binding to the cadherin receptor initiates a $Mg2^{+}$ dependent signal cascade pathway that includes a guanine nucleotide-binding protein, adenylyl cyclase, and protein kinase A which ultimately results in cell death.

Resistance of Lepidopteran Insects to Bt Toxins

More recently there have been reports of field resistance to Bt crops in pink bollworm (*Pectinophore gosspiella*), cotton bollworm (*Helicoverpa* spp), armyworm (*Spodoptera frugiperda*) and western corn rootworm (*Diabrotica virgifera virgifera*. Some insects collected from the field have Bt resistance that has been characterized in the laboratory. However, there is debate about the relevance of this laboratory resistance in the field. A decrease in field performance of Bt corn against *S. frugiperda* was observed in Puerto Rico and against *Busseola fusca* in South Africa. In southeastern US problems with control of *H. zea* on Bt cotton have also been reported.

The most common mechanism of resistance is the disruption of binding of Bt toxin to receptors in the midgut membrane. This disruption may be caused either by mutations in the receptor that blocks binding (reviewed in) or changes in expression of the receptors. Mutations in cadherin genes are responsible for Bt resistance in *Heliothis virescens, Helicoverpa armigera* and *Pectinophora gossypiella*. Another resistance mechanism associated with an ABC transporter locus has been reported in three lepidopteran spp (*H. virescens, P. xylostella and T. ni*). Resistance to Bt in *Ostrinia nubialis* is due to reduced midgut protease activity resulting in less activation of the protoxins.

Management of Resistance to Bt Crops

There are two main strategies for management of insect resistance to Bt crops: Refuge and pyramiding. The main approach for delaying evolution of resistance to Bt crops is the refuge strategy. Farmers are mandated to maintain an abundance of host non-Bt crops as a refuge surrounding their Bt crops. The theory behind this strategy is that any Bt resistant larvae that arise on the Bt crops will mate with susceptible individuals from neighbouring non-Bt crops. As long as inheritance of resistance remains recessive the offspring will be susceptible to Bt crops. This strategy is then combined with several other mandatory farming practices that include control of volunteer and ratoon plants that arise post-harvest, planting within a defined period of time to restrict the exposure of the Bt crop to the insect pests, restricted use of foliar Bt and the cultivation of crop residues. The other major strategy to combat the evolution of Bt resistance is gene pyramiding. For example, the development of second generation Bt cotton that has at least two Bt toxins such as the Monsanto Bollgard II cotton variety, but up to six Bt toxins. Another resistance management strategy which is still in the research phase of development is the use of insecticidal genes with completely different modes of action such as proteinase inhibitors. The success of combining multiple Bt genes for resistance management is contingent on the individual toxins having different targets to prevent cross resistance developing. Binding studies with various Cry toxins have been used to identify toxins with different binding sites in the lepidopteran midguts. This information can be used to design combinations of Cry toxins that complement each other to delay the development of resistance to Bt crops.

In addition to the resistance management plan for Bollgard cotton outlined above, farmers also use integrated pest management (IPM) systems as a sustainable approach to control all pests. IPM systems deploy a tactical combination of biotechnological, chemical, biological and cultural control methods to avoid pest problems. Some of the major IPM strategies and tools include maintenance of beneficial insect populations, ensuring healthy plant growth, managing weed hosts and monitoring pest populations and plant damage regularly. All these additional practices lead to better control of insect populations in general and therefore helps prevent the development of resistance in insect populations to Bt.

Protease Inhibitors for the Control of Insect Pests

Protease inhibitors are one component of a plant's natural defence mechanism against herbivores and pathogens. Plants protect themselves directly by constitutively expressing protease inhibitors and by inducing protease inhibitors in response to mechanical wounding or insect attack. They may also release volatile compounds after insect damage that function as potent attractants for predators of insect herbivores. The release of volatile compounds after wounding, such as methyl jasmonate also triggers the production of proteinase inhibitors in neighbouring unwounded plants essentially prearming the local population against insect attack.

Mechanism of Action of Protease Inhibitors on Lepidopteran Insects

Protease inhibitors when incorporated into artificial diets or expressed in transgenic plants increase mortality and reduce the growth and development of larvae from many insect pest species including Coleoptera, Orthoptera and Lepidoptera. The mechanisms by which ingested PIs mediate their effects on insect physiology differs between insect species. Proteinase inhibitors bind to insect digestive proteases, preventing proteolysis which blocks digestion of protein. This effectively starves the larvae of protein and essential amino acids required for insect growth, development and reproduction. To compensate for this inhibition, several insect species increase production of proteases to swamp the ingested PIs. This in turn can lead to a limitation in bioavailability of essential amino acids for protein synthesis, impairment of growth and development, and potentially death.

The loss of the sulphur-containing amino acids (cysteine and methionine) is critical because the sulfydryl content in trypsin and chymotrypsin is high and reprocurement of the sulphur-containing amino acids is difficult since cysteine and methionine are in relatively low concentrations in the diet, especially if the food source is plant material. Broadway and colleagues confirmed this hypothesis in bioassays with *Spodoptera exiqua* where the weight-reducing effects obtained with soybean trypsin inhibitor were eliminated when the diets were supplemented with methionine.

PIs in Transgenic Plants for Plant Protection: Success and Failure

Several groups have reported enhanced protection of plants against lepidopteran pests after transformation with genes encoding PIs. Despite this substantial body of work, defence strategies based on PI expression in plants have not resulted in any commercial application so far. This relates to two distinct problems:

(1) the levels of PI-expression in transgenic plants and

(2) the pest's capacity to react to PI consumption.

Most problems arise from the use of a single transgene producing a PI that targets only one protease or one class of protease in the insect midgut.

The first problem of inadequate levels of PI expression is best exemplified by studies with *P. xylostella*, the diamondback moth. When larvae of the diamondback moth consumed transgenic plants expressing the chymotrypsin and trypsin specific potato type II proteinase inhibitor, Pot II, they suffered lower growth rates. However, this did not confer an advantage to the plants because the larvae consumed more tissue to compensate for their decrease in metabolism. As a result, the insects maintained population growth rates similar to those of larvae on non-transgenic plants. Growth enhancement has been reported after PI ingestion in insects from a number of orders. Larvae that consumed tobacco leaves expressing low levels of mustard trypsin inhibitor 2 (MTI-2) developed faster, had an increased mean weight and caused more damage to leaves compared to control larvae on non-transgenic tobacco. The increase in leaf surface consumption observed with plants expressing low levels of MTI-2 may have resulted from a decrease in available protein due to the presence of MTI-2 and/or to an increase in gut proteolytic capacity induced by PI consumption.

The second problem, the pest's capacity to react to PI consumption, is exemplified by the observation that several PIs that are potent inhibitors of insect proteases *in vitro* fail to produce any deleterious effect when fed to larvae. Several mechanisms have been reported for this lack of effect. For example, the complement of proteolytic enzymes in the insect midgut can be altered after PI ingestion. This could involve a switch to enzymes of different substrate specificity, but the same mechanistic class. For example, production of a chymotrypsin-like enzyme rather than a trypsin-like protease. Another mechanism used to detoxify the PIs is degradation via endogenous proteases within the insect midgut. Insects that feed regularly on a particular host plant are generally not affected by the PIs produced by the host. For example the PIs from chickpea, a host plant for *H. armigera,* are rapidly degraded by the *H. armigera* gut proteases.

Similarly, single domain cystatins from potato multicystatin are degraded when fed to larvae of *Diabrotica* spp (Coleoptera). Sometimes non-host PIs are also rapidly degraded. Human stefin A, a potent inhibitor of human cysteine proteases, was degraded by cystatin-insensitive proteases in the gut of Colorado potato beetle (*Leptinotarsa decemlineata*) and black vine weevil (*Otiorynchus sulcatus*). Another anti-PI mechanism is the production of midgut inhibitor-resistant serine proteases. Some insect larvae adapt to the presence of PIs by replacing the inhibited enzymes with other PI-resistant proteases and can exhibit increased ingestion rates and faster development than larvae fed on control diets lacking PIs.

Some classic examples of this phenomenon are as follows. Soybean Kunitz trypsin inhibitor (SKTI) is normally an effective inhibitor of protease activity in gut extracts from *H. armigera* larvae, this insect is not seriously affected by ingestion of this PI because it responds to chronic ingestion of SKTI by increasing activity of an SKTI-resistant trypsin. Similarly, growth and development of *S. exigua* larvae was not impacted when fed leaves from tobacco plants transformed with the chymotrypsin/trypsin specific potato proteinase inhibitor II (Pot II).

Analysis of the trypsin activity in the gut of these insects demonstrated that only 18% of the trypsin activity of insects reared on these transgenic plants was inhibited by Pot II, whereas 78% of the trypsin activity in the gut of insects reared on control plants was Pot II-inhibitable. The larvae had compensated for the loss of the PI-inhibitable trypsin by a 2.5-fold induction of new activity that was resistant to inhibition by Pot II. Another observation of induction of PI-resistant enzymes was made by Markwick and coworkers who reported that the trypsin in three species of leaf rollers (Tortricidae) that had fed on diets containing SKTI was less inhibited by bovine pancreatic trypsin inhibitor (BPTI) compared to the trypsin in control larvae.

These responses have been reported for lepidopteran species that have ingested PIs in native plants, transgenic plants, and artificial diets. In summary, potent inhibition of an insect digestive enzyme *in vitro* by a particular PI is not a good prediction that the PI will be useful when expressed as a transgene for crop protection. That is, expression and regulation of midgut serine proteases in herbivorous insects is tightly regulated and is heavily influenced by the levels and the nature of ingested PIs. The mechanism by which changes in protease levels and protease isoforms is regulated in response to PI ingestion is still unknown for phytophagous insects.

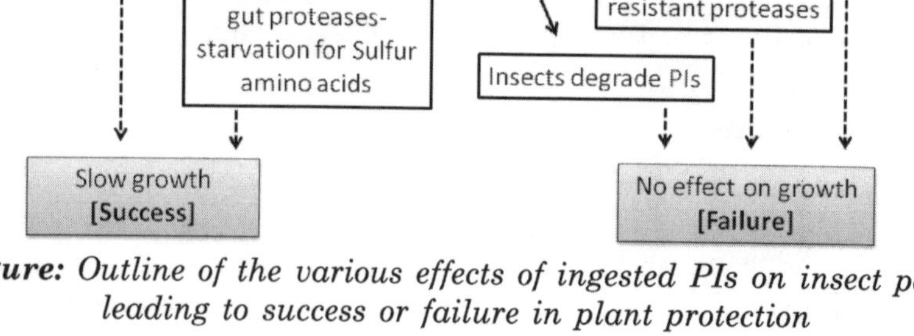

***Figure:** Outline of the various effects of ingested PIs on insect pests leading to success or failure in plant protection*

Proteinase Inhibitors from Nicotiana Alata as Defence Molecules against Insect Pests

Female reproductive tissues and wounded leaves of the ornamental tobacco, *Nicotiana alata* amass high levels of serine proteinase inhibitors for protection against insect pests and pathogens. These serine proteinase inhibitors (NaPI) belong to the Potato type II family (Merops family

I20) which have only been described in the Solanaceae. The NaPI precurser protein (NaProPI; 43 kDa), is composed of an ER signal peptide (29 amino acids), six repeated domains each with a potential PI-reactive site, and a 25 residue C-terminal domain that is essential for vacuolar targeting (VTS)

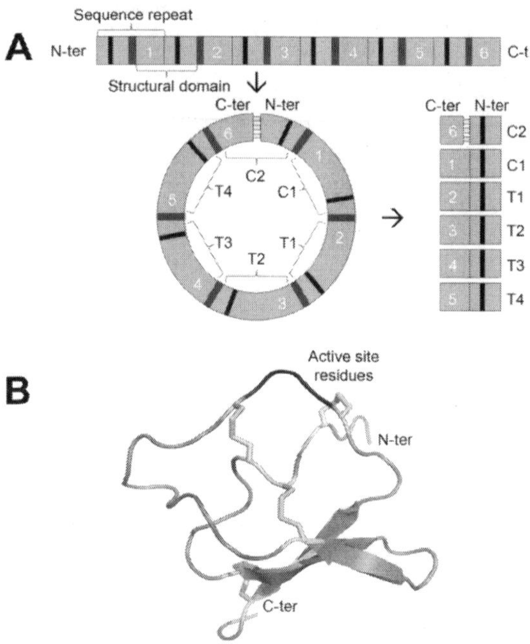

(A) The NaPI precursor protein, shown as a linear gene product, forms a circular 'bracelet' structure that is 'clasped' by three disulphide bonds (yellow) between the N- and C-terminal repeats. Each repeat (labelled 1-6) contains a protease-reactive site (black), which is specific for either chymotrypsin (C1 and C2) or trypsin (T1-4). The six linker regions (red), with sequence EEKKN, are cleaved to release the six active inhibitor domains. The N-terminal signal sequence and the C-terminal vacuolar targeting signal have been omitted for clarity. Figure adapted from Scanlon *et al.*. (B) Ribbon view of T1 showing the major secondary structural element, a triple stranded β-sheet (green) and the cysteines involved in disulfide bonds (yellow). The reactive site residues (black) are positioned between two cysteines that anchor the reactive loop to the central coil. The other five inhibitors have the same structure.

Processing of NaProPI in the secretory pathway removes the ER signal peptide and VTS, and releases six PIs. Processing of the six repeat NaProPI occurs at sites located within, rather than between, these repeated regions. Complete removal of the linker sequence (Glu-Glu-Lys-Lys-Asn) contained within each repeated region, generates five contiguous inhibitors, a chymotrypsin inhibitor (C1) and four trypsin inhibitors (T1-T4), and two flanking peptides from the N- and C-termini. The flanking peptides form a novel two-chain chymotrypsin inhibitor (C2) that can only be formed if NaPI adopts a circular structure. The peptides have very similar amino acid sequences. The three-dimensional structures of C1, C2, T1, T2, T3 and T4 have been determined by NMR spectroscopy. A triple stranded -sheet is the dominant secondary structural feature; several -turns and a short region -helix are also present. The reactive site is located on an exposed loop which has a higher degree of mobility than other regions of the protein. This is a common feature of PIs and is thought to allow the inhibitor to adapt to slightly different enzymes.

Atkinson and colleagues suggested NaPIs may be involved in deterring insects from feeding on stigmas or in protecting the stigma from pathogen invasion since the related type-II PIs from potato and tomato are effective against proteases of fungal, bacterial and insect origin. The PIs from *N. alata* inhibit the digestive gut proteases from five insect orders *in vitro* and display significant inhibitory activity against the midgut proteases of *H. punctigera* and *T. commodus*. Significant mortality was recorded when *H. punctigera* larvae were fed transgenic tobacco or transgenic peas expressing the NaPI precursor. More recently, the response of *Helicoverpa* larvae to ingestion of NaPI has been more thoroughly characterized. Following ingestion of NaPI, all surviving *Helicoverpa punctigera* larvae produced high levels of a chymotrypsin that was resistant to inhibition by NaPI. However this NaPI-resistant chymotrypsin was strongly inhibited by a potato type 1 inhibitor which is also produced by solanaceous plants, but belongs to a different class of serine proteinase inhibitors. When presented to *H. armigera* larvae in an artificial diet the combination of NaPI and the potato type I inhibitor had a much more dramatic effect on growth and development of the larvae compared to either of the inhibitors alone.

Neonates were transferred to cotton-leaf based artificial diets containing 0.3% of PIs (NaPI, StPin1A) and growth (mg) measured every 2nd day until day 11. Day 11, the % average weights compared to casein control are shown with representative larvae from each treatment.

7

Enhancing the Virulence of Baculovirus as Biopesticides with Wasp Parasitoid Polydnavirus Genes to Control Lepidopteran Insect Pests

The baculoviruses are a family of large rod-shaped viruses that can be divided to two genera: nucleopolyhedroviruses (NPV) and granuloviruses (GV). While GVs contain only one nucleocapsid per envelope, NPVs contain either single (SNPV) or multiple (MNPV) nucleocapsids per envelope. The enveloped virions are further occluded in granulin matrix in GVs and polyhedrin for NPVs. Moreover, GV have only single virion per granulin occlusion body while polyhedra can contain multiple embedded virions.

Baculoviruses have very species-specific tropisms among the invertebrates with over 600 host species having been described. Immature (larval) forms of moth species are the most common hosts, but these viruses have also been found infecting sawflies, mosquitoes, and shrimp. Although baculoviruses are capable of entering mammalian cells in culture they are not known to be capable of replication in mammalian or other vertebrate animal cells. Baculoviruses contain circular double-stranded genome ranging from 80–180 kbp.

The earliest records of baculoviruses can be found in the literature from as early as the sixteenth century in reports of "wilting disease" infecting silk-producing larva. Starting in the 1940s they were used and studied widely as biopesticides in crop fields. Since the 1990s they have been utilized for producing complex eukaryotic proteins in insect cell cultures. These recombinant proteins have been used in research and as vaccines in both human and veterinary medical treatments (for example, the most widely used vaccine for prevention of H5N1 avian influenza in chickens was produced in a baculovirus expression vector). More recently it has been found that baculoviruses can transduce mammalian cells with a suitable promoter.

These medical and potential medical uses have accelerated the number of publications on baculoviruses since 1995.

Baculovirus Life Cycle

The baculovirus life cycle involves two distinct forms of virus. Occlusion derived virus (ODV) is present in a protein matrix (polyhedrin or granulin) and is responsible for the primary infection of the host while the budded virus (BV) is released from the infected host cells later during the secondary infection.

Typically, the initial infection occurs when a susceptible host insect feeds on plants that are contaminated with the occluded form of the virus. The protein matrix dissolves in the alkaline environment of the host midgut (stomach), releasing ODV that then fuse to the columnar epithelial cell membrane of the host intestine and are taken into the cell in endosomes. Nucleocapsids escape from the endosomes and are transported to nucleus.

This step is possibly mediated by actin filaments. Viral transcription and replication occur in the cell nucleus and new BV particles are budded out from the basolateral side to spread the infection systemically. During budding, BV acquires a loosely fitting host cell membrane with expressed and displayed viral glycoproteins.

Baculovirus infection can be divided to three distinct phases, early (0–6 h post-infection), late (6–24 h p.i.) and very late phase (18–24 to 72 h p.i.). While BV is produced in the late phase, the ODV form is produced in the very late phase acquiring the envelope from host cell nucleus and embedded in the matrix of occlusion body protein. These occlusion bodies are released when cells lyse to further spread baculovirus infection to next host.

The extensive lysis of cells frequently causes the host insect to literally disintegrate, thus the reason for the historic name "wilting disease." the complete ODV-polyhedrin particles are resistant to heat and light inactivation, whereas the naked BV virion is more sensitive to environment.

Structure of the Virion

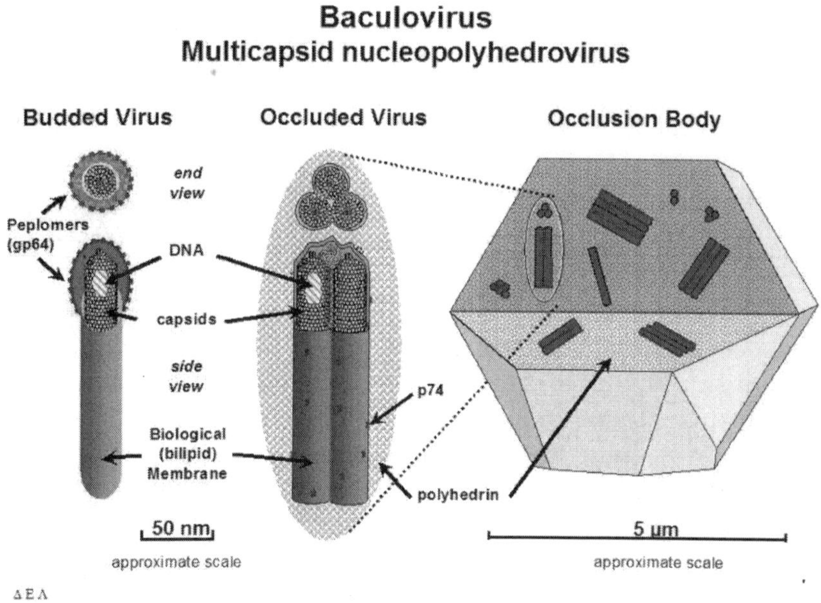

The most studied baculovirus is *Autographa californica multicapsid nucleopolyhedrovirus* (AcMNPV). The virus was originally isolated from the alfalfa looper (a lepidopteran) and contains a 134-kbp genome with 154 open reading frames (ORF). The major capsid protein VP39 together with some minor proteins forms the nucleocapsid (21 nm x 260 nm) that encloses the DNA with p6.9 protein.

BV acquires its envelope from the cell membrane and requires a glycoprotein, gp64, to be able to spread systemic infection. This protein forms structures called peplomers on one end of the budded virus particle but is not found on ODV (although several other proteins are only associated with the ODV form).

Some differences also exist in the lipid composition of the viral envelope of the two forms. While the BV envelope consists of phosphatidylserine, ODV contains phosphatidylcholine and phosphatidylethanolamine.

Major Envelope Glycoprotein gp64

During periods of evolution, the baculoviral envelope glycoproteins have undergone changes. Ld130, also known as baculovirus F-protein from *Lymantria dispar* (LdMNPV) is suggested to be an ancestral envelope fusion protein which has been replaced by non-orthologous gene replacement with gp64 in AcMNPV, *Bombyx mori* (BmNPV) and *Orgyia pseudotsugata* (OpMNPV) while they still retain the ld130 gene.

Gp64 is a homotrimeric membrane glycoprotein which is polarly present on the rod-shaped virion. It consists of 512 amino acids (aa) with four glycosylation sites at asparagine residues and has a N-terminal signal sequence (20 aa), oligomerization and fusion domain and a hydrophobic transmembrane domain near the C-terminus (7 aa).

It is produced in both early and late phases of the infection cycle with a maximal rate of synthesis occurring in 24–26 h p.i.. Trimerization with intermolecular cysteine-bonds seems to be a crucial step for protein transport to cell surface, since only 33% of synthesized protein reaches cell surface as monomeric gp64 is degraded within the cells.

Gp64 is essential for efficient budding of the virion and for the cell-to-cell transmission during the infection cycle as well as viral entry i.e. causing viral trophism and endosome mediated uptake to the cell. The major function of the gp64 envelope protein is to cause the pH-mediated envelope fusion to the endosome. Although gp64 has variety of essential functions, it has been reported that gp64-null baculoviruses can be substituted with other viral glycoproteins such as Ld130, G-protein of *Vesicular stomatitis* virus. These substitutions will result in functional virons.

Applications

Baculovirus expression in insect cells represents a robust method for producing recombinant glycoproteins. Baculovirus-produced proteins are currently under study as therapeutic cancer vaccines with several immunologic advantages over proteins derived from mammalian sources.

Biosafety

Baculoviruses are incapable of infecting mammals and plants. They have a restricted range of hosts that they can infect that is typically restricted to a limited number of closely related insect species. Because baculoviruses are not harmful to humans, they are considered a safe option for use in research applications.

Taxonomy

This family has been divided into four genera: Alphabaculovirus (lepidopteran-specific nucleopolyhedroviruses), Betabaculovirus (lepidopteran-specific Granuloviruses), Gammabaculovirus (hymenopteran-specific nucleopolyhedroviruses) and Deltabaculovirus (dipteran-specific nucleopolyhedroviruses).

Evolution

Baculoviruses are thought to have evolved from the Nudivirus family of viruses 310 million years ago.

Wasp Parasitoid Polydnavirus Genes

The Polydnaviruses, (PDV) are a family of insect viruses that contain two genera: Ichnoviruses (IV) and Bracoviruses (BV). The ichnoviruses occur in ichneumonid wasps species and bracoviruses in braconid wasps.

The genome of the virus is composed of multiple segments of double-stranded, superhelical DNA packaged in capsid proteins and a double layer (IV) or single layer (BV) envelope. Little or no sequence homology exists between BV and IV, suggesting that the two genera evolved independently.

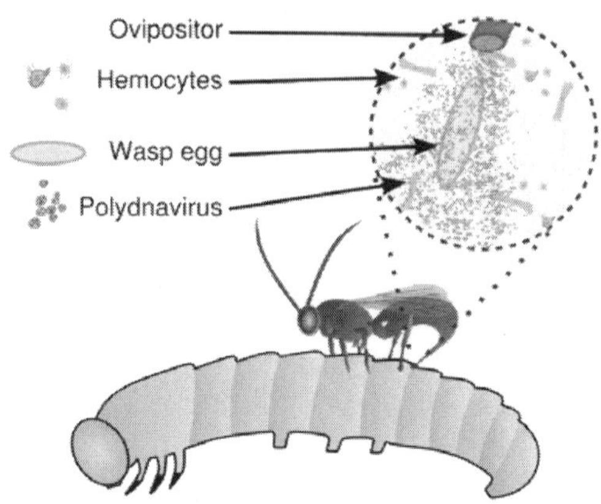

***Figure:** Diagram of a PDV host association*

These viruses are part of a unique biological system consisting of an endoparasitic wasp (parasitoid), an insect (usually lepidopteran) larva, and the virus. The full genome of the virus is integrated into the genome of the wasp and the virus only replicates in a particular part of the ovary, called the calyx, of pupal and adult female wasps.

The virus is injected along with the wasp egg into the body cavity of a lepidopteran host caterpillar and infects cells of the caterpillar. The infection does not lead to replication of new viruses, rather it affects the caterpillar's immune system.

Without the virus infection, phagocytic hemocytes (blood cells) will encapsulate and kill the wasp egg but the immune suppression caused by the virus allows for survival of the wasp egg, leading to hatching and complete development of the immature wasp in the caterpillar. Additionally, genes expressed from the polydnavirus in the parasitised host alter host development and metabolism to be beneficial for the growth and survival of the parasitoid larva. Thus the virus and wasp have a symbiotic (mutualistic) relationship.

Characteristics

Both genera of PDV share certain characteristics:

- the virus particles of each contain multiple segments of dsDNA (double-strand, or *normal* DNA, as contrasted with positive or negative-sense single-strand DNA or RNA, as found in some other viruses) with each segment containing only part of the full genome(much like chromosomes in eukaryotic organisms);
- the genome of the virus has eukaryotic characteristic such as the presence of introns and a low coding density
- the genome of each virus is integrated into the host wasp genome;
- the genome is organized in several multiple members genes families (which are different between Bracoviruses and Ichnoviruses).
- the virus particles are only replicated (produced) in specific cell types in the female wasp's reproductive organs.

However, the morphology of the two genera are different when observed by electron microscopy. Ichnoviruses tend to be ovoid (egg-shaped) while bracoviruses are short rods. The virions of Bracoviruses are released by cell lysis whereas virions of Ichnoviruses are released by budding.

Evolution

Nucleic acid analysis suggests a very long association of the viruses with the wasps (greater than 70 million years).

Two proposals have been advanced for how the wasp/virus association developed. The first suggests that the virus is derived from wasp genes. Many parasitoids that do not use PDVs inject proteins that provide many of the same functions, that is, a suppression of the immune response to the parasite egg. In this model, the braconid and ichneumonid wasps packaged genes for these functions into the viruses – essentially creating a gene-transfer system that results in the caterpillar producing the immune-suppressing factors. In this scenario, the PDV structural proteins (capsids) were probably "borrowed" from existing viruses.

The alternative proposal suggests that ancestral wasps developed a beneficial association with an existing virus that eventually led to the integration of the virus into the wasp's genome. Following integration, the genes responsible for virus replication and the capsids were (eventually) no longer included in the PDV genome. This hypothesis is supported by the distinct morphology differences between IV and BV, suggesting different ancestral viruses for the two families. The IV have remarkable similarities to ascoviruses while BV may be related to baculoviruses

The family *Polydnaviridae* may have evolved from the *Ascoviridae*. The ichnoviruses resemble the ascoviruses.

The brachivruses may have evolved from a nudivirus ~100 million years ago.

The two groups of viruses in the family are not in fact be phylogenetically related suggesting that this taxon may need revision.

Effect on Host Immunity

In the host, several mechanisms of the insect immune system can be triggered when the wasp lays its eggs and when the parasitic wasp is developing.

When a large body (wasp egg or small particle used experimentally) is introduced into an insect's body, the classic immune reaction is the encapsulation by the hematocytes. An encapsulated body can also be melanised in order to asphyxiate it, thanks to another type of hemocyte, which uses the phenoloxidase pathway to produce melanin. Small particles can be phagocytosed, and macrophage cells can then be also melanised in a nodule. Finally, insects can also respond with production of antiviral peptides.

PolyDNAvirus protect the hymenopteran larvae from the host immune system, acting at different levels.

- First they can disable or destroy hematocytes. The polyDNAvirus associated with *Cotesia rubecula*, code for a protein CrV1 that denatures actin filaments in hematocytes, so those cells become less able to move and adhere to the larvae. *Microplitis demolitor* Bracovirus (MdBV) induce apoptosis of hematocytes, thanks to its gene PTP-H2. It also decreases the adhesion capacity of hematocytes, thanks to its gene Glc1.8. The gene also inhibits phagocytosis.
- PolyDNAvirus can also act on melanisation, MdBV interferes with the production of phenoloxidase.
- Finally, polyDNAvirus can also produce viral ankyrins, that interfere with production of antiviral peptides. In some Ichnoviruses, Vankyrin can also prevent apoptosis, the extreme reaction of a cell to block viral propagation.
- The Ichnoviruses produce some proteins called vinnexins which have been recognized as homologous to the innexins of insects.

They are responsible for the encoding of the structural units of the gap-junctions. These proteins may alter the intercellular communication which could explain the disruption of the encapsidation process.

Virus Like Particles

Another strategy used by parasitoid Hymenoptera to protect their offspring is production of Virus Like Particles. VLPs are very close to viruses in their structure but they don't carry any nucleic acid. For example, *Venturia canescens* (Ichneumonidea) and *Leptopilina sp.* (Figitidaea) produce VLPs.

VLPs can be compared to PolyDNAvirus because they are secreted in the same way, and they both act to protect the larvae against the host's immune system. *V. canescens*-VLPs (VcVLP1, VcVLP2,VcNEP ...) are produced in the calix cells before they go to the oviducts. But they are composed of proteins that are closer in structure and function to Hymenoptera's proteins than to viral proteins. This suggests that they don't have a viral origin, unlike polyDNAvirus.

VLPs protect the Hymenoptera larvae locally, whereas polyDNAvirus can have a more global effect. VLPs allow the larvae to escape the immune system: the larva is not recognised as harmful by its host, or the immune cells can't interact with it thanks to the VLPs.

The wasp *Leptopilina heterotoma* secrete VLPs that are able to penetrate into the Lamellocytes, thanks to specific receptors, and then modify their shape and surface properties so they become inefficient and the larvae are safe from encapsulation.

Micro-RNA

MicroRNA are small RNA fragments produced in the host cells thanks to a specific enzymatic mechanism. They promote viral RNA destruction. MicroRNA attach to viral-RNA because they are complementary. Then the complex is recognised by an enzyme that destroys it. This phenomenon is known as PTGS (for post transcriptional gene silencing)

It is interesting to consider the microRNA phenomenon in the polyDNAvirus context. Many hypotheses can be formulated :

- Braconidae carry nudivirus-related genes in their genome, so they are may be able to produce microRNA against nudivirus, as an innate immunity.
- Wasps perhaps use microRNA to control the viral genes they carry.

- PolyDNAvirus can also use PTGS to interfere with the host's gene expression.
- PTGS is also used for organisms' development, using the same enzymes as antiviral gene silencing, so we can imagine that if the host uses PTGS against polyDNAvirus, perhaps it also affects its development.

Control of Lepidopteran Insect Pests

Insect pests are the main limiting factor for vegetable production in India and tropical Asia, amongst which Lepidopteran pests cause a significant amount of damage. Vegetables are some of the most profitable crops and farmers everywhere feel the need to protect such high value crops from any type of damage caused by insect pests. They often use synthetic insecticides indiscriminately, and insect resistance to insecticides is very common. Lepidopteran pests such as diamond-back moth (DBM) on cabbage (*Plutella xylostella*), fruit borer on tomato (*Helicoverpa armigera*), pod borer on chilli (*Spodoptera litura*), shoot and fruit borers on brinjal (*Leucinodes orbonalis*) and okra (*Earias fabia*) are among the major pests of vegetables.

To control Lepidopteran pests efficiently, some new pesticides, with novel modes of action, have been developed recently. To determine the effectiveness of the new pesticides compared with traditional chemicals (chlorpyrifos+cypermethrin), five field experiments were done on brinjal, okra tomato, cabbage and chilli, over two seasons. Altogether nine pesticides (flubendiamide, spinosad, emamectin benzoate, chlorfenapyr, *Bacillus thuringiensis*, novaluron, lufenuron, methoxyfenozide and mixed formulations of chlorpyrifos+cypermethrin) were used with untreated controls. All other agronomic practices were standard, and a blanket spray to control sucking pest insects and mites was applied when necessary.

Results showed that flubendiamide, spinosad and chlorfenapyr were highly effective in reducing the damage caused by diamond-back moth on cabbage, fruit borer on tomato, pod borer on chilli, shoot and fruit borer on brinjal and shoot and fruit borer on okra, and led to increases in yield. Novaluron was most effective against pod borer on chilli, closely followed by spinosad, flubendiamide and chlorfenapyr. In the case of fruit borer on tomato, spinosad was the most effective. Shoot and fruit borers on brinjal and okra were also controlled effectively by flubendiamide, spinosad and chlorfenapyr. Emamectin benzoate methoxyfenozide, and *Bacillus thuringiensis*, also performed well in reducing damage and increasing yield. *Bacillus thuringiensis* and methoxyfenozide were, however, less effective against shoot and fruit borer on brinjal and okra. The new pesticides, with novel modes of action and high selectivity, were highly effective against Lepidopteran pests. They are safer to non-target organisms and quickly degrade to non-toxic products. Lepidopteran (order Lepidoptera), any of more than 155,000 species of butterflies, moths, and skippers. This order of insects is second in size only to Coleoptera, the beetles.

Because of their day-flying habits and bright colours, the butterflies are more familiar than the chiefly night-flying and dull-coloured moths, but the latter are far more varied and abundant. The skippers are a worldwide group intermediate between butterflies and moths. With the exception of a few moths, all adult lepidopterans have two pairs of wings. The name Lepidoptera is derived from the Greek, meaning "scaly winged," and refers to the characteristic covering of microscopic dustlike scales on the wings.

The life cycle of lepidopterans consists of four stages: egg, larva (caterpillar), pupa (chrysalis), and adult (imago). The larvae do most of the eating, with the majority feeding on foliage, although many species eat stems, roots, fruits, or flowers. A number of moth and a few butterfly larvae are serious pests in agriculture and forestry. The adults of many species are important for their role in pollination, which occurs as they visit flowers for nectar. In most land environments the lepidopterans are ecologically important because they transform large amounts of plant matter into animal matter and in turn serve as food for many other groups of animals.

Many members of the order, especially the butterflies, have appealed to the human imagination for thousands of years as symbols of fragile and ephemeral beauty. References to them abound in literature, and they have been depicted in many paintings, have inspired the designs of jewellery, ornaments, and textiles, and have even occurred in many heraldic devices and on postage stamps.

Moths, butterflies, and skippers show great diversity in size and development rates. Some moths have wingspans as small as 4 mm (0.13 inch), whereas the largest moths and butterflies measure nearly 30 cm (about 1 foot). Fast-developing species may complete their development in as little as three weeks, while slower ones may require as long as two or even three years.

Lepidopterans live on every continent except Antarctica. Though they are far more numerous and diversified in the tropics, some species survive at the limits of polar vegetation. There are many successful species in nearly every environment, from arid deserts and high mountaintops to marshes and tropical rainforests. Most have become adapted for life in relatively restricted ecological niches and may be limited to only one or a small group of food plants, often to only a single part of one plant. They are therefore seldom abundant in more than one type of habitat. However, a few species have broader food habits and may occasionally reach peaks of abnormal abundance and defoliate large areas of deciduous forest or grassland. Some of the insects most injurious to human interests in agriculture and forestry are members of the Lepidoptera.

The large family-level groups and many smaller ones are worldwide in distribution, with owlet moths (family Noctuidae), measuring worm moths (family Geometridae), pyralid, or snout, moths (family Pyralidae), brush-footed butterflies (family Nymphalidae), and skippers (family Hesperiidae) being dominant elements of the insect fauna everywhere. A few families are characteristically more abundant in one faunal region than in others. Central and South America are characterized by great diversity of American false tiger moths (family Notodonidae) and ctenuchid moths (family Arctiidae). The North American (Nearctic) and Eurasian (Palaearctic) families show much evidence of close connections, chiefly between Asia and western North America. Each region has, however, many distinctive genera.

Many lepidopterans exist only in isolated colonies as relict (remnant) populations, cut off from relatives elsewhere by geologic or climatic changes. Australia and New Zealand have unusually diverse relict populations of the primitive mandibulate moths (family Micropterigidae) and swifts, or ghost moths (family Hepialidae). In North America, Europe, and Asia, many relict species have survived since the last ice age on isolated southern mountaintops.

Importance

Many hundreds of Lepidoptera injure plants useful to humans, including some of the most important sources of food, fabrics, fodder, and timber. The great majority of the injurious species are moths, and the detrimental life stage is always the larva. However, unlike members of other insect orders, lepidopterans do not act as carriers of plant diseases, nor are any of them parasites of or injurious to humans. However, some species feed on open wounds or bodily secretions of wild or domestic animals.

The list of valuable plants subject to damage by lepidopterans is a long one, including many grains, sugar beets and sugarcane, cotton, tobacco, some root crops and leaf crops, many fruits, and timber and shade trees. The damage may involve the leaves, stems, roots, or fruit.

Woolens, furs, silk, and even feathers are eaten by fungus moths of several genera (clothes moths). The greater wax moth (*Galleria mellonella*) causes considerable damage in beehives.

A few Lepidoptera are directly beneficial to humans. Nearly all silk is obtained from the domesticated silkworm (*Bombyx mori*), which is originally from China. Other silks such as shantung and tussah are the products of various Asiatic giant silkworm moths (family Saturniidae). The larvae and sometimes the adults of a few species are used for food. The larvae of one skipper (*Rhopalocampta libeon*) are collected in large quantities in the Congo, and the 10-cm (4-inch) caterpillars of giant skippers (family Megathymidae), known in Mexico as *gusanos de magüey*, are both consumed domestically and canned and exported for consumption as hors d'oeuvres. The South American cactus moth (*Cactoblastis cactorum*) has been highly beneficial in weed control, clearing more than 150 million ha (60 million acres) in Australia of alien prickly pear cactus. Doubtless, humans also benefit from much unrecognized weed eating by caterpillars and flower pollination by adults.

Many lepidopterans are valuable in biological research, including work in ecology, biogeography, systematics, genetics, and physiology. Much of the present knowledge of endocrine controls of insect development has come from studies of the silkworm moth and its relatives. The study of the British peppered moth (*Biston betularia*) has profoundly influenced ideas about rates of evolutionary change. An increase in the proportion of dark moths, a change thought to be brought on by airborne soot produced during the Industrial Revolution, has been termed *industrial melanism.*

Environmental Impact and Cost Benefit Analysis of Biological Control Application

Biological control proves to be very successful economically, and even when the method has been less successful, it still produces a benefit-to-cost ratio of 11:1. One study has estimated that a successful biocontrol programme returns £32 in benefits for each £1 invested in developing and implementing the programme, i.e., a 32:1 benefit-to-cost ratio. The same study had shown that an average chemical pesticide programme only returned profits in the ratio of 2.5:1.

Biological control can be defined as the use of an organism to reduce the population density of another organism and thus includes the control of animals, weeds and diseases. This definition establishes two of the main principles of biological control. Firstly, in nature, most organisms are consumed by other organisms, which in many cases leads to drastic reductions in the population of the prey species; in biological control, man exploits this 'natural control' to suppress the numbers of pest species. Secondly, biological control reduces rather than eradicates pests, such that the pest and natural enemy remain in the agro-ecosystem at low densities. A number of important pests can be kept at a low population density by biological control over long periods of time. In other cases, populations of pests are significantly reduced by natural enemies, but repeated releases or additional methods are needed to achieve an adequate level of control. These methods include, for example, resistant plants, cultural techniques, physical barriers, semiochemicals and, as a last resort, the use of selective chemicals; this is the fundamental philosophy of integrated pest management.

Many biological control schemes use predatory insects and mites, insects that parasitize other insects (parasitoids) or nematodes, targeted against insect and mite pests; these are the so-called 'macrobial' agents. There are also various 'microbial' agents (bacteria, viruses and fungi) that have been developed and applied in arthropod biological control. Herbivorous insects and mites have also been used in the biological control of weeds. Biological control schemes operate throughout the world as part of the management of pests in agriculture, forestry and greenhouse horticulture.

Although biological control has sometimes been introduced to combat arthropod pests that have developed resistance to insecticides and acaricides, the first biological control programmes pre-date the modern pesticide era. The intensification of agriculture through the twentieth century has been accompanied by increased international trade and the resultant transfer of pest species on plants and products across the globe. In addition, the introduction of new crops (such as soya in central Europe), and consumer demand for 'blemish-free' produce, have all contributed to an over-reliance on chemical control, with the well-described consequences of pest resistance, uneconomic production costs, bioaccumulation through food chains, environmental pollution, loss of biodiversity and risks for human health.

At the start of the twenty-first century, the opportunities and need for effective biological control are greater than ever: pest resistance continues to be a problem, pesticides are being withdrawn on environmental grounds without suitable replacements, and in some areas of the world, including Europe, there is uncertainty over the durability and public acceptance of genetically modified (GM) pest-resistant crops.

Types of Biological Control

There are different types of biological control, but both the exact number of types, and the definition of each type vary depending on the literature consulted.

There are three main techniques of biological control: *classical* (sometimes described as 'inoculative' biological control), *augmentative* (where a distinction can be made between 'inundation' and 'seasonal inoculation') and *conservation* control. Classical or inoculative control is used mainly against 'exotic' pests that have become established in new countries or regions of the world. Relatively, small numbers (usually less than 1000) of a certain species of natural enemy are collected from the country or region of origin of the pest, 'inoculated' into the new environment, and allowed to build up the level of control, which can be maintained over very long periods of time.

This type of biological control has been most successful with perennial crops (fruit plantations and forests), where the long-term nature of the ecosystem enables the interactions between pest and natural enemy to become fully established over a period of time: for example, the successful import and release of the predatory ladybird *Rodolia cardinalis* for control of the accidentally introduced citrus pest *Icerya purchasi* in Mediterranean Europe around 1900, and release of the parasitoid *Aphelinus mali* in the 1930s for control of the accidentally imported woolly apple aphid, *Eriosoma lanigerum*, in apple orchards throughout Europe.

Augmentation refers to all forms of biological control in which natural enemies are periodically introduced, and usually requires the commercial production of the released agents. Inundation involves the mass production and release of large numbers of the control agent, such as the *Trichogramma* egg parasitoids of various lepidopteran pests including the cotton bollworm *Heliothis virescens*, the sugar cane borer *Diatraea saccharalis* and the European corn borer *Ostrinia nubilalis*. The cultivation systems used for short-term annual crops destabilize the pest–natural enemy relationship, and thus often prevent the establishment of viable breeding populations of natural enemy species between the crop production cycles.

The aim of inundative releases is to create a massive ratio in favour of the natural enemy, analogous to the use of a pesticide, producing a rapid reduction or local extinction of the pest. Control is achieved mainly by the individuals that have been released rather than their offspring. However, the control is usually transient, and re-releases are required, sometimes more than once a year.

Seasonal inoculative control is a form of augmentation in which natural enemies are similarly mass reared in the laboratory and periodically released into short-term crops where many pest generations can occur in each growing season. As with augmentative control, relatively large numbers of natural enemies are released to obtain immediate control, but in addition, a build-up of the natural enemy population occurs through successive generations during the same growing season. This method can be applied when the cropping system prevents control extending over many years as with classical (inoculative) biological control. Examples include the control of whiteflies, leafminers, thrips, aphids and mites by parasitoids and predators in greenhouses.

Conservation control refers to the use of indigenous predators and parasitoids, usually against native pests. Various measures are implemented to enhance the abundance or activity of the natural enemies, including manipulation of the crop microclimate, creation of overwintering refuges (like 'beetle banks'), increasing the availability of alternative hosts and prey, and providing essential food resources such as flowers for adult parasitoids and aphidophagous hoverflies.

Landmark Events in Biological Control

The first historical record of biological control dates back to around AD 300 when predatory ants were used to control pests in citrus orchards. In the modern era, the control of the cottony-cushion scale *I. purchasi* on citrus crops in California in the 1880s by an imported ladybird (the vedalia beetle *R. cardinalis*) and dipteran parasitoid (*Cryptochaetum iceryae*), is widely regarded as representing the first major success of classical biological control.

The scale insect was accidentally introduced to the United States on *Acacia* plants imported from Australia. Two natural enemy species of *I. purchasi* were collected in Australia, released in small numbers in Californian citrus groves and achieved control of the scale in less than 2 years, the ladybird being the most effective agent. The success of this scheme, which has been repeated in other locations against the same pest worldwide, is attributable to a number of ecophysiological characteristics of the predator–prey relationship: both the adult and larval stages of *R. cardinalis* were active and could seek out their sedentary prey, the rate of development of the ladybird was more rapid than the scale, and the predator had no natural enemies of its own.

Another early example of classical control involved the coconut moth *Levuana iridescens* in Fiji, and established other important principles in the development of biological control. A tachinid parasitoid

Ptychomyia remota was known to attack the moth *Artona catoxantha* in Malaysia. The parasitoid was introduced to Fiji to control a 'new' host which it had not previously encountered (*L. iridescens*), and was successful across the whole island within 2 years. The stable Fijian climate allowed year-round reproduction of *L. iridescens*, providing a continuous source of prey for *P. remota*; additionally, the island ecosystem was small and isolated.

One of the most recent successes in biological control is a further example of classical control. Cassava (*Manihot esculenta*) is one of the staple diet crops for a number of African countries. Cassava was introduced to Africa from South America and had been grown for hundreds of years without any major pest problems, until the 1970s. The crop then became infested with an accidentally imported mealy bug from South America *Phenacoccus manihoti*, which spread rapidly over the next 15 years through many countries. A search in South America found the pest in Paraguay, together with one of its natural enemies, the parasitoid *Epidinocarsis lopezi*. The parasitoid showed good searching ability, apparently locating its host by attraction to damaged leaves, and was able to develop more quickly than the mealy bug. After its first release in Nigeria in 1981, *E. lopezi* spread rapidly through neighbouring African countries and is now regarded as one the most successful programmes in biological control with enormous economic benefits.

One of the first successes with inundative control on a field scale involved the citrophilus mealy bug, *Pseudococcus calceolariae* Fernald (equal to *gahani* Green), a pest of citrus in southern California. With the discovery and spread of the mealy bug in 1913, the citrus growers were faced with a serious problem that threatened the crop's viability. In response to this problem, the citrus industry established insectaries to mass-produce the predatory beetle *Cryptolaemus montrouzieri*. Mass production techniques were developed by the early 1920s and, for the next decade, the beetle was the principal means of suppressing this pest. In 1930, at the height of the *P. calceolariae* campaign, 16 insectaries had been established and were producing 20 million beetles annually for release in infested groves.

Inundative releases are currently used on a large scale against a range of pests. In Brazil, for example, *Cotesia* parasitoids are applied against sugar cane borer on 300 000 ha, *Anticarsia gemmatalis* multiple nucleopolyhedrovirus (AgMNPV) against soya bean caterpillar on more than 1 Mha, and egg parasitoids of soya bean bugs on 20 000 ha. Egg parasitoids of the genus *Trichogramma* are now released on more than 10 Mha worldwide.

The first success of seasonal inoculative biological control in protected cultivation involved the glasshouse whitefly *Trialeurodes vaporariorum* and the parasitic wasp *Encarsia formosa*. The whitefly and other greenhouse pests (spider mite, thrips and aphids) attack a range of vegetable crops (tomato, cucumber and peppers) and ornamental flowers. These pests reduce yields, lower the market value of products that are sold on appearance, and provide a breeding medium for secondary infections such as the sooty mould (*Cladosporium sphaerospermum*). Biological control of *T. vaporariorum* was established before 1930 in the UK and ran successfully until the introduction of DDT and other organochlorine insecticides in the 1940s. However, the development of resistance in the whitefly, together with other concerns about the overuse of pesticides led to the re-introduction of the biological control scheme in the 1970s. For both *T. vaporariorum* and the glasshouse spider mite *Tetranychus urticae* controlled by the predatory mite *Phytoseiulus persimilis*, sophisticated management schemes have now been developed, involving the commercial production of the control agents, flexible release plans (frequency and density), introduction of the pest species to maintain natural enemy populations when required, and manipulation of the climatic environment (temperature and humidity) to optimize control. Currently, all greenhouse vegetable pests can be managed with biological control agents (van Lenteren 2000*a*).

A major success in conservation biological control was achieved with pest management in European fruit orchards. A few years after the use of blanket sprays of synthetic pesticides there were marked declines in the abundance of several species of naturally occurring predators and parasitoids. A series of related studies investigated the occurrence of pests and natural enemies in sprayed and unsprayed orchards, and the relative toxicity of different pesticides on the beneficial species. This information was used to design an IPM programme based on the re-establishment of the natural enemy fauna (predatory insects, mites and parasitoids) to control key pests such as leaf rollers, leaf miners and spider mites, and the selective and local use of chemicals against other pest species (Gruys 1982).

Advantages and Limitations of Biological Control

The advantages and limitations of biological control are often expressed by comparisons with pesticides. Thus, predators and parasitoids are naturally occurring organisms and usually fairly specific in the range of prey that they will attack. Natural enemies actively seek out their prey and can increase the level of control over time. It is unlikely that resistance will develop to a control agent, and in many cases, the control can be self-perpetuating over long periods of time. The arguments against chemical pesticides are that they not only kill the pest organism, but also many non-target species, including natural enemy species, which in turn, may increase the pest status of species that were previously unimportant or easy to control. Furthermore, chemical control is limited to the area within which the pesticide is applied, frequent application may be required, and this selects for pest resistance.

The main limitation of biological control is that it is slower to suppress pest populations than most pesticides as parasitized organisms may take several days to die; and also, predators require a period of time to establish an economic level of pest suppression. Development costs of biological control are sometimes described as 'high', but these costs are much lower than for the equivalent synthesis, toxicological evaluation and marketing of a new pesticide, and substantial profits can be achieved from biocontrol with long term, effective natural enemies, of which *R. cardinalis*, *E. lopezi*, *E. formosa* and *P. persimilis* are good examples. The comparative advantages of pesticides are that control is rapid, and in the absence of resistance, a high and predictable level of mortality is normally guaranteed.

There are other reported limitations of biological control where pesticides seem to be the more advantageous approach, but the evidence for this conclusion is limited or absent (van Lenteren 1993*a*,*b*). For instance, the fact that biological control 'regulates' whereas pesticides 'eradicate' is viewed as disadvantageous, but this is a spurious argument. In biological control, pest populations are usually not eradicated, but maintained at very low densities; long-term suppression of pest species is the desirable aim. Eradication of a pest population after chemical treatment, even if this is achieved, occurs only on local scales, and the environment is then open to re-invasion, often with a much reduced natural enemy fauna.

Biological control has been described as 'unreliable' when compared with pesticides, but again the information for this view is equivocal. Some biological control schemes have been 'partial' rather than 'complete' successes, and in some instances, there have been periodic fluctuations in the level of control. But as a counter argument, pesticides also vary in their effectiveness and over time, the development of resistance can lead to the failure of a previously successful chemical.

It is now widely acknowledged that successful biological control depends on extensive preliminary studies to gain a comprehensive understanding of the biology and ecology of the pest and natural enemy complex, and of the environments from which they originate and have subsequently colonized, or into which they will be released. This research can be time consuming, but it is essential, and analogous to the period of time (estimated to be around 10 years) between the synthesis of a new molecule as a potential pesticide and placing a product on the market. Where all of the necessary 'efficacy studies' have been carried out, such as with *P. persimilis* and *E. formosa*, the control programmes have been consistently successful. In other situations, and under strong market pressure, control agents have sometimes been released without adequate evaluation, and control has been less successful and the pest problems have continued; for example, predatory mites released against *Frankliniella occidentalis*.

It has been suggested that the introduction of a biological control scheme against a pest, and the resultant withdrawal of broad-spectrum pesticides, can lead to new pest problems, but there is no evidence for this in several well-studied agro-ecosystems, including glasshouse systems. Research on biological control was undertaken from 1965 to 1975 to combat glasshouse pests (*P. persimilis* and *E. formosa*) that had become resistant to the then available pesticides (DDT, malathion and resmethrin), but the wide-scale introduction of natural enemies against these two species did not by itself lead to the occurrence of any new pests. The pest problems that have arisen in the European glasshouse industry since 1975 (*Spodoptera exigua*, *Liriomyza trifolii*, *Liriomyza huidobrensis*, *F. occidentalis* and *Bemisia tabaci*) have all been unintentional introductions that arrived with varying levels of resistance to most pesticides. These newly imported pests have threatened the biological control of existing pests because governments usually respond to pest invasions by initiating extermination programmes based on frequent application of pesticides, thereby killing the natural enemies of the 'old' pests. Therefore, it is crucial to identify effective natural enemies of new pests before or soon after invasion, in order to maintain stability in the commercial biological control of existing pests.

Pest control must be cost effective relative to the value of the crop. A comparison of the costs of both the development and the use of chemical and biological control indicates that, in both respects, natural enemies are more cost effective than pesticides.

Comparison of aspects related to the development and application of chemical and biological control.

Cost–benefit analyses suggest that research on biological control is more cost effective than on chemical control (30 : 1 and 5 : 1, respectively. Despite this advantage, the main reason that biological control is not used on a larger scale is attributable to problems associated with the production and distribution of parasitoids and predators; particularly, the limited shelf life (days or at most weeks) of most natural enemy species and the impossibility of patenting a naturally occurring, unmodified species.

It is evident that many more chemical compounds have been tested than species of natural enemy, though there are still vast numbers of predator and parasitoid species that remain to be screened for their use in biological control. The success rate for finding a successful natural enemy is much higher than for a chemical compound, which is mainly attributable to the 'directed search' that is used for natural enemies compared with the more random approach for chemical compounds, although pesticide discovery and design has become far more rational over time. Development costs are much higher for chemical pesticides, largely as a result of the very stringent requirements for ecotoxicological studies as part of the registration process.

Interestingly, the developmental time of an effective product, be it a chemical pesticide or a natural enemy species, is the same. It is common for insect and mite pests to develop resistance against chemical pesticides, whereas as resistance against natural enemies is unknown, at least to the extent that the control agent becomes ineffective.

When compared on the criterion of specificity, even the most selective chemical is likely to kill many species of arthropod, whereas natural enemy species used in biological control are usually highly specific, killing only one or a few related species of prey. When selective natural enemies are used in biological control programmes, harmful side effects do not occur, but with highly polyphagous species there may be some negative environmental effects.

The favourable economics of biological control relative to the use of pesticides also applies to the commercial application of the technique. In the 1980s, Ramakers (1982) estimated the cost (product and labour) of controlling glasshouse whitefly (*T. vaporariorum)* to be twice as expensive using insecticide compared with the parasitoid *E. formosa.* A similar difference in costs was found with the control of spider mites (*T. urticae*) by predatory mites compared with acaricides.

In the control of tomato and cucumber pests in the UK, Wardlow (1992) estimates chemical control to be three- or five times more expensive than using commercially produced natural enemies, and even where different biological control agents are used in combination (as with cucumber), the costs are still not higher than when using chemicals.

Periodic analyses of successes and failures of biological control have suggested a number of more generalized limitations that relate to the types of pests that can be more easily controlled by predators and parasitoids, features of the targeted ecosystem and suitability of climates. For classical biological control, some of the most effective schemes have been applied against relatively sedentary pests (such as scale insects and mealy bugs), living in island ecosystems and in year-round stable climates that allow continuous development and reproduction of the pest and natural enemy (DeBach 1964). This view gained support from the control of *I. purchasi* in the citrus groves ('ecological islands') of California, *L. iridescens* in Fiji and sugar cane pests in Hawaii. However, as the scientific understanding of biological control has increased, through both theoretical and practical studies, it is apparent that these limitations are not an insurmountable obstacle to success. For example, biological control schemes have been successfully applied against mobile pests such as winter moth (*Operophtera brumata*) in Canada (by the parasitoids *Cyzenis albicans* and *Agrypon flaveolatum* from Europe), where there is a strongly seasonal climate with severe winters. Nevertheless, it is evident that the seasonal temperate climate can be a barrier to the long-term survival of natural enemies originating from tropical or semi-tropical climates, such as the failure of the parasitoid *A. mali* to control the woolly aphid *E. lanigerum* in UK apple orchards in the 1920s and 1930s, through a lack of overwintering ability, even though the natural enemy had been successfully introduced into other countries. However, the main reason for the limited success of biological control in the field in northern Europe is not the weather, but the major changes that occurred in agricultural production systems after 1945 with the introduction of pesticide-dominated methods of pest control, and the loss of resistance to pests and diseases through plant breeding programmes that focused selectively on yield. Where IPM schemes have been established in orchards, and thus pesticide use is highly restricted (or terminated), a number of key pests can be well-controlled by naturally occurring beneficial insects (which may have to be locally reintroduced). There may be other pests that cannot be adequately controlled by naturally occurring natural enemies, and alternative methods will then be required (pheromones, host-plant resistance and hormonal control).

Selection of Successful Agents

In the early stages of development, biological control of necessity followed mainly a 'trial and error' approach. But now, with a history of over 100 years, there is a large and increasing database of information that can be analysed with the aim of identifying the causes of success and failure, and hence the features that are commonly associated with the most successful agents. These analyses suggest that successful biocontrol agents show strong searching ability to locate prey and identify areas of high pest density, and ideally should have higher potential rates of increase than their prey (faster development, more generations per year, and greater fecundity or a greater predation rate than the reproduction capacity of the pest). Control agents must be able to survive at low pest density, otherwise control will fail and repeated releases will be necessary. The target pest should be the preferred prey of the natural enemy, though use of alternative prey is desirable during periods when the pest occurs at a very low density, or is not in a stage of the life cycle that is parasitized or predated by the control agent. However, it is vitally important that the released agent should not attack non-target species, and this feature is a key element in the pre-licensing environment risk assessment that is now routine in many countries. In situations where the natural enemy is strongly host specific, it is important for the life cycles of the predator and prey or parasitoid and host to be synchronous, especially in seasonal climates where diapause and the timing of post-winter emergence can influence the respective phenologies from year to year. The ability of parasitoids to detect and avoid hosts that have already been parasitized is also an advantageous feature of successful agents. The similarity in climate between collection and release sites is a further important consideration that can have a major impact on successful establishment in new environments.

It is clear that such analyses are not a guarantee of success, and there are also contradictory views among scientists that question some of the fundamental theories of biological control upon which such analyses are based. For instance, it has been argued that the most desirable biological characteristics identified in this selection process are derived from theoretical models of the interactions between prey species and predators or parasitoids without any strong evidence that such theory actually increases the frequency of success or provides an explanation for failures (Gutierrez *et al.* 1994). Likewise, it has been suggested that parasitoids and predators co-evolve with their prey such that control agents would be expected to become less effective over time (Pimentel 1963). If true, this suggests that the concept of classical biological control in which exotic pests are 'reunited' with their former natural enemies collected from the country or area of origin of both species is likely to be successful for limited periods of time; and conversely, that species collected from different locations to the native area of the pest, or from other host or prey associations, may be more successful, or maintain control for longer periods of time (Hokkanen & Pimentel 1984). Analysis of databases of successes and failures in biological control have not produced any strong evidence to support these ideas (Waage & Greathead 1988), and some of the classical control schemes based on the collection of natural enemies associated with the pest from its country of origin (cottony-cushion scale in California and sugar cane leafhopper in Hawaii) have now been operating successfully for 50–100 years. It also has to be recognized that there are important differences in predator–prey relationships in natural environments and those occurring with pests and biological control agents in agricultural ecosystems, particularly in terms of the density and uniformity of the plant hosts; as such, observations and predictions from naturally co-evolved interactions may not be a reliable predictor of outcomes in cropping regimes.

Increasing the Success of Biological Control

Some biological control schemes have the potential to be successful but require additional inputs from man, either prior to release or during natural colonization of the crop environment or to encourage post-release establishment, and this is particularly true of conservation biological control. As examples, adult parasitoids and hoverflies require sources of nectar and pollen to mature their eggs (Wäckers 2003), and growers have been encouraged to plant selected nectar-producing flowers around field boundaries to increase the attractiveness of the crop environment to such natural enemies, with some experimental evidence that an increase in crop diversity with flowers can increase the natural control of pests.

An increase in plant diversity can also attract insect herbivores that act as alternative hosts or prey for the released agents. The parasitoid *Anagrus epos* parasitizes the eggs of grape leafhoppers (*Erythromeura* species) in California. The grape leafhopper overwinters in the adult stage, leading to a loss of synchrony in the host–parasitoid relationship. A different leafhopper (*Edwardsiana prunicola*) associated with prune trees provides an alternative host for *A. epos*, ensuring effective control of the grape leafhoppers at the start of each growing season. The importance of alternative prey has also been identified in the control of western flower thrips (*F. occidentalis*) by the anthocorid bug *Orius majusculus* and with the egg parasitoids of rice plant hoppers (Yu *et al.* 1996).

Sometimes, alternative prey, together with natural enemies, can be offered on non-crop plants adjacent to the crop, to enhance natural enemy populations by the so-called 'banker plant system'. As an example, one of the most effective methods of suppressing aphid populations in protective cultivation is to introduce the natural enemy species into the greenhouse even before any aphids have been discovered on the crop plants. This can be achieved by using open rearing units ('banker plants') consisting of wheat plants with wheat aphids (which cannot develop on the greenhouse crops) together with predators or parasitoids (van Steenis 1995). In other situations, slow-release systems are used to obtain a regular re-infestation of the crop with natural enemies as with the 'sachet-system' used for the control of thrips by predatory mites. Each sachet contains the predator with an alternative prey source; the mites consume the provided prey and emerge from the sachet progressively over several weeks and then predate the target pest.

Assistance from man may extend to more specialized release systems such as spray equipment and machinery that 'shoots' small containers loaded with natural enemies into the crop (*Trichogramma* in maize), release by model airplane (whitefly parasitoids in infested patches of cotton), and full field release by helicopter, ultralight or small airplane (release of *Trichogramma* in maize and other crops). One of the most important actions man can take is to aid the synchronous occurrence of pest and natural enemy by well-timed releases of the control agent.

Current use of Biological Control

Biological control can be defined as the use of an organism to reduce the population density of another organism. Natural biological control ensures that the Earth is 'green' and that plants can produce sufficient biomass to sustain other forms of life. Without biological control, the production of energy by plants would be a small fraction of current levels.

Natural biological control of pest organisms has occurred since the evolution of the first ecosystem some 500 Myr ago, and continues to the present day across 55.5 billion hectares of the world's terrestrial ecosystems and without human intervention. Its role in marine or freshwater ecosystems has not been quantified, but is certainly equally important as in terrestrial environments. In forest and agro-ecosystems, most of the potential arthropod pests (95% of the total or approximately 100 000 species; DeBach & Rosen 1991; J. C. van Lenteren 2000, unpublished data) are under natural biological control. All other modern methods of control are targeted at the remaining 5000 arthropod pest species. What is often unknown or easily ignored is that the maintenance of ecosystem functions achieved by natural biological control is estimated to have a minimum value of 400 billion US$ per year (Costanza *et al.* 1997), which is an enormous sum of money compared with the annual expenditure on insecticides of only 8.5 billion US$.

Classical biological control is estimated to be applied on 350 Mha worldwide (10% of land under culture; J. C. van Lenteren 2005, unpublished) and has very favourable cost–benefit ratios of 1 : 20–500. As soon as a natural enemy has been released and becomes effective, it contributes annually to benefits, while costs are minimal. As with natural biological control, the continuously increasing benefits of classical biological control programmes are often overlooked once a pest disappears as a result of the release of an efficient natural enemy. The comparative estimate for area usage of augmentative biological control is 16 Mha (0.046% of land under culture, van Lenteren & Bueno 2003), with a cost–benefit ratio of 1 : 2–5, which is similar to chemical pest control. More than 5000 introductions of approximately 2000 species of non-native arthropod biological control agents have been made over the past 120 years in classical biological control programmes targeted against arthropod pests in 196 countries or islands; and more than 150 species of natural enemies (parasitoids, predators and pathogens) are currently commercially available for augmentative forms of biological control.

Modern Biological Control in Sustainable Agriculture: Restoring the Ecosystem Function of Pest Management

Before the large-scale application of chemical pesticides, biological control was one of the main pest management methods embedded in a 'systems approach' to pest prevention and reduction, covering animals, weeds and pathogens. A farmer had to think about pest prevention before designing the next season's planting scheme and choice of crops. This concept generally made use of three methods of pest management: cultural control, host-plant resistance and biological control. Cultural methods such as crop rotation, cover crops and manipulation of sowing and harvesting dates were used to prevent the build-up of pest numbers. Plants that had a high degree of resistance or tolerance to pests were another cornerstone of pest prevention, and the third was natural, classical, inundative and conservation biological control.

From around 1950, these methods became redundant as almost all pests could be easily managed by the newly discovered pesticides. As a result, pest control research became a highly reductionist activity, and changed from an important decision-making exercise in pest management to a routine but initially successful fire brigade activity. Another effect was that plants were no longer selected for resistance to pests, but only for the highest production of biomass (food) or cosmetic appearance (flowers), produced under blanket applications of pesticides. This, in turn, has led to the current situation in which crops have become 'incubator plants', unable to survive without frequent pesticide applications and agro-ecosystems that have few, if any, natural enemy populations.

However, now that chemical pesticides are no longer seen as the major solution for sustainable pest management, it is not possible to simply return in a year or so to pre-pesticide methods of control: crops are too weak to survive without pesticides, natural enemy faunas are absent, and many farmers remain 'pesticide-addicted'. So, first there is a need to invest in the development of new crop cultivars with resistance to pests and diseases, and there is evidence that this is happening in some countries. For example, the Dutch plant breeding industry now invests 35% of its research funds in the development of pest resistance compared with only 5%, 20 years ago. At the same time, there is a need to restore previously used natural, classical, inundative and conservation biological control.

Further, several other alternatives to conventional chemical pest control can also be implemented such as mechanical, physical, genetical, pheromonal and semiochemical control. Also, the environment can be manipulated to make it more attractive and favourable for natural enemies. This strategy involves a range of methods, from manipulation of biotic and abiotic components of the environment such as modifying the climate (e.g. greenhouses and wind shields), to the application of chemicals that stimulate the activity of natural enemies. If natural enemies fail to establish (either due to agricultural practices or lack of adaptive ability in the natural enemy) or, if once established, fail to control the pest, manipulation of the natural enemy or its environment may lead to greater control. The local habitat may lack certain key requisites, which when provided may allow natural enemies to establish or become more effective. Manipulation of the environment is currently applied on a limited scale though there are many opportunities for implementation.

The successful introduction (or more accurately, 're-introduction') of these 'new' pest management strategies will require retraining of the extension and advisory service and of growers. This is easier said than done, because it is difficult to simply replace a certain pesticide with an alternative method of control. Instead, there is a need to return to the systems approach, where the influence of all factors affecting pest abundance is considered. An example of a successful systems approach is seen with the optimization of fertilizer use, where a considerable reduction in fertilizer application reduced the development of several pests including aphids, whiteflies and leaf miners. The aim of this approach is to create a system that is inherently resistant to many pests and, thus, requires fewer or no treatment with conventional pesticides. Here, it is important to recognize that maximizing net income is not synonymous with maximizing yield. High yields are obtained with excessively high inputs of fertilizers and pesticides. Reducing the level of inputs may lead to somewhat lower yields, but as the financial inputs are also lower, the net income may be the same or higher. In general, integrated farming takes a more complete account of the impacts of agriculture on ecosystems (pesticide and fertilizer pollution of soil and water, preservation of flora and fauna, quality and diversity of landscape, conservation of energy and non-renewable resources), as well as sociological considerations (employment, public health and well-being of persons associated with agriculture), than is the case with conventional farming practices.

Although research on integrated farming is still limited, this approach is gaining impetus in Europe. The practices which can be manipulated in integrated farming programmes are crop rotation, cultivation, fertilization, pesticide use, cultural control measures, biological control and other alternatives to conventional chemical control. The practical results obtained in a large, long-term project in The Netherlands indicate that environmental pollution can be reduced through a decrease in fertilizer use and the replacement of chemical pesticides by non-chemical measures. In integrated farming, artificial fertilizers tend to be replaced by organic manure, reducing the total amount of freely available nitrogen (N) in the system, and hence the crop is less attractive or suitable for pest and disease development. Weed, pest and disease problems are reduced in integrated farming through the use of weed-competitive or disease- and pest-resistant varieties, reduction of N-fertilization, adoption of specific sowing dates and plant spacing, mechanical weed control and natural control. Chemical pest control in integrated farming is based on accurate sampling of pest population densities linked to economic thresholds. Using this approach, a consistent reduction of more than 90% in pesticide use can be achieved with the same economic return as conventional farming (van Lenteren 1997*a*). The generally lower yields from the integrated system are compensated by reductions in cost from the lower input of pesticides and fertilizers.

Biological control will be a key feature of future sustainable crop production (van Lenteren 1998). The landscape in which agriculture currently takes place is characterized by (i) low species diversity, (ii) plants with little architectural complexity, and (iii) plants and animals that are short-lived with high fecundity, relatively good dispersal but poor competitive ability (Bukovinszky 2004). Further, many agro-ecosystems are dominated by weeds, insects and pathogens that are highly adapted for rapid colonization and population increase. Plants with simple architectures have fewer associated species of insects (pests and beneficials) than diverse and architecturally complex plant communities (Landis & Marino 1997). As a consequence of these low-diversity plant and herbivore communities, agro-ecosystems often have strongly impoverished natural enemy communities when compared with natural ecosystems (Landis *et al.* 2000). However, the areas adjacent to crop fields are usually less disturbed and architecturally more complex, with a richer and more stable natural enemy fauna that can provide source populations of beneficial arthropods to promote pest management. But it should be realized that these extra-field communities may also act as a reservoir for pest species (Winkler 2005).

Sustainable pest management must, therefore, be based on an appreciation of how the structure of the agricultural landscape can influence the interactions between extra-field and within-field processes. An understanding of the interchange of organisms between areas of the landscape, and the influence of landscape structure on these interchanges, is critical for predicting and managing pest populations in agricultural fields (Lewis *et al.* 1997). However, as a starting point, it might be more efficient to first concentrate on changes within cropping systems that could increase natural pest control. One approach would be the use of multi- or poly-cropping systems with (i) two or more crop species, (ii) a crop with an undersowing of an economically unimportant plant, or (iii) a multi-crop consisting of a crop species and herbaceous field margins. Although there are many publications which state that the natural enemy fauna is richer in multi-crops, and that natural biological control is greater than in monocultures, there is very little experimental, quantitative data to support this view. There is, for example, hardly any information on how natural enemies search for prey in multi-crop systems compared with searching behaviour in monocultures. However, one of the most frequently cited reasons for multi-cropping (which is applied on 60% of the world area used for food production) is protection from pests (Vandermeer 1989). Pest pressure is usually lower in multi-crops, though not always. The presence of associated plants in a multi-crop can reduce the damage to the main crop species in three ways, all involving a lower population growth rate of the pest. In the first situation, the associated plants are more attractive hosts for the pest than the crop (*host-plant quality hypothesis*). In the second response, the associated plants interfere directly with activities of the pest (*disruptive crop hypothesis*); and in the third, the associated plants change the environment to favour natural enemies of the pests (*natural enemies hypothesis*).

A series of studies have tried to identify the mechanisms by which the numbers of insect herbivores are reduced in multi-crop systems and concluded that in most cases the disruptive crop hypothesis is the best explanation for the data, but that the natural enemies hypothesis could often be supported (Risch *et al.* 1983). Some recent experimental studies indicate that all three hypotheses for pest reduction, the host-plant quality hypothesis, the disruptive crop hypothesis and the natural enemies hypothesis may be valid. Other reviews reveal similar results and data clearly show that plant diversity often results in higher natural enemy populations (e.g. Andov 1983).

An analysis of 51 recent studies of habitat manipulation to enhance conservation biological control (Gurr *et al.* 2000) showed that in the vast majority of cases there were significant benefits for natural enemies. However, a significantly beneficial effect on natural enemies did not always result in a greater reduction of pest populations or higher yields. Owing to the empirical approach that typifies many of these studies, the effects of agro-ecosystem diversification on the searching behaviour and success of arthropod natural enemies remains poorly understood. Studies in this area are a priority to inform the design and fine-tuning of farming schemes aimed at pest prevention.

Biological Control and GM Crops

The estimated global area of commercially grown GM crops in 2004 was 81 Mha, equivalent to approximately 5% of the global area available for cultivating crops and 3.2 times the total land area of the UK (James 2004). Four major GM crops (soya bean, cotton, maize and oilseed rape) were grown by 8.25 million farmers in 17 countries in 2004, and approximately 90% of the farmers were resource poor from developing countries. GM herbicide tolerance in these four crops occupied 58.6 Mha (72% of the total GM area), while GM insect resistance in maize and cotton was grown on 15.6 Mha (19%). Stacked genes for herbicide tolerance and insect resistance in maize and cotton occupied 6.8 Mha (9%). GM crops containing genes of *Bacillus thuringiensis* (Bt) expressing truncated Cry proteins which are toxic to specific insect groups (e.g. Lepidoptera, Coleoptera) are currently the only commercial insect resistant GM plants and are grown on 22.4 Mha (28%). Bt Cry1-expressing maize and cotton are protected from attack by lepidopteran pests including corn borers (mainly *O. nubilalis*) in maize and the budworm–bollworm complex (*H. virescens*, *Helicoverpa* spp., *Pectinophora gossypiella*) in cotton (Shelton *et al.* 2002). In 2003, Bt maize for control of *Diabrotica* spp. (corn rootworms) was commercialized, expressing the coleopteran-specific Cry3Bb toxin.

An advantage of insect-resistant transgenic Bt plants is the reduced need for conventional insecticides, providing benefits for human health and the environment (Shelton *et al.* 2002). For example in US cotton, the average number of insecticide applications used against the budworm–bollworm complex decreased from 4.6 in 1992–1995 to 0.8 in 1999–2001, largely due to the introduction of Bt cotton (FAO 2004). In China, the introduction of Bt cotton has led to a 60–80% decrease in the use of foliar insecticides (Fitt *et al.* 2004).

Detailed analysis of potential effects of GM crops on the environment and the crop ecosystem is crucial prior to commercial release (Dale *et al.* 2002). Key concerns are potential ecological consequences on non-target organisms, including the natural enemies of pests. For Bt toxin, direct effects can be expected only if it is ingested and the parasitoid/predator is susceptible. This requires that the toxin is acquired through direct plant feeding (e.g. pollen), or that it is passed on in a biologically active form by the host/prey. Indirect, host/prey-quality mediated effects can be expected if susceptible herbivores ingest the toxin. Affected herbivores are often smaller, develop slower, behave differently and/or have an altered tissue composition compared to healthy individuals. Such changes may influence the quantity of prey and their nutritional quality for the next tropic level. Effects on natural enemies include death, sublethal effects (e.g. prolonged development, reduced weight), altered behaviour (e.g. reduced parasitization rate, changes in prey choice), or species may be unaffected.

Over the past 10 years, a number of laboratory studies have investigated possible deleterious effects of Bt plants on the mortality, longevity and development of predators and parasitoids. To date, there is no indication of direct effects of Bt transgenic plants on any of the species studied so far, either in direct plant-feeding assays or when the natural enemies had been provided with non-susceptible prey or hosts containing the Bt (Cry) proteins (Romeis *et al.* 2006). Adverse effects have only been reported in studies using herbivores as prey or hosts that ingested the Bt toxin and were susceptible to it (Romeis *et al.* 2006). Such indirect host-quality mediated effects are to be expected because sublethally intoxicated prey or hosts are generally of lower nutritional quality. However, such indirect effects are not unique to Bt toxins, and are known to occur with other insecticides.

On a larger scale, more than 50 studies have been conducted in both experimental and commercial fields to evaluate the impact of Bt crops on natural enemies. Experimental field studies have revealed only minor, transient or inconsistent effects of Bt crops when compared to a non-Bt control. Exceptions were observed with specialist natural enemies which were virtually absent in Bt fields due to the lack of target pests as prey (Riddick *et al.* 1998) or hosts (Pilcher *et al.* 2005). A few studies in Bt crops revealed consistent reductions in the abundance of different generalist predators that were also associated with the reduced availability of sensitive (target) lepidopteran prey.

Some studies have compared the function of natural enemies (parasitization and predation) in Bt and conventional crops; parasitization rates of naturally occurring or sentinel larvae of sensitive (target) lepidopteran species in Bt crops have been reported to be lower compared to control plots. This reduction in parasitism is not surprising given that host populations were significantly decreased by the Bt crop. Predation rates on sentinel lepidopteran eggs or larvae did not differ between Bt and untreated non-Bt fields, but were significantly reduced by the application of broad-spectrum insecticides. Studies in commercially managed cotton fields revealed much higher predation rates in Bt cotton compared to non-Bt fields, where more insecticides were applied (Head *et al.* 2005).

A 6-year field study in Bt cotton on the abundance of 22 arthropod natural enemy taxa indicated that the exposure to the Bt toxin over multiple generations did not cause any chronic long-term effects (Naranjo 2005*a*). As yet, there is little evidence that secondary pest outbreaks in Bt crops have emerged as a problem requiring significant use of insecticides. This confirms that overall biological control is not negatively affected by the use of Bt plants.

The analysis of all published peer-reviewed studies on Bt crops and natural enemies (Romeis *et al.* 2006) has indicated that parasitoid and predator abundance and level of biological control are similar in Bt and untreated non-Bt crops, whereas broad-spectrum insecticides generally lead to a drastic reduction in natural enemy populations as well as a loss of biological control function. It seems, therefore, that in those Bt crops that have been cultivated in commercial production, the Bt technology can contribute to natural enemy conservation and be a useful tool in IPM. Furthermore, it appears that Bt crops and biological control could be used in combination, which may be particularly valuable where the GM-conferred insect resistance is not effective against other key pests. As GM technology evolves, it will be important to evaluate

GM traits conferring insect resistance on a case-by-case basis, to identify any adverse effects on the associated natural enemy fauna, and the compatibility of the GM crop with biological control.

Risks, Benefits and Regulation

Since pest problems in agriculture involve plants, plant-feeding organisms and their natural enemies, the regulation of biological control agents has usually been the responsibility of national plant quarantine services. For this reason, regulation over several decades focused on the need to ensure that introduced natural enemies would not become agricultural pests (Waage 1997). Concern about the risk which introduced biological control agents used against arthropod pests might pose to natural, non-agricultural ecosystems did not become a major issue until much more recently. In contrast, the potential for herbivorous arthropods to become pests when introduced for the biological control of weeds has been recognized for a long time, and pre-release screening procedures, such as host-specificity testing, have been in place for over 40 years (Harris & Zwoelfer 1968). Pre-release screening of entomophagous biological control agents lagged behind their weed counterparts until the warnings by Howarth and, which raised concerns about possible impacts on non-target species. Interestingly, an analysis of published and unpublished information found that only 1.5% of entomophagous biological control agents introduced before 1999 had undergone any pre-release host specificity testing (Lynch *et al.* 2001). Yet, despite the fact that non-target effects of entomophagous biological control agents were rarely considered prior to import and release, there are only a few known cases where any adverse effects can be attributed to the release of such organisms.

The increasing popularity of biological control as an alternative to pesticide-based programmes has resulted in more frequent import and release of exotic entomophagous organisms in many countries (van Lenteren 1997*b*). In the past, Europe has generally been a source rather than a recipient of invertebrate biological control agents, in comparison to other countries with greater experience in classical biological control such as Australia, Canada, New Zealand, South Africa and the USA (Greathead 1976). These countries had legislation and testing procedures in place relatively early to regulate imports and analyse the risks of exotic entomophagous biological control agents (Sheppard *et al.* 2003).

Increasing international trade in agricultural products together with accidental introductions of organisms related to tourism and global trade have become important sources for new imports of exotic pests as demonstrated by Bin & Bruni (1997) for Italy. Many of these introduced organisms are candidates for classical biological control, especially if they establish in conservation areas where they may threaten native species and communities. A significant event in the history of regulation of entomophagous biological control agents was the adoption of the Convention on Biological Diversity following the Rio Conference in 1992. Signatory countries have accepted the obligation to prevent the introduction and to control alien species that threaten indigenous ecosystems and habitats. A broad and ongoing discussion within the scientific community and in public fora was initiated by the publication of the Convention, and biodiversity issues made a sudden appearance on the agendas of policy makers. As a consequence, the public perception of threats to biodiversity has changed markedly in recent years. Biological control, which had previously been viewed as an environmentally friendly method of pest control was subjected to a more critical evaluation, the assumed safety was questioned, and the need for regulation of the import and release of non-native species has become a keenly debated issue.

Harmonization of Regulation of Entomophagous Biological Control Agents

Regulation of the introduction and release of entomophagous biological control agents differs between countries, and some have yet to establish guidelines and procedures. International laws and agreements require a harmonized and effective regulatory system between countries. In many cases, introductions of biological control agents are administered under regulations which were established for other purposes, such as plant quarantine, wildlife conservation and GM plants. The application of appropriate regulatory procedures is important in order to maintain public confidence in biological control and to facilitate introduction and use of non-native species.

A number of regulatory documents for the import and release of invertebrate biological control agents have been produced by international organizations; for example, the Food and Agricultural Organization of the United Nations, the European and Mediterranean Plant Protection Organization and the Organization for Economic Co-operation and Development. More recently, guidance on procedures and methods for environmental risk assessment (ERA) of non-native invertebrate biological control agents has been proposed. The biological control industry was concerned when the OECD guidance document was published, as the information requirements were considered to be too stringent, and manufacturers feared that national authorities would establish their own regulatory systems. As a consequence, the

International Biocontrol Manufacturers Association proposed that the International Organization for Biological Control of Noxious Animals and Plants/West Palaearctic Regional Section (IOBC/WPRS) should coordinate the harmonization of the various regulatory documents. This resulted in the establishment of a Commission of the IOBC/WPRS in 2003 with the aim of offering a platform to scientists, regulators and industry to develop coordinated guidelines for European countries. The recommendations of this Commission were recently published, and can now be adopted and implemented by national authorities.

The regulatory guidelines propose a two-step procedure in which the first stage is to license the import of an organism for research under contained conditions. This enables researchers to identify the key biological characteristics of the organism, such as host range and climatic requirements, which determine the ability of species to establish in new environments. This phase of the regulatory process should provide sufficient information for judgements to be made on the safety and efficacy of the organism. In the second step, companies submit an application for a licence to release the organism, with a dossier of information compiled in line with the guidelines, and including an evaluation of the risks and benefits arising from the proposed release. The risk–benefit analysis should include both positive and negative effects to the economy (farmer, producer and public), human/animal health and the environment.

In contrast to pesticides, biological control agents usually lack the potential to adversely affect human and animal health. There have been some rare cases of adverse effects on human health in allergic reactions to some organisms (for example, to wing scales of Lepidoptera) among personnel working in production facilities. Environmental risks and benefits of biological control extend to a comparative evaluation (with pesticides) of effects on water, soil, air, biodiversity and ecosystem functions. Invertebrate biological control agents do not pollute water, soil or air, and thus the assessment of potential environmental effects have focused on biodiversity issues. However, it is difficult to assign monetary values to the damage or loss of species or ecosystem functions, such that environmental effects have to be considered in a qualitative manner. As a generalization, a strong case can be made for the view that biological control contributes to a reduction in pesticide use with corresponding benefits to the environment. While it should not be assumed that introduced biological control agents pose no risk to native species and ecosystems, history suggests that there have been remarkably few problems, even through the many decades when regulation was virtually non-existent. The recently published regulatory guidelines are intended to produce a careful analysis of the establishment potential of an imported species in a new environment and any adverse effects on non-target organisms and ecosystem functions that may then arise.

Consequences of Regulation for the Use of Biological Control Agents in Sustainable Agriculture

Stakeholders in the area of biological control include scientists, industrial producers, retailers, growers, consumers, environmentalists and government regulators. Different stakeholders may have overlapping or conflicting interests; for example, the biological control industry and national organizations with responsibility for protecting the natural environment. It can be a time-consuming exercise to find a consensus between these differing views, with the net effect that innovative products are kept off the market for a long time, or that some species are never released. The costs of delaying product introduction to the market or the failure to licence a species at all can be calculated in terms of the economic loss for producers, retailers and growers, whereas the benefits of effective regulation can be expressed as 'public goods', like environmental protection. When seen in these terms, it is clearly difficult to make direct comparisons between the costs and benefits of regulation.

Markets for biological control products are usually small and cannot support or justify high levels of expenditure on the evaluation of environmental risks and preparation of the dossiers that are required for registration. The procedure is not only costly but also time consuming; the dossier for the registration of the bacterium *Pseudomonas chlororaphis* used for seed treatment against seed-borne diseases of barley and wheat was submitted in January 1996 under European Union (EU) directive 91/414 and was finally approved in April 2004. The regulatory process can be very costly and therefore may prevent small and medium-sized enterprises from developing biological control products. Only larger companies have the necessary finance to conduct research, produce a dossier and enter into the iterative discussions that are part of the registration procedure. As a consequence of overly stringent regulations, small and medium-sized businesses may be discouraged or prevented from identifying and assessing new species as potential biological control agents. Unnecessary over-regulation thus inhibits the development and introduction of innovative products for sustainable agriculture, and also impacts on start-up and spin-off companies in the biological control sector, as venture capital becomes even more scarce in the light of reduced returns on investment.

The current dilemma is not whether to regulate biological control agents, but how to regulate without introducing negative trade-offs which prevent innovation and limit the introduction of novel biological control agents in sustainable crop protection systems. In essence, the regulatory framework, and therefore associated costs, should be set at the minimum level that does not compromise environmental safety and human health.

An efficient and low-cost regulatory system is likely to be characterized by a number of key features:

(i) applicants and regulators should establish contact at an early stage in the regulatory process and exchange information openly thus preventing unnecessary data generation;

(ii) harmonized regulation among countries (e.g. across Europe) will be cost-effective because companies will be able to submit the same (or slightly modified) risk assessment dossier in different countries;

(iii) governments should support public–private partnership to gather information on risks and benefits of candidate agents; and

(iv) when previous releases in other countries or ecoregions with similar climatic conditions have proved to be effective and environmentally safe, a light-touch process of notification rather than regulation should be applied.

There have been a number of recent developments that hold out the prospect of a harmonized regulatory system in Europe. Firstly, scientists and representatives of the biological control production industry collaborated in an EU-funded project (ERBIC, 'Evaluating Environmental Risks of Biological Control Introductions into Europe') to develop a structure for a general ERA, consisting of

(i) characterization and identification of the biological control agent, and determination of

(ii) risks to human health,

(iii) environmental risks, and

(iv) efficacy.

The ERA was proposed to focus on five risk factors of natural enemies: host/prey range, establishment, dispersal, and direct and indirect effects of releases. It was recognized that the design and application of an ERA depends crucially on the availability of appropriate methods that are both realistic and cost-effective for the industry to apply. In this respect, the recent discovery that the outdoor winter survival (and hence establishment potential) of a number of insect and mite agents used in greenhouses is strongly correlated with their low-temperature survival under laboratory conditions is an important contribution to a risk assessment protocol.

The requirements of an ERA for non-native biological control agents have recently been addressed by Bigler *et al.*, with detailed background information and methodologies for the determination of host ranges, establishment, dispersal, and direct and indirect effects of releases on non-target organisms. Concurrently, the IOBC established a commission to review and amalgamate various risk assessments documents (from FAO, OECD and EPPO) and produce a harmonized regulatory framework for Europe. The current state of affairs with regard to the risks of releasing exotic natural enemies, variation in regulatory systems in different countries worldwide, and proposals for both comprehensive and quick-scan methods for ERA have been reviewed by van Lenteren *et al.*. The EU has recently expressed the view that the application and success of biological control (with macrobial and microbial agents, and semiochemicals) has been lower in Europe than in other parts of the world, and questioned whether this relative lack of success is attributable to a fragmented or over-cautious regulatory process in different European countries. A new EU project (Registration of Biological Control Agents, REBECA) will address these issues and bring forward recommendations for a Europe-wide regulatory framework for these different types of control agents.

Future Perspectives

It may not be an easy task at the present time to convince farmers worldwide to adopt a systems approach to pest management and make greater use of biological control. Yet in an agricultural industry that is still dominated by pesticides, biological control has found its place in the form of augmentative releases, particularly for the management of pests that are difficult to control with insecticides. Each pest species has tens to hundreds of associated natural enemy species (parasitoids, predators and pathogens), and thus thousands of natural enemies still await discovery. During the past 40 years, the identification and pre-release evaluation of natural enemies—including ERAs—has improved greatly, with more than 150 species of natural enemy now commercially available for augmentative biological control. Based on the criteria of current evaluation methods, inefficient or hazardous natural enemies can be identified quickly, and thus avoid unnecessary expenditure on further research. Improvement of networking among the world's biological control community and construction of easily accessible databases containing information on all studied natural enemies (with appropriate evaluation) will help to increase the rate of identification of new and efficient control agents. Once a good natural enemy has been found, it is important to train the extension service and farmers in its use, an aspect of biological control that is often neglected. The FAO Farmers Field School (FFS) projects in Asia and Africa have shown that, as a result of self-learning and experimenting, farmers are capable of quickly selecting the most appropriate pest management strategy for their crops, and rapidly move away from chemicals to cultural methods and biological control (Ooi & Kenmore 2005). An adaptation of the FFS approach might be considered as a means of achieving sustainable pest management in the developed world as well. But biological control practitioners will also have to spend time creating societal awareness about the benefits of sustainable and environmentally friendly pest management, otherwise conventional chemical control will continue to dominate control options.

Though it is evident that biological control programmes have been successfully implemented in a wide range of crop environments in all parts of the world, and the potential to increase the role of biological control is great (including in IPM), it remains the case that the adoption and implementation of this method of control is slow. Owing to technical and economic, but mainly attitudinal barriers, ecologically sound pest management has not been used for a wide variety of pests and diseases. Although the funding for research on biological control is limited, it is our conviction that the main constraints are related to the attitudes of advisory personnel and farmers, combined with the disinterest in anything other than chemical control among the pesticide industries, and the hollow endorsement of alternative methods made by policy makers in which words are yet to be matched by a realistic investment in research and development. We are confident, however, that future pest management will depend strongly on biological control because it is the most sustainable, cheapest and environmentally safest system of pest management, with additional benefits for growers and consumers. Biological control is expected to account for a significantly increased proportion of all crop protection methods by the year 2050.

Bibliography

Alfred Steferud: *Diseases of Fruits and Nuts*, Biotech Books, Delhi, 2005.

Anderson, E. F.: *The Cactus Family*, Timber Press, Portland, Oregon, 2001.

Andrews L.: *Citrus Production - Orange*, St. Augustine, Trinidad and Tobago, 1990

Arora, Chandralekha : *Advances in Horticulture Science Research*, Oxford Book Company, Delhi, 2009.

Benson, L. H.: *The Cacti of the United States and Canada*. Stanford University Press, Stanford, California, 2001.

Benu Singh: *A Modern Book on Forestry and Horticulture*, Vista International Pub, Delhi, 2010.

Bhandari, M.M.: *The Flora of the Indian Desert*. Scientific Publisher, Jodhpur, 1990.

Bhattacharjee, Supriya Kumar: *Advances in Ornamental Horticulture*, Pointer, Delhi, 2006.

Bhutani, R.C.: *Fruit and Vegetable Preservation*, Biotech Books, Delhi, 2003.

Chadha, K.L. : *Advances in Horticulture*, Malhotra Publishing House, Delhi, 2006.

Collymore L.: *Fruit Production in Barbados*, Port of Spain, Trinidad and Tobago, 1996.

Dabholkar, A.R. : *General Plant Breeding*, Concept, Delhi, 2006.

Doijode S. D.: *Seed Germination in Fruits*, New Delhi, Malhotra Publishers, 1993.

Goswami, C L : *Plant Physiology : Research Methods*, Scientific, Delhi, 2007.

Julia F.: *Fruits of Warm Climates*, Miami, Julia F. Morton Publisher, 1987.

Kumar Supriya : *Amenity Horticulture, Biotechnology and Post-Harvest Technology*, Pointer, Delhi, 2006.

Little, E.L.: *Common Fuelwood Crops*, Communi-Tech Assoc., Morgantown, West Virginia, 1983.

Malins A.: *Postharvest Handling of Pineapple and Mango*, Port of Spain, Trinidad and Tobago, 1992.

Mitra S.: *Postharvest Physiology and Storage of Tropical and Subtropical Fruits*, Oxon, CABI, 1997.

Morton, Julia F.: *Fruits of Warm Climates*. Wipf and Stock Publishers, 1987.

Pareek O. P.: *Advances in Horticulture: Fruit Crops,* New Delhi, Malhotra Publishing House, 1993.

Paul B. : *Food Biotechnology in Ethical Perspective*, Aspen, CO: Aspen Publishers, 1997.

Peter, K.V. : *Basics of Horticulture*, New India, Delhi, 2009.

Popenoe, W.: *Manual of Tropical and Subtropical Fruits*. Hafner Press, 1974.

Proctor F. J.: *Onions in Tropical Regions*, Kent, Natural Resources Institute, 1990.

Purohit, S.S. : *Organic Farming : Biocontrol and Biopesticide Technology*, Agrobios, Delhi, 2008.

Saini, M.L. : *Plant Breeding and Crop Improvement*, CBS, Delhi, 1997.

Saroj, P.L. : *Advances in Arid Horticulture*, International Book Dist, Delhi, 2004.

Sharma, V.K. : *Advances in Horticulture : Strategies, Production, Plant Protection and Value Addition*, Deep and Deep, Delhi, 2004.

Singh Z.: *Propagation and Rootstocks of Citrus*, New Delhi, Malhotra Publishers, 1993.

Thomas E.: *Fruit Production in St. Kitts and Nevis*, Port of Spain, IICA, 1996.

Tyagi, I.D. : *Plant Breeding and Genetics at a Glance*, South Asian, Delhi, 2005.

Index

❑❑❑

Handbook of Biological Control and Horticultural Crops

Handbook of Biological Control and Horticultural Crops

Vol. 2

Dr. J.S. Bohra

AGROTECH PRESS

JAIPUR ▪ NEW DELHI

AGROTECH PRESS

Regd. Off. "Bony Residency", Gate No. 2
Opp. Tilak Public School, Vishweswariya Nagar,
Gopalpura Road, Jaipur - 302018, India
Ph: 0141 - 2761280, Telefax: 2761381
E-mail : agrotechpress@gmail.com

Delhi Office: 102, 1st Floor, Satyam House,
4327/3, Ansari Road, Darya Ganj,
New Delhi - 110002, India
Ph: 011-45652440
visit us: www.agrotechpress.com

Distributed By
KEYA INTERNATIONAL PRESS LLP
E-mail: keyainternationalpressllp@gmail.com

Published in India, 2025

ISBN : 9789384568559 (3 Vols. Set)

Typeset by Thomson Press

Printed and bound in India by Thomson Press Ltd.,
New Delhi

Contents

Preface

The term biological control was first coined by Harry Smith of the University of California in relation to the biological control of insects. He defined biological control as the suppression of insect populations by the actions of their native or introduced enemies. Since then, many definitions have been suggested. Most of these are related to the biological control of insects, weeds or diseases. In generic terms, biological control can be defined as a population-levelling process in which the population of one species lowers the number of another species by mechanisms such as predation, parasitism, pathogenicity or competition. In other words, biological control broadly refers to the use of one living organism to curtail the growth and proliferation of another, undesirable one.

Biological control has become an alternate and most important of plant disease management as chemical pesticides possess adverse effect on environment and human health. Biocontrol agents are useful organisms that reduce the adverse effects of phytopathogens and enhance positive response in the plant. Judicious use of biocontrol agents can significantly reduce use and subsequently cost of pesticides and thereby contribute to sustainable development of agriculture. With growing concern for environmental pollution and presence of chemical residues in grains, vegetables, fruits and other food materials, biological disease management tactics have emerged as potential alternative to chemical application for containing crop diseases. Historically, research in biocontrol has yielded discoveries of both fundamental biology and solutions for practical problems in agriculture. Advances on both fronts will be furthered by a better understanding of the complex ecology that surrounds the disease-retardant interactions of microorganisms and plants. The emerging tools of modern biology afford increasingly sophisticated approaches to dissect the multichannel dialogue among the plants, pathogens, biological control agents, and microbial communities that provide the biological context for disease and its suppression. As these research avenues are pursued, new principles of organismal interactions and community function and new strategies for deployment of biocontrol agents will emerge.

Biological management of diseases of crops involves utilization of biotic and abiotic agents that act through one or more mechanisms to reduce the potential of the pathogen directly or indirectly by activating the host defence systems to reduce the disease incidence and/or intensity. Biotic biological control agents include living oomycetes, fungi, bacteria and viruses that have inhibitory effects on the microbial pathogens through various mechanisms of action such as antagonism, competition for nutrients and niches, prevention of colonization of host tissues by the pathogen and induction resistance in plants against the diseases. It is essential to assay the biocontrol potential of all species/isolates of fungal, bacterial and viral BCAs in in vitro, greenhouse, and under field and storage conditions, in addition to their precise identification by biological, immunological and nucleic acid-based assays. Abiotic biological control agents include solarization, physical and chemical agents and those derived from diverse organic and inorganic sources. Organic amendments such as composts, green manures, vegetable wastes, plant extracts and secondary metabolites like essential oils have been shown to have high level of disease-suppressive activity. Chitosan derived from the crab shell, synthetic organic compounds such as SA, ASM, BTH and BABA have been used for treating seeds and plants. Combination of biotic and abiotic agents leads to synergism and consequent improvement in the effectiveness of disease control. Some of the biotic abiotic agents have provided effective disease suppression, when tested under in vitro and field and storage conditions.

This text encompasses current trends in biological control of plant diseases, biomanagement of soil borne plant pathogens through fungal antagonists, management of soil borne diseases of cotton through ecofriendly approaches, biocontrol of fungal disease, biocontrol of post harvest diseases, ecofriendly disease management in spice crops using microbial agents, biological control of seed-borne fungal diseases of legumes, biological control of human pathogenic fungi etc. It also dwells upon the future research challenges in biocontrol move beyond the suppression of plant disease to include impacts on human health and registration processes. By most definitions, biocontrol is an applied field of research involving the human use of a microorganism to enhance crop productivity. But the research generated by the desire to use this practice in agriculture has revealed fascinating biology and led to copious fundamental discoveries. The opportunity to conduct such basic research is coupled with a responsibility to solve the practical problems that prevent the successful deployment of many biocontrol agents. Future trends in biocontrol research will unite fundamental biology with the quest for solutions that will make biocontrol integral to the safe and wise management of every agroecosystem.

Dr. J.S. Bohra

1

Plant Disease Concepts

Plant disease, an impairment of the normal state of a plant that interrupts or modifies its vital functions.

All species of plants, wild and cultivated alike, are subject to disease. Although each species is susceptible to characteristic diseases, these are, in each case, relatively few in number. The occurrence and prevalence of plant diseases vary from season to season, depending on the presence of the pathogen, environmental conditions, and the crops and varieties grown. Some plant varieties are particularly subject to outbreaks of diseases; others are more resistant to them.

Nature and Importance of Plant Diseases

Plant diseases are known from times preceding the earliest writings. Fossil evidence indicates that plants were affected by disease 250 million years ago. The Bible and other early writings mention diseases, such as rusts, mildews, blights, and blast, that have caused famine and other drastic changes in the economy of nations since the dawn of recorded history. Other plant disease outbreaks with similar far-reaching effects in more recent times include late blight of potato in Ireland (1845–60); powdery and downy mildews of grape in France (1851 and 1878); coffee rust in Ceylon (starting in the 1870s); *Fusarium* wilts of cotton and flax; southern bacterial wilt of tobacco (early 1900s); Sigatoka leaf spot and Panama disease of banana in Central America (1900–65); black stem rust of wheat (1916, 1935, 1953–54); and southern corn leaf blight (1970) in the United States.

Loss of crops from plant diseases may result in hunger and starvation, especially in less developed countries where access to disease-control methods is limited and annual losses of 30 to 50 percent are common for major crops. In some years, losses are much greater, producing catastrophic results for those who depend on the crop for food. Major disease outbreaks among food crops have led to famines and mass migrations throughout history.

The devastating outbreak of late blight of potato (*Phytophthora infestans*) that began in Europe in 1845 and brought about the Irish famine caused starvation, death, and mass migration of the Irish population. Of a population of eight million, approximately one million (about 12.5 percent) died of starvation and 1.5 million (almost 19 percent) emigrated, mostly to the United States, as refugees from the destructive blight. This fungus thus had a tremendous influence on the economic, political, and cultural development in Europe and the United States. During World War I, late blight damage to the potato crop in Germany may have helped end the war.

Losses from plant diseases also can have a significant economic impact, causing a reduction in income for crop producers and distributors and higher prices for consumers. In 1993 the United States lost more than one million acres (405,000 hectares) of crops to disease. More than 800,000 acres of wheat succumbed to disease, exacting a monetary loss in the millions of dollars.

Diseases—a normal Part of Nature

Plant diseases are a normal part of nature and one of many ecological factors that help keep the hundreds of thousands of living plants and animals in balance with one another. Plant cells contain special signalling pathways that enhance their defences against insects, animals, and pathogens. One such example involves a plant hormone called jasmonate (jasmonic acid). In the absence of harmful stimuli, jasmonate binds to special proteins, called JAZ proteins, to regulate plant growth, pollen production, and other processes. In the presence of harmful stimuli, however, jasmonate switches its signalling pathways, shifting instead to directing processes involved in boosting plant defence. Genes that produce jasmonate and JAZ proteins represent potential targets for genetic engineering to produce plant varieties with increased resistance to disease.

Humans have carefully selected and cultivated plants for food, clothing, shelter, fibre, and beauty for thousands of years. Disease is just one of many hazards that must be considered when plants are taken out of their natural environment and grown in pure stands under what are often abnormal conditions.

Many valuable crop and ornamental plants are very susceptible to disease and would have difficulty surviving in nature without human intervention. Cultivated plants are often more susceptible to disease than are their wild relatives.

This is because large numbers of the same species or variety, having a uniform genetic background, are grown close together, sometimes over many thousands of square kilometres. A pathogen may spread rapidly under these conditions.

Definitions of Plant Disease

In general, a plant becomes diseased when it is continuously disturbed by some causal agent that results in an abnormal physiological process that disrupts the plant's normal structure, growth, function, or other activities. This interference with one or more of a plant's essential physiological or biochemical systems elicits characteristic pathological conditions or symptoms.

Plant diseases can be broadly classified according to the nature of their primary causal agent, either infectious or noninfectious. Infectious plant diseases are caused by a pathogenic organism such as a fungus, bacterium, mycoplasma, virus, viroid, nematode, or parasitic flowering plant. An infectious agent is capable of reproducing within or on its host and spreading from one susceptible host to another.

Noninfectious plant diseases are caused by unfavourable growing conditions, including extremes of temperature, disadvantageous relationships between moisture and oxygen, toxic substances in the soil or atmosphere, and an excess or deficiency of an essential mineral. Because noninfectious causal agents are not organisms capable of reproducing within a host, they are not transmissible.

In nature, plants may be affected by more than one disease-causing agent at a time. A plant that must contend with a nutrient deficiency or an imbalance between soil moisture and oxygen is often more susceptible to infection by a pathogen; a plant infected by one pathogen is often prone to invasion by secondary pathogens. The combination of all disease-causing agents that affect a plant make up the disease complex. Knowledge of normal growth habits, varietal characteristics, and normal variability of plants within a species—as these relate to the conditions under which the plants are growing—is required for a disease to be recognized.

The study of plant diseases is called plant pathology. Pathology is derived from the two Greek words *pathos* (suffering, disease) and *logos* (discourse, study). Plant pathology thus means a study of plant diseases.

Disease Development and Transmission

Pathogenesis and Saprogenesis

Pathogenesis is the stage of disease in which the pathogen is in intimate association with living host tissue. Three fairly distinct stages are involved:

1. Inoculation: transfer of the pathogen to the infection court, or area in which invasion of the plant occurs (the infection court may be the unbroken plant surface, a variety of wounds, or natural openings—*e.g.,* stomates [microscopic pores in leaf surfaces], hydathodes [stomatelike openings that secrete water], or lenticels [small openings in tree bark])
2. Incubation: the period of time between the arrival of the pathogen in the infection court and the appearance of symptoms
3. Infection: the appearance of disease symptoms accompanied by the establishment and spread of the pathogen.

One of the important characteristics of pathogenic organisms, in terms of their ability to infect, is virulence. Many different properties of a pathogen contribute to its ability to spread through and to destroy the tissue. Among these virulence factors are toxins that kill cells, enzymes that destroy cell walls, extracellular polysaccharides that block the passage of fluid through the plant system, and substances that interfere with normal cell growth. Not all pathogenic species are equal in virulence—that is, they do not produce the same amounts of the substances that contribute to the invasion and destruction of plant tissue. Also, not all virulence factors are operative in a particular disease. For example, toxins that kill cells are important in necrotic diseases, and enzymes that destroy cell walls play a significant role in soft rot diseases.

Many pathogens, especially among the bacteria and fungi, spend part of their life cycles as pathogens and the remainder as saprophytes.

Saprogenesis is the part of the pathogen's life cycle when it is not in vital association with living host tissue and either continues to grow in dead host tissue or becomes dormant. During this stage, some fungi produce their sexual fruiting bodies; the apple scab (*Venturia inaequalis*), for example, produces perithecia, flask-shaped spore-producing structures, in fallen apple leaves. Other fungi produce compact resting bodies, such as the sclerotia formed by certain root- and stem-rotting fungi (*Rhizoctonia solani* and *Sclerotinia sclerotiorum*) or the ergot fungus (*Claviceps purpurea*). These resting bodies, which are resistant to extremes in temperature and moisture, enable the pathogen to survive for months or years in soil and plant debris in the absence of a living host.

Epiphytotics

When the number of individuals a disease affects increases dramatically, it is said to have become epidemic (meaning "on or among people"). A more precise term when speaking of plants, however, is epiphytotic ("on plants"); for animals, the corresponding term is epizootic. In contrast, endemic (enphytotic) diseases occur at relatively constant levels in the same area each year and generally cause little concern.

Epiphytotics affect a high percentage of the host plant population, sometimes across a wide area. They may be mild or destructive and local or regional in occurrence. Epiphytotics result from various combinations of factors, including the right combination of climatic conditions. An epiphytotic may occur when a pathogen is introduced into an area in which it had not previously existed. Examples of this condition include the downy mildews (*Sclerospora* species) and rusts (*Puccinia* species) of corn in Africa during the 1950s, the introduction of the coffee rust fungus into Brazil in the 1960s, and the entrance of the chestnut blight (*Endothia parasitica*) into the United States shortly after 1900.

Also, when new plant varieties are produced by plant breeders without regard for all enphytotic diseases that occur in the same area to some extent each year (but which are normally of minor importance), some of these varieties may prove very susceptible to previously unimportant pathogens. Examples of this situation include the development of oat varieties with Victoria parentage, which, although highly resistant to rusts (*Puccinia graminis avenae* and *P. coronata avenae*) and smuts (*Ustilago avenae, U. kolleri*), proved very susceptible to *Helminthosporium* blight (*H. victoriae*), formerly a minor disease of grasses. The destructiveness of this disease resulted in a major shift of oat varieties on 50 million acres in the United States in the mid-1940s. Corn (maize) with male-sterile cytoplasm (*i.e.,* plants with tassels that do not extrude anthers or pollen), grown on 60 million acres in the United States, was attacked in 1970 by a virulent new race of the southern corn leaf blight fungus (*Helminthosporium maydis* race T), resulting in a loss of about 700 million bushels of corn. More recently the new *Helminthosporium* race was widely disseminated and was reported from most continents. Finally, epiphytotics may occur when host plants are cultivated in large acreages where previously little or no land was devoted to that crop.

Epiphytotics may occur in cycles. When a plant disease first appears in a new area, it may grow rapidly to epiphytotic proportions. In time, the disease wanes, and, unless the host species has been completely wiped out, the disease subsides to a low level of incidence and becomes enphytotic. This balance may change dramatically by conditions that favour a renewed epiphytotic. Among such conditions are weather (primarily temperature and moisture), which may be very favourable for multiplication, spread, and infection by the pathogen; introduction of a new and more susceptible host; development of a very aggressive race of the pathogen; and changes in cultural practices that create a more favourable environment for the pathogen.

Environmental Factors Affecting Disease Development

Important environmental factors that may affect development of plant diseases and determine whether they become epiphytotic include temperature, relative humidity, soil moisture, soil pH, soil type, and soil fertility.

Temperature

Each pathogen has an optimum temperature for growth. In addition, different growth stages of the fungus, such as the production of spores (reproductive units), their germination, and the growth of the mycelium (the filamentous main fungus body), may have slightly different optimum temperatures. Storage temperatures for certain fruits, vegetables, and nursery stock are manipulated to control fungi and bacteria that cause storage decay, provided the temperature does not change the quality of the products. Little, except limited frost protection, can be done to control air temperature in fields, but greenhouse temperatures can be regulated to check disease development.

Knowledge of optimum temperatures, usually combined with optimum moisture conditions, permits forecasting, with a high degree of accuracy, the development of such diseases as blue mold of tobacco (*Peronospora tabacina*), downy mildews of vine crops (*Pseudoperonospora cubensis*) and lima beans (*Phytophthora phaseoli*), late blight of potato and tomato (*Phytophthora infestans*), leaf spot of sugar beets (*Cercospora beticola*), and leaf rust of wheat (*Puccinia recondita tritici*). Effects of temperature may mask symptoms of certain viral and mycoplasmal diseases, however, making them more difficult to detect.

Relative Humidity

Relative humidity is very critical in fungal spore germination and the development of storage rots. *Rhizopus* soft rot of sweet potato (*Rhizopus stolonifer*) is an example of a storage disease that does not develop if relative humidity is maintained at 85 to 90 percent, even if the storage temperature is optimum for growth of the pathogen. Under these conditions, the sweet potato root produces suberized (corky) tissues that wall off the *Rhizopus* fungus.

High humidity favours development of the great majority of leaf and fruit diseases caused by fungi and bacteria. Moisture is generally needed for fungal spore germination, the multiplication and penetration of bacteria, and the initiation of infection. Germination of powdery mildew spores occurs best at 90 to 95 percent relative humidity. Diseases in greenhouse crops—such as leaf mold of tomato (*Cladosporium fulvum*) and decay of flowers, leaves, stems, and seedlings of flowering plants, caused by *Botrytis* species—are controlled by lowering air humidity or by avoiding spraying plants with water.

Soil Moisture

High or low soil moisture may be a limiting factor in the development of certain root rot diseases. High soil-moisture levels favour development of destructive water mold fungi, such as species of *Aphanomyces*, *Pythium*, and *Phytophthora*. Excessive watering of houseplants is a common problem. Overwatering, by decreasing oxygen and raising carbon dioxide levels in the soil, makes roots more susceptible to root-rotting organisms.

Diseases such as take-all of cereals (*Ophiobolus graminis*); charcoal rot of corn, sorghum, and soybean (*Macrophomina phaseoli*); common scab of potato (*Streptomyces scabies*); and onion white rot (*Sclerotium cepivorum*) are most severe under low soil-moisture levels.

Plant Disease Epidemiology

Plant disease epidemiology is the study of disease in plant populations. Much like diseases of humans and other animals, plant diseases occur due to pathogens such as bacteria, viruses, fungi, oomycetes, nematodes, phytoplasmas, protozoa, and parasitic plants. Plant disease epidemiologists strive for an understanding of the cause and effects of disease and develop strategies to intervene in situations where crop losses may occur. Typically successful intervention will lead to a low enough level of disease to be acceptable, depending upon the value of the crop.

Plant disease epidemiology is often looked at from a multi-disciplinary approach, requiring biological, statistical, agronomic and ecological perspectives. Biology is necessary for understanding the pathogen and its life cycle. It is also necessary for understanding the physiology of the crop and how the pathogen is adversely affecting it.

Agronomic practices often influence disease incidence for better or for worse. Ecological influences are numerous. Native species of plants may serve as reservoirs for pathogens that cause disease in crops. Statistical models are often applied in order to summarize and describe the complexity of plant disease epidemiology, so that disease processes can be more readily understood. For example, comparisons between patterns of disease progress for different diseases, cultivars, management strategies, or environmental settings can help in determining how plant diseases may best be managed. Policy can be influential in the occurrence of diseases, through actions such as restrictions on imports from sources where a disease occurs.

In 1963 J. E. van der Plank published “Plant Diseases: Epidemics and Control”, a seminal work that created a theoretical framework for the study of the epidemiology of plant diseases. This book provides a theoretical framework based on experiments in many different host pathogen systems and moved the study of plant disease epidemiology forward rapidly, especially for fungal foliar pathogens. Using this framework we can now model and determine thresholds for epidemics that take place in a homogeneous environment such as a mono-cultural crop field.

Elements of an Epidemic

Disease epidemics in plants can cause huge losses in yield of crops as well threatening to wipe out an entire species such as was the case with Dutch Elm Disease and could occur with Sudden Oak Death. An epidemic of potato late blight, caused by *Phytophthora infestans*, led to the Great Irish Famine and the loss of many lives.

Commonly the elements of an epidemic are referred to as the “disease triangle”: a susceptible host, pathogen, and conducive environment. For disease to occur all three of these must be present. Below is an illustration of this point. Where all three items meet there is disease. The fourth element missing from this illustration for an epidemic to occur, is time. As long as all three of these elements are present disease can initiate, an epidemic will only ensue if all three continue to be present. Any one of the three might be removed from the equation though. The host might out-grow susceptibility as with high-temperature adult-plant resistance, the environment changes and is not conducive for the pathogen to cause disease, or the pathogen is controlled through a fungicide application for instance.

Sometimes a fourth factor of time is added as the time at which a particular infection occurs, and the length of time conditions remain viable for that infection, can also play an important role in epidemics.

The age of the plant species can also play a role, as certain species change in their levels of disease resistance as they mature; a process known as ontogenic resistance.

If all of the criteria are not met, such as a susceptible host and pathogen are present but the environment is not conducive to the pathogen infecting and causing disease, disease cannot occur. For example, corn is planted into a field with corn residue that has the fungus *Cercospora zea-maydis*, the causal agent of Grey leaf spot of corn, but if the weather is too dry and there is no leaf wetness the spores of the fungus in the residue cannot germinate and initiate infection.

Likewise, it stands to reason if the host is susceptible and the environment favours the development of disease but the pathogen is not present there is no disease. Taking the example above, the corn is planted into a ploughed field where there is no corn residue with the fungus *Cercospora zea-maydis*, the causal agent of Grey leaf spot of corn, present but the weather means long periods of leaf wetness, there is no infection initiated.

When a pathogen requires a vector to be spread then for an epidemic to occur the vector must be plentiful and active.

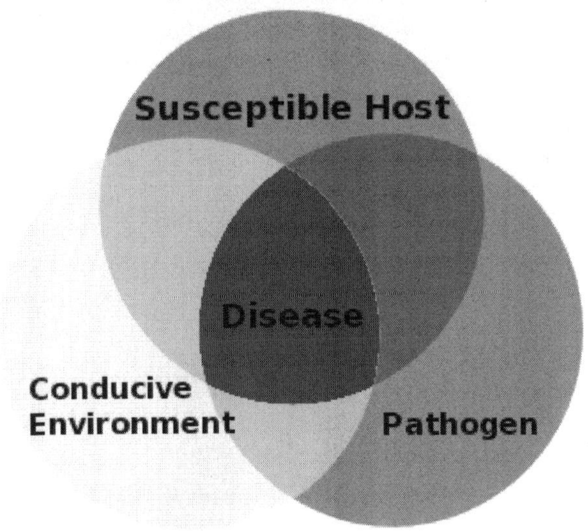

***Figure:** Plant disease triangle illustration*

Types of Epidemics

Monocyclic epidemics are caused by pathogens with a low birth rate and death rate meaning they only have one infection cycle per season. They are typical of soil born diseases such as Fusarium wilt of flax. Polycyclic epidemics are caused by pathogens capable of several infection cycles a season. These are most often caused by airborne diseases such as powdery mildew. Bimodal polycyclic epidemics can also occur. For example in brown rot of stone fruits the blossoms and the fruits are infected at different times.

For some diseases it is important to consider the disease occurrence over several growing seasons, especially if growing the crops in monoculture year after year or growing perennial plants. Such conditions can mean that the inoculum produced in one season can be carried over to the next leading to a build of an inoculum over the years. In the tropics there are no clear cut breaks between growing seasons as there are in temperate regions and this can lead to accumulation of innoculum.

Epidemics that occur under these conditions are referred to as polyetic epidemics and can be caused by both monocylcic and polycyclic pathogens. Apple powdery mildew is an example of a polyetic epidemic caused by a polycyclic pathogen and Dutch Elm disease a polyetic epidemic caused by a monocyclic pathogen.

Classification of Plant Diseases

Plant disease may be grouped as seed - borne, soil borne or air borne. When the disease spreads through seed, soil, or through wind.

1. *Endemic diseases:* Endemic means prevalent in and confined to a particular locality. These diseases are more or less constantly present in a particular area.
2. *Epidemic diseases:* These diseases occur incidentally and occasionally in a particular locality. The word epiphytotic is used particularly for plant diseases instead of epidemic.
3. *Sporadic diseases:* These occur at very irregular intervals and locations.

Modes of Spreading of Diseases

1. *Soil borne diseases:* Inoculums of the diseases causing pathogen remains in soil and penetrate the plant resulting in diseased condition e.g. Root rot, wilt.
2. *Seed borne diseases:* The micro organisms are carried along with seeds and cause diseases when congenial condition occurs. E.g. Damping off.
3. *Air borne diseases:* The micro-organisms are spread through air and attack the plants causing diseases. E.g. Blight, rust, powdery mildew.
4. *Diseases spread by insects:* The viral diseases are spread by insects. The insets which carry the viruses are known as vectors.

Majority of Plant Pathology

The vast majority of plant pathogens are fungi, however, plant diseases are also caused by insects, bacteria, nematodes, viruses and phytoplasmas. Disease-like symptoms can also be caused by abiotic factors, such as temperature, light, chemical agents and water or nutrient deficiencies.

Symptoms of disease include death and destruction of host tissue, wilting, abnormal growth and differentiation and discolouration of host tissue. Some parasites, called necrotrophs, secrete enzymes that kill host tissue, extract nutrients from the cells and then live in the dead tissue. The necrotic lesions caused by pathogens can be localised or extensive. Local necrotic lesions appear as discrete necrotic areas, while extensive, or spreading lesions spread until the whole organ or plant is killed. Wilting occurs when water loss is greater than water intake. It results from either: interference with water and nutrient absorption at the roots, interference with water conduction within the plant (i.e. infection of the vascular tissue), or loss of control of transpiration. Abnormal growth and differentiation results from deviation from the complex balance of interrelated reactions that take place in plants. Parasites can alter the hormonal balance in plants causing an abnormal increase in the size or number of cells, resulting in abnormal growth and differentiation, for example, the formation of galls. Discolouration of tissue is most commonly by chlorosis or mosaics of leaves, both of which can have a number of causes. Anything that interferes with the production of chlorophyll causes leaves to turn yellow, or chlorotic. Mosaicism is a symptom of many virus infections and is characterised by alternating light and dark green areas on the leaves.

Table: *Some of the most common terms used to describe symptoms of plant diseases.*

Blight	A disease characterised by widespread death of plant tissue.
Canker	A sunken necrotic lesion often of a main stem, branch or root.
Damping-off	Collapse and rot of seedlings near soil level before emergence or soon after emergence caused by *Pythium* spp., *Phytophthora* spp., *Fusarium* spp., and *Rhizoctonia* spp.
Dieback	Partial defoliation, twig and branch death and even complete death of plants.
Downy mildew	White or grey 'bloom' on leaves and stems caused by production of sporangiophores and sporangia by members of the Peronosporales (downy mildew fungi).

Contd...

Gall	An abnormal growth or swelling produced as a result of pathogenic invasion.
Mosaic	Patchy variation of normal green colour in leaves, usually light and dark green mosaic, symptomatic of many viral diseases.
Powdery mildew	White powdery 'bloom' on the plant surface caused by the production of fungal mycelium, conidiophores and conidia by members of the Erysiphales (powdery mildew fungi).
Pustule	A blister-like spore mass breaking through a plant epidermis.
Rot	Disintegration of tissue, often caused by enzymes or toxins produced by pathogens.
Rust	Rust-coloured pustules formed by members of the Uredinales (rust fungi).
Scab	A discrete, superficial roughened lesion.
Smut	A disease characterised by black spore masses on leaves, stems or inflorescences, caused by members of the Ustilaginales (smut fungi).
Vascular wilt	A disease in which the pathogen is confined to the vascular system of the host and in which wilting is a characteristic symptom; plants lose their turgidity and become flaccid, leaves collapse.

The amount of disease that develops in a plant community is dependent on properties of the host, the pathogen and the environment. The environment can affect both the susceptibility of the host (e.g. by creating stress in the plant) and the activity of the pathogen (e.g. providing moisture for spore germination). The pathogen and the host can affect each other's performance. The plant can also change its environment, by creating a microclimate around it.

***Table:** Factors that affect disease development.*

Pathogen	*Host*	*Environment*
Presence of pathogen		Temperature
Pathogenicity	Susceptibility	Rainfall / Dew
Adaptability period	Growth stage & form	Leaf wetness
Dispersal efficiency	Population density & structure	Soil properties
Survival efficiency	General health	Wind
Reproductive fitness		Fire history
		Air pollution
		Herbicide damage

The Pathogen

The presence or absence of a pathogen is the main factor that determines whether disease occurs. Introduction of a pathogen to an area from which it has previously been absent can cause major outbreaks of disease in plant communities. The amount of disease that develops is often determined by the pathogenicity of the main pathogen. The term pathogenicity relates to both the virulence (infection ability) and the aggressiveness (the vigour of the infection) of the pathogen. Pathogenicity is dependent on the pathogen's reproductive, dispersal and survival fitness.

The adaptability of the pathogen is also important in determining its ability to infect resistant hosts or to survive changed environmental conditions. Adaptability is determined by the pathogen's genetic flexibility and reproductive efficiency. The spread of a disease and the formation of epidemics is reliant on the pathogen's ability to disperse rapidly over long distances. The spores of cereal rusts, for instance, can be blown over vast distances in a few days, while soil-borne pathogens have little scope for extensive spread. For a pathogen to cause disease in successive seasons, it must be able to survive the intervening time. Some pathogens form spores or sclerotia that can survive in the soil for years, while others colonise alternative plant species until the season of their primary host comes around again. Beyond the mere presence of a pathogen, the number of infective propagules available to infect plants is a crucial factor in determining the amount of disease that develops. Generally, as the number of propagules increases, the level of disease increases, levelling off when the amount of disease reaches very high levels and there are few uninfected plants available. The survival of propagules, and therefore the number of propagules available to cause disease, is heavily influenced by environmental factors.

The Host

The development of disease in a plant community relies on the presence of individual hosts that are susceptible to that particular pathogen. If the majority of the population is susceptible to the pathotypes of a pathogen in the vicinity, an epidemic can occur.The best way of controlling disease is by planting species or cultivars that are not susceptible to pathogens of that area. The occurrence of disease can also be influenced by the host plant's growth stage and form. Some diseases are common in seedlings, while others are typical of mature plants. The growth stage of the population can also affect the microclimate around the plants; for example, the humidity and sunlight levels under the canopy. The population structure and density will also affect the development of disease in a plant community. The density of the main host species and the proportion of other plants that are not hosts within the community will determine the rate and extent of epidemic development. Crop plants tend to be densely planted, with no other species in amongst them, making them more susceptible to rapid spread of disease. Extensive, dense plantations can host spectacular epidemics, particularly if a new pathogen is introduced to the area. In addition, the general health of the host plant before infection is important in determining the success of a disease. Necrotrophs do well on poorly growing plants, while biotrophs thrive on a healthy host plant.

The Environment

The presence of a pathogen against a particular plant will generally not cause serious disease unless the environmental conditions are favourable. This includes the aerial environment and the soil (*edaphic*) environment. Human attempts at controlling disease usually involve manipulating the environment in some way. For example, breeding wheat cultivars to tolerate dry conditions allows Australian farmers to plant the crop in areas that are not favourable for pathogens such as powdery mildew and leaf rust. Properties of the aerial environment that influence disease development include moisture levels, temperature and pollution.

Moisture is particularly important to pathogenic bacteria and fungi. Rain splash plays an important role in the dispersal of some fungi and nearly all bacteria, and a period of leaf wetness is necessary for the germination of most airborne spores. By using water for dispersal, propagules are dispersed at a time when they are likely to be able to germinate as well. Because the process of germination and infection takes time, the duration of leaf wetness also influences the success of the infection. The duration necessary for infection varies with temperature. Usually, a longer period of leaf wetness is needed to establish an infection in cooler temperatures, as germination and infection are generally accelerated in warmer conditions.

Temperature also affects the incubation, or latent, period (the time between infection and the appearance of disease symptoms), the generation time (the time between infection and sporulation), and the infectious period (the time during which the pathogen keeps producing propagules). The disease cycle speeds up at higher temperatures, resulting in faster development of epidemics. The period of leaf wetness, combined with temperature information can be used to predict outbreaks of some diseases (infection periods) and be used to time preventative treatments, such as spraying. A recently recognised aspect of the aerial environment that can influence disease in plants is air pollution. A high concentration of pollutants can affect disease development and, in extreme cases, damage the plants directly by causing acid rain.

The edaphic (soil) environment affects soil-borne diseases, largely by determining the amount of moisture available to pathogens for germination, survival and motility. Germination and infection success also rely on the temperature of the soil. The fertility and organic matter content of the soil can affect the development of disease. Plant defences are weakened by nutrient deficiency, although some pathogens, such as rusts and powdery mildews, thrive on well-nourished plants. Other diseases thrive in soils that are specifically low in organic matter.

Interaction between Factors

The pathogen, the host and the environment interact, usually in ways that are difficult to quantify and predict. Control measures can include sowing of a crop species early, to avoid exposing seedlings to a disease during the time of year that provides the best environmental conditions for the pathogen.

Recognising Infectious Plant Diseases

A plant pathologist or a student taking plant pathology is often asked by friends or associates the following questions. What is wrong with my plant; followed by, what can I do to get rid of the problem? It may be too late to help the specific plant when the question is asked, but proper diagnosis may be extremely important in preventing the problem on other plants or in preventing the problem in the future.

How does a plant pathologist or a plant pathology student go about diagnosing plant problems? The diagnostician must have very good observation skills, and s/he also needs to be a good detective. It is important to keep an open mind until all of the facts related to the problem can be collected. The possibility of multiple causal factors must also be considered.

Control measures depend on proper identification of diseases and of the causal agents. Therefore, diagnosis is one of the most important aspects of a plant pathologist's training. Without proper identification of the disease and the disease-causing agent, disease control measures can be a waste of time and money and can lead to further plant losses. Proper disease diagnosis is therefore vital.

Often, plant pathologists have to rely on symptoms for the identification of a disease problem. For example, Dr. Shoe is asked by Ms. Green to examine azaleas in her nursery. When Dr. Shoe gets to the nursery, he observes that the azaleas in Greenhouse 1 are wilted. When he removes plants from their pots, the roots appear to be rotten. Ms. Green wants to know right away what she should do with the azaleas in Greenhouse 2 where no wilting is currently being observed. Dr. Shoe is asked to make recommendations even before he has looked at plants in the second greenhouse. Because similar symptoms can be produced in response to different causal agents, the use of symptoms alone is often an inadequate method for disease identification. The identification of the disease-causing agent may take a week or more. What can Dr. Shoe do for Ms. Green now?

One of the most important things is for Dr. Shoe to use his powers of observation. He needs to ask many questions related to the azalea's care and culture in order to eliminate or identify possible causes of the problem. He also needs to consider various environmental and cultural factors. As a result of his questions and observations he may:

- Be able to identify a disease and disease-causing agent,
- Be able to narrow the problem down to several possibilities which will require further study in the laboratory before he can make a final diagnosis, or
- Be completely baffled by the problem.

Regardless of the outcome, Ms. Green still expects a recommendation as to what she should do now.

This article presents the various steps/activities which are associated with accurate plant disease diagnosis. The process may vary with different diseases and conditions but the overall process is relatively consistent. The steps all require careful observations and questions. The steps include:

Know What is Normal

Proper plant identification. Identification of affected plants is one of the first steps in diagnosing a plant disease. Both scientific and common names of the plant should be noted. Common names should not be relied upon since some distinctly different plant species may have the same common name, and the common name used in one area may be used for a completely different species in another area. The common name "vinca" has been used to describe plants belonging to two different genera, *Vinca*, a perennial, and *Catharanthus*, an annual. Another example is "monkey grass" which is used to describe *Liriope* and *Ophiopogon* (mondo grass). An example from forestry is "cedar" which is used to describe eastern red cedar (*Juniperus*), western red cedar (*Thuja*), Port Orford cedar (*Chamaecyparis*), incense cedar

(*Libocedrus*), and Atlas cedar (*Cedrus*). Obviously the use of common names can cause confusion in identification and recognition of problems.

In addition to knowing the common and scientific names of an affected plant, it is important to know the specific variety or cultivar, whenever possible. A great variation in susceptibility to a specific disease may occur within different cultivars of a plant species. For example, when we look at the susceptibility of wheat to wheat stem rust caused by *Puccinia graminis* f. sp. *tritici*, we know that all wheat cultivars are not susceptible to all races of *P. graminis*. The major control measure for this disease is based on planting wheat cultivars each year that are resistant to the pathogen races that are predicted to be present during the growing season. Tomato cultivars having the "Better Boy" genetic background are generally resistant to root-knot nematodes while those with the genetic background of the variety "Rutgers" are susceptible, so knowing the genetic background of a cultivar can be important. Knowing the cultivar and its susceptibility to various diseases can narrow down the possible disease agents to consider.

Knowing the identity of the plant species affected allows the pathologist to utilize various resources that contain lists of plant diseases associated with specific plants. These lists are very helpful in suggesting possible pathogenic agents. An example of such a list is found on the American Phytopathological Society (APS) web site as a part of its the online resources. Once you select the plant of interest, you will see a list of bacterial, fungal, nematode, parasitic plants and viral diseases associated with the specific plant. *Westcott's Plant Disease Handbook* is useful because specific symptoms are associated with each disease.

Recognize healthy plant appearance. It is important to know the normal appearance of the plant species you are investigating. Each plant species has special growth habits, colours and growth rates. If you do not know what to expect of the plant you cannot recognize when something is wrong. Does the plant normally have new foliage that is yellow or red and becomes darker green as the foliage ages? Many ornamental shrubs have been developed and marketed for the ornamental value of such brightly coloured new growth. These plants are highly prized for this colouration; however, if an individual does not know that this colouration is the normal appearance of the plant, s/he may think that the plant is diseased. It is important to know what the normal appearance of a plant is before you decide there is a problem. It is also important to remember that appearance can vary with different cultivars. Some plant cultivars have naturally yellow to pale green leaves (e. g. new hosta cultivars, herbs like golden oregano, and coleus varieties) which at first glance appear to have symptoms of under-fertilization, root stress or soil pH problems.

Once the "normal" appearance of the specific plant is determined, several comparisons can be made between the problem plants and healthy plants. Compare characteristics such as overall size, shape, and colouration; leaf shape, size, colouration, and distribution; root distribution and colouration; and bark, stem or trunk texture and colouration. It is also important to note normal events, such as leaf drop, that may occur in a healthy plant. For example, some holly species normally drop leaves in the spring.

The affected parts of the plants should also be noted. Are there symptoms on the roots, leaves, stems, flowers, or fruit? Is the entire plant involved? Is only one limb or side of a plant involved? Answers to these questions can assist in the identification of the problem.

Check for Symptoms and Signs

Identify characteristic symptoms. Describing the characteristic symptoms exhibited by a specimen can be very difficult to do accurately. Because of this, it is often difficult, if not impossible, to determine what is wrong with a plant when a person is describing symptoms over the phone. As a test of this you may want to take a plant exhibiting symptoms and have three different individuals describe the symptoms that they observe on a sheet of paper. Next, compare the descriptions. Do the descriptions vary significantly? Could you visualize the symptoms by the way any one of the individuals described the diseased plant? Symptoms can often be grouped as follows;

- Underdevelopment of tissues or organs. Examples include such symptoms as stunting of plants, shortened internodes, inadequate development of roots, malformation of leaves, inadequate production of chlorophyll and other pigments, and failure of fruits and flowers to develop.
- Overdevelopment of tissues or organs. Examples include: galls on roots, stems, or leaves, witches' brooms, and profuse flowering.
- Necrosis or death of plant parts. These may be some of the most noticeable symptoms, especially when they affect the entire plant, such as wilts or diebacks. Other examples include shoot or leaf blights, leaf spots, and fruit rots.
- Alteration of normal appearance. Examples include mosaic patterns of light and dark green on leaves, and altered colouration in leaves and flowers.

Diseases also involve a progression of symptoms that can vary significantly. The progression of symptoms is one of the most important characteristics associated with problems caused by biotic agents. Diseases can result in primary and secondary symptoms. For example, decayed roots on a tree may be a primary symptom while the toppling over of the tree or windthrow is a secondary symptom.

At later stages of a disease, secondary invaders may also obscure the original disease symptoms so that symptoms observed at the later stages of the disease are not typical of the symptoms developed in response to the original pathogen. It is important to look for a progression of disease symptoms in plants exhibiting problems. In some cases, such as improper herbicide usage, symptoms observed may be similar to spots present as a result of an infectious agent. The difference is that with herbicide injury, the symptoms usually appear suddenly and there is no observable progression of symptoms. The spots may also follow spray patterns of the herbicide. Herbicides, such as 2,4-D, can cause leaf distortion which may be confused with viral diseases. However, when new leaves form, they will generally be free of symptoms, indicating a lack of symptom progression.

Figure: *Peach seedlings infected with various viruses alone or in combination. Peach seedlings infected with both* Prune dwarf virus *and* Prunus necrotic ringspot virus *(seedling on left), infected with* Prune dwarf virus *(seedling in middle) and infected with* Prunus necrotic ringspot virus *(seedling on right). (Used by permission of S. Scott)*

Identify symptom variability. Variations in symptoms expressed by diseased plants may lead to an improper diagnosis. These variations can result from a couple of factors. It is possible that there is more than one problem present, and in some cases there may be more than one pathogen infecting a plant. Symptoms associated with these infected plants may be significantly different from the symptoms expressed in response to each of the different pathogens acting separately. The disease symptoms exhibited by multiple pathogens infecting a plant may be either more severe or less severe than if the plant were infected with just one of the pathogens. This is commonly seen in multiple infections due to viruses. An example of this is shown in figure which shows peach seedlings infected with single or multiple viruses. The seedling on the left is infected with both *Prune dwarf virus* and *Prunus necrotic ringspot virus*. The seedling in the middle is infected with *Prune dwarf virus* alone and the seedling on the right is infected with *Prunus necrotic ringspot virus* alone.

Look for signs of biotic causal agents. Signs of plant disease agents are the observable evidence of the actual disease-causing agent. Signs may include the mycelia of a fungal agent, fungal spores, and spore-producing bodies. Indications of insects causing problems may include the actual insect, insect frass, mite webbing, and insect eggs. Signs are much more specific to disease-causing agents than are symptoms and are extremely useful in the diagnosis of a disease and identification of the agent causing the disease. The use of a hand lens and a knife can be valuable for a diagnostician in the field. Cutting into the bark of ornamental plants and trees at the soil surface may lead to the observation of mycelial mats of root rot fungi such as *Armillaria* spp. Bacterial ooze can be observed by cutting stems and placing them in water. Masses of different spores such as rust spores on leaves can also be important in disease diagnosis. Powdery mildews are typically diagnosed by the observation of the gray to white mycelia and conidia observed on the surface of leaves and flowers

Dissecting and compound microscopes are useful for the observation of specific spores and spore structures, and can lead to further identification of possible disease agents. Knowledge concerning the use of microscopes and a hand lens is vitally important to the diagnostician. Signs of plant disease agents can often be overlooked unless careful observations are conducted. Signs are not visible when taking a quick ride by plants looking through the windshield of a truck and may not even be visible to the naked eye.

Identify Plant Part Affected - Are symptoms associated with specific plant parts?

It is important to note if the symptoms observed are associated with specific plant parts. For example, is a wilt observed correlated with a disruption of the vascular system which may be indicated by browning of the vascular system or are the roots of the plants abnormal including rots, decreased feeder roots, etc.; are necrotic lesions observed strictly on younger leaves? The symptoms of some diseases are most commonly seen on specific plant parts and this observation can be important in diagnosis.

Observe Patterns

Check distribution of symptoms. One of the first things that a diagnostician should note is how the diseased plants are distributed over the affected area. Are they distributed uniformly across an area or are they localized? Is there a definite pattern to the distribution? For example, does it occur only along the edges of a greenhouse near open windows, next to roadways or driveways, in low spots of a field, along a planted row, or is it affecting plants at random in a field? This distribution can be especially important in looking at the possibility of non-infectious problems, such as improper herbicide use or various soil factors. A uniform pattern on an individual plant and uniform damage patterns over a large area are generally not associated with biotic agents, but are usually due to abiotic agents.

How prevalent is the problem? Are all plants affected? Infectious problems generally occur over time and there is a progression of symptoms. Rarely will all of the plants be affected. Generally, disease problems caused by biotic agents will be observed when they are causing problems on a low percentage of plants at least at the start of the disease, unless there were extenuating circumstances, such as the use of infected seeds. Even then, rarely will 100% infection be observed. When a problem appears in 100% of the plants, it more commonly results from factors such as soil conditions (deficiencies or toxicities), adverse climatic factors (cold temperatures, hail, drought, etc.), or toxic chemicals (improper pesticide use, growth regulators, air pollutants, such as ozone, etc.).

What has been the progression of symptoms on plants in the affected area? If the symptoms all appeared at the same time and there has been no further development of symptoms, this would indicate a possible episodic event such as a change in temperature or possible improper chemical usage. If however, the symptoms started in one area and slowly spread to other areas and the severity of disease symptoms changed over time, this would be more indicative of the presence of a biotic agent. Biotic agents can also include insects and mammals, such as voles, which may be feeding on plants in an area.

Check for host specificity. Is the problem occurring in only one plant species or are different plant species affected? If different plant species are affected, this suggests the possibility of a non-infectious problem which could be related to cultural or environmental problems. However, Phytophthora and Pythium root rots can cause problems on many different plant species; therefore, the fact that more than one plant species is affected does not completely eliminate infectious agents. If there is more than one species of plant involved, are these plants closely related and can they be infected by a common pathogen?

Ask Questions

Review the cultural practices and growing environment. It is vital that a diagnostician question the activities that have been conducted around the affected plants. The problem may not be due to anything that the grower has done; the problem could be related to what his/her neighbour has done. Information pertaining to the growing environment to which the affected plant has been exposed is a vital piece of the puzzle. It is especially important to document changes in the environment. Environmental factors to consider include: extreme temperatures (freezing and heat), rainfall, hail, lightning, prolonged drought, temperature inversions (important in possible air pollutant damage and pesticide drift) and prevailing winds. All of these abiotic factors can be important to the problem. Site factors such as soil type, possible drainage problems, and soil pH should also be evaluated.

Cultural and maintenance activities can be significant. What pesticides or other chemicals have been applied? At what rate and when were they applied? Who applied the chemicals? What equipment was used in its application? What other activities have occurred? Has someone been mowing in the area? Has the highway department been working along the roadway, possibly applying herbicides? Have any unusual occurrences or weather patterns been noted? Many times careful investigation by the diagnostician is required because, in some cases, someone may have done something improperly and may be unwilling to admit their error.

Laboratory Tests

Sometimes neither symptoms nor signs provide enough specific or characteristic information to decide the cause of an infectious plant disease. In such cases, it may be necessary to bring a sample back to the laboratory for further tests to isolate and identify the causal agent. This can be a time-consuming and labour-intensive process that takes specialized skills.

Incubation of plant material. One of the first steps when getting back to the laboratory may be to place a sample of the diseased tissue under conditions that will allow an infectious agent to grow and possibly induce sporulation. This can be accomplished by placing a leaf in a moist chamber. A moist chamber can be a sterile petri dish containing a wet filter paper in the bottom of the dish and a triangle of glass tubing on which the sample is placed so that the sample is not directly on the wet filter paper but is exposed to humid conditions. This type of moist chamber will work for small and relatively flat specimens such as leaves. Plastic bags or boxes may be necessary for larger specimens. Saprophytes that are present on the specimen can also be encouraged to grow in a moist chamber and a brief surface swab with 70% isopropanol or 0.1-1% sodium hypochlorite may be useful in reducing these saprophytes. Moist chambers are generally incubated at room temperature.

Isolation and identification of biotic plant disease causal agents. Isolation of fungi usually requires that pieces of infected plant tissue be placed on various nutrient media. The organism that grows out of this tissue is then isolated in pure culture. Bacteria are often isolated by chopping up infected tissue in a small amount of sterile water. This water:bacteria suspension is then streaked onto a bacteriological medium such as nutrient agar. Several problems can occur when trying to isolate the plant pathogenic agent. The infected plant tissue may contain one or more saprophytes which have moved into the infected tissue. These saprophytes may outgrow the plant pathogen on the nutrient medium, obstructing accurate identification of the pathogen. In some cases where a specific plant pathogen is suspected, a medium selective for the suspected pathogen may be utilized. It is also beneficial to attempt to isolate the plant pathogen from the margins of the diseased tissue where the pathogen is more numerous or more active than saprophytes that quickly colonize the recently killed tissue.

Once an organism is isolated, is that organism the true cause of the problem? Conducting Koch's postulates, which involves the inoculation of healthy plants, may be necessary to conclusively answer this question, especially if the organism has not been previously reported as a plant pathogen on that host. Koch's postulates are seldom conducted for routine diagnoses, but may be extremely important for new diseases and for research. Inoculation of a healthy host and obtaining the symptoms originally observed in the field may be difficult. This may be due to problems in replicating the conditions through which the host was inoculated and also in reproducing the environmental conditions present when the host became infected. It is often impossible to replicate in the laboratory the original conditions present during the disease development.

Once an infectious plant pathogen is successfully isolated, the organism must be identified. There are estimated to be some 1.6 million fungal species, most of which are not infectious pathogens. Many fungi and bacteria have never been isolated and identified. The characteristics upon which their identification is based are often complex and specialized training is necessary to be able to identify these fungi and bacteria. Diagnosticians with experience are often able to identify the most commonly isolated organisms. Identification of plant pathogenic nematodes also requires a trained individual.

Diagnostic tests for identification of biotic causal agents. A major problem in identification of biotic causal agents is the inability of some infectious pathogens to grow on artificial media. Viruses, as well as some fungi (e.g. powdery and downy mildew causing agents) and some prokaryotes (e.g. phytoplasmas), require a living host in order to grow. In cases where the plant pathogen is difficult or impossible to grow on artificial media, other methods may be used for their detection, such as the use of serological tests for viruses. Viral identification is often accomplished utilizing ELISA (enzyme-linked immunosorbent assay) which is based on the binding of an antibody produced to a specific virus with the virus in the infected plant material. More tests are currently being developed using the polymerase chain reaction (PCR) for detection of specific organisms. These types of reactions take specialized equipment and reagents, and the tests are not commonly done outside diagnostic and research laboratories. Other techniques used for the identification of viruses include negative staining and electron microscopy to view the viral particles in plant tissue or suspensions.

PCR and ELISA tests, as well as other laboratory tests, may be used for organisms that will grow on artificial media. Additional tests may include analysis of fatty acids of organisms, carbohydrate utilization (i.e. BIOLOG test), and enzyme activity testing (i.e. pectinase, isozyme patterns).

Diagnostic tests for identification of abiotic plant disease causal agents. It is extremely important to look for abiotic factors that may be important in observed symptoms. Soil and water tests may be necessary to determine pH, nutrient composition, salinity, and other factors such as pesticide residues that may induce various symptoms. It may also be important to get samples of plant tissue analyzed for nutrient content to determine if there are macro- or micronutrient deficiencies or toxicities.

Final Diagnosis

Diagnosis is a form of hypothesis testing, where the hypothesis is simply the identity of the disease, and a good diagnostician goes through multiple iterations of the scientific method (seeking evidence through testing that supports or refutes the hypothesis that s/he generates). These hypotheses are generated through observations of the plant, environment, and information from the grower. When all of the information is successfully collected, literature sources should be consulted to determine what is already known about diseases and disease-causing agents associated with the identified plant. Information can be obtained from published resources including plant disease compendia, plant disease indexes, technical notes, commodity newsletters, online resources, and personal communication with plant disease experts. When no information is available on the specific plant, information on diseases and disease causing-agents of similar plants may be useful. There may also be rare cases where no information is available related to the disease. Then, extensive testing may be necessary to determine the identification of the plant pathogen. When this type of testing is required, it may take a long time to develop research-based control recommendations and control measures may have to be based on diseases of similar etiology. If these diseases have occurred in other areas of the world, control measures that have been previously developed in other areas may be useful.

Causal agent identification and diagnosis of plant problems is just like a detective investigating an assault or murder case, only in this case, the victim is a plant. All clues should be investigated. Some clues may lead down blind alleys while others will lead down the correct path. It is important to note that there are exceptions that exist and these exceptions must be considered. It is the compilation of the information and clues that will ultimately lead to the most accurate diagnosis.

The figures that follow illustrate some of the common symptoms that may be produced by different types of problems. Images such as these symptoms and signs are often used in the diagnostic process. Studying these images may assist the diagnostician in narrowing down the possible diseases to consider and others which can be eliminated.

Commonly Observed Symptoms and Signs

Fungal leaf spots - spots usually vary in size. Generally are round and occasionally elongate on stems. Zones of different colour or texture may develop giving the spot a "bull's eye" effect. Spots are not limited by leaf veins.

Bacterial leaf spots - spots are often angular due to limitation by leaf veins. Colour is usually uniform and no signs of plant pathogen are evident. Tissue may appear initially as being water soaked but may become papery as it dries.

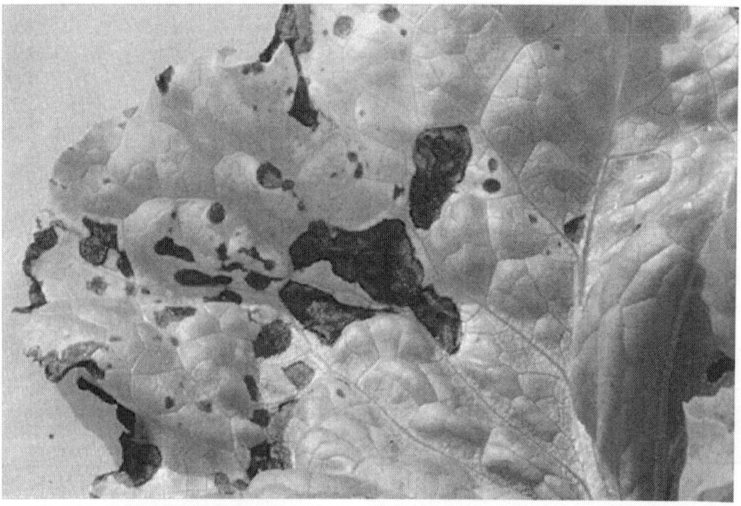

Figure: *Bacterial leaf spot on greenleaf lettuce caused by* Xanthomonas campestris *pv.* vitians.

Vein banding - Vein banding occurs when there is a band of yellow tissue along the larger veins of the leaf. This symptom is observed with viral diseases and is in contrast with nutrient deficiencies which may cause a dark green band along leaf veins.

Mosaic and Ringspot - Mosaic and ringspot are used to describe an irregular patchwork of green and yellow areas over the surface of a leaf. In some cases leaves may also become distorted. Often these symptoms are associated with viral pathogens. There is not a sharp margin between the affected and healthy areas. Distinct margins may indicate a nutritional problem or genetic variegation.

Leaf Distortion - Leaves of the infected plant may be distorted from their normal shape and size. Leaves may be elongated, smaller size, or thickened. This type of symptom can be associated with viral, fungal or bacterial infections as well as insect and mite infestations.

Powdery mildew - can affect leaves, stems, flowers and fruits with a white to gray surface coating of mycelia which can be rubbed off. Black specks may later develop in the mycelia. These specks are mature cleistothecia, the overwintering fungal structures which contain ascospores. Tissue may turn yellow, reddish or remain green under the mycelia and some leaf distortion may be observed especially on actively growing tissues.

Presence of Spores/Spore Structures - Several fungal diseases can be easily identified based on the presence of spores or spore structures on the leaf surface. Some examples of this are rusts which are often recognized by the rusty brown to black spores and smuts which are identified by the black spores which often replace the seed structure.

Cankers - Cankers are localized necrotic lesions which are often sunken in appearance. Cankers can result from mechanical injury (e.g. trees which have been damaged by collisions with cars or lawnmowers), and various fungi or bacteria. In the spring, ooze may be observed on the surface of bacterial cankers and fruiting bodies may be observed on the surface of fungal cankers.

Fruit Decays and Rots - Various fungi and bacteria can cause rots of fruit. These are often distinguished by the colour, lack of firmness of tissue, and signs of spores or fruiting bodies.

Fruit Discolouration - Discolouration of fruit is often associated with viral infections. This discolouration may be similar to mosaic and ringspot symptoms observed on leaves.

Wilts - Wilts are characterized by a general loss of turgidity of leaves or possibly entire plants due to the loss of water. The loss is most often caused by a blocking of the water flow through the xylem. This blockage can be caused by the presence of various bacteria (*Erwinia, Ralstonia*) and fungi (*Fusarium, Verticillium*) in the xylem. Wilts may also be observed when there is a destruction of the root system due to nematodes or root-rotting fungi (*Armillaria, Phytophthora, Pythium*) or an acute water shortage in the soil.

Shoot dieback or blights - Sudden dieback of a shoot usually indicates climatic or chemical injury rather than parasitic problems. If the line between affected and healthy bark is sharp, a soil chemical should be suspected. If dieback is somewhat more gradual and there is a cracking or splitting of the bark and wood, cold injury should be suspected along with bacterial blights caused by *Pseudomonas* or *Erwinia.* A bacterial streaming test with phase and compound microscope and isolation may be required to determine if the blights/dieback is due to a bacterial agent. Gradual decline of shoots and retention of dead leaves is more indicative of a parasitic disease. The margin between the affected and healthy tissue is often irregular and sunken. There may also be small pin-like projections or bumps over the surface of dead bark. These bumps are spore-producing structures of the fungal causal agent.

Dying branches of trees and shrubs - If scattered branches in a tree or shrub start to decline and eventually die, a canker disease or a shoot blight should be suspected. If branches die suddenly and, especially if affected branches are concentrated on one side of tree, weather conditions should be suspected (wind, snow, etc.) or animal or mechanical damage at the trunk base; however, this is not always the case. If symptoms develop over time on one side of a tree or plant then damage of the roots associated with one side may be suspected such as root rots due to *Phytophthora* spp.

Death of tree and shrub top - If all or a major portion of a tree or shrub dies over a period of time, the diagnostician should suspect a problem with the roots. Examples are diseases caused by Armillaria root rot and Verticillium wilt. The decline may be gradual and may eventually affect the entire tree, but in some cases the death may occur on one side of the plant initially. If the decline is sudden, a toxic chemical in soil or weather extremes such as freezing or drought should be suspected.

Overall Stunting or Decline - These symptoms can be caused by several very different factors. Systemic viral infections can result in stunting or decline, but such viral infections are often accompanied by other aboveground symptoms such as shortened internodes. In many cases, overall stunting of a plant may be due to problems associated with the root system. The roots should be examined for rotting and possible mycelial growth, reduction in roots especially feeder roots, and the presence of galls. Root galls can result from fungal and fungallike agents (*Plasmodiophora brassicae)*, nematodes (*Meloidogyne* spp. - root-knot), and bacteria (*Agrobacterium* sp.). Abiotic factors such as nutritional deficiencies, soil compaction and herbicide residues can also result in overall stunting or decline.

Damping-off - This term describes the rapid death and collapse of young seedlings. Often the seedlings will appear to be almost broken at the soil line. It may be observed in flats of plants begun in greenhouses and can result from infection of the seedlings by the fungal organisms *Fusarium, Phytophthora, Pythium, Rhizoctonia*, or *Thielaviopsis*.

Pattern Recognition Receptor

Pattern recognition receptors (PRRs) are a primitive part of the immune system. They are proteins expressed by cells of the innate immune system to identify pathogen-associated molecular patterns (PAMPs), which are associated with microbial pathogens or cellular stress, as well as damage-associated molecular patterns (DAMPs), which are associated with cell components released during cell damage. They are also called pathogen recognition receptors or primitive pattern recognition receptors because they evolved before other parts of the immune system, particularly before adaptive immunity.

Molecules Recognized

The microbe-specific molecules that are recognized by a given PRR are called pathogen-associated molecular patterns (PAMPs) and include bacterial carbohydrates (such as lipopolysaccharide or LPS, mannose), nucleic acids (such as bacterial or viral DNA or RNA), bacterial peptides (flagellin, ax21), peptidoglycans and lipoteichoic acids (from Gram positive bacteria), *N*-formylmethionine, lipoproteins and fungal glucans.

Endogenous stress signals are called danger-associated molecular patterns (DAMPs) and include uric acid.

Classification

PRRs are classified according to their ligand specificity, function, localization and/or evolutionary relationships. On the basis of function, PRRs may be divided into endocytic PRRs or signalling PRRs.

- *Signalling PRRs* include the large families of membrane-bound Toll-like receptors and cytoplasmic NOD-like receptors.
- *Endocytic PRRs* promote the attachment, engulfment and destruction of microorganisms by phagocytes, without relaying an intracellular signal. These PRRs recognize carbohydrates and include mannose receptors of macrophages, glucan receptors present on all phagocytes and scavenger receptors that recognize charged ligands, are found on all phagocytes and mediate removal of apoptotic cells.

PRR Types and Localization

Membrane-bound PRRs

Receptor Kinases: PRRs were first discovered in plants. Since that time many plant PRRs have been predicted by genomic analysis (370 in rice; 47 in Arabidopsis). Unlike animal PRRs, which associated with intracellular kinases via adaptor proteins, plant PRRs are composed of an extracellular domain, transmembrane domain, juxtamembrane domain and intracellular kinase domain as part of a single protein.

Toll-like Receptors (TLR)

Recognition of extracellular or endosomal pathogen-associated molecular patterns is mediated by transmembrane proteins known as toll-like receptors (TLRs). Toll-like receptors were first discovered in *Drosophila* and trigger the synthesis and secretion of cytokines and activation of other host defence programmes that are necessary for innate or adaptive immune responses. TLRs have been found in many species. In mammals, these receptors have been assigned numbers 1 to 11 (TLR1-TLR11). Interaction of TLRs with their specific PAMP induces NF-êB signalling and the MAP kinase pathway and therefore the secretion of pro-inflammatory cytokines and co-stimulatory molecules. Molecules released following TLR activation signal to other cells of the immune system making TLRs key elements of innate immunity and adaptive immunity.

C-type Lectin Receptors (CLR)

Many different cells involved in innate immune system starting from cells as complex as immature dendritic cells to cells as small as platelets express myriad CLRs which shape innate immunity by virtue of their pattern recognition ability. Most classes of human pathogens are covered by CLRs: e.g. among external or infectious pathogens mannose is the recognition motif for many viruses, fungi and mycobacteria; similarly fucose for certain bacteria and helminths; and glucans are present on mycobacteria and fungi. In addition, many of acquired nonself surfaces e.g. carcinoembryonic/oncofetal type neoantigens carrying "internal danger source"/"self turned nonself" type pathogen pattern are also identified and destroyed (e.g. by complement fixation or other cytotoxic attacks) or sequestered (phagocytosed or ensheathed) by the immune system by virtue of the CLRs. The name lectin is also a bit misleading because the family includes proteins with at least one C-type lectin domain (CTLD) which is a specific type of carbohydrate recognition domain. CTLD is a ligand binding motif found in more than 1000 known proteins (more than 100 in humans) and the ligands are often not sugars. If and when the ligand is sugar they need Ca2+ - hence the name "C-type", but many of them don't even have a known sugar ligand thus despite carrying a lectin type fold structure, some of them are technically not "lectin" in function.

CLRs are often endocytic receptors but may be also signalling as well, and depending on the ligand their signalling and trafficking fate might be different. They may have direct signalling including MAP kinase and NFkB activation or may act indirectly by modulating TLRs. CLRs are a very big category and includes soluble PRRs as well that serve many important immune functions.

Membrane receptor CLRs are broadly grouped into:

Group I CLRs: The Mannose Receptors

The mannose receptor (MR) is a PRR primarily present on the surface of macrophages and dendritic cells. The MR belongs to the multilectin receptor protein group and, like the TLRs, provides a link between innate and adaptive immunity. It recognizes and binds to repeated mannose units on the surfaces of infectious agents and its activation triggers endocytosis and phagocytosis of the microbe via the complement system. Specifically, mannose binding triggers recruitment of MBL-associated serine proteases (MASPs). The serine proteases activate themselves in a cascade, amplifying the immune response: MBL interacts with C4, binding the C4b subunit and releasing C4a into the bloodstream; similarly, binding of C2 causes release of C2b. Together, MBL, C4b and C2a are known as the C3 convertase. C3 is cleaved into its a and b subunits, and C3b binds the convertase. These together are called the C5 convertase. Similarly again, C5b is bound and C5a is released. C5b recruits C6, C7, C8 and multiple C9s. C5, C6, C7, C8 and C9 form the membrane attack complex (MAC).

Group II CLRs: Asialoglycoprotein Receptor Family

This is another large superfamily of CLRs that includes molecules like

1. The classic asialoglycoprotein receptor macrophage galactose-type lectin (MGL)
2. DC-SIGN (CLEC4L)
3. Langerin (CLEC4K)
4. Myeloid DAP12 associating lectin (MDL) 1 (CLEC5A)
5. DC associated C type lectin 1 (Dectin1) subfamily which includes
1. dectin 1/CLEC7A
2. DNGR1/CLEC9A
3. Myeloid C type lectin like receptor (MICL) (CLEC12A)
4. CLEC2 (also called CLEC1B)- the platelet activation receptor for podoplanin on lymphatic endothelial cells and invading front of some carcinomas.
5. CLEC12B
6. DC immunoreceptor (DCIR) subfamily which includes:
1. DCIR/CLEC4A
2. Dectin 2/CLEC6A
3. Blood DC antigen 2 (BDCA2) (CLEC4C)
4. Mincle i.e. macrophage inducible C type lectin (CLEC4E)

The nomenclature (mannose versus asialoglycoprotein) is a bit misleading as these the asialoglycoprotein receptors are not necessarily galactose (one of the commonest outer residues of asialo-glycoprotein) specific receptors and even many of this family members can also bind to mannose after which the other group is named !

Cytoplasmic PRRs

NOD-like Receptors (NLR): The NOD-like receptors (NLRs) are cytoplasmic proteins that regulate inflammatory and apoptotic responses. Approximately 20 of these proteins have been found in the mammalian genome and include two major subfamilies called NODs and NALPs, the MHC Class II transactivator (CIITA), IPAF, BIRC1, and other molecules. This family of proteins is greatly expanded in plants, and constitutes a core component of plant immune systems. Some of these proteins recognize endogenous or microbial molecules or stress responses and form oligomers that, in animals, activate inflammatory caspases (e.g. caspase 1) causing cleavage and activation of important inflammatory cytokines such as IL-1, and/or activate the NF-êB signalling pathway to induce production of inflammatory molecules. The NLR family is known under several different names, including the CATERPILLER (or CLR) or NOD-LRR family.

NODs: The ligands are currently known for NOD1 and NOD2. NOD1 recognizes a molecule called meso-DAP, which is a peptidoglycan constituent only of Gram negative bacteria. NOD2 proteins recognize intracellular MDP (muramyl dipeptide), which is a peptidoglycan constituent of both Gram positive and Gram negative bacteria. NODs transduce signals in the pathway of NF-êB and MAP kinases via the serine-threonine kinase called RIP2. NOD proteins are so named because they contain a nucleotide-binding oligomerization domain which binds nucleoside triphosphate. NODs signal via N-terminal CARD domains to activate downstream gene induction events, and interact with microbial molecules by means of a C-terminal leucine-rich repeat (LRR) region.

NALPs: Like NODs, these proteins contain C-terminal LRRs, which appear to act as a regulatory domain and may be involved in the recognition of microbial pathogens. Also like NODs, these proteins also contain a nucleotide binding site (NBS) for nucleoside triphosphates. Interaction with other proteins (e.g. the adaptor molecule ASC) is mediated via N-terminal pyrin (PYD) domain. There are 14 members of this subfamily in humans (called NALP1 to NALP14). Mutations in NALP3 are responsible for the autoinflammatory diseases familial cold autoinflammatory syndrome, Muckle-Wells syndrome and neonatal onset multisystem inflammatory disease. Activators of NALP3 include muramyl dipeptide, bacterial DNA, ATP, toxins, double stranded RNA, paramyxoviruses and uric acid crystals. Although these specific molecules have been shown to activate NALP3, it remains unclear whether this is due to direct binding or due to cellular stress induced by these agents.

Other NLRs: Other NLRs such as IPAF and NAIP5/Birc1e have also been shown to activate caspase-1 in response to Salmonella and Legionella.

RIG-I-like Receptors (RLR)

Further Information: RIG-I-like Receptor: Intracellular recognition of viral double-stranded (ds) and single stranded RNA has been shown to be mediated by a group of RNA Helicases which in turn recruit factors via twin N-terminal CARD domains to activate antiviral gene programmes. Three such helicases have been described in mammals: RIG-I and MDA5 (recognizing 5'triphosphate-RNA and dsRNA, respectively), which activate antiviral signalling, and LGP2, which appears to act as a dominant-negative inhibitor.

Plant PRRs

Plants contain a significant number of PRRs that share remarkable structural and functional similarity with drosophila TOLL and mammalian TLRs. The first PRR identified in plants or animals was the XA21 protein, conferring resistance to the Gram-negative bacterial pathogen Xanthomonas oryzae pv. oryzae. Since that time two other plants PRRs, Arabidopsis FLS2 (flagellin) and EFR (elongation factor Tu receptor)have been isolated. The corresponding PAMPs for XA21, FLS2 and EFR have all been identified. Upon ligand recognition, the plant PRRs transduce "PAMP-triggered immunity" (PTI). Plant immune systems also encode resistance proteins that resemble NOD-like receptors, that feature NBS and LRR domains and can also carry other conserved interaction domains such as the TIR cytoplasmic domain found in Toll and Interleukin Receptors. The NBS-LRR proteins are required for effector triggered immunity (ETI).

Secreted PRRs

A number of PRRs do not remain associated with the cell that produces them. Complement receptors, collectins, ficolins, pentraxins such as serum amyloid and C-reactive protein, lipid transferases, peptidoglycan recognition proteins (PGRs) and the LRR, XA21D are all secreted proteins. One very important collectin is mannan-binding lectin (MBL), a major PRR of the innate immune system that binds to a wide range of bacteria, viruses, fungi and protozoa. MBL predominantly recognizes certain sugar groups on the surface of microorganisms but also binds phospholipids, nucleic acids and nonglycosylated proteins. Once bound to the ligands MBL and Ficolin oligomers recruit MASP1 and MASP2 and initiate the lectin pathway of complement activation which is somewhat similar to the classical complement pathway.

Signalling via PRR

NonRD Kinases: PRRs commonly associate with or contain members of a monophyletic group of kinases called the interleukin-1 receptor-associated kinase (IRAK) family that include Drosophila Pelle, human IRAKs, rice XA21 and Arabidopsis FLS2. In mammals, PRRs can also associate with members of the receptor-interacting protein (RIP) kinase family, distant relatives to the IRAK family. Some IRAK and RIP family kinases fall into a small functional class of kinases termed non-RD, many of which do not autophosphorylate the activation loop. A survey of the yeast, fly, worm, human, Arabidopsis, and rice kinomes (3,723 kinases) revealed that despite the small number of non-RD kinases in these genomes (9%–29%), 12 of 15 kinases known or predicted to function in PRR signalling fall into the non-RD class. In plants, all PRRs characterized to date belong to the non-RD class. These data indicate that kinases associated with PRRs can largely be predicted by the lack of a single conserved residue and reveal new potential plant PRR subfamilies.

Disease Assessment

Critical information in the assessment of disease is the amount of disease that is present. This can be measured as the proportion of a plant community that is diseased (disease incidence) or as the proportion of plant area that is affected (disease severity). Often, disease has to exceed a certain threshold before it reduces the yield of a crop, but it is usually difficult to accurately estimate the yield reduction caused by a specific disease. For example, many diseases occur on senescing tissue that would not have contributed to the yield anyway. Easier diseases to assess are those that kill whole trees in orchards or plantations, and those that destroy the actual harvested product, such as fruit or grain.

Disease and crop loss assessments are necessary for evaluating the economic impact of a disease and the benefit of particular control strategies. There is no point in implementing a control measure if it will cost more that the increased crop yield will return. The growth of the crop, its yield potential, the development of the disease and its impact on yield all have to be measured to predict the impact on yield of particular levels of disease. This information can be combined with predictions of likely disease levels to determine whether preventative treatments are worthwhile.

In the past, most crop loss assessment has been qualitative, producing vague, inaccurate and sometimes misleading data. One major problem with this is the complex nature of disease development. Rarely can disease be attributed to just one factor. Assessment of the effect of disease on crop yield normally involves five steps:

- developing a descriptive growth stage key for the particular crop species
- developing methods to assess the incidence and severity of disease
- developing statistically sound methods of sampling crop populations for assessment of the amount of disease
- estimating the negative impact of particular levels of the disease on crop yield and quality, and
- evaluating the economic benefit from various methods available for reducing the amount of disease.

Assessment of Crop Growth and Development

The first step to quantify the effect of disease is to develop a key that describes the growth and development of healthy plants during the growing season. It should describe development, either from sowing to harvest, in the case of annual plants, or from season to season in perennial plants. Details drawings or photographs, showing characteristics of the various stages of development, including leaf formation, flowering, fruiting and senescence are needed. Standardised growth keys have been developed for a number of crop plants, enabling comparison between different countries and different conditions.

Assessment of Disease Incidence and Severity

Disease assessment methods need to easily provide objective measurements for a specific growth stage of a crop so that data from different sources is comparable, and provide an adequate sample of the crop for assessment. Whether it is disease incidence, or disease severity, or both, that are measured, depends on the nature of the disease. An "all or nothing" disease, for example, that inevitably kills any plant it infects, could be measured just by counting the number of plants that infected (disease incidence). However, in the case of a disease that causes varying degrees of damage to plants throughout the crop, a more complex measurement is needed, that assesses disease severity. The disease incidence for biotrophic pathogens can be measured by counting the number of plants, leaves, flowers etc. that are infected, but the disease severity is assessed by estimating the proportion of total photosynthetic area that is diseased. While this is generally less precise and less controllable than counting individual plants, it is usually a better predictor of crop loss. Because judging the proportion of diseased leaf by eye is unreliable, disease assessment keys, showing different disease severities as blackened areas, have been devised for various crops.

To produce a disease assessment key, the development of disease over the whole disease cycle and at different stages of plant growth must be studied to make prototype standard diagrams and/or descriptions. The accuracy of the key then needs to be tested, by assessing disease severity in the field using the key, and then assessing the same samples using accurate measurement techniques in the laboratory. There are also computer-programmes designed to train observers in disease severity assessment, by presenting images of diseased leaves, which the observer assesses, and comparing their result with the known level of disease in the diagram. This aims to reduce variation in results caused by different observers.

These disease assessment methods will only be accurate if performed on a representative sample of the crop. Samples of crop units (plants, leaves, fruit etc.) can be taken randomly from a crop, or standard quadrats can be placed in the crop and all plants within the quadrat assessed. In a test plot, a part of the plot is usually assessed for disease. Taking samples only from the edge of a plantation will not necessarily be representative of the bulk of the crop. To determine how many samples need to be taken, it is possible to sample a number of times with progressively more samples, and find the point at which the standard error is low. A disease that is uniformly spread throughout a crop will require fewer samples for an accurate assessment than a disease that has a patchy distribution throughout the crop.

Assessment of Crop Losses

Once the amount of disease has been determined, the next step is to assess, either experimentally or statistically, the effect of different levels of disease on the crop yield. Experimental assessment involves setting up test crops in which the level of disease is controlled. Monitoring crops with different levels of disease allows the comparison of epidemic progress and crop yield under different disease conditions. A relationship between disease parametres and yield can then be formulated, allowing prediction of crop loss for a certain level of disease at a particular point in the crop's growth. This method is dependent upon the assumption that the treatments used to keep disease at a certain level have no effect on crop yield themselves. This might not always be the case. Similarly, assessments using susceptible and resistant cultivars rely on both cultivars having a similar yield to start with. Again, this might not be true, and must be determined fist in a disease-free environment. All aspects of experimental design need to be carefully considered in order to gain an accurate assessment of the effect of disease on crop yield. For example, harvesting methods under experimental conditions are often more efficient than under field conditions, giving the appearance of a higher crop yield. From crop loss assessment studies, models can be devised. Crop loss models are usually based on one of three types of disease assessment: disease at a critical point in development, disease at multiple points in development, and disease throughout crop development.

The statistical approach to assessing disease involves statistical analysis of crop yields under different levels of disease that occur naturally in the field. The levels of disease and the crop yields are monitored, but not manipulated, and then yields from different seasons or areas with different levels of disease can be compared to determine the effect of disease on crop yield. The advantage of this method is the use of outcomes of real cropping situations in the field, not experimentally manipulated crops. The disadvantage of this approach is that in the field, the conditions are not controlled, and there are many factors that could affect yield besides disease. An alternative statistical approach to crop loss assessment is the use of questionnaires, filled in by the farmers, that allow disease and pest incidence and severity to be related to the yields they produce.

Biological Control of Plant Pathogens

Pathogenic microorganisms affecting plant health are a major and chronic threat to food production and ecosystem stability worldwide. As agricultural production intensified over the past few decades, producers became more and more dependent on agrochemicals as a relatively reliable method of crop protection helping with economic stability of their operations. However, increasing use of chemical inputs causes several negative effects, i.e., development of pathogen resistance to the applied agents and their nontarget environmental impacts. Furthermore, the growing cost of pesticides, particularly in less-affluent regions of the world, and consumer demand for pesticide-free food has led to a search for substitutes for these products. There are also a number of fastidious diseases for which chemical solutions are few, ineffective, or nonexistent. Biological control is thus being considered as an alternative or a supplemental way of reducing the use of chemicals in agriculture

There has been a large body of literature describing potential uses of plant associated bacteria as agents stimulating plant growth and managing soil and plant health.

Plant growth-promoting bacteria (PGPB) are associated with many, if not all, plant species and are commonly present in many environments. The most widely studied group of PGPB are plant growth-promoting rhizobacteria (PGPR) colonizing the root surfaces and the closely adhering soil interface, the rhizosphere. As reviewed by Kloepper et al. or, more recently, by Gray and Smith, some of these PGPR can also enter root interior and establish endophytic populations. Many of them are able to transcend the endodermis barrier, crossing from the root cortex to the vascular system, and subsequently thrive as endophytes in stem, leaves, tubers, and other organs.

The extent of endophytic colonization of host plant organs and tissues reflects the ability of bacteria to selectively adapt to these specific ecological niches. Consequently, intimate associations between bacteria and host plants can be formed without harming the plant. Although, it is generally assumed that many bacterial endophyte communities are the product of a colonizing process initiated in the root zone, they may also originate from other source than the rhizosphere, such as the phyllosphere, the anthosphere, or the spermosphere.

Despite their different ecological niches, free-living rhizobacteria and endophytic bacteria use some of the same mechanisms to promote plant growth and control phytopathogens.

The widely recognized mechanisms of biocontrol mediated by PGPB are competition for an ecological niche or a substrate, production of inhibitory allelochemicals, and induction of systemic resistance (ISR) in host plants to a broad spectrum of pathogens and/or abiotic stresses. This review surveys the advances of plant-PGPB interaction research focusing on the principles and mechanisms of action of PGPB, both free-living and endophytic bacteria, and their use or potential use for the biological control of plant diseases.

Fungicides and Germicides for Plant Diseases

Fungicides are biocidal chemical compounds or biological organisms used to kill or inhibit fungi or fungal spores. Fungi can cause serious damage in agriculture, resulting in critical losses of yield, quality, and profit. Fungicides are used both in agriculture and to fight fungal infections in animals. Chemicals used to control oomycetes, which are not fungi, are also referred to as fungicides as oomycetes use the same mechanisms as fungi to infect plants.

Fungicides can either be contact, translaminar or systemic. Contact fungicides are not taken up into the plant tissue, and protect only the plant where the spray is deposited; translaminar fungicides redistribute the fungicide from the upper, sprayed leaf surface to the lower, unsprayed surface; systemic fungicides are taken up and redistributed through the xylem vessels. Few fungicides move to all parts of a plant. Some are locally systemic, and some move upwardly.

Most fungicides that can be bought retail are sold in a liquid form. A very common active ingredient is sulfur, present at 0.08% in weaker concentrates, and as high as 0.5% for more potent fungicides.

Fungicides in powdered form are usually around 90% sulfur and are very toxic. Other active ingredients in fungicides include neem oil, rosemary oil, jojoba oil, the bacterium *Bacillus subtilis*, and the beneficial fungus *Ulocladium oudemansii.*

Fungicide residues have been found on food for human consumption, mostly from post-harvest treatments. Some fungicides are dangerous to human health, such as vinclozolin, which has now been removed from use. A number of fungicides are also used in human health care.

Natural Fungicides

Plants and other organisms have chemical defences that give them an advantage against microorganisms such as fungi. Some of these compounds can be used as fungicides:

- Tea tree oil
- Cinnamaldehyde
- Cinnamon essential oil
- Jojoba oil
- Neem oil
- Oregano oil
- Rosemary oil
- Monocerin
- Milk
- Ampelomyces quisqualis AQ10, CNCM I-807

Whole live or dead organisms that are efficient at killing or inhibiting fungi can sometimes be used as fungicides:

- The bacterium *Bacillus subtilis*
- The fungus *Ulocladium oudemansii*
- Kelp (powdered dried kelp is fed to cattle to protect them from fungi in grass)

Resistance

Pathogens respond to the use of fungicides by evolving resistance. In the field several mechanisms of resistance have been identified. The evolution of fungicide resistance can be gradual or sudden. In qualitative or discrete resistance, a mutation (normally to a single gene) produces a race of a fungus with a high degree of resistance. Such resistant varieties also tend to show stability, persisting after the fungicide has been removed from the market. For example, sugar beet leaf blotch remains resistant to azoles years after they were no longer used for control of the disease. This is because such mutations often have a high selection pressure when the fungicide is used, but there is low selection pressure to remove them in the absence of the fungicide.

In instances where resistance occurs more gradually, a shift in sensitivity in the pathogen to the fungicide can be seen. Such resistance is polygenic – an accumulation of many mutation in different genes each having a small additive effect. This type of resistance is known as quantitative or continuous resistance. In this kind of resistance, the pathogen population will revert to a sensitive state if the fungicide is no longer applied.

Little is known about how variations in fungicide treatment affect the selection pressure to evolve resistance to that fungicide. Evidence shows that the doses that provide the most control of the disease also provide the largest selection pressure to acquire resistance, and that lower doses decrease the selection pressure.

In some cases when a pathogen evolves resistance to one fungicide, it automatically obtains resistance to others – a phenomenon known as cross resistance. These additional fungicides are normally of the same chemical family or have the same mode of action, or can be detoxified by the same mechanism. Sometimes negative cross resistance occurs, where resistance to one chemical class of fungicides leads to an increase in sensitivity to a different chemical class of fungicides. This has been seen with carbendazim and diethofencarb.

There are also recorded incidences of the evolution of multiple drug resistance by pathogens – resistance to two chemically different fungicides by separate mutation events. For example, *Botrytis cinerea* is resistant to both azoles and dicarboximide fungicides.

There are several routes by which pathogens can evolve fungicide resistance. The most common mechanism appears to be alteration of the target site, in particular as a defence against single site of action fungicides. For example, Black Sigatoka, an economically important pathogen of banana, is resistant to the QoI fungicides, due to a single nucleotide change resulting in the replacement of one amino acid (glycine) by another (alanine) in the target protein of the QoI fungicides, cytochrome b. It is presumed that this disrupts the binding of the fungicide to the protein, rendering the fungicide ineffective.

Upregulation of target genes can also render the fungicide ineffective. This is seen in DMI-resistant strains of *Venturia inaequalis*.

Resistance to fungicides can also be developed by efficient efflux of the fungicide out of the cell. *Septoria tritici* has developed multiple drug resistance using this mechanism. The pathogen had 5 ABC-type transporters with overlapping substrate specificities that together work to pump toxic chemicals out of the cell.

In addition to the mechanisms outlined above, fungi may also develop metabolic pathways that circumvent the target protein, or acquire enzymes that enable metabolism of the fungicide to a harmless substance.

Fungicide Resistance Management

The fungicide resistance action committee (FRAC) has several recommended practices to try to avoid the development of fungicide resistance, especially in at-risk fungicides including *Strobilurins* such as azoxystrobin.

Products should not be used in isolation but rather as mixture, or alternate sprays, with another fungicide with a different mechanism of action. The likelihood of the pathogen's developing resistance is greatly decreased by the fact that any resistant isolates to one fungicide will be killed by the other; in other words, two mutations would be required rather than just one. The effectiveness of this technique can be demonstrated by Metalaxyl, a phenylamide fungicide. When used as the sole product in Ireland to control potato blight (*Phytophthora infestans*), resistance developed within one growing season. However, in countries like the UK where it was marketed only as a mixture, resistance problems developed more slowly.

Fungicides should be applied only when absolutely necessary, especially if they are in an at-risk group. Lowering the amount of fungicide in the environment lowers the selection pressure for resistance to develop.

Manufacturers' doses should always be followed. These doses are normally designed to give the right balance between controlling the disease and limiting the risk of resistance development. Higher doses increase the selection pressure for single-site mutations that confer resistance, as all strains but those that carry the mutation will be eliminated, and thus the resistant strain will propagate. Lower doses greatly increase the risk of polygenic resistance, as strains that are slightly less sensitive to the fungicide may survive.

It is also recommended that where possible fungicides are used only in a protective manner, rather than to try to cure already-infected crops. Far fewer fungicides have curative/eradicative ability than protectant. Thus, fungicide preparations advertised as having curative action may have only one active chemical; a single fungicide acting in isolation increases the risk of fungicide resistance. It is better to use an integrative pest management approach to disease control rather than relying on fungicides alone. This involves the use of resistant varieties and hygienic practices, such as the removal of potato discard piles and stubble on which the pathogen can overwinter, greatly reducing the titre of the pathogen and thus the risk of fungicide resistance development.

A satisfactory fungicide must be one that does not injure the plants and at the same time is effective against the parasite. For spraying, additional requirements are imposed; it should not dissolve readily in rain water; it should adhere to foliage and fruit; in some cases it should be colourless in order not to make ornamentals more unsightly than when diseased. The fungicide which has been used most for general purposes is bordeaux mixture. Lately some other preparations, particularly lime-sulfur combinations, have come into use, and in many cases are supplanting bordeaux. There are in addition a large number of other substances which have fungicidal value and are in more limited use for specific cases.

Practices

Destroying affected parts. - It is important that all affected parts should be removed and burned, if possible. In the fall all leaves and fruit that have been attacked by fungi should be raked up and burned. Diseased branches should be severed at some distance below the lowest visible point of attack. Fungous diseases often spread rapidly, and prompt action is usually necessary. Practice clean and tidy culture.

Rotation of crops. - One of the most effective and practical means of heading off fungous diseases. Especially applicable to diseases of roots or root-crops, but also to many other diseases of annual plants.

Sterilizing by steam. - An effective fungicidal practice for several soil-inhabiting organisms which attack roots and stems. This includes nematode worms. It is especially applicable in the greenhouse, where it may be applied (a) through sub-irrigation tile or through specially laid perforated steam pipes in the bottom of the bed. Cover the beds with blankets, introduce steam under pressure of 40 to 80 pounds for two hours. Insert thermometres at various places to see that the soil is being uniformly heated. (6) A large galvanized iron tight box may be constructed with finely perforated trays 4 to 6 inches in depth. Soil placed in these trays and steamed for two hours as above will be freed from parasitic organisms. In this case the frames should be sprayed with a solution of formalin, 1 pint in 10 gallons of water.

Steam sterilization of soil may be used on intensively cultivated areas or extensive seed-beds. A portable boiler is necessary. The beds are sterilized after they have been prepared for seed, and just before the seed is sown. A galvanized pan 10 by 6 feet and 6 inches deep is inverted, and the edges are pushed down into the soil one or two inches. The pan is connected with the steam boiler by means of a steam hose and live steam is run into the pan from about forty minutes under a pressure of 100 pounds and up. The higher the pressure the more thoroughly the soil will be sterilized.

The cost of sterilizing is approximately three-fourths of a cent the square foot. It should be noted that soil sterilization has an invigorating effect on the plants, and it will be necessary to run greenhouses at a lower temperature (5°-10°) both night and day. Field sterilization also kills weed seeds, and with the reduction of the cost of weeding makes the process practicable.

Substances

Bordeaux mixture. - A bluish-green copper compound that settles out when freshly slaked lime and a solution of copper sulfate (blue254 FUNGICIDES AND GERMICIDES FOR PLANT DISEASES stone) are mixed. Many formulas have been recommended and used. The 5-5-50 formula may be regarded as standard. In such a formula the first figure refers to the number of pounds of copper sulfate, the second to the stone or hydrated lime, and the third to the number of gallons of water. Bordeaux must often be used as weak as 2-2-50, on account of injury to some plants.

Bacterial Diseases in Plants - Control & Prevention

Just like people contracting bacterial disease (strep throat, TB, salmonella poisoning), plants can also be effected by bacterial disease. Bacterial diseases in plants may affect stems, leaves or roots or be carried internally. Generally, they belong to the genera *Erwinia, Pectobacterium, Pantoea, Agrobacterium, Pseudomonas, Ralstonia, Burkholderia, Acidovorax, Xanthomonas, Clavibacter, Streptomyces, Xyllella, Spiroplasma,* and *Phytoplasma.*

A bacterial disease may cause a variety of symptoms: blights, cankers, galls, leaf spots, overgrowths, specks, scabs, or wilts. Generally, the common name of the disease is a combination of the symptom or appearance and its location on the plant, like Bacterial Leaf Spot caused by *Pseudomonas cichorii* and fireblight in pears and apples, caused by *Erwinia amylovora.* In contrast to viruses that live inside plant cells, bacteria grow in the spaces between cells, producing toxins, special proteins or enzymes that damage the plant cells. *Agrobacterium* causes cells to genetically modify, producing cancer-like growths called galls.

Bacterial diseases are spread in many ways – rain, wind, birds or insects. People can also spread bacterial diseases by using infected pruning tools, by improper disposal of infected plant material, improperly managing plants in the winter, or introducing infected plants in an area. Bacteria require a wound or natural opening like stomata to get inside a plant to cause damage. Once inside, they kill host cells. Bacteria are hardy – if when spread, they find no ready host, they can go dormant until a suitable host is found.

Control of Bacterial Disease

Bacterial diseases are difficult to control; usually it is better to prevent the spread rather than cure a plant. Of course, in agricultural applications it is sometimes necessary to both prevent and treat on a large-scale basis.

- Choose plants that are resistant; use disease-free seed or plants.
- Disinfect pruning tools. Botani-Wipes are handy for this.
- Plan crop rotations to eliminate over-wintering.
- Prevent surface wounds on plants.
- Expose plant material you wish to compost to dry air, heat and sunlight to kill any disease-causing bacteria.
- After treating with any substance that kills bacteria, always replace them with good bacteria such as those in EM-1 or Compost Tea – otherwise any leftover disease causing bacteria will take over again!
- Control insects that might vector bacterial diseases.
- Water correctly; avoid wetting leaves on indoor plants. Misting may cause bacterial growth on leaves.
- Give plants plenty of room, whether in a pot or in the ground.
- Pay attention to indoor plant care in the winter because of lower light intensity.

- Isolate the diseased plant if possible and prune infected leaves. Do not prune more than 1/3 of the leaves of a plant.
- If the disease is systemic, affecting the stem as well as leaves, the plant cannot recover. Destroy it to prevent the spread of the disease to other plants.

Make sure the soil is healthy with plenty of organic nutrients and beneficial microorganisms.

Insects Affected Plants and Their Control

Insects are responsible for two major kinds of damage to growing crops. First is direct injury done to the plant by the feeding insect, which eats leaves or burrows in stems, fruit, or roots. There are hundreds of pest species of this type, both in larvae and adults, among orthopterans, homopterans, heteropterans, coleopterans, lepidopterans, and dipterans. The second type is indirect damage in which the insect itself does little or no harm but transmits a bacterial, viral, or fungal infection into a crop. Examples include the viral diseases of sugar beets and potatoes, carried from plant to plant by aphids.

Although most insects grow and multiply in the crop they damage, certain grasshoppers are well-known exceptions. They can exist in a relatively harmless solitary phase for a number of years, during which time their numbers may increase. They then enter a gregarious phase, forming gigantic migratory swarms, which are transported by winds or flight for hundreds or thousands of miles. These swarms may completely destroy crops in an invaded region. The desert locust (*Schistocerca gregaria*) and migratory locust (*Locusta migratoria*) are two examples of this type of life cycle.

Medical Significance

Insect damage to man and livestock also may be direct or indirect. Direct human injury by insect stings and bites is of relatively minor importance, although swarms of biting flies and mosquitoes often make life almost intolerable, as do biting midges (sand flies) and salt-marsh mosquitoes. Persistent irritation by biting flies can cause deterioration in the health of cattle. Some blowflies, in addition to depositing their eggs in carcasses, also invade the tissue of living animals including humans, a condition known as myiasis. An example of an insect that causes this condition is the screwworm fly (*Cochliomyia*) of the southern U.S. and Central America. In many parts of the world, various blowflies infest the fleece and skin of living sheep. This infestation, called sheep-strike, causes severe economic damage.

Many major human diseases are produced by microorganisms conveyed by insects, which serve as vectors of pathogens. Malaria is caused by the protozoan *Plasmodium*, which spends part of its developmental cycle in *Anopheles* mosquitoes. Epidemic relapsing fever, caused by spirochetes, is transmitted by the louse *Pediculus*. Leishmaniasis, caused by the protozoan *Leishmania*, is carried by the sand fly *Phlebotomus*. Sleeping sickness in humans and a group of cattle diseases that are widespread in Africa and known as nagana are caused by protozoan trypanosomes transmitted by the bites of tsetse flies (*Glossina*). Under nonsanitary conditions the common housefly *Musca* can play an incidental role in the spread of human intestinal infections (e.g., typhoid, bacillary and amebic dysentery) by contamination of food. The tularemia bacillus can be spread by deerfly bites, the bubonic plague bacillus by fleas, and the epidemic typhus rickettsia by the louse *Pediculus*. Various mosquitoes spread viral diseases (e.g., several encephalitis diseases; dengue and yellow fever in humans and other animals).

The relationships among the various organisms are complex. Malaria, for example, has a different epidemiology in almost every country in which it occurs, with different *Anopheles* species responsible for its spread. These same complexities affect the spread of sleeping sickness. Some relationships are indirect. Plague, a disease of rodents transmitted by flea bites, is dangerous to humans only when heavy mortality among domestic rats forces their infected fleas to attack people, thereby causing an outbreak of plague. Typhus, tularemia, encephalitis, and yellow fever also are maintained in animal reservoirs and spread occasionally to humans.

2

Garden, Lawn and Landscape Pests Diseases

Disease Control in the Landscape

Landscape plants represent a substantial investment in the aesthetic appearance of the home and commercial landscape. Plant diseases can affect the survival or aesthetic value of these plants. Since landscape plants are generally costly, landscape managers and homeowners need to be aware that preventive maintenance to control diseases, or plant health care is less expensive than buying replacement plants.

Understanding Plant Diseases

The causes of plant diseases in the landscape are often misunderstood. This is because they may be incited by microbes (pathogens) that are not easily seen, and because the diseases result from complex interactions between the landscape plants, the disease-causing pathogens, and the environment affecting the interactions. Homeowners and landscape managers need to be aware of several important concepts regarding plant diseases.

Landscape plants differ in their disease susceptibility or resistance. Many nurseries label disease resistant cultivars (e.g., flowering crabapples) for the customer's information. Furthermore, susceptibility of many landscape plants to disease can be altered by the growing conditions. Plants having low energy reserves due to drought, injuries, and poor nutrition may have little energy left to fight diseases.

Many pathogens such as fungi and bacteria, are living microbes, and cause infections and disease. However, some "diseases" such as iron deficiency chlorosis, limb dieback, and tree decline can be caused by adverse growing conditions such as soil nutrient or pH imbalance, construction injuries, or soil compaction.

- Disease-causing microbes often exist unseen in a dormant form, waiting for the right conditions to occur before beginning an infection. Some microbes that cause disease are almost always present, no matter how the landscape is managed (e.g., rose black spot), while others can be kept out of the landscape by wise management (e.g., crown gall).
- Management of the landscape environ-ment can have a profound effect on whether or not a disease will occur, and how damaging it will be. In general, rainy, foggy weather and poor drying conditions favour foliar diseases, and wet soils favour most root decay diseases . Almost any change in the way the landscape is managed will alter the disease situation - sometimes for benefit, and sometimes for harm.
- Many cultural practices are useful as disease management tools, but biological methods for disease control are limited.

There are good reasons why homeowners and landscape managers prefer biological and cultural practices for andscape disease management. Reduced uses of chemical pesticides in urban landscapes have evident environmental, worker safety, and public health value. Landscape diversity lends itself to custom tailoring of site specific disease control methods, which often favour biological and cultural techniques. Some cultural practices provide broad-spectrum disease management.

- The landscape environment can be modified to control diseases. Of the three elements needed for diseasedevelopment (susceptible host, virulent pathogen, and favourable environment), changing a disease-favourable environment is often overlooked as a disease control procedure. A good plant-growing environ-ment unfavourable for disease development should be created prior to planting. However, even the environment of mature landscapes can be altered to reduce diseases of landscape plants.
- There are many ways to decrease the pathogen in the landscape to control diseases.

When a disease is found on plants in the landscape, the normal first reaction of homeowners and landscape managers is to do something to directly overpower the pathogen, the microbe that is causing the infection. In many cases, after the infection has already occurred, it is too late to harm the pathogen. There are biological and cultural activities that can be accomplished in the land-scape to decrease or reduce the spread of the pathogen before it has a chance to cause disease.

- Chemical control for diseases should be used only when needed. Most diseases of landscape plants do not require annual or regular chemical applications.

Consider a routine chemical application for disease control only when non-chemical control measures are inadequate and when the disease is a known threat to irreplaceable plants in your landscape. A known landscape plant disease threat usually fits into one of the following categories:

- o It has occurred on your plants in previous seasons, e.g., anthracnose diseases of shade trees or juniper tip blight.
- o It is a lethal or ruinous disease present in nearby landscapes, e.g., Dutch elm disease or pine wilt disease.
- o It is so devastating and common that routine protection is essential on susceptible cultivars, e.g., rose black spot or flowering crabapple scab.
- o Best cultural practices have been used but the disease is still active.

Many problems of woody plants result from poor growing practices and not from diseases. For example, girdling roots that develop due to planting too deeply or piling mulch high against the trunk are the result of poor growing practices. Attempting to grow plants not suited to Kentucky climate or in poor growing sites will result in sick plants due to lack of knowledge or judgement. Careless use of lawn or weed trimming equipment and construction activity can cause injuries which impact plant health.

Plant Disease Forecasting

Plant disease forecasting is a management system used to predict the occurrence or change in severity of plant diseases. At the field scale, these systems are used by growers to make economic decisions about disease treatments for control. Often the systems ask the grower a series of questions about the susceptibility of the host crop, and incorporate current and forecast weather conditions to make a recommendation. Typically a recommendation is made about whether disease treatment is necessary or not. Usually treatment is a pesticide application.

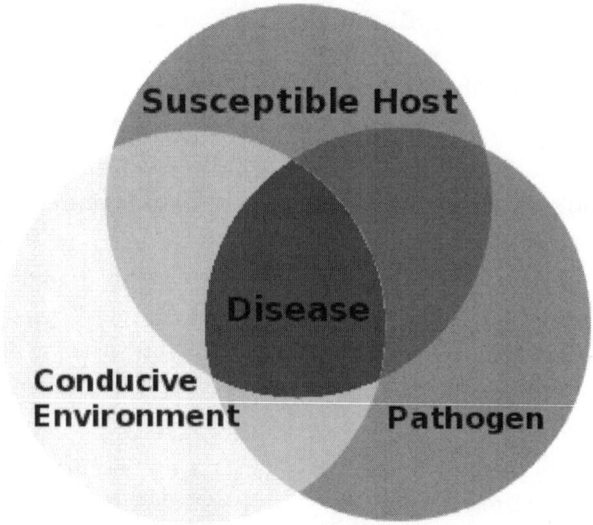

Figure: *The plant disease triangle represents the factors necessary for disease to occur*

Forecasting systems are based on assumptions about the pathogen's interactions with the host and environment, the disease triangle. The objective is to accurately predict when the three factors - host, environment, and pathogen - all interact in such a fashion that disease can occur and cause economic losses.

In most cases the host can be suitably defined as resistant or susceptible, and the presence of the pathogen may often be reasonably ascertained based on previous cropping history or perhaps survey data. The environment is usually the factor that controls whether disease develops or not.

Environmental conditions may determine the presence of the pathogen in a particular season through their effects on processes such as overwintering. Environmental conditions also affect the ability of the pathogen to cause disease, e.g. a minimum leaf wetness duration is required for grey leaf spot of corn to occur. In these cases a disease forecasting system attempts to define when the environment will be conducive to disease development.

Good disease forecasting systems must be reliable, simple, cost-effective and applicable to many diseases. As such they are normally only designed for diseases that are irregular enough to warrant a prediction system, rather than diseases that occur every year for which regular treatment should be employed.Forecasting systems can only be designed if there is also an understanding on the actual disease triangle parametres.

Examples of Disease Forecasting Systems

Forecasting systems may use one of several parametres in order to work out disease risk, or a combination of factors. One of the first forecasting systems designed was for Stewart's Wilt and based on winter temperature index as low temperatures would kill the vector of the disease so there would be no outbreak.

An example of a multiple disease/pest forecasting system is the EPIdemiology, PREdiction, and PREvention (EPIPRE) system developed in the Netherlands for winter wheat that focused on multiple pathogens. USPEST.org graphs risks of various plants diseases based on weather forecasts with hourly resolution of leaf wetness. Forecasting models are often based on a relationship like simple linear regression where x is used to predict y. Other relationships can be modelled using population growth curves. The growth curve that is used will depend on the nature of the epidemic. Polycyclic epidemics such as potato late blight are usually best modelled by using the logistic model, whereas monocyclic epidemics may be best modelled using the monomolecular model. Correct choice of a model is essential for a disease forecasting system to be useful.

Plant disease forecasting models must be thoroughly tested and validated after being developed. Interest has arisen lately in model validation through the quantification of the economic costs of false positives and false negatives, where disease prevention measures may be used when unnecessary or not applied when needed respectively. The costs of these two types of errors need to be weighed carefully before deciding to use a disease forecasting system.

Future Developments

In the future, disease forecasting systems may become more useful as computing power increases and the amount of data that is available to plant pathologists to construct models increases. Good forecasting systems also may become increasingly important with climate change. It will be important to be able to accurately predict where disease outbreaks may occur, since they may not be in the historically known areas.

Disease Pest Identification

Identifying plant disease in the garden, lawn, and landscape begins by investigating the host plant and looking for clues. The following resources provide information, many through pictures, that aid in identifying the disease pest.

Garden Disease Identification

- Vegetable Disease Problems by Region
- Fruit Disease Pests by Region

Lawn Disease Identification

- Turfgrass Diseases in Georgia: Identification and Control (University of Georgia Cooperative Extension)

Landscape (Tree and Shrub) Disease Identification

- Tree and Shrub Disease Problems by Region

Image Resources

Pictures can be an important tool in helping you diagnose problems in the landscape.

- Oregon State University Image Index (Oregon State University Extension) provides a searchable index of plant disease images
- Integrated Pest Management Images (ipmimages.org) offers an extensive collection of images on many topic, including diseases, insects, and other pests
- Invasive Diseases (www.invasive.org) documents many of the most troublesome diseases that originated in other parts of the world

Local Diagnostic Laboratories

If you cannot identify your disease problem with the resources provided, you may be able to submit information to your state diagnostic plant laboratory using:

- Plant clinic listings @ APSNet - University Diagnosticians (Maintained by the American Phytopathological Society)

Fungicide Use and Products

While plant diseases may be caused by a number of organisms, such as bacteria, fungi, viruses, or phytoplasmas, only fungal plant diseases can be effectively controlled through the use of chemical pesticides. A fungal pesticide is called a fungicide. Fungicides and their use vary by state.

Biological Control of Rice Blast Disease (Piricularia Oryzae)

Magnaporthe grisea, also known as rice blast fungus, rice rotten neck, rice seedling blight, blast of rice, oval leaf spot of graminea, pitting disease, ryegrass blast, and Johnson spot, is a plant-pathogenic fungus that causes a serious disease affecting rice. It is now known that

M. grisea consists of a cryptic species complex containing at least two biological species that have clear genetic differences and do not interbreed. Complex members isolated from *Digitaria* have been more narrowly defined as *M. grisea.* The remaining members of the complex isolated from rice and a variety of other hosts have been renamed *Magnaporthe oryzae.* Confusion on which of these two names to use for the rice blast pathogen remains, as both are now used by different authors.

Members of the *Magnaporthe grisea* complex can also infect other agriculturally important cereals including wheat, rye, barley, and pearl millet causing diseases called blast disease or blight disease. Rice blast causes economically significant crop losses annually. Each year it is estimated to destroy enough rice to feed more than 60 million people. The fungus is known to occur in 85 countries worldwide.

Hosts and Symptoms

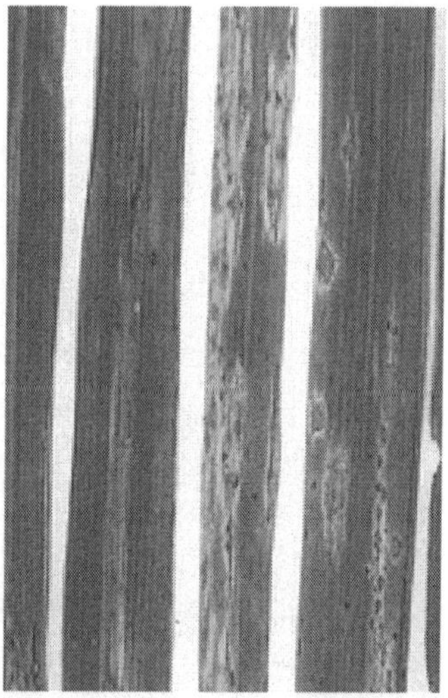

Figure: *Lesions on rice leaves caused by infection with* M. grisea

M. grisea is an ascomycete fungus. It is an extremely effective plant pathogen as it can reproduce both sexually and asexually to produce specialized infectious structures known as appressoria that infect aerial tissues and hyphae that can infect root tissues.

Rice blast has been observed on rice strains M-201, M-202, M-204, M-205, M-103, M-104, S-102, L-204, Calmochi-101, with M-201 being the most vulnerable. Initial symptoms are white to gray-green lesions or spots with darker borders produced on all parts of the shoot, while older lesions are elliptical or spindle-shaped and whitish to gray with necrotic borders. Lesions may enlarge and coalesce to kill the entire leaf. Symptoms are observed on all above-ground parts of the plant. Lesions can be seen on the leaf collar, culm, culm nodes, and panicle neck node. Internodal infection of the culm occurs in a banded pattern. Nodal infection causes the culm to break at the infected node (rotten neck). It also affects reproduction by causing the host to produce fewer seeds. This is caused by the disease preventing maturation of the actual grain.

Disease Cycle

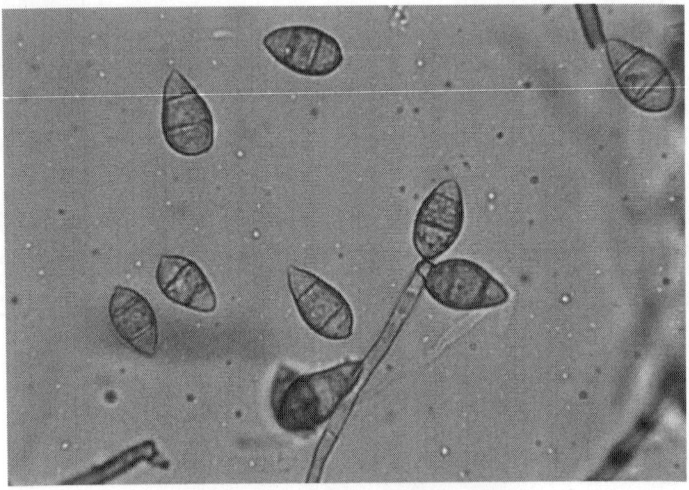

Figure: *Spores of* M. grisea

The pathogen infects as a spore that produces lesions or spots on parts of the rice plant such as the leaf, leaf collar, panicle, culm and culm nodes. Using a structure called an appressorium, the pathogen penetrates the plant. *M. grisea* then sporulates from the diseased rice tissue to be dispersed as conidiospores. After overwintering in sources such as rice straw and stubble, the cycle repeats.

A single cycle can be completed in about a week under favourable conditions where one lesion can generate up to thousands of spores in a single night. With the ability to continue to produce the spores for over 20 days, rice blast lesions can be devastating to susceptible rice crops.

Environment

Rice blast is a significant problem in temperate regions and can be found in areas such as irrigated lowland and upland. Conditions conducive for rice blast include long periods of free moisture where leaf wetness is required for infection and high humidity is common. Sporulation increases with high relative humidity and at 77-82 degrees F, spore germination, lesion formation, and sporulation are at optimum levels.

In terms of control, excessive use of nitrogen fertilization as well as drought stress increase rice susceptibility to the pathogen as the plant is placed in a weakened state and its defences are low. Extended drain periods also favour infection as they aerate the soil, converting ammonium to nitrate and thus causing stress to rice crops, as well.

Management

The fungus has been able to establish resistance to both chemical treatments and genetic resistance in some types of rice developed by plant breeders. It is thought that the fungus can achieve this by genetic change through mutation. In order to most effectively control infection by *M. grisea*, an integrated management programme should be implemented to avoid overuse of a single control method and fight against genetic resistance. For example, eliminating crop residue could reduce the occurrence of overwintering and discourage inoculation in subsequent seasons. Another strategy would be to plant resistant rice varieties that are not as susceptible to infection by *M. grisea*. Knowledge of the pathogenicity of *M. grisea* and its need for free moisture suggest other control strategies such as regulated irrigation and a combination of chemical treatments with different modes of action. Managing the amount of water supplied to the crops limits spore mobility thus dampening the opportunity for infection. Chemical controls such as Carpropamid have been shown to prevent penetration of the appressoria into rice epidermal cells, leaving the grain unaffected.Pseudomonadaceae

The Pseudomonadaceae are a family of bacteria that includes the genera *Azomonas*, *Azomonotrichon*, *Azorhizophilus*, *Azotobacter*, *Cellvibrio*, *Mesophilobacter*, *Pseudomonas* (the type genus), *Rhizobacter*, *Rugamonas*, and *Serpens*. The Azotobacteriaceae were recently published as belonging in this family as well.

History

Pseudomonad literally means 'false unit', being derived from the Greek *pseudo* ('false') and *monas* (ìïíïò - 'a single unit'). The term "monad" was used in the early history of microbiology to denote single-celled organisms. Because of their widespread occurrence in nature, the pseudomonads were observed early in the history of microbiology. The generic name *Pseudomonas* created for these organisms was defined in rather vague terms in 1894 as a genus of Gram-negative, rod-shaped and polar-flagella bacteria. Soon afterwards, a very large number of species was assigned to the genus. Pseudomonads were isolated from many natural niches and a large number of species names was originally assigned to the genus. New methodology and the inclusion of approaches based on the studies of conservative macromolecules have reclassified many strains.

Pseudomonas aeruginosa is increasingly recognized as an emerging opportunistic pathogen of clinical relevance. Studies also suggest the emergence of antibiotic resistance in *P. aeruginosa*.

In 2000, the complete genome sequence of a *Pseudomonas* species was determined; more recently, the sequences of other strains have been determined, including *P. aeruginosa* strains PAO1 (2000), *P. putida* KT2440 (2002), *P. fluorescens* Pf-5 (2005), *P. fluorescens* PfO-1 and *P. entomophila* L48. Several pathovars of *Pseudomonas syringae* have been sequenced, including pathovar tomato DC3000 (2003), pathovar *syringae* B728a (2005), and pathovar *phaseolica* 1448A (2005).

Distinguishing Characteristics

- Oxidase positive - this is due to the presence of cytochrome c oxidase enzyme
- Non-fermentative
- Many metabolise glucose by the Entner Doudoroff pathway mediated by 6-phosphoglyceraldehyde dehydrogenase and aldolase
- Polar flagella, enabling motility
- Many members produce derivatives of the fluorescent pigment pyoverdin

The presence of oxidase and polar flagella and inability to carry out fermentation differentiate pseudomonads from Enterobacteriaceae.

Cultural Control

Blast is favoured by excessive nitrogen fertilization, aerobic soils, and drought stress. High nitrogen rates and nitrate nitrogen increase rice susceptibility to the disease. Extended drain periods may also encourage the disease by aerating the soil, by converting ammonium to nitrate, and by causing drought stress to rice.

Use proper seed sampling and testing to identify and avoid the use of blast-infested seed in areas where blast is not a problem. This may help limit the introduction of the disease into noninfested areas. Water seeding is recommended to reduce or eliminate disease transmission from seed to seedlings. Drill seeding is not recommended because it may allow seed transmission, nitrate formation, and result in drought stress.

Continuous flooding is recommended to limit blast development. Avoid field drainage, especially for extended periods because it allows the formation of nitrate and may cause drought stress. Some studies in other areas suggest that shallow water is more favourable to blast development than deeper water.

Importance

Rice blast is the most important disease concerning the rice crop in the world. Since rice is an important food source for much of the world, its effects have a broad range. It has been found in over 85 countries across the world and reached the United States in 1996. Every year the amount of crops lost to rice blast could feed 60 million people. Although there are some resistant strains of rice, the disease persists wherever rice is grown. The disease has never been eradicated from a region.

Natural Rice Rhizospheric Microbes Suppress Rice Blast Infections

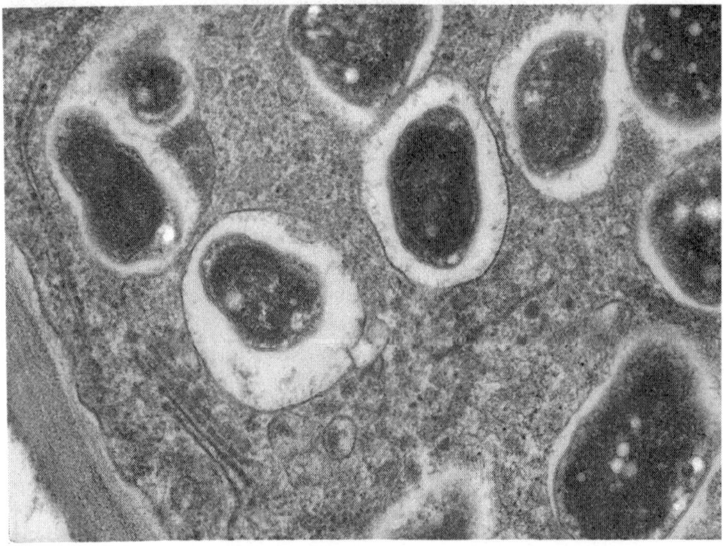

Figure: *Cross section though a soybean (*Glycine max *'Essex') root nodule. The Rhizobacteria,* Bradyrhizobium japonicum, *colonizes the roots and establishes a nitrogen fixing symbiosis.*

Rhizobacteria are root-colonizing bacteria that form symbiotic relationships with many plants. The name comes from the Greek *rhiza*, meaning *root*. Though parasitic varieties of rhizobacteria exist, the term usually refers to bacteria that form a relationship beneficial for both parties (mutualism). They are an important group of microorganisms used in Biofertilizer. Biofertilization accounts for approximately 65% of the nitrogen supply to crops worldwide. Rhizobacteria are often referred to as plant growth-promoting rhizobacteria, or PGPRs. The term PGPRs was first used by Joseph W.

Kloepper in the late 1970s and has become commonly used in scientific literature. PGPRs have different relationships with different species of host plants. The two major classes of relationships are rhizospheric and Endophytic. Rhizospheric relationships consist of the PGPRs that colonize the surface of the root, or superficial intercellular spaces of the host plant, often forming Root nodules. The dominant species found in the rhizosphere is a microbe from the genus *Azospirillum*. Endophytic relationships involve the PGPRs residing and growing within the host plant in the apoplastic space.

Nitrogen Fixation

Nitrogen Fixation is one of the most beneficial processes performed by rhizobacteria. Nitrogen is a vital nutrient to plants and gaseous nitrogen (N_2) is not available to them due to the high energy required to break the triple bonds between the two molecules. Rhizobacteria through nitrogen fixation are able to convert gaseous nitrogen (N_2) to ammonia (NH_3) making it an available nutrient to the host plant which can support and enhance plant growth. The host plant provides the bacteria with amino acids so they do not need to assimilate ammonia. The amino acids are then shuttled back to the plant with newly fixed nitrogen. Nitrogenase is an enzyme involved in nitrogen fixation and requires anaerobic conditions. Membranes within root nodules are able to provide these conditions. The rhizobacteria require oxygen to metabolize therefore oxygen is provided by a hemoglobin protein called Leghemoglobin which is produced within the nodules. Legumes are well known nitrogen fixing crops and have been utilized for centuries in crop rotation to maintain the health of the soil.

Symbiotic Relationships

The symbiotic relationship between rhizobacteria and its host plant is not without costs. For the plant to be able to benefit from the added available nutrients provided by the rhizobacteria, it needs to provide a place and the proper conditions for the rhizobacteria to live. Creating and maintaining root nodules for rhizobacteria can cost between 12%-25% of the plants total photosynthetic output. Legumes are often able to colonize early successional environments due to the unavailability of nutrients. Once colonized though the rhizobacteria make the soil surrounding the plant more nutrient rich, which in turn can lead to competition with other plants. The symbiotic relationship in short can lead to increased competition.

PGPRs increase the availability of nutrients through the solubilization of unavailable forms of nutrients and by the production of siderophores which aids in the facilitating of nutrient transport. Phosphorus, a limiting nutrient for plant growth, can be plentiful in soil but is most commonly found in insoluble forms. Organic acids and phosphotases released by rhizobacteria found in plant rhizospheres facilitate the conversion of insoluble forms of phosphorus to plant available forms such as $H_2PO_4^-$. PGPR bacteria include *Pseudomonas putida, Azospirillum fluorescens,* and *Azospirillum lipoferum* and notable nitrogen fixing bacteria associated with legumes includes *Allorhizobium, Azorhizobium, Bradyrhizobium,* and *Rhizobium.*

Though microbial inoculants can be beneficial for crops, they are not widely used in industrial agriculture, as large-scale application techniques have yet to become economically viable. A notable exception is the use of rhizobial inoculants for legumes such as peas. Inoculation with PGPRs ensure efficient nitrogen fixation, and they have been employed in North American agriculture for over 100 years.

Plant Growth-promoting Rhizobacteria

Plant growth-promoting rhizobacteria *(PGPR)* was first defined by Kloepper and Schroth to describe soil bacteria that colonize the roots of plants following inoculation onto seed and that enhance plant growth. The following are implicit in the colonization process: ability to survive inoculation onto seed, to multiply in the spermosphere (region surrounding the seed) in response to seed exudates, to attach to the root surface, and to colonize the developing root system. The ineffectiveness of PGPR in the field has often been attributed to their inability to colonize plant roots.

A variety of bacterial traits and specific genes contribute to this process, but only a few have been identified. These include motility, chemotaxis to seed and root exudates, production of pili or fimbriae, production of specific cell surface components, ability to use specific components of root exudates, protein secretion, and quorum sensing. The generation of mutants altered in expression of these traits is aiding our understanding of the precise role each one plays in the colonization process. Progress in the identification of new, previously uncharacterized genes is being made using nonbiased screening strategies that rely on gene fusion technologies. These strategies employ reporter transposons and in vitro expression technology (IVET) to detect genes expressed during colonization.

Using molecular markers such as green fluorescent protein or fluorescent antibodies it is possible to monitor the location of individual rhizobacteria on the root using confocal laser scanning microscopy. This approach has also been combined with an rRNA-targeting probe to monitor the metabolic activity of a rhizobacterial strain in the rhizosphere and showed that bacteria located at the root tip were most active

Mechanisms of Action

PGPR enhance plant growth by direct and indirect means, but the specific mechanisms involved have not all been well-characterized, Direct mechanisms of plant growth promotion by PGPR can be demonstrated in the absence of plant pathogens or other rhizosphere microorganisms, while indirect mechanisms involve the ability of PGPR to reduce the harmful effects of plant pathogens on crop yield. PGPR have been reported to directly enhance plant growth by a variety of mechanisms: fixation of atmospheric nitrogen that is transferred to the plant, production of siderophores that chelate iron and make it available to the plant root, solubilization of minerals such as phosphorus, and synthesis of phytohormones. Direct enhancement of mineral uptake due to increases in specific ion fluxes at the root surface in the presence of PGPR has also been reported . PGPR strains may use one or more of these mechanisms in the rhizosphere. Molecular approaches using microbial and plant mutants altered in their ability to synthesize or respond to specific phytohormones have increased our understanding of the role of phytohormone synthesis as a direct mechanism of plant growth enhancement by PGPR PGPR that synthesize auxins and cytokinins or that interfere with plant ethylene synthesis have been identified.

Pathogenic Roles

Studies conducted on sugar beet crops found that some root-colonizing bacteria were deleterious rhizobacteria (DRB). It was found that sugar beet seeds inoculated with DRB had reduced germination rates, root lesions, reduced root elongation, root distortions, increased fungi infection and decreased plant growth. In one trial the sugar beet yield was reduced by 48%.

Six strains of rhizobacteria have been identified as being DRB. The strains are in the genera *Enterobacter*, *Klebsiella*, *Citrobacter*, *Flavobacterium*, *Achromobacter*, *Arthrobacter*. Due to a large number of taxonomic species yet to be described complete characterization has not been possible as DRB are highly variable.

The presence of PGPRs has proven to reduce and inhibit the colonization of DRB on sugar beet roots. Plots inoculated with PGPRs and DRBs had an increase in production of 39% while plots only treated with DRbs had a reduction in production of 30%.

Biocontrol

Rhizobacteria are also able to control plant diseases that are caused by other bacteria and fungi. Disease is suppressed through induced systematic resistance and through the production of anti-fungal metabolites. *Pseudomonas* Biocontrol strains have been genetically modified to improve plant growth and improve the disease resistance of agricultural crops. In agriculture, inoculant bacteria are often applied to the seed coat of seeds prior to being sown. Inoculated seeds are more likely to establish large enough rhizobacteria populations within the rhizosphere to produce notable beneficial effects on the crop.

Plant Disease Management Through Diversified Cropping Systems: an Eco-friendly Approach

Diversifying rice systems in rotation with other cereals like wheat or maize, high-value crops like potato, legumes, or fodder crops is one way rice farmers can optimize their use of resources. Diversified cropping systems broaden the source of a farmer's food and income, increases their land productivity, and minimizes unpredictable risks such as the build-up of pest and diseases common in rice monoculture.

Owing to market forces, rice-based cropping systems in irrigated and otherwise favourable environments in Asia and Africa – those are areas with sufficient rainfall, good soils, and good market access – are continuously being intensified to include wheat and maize. Rice-wheat systems presently occupy 13.5 million hectares in South Asia's Indo-Gangetic Plains and another 3.5 million hectares in China. Rice-maize systems are also gaining importance in South and Southeast Asia.

The increasing demand for maize in many Asian countries has led to diversifying rice-rice or rice-wheat systems into rice-maize systems that may also occupy up to 3 million hectares at present. In addition, rice is also grown with other crops like potato, legumes, or vegetables in about 8 million hectares.

Agricultural Diversification

In the agricultural context, diversification can be regarded as the re-allocation of some of a farm's productive resources, such as land, capital, farm equipment and paid labour, into new activities. These can be new crops or livestock products, value-adding activities, provision of services to other farmers and, particularly in richer countries, non-farming activities such as restaurants and shops. Factors leading to decisions to diversify are many, but include; reducing risk, responding to changing consumer demands or changing government policy, responding to external shocks and, more recently, as a consequence of climate change.

Definitions of Diversification

While most definitions of diversification in developing countries work on the assumption that diversification primarily involves a substitution of one crop or other agricultural product for another, or an increase in the number of enterprises, or activities, carried out by a particular farm, the definition used in developed countries sometimes relates more to the development of activities on the farm that do not involve agricultural production. For example, one section of the British Department for Environment, Food and Rural Affairs (DEFRA) defines diversification as "the entrepreneurial use of farm resources for a non-agricultural purpose for commercial gain". Using this definition DEFRA found that 56% of UK farms had diversified in 2003. The great majority of diversification activities simply involved the renting out of farm buildings for non-farming use, but 9% of farms had become involved with processing or retailing, 3% with provision of tourist accommodation or catering, and 7% with sport or recreational activities. Others adopt a broader definition, which may include development of new marketing opportunities.

In developing countries such as India, which has been one of the leaders in promoting diversification, the concept is applied both to individual farmers and to different regions, with government programmes being aimed at promoting widespread diversification. The concept in India is seen as referring to the "shift from the regional dominance of one crop to regional production of a number of crops ...... (which takes into account)..... the economic returns from different value-added crops... with complementary marketing opportunities".

Drivers of Diversification

Diversification can be a response to both opportunities and threats.

Opportunities

- Changing consumer demand. As consumers in developing countries become richer, food consumption patterns change noticeably. People move away from a diet based on staples to one with a greater content of animal products (meat, eggs, and dairy) and fruits and vegetables. In turn, more dynamic farmers are able to diversify to meet these needs.
- Changing demographics. Rapid urbanization in developing countries has an impact on consumption patterns. Moreover, a

smaller number of farmers, in percentage terms at least, has to supply a larger number of consumers. While this may not imply diversification it does require adaptation to new farming techniques to meet higher demand.

- Export potential. Developing country farmers have had considerable success by diversifying into crops that can meet export market demand. While concern about food miles, as well as the cost of complying with supermarket certification requirements such as for GlobalGAP may jeopardize this success in the long run, there remains much potential to diversify to meet export markets.
- Adding value. The pattern witnessed in the West, and now becoming widespread in developing countries, is for consumers to devote less and less time to food preparation. They increasingly require ready-prepared meals and labour-saving packaging, such as pre-cut salads. This provides the opportunity for farmers to diversify into value addition, particularly in countries where supermarkets play a major role in retailing.
- Changing marketing opportunities. The changing of government policies that control the way in which farmers can link to markets can open up new diversification possibilities. For example, in India, policy changes to remove the monopoly of state "regulated markets" to handle all transactions made it possible for farmers to establish direct contracts with buyers for new products.
- Improving nutrition. Diversifying from the monoculture of traditional staples can have important nutritional benefits for farmers in developing countries.

Threats

- Urbanization. This is both an opportunity and a threat, in that the expansion of cities places pressure on land resources and puts up the value of the land. If farmers are to remain on the land they need to generate greater income from that land than they could by growing basic staples. This fact, and the proximity of markets, explains why farmers close to urban areas tend to diversify into high-value crops.
- Risk. Farmers face risk from bad weather and from fluctuating prices. Diversification is a logical response to both. For example, some crops are more drought-resistant than others, but may offer poorer economic returns. A diversified portfolio of products should ensure that farmers do not suffer complete ruin when

the weather is bad. Similarly, diversification can manage price risk, on the assumption that not all products will suffer low prices at the same time. In fact, farmers often do the opposite of diversification by planting products that have a high price in one year, only to see the price collapse in the next, as explained by the cobweb theory.

- External threats. Farmers who are dependent on exports run the risk that conditions will change in their markets, not because of a change in consumer demand but because of policy changes. A classic example is the Caribbean banana industry, which collapsed as a result of the removal of quota protection on EU markets, necessitating diversification by the region's farmers.
- Domestic policy threats. Agricultural production is sometimes undertaken as a consequence of government subsidies, rather than because it is inherently profitable. The reduction or removal of those subsidies, whether direct or indirect, can have a major impact on farmers and provide a significant incentive for diversification or, in some cases, for returning to production of crops grown prior to the introduction of subsidies.
- Climate change. The type of crop that can be grown is affected by changes in temperatures and the length of the growing season. Climate change could also modify the availability of water for production. Farmers in several countries, including Canada, India, Kenya, Mozambique, and Sri Lanka have already initiated diversification as a response to climate change. Government policy in Kenya to promote crop diversification has included the removal of subsidies for some crops, encouraging land-use zoning and introducing differential land tax systems.

Opportunities for Diversification

In making decisions about diversification farmers need to consider whether income generated by new farm enterprises will be greater than the existing activities, with similar or less risk. While growing new crops or raising animals may be technically possible, these may not be suitable for many farmers in terms of their land, labour and capital resources. Moreover, markets for the products may be lacking. The United Nations Food and Agriculture Organization (FAO) has been one of the development organizations promoting diversification by small farmers and has produced booklets identifying beekeeping, mushroom farming, milk production, fish ponds and sheep and goats, among others, as diversification possibilities.

Measures of Diversification

Agricultural diversification is measured in a number of ways throughout the world. For example, one such measure is the *index of maximum proportion*, which is "defined as the ratio (proportion) of the farm's primary activity to its total activities".

Biotechnological Approaches for Disease Management in Forest trees

The general objective of a genetic improvement programme should be the *sustainable* management of genetic variation to generate, identify and multiply for operational planting high-yielding and well adapted genotypes. For an outcrossing species, breeding typically incorporates:

i. *Establishment of an initial base population*: In its broadest sense, for forest species this includes species and provenance testing, and the development of breeding and gene conservation populations.

ii. *Population improvement*: For forest species this typically includes recurrent cycles of selection and recombination.

iii. *The derivation and multiplication of strains (e.g. families or clones) to be used operationally.*

This is the step in which breeding gains are captured and transferred to production populations.

The above are incorporated into a breeding strategy. A breeding strategy should permit continued improvement, and must allow for future alterations in breeding objectives. A sound breeding strategy also incorporates the early collection of appropriate biological and genetic data concerning the species, and should include measures to maintain a broad genetic base and to minimize inbreeding.

There exists a virtually unlimited number of alternative breeding strategies, varying with respect to, for example:

- number of populations used
- approach to selection (e.g. phenotypic, progeny test, clonal test, family)
- mating design for recombination and the estimation of genetic parametres (e.g. open-pollinated, pair mating, factorials, diallels)
- approach to handling the problem of minimization of inbreeding (e.g. pedigree control and sublining)
- approach to the capture of genetic gain (e.g. open pollinated general combiner orchards, full-sib family options, clones)

- method for integrating all of the above, and for balancing the frequently conflicting objectives - e.g. the problem of maximizing selection intensity while maintaining variation.

Timing of the different phases is an important component of a breeding strategy. Thorough species and provenance testing and biological studies are important early objectives. The desirability of early availability of improved genotypes for planting dictates, however, that some selection work will precede completion of the above. Some plantation programmes for certain species may not warrant the expense of an intensive breeding programme, and a genetic improvement programme may include little more than species testing and development of appropriate propagation methods.

The Status of Tree Improvement

It has been estimated recently that, of a total of at least 50 000 tree species, some 400 have been the subject of formal breeding or testing activities of some description, 140 with at least one generation of selection and mating, and 35 of these involving at least 20 seed parents (Committee on Managing Global Genetic Resources 1991). The following is a brief survey of the current status of improvement for a representative sample of these species.

Industrial Dpecies

For convenience, the term "industrial" is applied here to species used predominantly in large scale plantations aimed mainly at the commercial production of timber. There are currently about 100 million hectares of industrial plantations worldwide (Gauthier 1991), and it has been estimated that projected demands for industrial timber early next century might be met by a total estate of 100–200 million hectares (Mather 1990). For convenience, industrial plantation species are grouped into broad classes in the following survey. It is admitted though that some groups are quite heterogeneous, and programmes for many species straddle climatic and rotation length classes.

Tropical and sub-tropical Eucalypts

These currently comprise less than 5% of established industrial plantings, but this is a group of rapidly increasing importance due to the potential which exists for expanded plantings in many tropical countries. The most important taxa are *E. grandis, E. tereticornis* and *E. camaldulensis*, and notably hybrids between *E. grandis* and *E. urophylla* and the other two species. *E. grandis* is widely planted in moister areas, particularly of South America and Africa, due to its fast growth, good silvicultural characteristics, and good pulping qualities.

E. camaldulensis and *E. tereticornis*, more drought hardy than grandis but without the same ability to capture the site, are very widely planted, particularly in India and Africa (Vivekanandan 1985, Koyo 1989). Other species with potential in tropical Africa, Asia and the Pacific, but much less widely used, include *E. microcorys, E. pilularis, E. cloeziana, E. citriodora, E. pellita,* and possibly *E. deglupta* (Burley & Barnes 1989). Private companies have featured prominently in the more well known plantings, but government enterprises are also important.

By the standards of most industrial forest plantation species, the tropical eucalypts are relatively easy to breed:

- many flower abundantly and precociously
- they are generally characterised by short rotations, permitting early selection and short generations (e.g. generations of 4–7 years for *E. grandis*)
- although not as easy as for pines, controlled pollination is still quite achievable
- most coppice, and can be propagated reasonably easily by cuttings
- seeds can be stored satisfactorily.

A few very advanced breeding programmes exist, undertaken by both government institutes and private growers. Four generations of selection and open-pollinated recombination have yielded large gains in the Florida (USA) programme (Squillace 1989). Spectacular and well-publicized successes have been recorded with clonal propagation of hybrids in Brazil and the Congo, although the breeding programmes underlying these clonal selection programmes are not as advanced (Campinhos & Ikemori 1989, Vigneron 1989). Clonal programmes are underway also for *E. grandis* in other tropical countries, but are perhaps unlikely to result in the large gains reported for the hybrids. These advanced programmes represent a very small proportion of the world plantation estate in tropical eucalypts.

Provenance and progeny trials have been planted in many areas, some selections made, and seedling seed orchards established, but much of the planting is still conducted with material subjected to little if any improvement. Vigour, form and wood quality are typically the most important selection criteria. Cold tolerance has been an important criterion in the *E. grandis* programme in Florida (Meskimen 1983, Rockwood *et al.* 1989), although some freezes probably exceed the levels which can be tolerated by naturally available variants (Meskimen 1983). Most programmes are characterized by open pollinated approaches to recombination, testing and capture of genetic gain (e.g. Geary *et al.*

1983). Genetic variation of the tropical eucalypts is fairly well preserved in *ex situ* and *in situ* stands.

To summarize, genetic improvement programmes offer clear benefits for industrial plantation forestry with tropical eucalypts, and no major biological impediments to the breeding of these species exist. In terms of practical priorities, there is a lot to be gained by the extension of existing technologies to additional programmes. For the pure species, open-pollinated approaches are relatively simple and give good gains. Insufficient understanding and application of sound breeding strategies is a limitation in some programmes.

For the more advanced programmes, the availability of greater frost tolerance would be advantageous in some areas. The availability of better selection methods would circumvent the problem of the expense of clonal testing, and permit the imposition of higher selection intensities in the clonal programmes. Although generation intervals in these species are relatively short, large gains would be available through even earlier selection and flowering.

Cold-tolerant Eucalypts

Although comprising no more than 2–3% of currently existing industrial plantations, this group is of expanding importance, particularly in plantation programmes by private companies in Chile, Africa and Europe. Around 50 000 ha of *Eucalyptus nitens* plantations have been established worldwide, while *E. globulus* plantations are much more extensive - over 1 000 000 ha in Portugal alone (Eldridge & Griffin 1990). Species for which plantings are more limited include *E. dunnii, E. smithii, E. radiata, E. fraxinoides, E. McArthurii* and *E. fastigata.*

Although generally grown on short rotations and selectable at a similar age to the tropical eucalypts, these species are generally more difficult subjects for breeding:

- flowering of *E. globulus* is poor in some locations, and *E. nitens, E. dunnii and E. smithii* are all shy flowerers
- these four species are difficult to root from cuttings.

E. globulus is characterized by sophisticated and well-planned breeding programmes which have been established in recent years, e.g. the “nucleus breeding” approach in Portugal, and the “multiple population” programmes in South Africa and Chile. Clonal and seedling seed orchards have been established for these species, but are yet to produce seed in required quantities. Selection criteria are principally vigour, stem form, and wood quality. Resistance to winter freezing is of some interest in relation to possible wider planting of *E. globulus*

(Tibbets *et al.* 1991). Genetic resources of these species are relatively secure.

Addressing the flowering and rooting problems should be major priorities for more effective breeding of these species. Breeding programmes for these species would also benefit from the availability of greater frost tolerance, better selection methods, and even shorter generations, as described above.

Other Tropical Hardwoods

Although currently comprising less than 5% of established industrial plantings, this heterogeneous group is of rapidly increasing importance with considerable potential in many tropical countries. Some species, in particular *Tectona grandis, Gmelina* and the tropical acacias, are already popular species in both private and public plantation operations.

Total plantings of *Gmelina arborea* probably stand at over 200 000 ha, with Brazil, West Africa, the Philippines and Malaysia the major centres. Potential exists in many other areas of the tropics, although site sensitivity is a limitation. The tree provides a very useful general purpose hardwood, and high quality pulp. With short rotations (7–12 years in some programmes), early flowering (at 3–5 years), reliable seed storage (under the right conditions), ready coppicing, and good rooting of cuttings, the species lends itself to rapid breeding, although only limited improvement has been undertaken. Controlled pollination is considerably more difficult than for the tropical eucalypts. Provenances have been quite widely tested, and seed orchards have been established in a few locations, e.g. Brazil, Malaysia, and the Philippines. Vegetative propagation of phenotypic selections is practised in Malaysia. Selection criteria are mainly stem straightness, cylindricity and volume.

Tectona grandis is widely planted in Indonesia (over 1 000 000 ha), Thailand, India, several other Asian countries, South and Central America and in many parts of Africa (where it is the most widely planted exotic) (White 1991). The species is moderately amenable to improvement - seedlings flower at age six to nine (age of first flowering is positively correlated with bole length to forking), yield of viable seed is often poor, seed can be stored, trees coppice, and juvenile cuttings root readily. Control pollination is feasible, but not easy. Long rotations, and a consequent long selection interval, is a limitation on rapid breeding. Trials in several places demonstrate wide provenance variation. Breeding programmes are in existence in Thailand, India and elsewhere, and many clonal seed orchards have been established (Burley and Barnes 1989, Kaosa-ard 1989, White 1991).

Acacia mangium has been planted extensively in Malaysia, and large plantings are continuing also in Indonesia. Other species such as *A. auriculiformis, A. crassicarpa* and *A. aulacocarpa* are also of interest. Seed production commences relatively early, seed can be stored satisfactorily, controlled pollination is feasible although not easy (Sedgley *et al.* 1992), and juvenile cuttings root well (Wong & Haines 1992). Broad provenance collections have been made and trials planted, and seedling seed orchards established at several centres. Gene pools are relatively secure.

Cordia alliodora is planted in Central America, Colombia, Venezuela and Ecuador, in plantations as well as an agroforestry crop (Newton *et al.* 1991). Rotations are about 20 years, and timber is of high quality. Heavy flowering commencing as early as two years and reliable seed storage (Greaves & McCarter 1990) are conducive to rapid breeding. Seed collections have been made and international provenance/progeny trials established. Seedling seed orchards have been established in Central America. The species is quite variable (Greaves and McCarter 1990), and some gene pools are endangered (Newton *et al.* 1991).

Triplochiton scleroxylon and *Terminalia superba* have been the subject of provenance/progeny tests and clonal trials in Africa, and plantations are being established. Selections are propagated by cuttings from stump coppice in the case of *Triplochiton*, and shoots from mature scions grafted onto young seedlings in the case of *Terminalia* (Leakey 1987). Rotation lengths and selection intervals are longer than for tropical eucalypts. *Paraserianthes falcataria* is planted in the Philippines and Malaysia, provenance/progeny trials have been established, and seedling seed orchards established. Gene conservation collections are being undertaken for *Bombacopsis quinata, Sterculia apetala, Alnus acuminata*, and *Vochysia hondurensis*.

All have populations in danger of extinction or genetic erosion (Newton *et al.* 1991). Seedling seed orchards of *Bombacopsis quinata* have been established several countries in Central and South America including Honduras and Venezuela. Several members of the Meliaceae are potentially very valuable plantation species, e.g. *Swietenia macrophylla, S. humilis, S. mahogani* and *Cedrela odorata*. Damage by shoot borers, however, severely limits the extent to which plantations can be established. International provenance and progeny trials have been established of some of these species. Intraspecific variation in levels of resistance is evident, but the extent to which effective levels are available not determined (Newton *et al.* 1991, Newton *et al.* 1993). Some of these species have suffered severe genetic erosion. The use of valuable species of the Dipterocarpaceae (*Dipterocarpus elatus, Hopea* and *Shorea*) in plantations has been limited by flowering problems, seed recalcitrance, and silvicultural difficulties. Some of these species are also threatened by erosion of gene pools. A large number of other species, as yet untested or poorly known, are of potential use, including species of *Octomeles, Cleistopholis, Pterocarpus and Canarium.*

To summarize then, some tropical hardwoods compare well with tropical eucalypts as subjects for breeding, while some are a little more difficult. In general though, breeding is not yet as advanced. Broader implementation of good improvement programmes is an important priority for these species. Some other species pose major problems: e.g. insect susceptibility in the Meliaceae, flowering and seed problems in Dipterocarpaceae. Many other potentially valuable species are simply not well known, in terms of their adaptation and biological and genetic features, and gene pools are probably under threat in parts of the species ranges. The testing of potentially useful species, characterization of mating systems, provenance collection, establishment of trials, implementation of gene conservation measures and commencement of other breeding work poses a very large and urgent task.

Poplars and Willows

Comprising perhaps 5–10% of established industrial plantations, poplars and willows are important plantation trees in many countries, including in Europe, North America, China, Argentina, India, Africa, and Australia (Pryor 1992). Small private plantings feature strongly. The group includes a large number of species and subspecies.

Relatively short rotations, early and prolific flowering and ease of vegetative propagation are features which render the poplars amenable to rapid breeding and good capture of genetic gain, although seed storage is a problem. Apart from the long history of improvement, two features of poplar breeding programmes distinguish them from those for most other forest tree species (and present similarities to some fruit tree breeding programmes):

- The emphasis on between and within species hybridization, frequently involving well characterized clones, followed by clonal testing (Ostry & McNabb 1986, Carter *et al.* 1988, Herpka 1987). Most commercial plantings comprise tested clones, frequently with cultivar names as for ornamental horticulture.
- The emphasis on disease resistance (e.g. to *Melampsora, Marssonina* and *Septoria*) as the major selection criterion (Herpka 1987, Ostry & McNabb 1986, McNabb *et al.* 1990, Pryor 1992),

with maintenance of growth characteristics a secondary objective. Disease resistance breeding has emerged in response to several clone-specific calamities.

The importance of the disease problem in poplars, and the long history of wide deployment of individual clones, may not be coincidental, and may provide a lesson for more considered and cautious deployment in clonal programmes with other species.

An important factor limiting improvement in many poplar programmes is the availability of required levels of disease resistance. As above, large benefits would accrue through the development of better selection methods and shorter generation intervals.

Other Temperate Hardwoods

Plantations of species of this heterogeneous group, located in particular in Europe, North America and China, comprise perhaps 5–10% of the current industrial forest plantation estate.

Although plantings of long rotation, high value species such as *Juglans* and *Prunus* are currently insignificant in total extent, it has been suggested that small private plantings of these could assume greater importance in the future on disused agricultural land (Von Althen 1991). For these species, small breeding programmes, by European and North American government institutes and cooperatives, are in existence. The long generation interval (related to both the long selection interval and the long delay to flowering), difficulty in controlled pollination, intermittent and light seed crops, and seed recalcitrance (Rink & Stelzer 1982, Beineke 1982, Robinson & Overton 1989) are limitations to the breeding of walnut. Orchards established many years ago are only now producing seed. Vigour and stem form are important selection criteria, but wood quality is also of particular importance. High values of the product have led to some use of grafted plantations, and patented clones are available (Beineke 1982, Beineke 1989, Robinson & Overton 1989). Unmanaged exploitation has led to scarcity of the species in many areas of its natural range (Robinson & Overton 1989). Breeding programmes incorporating clonal testing are underway for *Prunus* in Europe. With restricted current plantings and long generation intervals, breeding of these species beyond selection and either seed orchard establishment or propagation of superior genotypes is not readily justified. This is a group for which effective early selection would be particularly valuable.

Robinia pseudoacacia is a leguminous tree of considerable value in fast growing biomass plantations. Plantings stand at over 3 250 000 ha, with large areas in China, South Korea, and Hungary (Keresztesi 1983, Hanover *et al.* 1991). A precocious flowerer, regular and abundant seeder, with seed readily stored, and easily propagated from shoot or root cuttings, the species is very amenable to breeding. Hungarian breeding programmes commenced several decades ago have been based on crossing (including with related species) and clonal selection, and many registered cultivars are available (Keresztesi 1983). Selection criteria are wood quality (in particular), frost tolerance, growth and form. The origin and genetic base of many plantings, however, are unknown. Breeding programmes incorporating provenance and progeny testing have been commenced in North America (Mebrahtu and Hanover 1989, Hanover *et al.* 1991). Greater frost resistance in the species would extend its range.

Acacia mearnsii is widely planted for bark (tannin), mining timber, poles and fuelwood especially in southern and eastern Africa, and China. Difficult to control pollinate and to propagate vegetatively, breeding programmes are based on the open-pollinated progeny trial/ seeding seed orchard approach (Raymond 1987, Hillis 1989). Alnus glutinosa is a very precocious (as early as two to three years) and heavy seed producer. Provenance/progeny tests have been established and seed orchards have been planned (Carter *et al.* 1988). Betula species similarly flowerer precociously and abundantly, and greenhouse seed orchards in Finland are now producing seed (Carter *et al.* 1988). Species involved in these programmes are B. verrucosa and B. pubescens. A plus tree selection programme for Betula verrucosa commenced in Sweden in 1988, and progeny trials and clonal tests have been established (Danell 1991). Small breeding programmes are in existence, and orchards have been established, for a number of other North American hardwoods, e.g. *Platanus occidentalis, Fraxinus pennsylvanica, Liquidambar styraciflua, Quercus rubra and Liriodendron tulipifera* (Squillace 1989).

Medium Rotation Conifers

Species of this group make up 35–40% of the world plantation estate, and potential exists for further plantings in many developing countries. Included are the important *Pinus* species *P. taeda, P. elliottii, P. caribaea, P. radiata* and *P. pinaster. P. taeda* comprises about 80% of some 10 000 000 ha of southern pine plantations in south-eastern USA (Lantz & Kraus 1987), and substantial areas exist also in China (220 000 ha) and Brazil. *P. elliottii* makes up the major part of the remainder of the USA plantations, and is important also in China (880 000 ha), Brazil (several hundred thousand hectares at least) and South Africa (some 150 000 ha). Substantial plantations of P. radiata exist in Chile, New Zealand and Australia (each more or less 1 000 000 ha). *P. pinaster* is an important plantation species in France (1 200 000 ha) and Portugal (900 000 ha), and is planted also in Chile and Australia (Destremau *et al.* 1982). World plantings of *P. caribaea* are over 500 000 ha, mostly in Venezuela although the species is important also in Australia, Fiji and Brazil, and potentially so in many other tropical countries. Several other pine species are important regionally, e.g. *P. patula, P. rigida, P. palustris, P. echinata, P. virginiana*, and *P. kesiya*. Some other pine species, e.g. *P. oocarpa* and *P. tecunumanii*, have potential in tropical regions.

Although the seed development cycle is longer than for many hardwoods, most of these pine species have some features which facilitate breeding:

- many flower reliably and well (although, for example, *P. caribaea* and *P. merkusii* flower poorly in some areas to which they have been introduced)
- controlled pollination is relatively easy
- seed can be stored readily
- grafting is easy

Cuttings can be propagated from young seedlings of many species, but propagation of cuttings from older trees generally is difficult.

Well established breeding programmes are in existence, at or near the third generation of selections in the USA *P. taeda*, New Zealand *P. radiata*, and Queensland *P. caribaea* programmes. (NCSU- I.CTIP 1990, Shelbourne 1986, Kanowski & Nikles 1989). A substantial proportion of the world's pine plantings are conducted with genetically improved material derived from the breeding programmes. The ease of controlled pollination has led to its frequent use in mating programmes. The large USA breeding programmes are typified by sophisticated mating designs for recombination and testing, e.g. diallels and factorials. Pair mating, polycrossing, or open pollination is more common elsewhere. Selection criteria are mainly vigour, stem form and wood quality, with disease resistance of some interest also, particularly for *P. taeda, P. elliottii* and *P. radiata*. The most advanced programmes employ controlled pollination also for the capture of genetic gain - to produce superior full-sib families, sometimes with multiplication by cuttings. The latter applies in particular to *P. radiata* in New Zealand and Australia, and to the hybrid between *P. elliottii* and *P. caribaea* in Queensland. Open pollination, however, remains the most common approach to the capture of genetic gain, and pollen contamination in seed orchards (50% even in well isolated orchards according to Hodge *et al.* 1991), is undoubtedly a cause of substantial leakage of this genetic gain in many programmes.

Major limitations to improvement are the long intervals to flowering and selection (resulting in a long generation interval), inaccurate selection, and the poor rooting of cuttings from older trees. Gene pools for most species are reasonably well preserved in *ex situ* plantings, although some provenances of the Central American pines are endangered.

Also included in this group are *Cryptomeria japonica* and *Chamaecyparis obtusa*, which make up the greater part of the Japanese plantations of some 10 000 000 ha. Historically, the Japanese programmes have been based on the vegetative propagation of selected varieties, each comprising a small number of phenotypically similar clones. Provenance and progeny tests have been established and crossing programmes implemented in more recent times.

Araucaria cunninghamii is a valuable timber tree represented by 50 000 ha in Australia but with considerable potential elsewhere in the tropics. Mature trees coppice readily, and cuttings root well, but a rigid orthotropic and plagiotropic branching system places restrictions on vegetative multiplication rates with cuttings. Irregular flowering, and long delays to the production of male strobili, place further constraints on the breeding of the species, although an active breeding programme tailored to accommodate these characteristics has resulted in substantial gains (Nikles *et al.* 1988). Another fairly widely planted species of this genus is *A. angustifolia*, in South America. For this species, however, little improvement has been undertaken, and the species is endangered in parts of its natural range.

Genetic improvement offers clear benefits for planting programmes with species of this group. Broader application of good improvement programmes is an important priority, particularly in tropical countries. The broader development of technologies for the mass production of superior full-sib families would also be highly beneficial. In terms of research objectives in advanced breeding programmes, major benefits are likely to accrue from more effective control of the maturation state (particularly in relation to the rooting of cuttings), more effective selection, and shorter generation intervals.

Long Rotation Temperate Conifers

This group, comprises 35–40% of global industrial plantations. Important species included here are *Pinus sylvestris* (annual plantings of 90 000 ha, 90 000 ha, 100 000 ha, 40 000–50 000 ha in Finland, Sweden, China and Poland respectively, and with additional large programmes in countries of the former USSR), *Picea abies* (a major plantation species in Europe and North America), other *Picea* species, e.g. *P. sitchensis, P. mariana P. glauca* and *P. engelmannii* (important in U.K., Canada and north-eastern USA), *Pinus contorta* (at least 350 000 ha established in Sweden (Fries 1987), and large areas also in Canada (Ying *et al.* 1985, Konishi 1986)), *Pseudotsuga menziesii* (a very important plantation species in north-western USA and Canada, but represented by some 350 000 ha also in Europe), *Pinus banksiana* (important in eastern Canada), *P. strobus* (USA), *Larix* species (extensive plantings in South Korea and grown also in north-eastern USA and Europe). Some large planting programmes are undertaken by government bodies, e.g. in the UK and Canada. A prominent feature of the industry in several European countries is the large number of small growers.

Pinus contorta (Fries 1987) and *P. banksiana* (Park *et al.* 1989) are reasonably precocious and reliable flowerers, but several of the other species mentioned are characterized by light and/or irregular flowering, e.g. *P. sylvestris* (Matyas 1991, Koski 1991), *Picea abies* (Wellendorf 1989, Dietrichson 1989, Ruotsalainen and Nikkanen 1989) and *P. engelmannii* (Konishi 1986). The major obstacle to improvement, however, and the main difference from the medium rotation conifers, is the longer selection interval, rendering traditional breeding very slow.

Well established breeding programmes, largely the responsibility of government institutes, exist for these species. Most testing and recombination programmes are based on open-pollination (Lindgren 1991. Fries 1987, Morgenstern and Park 1991) although some pair mating is used, and more sophisticated mating designs for Douglas fir. Interspecific hybridization is of significance in *Larix* (Paques 1989). Some of these species display marked provenance X planting site interaction, particularly in relation to temperature sensitivity, and definition of breeding zones is an important feature of some programmes, e.g. for *Pinus sylvestris* (Danell 1991), *P. banksiana* (Miller 1984), and *Picea abies* (Dietrichson 1989). Major selection criteria are vigour, straightness and wood properties, frost resistance in *Pinus sylvestris* (Oleksyn 1991) and *Picea abies* (Van de Sype 1989), disease resistance in *P. contorta* (Ying *et al.* 1985), and insect resistance in *Picea abies.* The open-pollinated orchard, of which large areas have been established, remains the main approach to the capture of genetic gain, but most orchards are not producing enough seed to make much impact on the requirement for planting stock. Approximately 200 ha of clonal tests have been established in Germany (Kleinschmit & Svolba 1989), and over 10 000 clones are under test in Sweden (Danell 1991). Commercial use of clones, however, remains very limited, due partly to a reluctance of growers to pay higher prices for cuttings (P. Monchaux pers. com.) and to doubts concerning the capacity to retain sufficient levels of juvenility through a lengthy sequential propagation programme (Kleinschmit & Svolba 1989). Most gene pools are reasonably well preserved. The broader development of technologies for the mass production of superior full-sib families would be useful for these species. Some programmes would benefit from a greater capacity to manipulate frost tolerance, and some from the ability to manipulate flowering. Breeding of the long rotation species would benefit particularly from reductions of the generation interval through early selection. It should be noted that in some countries, plantations of these species are used for many purposes. Intensive breeding for enhanced timber production in such plantation programmes may not be desirable, particularly where reduction of genetic diversity is involved. Legislation reinforces this view in some countries.

Non-industrial Species

There is a large requirement worldwide, mainly in developing countries, for fuelwood plantations, plantings to arrest and reverse land degradation, and, most importantly, tree plantings integrated into farming systems to provide mulch, fodder, shade, fuelwood, timber, fruit, and soil improvement and protection. Many of the species discussed in this category can be used in industrial plantations (as defined above), and identification of a "non-industrial" group of species is a demarcation which is more convenient than clearly defined.

Large fuelwood plantations, e.g. of tropical eucalypts or casuarinas, located near cities are one approach to the urban fuelwood crisis. Operated on a commercial basis, these can be viewed as industrial species. Current plantings are not widely based on improved material, but simple improvement programmes would be justified in some of these.

Major rehabilitation plantings, e.g. with casuarinas or acacias, are publicly funded. Improvement beyond species and provenance selection may be difficult to justify on purely economic grounds, and may furthermore be of limited desirability in those cases where maintenance of diversity in such plantings is of major importance.

Examples of integration of trees into farming systems include:

- the alley farming systems, incorporating the use of trees such as *Gliricidia* or *Leucaena* for protection and mulch production.

- shade tree systems, e.g. *Inga* and *Erythrina* species grown as shade over coffee in Latin America.
- trees grown principally for fodder and fuelwood in agroforestry systems, e.g. *Prosopis* and arid zone acacias.
- trees grown in association with crops mainly for timber, e.g. *Paulownia* in China and *Dalbergia* in Pakistan.
- windbreak plantings, e.g. of *Casuarina.*

Gliricidia is a fast growing tree used widely in the Philippines, India, Sri Lanka, Indonesia, East Africa and its native Central and South America for fodder, fuelwood, live fences, construction wood and mulch. Subjected to agricultural use for centuries, its true natural range is uncertain, and many land races have been created (Hughes 1987). Flowers are produced within the first year or two, seed production is prolific, and cuttings are easily propagated. Large collections have been made and some provenance comparisons carried out, but breeding is not very advanced. A breeding seed orchard of a particular desirable provenance will be established at the ICRAF research station in Zambia (Simons 1992b). Cuttings of very limited numbers of locally selected clones are the planting stock in some areas. Genetic variation with respect to a range of characters is substantial (Simons 1992b). Selection criteria vary with use, but timing and reliability of leaf production are important parametres for fodder production.

Leucaena is a long cultivated tree of importance in South East Asia, Nepal, India, Africa and Latin America. It is nitrogen fixer which provides good forage, green manure, firewood and small timber. Frequently used with other crops, it is also the major component of energy plantations established in the Philippines (Durst 1987). *L. leucocephala* seeds very precociously and prolifically, but is not easily propagated by cuttings. Unlike other tree species reviewed here, *L. leucocephala* is frequently self-pollinating, although this is not a feature of other species of the genus. With respect to genetic improvement, *Leucaena* has received more attention than most other non-industrial trees. Two directions have been followed in the Hawaiian programme of Dr Brewbaker:

- testing of self-pollinated lines
- interspecific hybridization

In most areas, the psyllid *Heteropsylla cubana* causes serious damage (McDicken & Taylor 1988, Durst 1987), and resistance to this pest is an important selection criterion. Others include biomass, cold tolerance, seedlessness, fodder quality, and dry season leaf retention. Although yield increases of well over 100% have been achieved through simple selection (MacDicken & Mehl 1990), breeding is yet to have a major impact in planting programmes, many of which have suffered from off-site planting and a very narrow genetic base. For *L. leucocephala*, locally produced seed is what is generally used. Some *Leucaena* species are under threat, and collections are underway.

Calliandra calothyrsus is a nitrogen fixing, thornless legume producing high quality fuelwood and fodder, with considerable potential in site amelioration, particularly sites too wet for *Gliricidia* and *Leucaena*. The species flowers in its second year and coppices and roots well. Provenance and progeny tests have been planned (Burley & Barnes 1989, MacQueen 1992).

Prosopis is an arid zone genus of about 45 species, highly tolerant of drought and of difficult soils, grown widely on marginal land in India and countries in South America and with potential in many other countries. The trees provide fuelwood, construction timber and fodder. Coppicing is prolific (Walker 1988). A range of provenance trials coordinated *i a* by FAO, involving several species, are in existence (Cossalter *et al.* 1987, FAO 1988a, Matheson 1990, Dunsdon *et al.* 1991, Bach 1992, Graudal & Thomsen 1992). Objectives of breeding include growth increases, reduction of thorniness, and improvement of fodder quality.

A mimosoid legume genus of about 400 species, *Inga* is planted widely in Latin America to provide shade over coffee, wood, fuelwood, nitrogen fixation, green manure and fodder. Seed recalcitrance has limited plantings in other regions. Little collection work has been undertaken.

There are about 135 African acacias, of which *A. nilotica, A. tortilis, A. senegal*, and *Faidherbia alba* are particularly important. Very drought resistant, these species provide fodder, timber, charcoal, shade, honey, gum arabic, tannins and dune stabilization in arid areas of Africa and the Indian subcontinent. Some species display salt tolerance. The species tend to be prolific seeders, seeds can be stored, and stumps produce coppice (Fagg & Barnes 1990). Species are characterized by considerable genetic variation, and some collections have been undertaken and provenance trials established under the auspices of FAO (Cossalter *et al.* 1987, FAO 1988a, Koyo 1989, Bach 1992, Graudal & Thomsen 1992). *A. nilotica* is regarded as a weed in Indonesia and Australia.

Erythrina poeppigiana and *E. fusca* are used as pollarded trees providing shade and mulch for coffee in Central America. *E. berteroana* is used in living fences. The species flower precociously (at three years) and abundantly. Cuttings root well, and direct planting of unrooted stakes is common. Improvement work conducted at CATIE is along poplar lines, involving the clonal testing of individuals from natural and planted stands.

The genus *Paulownia* is represented by about nine species native to China, where there are reportedly 1 300 000 ha planted in mixed cultivation with crops (Chinese Acad. For. 1986). The species is suited to intercropping due to late leaf emergence and leaf fall, and a deep root system (Chinese Acad. Forestry 1986), and produces timber within five to six years. *P.tomentosa* and *P.kawakamii* usually flower in the second year after planting, and *P.fortunei* and *P. catalpifolia* in the fifth to sixth year. Seeds can be stored satisfactorily, and cuttings can be propagated from both young seedlings and the roots of mature trees. In China, the approach to breeding envisaged is controlled pollination (in particular interspecific hybridization) and then clonal testing (Chinese Acad. For. 1986). Selection criteria are height (to minimize shading of crops) and freedom from the many diseases and pests. Planting stock currently used comprises selected clones of some good hybrids. The North American breeding programme is at the species and provenance evaluation stage (NCSU-Industry Hardwood Research Co-op. 1990).

Dalbergia sissoo has long been valued in Pakistan, India and Nepal for its high value cabinet timber, fodder, fuelwood, charcoal, honey, medicinal properties, windbreak quality and capacity to fix nitrogen (White 1990). It is used both with crops and as an industrial plantation species (Sheikh 1988). The species flowers abundantly and regularly from age four, and good rooting of cuttings has led to their use as the most common method of propagation (Sheikh 1988, White 1990). The breeding system of the species is not well known (White 1990). Seed orchards were established in Pakistan in 1975 but improvement is not very advanced (White 1990). Selection criteria are wood yield and quality, but with consideration given also to fodder and honey properties.

Casuarina equisetifolia and *C. cunninghamii* are used widely for windbreak plantings and fuelwood production in areas of the tropics. *C. equisetifolia* comprises the major component of a shelterbelt covering over 1 000 000 ha in China (National Research Council 1984), Coastal sand dune fixation is one of the most successful applications of this species. Both species are dioecious. For industrial plantations, provenance trial establishment, plus tree selection, and the establishment of Breeding Seed Orchards using the multiple population approach has been recommended (Burley & Barnes 1989). Breeding of *Casuarina* in Egypt incorporates selection for drought resistance (Burley 1985). A sterile hybrid between *C.junghuhniana* and *C.equisetifolia* is propagated vegetatively and widely planted in Thailand (Willan *et al.* 1990).

Azadirachta indica is a promising species of uncertain natural distribution due to centuries of traditional use (Willan *et al.* 1990). Poor seed viability is a serious problem. Seed collections have been made and provenances are under test (F/FRED 1988, Nikles 1992). Growth rate, insect resistance and azadirachtin content will be important selection criteria (Nikles 1992).

A large number of other species are of potential value in non-industrial plantings, most remaining poorly studied. These include species of *Sesbania, Pithecellobium, Albizia, Caesalpinia* and *Parkinsonia* (MacDicken & Mehl 1990, Dunsdon *et al.* 1991).

To summarize for the non-industrial species then, moderately comprehensive data on biological features influencing breeding are available for only a handful of species. Several of these taxa are highly variable and display features which render them particularly amenable to very rapid improvement by traditional means - precocious and abundant seeding, coppicing, and ease of vegetative propagation. Biological features of many potentially valuable non-industrial species remain largely unknown, and collection of data of this type will be an important component of the initial phase of any breeding programmes. Although not well studied, some gene pools are under threat, and some species display unpredictable patterns of variation as a result of human disturbance, rendering sampling strategies more difficult to design.

Considered collectively, non-industrial trees differ markedly from industrial trees in the wide variety of end uses and thus selection criteria - e.g. fuelwood quality, foliage yield, fodder quality, coppicing ability, phenological characteristics, timber quality, nitrogen fixation, root features, tree form and resistance to insects, disease, drought and other stresses. Some of these, e.g. fuelwood quality, are very complex. For some farmers, reliability of production may be more important than mean annual yield (Simons 1992a). In agroforestry systems, the interactive effect of the tree component with the associated crop is important (Willan *et al.* 1990). Desired features of agroforestry species have sometimes been integrated in terms of "ideotypes" (Glover 1990). Although selection work has been undertaken in a few programmes, and clonal propagation of selected clones is used in *Paulownia* and *Erythrina,* most non-industrial species remain at the species testing stage. The work involved in just selecting the most promising of perhaps thousands of other potential valuable but as yet untested species is formidable.

Improvement beyond species and provenance testing may not be warranted in many programmes with non-industrial species:

- the value of the product may not warrant the effort, particularly where the tree crop is the lesser component of the farm product.
- selection criteria for a particular species may vary considerably with farming system and region.
- there may be some difficulties in transferring the results of central breeding programmes to the growers.

Major emphases of improvement for non-industrial species are likely to be taxonomic studies of variation, species and provenance testing, the assessment of some features of reproductive biology, and conservation activities. Where regionally consistent selection criteria can be defined, some clonal selection may be important for species easily propagated vegetatively, and perhaps seed tree selection for species propagated by seed. Sophisticated breeding programmes are unlikely to be warranted for most of these species. Interspecific hybridization is of proven value for some taxa, and holds promise also for others, but must be used in the context of a sound improvement programme.

Tree Improvement in Developed Countries

Tree improvement in developed countries mainly concerns industrial species, although agroforestry is becoming more important. Some form of genetic improvement programme is a feature of most plantation programmes for industrial species. Most are based on recurrent cycles of phenotypic selection (frequently followed by testing of progeny) and recombination among selected individuals. Exceptions are the poplars and cryptomeria, for which clonal selection has been the traditional approach. Vigour, stem form and wood quality are the most important selection criteria in most programmes. Resistance to disease or insects in particular, and also to cold and drought, are significant in several programmes. The importance of disease resistance in the poplars is worthy of special note. Most characters of major importance are under polygenic control. Although there are some clonal programmes, and programmes based on the propagation of full-sib families, the open-pollinated orchard is the most common approach to the capture of genetic gain. Gene pools are reasonably well preserved for most of these species, although locally adapted populations may be in danger of depletion or genetic pollution in some cases.

Breeding activities (e.g. crossing programmes, progeny testing, selection) are generally carried out under the guidance of personnel with expertise in quantitative plant breeding, supported by good systems of field experimentation and data base management. The personnel involved are frequently employees of the grower organizations, although independent research institutes are involved in some cases. Breeding cooperatives, sharing genetic material, knowledge and work, are particularly important for *Pinus taeda, P. elliottii, P. radiata,* and some north American hardwoods (e.g. the North Carolina, Florida and North Central Hardwood cooperatives, and the Southern Tree Breeding Association).

Supportive research (e.g. assessment of reproductive behaviour, mating patterns, genetic parametres) is conducted frequently by the above groups, but also by universities and other research organizations. Strategic research (aimed at the development of better breeding methods) is undertaken primarily by universities and other research organizations, although several of the larger private growers also conduct long term strategic research. As pointed out by Kanowski (1993), research in tree improvement is greatly under-resourced, in both absolute terms and relative to agriculture.

Significant genetic gains are being achieved in these breeding programmes but, in particular for the long rotation species, there has been only a minor impact to date on the genetic quality of plantations.

Major limitations to rapid improvement with most of these species are:

- Long generation intervals, related to:
  - o poor juvenile-mature correlations. Although there is some debate about this, in general, for many characters, these correlations are not high enough to permit very early selection. For *Pinus elliottii,* for example, it has been suggested that growth up to five years involves different physiological mechanisms to subsequent growth (Hodge *et al.* 1991). It should be noted that selection intervals for many of the breeding programmes discussed above are shorter than those recommended by some authorities. Zobel & Talbert (1984), for example, advocate half rotation age as the appropriate selection interval.
  - o long juvenile phase with respect to flowering.
- Low effectiveness of selection for many characters. Most major selection criteria are of low heritability, such that progeny testing is necessary. Some characters, e.g. resistance to frost, drought and insects, are difficult to test in a systematic way.

- Through the use of open-pollinated orchards, the exploitation of only a part of the genetic variation available.

Major research priorities for these species should be the broader development of methods for the propagation of full-sib families and clones, and the development of methods for early and more accurate selection and the promotion of precocious flowering.

It is of some interest to note that an apparently slow rate of advancement with respect to genetic quality of plantations is not unique to forest tree species. The avocado varieties Fuerte and Hass, for example, were introduced to the industry over 60 years ago, but remain the major commercial varieties. Similarly, the major kiwifruit varieties were selected several decades ago.

The advances which have taken place with these crops have been horticultural, not genetic. Silvicultural advances are critical also for forest tree species, but it is unlikely that the intensity of horticulture given to fruit tree crops would be affordable. For some agricultural species, the genetic superiority of modern varieties requires intensive agriculture, including comprehensive pest and disease management, for its expression. Modern high yielding varieties tend to lack much of the natural resistance of old land races or wild relatives (Harms 1992). The lower intensity of silviculture affordable renders desirable, then, the more careful management of genetic improvement for forest tree species.

Biological Control of Pigeon Pea Wilt

The pigeon pea (*Cajanus cajan*) is a perennial legume from the family Fabaceae. Since its domestication in South Asia at least 3 500 years ago, its seeds have become a common food grain in Asia, Africa, and Latin America.

The cultivation of the pigeon pea goes back at least 3,500 years. The centre of origin is the eastern part of peninsular India, including the state of Odisha, where the closest wild relatives (*Mansi*) occur in tropical deciduous woodlands.

Archaeological finds of pigeon pea include those from two Neolithic sites in Odisha, Gopalpur and Golbai Sassan dating between 3,400 and 3,000 years ago, and sites in South India, Sanganakallu and Tuljapur Garhi, also dating back to 3,400 years ago. From India it travelled to East Africa and West Africa. There, it was first encountered by Europeans, so it obtained the name Congo Pea. By means of the slave trade it came to the American continent, probably in the 17th century.

Cultivation

Today, pigeon peas are widely cultivated in all tropical and semitropical regions of both the Old and the New Worlds. Pigeon peas can be of a perennial variety, in which the crop can last three to five years (although the seed yield drops considerably after the first two years), or an annual variety more suitable for seed production.

Pigeon peas are an important legume crop of rainfed agriculture in the semiarid tropics. The Indian subcontinent, eastern Africa and Central America, in that order, are the world's three main pigeon pea-producing regions. Pigeon peas are cultivated in more than 25 tropical and subtropical countries, either as a sole crop or intermixed with cereals, such as sorghum (*Sorghum bicolour*), pearl millet (*Pennisetium glaucum*), or maize (*Zea mays*), or with other legumes, such as peanuts (*Arachis hypogaea*). Being a legume, the pigeon pea enriches soil through symbiotic nitrogen fixation.

The crop is cultivated on marginal land by resource-poor farmers, who commonly grow traditional medium- and long-duration (5–11 months) landraces. Short-duration pigeon peas (3–4 months) suitable for multiple cropping have recently been developed. Traditionally, the use of such input as fertilizers, weeding, irrigation, and pesticides is minimal, so present yield levels are low (average = 700 kg/ha). Greater attention is now being given to managing the crop because it is in high demand at remunerative prices.

Pigeon peas are very drought resistant, so can be grown in areas with less than 650 mm annual rainfall. With the maize crop failing three out of five years in drought-prone areas of Kenya, a consortium led by the International Crops Research Institute for the Semi-Arid Tropics (ICRISAT) aimed to promote the pigeonpea as a drought-resistant, nutritious alternative crop. Successive projects encouraged commercialization of legumes, by stimulating the growth of local seed production and agro-dealer networks for distribution and marketing. This work, which included linking producers to wholesalers, helped to increase local producer prices by 20–25% in Nairobi and Mombasa. The commercialization of pigeonpea is now enabling farmers to buy assets, ranging from mobile phones to productive land and livestock, and is opening pathways for them to move out of poverty.

World production of pigeon peas is estimated at 4.3 million tons[2]. About 82% of this is grown in India. These days it is the most essential ingredient of animal feed used in West Africa, most especially in Nigeria, where it is also grown.

Uses

Pigeon peas are both a food crop (dried peas, flour, or green vegetable peas) and a forage/cover crop. In combination with cereals, pigeon peas make a well-balanced human food. The dried peas may be sprouted briefly, then cooked, for a flavor different from the green or dried peas. Sprouting also enhances the digestibility of dried pigeon peas via the reduction of indigestible sugars that would otherwise remain in the cooked dried peas.

In India, split pigeon peas (*toor dal*) also called Kandhi pappu in Telugu, Togari bele in Kannada and Tuvaram paruppu in Tamil are one of the most popular pulses, being an important source of protein in a mostly vegetarian diet. In regions where it grows, fresh young pods are eaten as a vegetable in dishes such as *sambar*. In Ethiopia, not only the pods, but also the young shoots and leaves are cooked and eaten.

In some places, such as the Caribbean coast of Colombia, Dominican Republic, Panama and Hawaii, pigeon peas are grown for canning and consumption. A dish made of rice and green pigeon peas (called *moro de guandules*) is a traditional food in the Dominican Republic. Pigeon peas are also made as a stew, with plantain balls. In Puerto Rico *arroz con gandules* is made with rice and pigeon peas and is a typical dish. Trinidad and Tobago and Grenada have their own variant, called *pelau*, which includes either beef or chicken, and occasionally pumpkin and pieces of cured pig tail. In the Atlantico department of Colombia the sopa de guandú con carne salada (or simply "guandules") is made with pigeon peas.

In Thailand, pigeon peas are grown as a host for scale insects which produce lac. Pigeon peas are in some areas an important crop for green manure, providing up to 90 kg nitrogen per hectare (Adu-Gyamfi et al., 2007) . The woody stems of pigeon peas can also be used as firewood, fencing and thatch.

Genome Sequence

The pigeon pea is the first seed legume plant to have its complete genome sequenced. The first draft was done by a group of 31 Indian scientists from the Indian Council of Agricultural Research under the leadership of Nagendra Kumar Singh.

In contrast to the mature seeds, the immature seeds are generally lower in all nutritional values, however they contain a significant amount of vitamin C (39 mg per 100 g serving) and have a slightly higher fat content. Research has shown that the protein content of the immature seeds is of a higher quality.

Pigeon pea plants are susceptible to infection by the wilt pathogen at all stages of development. The main symptom of the disease is wilting characterized by gradual, sometimes sudden, yellowing, withering and drying of leaves followed by drying of entire plant or some of its branches. Such plants exhibit loss of leaf turgidity, interveinal clearing and chlorosis before death. Isolated wilted plants may appear about a month after sowing and patches of dead plants in the field, usually at the flowering and podding stage, are the first indication of wilt. The characteristic symptom of the disease in the adult plants is a purple band extending upwards from the base of the main stem. The band can clearly be seen in pigeon pea, when the green stems of healthy plants are compared to the diseased stems with coloured lines. Partial wilting is quite common and distinguishes the disease from termite damage, drought and Phytophthora blight, which also kill the plants. Partial wilting is due to infection of lateral roots, while total wilt is a result of tap root infection. If an infected plant is split open below the purple band, browning of the stem and brown to black discolouration of xylem vessels is visible. Sometimes lower branches show die-back with the purple band extending from the tip downward and intensive xylem blackening.

Prevention and control

Resistant Crop Cultivars

There has been considerable research on the control of this disease. The major emphasis has been on the identification of resistance and the production of wilt-resistant cultivars. Pot and field screening techniques have been developed for the identification of resistance. For a pot-screening technique, autoclaved pigeon pea stem pieces are mixed with non-autoclaved alfisol in pots. The soil is inoculated with a F. udum culture multiplied on sand-pigeon pea flour (9:1) medium. A susceptible cultivar is sown and the wilted plant material is re-incorporated in the pots for three cycles.

Wilt-diseased plots have been used to screen crops against several vascular wilts.

Multilocation tests conducted in India and eastern Africa helped to identify several lines resistant to wilt at several locations. Moderately resistant lines are available in all maturity groups. Some of these lines also show resistance across locations and seasons. These lines includes ICP 8863 (Maruti), ICP 9145, ICP 9174, ICP 12745, ICPL 333, ICPL 8363, ICPL 88047, BWR 37. DPPA 85-2, DPPA 85-3, DPPA 85-8, DPPA 85-13, DPPA 85-14, Bandpalera, ICP 47769, ICP 9168, ICP 10958, ICP 11299, C11 (ICP 7118), BDN1 (ICP 7182). ICP 8864 and ICP 9145 have been released for commercial cultivation in Malawi and have become very popular (Reddy et al., 1995). The lines that showed resistance in Kenya were ICP 8869, ICP 9145 and ICP 10960. The lines that were resistant in Malawi are ICP 7855, ICP 9145, ICP 9154, ICP 9174, ICP 9177, ICP 10958, ICP 11297, ICP 11299 and ICP 12738. Maruti (ICP 8863), which has been released for commercial cultivation in India, has become very popular in peninsular India.

Lines which combine wilt resistance and resistance to other diseases have been identified. For example, ICPL 87 and C11 possess resistance to wilt, sterility mosaic and Phytophthora blight; ICP 7667, ICP 8861, ICP 8662 (Hy3C) to wilt, sterility mosaic and powdery mildew; ICPL 81 to wilt and halo blight; BDN1 to wilt, Phytophthora blight and halo blight; ICP 8661, ICP 8662, ICP 8867, ICP 8869, ICP 10962 to wilt and Alternaria blight.

Inheritance of Resistance

Limited information is available on the inheritance of resistance to the disease. In a cross between one resistant (ICP 8863) and two susceptible (ICP 2376 and LRG 3C) lines, resistance was found to be controlled by a single recessive gene. The gene was designated pwr1.

Nature of Resistance

The nature of resistance was analysed in cultivars that were resistant and susceptible to wilt. The extract from a resistant cultivar (C-11-6) inhibited spore germination and the growth of the germtube. The inhibitory compounds in the resistant cultivar were identified as chlorogenic acid and caffeic acid.

Cultural Practices

Pigeon pea is generally grown in inter- and mixed-cropping systems in rotation with other crops. However, since the fungus survives on deep-seated roots of the host, below the depth of ordinary cultivation, the success of rotation will depend upon the field sanitation (removal of affected plants with their roots), hot weather cultivation etc. A 4-5-year rotation has been found to free the field completely of the wilt pathogen. Field studies conducted at ICRISAT Centre have shown that crops such as sorghum, castor, maize and groundnut inhibit the soil population of F. udum.

One-year breaks with either sorghum or fallow reduced wilt in the following pigeon pea crop from 60-90% to 16 and 31%, respectively. Pigeon pea rotation with tobacco has been recommended as a possible means of control because of the adverse effect of tobacco root exudates on the pathogen. Wilt incidence was increased with increasing pigeon pea biomass .

Root exudates from a range of crops, which are frequently intercropped with pigeon pea, and green manuring reduce wilt incidence either by reducing population of F. udum or by increasing the antagonistic activity of the microbial population in the soil.

A higher incidence of wilt in sterilized soil than in unsterilized soil was attributed to the action of antagonistic organisms such as Aspergillus niger, Rhizopus nigricans and Bacillus subtilis.

The soil composition and certain cultural practices can affect wilt incidence.

3

Advances in Biological Control of Some Pre and Post Harvest Crops

Postharvest Diseases of Fruits

Stored produce is subject to a variety of rots and decay caused by fungi or bacteria. These organisms may cause soft spots or light brown lesions on fruits and vegetables. Fungal growth, in a variety of colours, may also be apparent on the surface of infected produce. In time, the entire fruit or vegetable can become dry and mummified, or, under moist conditions, a soft, wet mass.

Postharvest diseases may start before or after harvesting. Plants or fruits infected in the field may not develop symptoms until stored. Once in storage, infections continue to develop on the fruits and vegetables. Wounds, cuts, or bruises caused during harvesting are common entry points for bacteria and fungi. Penetration can also occur during storage through natural openings, such as lenticels, or directly through the cuticle and epidermis.

Spread of the infection usually requires the presence of warm temperatures and high moisture, although some storage rots can continue to develop at low temperatures. *Penicillium*, a fungus which causes blue and green mold on fruits, can continue to grow slowly even at temperatures near freezing. Once produce is infected, the disease can spread to healthy produce by direct contact. Insects can also be involved in spreading disease. Fruit flies, attracted to fruits and vegetables infected with "sour rot" carry spores of the fungus *Geotrichum* from infected produce to healthy produce.

To prevent postharvest diseases, monitor and manage diseases during the growing season. Stake tomatoes to keep fruits from contacting the soil. Pick fruits and vegetables during cool, dry weather and avoid bruising or wounding them.

Biological Control-A new Climate

Biological control agents have had a difficult time making it from the laboratory to the marketplace. This difficulty has been largely due to problems of ineffectiveness when the biocontrol agents are subjected to the "uncontrolled" environment of the field and to the lack of economic incentive to develop the technology necessary for their effective use. Lack of support for widespread, sustained research in this area has been largely due to the effectiveness and perceived safety of chemical fungicides. However, because of the changing socioeconomic climate and recent advances in genetic engineering, interest in biological control as a meaningful approach to pest and disease management has been rejuvenated.

Several venture capital companies or subsidiaries of well-known agrochemical companies, with a focus on biological control, have arisen in response to this new climate. Therefore, the outlook for developing economically viable biological control products looks very promising.

Postharvest Biological Control-A unique Environment

Two basic approaches are available for using microorganisms to control postharvest diseases: use and management of the beneficial micro flora that already exist on fruit and vegetable surfaces or the artificial introduction of antagonists against postharvest pathogens. Our knowledge of methods to manipulate naturally occurring populations of mixed species of microorganisms in a beneficial manner, however, is meager, and the greatest use of biological control (pre- and postharvest) has come through the artificial introduction of large numbers of a known antagonist, Several factors indicate that postharvest biological control with the use of artificially introduced antagonists may prove to be an effective technology. First, environments for the storage of harvested commodities are often controlled and maintained. This control should lessen the problem of introducing the bio-control agent into an unpredictable and highly variable environment, which previously has been a limiting factor in field-released biocontrol measures. Second, the ability to target the biocontrol agent to the site needed for activity is enhanced in postharvest application. Third, the high value of harvested commodities may make the application of elaborate biological control procedures more cost-effective than similar procedures in the field. Fourth, for some commodities harvested for fresh-market consumption, protection from postharvest diseases is only needed for a short duration.

Although it appears that the postharvest environment may be especially favourable f the development of biocontrol agents, a considerable investment of time and money is required to establish whether any particular organism has commercial potential. There fore, the isolation, screening, and selection of potential antagonists should receive careful deliberation. As indicated by Roberts (1991), any potential antagonist must have the ability to colonize and persist on the commodity at effective levels; be compatible with other postharvest practices, processes, and chemicals; and be effective under cold and, in some cases, controlled-atmosphere conditions. Additionally, the organism must be amenable to large-scale production using low-cost substrates. Given these constraints, several antagonists for control of postharvest diseases of fruit have been identified for their commercial potential.

Research around the world has proven the effectiveness of biological control of postharvest diseases and, at the same time, identified its limitations. Many of these limitations result from knowledge gaps that can be addressed by additional research to better understand the mechanisms of biocontrol on fruit, microbial ecology of the fruit surfaces, and the survival of antagonists under adverse conditions. Additional research is also needed to prolong the shelf life of antagonists, determine the compatibility of biocontrol with other alternative treatments, and adapt biocontrol to various postharvest handling practices, and to new production technologies, such as mechanical harvesting, that may predispose fruit to decay.

The greater use of biological control of postharvest diseases can be achieved by expanding the application of available products to new commodities and different pathogens. The products currently available were developed to control decays originating from wound infections; however, significant postharvest losses on various fruits may also occur from decays originating from latent infections that develop before harvest. Controlling these decays has become the new frontier for biological control. New approaches utilizing natural fruit wax and artificial membranes were developed to find antagonists effective in controlling latent infections under laboratory conditions. The high cost of tests required for registration prohibited the commercial development of several effective biocontrol agents. A new framework is needed for streamlining the registration of low risk biocontrol agents without known deleterious effects to human health. This is especially relevant for those agents that are consumed with foods in large quantities (e.g., in apple cider, fruit juices, or with various cultured food products (probiotics)). Biological control of postharvest diseases has been accepted by the fruit industry as a viable alternative to synthetic fungicides, and products registered in the United States can be used on organically-labellled fruit.

Biocontrol of Plant Diseases Through Microbes and Endomycorrhizae

A mycorrhiza is a symbiotic (generally mutualistic, but occasionally weakly pathogenic) association between a fungus and the roots of a vascular plant. In a mycorrhizal association, the fungus colonizes the host plant's roots, either intracellularly as in arbuscular mycorrhizal fungi (AMF or AM), or extracellularly as in ectomycorrhizal fungi. They are an important component of soil life and soil chemistry.

Mutualist Dynamics

Mycorrhizae form a mutualistic relationship with the roots of most plant species. While only a small proportion of all species has been examined, 95% of those plant families are predominantly mycorrhizal. They are named after their presence in the plant's rhizosphere (root system).

Recent research with ectomycorrhizal in boreal forests, has indicated mycorrhizae and plants have a relationship that may be more complex than simply mutualistic. This relationship was noted when mycorrhizae where unexpectedly found hoarding nitrogen from plant roots in times of nitrogen scarcity. Researchers argue that some mycorrhizae distribute nutrients based upon the environment with surrounding plants and other mycorrhizae, in a manner similar to a capitalistic market. Researchers go on to explain how this updated model explains why mycorrhizae do not alleviate plant nitrogen limitation, and why plants can switch abruptly from a mixed strategy with both mycorrhizal and nonmycorrhizal roots to a purely mycorrhizal strategy as soil nitrogen availability declines.

Sugar-water/mineral Exchange

This mutualistic association provides the fungus with relatively constant and direct access to carbohydrates, such as glucose and sucrose. The carbohydrates are translocated from their source (usually leaves) to root tissue and on to the plant's fungal partners. In return, the plant gains the benefits of the mycelium's higher absorptive capacity for water and mineral nutrients due to the comparatively large surface area of mycelium: root ratio, thus improving the plant's mineral absorption capabilities.

Plant roots alone may be incapable of taking up phosphate ions that are demineralized in soils with a basic pH. The mycelium of the mycorrhizal fungus can, however, access these phosphorus sources, and make them available to the plants they colonize. Thus many plants are able to obtain phosphate, without using soil as a source. For example, in some dystrophic forests large amounts of phosphate are taken up by mycorrhizal hyphae acting directly on leaf litter, bypassing the need for soil uptake. *Inga alley cropping*, proposed as an alternative to slash and burn rainforest destruction, relies upon mycorrhiza within the Inga Tree root system to prevent the rain from washing phosphorus out of the soil. In some cases, the transport of water, carbon, and nutrients could be done directly from plant to plant through mycorrhizal networks that are underground hyphal networks created by mycorrhizal fungi that connect individual plants together.

Suillus tomentosus, a basidiomycete fungus, produces specialized structures known as tuberculate ectomycorrhizae with its plant host lodgepole pine (*Pinus contorta* var. *latifolia*). These structures have been shown to host nitrogen fixing bacteria which contribute a significant amount of nitrogen and allow the pines to colonize nutrient-poor sites.

Mechanisms

The mechanisms of increased absorption are both physical and chemical. Mycorrhizal mycelia are much smaller in diametre than the smallest root, and thus can explore a greater volume of soil, providing a larger surface area for absorption. Also, the cell membrane chemistry of fungi is different from that of plants (including organic acid excretion which aids in ion displacement). Mycorrhizas are especially beneficial for the plant partner in nutrient-poor soils.

Disease and Drought Resistance and its Correlation to Mycorrhizae

Mycorrhizal plants are often more resistant to diseases, such as those caused by microbial soil-borne pathogens. AMF was also significantly correlated with soil biological fertility variables such as soil fungi and soil bacteria, including soil disease. Furthermore, AMF was significantly correlated with soil physical variable, but only with water level and not with aggregate stability. and are also more resistant to the effects of drought.

Colonization of Barren Soil

Plants grown in sterile soils and growth media often perform poorly without the addition of spores or hyphae of mycorrhizal fungi to colonise the plant roots and aid in the uptake of soil mineral nutrients. The absence of mycorrhizal fungi can also slow plant growth in early succession or on degraded landscapes. The introduction of alien mycorrhizal plants to nutrient-deficient ecosystems puts indigenous non-mycorrhizal plants at a competitive disadvantage.

Resistance to Toxicity

Fungi have been found to have a protective role for plants rooted in soils with high metal concentrations, such as acidic and contaminated soils. Pine trees inoculated with *Pisolithus tinctorius* planted in several contaminated sites displayed high tolerance to the prevailing contaminant, survivorship and growth.

One study discovered the existence of *Suillus luteus* strains with varying tolerance of zinc. Another study discovered that zinc-tolerant strains of *Suillus bovinus* conferred resistance to plants of *Pinus sylvestris*. This was probably due to binding of the metal to the extramatricial mycelium of the fungus, without affecting the exchange of beneficial substances.

Occurrence of Mycorrhizal Associations

At around 400 million years old, the Rhynie chert contains the earliest fossil assemblage yielding plants preserved in sufficient detail to detect mycorrhizas - and they are indeed observed in the stems of *Aglaophyton major*.

Mycorrhizas are present in 92% of plant families studied (80% of species), with arbuscular mycorrhizas being the ancestral and predominant form, and indeed the most prevalent symbiotic association found in the plant kingdom. The structure of arbuscular mycorrhizas has been highly conserved since their first appearance in the fossil record, with both the development of ectomycorrhizas, and the loss of mycorrhizas, evolving convergently on multiple occasions.

Types of Mycorrhiza

Mycorrhizas are commonly divided into *ectomycorrhizas* and *endomycorrhizas*. The two types are differentiated by the fact that the hyphae of ectomycorrhizal fungi do not penetrate individual cells within the root, while the hyphae of endomycorrhizal fungi penetrate the cell wall and invaginate the cell membrane. Additionally, many plants in the order Ericales form a third type, ericoid mycorrhizas, while some members of the Ericales form arbutoid and monotropoid mycorrhizas. All orchids are myco-heterotrophic at some stage during their lifecycle and form orchid mycorrhizas with a range of basidiomycete fungi.

Endomycorrhiza

Endomycorrhizas are variable and have been further classified as arbuscular, ericoid, arbutoid, monotropoid, and orchid mycorrhizas. Arbuscular mycorrhizas, or AM (formerly known as vesicular-arbuscular mycorrhizas, or VAM), are mycorrhizas whose hyphae enter into the plant cells, producing structures that are either balloon-like (vesicles) or dichotomously branching invaginations (arbuscules).

The fungal hyphae do not in fact penetrate the protoplast (i.e. the interior of the cell), but invaginate the cell membrane. The structure of the arbuscules greatly increases the contact surface area between the hypha and the cell cytoplasm to facilitate the transfer of nutrients between them. Arbuscular mycorrhizas are formed only by fungi in the division Glomeromycota. Fossil evidence and DNA sequence analysis suggest that this mutualism appeared 400-460 million years ago, when the first plants were colonizing land. Arbuscular mycorrhizas are found in 85% of all plant families, and occur in many crop species.

The hyphae of arbuscular mycorrhizal fungi produce the glycoprotein glomalin, which may be one of the major stores of carbon in the soil. Arbuscular mycorrhizal fungi have (possibly) been asexual for many millions of years and, unusually, individuals can contain many genetically different nuclei (a phenomenon called heterokaryosis).

Ectomycorrhiza

Figure: *Ectomycorrhizal beech*

Ectomycorrhizas, or EcM, are typically formed between the roots of around 10% of plant families, mostly woody plants including the birch, dipterocarp, eucalyptus, oak, pine, and rose families, orchids, and fungi belonging to the Basidiomycota, Ascomycota, and Zygomycota. Some EcM fungi, such as many *Leccinum* and *Suillus*, are symbiotic with only one particular genus of plant, while other fungi, such as the *Amanita*, are generalists that form mycorrhizas with many different plants. An individual tree may have 15 or more different fungal EcM partners at one time. Thousands of ectomycorrhizal fungal species exist, hosted in over 200 genera. A recent study has conservatively estimated global ectomycorrhizal fungal species richness at approximately 7750 species. Although, on the basis of estimates of knowns and unknowns in macromycete diversity, a final estimate of ECM species richness would likely be between 20000 and 25000.

Ectomycorrhizas consist of a hyphal sheath, or mantle, covering the root tip and a Hartig net of hyphae surrounding the plant cells within the root cortex. In some cases the hyphae may also penetrate the plant cells, in which case the mycorrhiza is called an ectendomycorrhiza. Outside the root, Ectomycorrhizal extramatrical mycelium forms an extensive network within the soil and leaf litter.

Nutrients can be shown to move between different plants through the fungal network. Carbon has been shown to move from paper birch trees into Douglas-fir trees thereby promoting succession in ecosystems. The ectomycorrhizal fungus *Laccaria bicolour* has been found to lure and kill springtails to obtain nitrogen, some of which may then be transferred to the mycorrhizal host plant. In a study by Klironomos and Hart, Eastern White Pine inoculated with *L. bicolour* was able to derive up to 25% of its nitrogen from springtails.

The first genomic sequence for a representative of symbiotic fungi, the ectomycorrhizal basidiomycete *Laccaria bicolour*, has been published. An expansion of several multigene families occurred in this fungus, suggesting that adaptation to symbiosis proceeded by gene duplication. Within lineage-specific genes those coding for symbiosis-regulated secreted proteins showed an up-regulated expression in ectomycorrhizal root tips suggesting a role in the partner communication. *Laccaria bicolour* is lacking enzymes involved in the degradation of plant cell wall components (cellulose, hemicellulose, pectins and pectates), preventing the symbiont from degrading host cells during the root colonisation. By contrast, *Laccaria bicolour* possesses expanded multigene families associated with hydrolysis of bacterial and microfauna polysaccharides and proteins. This genome analysis revealed the dual saprotrophic and biotrophic lifestyle of the mycorrhizal fungus that enables it to grow within both soil and living plant roots.

Ericoid Mycorrhiza

Ericoid mycorrhizas are the third of the three more ecologically important types. They have a simple intraradical (grow in cells) phase, consisting of dense coils of hyphae in the outermost layer of root cells. There is no periradical phase and the extraradical phase consists of sparse hyphae that don't extend very far into the surrounding soil. They might form sporocarps (probably in the form of small cups), but their reproductive biology is little understood.

Ericoid mycorrhizas have also been shown to have considerable saprotrophic capabilities, which would enable plants to receive nutrients from not-yet-decomposed materials via the decomposing actions of their ericoid partners.

Jatropha Plant Diseases and Their Biological Control

Jatropha is a genus of flowering plants in the spurge family, Euphorbiaceae. The name is derived from the Greek words 0áôñüò (*iatros*), meaning "physician," and ôñïöÞ (*trophe*), meaning "nutrition," hence the common name physic nut. It contains approximately 170 species of succulent plants, shrubs and trees (some are deciduous, like *Jatropha curcas*). Most of these are native to the Americas, with 66 species found in the Old World. Mature plants produce separate male and female flowers. As with many members of the family Euphorbiaceae, *Jatropha* contains compounds that are highly toxic.

In 2007 Goldman Sachs cited *Jatropha curcas* as one of the best candidates for future biodiesel production. It is resistant to drought and pests, and produces seeds containing 27-40% oil, averaging 34.4%. The remaining press cake of jatropha seeds after oil extraction could also be considered for energy production. However, despite their abundance and use as oil and reclamation plants, none of the *Jatropha* species have been properly domesticated and, as a result, their productivity is variable, and the long-term impact of their large-scale use on soil quality and the environment is unknown. Igbinosa and colleagues (2009) demonstrated potential broad spectrum antimicrobial activity of *J. curcas*.

The stems of haat (*Jatropha cuneata*) are used for basketmaking by the Seri people in Sonora, Mexico. The stems are roasted, split and soaked through an elaborate process. The reddish dye that is often used is made from the root of another plant species, *Krameria grayi*. Spicy jatropha (*J. integerrima*) is cultivated as an ornamental in the tropics for its continuously blooming crimson flowers. Buddha belly plant (*J. podagrica*) was used to tan leather and produce a red dye in Mexico and the southwestern United States. It is also used as a house plant.

The oil from Jatropha curcas is mainly converted into biodiesel for use in diesel engines. The cake can be used for fish or animal feed (if detoxified), biomass feedstock to power electricity plants, or as biogas or high-quality organic fertilizer. It can also be used as a bio-pesticide and for medicinal purposes. Furthermore, it has been found that Jatropha curcas can be planted in arid and hot regions such as the desert areas of Egypt, India, and Madagascar, and contribute a reduction of up to 25 t of CO2 per hectare per year from the atmosphere (over a 20 yr period), while still producing bio fuel and also the dry cakes from the oil extraction. Currently, research plantations are being planted to test the results and see the viability of this.

Toxicity

Much like other members of the family Euphorbiaceae, members of the genus *Jatropha* contain several toxic compounds. The seeds of *Jatropha curcis* contain the highly poisonous toxalbumin curcin, a lectin dimer. They also contain carcinogenic phorbol. Despite this, the seeds are occasionally eaten after roasting, which reduces some of the toxicity. Its sap is a skin irritant, and ingesting as few as three untreated seeds can be fatal to humans. In 2005 Western Australia banned *Jatropha gossypiifolia* as invasive and highly toxic to people and animals

Major Diseases of the Biofuel Plant, Physic Nut (*Jatropha Curcas*)

Worldwide, concern over the consequences of global warming has resulted in intensified searches for potential plants that couldsupply raw materials for producing renewable fuels. Therein, physic nut (*Jatropha curcas* L.) has gained attention as a perennial culture that produces seeds with high oil content and excellent properties. In addition to these attributes, many studies have describedphysic nut as a culture resistant to pests and disease. However, in recent years, the expansion of areas under cultivation has been accompanied by the appearance of various diseases. Thus, this chapter aims to provide information about the main diseases that occur in physic nut and their diagnosis and to encourage further research on disease control.

Diseases

Although several descriptions of fungi exist, this chapter will discuss the most common and damaging diseases that affect physic nut, and draws on the following descriptions:

Anthracnose

Colletotrichum gloeosporioides (Penz.)Sacc.

Colletotrichum capsici (Syd.)Butl.and Bisby

The most commonly observed symptoms are brown to black necrotic lesions that are irregularly shaped and appear on the edges and centre of the leaf and which may contain a yellow halo. The lesions appear in the form of small, isolated points that coalesce and subsequently cause the complete destruction of the leaves. The fruit can also become infected, which leads to the appearance of dark brown lesions.

In addition to these symptoms, research in Mexico has indicated that the fungus *Colletotrichum capsici* caused stem canker and apical death of seedlings.

Colletotrichum is a fungus anamorph of the phylum Ascomycota and telemorph genus *Glomerella*. The species of this genus have the following characteristics: conidiomata that are acervular, subcuticular or epidermal, and may contain setae; conidiophores that are hyaline to brown; conidiogenous cells that are enteroblastic, phialidic and hyaline; conidia that arehyaline, aseptate (except prior to germination), straight or falcate, smooth and thin-walled; and appressoria that are brown, entirely or with crenate to irregular margins produced with germination of conidia.

Colletotrichum spp. is known to infect a large range of hosts and to cause various symptoms, the most common of which is anthracnose. This fungus can survive in seeds, crop residues, infected plants, and in soil as saprophytes. Although the disease occurs in various regions of the world, it is more severe in regions with a hot and humid climate.

So far, there are no recommendations for controlling this disease. Because of the damage it can cause to physic nut, this disease should be studied further.

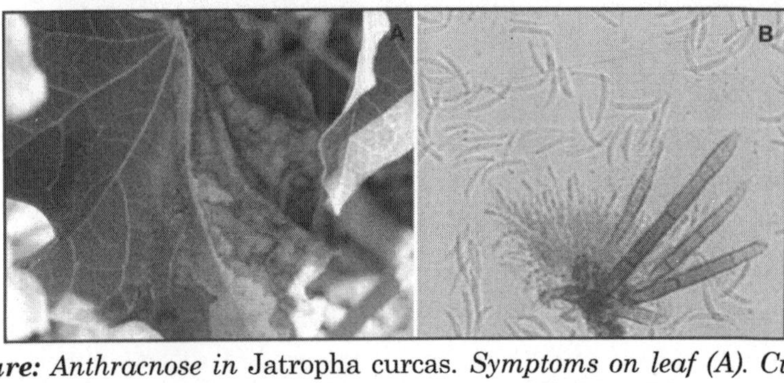

Figure: *Anthracnose in* Jatropha curcas. *Symptoms on leaf (A). Curved conidia, dense conidiophores and septate setae of* Colletotrichum capsici *(B).*

Passalora Leaf Spot

Passalora ajrekari (Syd.) U. Braun

Passalora jatrophigena U. Braun & F.O. Freire

This disease was first described in Brazil by Braun & Freire , and later by Freire & Parente (2006) in leaves of *Jatropha curcas* and *Jatropha podagrica*, and in others countries by Crous & Braun (2003).

The primary symptoms of this disease are rounded leaf lesions that are creamy to light brown in colour, with a narrow dark brown halo, and later become limited by leaf veins and darken. Lesions measure 1-2 cm in diametre and rarely coalesce.

The genus *Passalora* is a cercosporoid fungus, previously included in the genus *Cercospora* that has as its teleomorph the *Mycosphaerella.* Species share taxonomic characteristics such as branched, septate, smooth, hyaline to pigmentedhyphae; absent to well-developed stromata; solitary or fasciculate to synnematous conidiomataconidiophores, arising from stromata or hyphae, internal or superficial, pluriseptate, subhyaline to pigmented; conspicuous conidiogenous loci, with scars that are somewhat thickened and darkened; conidia that are solitary to catenate in simple or branched chains, amerosporous to scolecosporous, aseptate to pluriseptate, and pale to distinctly pigmented and hila that are somewhat thickened and darkened.

Although it has been reported in several countries, to date this disease has not presented risk to physic nut cultivation.

*Cercospora/Pseudocercospora*leaf spot

*Cercospora jatrophicola (*Speg.) Chupp,

Cercospora jatrophigena U. Braun

Pseudocercospora jatrophae-curcas (J.M. Yen) Deighton

Pseudocercospora jatrophae (G.F. Atk.) A.K. Das & Chattopadh.

Pseudocercospora jatropharum (Speg.) U. Braun

This disease manifests in the form of leaf spots that consist of well-delimited brown irregular necrotic spots.

The genera mentioned above have the following taxonomic characteristics:

The genus *Cercospora* groups anamorphs of *Mycosphaerella* with hyphae that are colourless or near-colourous to pigmented, branched, septate, and smooth to faintly rough-walls. Stromata are lacking to well-developed, subhyaline to usually pigmented. Conidiophores are solitary to fasciculate, arising from internal hyphae or stromata, erect, subhyaline to pigmented. Conidiogenous loci (scars) are conspicuous, thickened and darkened. Conidia are solitary, scolecosporous, cylindrical-filiform, hyaline or subhyaline, mostly pluriseptate, and smooth and hila are thickened and darkened.

The genus *Pseudocercospora*groups anamorphs of *Mycosphaerella* with basically pigmented conidiophores and inconspicuous, unthickened, not darkened conidiogenous loci; solitary, or catenulateconidia, aseptate to pluriseptate with conidial scars that are inconspicuous and not thickened.

Crous & Brauncite the occurrence of five species of cercosporoid, indicated above, in the culture of physic nut. However, few studies have examined fungi cercosporoid in this crop. As a result, there is no information about favourable conditions, symptoms or disease control. To date, this disease has not presented risk to the cultivation of physic nut.

Powdery Mildew

The powdery mildew caused by the fungus *Pseudoidium jatrophae* was previously described as *Oidium heveae* Steim by Viégas (1961) in Braziland *Oidium jatrophae* Hosag., Siddappa, Vijay. & Udaiyan in India. This disease occurs commonly in physic nut plantations and it has been frequently observed in various regions of Brazil and the rest of world.

The most common symptoms of the disease are the production of abundant white or gray mycelia in leaves, petioles, stems, flowers and fruits. With the evolution of the disease, infected plants may show necrotic lesions, which cause leaf fall, underdevelopment, death of buds and young fruit deformation.

The fungus that causes this disease is a typical biotrophic pathogen of the phylum Ascomycota, order Erysiphales. This pathogen may be characterized by white or grayish colonies, septate and branched mycelia; conidiophores that are erect or ascending, cylindrical, hyaline, septate and forming conidia singly; conidia that are usually large in proportion to the diametre of the conidiophores, simple, smooth, ellipsoid-ovoid doliiform, hyaline, single-celled

The disease generally favours warm temperatures, humidity of 75-80% and reduced light. Heavy rains are generally unfavourable to the pathogen (Furtado &Trindade 2005). In Brazil, the disease usually occurs in the dry season, apparently without causing extensive losses, because its occurrence coincides with the plants' period of natural defoliation (Saturnino et al. 2005).

Currently, there are no fungicides recommended for culture, but some studies cite that spraying sulfur fungicides works to control this fungus. Another measure is to control alternative hosts, especially plants of the family Euphorbiaceae (Furtado & Trindade 2005; Saturnino et al. 2005; Dias et al. 2007).

Rust

The first report of this disease in *Jatropha curcas*, described its cause as *Uredo jatrophicola* Arthur(Arthur 1915). In Brazil, this disease was first found in 1936 in São Paulo (Viégas 1945). Currently, it is widely distributed throughout Brazil (Dias et al. 2007) and several other countries.

The fungus that causes this disease was previously classified as *Phakopsora jatrophicola* (Arthur) Cummins; however, it was reclassified as *Phakopsora arthuriana* Buriticá & Hennen.

The symptoms manifest in the leaves, initially in the form of small chlorotic points on the upper surface, which correspond to the underside of the leaf, and then small protruding pustules, which after breaking, release a powdery mass of uredospores of orange colour, giving a ferruginous aspect. In severe infections, pustules coalesce to form necrotic spots, which are reddish brown and irregularly shaped and can destroy the leaf.

The *Phakopsora arthuriana* belongs to the phylum Basidiomycota, class Pucciniomycetes. It is characterized by uredinia hypophyllous, occasionally epiphyllous, in small groups opening by a pore, surrounded by numerous not septate paraphyses that project outside the host; urediniospores, ellipsoid, to obovoid, sessile, closely and finely echinulate, germ pores obscure; telia hypophyllous, suberpidermal in origin, closely around the uredinia; teliospores irregularly arranged, cuboid, ellipsoid to polygonal.

Currently there are no fungicides recommended for this culture.

Stem Canker and Dieback

The disease manifests in the form of dieback that can progress until it takes over the trunk of the plant. Stem cankers have also been observed, causing necrotic lesions on branches and vascular discolouration. In Malaysia, disease incidence can be as high as 80% of a plantation

Characteristics of the *Lasiodiplodia* species commonly include the presence of paraphyses within the conidiomata pycnidial and conidia that are initially hyaline and aseptate. But in maturity, one median septum is formed, and the walls become dark brown with the formation of longitudinal striations due the deposition of melanin granules on the inner surface of the wall.

The identification of the *Lasiodiplodia* species based solely on morphological characteristics is not easy. Currently, it is known that what was initially identified as *Lasiodiplodia theobromae* is in fact a species complex (Alves et al. 2008). Thus, molecular studies are needed to correctlyidentify the pathogen, as was done by Thanarajoo & Sulaiman (2012).

Lasiodiplodia spp. is a fungus of the phylum Ascomycota, family Botryosphaeriaceae. Fungi in this family are known to survive as endophytes and demonstrate symptoms when plants are under some stress (Slippers & Wingfield 2007). Thus, many researchers see them as opportunistic pathogens.

Control of this disease can be achieved by pruning and destroying affected branches. Later plants should be brushed with copper fungicides or thiophanate methyl for injuries (Furtado & Trindade 2005). Additionally, balanced fertilization, soil analysis and sufficient levels of irrigation in regions with long periods of drought can aid in disease control.

Collar and Root Rot

Fusarium solani(Martius) Appel & Wollenweber *Lasiodiplodia theobromae*(Pat.) Griffon & Maubl *Neoscytalidium dimidiatum(Penz.) Crous & Slippers Macrophomina phaseolina*(Tassi) Goid.

The first report of this disease in Brazil was made by Pereira et al. (2009), who identified it as being caused by *Lasiodiplodia theobromae.* In India, this same pathogen was reported by Latha et al. (2009), and *Macrophomina phaseolina*was reported by Patel et al. (2008). In China, Yue-Kai et al. (2011) identified the fungus *Fusarium solani*, and Machado et al. (in press) made the first description of *Neoscytalidium dimidiatum(Penz.). Crous & Slippers associated this pathogen with collar and root rot in physic nut in Brazil.*

All the pathogens mentioned above are typical soil fungi. They occur in a wide range of hosts, can be spread by seeds, and survive as parasites, saprophytes, endophytes or resistant structures, such as chlamydospores in Fusarium and Neoscytalidium or sclerotia in Macrophomina.

This disease has acquired great importance, because it can reduce productivity by causing the sudden death of plants and making cultivation areas unviable. The symptoms most commonly observed are wilting, leaf yellowing with subsequent fall, and cracks in the collar region. In the collar region, the appearance of black fungal structures in the bark of the plant has been consistently observed. Upon being removed from the soil, plant roots rot and the vascular system is affected by necrotic symptoms, ranging from light brown to black. Due to loss of support, the plants have often already fallen due to the wind.

The genus *Fusarium*has the following general characteristics: bright aerial mycelium, hyphae septate, conidiophores variable, single or grouped in sporodochia; conidia hyaline variable, principally of two kinds -multicellularmacroconidia, slightly curved or bent at the pointed ends and typically canoe-shaped; unicellular, ovoid or oblongmicroconidia, borne singly or in chains and also grouped in false heads, formed in mono or polyphialides; thick-walled chlamydospores are common in some species.

The common characteristics of *Lasiodiplodia* species include the presence of paraphyses within the conidiomata pycnidial and initially hyaline and aseptateconidia. However, in maturity, one median septum is formed, and the walls become dark brown with the formation of longitudinal striations, due the deposition of melanin granules on the inner surface of the wall.

The genus *Neoscytalidium*is a group of fungi that produces synanamorph *Scytalidium*-like with septate and oblong to globosearthroconidia formed from aerial mycelia. Initially hyaline, with age, the arthroconidia become brown and with a thick wall. Commonly observed are pycnidia that are dark and globose immersed or superficially in a stroma that produces *Fusicoccum*-like conidia that are hyaline and ellipsoid to nearly fusiform. Dark septate conidia can also be observed.

Characteristics of the *Macrophomina* spp. generally include the formation of dark mycelia and abundant production of sclerotia in PDA. Eventually, the formation of conidiomata pycnidial can be observed, with the release of hyaline conidia with apical mucoid appendages.

In areas prone to prolonged dry seasons, a higher incidence of collar and root rot has been observed. Therefore, it is believed that the water stress is the main factor that predisposes plants to this disease.

The above-mentioned pathogens are difficult to control, due to the fact that they survive in soil.

Therefore, to reduce disease incidence, it is first necessary to provide water and fertilizer balanced for proper plant development. When transplanting seedlings to the field, all forms of injury should be avoided. Another control measure would be to use healthy propagative material as well as seed treatments.

Yellow Mosaic

In addition to the several fungal diseases mentioned, there is also yellow mosaic, a disease caused by a strain of the virus *Indian Cassava Mosaic Virus*. This disease, detected in physic nut plantations in India, causes mosaic, reduced leaf size, leaf distortion, blistering and stunting of diseased plants. The disease is transmitted by the vector *Bemisia tabaci* in a non-persistent manner, but not through mechanical inoculation or seeds.

Seed Associated Fungi

Seeds propagate the majority of cultures worldwide. These cultures are vulnerable to infection by several pathogens that can survive in seeds. These pathogens may cause reduction of seed germination, as well as deformation, discolouration, reductions in size and weight, and deterioration during storage. They can further contribute to rotting roots, damping-off, necrosis in leaves, and the spread of diseases across long distances. Consequently, these diseases cause losses valued at billions of dollars. To date, few studies have addressed the seed pathology of physic nut, and there is no information available about the losses that seed pathogens cause in this culture. But, follows below the major pathogens and saprophytic fungi associated with seeds.

Disease Management in Indian Mustard: Molecular Advances

Rapeseed and mustard are most important rabi oilseed crops of our country. This group of crops comprises toria, brown sarson, yellow sarson, gobhi sarson, raya or Indian mustard, black mustard, rai and taramira. India is the second largest producer of rapeseed-mustard in the world after China. Major bottleneck in its production is reduction of yield due to pest attacks. Insects and diseases are appears important limiting factors which restrict the fast expansion of cultivation and productivity of these crops. One of the major concerns in enhancing and stabilizing the yield of rapeseed mustard is the incidence of insect and diseases, which are causing the damage to crop at different stages and responsible for huge yield losses to a extent ranging from 10 to 90 per cent. In present article important pests of mustard and their integrated management has been described in detail.

A. Major Diseases

i. White rust (*Albugo candida*): It is a serious problem of mustard causes leaf phase (local) and stag head (systemic) infections. The symptom develops on all aerial parts except root as light brown water soaked lesion covered with white growth of mycelium.

Prominent white or creamy yellow pustules appear on lower surface of leaves and clearly visible during February - March. The small size pustules may also appear on upper surface of leaves. The infected inflorescences result in deformed floral parts (staghead) usually on top of branches. Flowers get malformed and become sterile. The seeds may become smaller and shriveled.

ii. *Alternaria* Blight (*A.brassicae* & *A.brassicicola*): Small dark brown- blackish sooty and velvety spot with round concentric rings appeared on stem, leaves and pods. Centre of spots soon dries and drops out. In severe attack leaves, pods and upper parts of stem withers. Pod infection results in loss of seed yield and germ inability. The pathogen remains viable in soil debris.

iii. Bacterial Leaf spot/ blight: The Causal organism is *Xanthomonas campestris* pv. *campestris*. V shaped lesion are produced. It also leads to collapse of interveinal tissues. Mostly plants are attacked from early seedling stage. The tissue turns yellow and chlorosis reached towards the centre of the leaf and forming V shaped area with base of V towards the midrib. The veins show brown to black discolouration. There is premature fall of the leaves is observed.

iv. Downy mildew (*Peronospora parasitica*): This disease alone or in combination with white rust is responsible for severe losses in yield. Symptoms appear on all above ground plant parts particularly on leaves and inflorescence. Purplish-brown spots appear on underside of the leaves. The most conspicuous symptom is thickening of inflorescence (staghead).Fungus stimulate hypertrophy and hyperplasia consequently results in deformation of various floral organs. The affected inflorescence does not bear siliquae or produce abnormal curled siliquae without seeds.

Integrated Disease Management

- Plough the field in summer to reduce the soil borne pathogen and weeds.
- Sowing of the crop should be done in first fort night of October as it escapes the disease. In delayed planting, the incidence of white rust, blight, is more.
- Adopt line sowing with 30x 10-15 cm row and plant distances.

- Removal of 50 per cent lower leaves at 60 days of crop age is advocated as it reduces the secondary spread of diseases and enhancing yield.
- Removal and destruction of stagheads (White rust& Downy mildew) from affected plants should be practiced.
- Seed treatment with Metalaxyl (Apron 35 SD) @ 6.0 g/kg, thiram or carbendazim @ 2.0 g/kg seed before sowing.
- Appliance of balanced (NPK-100:40:40) fertilizers for effective disease management.
- Application of 25-40 kg sulphur along with NPK increases the yield and disease resistance.
- Prophylactic spray of aqueous extract of garlic bulb or Eucalyptus leaf extract (2.0 % w/v) or mancozeb 75 WP @ 0.2 per cent at 50 and 70 days after sowing is effective against blight, white rust etc.

Integrated Management of Rhizome Rot of Ginger

Ginger is one of the most promising spice crop grown in North Eastern States particularly in the state of Arunachal Pradesh, Meghalaya, Mizoram, Nagaland etc. It is estimated that 50% of the national production ginger comes from the North Eastern States. India one of the leading producer and exporter of ginger in the world. India's production of ginger constitutes about 50% of the total world's production of ginger. Rhizome or soft rot is one of the most devastating diseases of ginger causing heavy yield loss, wherever this crop is grown.

Due to this disease more than 50% yield loss has been reported. This disease is mainly caused by the species of Pythium but associated with this disease are the fungus *Fusarium* spp. and the root knot nematode *Meloidogyne* sp. occasionally, the fungi *Rhizoctonia solani*, along with the bacterium *Pseudomonas* sp. have been isolated from diseased rhizomes.

Symptoms

- Initial symptom of the disease appear as light yellowing of the tips of lower leaves which gradually spreads down to the leaf blade and leaf sheath along the margin.
- In the early stages, the middle portion of the lamina remains green while the margins become yellow.
- The yellowing spreads to all the leaves of plant from bottom upwards and is followed by drooping, withering and drying.

- The collar region of the pseudo-stem shows pale translucent brown colour which becomes water-soaked, due to destruction of parenchyma tissues.
- The infected shoot can be easily pulled out from the soil; the infection from the collar spreads to the rhizome gradually.

Causal organism: *Fusarium oxysporium* f.sp *zingiberi, Pythium aphanidermatum,* and *Pythium myriotylum*

Disease cycle and Environmental Relations:

- Infected rhizomes used for seed are the main agents of introducing the fungus in the field.
- Diseased plant debris containing oospores is the main source of soil-borne primary inoculums.
- Low-lying fields and water-logging are conducive to disease development.

Management Practices

Cultural Management:

- Use of healthy planting materials is one of the most important measures for management of rhizome or soft rot of ginger.
- Water-logging in the field must be avoided.
- Provide proper drainage and keep land free from weeds at all times.
- Collect the diseased material as and when the disease is noticed and burn them.
- Early planting during April month.

Physical Management:

- Treat the rhizomes with hot water at 51° C for 10 minutes.

Biological Management:

- Spraying of the soil surface with Bordeaux mixture @ of 7 litre / square foot, once before sowing (6:6:50 conc. Means 6kg lime dissolve in 25 litre water and remaining 25 litre water dissolve 6kg copper sulphate therefore, mixed both solutions) and then after germination at 2-3 weeks interval (5:5:50 conc.) give partial control of the disease.
- Rhizomes treated with (*Trichoderma* bio-control agent) @ 5g / kg rhizome for 30 minutes
- Soil application of *Trichoderma* bio-control agent (2.5 kg mixed with 50 kg FYM) 10-15 days before sowing.

- Application of oil cakes made from *Azadirachta indica, Calophyllum inophyllum, Pongamia glabra Hibiscus sabdariffa* and *Brassica compestris* are recommended.

Marker Assisted Selection for Resistance Improvement in Potato

For a new cultivar to be successful, it must have an advantage over existing varieties that can be readily observed and jointly agreed upon by both growers and processors (for processing varieties) and by growers and consumers (for fresh market and novelty varieties). For breeders, this means not only selecting for numerous traits, but combining those traits into one variety. Another challenge for breeders is to be able to identify and breed for traits before they are demanded by growers, processors, and consumers. Therefore, the goal of most potato breeding programmes is twofold: develop improved varieties and increase the frequency of important alleles in the programme germplasm.

Approximately 85% of the potatoes grown in Michigan are utilized by the potato chip industry. As a result, the potato breeding programme at Michigan State University is dedicated to developing new potato chip varieties. Currently, the emphasis is on varieties that meet or exceed current agronomic standards, resist cold-sweetening, produce acceptable chips with low acrylamide levels, and have scab (*Streptomyces scabies*) resistance. Additional traits such as resistance to late blight (*Phytophthora infestans*), potato virus Y (PVY), Verticillium wilt (*Verticillium dahlia*), and golden nematode (*Globodera rostochiensis*) are being incorporated into early generation breeding materials. Although these additional traits are not presently required by growers, processors, or consumers, and are not currently viewed as market limiting traits in Michigan, they represent potential challenges to our potato industry and may already be challenges in other regions of potato production.

Golden nematode (GN) is a soil-borne, potato cyst nematode (PCN) that is a concern in some US potato growing regions and has been recognized in Europe as a significant pest of potatoes for many years. Heavy infestations of this pest can cause significant yield losses and limit the choice of potato varieties that can be grown effectively. The first reported identification of GN in the United States was in New York in 1941. Subsequent Federal and State quarantines have limited the spread of this pest to 9 counties in New York. In addition to limiting the spread of this pest, the best management practices include: rotating fields with non-host crops (corn, soybeans and wheat) and growing GN resistant potato varieties (Animal and Plant Health Inspection Service [APHIS], 2008).

Phenotypic Screening for Golden Nematode Resistance

Phenotypic screening for GN resistance is done in replicated greenhouse tests. Briefly, tubers are planted in pots and inoculated with *G. rostochiensis* eggs. Eight weeks after emergence, the plants are uprooted and the number of cysts are counted (0-5 = resistant; >5 = susceptible) (Brodie and Plaisted, 1987). The time and cost involved in this assay are prohibitive to screening thousands of early generation breeding lines for GN resistance. Therefore, when limited to phenotypic screening, only a small number of later generation breeding lines can be screened for GN resistance based on a pedigree suggestive of the potential for resistance. An alternative to phenotypic screening is marker-assisted selection (MAS).

MAS for H1-based Golden Nematode Resistance

In the Michigan State University Potato Breeding and Genetics programme, MAS for golden nematode resistance is conducted after the 3rd year of a breeding cycle. At harvest, the third-year (12-hill) plots are evaluated and approximately 25% of the lines (roughly 500) are selected for advancement. Of the selected lines, those with the potential (based on pedigree) to harbor the *H1* resistance locus are genotyped. For genotyping, template DNA is extracted from leaf tissue that is collected from the plants growing in the field or from post-harvest tuber tissue. A high-throughput, 96-well DNA extraction protocol is used to increase the efficiency and cost effectiveness of the process. Following PCR, using the template DNA and the *H1* locus primers, and electrophoretic separation on an agarose gel (2%), the individual clones are scored for the presence or absence of the 119 bp band linked to the *H1* locus. Breeding lines that are positive for the *H1* marker band are suspected to be resistant to GN. Although market-limiting traits are given priority, the presence of the *H1* marker band helps us make more informed decisions during future rounds of selection. Furthermore, lines that have the *H1* marker but are not selected to advance towards commercialization are often used as parents for subsequent rounds of crossing and selection. This increases the frequency of this valuable allele in the programme germplasm. Breeding lines that are being considered for commercialization and test positive for the *H1* marker are phenotyped to ensure that they are GN resistant.

Potential Application of Trichoderma as Biocontrol Agents, their Molecular Characterization and Diversity Analysis

Trichoderma is a genus of fungi that is present in all soils, where they are the most prevalent culturable fungi. Many species in this genus can be characterized as opportunistic avirulent plant symbionts. This refers to the ability of several *Trichoderma* species to form mutualistic endophytic relationships with several plant species. The genomes of several *Trichoderma* species have been sequenced and are publicly available from the JGI.

Characteristics

Cultures are typically fast growing at 25–30°C, but some species of *Trichoderma* will grow at 45°C. Colonies are transparent at first on media such as cornmeal dextrose agar (CMD) or white on richer media such as potato dextrose agar (PDA). Mycelium are not typically obvious on CMD, conidia typically form within one week in compact or loose tufts in shades of green or yellow or less frequently white. A yellow pigment may be secreted into the agar, especially on PDA. Some species produce a characteristic sweet or 'coconut' odour.

Conidiophores are highly branched and thus difficult to define or measure, loosely or compactly tufted, often formed in distinct concentric rings or borne along the scant aerial hyphae. Main branches of the conidiophores produce lateral side branches that may be paired or not, the longest branches distant from the tip and often phialides arising directly from the main axis near the tip. The branches may rebranch, with the secondary branches often paired and longest secondary branches being closest to the main axis. All primary and secondary branches arise at or near 90° with respect to the main axis. The typical *Trichoderma* conidiophore, with paired branches assumes a pyramidal aspect. Typically the conidiophore terminates in one or a few phialides. In some species (e.g. *T. polysporum*) the main branches are terminated by long, simple or branched, hooked, straight or sinuous, septate, thin-walled, sterile or terminally fertile elongations. The main axis may be the same width as the base of the phialide or it may be much wider.

Phialides are typically enlarged in the middle but may be cylindrical or nearly subglobose. Phialides may be held in whorls, at an angle of 90° with respect to other members of the whorl, or they may be variously penicillate (gliocladium-like). Phialides may be densely clustered on wide main axis (e.g. *T. polysporum*, *T. hamatum*) or they may be solitary (e.g. *T. longibrachiatum*).

Conidia typically appear dry but in some species they may be held in drops of clear green or yellow liquid (e.g. *T. virens*, *T. flavofuscum*). Conidia of most species are ellipsoidal, 3–5 x 2–4 μm (L/W = > 1.3); globose conidia (L/W < 1.3) are rare. Conidia are typically smooth but tuberculate to finely warted conidia are known in a few species.

Synanamorphs are formed by some species that also have typical *Trichoderma* pustules. Synanamorphs are recognized by their solitary conidiophores that are verticillately branched and that bear conidia in a drop of clear green liquid at the tip of each phialide.

Chlamydospores may be produced by all species, but not all species produce chlamydospores on CMD at 20°C within 10 days. Chlamydospores are typically unicellular subglobose and terminate short hyphae; they may also be formed within hyphal cells. Chlamydospores of some species are multicellular (e.g. *T. stromaticum*).

Trichoderma genomes appear to be in the 30–40 Mb range, with approximately 12,000 genes being identifiable.

Teleomorph

Teleomorphs of *Trichoderma* are species of the ascomycete genus *Hypocrea*. These are characterized by the formation of fleshy, stromata in shades of light or dark brown, yellow or orange. Typically the stroma is discoidal to pulvinate and limited in extent but stromata of some species are effused, sometimes covering extensive areas. Stromata of some species (Podostroma) are clavate or turbinate. Perithecia are completely immersed. Ascospores are bicellular but disarticulate at the septum early in development into 16 part-ascospores so that the ascus appears to contain 16 ascospores. Ascospores are hyaline or green and typically spinulose. More than 200 species of *Hypocrea* have been described but few have been grown in pure culture and even fewer have been described in modern terms.

Habitat

***Figure:** Trichoderma colony in nature*

Trichoderma species are frequently isolated from forest or agricultural soils at all latitudes. *Hypocrea* species are most frequently found on bark or on decorticated wood but many species grow on bracket fungi (e.g. *H. pulvinata*), Exidia (*H. sulphurea*) or bird's nest fungi (*H. latizonata*) or agarics (*H. avellanea*).

Biocontrol Agent

Several strains of *Trichoderma* have been developed as biocontrol agents against fungal diseases of plants. The various mechanisms include antibiosis, parasitism, inducing host-plant resistance, and competition.

Most biocontrol agents are from the species *T. harzianum*, *T. viride* and *T. hamatum*. The biocontrol agent generally grows in its natural habitat on the root surface, and so affects root disease in particular, but can also be effective against foliar diseases.

Causal Agent of Disease

T. aggressivum (formerly T. harzianum biotype 4) is the causal agent of green mold, a disease of cultivated button mushrooms. . *Trichoderma viride* is the causal agent of green mold rot of onion.

Toxic House Mould

The common house mould, *Trichoderma longibrachiatum*, produces small toxic peptides containing amino acids not found in common proteins, like alpha-aminoisobutyric acid, called trilongins (up to 10% w/w). Their toxicity is due to absorption into cells and production of nano-channels that obstruct vital ion channels that ferry potassium and sodium ions across the cell membrane. This affects in the cells action potential profile, as seen in cardiomyocytes, pneumocytes and neurons leading to conduction defects. Trilongins are highly resistant to heat and antimicrobials making primary prevention the only management option.

Medical Uses

Cyclosporine A (CsA), a calcineurin inhibitor produced by the fungi *Tolypocladium inflatum* and *Cylindrocarpon lucidum*, is an immunosuppressant prescribed in organ transplants to prevent rejection.

Industrial Use

Trichoderma, being a saprophyte adapted to thrive in diverse situations, produces a wide array of enzymes. By selecting strains that produce a particular kind of enzyme, and culturing these in suspension, industrial quantities of enzyme can be produced.

- *T. reesei* is used to produce cellulase and hemicellulase
- *T. longibratum* is used to produce xylanase
- *T. harzianum* is used to produce chitinase.

Development of Pseudomonas Fluorescens Based Biopesticides Biofor-pf and Management of Vegetable Diseases

Pseudomonas fluorescens is a common Gram-negative, rod-shaped bacterium. It belongs to the *Pseudomonas* genus; 16S rRNA analysis has placed *P. fluorescens* in the *P. fluorescens* group within the genus, to which it lends its name.

General Characteristics

P. fluorescens has multiple flagella. It has an extremely versatile metabolism, and can be found in the soil and in water. It is an obligate aerobe, but certain strains are capable of using nitrate instead of oxygen as a final electron acceptor during cellular respiration.

Optimal temperatures for growth of *Pseudomonas fluorescens* are 25-30 degrees Celsius. It tests positive for the oxidase test. Pseudomonas fluorescens is also a nonsaccharolytic bacteria.

Heat-stable lipases and proteases are produced by *Pseudomonas fluorescens* and other similar pseudomonads.

These enzymes cause milk to spoil, by causing bitterness, casein breakdown, and ropiness due to production of slime and coagulation of proteins.

The Name

The word *Pseudomonas* means 'false unit', being derived from the Greek words *pseudo* and *monas*. The word was used early in the history of microbiology to refer to germs. The name 'fluorescens' refers to the microbe's secretion of a soluble fluorescent pigment called pyoverdin, which is a type of siderophore.

Genome Sequencing Projects

The genomes of *P. fluorescens* strains SBW25, Pf-5 and PfO-1 have been sequenced.

Biocontrol Properties

Some *P. fluorescens* strains (CHA0 or Pf-5, for example) present biocontrol properties, protecting the roots of some plant species against parasitic fungi such as *Fusarium* or *Pythium*, as well as some phytophagous nematodes.

It is not clear exactly how the plant growth-promoting properties of *P. fluorescens* are achieved; theories include:

- the bacteria might induce systemic resistance in the host plant, so it can better resist attack by a true pathogen
- the bacteria might outcompete other (pathogenic) soil microbes, e.g., by siderophores, giving a competitive advantage at scavenging for iron
- the bacteria might produce compounds antagonistic to other soil microbes, such as phenazine-type antibiotics or hydrogen cyanide

To be specific, certain *P. fluorescens* isolates produce the secondary metabolite 2,4-diacetylphloroglucinol (2,4-DAPG), the compound found to be responsible for antiphytopathogenic and biocontrol properties in these strains. The *phl* gene cluster encodes factors for 2,4-DAPG biosynthesis, regulation, export, and degradation. Eight genes, *phlHGFACBDE*, are annotated in this cluster and conserved organizationally in 2,4-DAPG-producing strains of *P. fluorescens.*

Of these genes, *phlD* encodes a type III polyketide synthase, representing the key biosynthetic factor for 2,4-DAPG production. *PhlD* shows similarity to plant chalcone synthases and has been theorized to originate from horizontal gene transfer. But phylogenetic and genomic analysis has revealed that the entire *phl* gene cluster is ancestral to *P. fluorescens*, many strains have lost the capacity, and it exists on different genomic regions among strains.

There is experimental evidence to support all of these theories, in certain conditions; a good review of the topic is written by Haas and Defago.

The strain referred to as Pf-CL145A has proved itself a promising solution to the invasive Dreissena (zebra and quagga) mussels. The strain of the bacteria produces toxins that destroy the digestive system of the mussels and produces a >90% kill rate.

Several strains of "P. Fluorescens", such as Pf-5 and JL3985, have developed a natural resistance to ampicillin and streptomycin. These antibiotics are regularly used in biological research as a selective pressure tool to promote plasmid expression.

Medical Properties

By culturing *Pseudomonas fluorescens*, Mupirocin (an antibiotic) can be produced, which has been found to be useful in treating skin, ear, and eye disorders. Mupirocin free acid and its salts and esters are agents currently used in creams, ointments, and sprays as a treatment of Methicillin-resistant *Staphylococcus aureus* (MRSA) infection.

P. fluorescens demonstrates hemolytic activity and, as a result, has been known to infect blood transfusions.

It is also used in milk to make yogurt.

United States Patents: 6489358, 4873012, 6156792

Disease

P. fluorescens is an unusual cause of disease in humans, and usually affects patients with compromised immune systems (e.g., patients on cancer treatment). From 2004 to 2006, there was an outbreak of *P. fluorescens* in the United States, involving 80 patients in six states. The source of the infection was contaminated heparinized saline flushes being used with cancer patients.

Plant Protection

Plant Protection continues to play a significant role in achieving targets of crop production. The major thrust areas of plant protection are promotion of ensuring availability of safe and quality pesticides for sustaining crop production from the ravages of pests and diseases, streamlining the quarantine measures for accelerating the introduction of new high yielding crop verities, besides eliminating the chances of entry of exotic pests and for human resource development including empowerment of women in plant protection skills.

Keeping in view the ill effects of chemical pesticides such as development of pest resistance, pest resurgence, outbreak of secondary pests, pesticide residues in food, fodder, soil, air and water resulting in human health hazards and ecological imbalances, Integrated Pest Management (IPM)as the cardinal principle and main plank of plant protection strategy has been emphasized throughout the country. IPM is an eco-friendly approach aimed at minimum use of chemical pesticides by employing available alternative methods for pest control like cultural, mechanical and biological and use of bio-pesticides.

BIOFOR PF-2: Safe Organic Pesticide for Crop Disease Management

A living soil is the basis of organic farming. A live healthy soil with proper cropping patterns, crop residue management and effective crop rotation can sustain optimum productivity over the years without any loss of fertility. Organic farming is the holistic production management system that avoids use of synthetic fertilizers and pesticides, minimizes pollution of air, soil and water and optimizes the health and productivity of independent communities of life, plants and animals and people. It envisages a comprehensive management approach to improve soil health, ecosystem of the region and quality of the produce. Advantages of organic farming:

- Helps in maintaining environment health by reducing the level of pollution.
- Reduces human and animals hazards by decreasing the level of residues in the product.
- Ensures optimum utilization of natural resources for short term benefits and helps in conserving them for future generation.
- Makes agriculture sustainable.
- Improves the physical and chemical properties of soil.
- Helps in conserving natural resources for future generation.

BIOFOR PF-2 is an organic biopesticide containing beneficial microbes-Pseudomonas fluorescence & Trichoderma harzianum . It has ability to biologically control Ginger rot and wilt diseases of Tomato, Brinjal, Potato & Chilli.

Benefits of BIOFOR PF-2: It is safe to health and environment as there is no residual or environmental hazards; It is being plant-growth-promoting, beneficial to the plant, soil and environment; The delivery system is very easy and does not need sophisticated equipments; It improves soil health as it has the ability to convert unavailable phosphatic nutrients; It increases soil nutrients, rejuvenate soil, enhances plant growth and combat soil borne diseases.It is economic compared to application of commercial chemicals and antibiotics.

Application Methods: Seed treatment: Make a paste of Biofor with water and sticker (Rice gruel/Revive) and treat the seeds for 1 hour.

Requirement:

For Tomato, Brinjal and Chilli	100g/kg seed
For Potato & Ginger	10kg/qt seed or rhizome

OFT on Soft rot management in Ginger

Seedling treatment : (For Tomato, Brinjal and Chilli) The roots of seedlings are dipped in the paste of Biofor pf-2 for 1 hour before transplanting.

Requirement : 1 Kg Biofor for 1000 seedlings.

Soil treatment : Mix 1 Kg Biofor with 10 kg dry cowdung or Vermicompost and apply the mixture @ 100 g/plant.

Requirement : 100 Kg Biofor/ha

Soft rot is the common field disease of ginger in the district. Therefore, an On Farm Trial has been carried out on management of soft rot of ginger in Gidemi village using Biofor-PF2 in farmers field.

Bio control of Chilli leaf Curl Disease-a Review

Chilli leaf curl disease is important and causes heavy losses in the yield and quality of fruits. In the initial stages of growth resulting in no fruit formation. In case any fruit is formed that remain deformed and undersized.

Symptoms

1. Curling of the leaves, and thickening and swelling of veins.
2. Abaxial and adaxial curling of leaves accompanied by puckering and blistering of interveinal areas and thickening and swelling of veins.
3. Stunted plant growth due to shortened internodes and leaves greatly reduced in size.
4. Flower buds abcise before attaining full size and anthers do not contain pollen grains.

Causal Organism: Gemini Virus, Begomo Virus

Transmission: Bamici Tabaci, Grafting:

Host range: Annual and Perennial plants.

Anthracnose and Fruit Rot of Chilli and Their Management

The sustainability of chilli-based agriculture is threatened by a number of factors. Main biotic stresses such as bacterial wilt, anthracnose, viruses and several insect pests have been reported to impair the crop productivity (Isaac, 1992). Anthracnose disease is a major problem in India and one of the more significant economic constraints to chilli production worldwide, especially in tropical and subtropical regions.

The disease causes both pre- and post-harvest fruit decay. Chilli anthracnose usually develops under high humid conditions when rain occurs after the fruits have started to ripen with reported losses of up to 84%. Economic losses caused by the disease are mainly attributed to lower fruit quality and marketability. Although infected fruits are not toxic to humans or animals, severely affected fruits showing blemishes are generally considered unfit for human consumption. This is because the anthracnose causes an unpleasant colour and taste in chilli products. Management of the disease under the prevailing farming systems in India has become a recurrent problem to chilli growers.

This technical bulletin has been produced as an overview of a body of work and aims to mitigate the threat of chilli anthracnose. Producers, seed companies, field inspectors, seed analysts, research workers and students of plant pathology may find useful information in this publication.

Distribution

Chilli anthracnose was first reported in India on chillies from the Coimbatore of Madras Presidency. The disease has been identified in all the chilli producing regions of the world and has become a serious constrain in chilli production whenever the crop is grown. Different species of Colletotrichum, namely C. capsici, C. gloeosporioides and C. acutatum are known to cause anthracnose in chilli in India. Recent surveys conducted revealed C. capsicias the most predominant species in the major chilli growing states namely Karnataka and Andhra Pradesh in India.

Seed-borne infections appear to be common in seed samples collected from affected fields. The most serious infection of C. capsici was reported from seed samples of the Mulabagilu village of the Kolar district of Karnataka.

Damage Symptoms

Anthracnose, deriving from a Greek word meaning 'coal', is the common name for plant diseases characterised by very dark, sunken lesions, containing spores. The disease appears as small circular spots that coalesce to form large elliptical spots on fruits and leaves. Under severe conditions, defoliation of affected plants occurs. Kim et al. reported that different species cause diseases of different parts of the chilli plant.

The disease has been observed to occur in three phases, they are: (i) Seedling blight or damping off, ii) Leaf spot and die back, and (iii) Anthracnose or fruit rot. Capsicum fruit rot reduces fruit dry weight and quantities of capsacin and oleoresin. In anthracnose or fruit rot caused by C. capsici, the ripe fruits turning red are mostly affected.

The disease is characterised by the appearance of small black circular spots on the skin of the fruits that spread in the direction of the long axis, thus becoming more or less elliptical. As the infection progresses, the spots get either diffused and black, greenish or dirty grey in colour or they are markedly delimited by a thick and sharp black outline enclosing a lighter black or straw coloured area. In some cases, the lesions are brown, and then turn black from the formation of setae and sclerotia. Severe infection results in change of fruit colour from red to straw or white. Numerous acervuli are scattered on the discoloured area of the infected fruit. When a diseased fruit is cut open, the lower surface of the skin is covered with minute, spherical, black stromatic masses or sclerotia of the fungus. A mat of fungal hyphae covers the seeds. Such seeds turn rusty in colour. Affected fruits are deformed, white in colour and lose their pungency. Ultimately, the diseased fruit shrivels and dry up.

Pathogen Survival and Dissemination

Either chilli growers produce their own seed for planting, exchange with others, or buy it at the markets. Chilli seeds and transplants in these production systems are commonly infested with the pathogen Colletotrichum capsici. The fungus survives in and on seeds.

Anthracnose is introduced into the field on infected transplants or survives between seasons in plant debris or on weed hosts.

Alternative hosts include other solanaceae (tomato, potato, eggplant). Fruits are infected when spores of the fungus or infested debris are rain splashed onto chilli plants. New spores are produced within the infected tissue and are then dispersed to other fruits. Workers may also move spores with equipment or during handling of infected plants.

Infection usually occurs during warm, wet weather. Temperatures around 27 °C are optimum for disease development, although infection occurs at both higher and lower temperatures. Severe losses occur during rainy weather because the spores are washed or splashed to other fruit resulting in more infections. The disease is more likely to develop on mature fruits, although it can occur on immature fruits as well.

Detection of the Pathogen from Seeds

Colletotrichum capsici is both internally and externally seed-borne. Such infected seeds and infected seedlings act as the primary sources of inoculum. Detection methods applied at the Asian Seed Health Centre: The blotter method, pathogenicity and PCR tests are described as follows:

Blotter Method

Seed-borne C. capsicican be detected by subjecting chilli seed samples to the standard blotter test. Individual samples of 400 seeds are plated in four replicates of 100 seeds each.

Twenty-five seeds per plate are placed on three-layered moist blotter paper plates. The thus plated seeds are then incubated at 24±1 °C under alternating cycles of 12/12h of near ultra violet light and darkness for seven days. On the eighth day of incubation, the seeds are observed for the presence of the fungal growth with the help of the binocular microscope (Leica, Germany).

Colletotrichum capsiciis identified based on morphological characters of the acervuli and conidia growing on the seed and on the blotter papers and by pathogenicity tests of the suspected organism.

Pathogenicity Tests

The isolates are used in the inoculation of chilli seedlings and fruits by the detached leaf assay procedure. (i) Conidial suspension (1 x 106 conidia per ml) of Twelve day old PDA grown cultures is sprayed on one-month-old chilli plants. (ii) The inoculated plants are covered with plastic bags for two days to maintain humidity. (iii) The plants are assayed for disease seven days after inoculation and continued to be so for up to 20 days; (iv) The presence of the pathogen is further confirmed by incubating the leaves in moist chambers for 5-7 days at 22 ± 2°C and observed for the development of fungal growth.

PCR Detection

The pathogen can also be detected by PCR-based methods using specific primers. This method of detection is useful for seed analysts working in seed testing and plant quarantine laboratories and in Germplasm banks where the sample size is limited and precious and to obtain fast and accurate test results.

Preparation of DNA From Seeds:

The seeds are macerated into fine powder in liquid nitrogen using a mortar and pestle. Genomic DNA is extracted using DNeasy Plant Minikit (Qiagen, France) as per the manufacturer's instructions.

From Agar Cultures:

The mycelia mat (100 mg) of a seven-day-old culture of the fungus grown on potato-dextrose broth medium is ground in liquid nitrogen for total genomic DNA isolation using Qiagen DNA extraction kit following the instructions of the manufacturer.

Specific primers PCR detection assay for C. capsici is conducted using species specific primers: CC1F1-5'ACCTAACTGTTGCTTCGGCG3' and CC2R2-5' AAAT TTGGGGGTTTTACGGC 3' with an expected amplicon size of 447 bp.

Management Strategies

Rot and die back caused by Colletotrichum capsici and bacterial wilt caused by Ralstonia solanacearum Smith and insect pests are the major diseases of chillies in the Karnataka region. The integrated management techniques are recommended, as no single specific management programme including crop rotation can eliminate chilli anthracnose.

Cultural Control

- Production of pathogen-free planting materials is the key control measure used to manage the disease. Planting of contaminated seed or transplants facilitates disease spread. Use of healthy pathogen-free chilli seed or transplants should be adopted
- Early removal of affected plants will control the spread of the diseases
- Transplants should be kept clean by controlling weeds and solanaceous volunteers in the vicinity of the transplant houses
- Stagnation of water should not be allowed in nursery beds and fields in order to avoid fungal infection. The field should have good drainage and be free from infected plant debris.

Use of Resistant Cultivars

- Use of resistant varieties that will eliminate losses from diseases, use of chemicals and expenses incurred in disease control
- IIHR, Bangalore, TNAU, Coimbatore, and PAU, Ludhiana, have developed resistant varieties for anthracnose disease viz., IIHR 275-13-5, IIHR345-6, IIHR 332 -109, CC4, Ujwala, CA 87-4. S-20-1, Lorai and BG-1.

Screening for Resistance Against Anthracnose

Plant breeder material and introduced varieties can be screened for resistance to the disease before adoption by the growers.

Preparation of Inoculum

- Colletotrichum capsici isolated from infected seeds/fruits are maintained on potato-dextrose agar at 12/12h cycles of NUV/ dark conditions for 10 days
- Conidial mass is scrapped from the plate with sterile water using a sterile glass slide
- Mycelial mat and conidia are separated using double layered cheese cloth
- The conidial concentration is adjusted with sterile, distilled water using a haemocytometre.

Fruit Injection

- The spore suspension of C. capsici(H”500 spores.ìl-1) are injected at two equidistant points of the fruits and incubated for 6 days at a temperature of 25°C and 95% RH

- Evaluate fruits for anthracnose symptoms and score for disease development.

Fruit Spray

- The C. capsicisuspension (107 conidia.ml-1) is applied at 2 kg.cm-2 pressure over the entire surface of the fruit and incubated for 10 days at a temperature of 25°C and 95% RH
- Evaluate fruits for disease symptoms.

Biological Control

- Antagonistic Pseudomonas fluorescens as seed treatment and as well as spray treatment @108 CFU.g-1 were found to be effective against C. capsici.
- Trichoderma species are able to effectively control C. capsici infection in chilli.
- Other biological control agents that have been tested for efficacy against Colletotrichum include Bacillus subtilis and Candida oleophila.

Use of Plant Extracts

- Plant products have been tested in many laboratories. Seed and spray treatment showed that the crude extract from rhizome, leaves and creeping branches of sweetflag (Acorus calamus L.), palmorosa (Cymbopogon martinii) oil, Ocimum sanctum leaf extract, and neem (Azadirachta indica) oil could restrict growth of the anthracnose fungus.

Fungal Diseases of Apple and their Management

Apple Crown Rot

Crown rot continues to be a major cause of tree death in Pennsylvania orchards. It often is observed on 3- to 8-year-old trees. Certain rootstocks are more susceptible to the pathogen than others. The disease often occurs in low-lying areas of orchards with heavy, poorly drained soils. The incidence of this disease has increased with the introduction of more dwarfing rootstocks.

Symptoms: The first symptoms to appear in the spring are delayed bud break, leaf discolouration, and twig dieback. These symptoms indicate that crown infection is advanced. Although infected trees might survive the growing season, they show symptoms of leaf and bark discolouration and premature leaf drop in the fall.

The most obvious symptom found on affected trees is a partial or complete girdling of the trunk. Infected bark becomes brown and is often slimy when wet. Close examination of the roots often reveals reddish-brown, water-soaked areas of necrotic tissue located at the base of the root where it attaches to the rootstock. The entire underground portion of the stem is usually water soaked and brown, and the necrotic area usually extends upward to the graft union.

Disease Cycle: The disease is caused by fungi in the genus *Phytophthora*, which belongs to a group of fungi known as the water molds. The fungus survives in the soil for several years as spores that are resistant to drought and somewhat resistant to chemicals. The fungus requires high levels of moisture and cool temperatures for growth and reproduction, and grows best at temperatures around 56°F. Trees, therefore, are attacked at about blossom time (April) and during the onset of dormancy (September). The fungus can infect apple trees in the following ways:

1. collar rot—infection above the tree union,
2. crown rot—infection of the lower trunk and root bases, and
3. root rot—infection of the lateral and fibrous root system.

Disease Management: The following techniques are useful in managing apple crown and collar rot:

- Rootstock selection—Of the rootstocks preferred by growers, none is completely resistant to crown rot. The rootstocks M.7 and MM.106 have appeared to be the most susceptible. The most resistant rootstock is M.9.
- Orchard site selection—Avoid planting orchards in heavy, poorly drained soils. These sites favour fungal growth and development. Crown rot prevention is difficult, and eradication almost impossible in low-lying, poorly drained sites.
- Horticultural—If the tree has not been girdled completely, remove the soil from the base of the tree; scrape the surface of the discoloured area and leave it exposed to dry. Drying often stops crown rot from progressing further.

Apple Scab

Apple scab is Pennsylvania's most important apple disease, attacking wild and cultivated apple and crabapple. Early season disease management is directed primarily at controlling apple scab.

Symptoms: The first infections often occur on the leaves surrounding flower buds. Dull, olive-green areas visible on the undersides of leaves are the first evidence of the disease. As the lesions (infected areas) become older, they assume a definite outline as olive-green or brown circular spots. Leaves are susceptible to infection for about 5 days after they unfold. Severe early leaf infection can result in dwarfed, twisted leaves, which might drop later in the season.

Early infection can occur on the calyx (blossom end of the fruit) or on the pedicel (fruit stem). Severe pedicel infection results in fruit drop. Fruit can become infected at any time in its development. Typical fruit lesions are distinct, almost circular, rough-surfaced, olive-green spots. Heavily infected fruits are usually misshapen and might crack and drop prematurely. When leaf infection is active just before harvest, the fruit might become infected.

These spots do not show at harvest time, but develop slowly while the apples are in storage. This phase of apple scab disease is termed storage scab.

Disease Cycle: Apple scab is caused by the fungus *Venturia inaequalis*, which overwinters in infected leaves that have fallen to the ground. Fruiting bodies are produced within the dead leaf tissue. As spring approaches, these begin to mature and produce spores (ascospores) that are discharged into air currents and carried to developing apple buds. The fruiting bodies in the fallen leaves must be wet for the spores to discharge. The ascospores are not all discharged with the first spring rains since they mature over a 4- to 6-week period.

If the spores remain wet for a few hours after landing on wet apple buds, leaves, or fruit, they germinate and grow into the apple tissue. The time required for germination and penetration depends on the temperature and the presence of a wet surface. After the fungus has penetrated, it continues to grow beneath the cuticle. After 8 to 18 days (development occurs most rapidly at high temperatures), a visible scab lesion is produced.

On its surface appear more spores (conidia), which are easily dislodged when the lesions are wet. The spores are splashed around by rain and blown by wind to new leaf and fruit surfaces within the tree. They germinate on wet surfaces, infect the tissue, and produce a new lesion. In this manner, several secondary infection cycles can occur in the course of a growing season.

Disease Management: Scab infections may be prevented by applying fungicides at regular intervals throughout the growing season. The objective is to provide a protective coating that will inactivate any spores landing on the fruit and foliage. It is critical to control scab early in the season, from bud emergence through the second spray after blossom petals fall (second cover period). If scab infection can be prevented during the time all the ascospores are discharged from the fruiting bodies in the fallen leaves, the disease cycle is broken and no further source of infection remains for the rest of the season. If the cycle is not controlled, however, and leaf and fruit infection do occur, then conidia are produced on these lesions and scab will remain a constant threat all season whenever wet weather occurs.

Bitter Rot

Bitter rot, Glomerella cingulata, is an important disease in the southern states and is becoming more important in Pennsylvania. Its hosts are apple and pear trees. On peach and nectarine the same fungus causes a disease known as anthracnose; on grape it causes ripe rot; and on chestnut it causes blossom-end rot of green burrs. The discussion below is limited to the disease as it affects apple and pear trees.

Symptoms: Bitter rot occurs only on fruit. Cankers can form on twigs, but they are rare. The fungus is one of the few fruit rot organisms that can penetrate the unbroken skin of the fruit. The disease first appears during midsummer or later as a small, light-brown, circular spot. One or many spots might appear; if temperatures are high, they enlarge quite rapidly and soon change to a dark brown. By the time the spots are 1/8 to 1/4 inch in diametre, they are distinctly sunken or saucer shaped. When they reach 1/2 inch in diametre, small, black dots—the fruiting bodies of the fungus—appear in the sunken lesion. These might be arranged in concentric rings. Later, they ooze a gelatinous, salmon-pink mass of spores, which is washed by rains onto other fruit. Beneath the surface of the spot, the flesh is light brown and watery in a cone-shaped area, with the small end of the cone towards the fruit centre. As the fruit ripens, it decays rapidly and finally shrivels into a mummy.

Disease Cycle: The fungus overwinters in mummified fruit and in cracks and crevices in the bark. Jagged ends of broken limbs are ideal sites. With the advent of warm weather, the fungus produces spores washed by rains onto developing fruit. Often, the first infections appear as a cone-shaped area within the tree and can be traced to a source of spores at the tip of the cone. The disease develops optimally in rainy conditions, with a relative humidity of 80 to 100 percent and a temperature of 85°F.

Disease Management: Routine fungicide sprays normally control bitter rot in Pennsylvania. Summer fungicide applications should not be extended beyond 14-day intervals. Sanitation practices, such as removing mummified fruit, are important in controlling this disease.

Black Rot

The black rot fungus, Botryosphaeria obtusa, covers a wide geographical range, attacking the fruit, leaves, and bark of apple trees and other pomaceous plants. The fungus is a vigorous saprophyte and may colonize the dead tissue of many other hosts. However, its parasitic activities are confined mainly to pome fruits.

Symptoms: The first signs of black rot are small, purple spots appearing on the upper surfaces of leaves 1 to 3 weeks after petal fall. Leaf margins remain purple, while the centres turn brown, tan, or yellowish brown. After a few weeks, secondary enlargement of leaf spots occurs. Because this is not a uniform expansion, the spots become irregular or lobed in shape, at which time they assume a characteristic "frog-eye" appearance: a purple margin with a zone of dark brown surrounding the tan-to-gray centre. Small, black pycnidia (pimple-like fruiting bodies of the fungus) might appear in the centres. Leaves that are heavily infected will drop from the tree.

Infected areas of branches and limbs are reddish brown and sunken slightly below the level of the surrounding healthy bark. These cankers might expand each year, with a few eventually reaching several feet in length. Pycnidia are produced abundantly on limb cankers. Fruit rot usually appears at the calyx end of the fruit. It can originate at any wound that penetrates the epidermis, including insect injuries. Usually only one spot occurs per fruit—a characteristic that distinguishes black rot from bitter rot. Initially, the infected area becomes brown and stays brown or turns black as it increases in size. As the rotted area increases, a series of concentric rings often forms. The flesh of the decayed area remains firm and leathery. Eventually, the apple completely decays, dries, and shrivels into a mummy. Pycnidia that contain spores of the black rot fungus appear on the surface of the rotted tissue.

Disease Cycle: The fungus overwinters in dead bark, dead twigs, cankers, and mummified fruit. It can invade almost any dead, woody tissue and frequently is found in tissue killed by fire blight. Early leaf infections often are visible as a cone-shaped area on the tree, with a dead twig or mummified fruit at the apex.

In the spring, spores (conidia and ascospores) are released during rainfall throughout the season. Conidia might continue to be produced during wet periods throughout the summer and can remain viable for long periods. Disseminated by splashing rains, wind, and insects, these spores can infect leaves, the calyxes of blossoms, tiny fruit, and wounds in twigs and limbs. Leaf infection develops during petal fall, at which time conidia attach, germinate, and penetrate through stomata or wounds. Infections of fruit and wood might not become visible for several weeks. Initial fruit infections occur during the bloom period but are not usually apparent until midsummer as the apple approaches maturity. Throughout the growing season, infections occur through wounds.

Disease Management: Cultural control strategies can affect the level of control achieved for black rot. Management programmes based on sanitation to reduce inoculum levels in the orchard are the primary means of control.

- Carefully prune and dispose of dead wood. This should be an important component of both current-season and long-range management.
- Prune and remove cankers; properly dispose of prunings by burial or burning.
- Remove all mummified fruit.
- Control fire blight by pruning out infected wood or controlling insect vectors.

Blue Mold

Blue mold of apple is a common rot of stored apples and pears and is caused by the fungus Penicillium expansum. Other names for the disease are soft rot and Penicillium rot. Blue mold is a disease of ripe fruit and develops mostly on apples that are picked before they are mature. Firm fruit in the same container as decaying fruit may absorb a moldy odour and flavor.

Symptoms: Soft rot appears as soft, light-brown, watery spots that begin around injuries or lenticels on the outer surface of fruit. Rotted fruits have a characteristically moldy odour and flavor. When the relative humidity is high, grayish-blue masses of spores might appear on the fruit surface. Under favourable conditions, the entire fruit can rot in 2 weeks.

Disease Cycle: Spores of the soft rot fungus are present almost everywhere and can survive long periods of unfavourable conditions. Injuries to fruit, especially during picking and handling operations, are the primary points of entry. At ordinary temperatures, infected fruit can rot in 2 weeks or less.

Disease Management: To control blue mold, it is important to prevent fruit from becoming bruised during picking and handling. Fruit should be harvested at optimum maturity. It is also essential to move harvested fruit into cold storage as rapidly as possible.

Brooks Fruit Spot

Brooks fruit spot is a minor disease that occurs throughout the northeastern and mid-Atlantic apple growing regions in the United States. The disease is caused by the fungus Mycosphaerella pomi, Brooks fruit spot is also known as Phoma fruit spot. The disease attacks apple and crabapple trees.

Symptoms: Brooks spot appears on immature fruit in late June and July. Spots on fruit are about 1/4 inch in diametre, irregular in shape, slightly sunken, and usually most numerous on the calyx end. On red-skinned varieties, lesions are dark red to purple. On light-skinned varieties, lesions remain dark green. Unless infected fruit is placed in cold storage immediately after harvest, the spots increase in size and become more sunken, thus more visible.

Disease Cycle: The fungus overwinters in apple leaves on the orchard floor. About the time of petal fall, ascospores are discharged from fallen leaves. How the fungus gets on the leaves is not known since there is no evidence of the disease on leaves while they remain on the trees. Rain and high humidity favour spore discharge and infection of fruit. Infections continue until midsummer, although they decrease as the season progresses.

Disease Management: Routine fungicide applications normally control this disease in Pennsylvania. Summer fungicide applications should not be extended beyond 14-day intervals. Golden Delicious, Jonathan, Rome Beauty, and Stayman, are susceptible. Delicious is somewhat resistant.

Crown Gall

Crown gall occurs on a wide range of herbaceous and woody plant species including pome and stone fruit trees. The disease occurs worldwide and is especially troublesome in nurseries. Losses in orchards are sporadic. The disease is caused by the bacterium Agrobacterium tumefaciens.

Symptoms: The formation of galls on roots and crowns of plants are characteristic of crown gall. Galls are often located at pruning wounds, especially on apple. Small galls are initially smooth on the surface. As they enlarge they become dark, hard, woody tumors with gnarled, irregular surfaces. Gall shapes and sizes vary. The same root or crown may contain numerous galls. Secondary fungi and insects are attracted to galls as they age.

Disease Cycle: The bacterium causing crown gall is spread from infected nursery stock, which can contaminate orchard soils. These bacteria enter the roots and crowns of plants through wounds. The development of galls can occur in a few weeks or a few years, depending on various environmental factors (temperature) and host affected. The bacteria can escape from these galls and infect healthy roots and contaminate surrounding soil. The pathogen spreads from this contaminated soil to new sites and plants by irrigation water, splashing rain, tools, wind, insects, and plant parts used for propagation.

The development of crown gall is influenced by planting site and rootstock type. The most susceptible rootstock to crown gall is M.7 followed by M.9 and M.26. The disease is reported to be more severe on apple trees in sites previously planted in nursery crops such as peach, grape, raspberry, and rose. Planting trees in poorly drained soils seem to contribute to the incidence of crown gall.

Disease Management: Good sanitation and cultural practices are necessary for the control of crown gall. Planting disease-free nursery stock is essential to avoid the introduction of this disease. Choose rootstocks that are not susceptible. Adopt management practices that minimize wounding. Plant trees in well-drained soils.

In recent years, the use of a bacterial antagonist has been shown to be very effective in preventing crown gall on all tree fruit species, except apple. This antagonist is used on seeds, roots and stems of propagation material.

Fumigation is ineffective.

Crown and Collar Rot

Crown rot continues to be a major cause of tree death in Pennsylvania orchards. It often is observed on 3- to 8-year-old trees. Certain rootstocks are more susceptible to the pathogen than others. The disease often occurs in low-lying areas of orchards with heavy, poorly drained soils. The incidence of this disease has increased with the introduction of more dwarfing rootstocks.

Symptoms: The first symptoms to appear in the spring are delayed bud break, leaf discolouration, and twig dieback. These symptoms indicate that crown infection is advanced. Although infected trees might survive the growing season, they show symptoms of leaf and bark discolouration and premature leaf drop in the fall.

The most obvious symptom found on affected trees is a partial or complete girdling of the trunk. Infected bark becomes brown and is often slimy when wet. Close examination of the roots often reveals reddish-brown, water-soaked areas of necrotic tissue located at the base of the root where it attaches to the rootstock. The entire underground portion of the stem is usually water soaked and brown, and the necrotic area usually extends upward to the graft union.

Disease Cycle: The disease is caused by fungi in the genus *Phytophthora*, which belongs to a group of fungi known as the water molds. The fungus survives in the soil for several years as spores that are resistant to drought and somewhat resistant to chemicals. The fungus requires high levels of moisture and cool temperatures for growth and reproduction, and grows best at temperatures around 56°F. Trees, therefore, are attacked at about blossom time (April) and during the onset of dormancy (September). The fungus can infect apple trees in the following ways:

1. collar rot—infection above the tree union,
2. crown rot—infection of the lower trunk and root bases, and
3. root rot—infection of the lateral and fibrous root system.

Disease Management: The following techniques are useful in managing apple crown and collar rot:

- Rootstock selection—Of the rootstocks preferred by growers, none is completely resistant to crown rot. The rootstocks M.7 and MM.106 have appeared to be the most susceptible. The most resistant rootstock is M.9.
- Orchard site selection—Avoid planting orchards in heavy, poorly drained soils. These sites favour fungal growth and development. Crown rot prevention is difficult, and eradication almost impossible in low-lying, poorly drained sites.
- Horticultural—If the tree has not been girdled completely, remove the soil from the base of the tree; scrape the surface of the discoloured area and leave it exposed to dry. Drying often stops crown rot from progressing further.

Mucor Rot

Mucor rot is a fungal disease of apples and pears. The disease is a postharvest storage problem. It does not occur as frequently as blue mold, however, losses due to Mucor infection can be serious.

Symptoms: Mucor rot develops at the calyx end of fruit or at wound punctures in the skin. Infected fruit completely decay after about two months in cold storage. Decaying fruit become very "juicy" and within this juice are abundant spores of the fungus . This is how secondary spread occurs, which is more common in pears but not in apples.

Disease Cycle: Mucor rot is caused by the fungus *Mucor piriformis.* The fungus survives in the soil as spores. Populations of these spores in the soil increase one to two months after fruit have been harvested. Spore populations severely decline during winter. Fruit that have fallen on the orchard floor are infected through contact with infested soil. Throughout the fall and winter, spores are dislodged from the decaying fruit into the soil. Spores are also dispersed by mowing, which scatters pieces of infected fruit. The fungus survives best in cool, dry soils. Fruit become most successful to infection one month before harvest. Fruit that are overmature are more susceptible to infection. The fungus enters the packing house in soil that adheres to fruit bins. Fruit infection occurs during harvest or in the dump tank with rot developing in storage.

Control:

- Remove fallen fruit from the orchard floor and destroy.
- Do not put fallen fruit in bins with harvested fruit.
- Harvest fruit in dry weather.
- Rinse fruit with fresh water to remove spores
- Dry fruit before placing in cold storage.
- Keep soil and debris off the underside of fruit bins.
- Disinfect dump tank solutions.
- Individual fruit wraps impregnated with copper helps reduce rot.

Fire Blight

Fire blight is destructive to apple and quince trees and is the most serious pear disease in the eastern United States. In recent years significant losses to the apple industry have resulted due to this disease. Caused by the bacterium Erwinia amylovora, the disease can attack some 75 species of plants in the rose family. Fire blight also occurs frequently on pyracantha, spirea, hawthorn, and mountain ash. In fruit trees, the disease can kill blossoms, fruit, shoots, limbs, and tree trunks. Certain varieties of apple are more susceptible than others. Susceptible varieties include Gala, Ginger Gold, Braeburn, Jonathan, Rome, Yellow Transparent, and Idared.

Nectria Twig Blight

Nectria twig blight, caused by the fungus Nectria galligena, is a minor disease that breaks out occasionally in Pennsylvania. This disease is a problem, however, in areas of coastal California where temperatures are more moderate. Because its symptoms are similar to those of fire blight, growers need to be able to recognize it. The chemical controls used for fire blight would not be used on nectria twig blight.

Symptoms: In early summer, leaves and shoots of infected twigs wilt and turn brown. Close examination will show that a canker has girdled the twig at the point where shoots begin to grow. Most often, this spot is located at the base of the previous season's cluster bud. Rome Beauty and close relatives such as Gallia Beauty, characterized by enlarged cluster-bud bases, are very susceptible. In midsummer, a few pink or coral fruiting structures of the fungus might appear in the cankered area.

Disease Management: No control is necessary.

Powdery Mildew

Powdery mildew, caused by the fungus Podosphaera leucotricha, attacks buds, blossoms, leaves, new shoots, and fruit of wild and cultivated apples and crabapples. It interferes with the proper functioning of leaves, reduces shoot growth, reduces fruit set, and produces a netlike russet on the fruit of some cultivars. It is often a serious problem in apple nurseries.

Symptoms: The first sign of powdery mildew in spring is a 3- to 4-day delay in the opening of infected buds. Leaves and blossoms of these buds soon become covered with a white to light-gray powder—the spores of the powdery mildew fungus. Flowers do not develop normally, are likely to be greenish white, and produce no fruit. On leaves of new shoot growth, symptoms of powdery mildew appear as felt-like, white patches on the margins and lower surfaces. Infected leaves curl upward and soon become covered with a powdery coating of spores. New infections of succulent leaves and growing shoots reduce the size of the entire shoot. By midsummer, leaves and shoots can turn brown.

Disease Cycle: The fungus overwinters as mycelium (the body of the fungus) inside infected buds. As these buds open in spring, all of their parts become covered with a powdery coating of spores. The spores are easily windblown and infect new leaves, fruit, and shoots. Fruit infection takes place during and shortly after the blossom period. Leaf and shoot infection can continue as long as shoot growth continues. Buds can become infected as they begin to form, until they are matured for overwintering. Infections occur at temperatures of 65 to 80°F when relative humidity is high. Mildew-susceptible varieties include Cortland, Idared, Jonathan, Monroe, and Rome Beauty.

Disease Management Protective mildew sprays should begin at the tight cluster bud stage. Where the disease is severe, mildew fungicides should be continued until new shoots stop growing, or around the fourth cover spray. Choose varieties that are resistant to powdery mildew. Some scab-resistant apple varieties are resistant to this disease

Rust Diseases

There are three rust diseases: cedar-apple rust, hawthorn rust, and quince rust. All three fungi spend part of their life cycle on the eastern red cedar and are problems only when red cedar is found close to the orchard. The most common is cedar-apple rust. These diseases can cause economic losses in several ways. Severe leaf infection and defoliation may make trees susceptible to winter injury. Severe defoliation reduces fruit size and quality, and infected fruit is deformed, sometimes very seriously. Cedar-apple rust occurs on leaves and fruit of apple and crabapple trees. Hawthorn rust occurs on leaves of pear, hawthorn, apple, and crabapple. Quince rust occurs on the leaves and fruit of quince and the fruit of pear, apple, and crabapple.

Symptoms: Cedar-apple rust is caused by the fungus *Gymnosporangium juniperi-virginianae*. Symptoms first appear as small, pale-yellow spots on the upper surfaces of leaves. The spots enlarge, and eventually tiny, black, fruiting bodies (pycnia) become visible. Often, a number of orange-yellow pustules, called aecia, are produced in each spot on the leaf. Infected leaves may remain on the tree or turn yellow and drop. Quince rust, caused by the related fungus *Gymnosporangium clavipes*, does not infect apple leaves but does infect leaves and fruit of quince and hawthorn and can be observed in the spring on cedar branches.

Fruit lesions caused by the cedar-apple rust fungus are somewhat like leaf lesions, they but are much larger and often cause fruit to become disfigured or to develop unevenly. Light-brown to reddish-brown galls form on the branches of red cedar. When they are dry and hard, they may be 1/2 to 2 inches in diametre and are known as "cedar apples." The surfaces of the galls are covered with depressions much like those on a golf ball. In the spring, when the "cedar apples" become wet, a yellow-orange, gelatinous horn up to 2 inches long protrudes from each depression. Fruit infected by quince rust is usually puckered and distorted at the calyx end.

Disease Cycle: Spores discharged from these gelatinous horns on red cedar are easily wind-borne and infect apple leaves and fruit. Spore discharge begins at about the pink stage of apple bloom and is usually completed in a few weeks. Following a few wet periods, the cedar galls die. Spots on apple leaves appear about 10 days after infection. Visible fruit infections require a somewhat longer time. Aecia on the undersides of apple leaves or on fruit lesions themselves produce spores. These spores can be carried back to the red cedar by wind and rain. After lodging in leaf axils or in crevices on cedar twigs, they germinate, infect the twig, and produce tiny galls the following spring. One year later, these galls become able to produce gelatinous horns bearing spores that can reinfect apple trees.

Disease Management: Remove red cedar trees from the vicinity of apple trees. Plant scab-resistant apple varieties. Some scab-resistant varieties are immune to cedar-apple rust.

Apply fungicides at the pink bud stage of apple.

Union Necrosis and Decline

Apple union necrosis and decline is primarily a problem on trees propagated on MM106 rootstock. It is especially serious on red delicious trees, which are on MM106. The disease is caused by the same virus, which is vectored by the dagger nematode causing Prunus stem pitting.

Symptoms: Affected trees begin to exhibit symptoms when they reach bearing age. Leaves are small and sparse and their colour is a dull pale green. Other symptoms include a reduction in terminal shoot growth and shortened stem internodes. Infected trees flower heavily and set large crops of small, highly coloured fruit. On severely infected trees, partial to complete separation of the graft union is common. Removal of bark above and below the graft union reveals abnormally thick, spongy bark and a distinct necrotic line at the graft union. Symptoms on apple are similar to those on peach.

Symptoms at the graft union may differ depending on the cultivar. The severity of apple union necrosis is influenced by the cultivar and rootstock combination. There has been some reported resistance of cultivars and rootstocks to this disease.

Disease Cycle and Management: Apple union necrosis and decline are caused by the tomato ringspot virus, which is vectored by the dagger nematode. The disease cycle and disease management tactics are the same as that for Prunus stem pitting.

Sooty Blotch and Flyspeck

Sooty blotch and flyspeck of apple are separate diseases affecting apple, crabapple, and pear trees. Oftentimes both diseases are normally present on the same fruit. They cause only surface blemishes that detract from fruit appearance and lower fruit quality and market value. Sooty blotch also shortens fruit storage life because of increased water loss. Sooty blotch is a disease complex caused by several unrelated fungi. Flyspeck is caused by the fungus Zygophiala jamaicensis.

Symptoms: Sooty blotch appears on fruit surfaces as sooty or cloudy blotches with indefinite borders. These blotches are olive green to black and can be removed by rubbing vigorously. Flyspeck looks like true "flyspecks," characterized by sharply defined, small, black, shiny dots in groups of a few to nearly 100 or more.

Disease Cycle: Both fungi overwinter on the twigs of many woody plants, as well as on apple and pear trees. The diseases are spread by these overwintering hosts. Spores of the fungi are windblown into and throughout the orchard, and fruit infection can occur anytime after petal fall. It is most prevalent in mid- to late summer. Disease outbreaks are favoured by extended periods of above-normal summer temperatures, combined with frequent rainfall and high humidity. These diseases usually appear on fruit late in the season.

Disease Management: Routine fungicide sprays normally control these diseases in Pennsylvania. Cultural controls include the removal of alternate hosts such as brambles from the orchard and surrounding hedgerows. Dormant-season and summer pruning that opens up the tree canopy and facilitates air movement and the drying of fruit after a rain period will help control these diseases. Thinning to separate the fruit clusters also will help prevent disease.

White Rot

The white rot fungus, Botryosphaeria dothidea, often referred to as "Bot rot" or Botryosphaeria rot, is most important on apple trees, but also attacks crabapple, pear, grape, and chestnut. On apple trees it can be observed as a distinct canker on twigs, limbs, and trunks. However, leaf infections do not occur. Losses from fruit rot can be considerable, especially in southeastern regions of the United States. Drought stress and winter injury have been associated with an increase in infection and canker expansion in northern states.

Diseases of Coriander and Their Management

Coriander, *Coriandrum sativum*, is an erect annual herb in the family Apiaceae. The leaves of the plant are variable in shape, broadly lobed at the base of the plant, and slender and feathery higher on the flowering stems. It is a soft, hairless plant. The flowers are produced in small umbels and are white or very pale pink in colour with the petals pointing away from the centre of the umbel longer than those pointing towards it. Coriander is an annual plant, surviving only one growing season and reaches up to 50 cm (19.7 in). Coriander may also be referred to as cilantro, chinese parsley or dhania and originates from the Near East.

Figure: *Coriander plant*

Uses

All parts of the coriander plant are edible, but the fresh leaves and the dried seeds are most commonly used. Leaves and seeds are used fresh or dried as a herb in cooking.

Propagation

Coriander is propagated from seed and prefers to be grown in full sun or partial shade. The plant grows optimally in areas with damp, cool springs and hot, dry summers at temperatures between 17 and 27°C (62.6–80.6°F) depending on the variety. The plant prefers a well-draining soil and can be grown in sandy loam, loam and clay soils as long as there is sufficient drainage. The plants should be spaced at least 10 cm (4 in) apart at a density of about 75 plants per square metre.

Disease Management: Diseases can also be a serious problem in a coriander crop. The main diseases of ginger are noted below:

Stem Galls of Coriander (Protomyces macrosporus)

This is a common and widespread disease of coriander causing heavy damage to the crop, especially when associated with wilt. The mean loss in fruit yield is estimated at 15% when about 23% plants are infected.

Symptoms: The disease appears in the form of tumor-like swellings of leaf veins, leaf stalks, peduncles, stems as well as fruits. The swellings of the veins give a swollen hanging appearance to the leaves. The tumors are at first glossy but later rupture and become rough. They are about 3 mm broad and up to 12.5 mm long. Badly affected plants may be killed. In the presence of excessive soil moisture, especially under shaded conditions, when the stem fails to harden and remain succulent, the tumors are numerous.

Control Measures: Use of clean, healthy seeds is the best remedy.

Field should be kept clean.

Soil solarization and use of resistant varieties RCr 41 can also be followed.

Crop should be rotated for long term with other crops belonging to other family to reduce the primary inoculum.

Wilt disease Causal Organism

Fusarium oxysporum f.sp. coanderii

Symptoms: The fungus which causes wilt affects root system of the plants.

Control Measures: To prevent the crop from infection by the fungus, deep ploughing should be done during summer season. Crop rotation may also be followed.

In those plots, where effect of this disease has been noticed, coriander crop should not be taken for 2-3 years.

The occurrence of the disease can be reduced by applying oil cakes in the field with high soil pH 8.2. Extreme dry climatic conditions should be avoided as it favours the disease.

Bio-control:

- The antagonistic fungi Trichderma viride @ 4 gm/kg seed treatment is effective.
- Seed treatment with Pseudomonas fluorescens @ 10 gm/kg seed + soil application of neem cake @ 150 kg/ha is effective bio-control method.

Powdery Mildew (Erysiphe Polygoni)

Symptoms: Slightly discoloured specks from which white powdery spots spread and overrun the leaf and pods.

Control Measures: Rouging of infected plants as soon as noticed in the field and subsequent burning.

Grow resistant/tolerant varieties.

Foliar spray of onion extract (5%) is effective in reducing severity.

The incidence is more in February-March, so the sowing should be completed in first week of October.

Blight Disease of Coriander

Coriander is sometimes affected by a blight that can cause severe losses in the crop.

Causal Organism

The most important cause of blight in coriander is Aureobasidium sp. & Ascochyta sp. Additional fungal species can cause the same disease on this crop and these are: Fusarium avenaceum, F. poae, F. culmorum, F. equiseti, F. sporotrichioides, F. graminearum, Botrytis cinerea and Sclerotinia sclerotiorum. The occurrence of persistent wet conditions during flowering could result in any one of these fungi causing blight.

Symptom: The emerging umbels containing the young flowers that have been killed. Later in the season when plants are beginning to mature it may be difficult to recognise a diseased field aside from the reduced seed production.

Control Measures: There is no simple method for the control of blight. The currently grown types of cultivars show no resistance to blight. Once blight is established in a field or on a farm it will be there for a long time. The following suggestions may lower the risk of severe damage. Weather remains an unpredictable but very important factor, as a prolonged moist period during flowering is very conducive to disease development. The ways to check/control the diseases are:

Avoiding excessive irrigation and high humidity.

Using disease-free seeds.

Adopting field sanitation measures.

Using resistant varieties.

Physiological Disorders: The coriander crop is most vulnerable to frost damage at flowering and early seed formation stages. The damage due to frost can be minimized by adopting suitable measures as given for frost in fennel.

Aphids: Aphid infestation occurs at the time of flowing or after flowering in coriander crop. It sucks the sap from tender parts of the plants resulting in heavy loss.

Control Measures: Clip off the heavily infested plant parts.

Spray the crop with pressurized water.

Application of fish oil, resin soap or neem seed kernel extract (3%), neem oil(2%) or tobacco decoction (0.05%) is found to be effective measure.

Spray of neem oil (1%) or neem kernel extract (3%) or 15 litre cow urine + 1.5 kg neem kernel extract/ha) are effective in controlling aphids. When sprayed three times at 15 days interval.

Grasshopper: Grasshoppers can seriously damage a coriander crop and may result in downgrading the price because of body parts in the sample that are hard to remove when cleaning.

Leafhopper: Leafhoppers also may cause problems as the vector for aster yellows. No insecticides are registered for use on the crop.

Cutworm

It is a caterpillar, brownish in colour, which cuts the plants from ground level and makes them to fall down. Infestation of this pest starts at the initial stage of plants resulting in heavy loss to the crop.

Mites: The mites frequently attack the coriander crop and whole plant becomes whitish yellow and appears sickly. The insect mostly feeds on young leaves and the infestation is more severe on young inflorescence. Small mites are seen on the lower side of the leaves and when serous, cause webbing and feed within the web. Plants get stunted at severe infestation.

Control Measures: Spraying of Neem oil (1%) is effective to check the population of mite in the field.

Neem cake can also be applied in the soil before sowing @ 2 t/ha.

Mango Malformation: Epidemiology and Management

The genus *Mangifera* is one among the 73 genera belonging to the family Anacardiaceae in the order Sapindales. Mango (*Mangifera indica*) has been cultivated for more than 5000 years and a wide genetic diversity exist in this crop. The largest number of Mangifera species are found in the Malay Peninsula, the Indonesian archipelago, Thailand, Indo-China and the Philippines. Edible fruits are produced by at least 27 species in the genus.

Mangoes are native to Southern Asia, especially eastern India, Burma and Andaman Islands. It's production predominates in dry and wet tropical low land areas 23° 26' North and South of the equator, on the Indian subcontinent, Southeast Asia and Central and South America. Mango the 'King of Fruits' is most popular fruit among millions of people in India and abroad. In terms of total fruit production on a global basis mango is next to banana. For mango, India ranks first in area and production in the world, accounting 41.0% of the total world mango production i.e.,, 19.22 million tones(mt). India's contribution to the world's mango production is highest i.e.,, 13,692,000 mt from 2,205,000 ha and about 54531 mt of mango is exported of approx. value of Rs. 127.42 crores. In spite of the highest area (22, 05,000 ha) and production of mango in the world, the mango productivity is very low i.e.,, 6.3 mt/ha. This is because of the several abiotic and biotic factors, of which malformation is one of the most important problems.

Mango malformation disease (MMD) is caused by one or more species of the fungus Fusarium.

The disease spreads on a tree very slowly, but if left unchecked, can reduce yields in orchards.

The main spread of MMD to new areas is by infected pruning equipment or vegetative planting material.Mango is the only known host of the disease. It may be associated with the bud mite, Aceria mangiferae but the mites have been shown to spread the disease within a tree and not between trees.

Tree with the disease developed galls in the leaf axils of flush tissue that had been produced at least six months earlier. The galls were roughly circular, 20–50 mm in diametre and consisted of a proliferation of shoots which grew to a length of 10–30 mm,before dying and promoting the growth of new shoots.

The MMD symptoms are a result ofa hormonal imbalance in the trees, induced by theFusarium infections, and associated with bud mite infestations.

Abnormal, compact development of shoots and flowers are common signs of mango malformation disease. Both normal growth and mango malformation disease-affected growth may be present on a plant at the same time.

Growing points such as leaf and stem buds produce misshapen shoots with short internodes and brittle leaves. The leaves are significantly smaller than those of healthy plants and re-curve towards the stem giving a squat, bunchy-top appearance.

Shoot damage occurs in mature trees but symptoms are particularly serious for young plants, which become severely stunted.

Affected flower stems or panicles are thickened and highly branched, producing up to three times the normal number of flowers.

The flowers are enlarged, sterile and do not bear fruit. There are often high numbers of male flowers compared with perfect flowers. Panicles may also form dwarfed and distorted leaves.

The severity of the disease varies from variety to variety and tree to tree in the same variety. Seasonal variations in the occurrence and severity of problem correlate with ambient temperature at flowering

Protecting Your Mango Treesproperty from Mango Malformation Disease(MMD)

Good orchard management occupies paramount importance in this context and plays a vital role in checking the disorder.

To avoid introducing mango malformation disease on to your trees, establish new plantings with pest-free nursery stock. Grafting or marcot material should never be taken from an infected orchard. Nurseries should not be established in orchards affected by mango malformation disease.

Thorough cleaning of pruning equipment between trees reduces the spread of pests and disease.

Make sure that you and your farm workers are familiar with mango malformation disease symptoms and other mango pests.

Ensure that workers, visitors, vehicles and equipment are decontaminated before they enter and leave your farm.

Infestation Control

Different measures like application of growth regulators, urea, nutrients, acaricides and fungicides are suggested to control it. Of these the most simple, effective, economically viable and eco-friendly approach that has been tested at IARI to control this malady is the use of the leaf extracts of common weeds namely Datura stramonium, Calotropis gigantean, and neem tree (Azadirachta indica).

The biological constituents of these weeds possessing medicinal properties are alkaloids from datura, azadirachtin in neem and sugars, flavonoids, flavonol glycosides, oxypregnane-oligoglycosides, terpenes, terpene derivatives, pentacyclic triterpenoids and triterpenoids from C.gigantea.

Control of Malformation: Since there are different schools of thoughts regarding the causes of this malady. attempts have been made by various workers to control it by application of plant growth regulators, nutrients, pesticides, phenolics, anti-malformins and malformin-antagonists. deblossoming with or without growth regulators and pruning without or in combination with pesticides.

Application of Plant Growth Regulators and Phenolic Compounds

Spray of NAA (100 to 200 ppm) during the first week of October has been worked out to reduce floral malformation.

Later, four sprays of 250 ppm NAA at weekly intervals from 20th October onwards have been suggested to reduce the intensity of malformation to the maximum possible.

Hence. spray of 100 ppm NAA has been suggested instead of 200 ppm NAA.

They also observed that spraying 200 ppm NAA in first week of October followed by spraying of 500 ppm of ethrel at bud inception stage during February was highly effective in reducing floral malformation.

Spray of 400 ppm of Ethephon at bud inception stage reduced the floral malformation significantly.

Increase in floral branches and yield was observed by the spraying of ethrel (250 or 500 ppm). or uniconazole 500 ppm in the month of December.

The gibberellic acid at 50 ppm causes delayed panicle emergence, increased number of perfect flowers and increased pollen viability.

Sprays in the months of October and November cause 50% reduction in malformation.

Spraying of 1000 ppm Paclobutrazol (10-60 gm/tree), prior to FBD, during the first week of October reduced malformation, increased number of healthy flowers and increased the yield.

Deblossoming

Deblossoming at the bud burst stage alone and in combination with the spraying of 200 ppm NAA was reported very effective.

Deblossoming alone at bud burst stage gives substantial reduction in malformation; and alleviate panicle malformation however there is no effect of deblossoming in reducing the intensity of malformation.

Malformed and healthy buds prior to deblossoming and found that only those buds which are free from malformation produced normal panicles and those having malformation produced diseased inflorescences and reported that there is no effect of deblossoming.

The process of deblossoming is very cumbersome and hence it is advisable to develop a chemical for deblossoming purpose.

The effect of abscission promoting chemicals e.g. ethephon (500 and 1000 ppm). M and B 25105 (250. 500. 1000 and 2000 ppm) and carbaryl (5000 and 10.000 ppm) on chemical deblossoming was studied in Dashehari and Chausa cvs. of mango.

Ethephon affected almost a total kill of panicles irrespective of initial size at both the concentrations In Dashehari and panicles were completely shed within seven days of the treatment.

M and B 25105 was slightly effective at lower concentrations while carbaryl was ineffective.

In case of Chausa various treatments were found to be ineffective in inducing deblossoming.

However, ethephon caused deblossoming of only smaller panicles.

Different chemicals were also found to exert profound inhibitory effect on the growth of mango panicles as compared to the control to deblossom chemically.

The application of 200 and 500 ppm ethrel to get complete control.

The panicle can be deblossomied by applying 250 ppm of cycloheximide.

Application of Nutrients

Soil application of N:P2O5:K2O (9:3:3) causes high N and low P and K reduced floral malformation.

Soil application of NPK + FeSO4 and is noticed a reduction in malformation and increase in yield.

The N reduced the floral malformation whereas high P and K had similar effect.

By the spraying of mangiferin Zn2+ and Cu2+, the partial control is achieved and good control by injecting phosetyl-Al+Zn+B in the tree trunk.

Reduction in floral malformation is achieved by cobalt sulphate (1000 ppm) when sprayed in first week of October

Application of Pesticides

The application of 0.04% diazinon controls malformation.

Two sprays of 0.32% diazinon minimize the malformation.

99% control of the disease is attained by excising the desired shoots 300-600 mm below the inflorescence and burning them and spraying trees with 0.1% parathion.

0.1% thiometon after every three weeks.

The simple removal of shoots In July and August had given 98% control with no disease for next two years.

Fumigating of the grafted plants with methyl bromide (30-40 mg/ litre) for 1-3 hour kills the adults and nymphs of mite.

The recovery from malformation saplings is possible after pruning the seedling shoots except a terminal bud and few axillary buds and sprayed with 0.1% diazinon.

Aphidon chlordane. diazinon. fenitrothion. mecarbam and disulfoton gàve 100% kill at 0.03 and fenthion. phosphamidon.

Phenthoate, azinphos-ethyl, methyl demeton and aldrin at 0.1% were equally effective against mite.

The percentage of malformed panicles is recorded with lowest (12.9) in swabbing treatment of the out ends with diazinon and highest (40.38) with the pruning alone.

Aldicarb. apocarb. dImethoate and phorate were most effec- tive against mite at 0.05 9 per plant.

Pruning is coupled with sprays of an acaricide and a fungicide have shown some promise in reducing the malformation.

It was found that 0.1% of diazinon, monocrotophos and phosphamidon were comparatively better in checking the infestation of bud mite, Acelia mangiferae.

It was found that diazinoll was the best followed closely by 0.1% monocrotophos and chlorophenamidine, 0.3% phosphamldon, 0.1% parathion and methyl parathion and 0.25% WP sulphur in reducing the malformation, if sprayed from July to December at 21 days of interval.

The complete control of the malady is achieved with the spraying of Fytolan (0.2%) once before inoculation followed by two sprays at 5 and 8 days intervals.

Management of Alternaria Leaf Spot Disease

Causes numerous dark brown to black spots, which may develop in concentric rings or appear targetlike, on infected leaves. The plant may appear chlorotic and leaves may dry or drop prematurely.

Stem lesions or cankers may develop, sometimes girdling and killing stems. Alternaria species persist in infected crops, seeds and plant debris. Spores are carried mainly by the wind, but can also be spread by watersplash to neighbouring plants.

Cultural Controls

To help control Alternaria species, use pathogen-free stock, keep foliage dry and use good sanitation. Certain Alternaria species in the greenhouse have been drastically reduced by covering the structure with special film that absorbs UV light, inhibiting spore formation.

Management of Fusarium Solani f.sp. Melongenae Causing Wilt Disease in Egg Plants

Fusarium solani is a filamentous fungus in the genus Fusarium, and the anamorph of Haematonectria haematococca. Commonly isolated from soil and plant debris. The fungus has a worldwide distribution, but its frequency as a medically important pathogen is not fully known. Aside from keratitis, it is an infrequent cause of fungal infections but remains the most common disease-causing fungus in its genus.

Brinjal wilt caused by Fusarium solani f.sp. melongenae is an important disease which causes huge yield loss. Use of fungicides, cultural and biological control is some common management practices for this disease. Due to the harmful effect of chemical pesticides on the environment, biological control is emerging as an alternative management option.

Resistance in Plants: Ecofriendly Disease Management

The relationship between a plant and a pest[1] is very complex. The ability of a pest to cause disease in a plant depends on environmental conditions, the properties of the organism itself and the capacity of the plant to defend itself. Varieties within a plant species can differ in their ability to defend themselves. Under different climatic conditions the interaction between the same plant and pest may have different outcomes.

Pathogens are known to develop and form new biotypes, pathotypes, races or strains that can cause damage to plants, which remain unaffected by the original form of the pathogen. Resistance is the ability of a plant variety to restrict the growth and development of a specified pathogen or the damage they cause when compared to susceptible plant varieties under similar environmental conditions and pathogen pressure. Resistant varieties may exhibit some disease symptoms or damage under heavy pathogen pressure.

Consistency in Terminology and Nomenclature

To promote consistency in the terms used to describe the reaction of a plant to a pest, the ISF Vegetable and Ornamental Crops Section has adopted definitions for terms describing reactions of plants to pests and recommends their use to vegetable seed companies around the world. The paper defines two levels of resistance. Company claims on the level of resistance are based on tests carried out with well-characterized isolates of a pest in controlled environmental conditions. An isolate is a population of (micro)organisms that has been obtained in pure culture from a field, from another location or in a laboratory, and that has been characterised using differential hosts. A characterised isolate is representative of a biotype, pathotype, race or strain of a pest and will not have any characteristic differentiating it from another isolate of the same biotype, pathotype, race or strain.

Seed companies are also encouraged to use uniform codes for pests affecting vegetable and cereal crops in catalogues and other communication with customers.

Disease resistance is a major goal in breeding new varieties and plays a key role in vegetable crop production and integrated pest management practices. It is also carefully described to differentiate new varieties from older ones on the market. Resistance genes may be effective against all or some biotypes, pathotypes, races or strains of a pest, and the emergence of new biotypes, pathotypes, races or strains is not uncommon.

To identify and distinguish different isolates within a species plant pathologists use different methods including 'differential hosts'. Isolates that after characterization do not match an already described pathotype, biotype, race or strain of the pest are designated as a new biotype, pathotype, race or strain following proof that it is established in nature in a scientific publication.

A consistent naming of pathogen races/strains builds stakeholder confidence in product performance under disease pressure. However, the system of naming pathogen races and strains varies from region to region. In some instances there is a single system in use everywhere, as in the case of leaf mould in tomato (Ff) and near wilt in pea (Fop). Different naming systems cause no problem when there is no overlap in strain/race names, as in anthracnose in bean (Cl) in Europe and USA, or where through the use of joint host differential sets the systems are 'linked', as is the case with downy mildew in lettuce (Bl). The most problematic case is when the use of differing systems causes an overlap in race/strain names, e.g. in Fusarium wilt in tomato (Fol 0, 1 and 2 in Europe, Fol 1, 2 and 3 in USA) and downy mildew in spinach (Pfs).

Together with representatives from public research institutions and regulatory organizations, the vegetable seed industry is engaged in developing a system to standardize identification of pathogen strains/races.

Tolerance to Infections

Tolerance to infections, or disease tolerance, is one of the mechanisms host organisms can fight against parasites, pathogens or herbivores that attack the host. Tolerance to infections is defined as the ability of a host to limit the impact of parasites, pathogens or herbivores on host health, performance, and ultimately on fitness. Tolerance is not equivalent to resistance. Disease resistance is the host trait that prevents infection or reduces the number of pathogens, parasites and herbivores within or on a host.

Tolerance to infections can be illustrated as the reaction of host performance along with increasing pathogen, parasite or herbivore load. This is a reaction norm in which host performance (on y-axis) is regressed against an increasing pathogen, parasite or herbivore burden (on x-axis). The slope of the reaction norm defines the degree of tolerance. High tolerance is indicated as a flat slope, i.e., host performance is not influenced by increasing number of pathogens, parasites or herbivores. Steep downward slope indicates low tolerance; host performance is strongly reduced with increasing pathogen burden. An upward slope indicates overcompensation; a host can increase its performance with increasing pathogen burden.

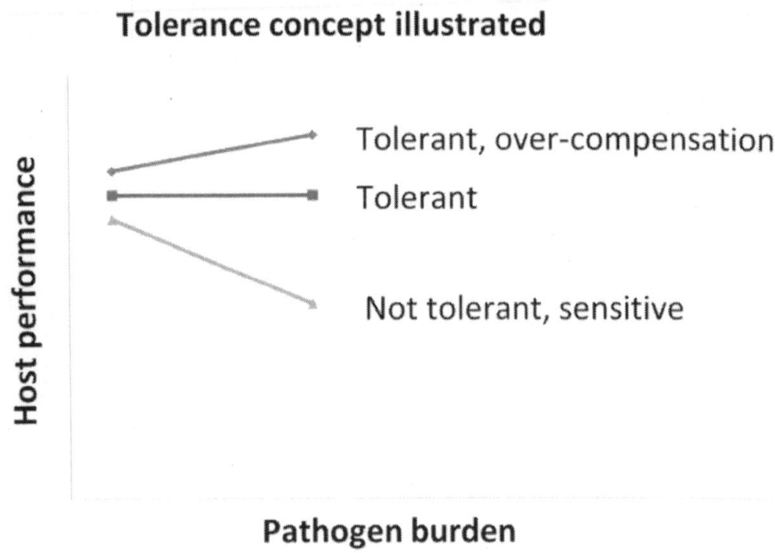

Figure: *Tolerance to infections concept illustrated as the host performance response to an increasing pathogen load*

In farm animal science, tolerance to infections is sometimes termed disease resilience.

Ecofriendly Management of Viral Diseases in Some Vegetables

Insect pests and diseases are the major biotic constraints to vegetable production. The crop losses in the country due to various pests range on an average from10-30 per cent. Apart from causing direct damage, many of them also act as vectors for several viral diseases which have aggravated the problem still further. In recent past, with changes in the cropping pattern, ecosystems and habitat, climate and wider use of high input intensive vegetable varieties/hybrids, a shift in status of pests and diseases has been realized in time and space. Most of the plant protection recommendations in vegetables so far indicated the calendar based application of pesticides. This has become a common practice over the years by most of the farmers growing vegetables in the country. Chemical method of management has its own demerits like ill effects on environment and human and animal healths. Many pests have adapted new hosts, developed resistance to pesticides and often there are secondary out breaks. Insect pests like serpentine leaf miner (Liriomyza trifolii) on tomato and cucurbitaceous crops, gall midge on brinjal, chilli and capsicum; stem fly on okra, leafhopper (Empoasca motti) on bitter gourd; head borer (Hellula undalis) on cabbage, red spider mite (Tetranychusspp.) on okra, brinjal, cowpea and Indian bean; yellow mite (Polyphagotarsonemus latus) on chilli and more recently mealy bug (Pheonacoccus solenopsis) on okra, chilli, brinjal and tomato specially in protected conditions have intensified the severity of occurrence in different parts of the country. Similarly, many soil borne aggressive pathogens such

as Rhizoctonia, Sclerotium, Pythium and plant parasitic nematodes are noted to threat the vegetable production system. Apart from these, the aerial pathogens such as fungal (Alternaria, Cercospora, Colletotrichum and mildews), bacterial (Xanthomonas, Corynebacteria, Pseudomonaes and Ralstonia) and different viruses (leaf curls, yellow mosaic) are noted to be major limiting factors for vegetable production in India.

Success of management of insect pests/diseases highly depends on correct identification and selection of appropriate control measures. Among the various methods of pest control, chemical method still enjoys first choice because of its quick action and easy availability. Consequently, vegetables consume about 13-14 per cent of the total pesticide consumption in India.

Presently, the maximum pesticide usage is in chilli (5.13 kg a.i. /ha) followed by brinjal (4.60 kg a.i. /ha), cole crops (3.73 kg a.i. /ha) and okra (2-3 kg a.i./ha). Unlike cereals, the green pods and fruits harvested at shorter intervals and used as vegetables are prone to retain pesticide residue.The residue problem becomes more severe when the Good Agricultural Practices (GAPs) are not followed. In the present context of food and environmental safety, biological and non-chemical approaches for the management of pests/ diseases are being emphasized in developing bio-intensive IPM/IDM technologies to reduce the toxic residues in the produce and cost of inputs.

However, field efficacy of bio-control agents in many cases has failed to combat the pests/disease due to evolution of new biotypes/ stains/pathotypes/races in the insect/pathogen system. Hence, it is imperative to develop, validate and implement multipronged approaches consisting of biological, host plant resistance/tolerance, cultural practices, etc. to manage the pests and diseases.

In the last couple of decades, researches on vegetable pest and disease management have been reoriented in this direction that has resulted in the generation of extensive information on these aspects.

Virus Diseases of Chilli Capsicum Annuum L. Commonly Occurring in Jammu Division

Diseases of Strawberry and their Management

Strawberry cultivars vary widely in size, colour, flavor, shape, degree of fertility, season of ripening, liability to disease and constitution of plant. Some vary in foliage, and some vary materially in the relative development of their sexual organs. In most cases, the flowers appear hermaphroditic in structure, but function as either male or female. For purposes of commercial production, plants are propagated from runners and, in general, distributed as either bare root plants or plugs. Cultivation follows one of two general models—annual plasticulture, or a perennial system of matted rows or mounds. A small amount of strawberries are produced in greenhouses during the off season.

Figure: *A field using the plasticulture method*

The bulk of modern commercial production uses the plasticulture system. In this method, raised beds are formed each year, fumigated, and covered with plastic to prevent weed growth and erosion. Plants, usually obtained from northern nurseries, are planted through holes punched in this covering, and irrigation tubing is run underneath. Runners are removed from the plants as they appear, in order to encourage the plants to put most of their energy into fruit development. At the end of the harvest season, the plastic is removed and the plants are plowed into the ground. Because strawberry plants more than a year or two old begin to decline in productivity and fruit quality, this system of replacing the plants each year allows for improved yields and denser plantings. However, because it requires a longer growing season to allow for establishment of the plants each year, and because of the increased costs in terms of forming and covering the mounds and purchasing plants each year, it is not always practical in all areas.

The other major method, which uses the same plants from year to year growing in rows or on mounds, is most common in colder climates. It has lower investment costs, and lower overall maintenance requirements. Yields are typically lower than in plasticulture.

A third method uses a compost sock. Plants grown in compost socks have been shown to produce significantly higher oxygen radical absorbance capacity (ORAC), flavonoids, anthocyanins, fructose, glucose, sucrose, malic acid, and citric acid than fruit produced in the black plastic mulch or matted row systems. Similar results in an earlier 2003 study conducted by the US Dept of Agriculture, at the Agricultural Research Service, in Beltsville Maryland, confirms how compost plays a role in the bioactive qualities of two strawberry cultivars.

Strawberries are often grouped according to their flowering habit. Traditionally, this has consisted of a division between "June-bearing" strawberries, which bear their fruit in the early summer and "ever-bearing" strawberries, which often bear several crops of fruit throughout the season. Research published in 2001 showed that strawberries actually occur in three basic flowering habits: short-day, long-day, and day-neutral. These refer to the day-length sensitivity of the plant and the type of photoperiod that induces flower formation. Day-neutral cultivars produce flowers regardless of the photoperiod. Strawberries may also be propagated by seed, though this is primarily a hobby activity, and is not widely practiced commercially. A few seed-propagated cultivars have been developed for home use, and research into growing from seed commercially is ongoing. Seeds (achenes) are acquired either via commercial seed suppliers, or by collecting and saving them from the fruit.

Strawberries can also be grown indoors in strawberry pots.

Kashubian strawberry (*Truskawka kaszubska* or *Kaszëbskô malëna*) are the first Polish fruit to be given commercial protection under EU law. They are produced in Kartuzy, Koœcierzyna and Bytów counties and in the municipalities of Przywidz, Wejherowo, Luzino, Szemud, Linia, £êczyce and Cewice in Kashubia. Only the following varieties may be sold as *kaszëbskô malëna*: Senga Sengana, Elsanta, Honeoye that have been graded as Extra or Class I.

Manuring and Harvesting

Most strawberry plants are now fed with artificial fertilizers, both before and after harvesting, and often before planting in plasticulture.

To maintain top quality, berries are harvested at least every other day. The berries are picked with the caps still attached and with at least half an inch of stem left. Strawberries need to remain on the plant to fully ripen because they do not continue to ripen after being picked. Rotted and overripe berries are removed to minimize insect and disease problems. The berries do not get washed until just before consumption. They are covered in a shallow pan and refrigerated when storing.

Soil test information and plant analysis results are used to determine fertility practices. Nitrogen fertilizer is needed at the beginning of every planting year. There are normally adequate levels of phosphorus and potash when fields have been fertilized for top yields. In order to provide more organic matter a cover crop of wheat or rye is planted in the winter the year before planting the strawberries. Strawberries prefer a pH from 5.5 to 6.5 so lime is usually not applied.

The harvesting and cleaning process has not changed substantially over time. The delicate strawberries are still harvested by hand. Grading and packing often occurs in the field, rather than in a processing facility. In large operations, strawberries are cleaned by means of water streams and shaking conveyor belts.

Diseases

Strawberry plants can fall victim to a number of diseases. The leaves may be infected by powdery mildew, leaf spot (caused by the fungus *Sphaerella fragariae*), leaf blight (caused by the fungus *Phomopsis obscurans*), and by a variety of slime molds. The crown and roots may fall victim to red stele, verticillium wilt, black root rot, and nematodes. The fruits are subject to damage from gray mold, rhizopus rot, and leather rot. To prevent root-rotting, strawberries should be planted every four to five years in a new bed, at a different site.

The plants can also develop disease from temperature extremes during winter. When watering strawberries, advice has been given to water only the roots and not the leaves, as moisture on the leaves encourages growth of fungus.

Botrytis Cinerea

***Figure:** Botrytis cinerea infection on strawberry.*

Botrytis cinerea ("botrytis" from Ancient Greek *botrys (âüôñõò)* meaning "grapes" plus the Neolatin suffix *-itis* for disease) is a necrotrophic fungus that affects many plant species, although its most notable hosts may be wine grapes. In viticulture, it is commonly known as *botrytis bunch rot*; in horticulture, it is usually called *grey mould* or *gray mold.*

The fungus gives rise to two different kinds of infections on grapes. The first, grey rot, is the result of consistently wet or humid conditions, and typically results in the loss of the affected bunches. The second, noble rot, occurs when drier conditions follow wetter, and can result in distinctive sweet dessert wines, such as Sauternes or the Aszú of Tokaji. The species name *Botrytis cinerea* is derived from the Latin for "grapes like ashes"; although poetic, the "grapes" refers to the bunching of the fungal spores on their conidiophores, and "ashes" just refers to the greyish colour of the spores *en masse*. The fungus is usually referred to by its anamorph (asexual form) name, because the sexual phase is rarely observed. The teleomorph (sexual form) is an ascomycete, *Botryotinia fuckeliana*, also known as *Botryotinia cinerea*

Biology

Botrytis cinerea is characterized by abundant hyaline conida (asexual spores) borne on grey, branching tree-like conidiophores. The fungus also produces highly resistant sclerotia as survival structures in older cultures. It overwinters as sclerotia or intact mycelia, both of which germinate in spring to produce conidiophores. The conidia are dispersed by wind and rain-water and cause new infections.

A considerable genetic variability has been observed in different *Botrytis cinerea* strains (polyploidy).

Gliocladium roseum is a fungal parasite of *Botrytis cinerea*.

Horticulture

Botrytis cinerea affects many other plants. It is economically important on soft fruits such as strawberries and bulb crops. Unlike wine grapes, the affected strawberries are not edible and are discarded. To minimize infection in strawberry fields, good ventilation around the berries is important to prevent moisture being trapped among leaves and berries. A number of bacteria have been proven to act as natural antagonists to *B. cinerea* in controlled studies.

In greenhouse horticulture, *Botrytis cinerea* is well known as a cause of considerable damage in tomatoes.

The infection also affects rhubarb, snowdrops and white meadowfoam.

Mycoviruses of Botrytis Cinerea

Botrytis cinerea not only infects plants, it also hosts several mycoviruses itself .

A range of phenotypic alterations due to the mycoviral infection have been observed from symptomless to mild impact, or more severe phenotypic changes including reduction in growth/suppression of mycelia, sporulation and sclerotia production, formation of abnormal colony sectors and virulence.

4

Soil Borne Diseases and their Management

Soils contain diverse communities of microscopic organisms that are capable of damaging plants. A detrimental interaction between a soil organism and a plant is often highly specific. For example, a fungus that causes root-rot of wheat may have no effect on the roots of another plant growing in the same soil. Highly specialised interactions between soil organisms and plants can kill seedlings and even adult trees. Many organisms target younger plants but others appear as problems at later stages in the life of the plant. Other pathogens are able to cause disease in many different plant species.

The soil organisms that have the potential to be plant pathogens include fungi, bacteria, viruses, nematodes and protozoa. Some pathogens of the above ground parts of plants (leaves, stems) survive in the soil at various stages in their life cycles. Therefore, a soil phase of a plant pathogen may be important, even if the organism does not infect roots.

In spite of the potential for severe damage to be inflicted on plants by soil organisms, most plants do not display serious symptoms of disease. Disease usually occurs when conditions are particularly unfavourable, or when a soil organism is accidentally introduced where a highly susceptible plant species occurs. Intensive production in agriculture, horticulture or forestry increases the opportunities for diseases to develop compared with undisturbed natural ecosystems. Planting of similar plant species together in monoculture increases the probability of disease outbreak. In contrast, the damage caused by the fungus *Phytophthora cinnamomi* in many different plant species in diverse natural ecosystems in Australia demonstrates the damage that can be caused by a pathogen that infects the roots of many unrelated plants.

Some plant pathogens depend on their host plant for survival and are unable to complete their life cycle without infecting their host plant. Biotrophic organisms of this type are often difficult to grow in laboratory media.

Disease-causing microorganisms and soil animals are a natural component of the soil community. The organisms are normally present in relatively low numbers. An outbreak of disease commonly follows either an increase in the abundance of the pathogen or a change in the susceptibility of the host to the pathogen. Important questions are:

- Why do outbreaks of disease occur?
- What are the soil conditions that allow disease to develop or that suppress disease?
- What are the soil conditions that suppress disease?
- How do management practices influence the soil as a habitat for plant pathogens?
- How do plant pathogens survive in the absence of their specific host plant?

What is Plant Disease?

Plant disease is the result of a complex interaction between the host plant, the pathogen and the environmental conditions including those in the soil. Soil conditions that influence the development of disease are pH, moisture, oxygen, and nutrients. In addition, other soil organisms interact with the pathogen. The level of this interaction is also influenced by soil conditions, which affects the growth and activity of the non-pathogenic organisms. Consequently, whenever the soil environment changes, there is potential for either a positive or negative influence on soil-borne plant pathogens.

Plant disease reduces plant survival and productivity, but the symptoms of plant disease are often difficult to recognise. One reason for this is that they are not necessarily the same in plants of different ages. The amount of a potential pathogen in a soil and the environmental conditions can also alter the symptoms displayed by a plant. Furthermore, symptoms may be related to infection by more than one organism. Interactions between organisms can produce symptoms that are not the same as those that occur when each type of organism is present alone. Similar symptoms can be caused by different pathogens and furthermore, the symptoms of some pathogens resemble those of various nutrient deficiencies. Therefore, it is not always easy to identify a pathogen responsible for a disease simply from the symptoms on the plant. A great deal of experience and knowledge of the range of symptoms produced by different pathogens and knowledge of symptoms of nutrient deficiency is essential for an accurate identification of most pathogens.

Examples of major soil-borne plant diseases:

(i) Disease caused by *Phytophthora cinnamomi*: *Phytophthora cinnamomi* causes a serious disease that threatens forests and other ecosystems especially in south-eastern and south-western Australia (Shea *et al.* 1984). Hundreds of different plant species are killed by this introduced pathogen. Vehicles are often responsible for the widespread distribution of the pathogen by disturbing and transporting infected soil. There is no simple solution to *Phytophthora* disease in the forest. Quarantine methods have been introduced to limit the spread of the fungus and cleaning of vehicles is mandatory.

    P. cinnamomi and related fungi also cause disease of horticultural plants such as azalea, pineapple and avocado. In horticultural systems, organic mulches have been used to stimulate the community of soil organisms and reduce the negative impact of the pathogens.

(ii) Take-all Disease of Wheat: Take-all disease is caused by the fungus *Gaeumannomyces graminis* var. *tritici*. This pathogen infects the vascular tissue of wheat roots and restricts the transport of water and nutrients within the plant. Severely infected plants have stunted root systems. In addition to root rot, a severe symptom is 'white heads' which occurs if plants survive seedling damage and grow to maturity. Such plants form seed heads with poor grain development that are characteristically white. The fungus survives in the soil on decaying plant material and relies on this material as a carbon source to sustain it until it is able to infect new roots in the following wheat crop or alternate hosts.

    The take-all fungus is a major root disease world-wide and is estimated to cause millions of dollars in lost wheat yield each year. The main method for control is by crop rotation with non-host plants and removal of weeds that may act as alternate hosts. There has been little success in breeding for resistance in wheat to the fungus *Gaeumannomyces graminis* var. *tritici*. Under some conditions, soils become naturally suppressive to the fungus. This has been observed in long-term monocultures of wheat. Reduced disease is likely to be due to changes in activity of other soil microorganisms.

(iv) Crown Gall: Crown gall occurs on many genera of plants and is characterised by the formation of root tumours caused by the bacterium *Agrobacterium tumefaciens*. The bacteria infect the root and induce plant cells to divide; a tumour-like swelling is formed that contains infected cells around its outer surface. Bacterial DNA is transferred to the host plant.

(v) Root Knot Nematode Disease: Root knot nematodes cause disease on hundreds of plant species, especially horticultural species, in warmer climatic zones. Species of nematodes in the genus *Meloidogone* induce the formation of numerous galls throughout the root system. The damaged roots also have malfunctioning root tips which reduce root growth, resulting in considerable yield losses.

How is the Source of Plant Disease Identified?

The fundamental approach to identifying the organism or organisms responsible for a disease is to follow the protocol known as Koch's Postulates:

Step 1. Visual inspection of the plant to identify signs of the presence of a pathogen and symptoms on the host (observed without the aid of a microscope.

Step 2. Observation of the diseased portion of the plant using a microscope.

Step 3. Isolation and purification of the pathogen (growth of the organism on artificial media).

Step 4. Inoculation of plants of the same type with organisms isolated from the diseased plant to observe symptoms of disease, for comparison with the original observations in Step 1.

The purpose of using Koch's Postulates is to identify the organism responsible for disease in a rigorous manner. This is the first step in working out how to overcome the problem. Not all organisms can be isolated from plants and grown in culture and for many diseases it is difficult to identify the responsible organism.

Once a potential pathogen has been identified, a variety of procedures can be used to determine the conditions under which the disease is likely to be most severe. An example of this approach was used to identify the causal organism of brown spot and root rot of lupin in Western Australia as described below.

Case Study: Pleiochaeta Root Rot of Lupin

The pathogen responsible for a serious root disease of lupin in Western Australia was not immediately identified when it was first detected. In fact, several species of fungi were commonly reported from diseased lupin roots, including species of *Fusarium*, *Pythium*, *Rhizoctonia* and *Pleiochaeta* further hindering the control of the disease.

The protocol used by Sweetingham to identify the causal pathogen was to

- Observe and describe the disease symptoms and occurrence in the field.
- Identify the fungi isolated from diseased roots. More than 10 species of fungi were isolated, but none were clearly identified as the possible pathogen.
- Perform pathogenicity tests to identify the degree of disease caused by each fungus when it was present by itself on lupin roots. Some fungi were more pathogenic than others, but none of the fungi, when present alone, was distinctly more pathogenic than any other.
- Test the effect of selected fungicides on the presence of different fungi and on the extent of disease using field soil. This technique was most successful in identifying the likely cause of the root rot.

The use of fungicides that targeted the major fungi present in soil where disease was observed demonstrated that *Pleiochaeta* was the most likely pathogen. The fungicides Ridomil, Benlate, Rovral and a mixture of all three were used. Ridomil eliminated *Pythium* without causing a major decline in the disease rating of roots. Benlate had little effect on *Pythium*, but reduced *Pleiochaeta* and *Fusarium* in the roots leading to a reduced disease rating. Rovral eliminated *Pleiochaeta* but left *Furarium* and *Pythium* and these roots had a much reduced disease rating. This set of fungicide treatments was successful in discriminating among the fungi that were present in the diseased roots.

This case study illustrates the complexity involved in identifying the causal agent of root disease, especially when more than one fungus may be involved. It cannot be presumed that the presence of an organism implicates it as a causal agent of disease. Non-pathogenic fungi often enter damaged roots and therefore a number of organisms may be present in diseased roots without being responsible for the disease.

How are Disease-causing Organisms Dispersed?

Disease occurs under the following conditions:

(i) if there are enough infective hyphae, spores or other fungal structures such as sclerotia of a pathogen present in the soil,

(ii) if there is a suitable environment for growth of the organisms, and

(iii) if the plant is susceptible to infection.

If the organism is present in low numbers, it may have little impact on the plant. If soil conditions are suitable for the multiplication of the pathogen, there will be a greater chance of disease if other factors are also appropriate for disease development. In general, the more organisms present, the more severe the disease (Sweetingham 1991), but plant and soil conditions are also important.

The hyphae of some fungal pathogens can grow long distances through the soil. For example, hyphae of the fungus *Armillaria* extend many metres horizontally in soil and simultaneously infect the roots of several plants. Some fungi, for example species of *Phytophthora*, form narrow and short hyphae that do not spread the fungus through the soil. Often these fungi are disseminated in water or in soil during erosion and by other mechanisms including transport of soil on vehicles or footwear. Spores of fungi that are pathogens of leaves and stems may have a soil phase during which spores remain dormant. Once the dormancy period ceases, spores survive in the soil until conditions are appropriate for their germination and infection of a new plant. Growth of some pathogens is stimulated by the presence of roots which release exudates or emit other signals.

Spores of some soil-borne fungi are dispersed by wind. Dispersal of bacterial plant pathogens is common when soil is transported from one place to another. Some disease-causing organisms are dispersed on plant debris. *Pleiochaeta setosa* spores are dispersed in soil via dead organic material from the previous crop.

How do Pathogens Enter Roots?

Pathogenic invasion of plant roots involves several steps. The first of these is recognition between the pathogen and the plant, which involves complex molecular mechanisms. Recognition occurs in the diffuse interface between the root and the soil created by root exudates and sloughed off root cells. This interface between soil and root may be a very suitable habitat for growth of the pathogen before it enters the root.

This initial interaction can also be the stimulus for the fungus to penetrate the plant cell wall. Some organisms produce enzymes that dissolve cell wall layers, allowing the hyphae to enter. Pathogens can also enter roots at points damaged by soil animals or other disturbances. A plant may react to the presence of a pathogen by producing chemical or physical barriers that prevent entry or confine potential pathogens to a small part of the root.

Once inside the root, pathogens occupy specific cell types. Some fungi enter the cells that conduct nutrients, carbon and water throughout the plant. If these cells are damaged, important processes that control the life of the plant are disrupted. Other fungi predominantly colonise the cortical root cells that are important in nutrient uptake from the soil.

How do Roots React to Pathogens?

Plants differ in their susceptibility to plant pathogens. Even cultivars of one plant species can be more or less resistant to a pathogen. Similarly, species or strains of pathogens differ in their capacity to infect a plant. Differences in the degree of disease can therefore be related to:

(i) the susceptibility of the plant;

(ii) the virulence of the pathogen;

(iii) the stage of development of the plant;

(iv) the abundance of the pathogen; and

(v) the environmental conditions.

How Does Plant Nutrition Affect Disease?

The nutritional status of a plant influences many of its physiological processes and this can alter the response that plants have to the presence of pathogens. Changes in the concentration of nutrients such as phosphorus in a plant can alter root and shoot growth as well as the movement of carbohydrate and other molecules within the plant. These changes have the potential to influence the development of a disease.

Fertilisers can also alter plant susceptibility to disease. In this study, in the absence of manganese fertiliser, barley grew less well if nematodes were present in the soil than when manganese was added at sufficient levels to meet the requirements for growth of the plant This occurred even when the numbers of nematodes in both treatments were similar.

What is 'Biological Control' of Plant Disease?

The control of pathogens and prevention of plant disease is a natural soil biological process. Indeed, in most situations, plant disease is not strongly evident even when potentially pathogenic fungi are present in a soil. This suppressive characteristic of soil is common in natural ecosystems, but becomes less apparent with high levels of disturbance and the introduction of plant monocultures. Disturbance creates soil conditions and a high density of susceptible roots that encourages the multiplication of pathogens. Once potentially damaging organisms become present in high numbers in a soil, they may be difficult to eradicate. Management practices are required that create conditions in the soil that are not favourable to pathogens so that their growth is limitedand therefore, disease it restricted.

There are many demonstrations of the potential for soil organisms to reduce disease. As demonstrated in this study, when apple trees were replanted in a soil previously prone to apple replant disease, increased tree growth occurred when various strains of bacteria were added to the soil, indicating a beneficial effect (Utkhede and Li 1989). Similar effects were observed at two sites, but because different quantities of the bacteria were added in the inoculum it is difficult to compare the effectiveness of each organism.

In some environments, take-all disease of wheat has declined naturally. These areas of natural suppression of disease have been investigated extensively to determine how the pathogen is suppressed. Specific bacteria that are capable of limiting the growth of the fungus, including *Pseudomonas* species, have been isolated from disease suppressive soils. But it is still unclear whether these bacteria are responsible for the decline in take-all disease.

Physiological Plant Disorder

Physiological plant disorders are caused by non-pathological disorders such as poor light, weather damage, water-logging or a lack of nutrients, and affect the functioning of the plant system. Physiological disorder are distinguished from plant diseases caused by pathogens, such as a virus or fungus. While the symptoms of physiological disorders may appear disease-like, they can usually be prevented by altering environmental conditions. However, once a plant shows symptoms of nutrient deficiency it is likely that that season's yields will be reduced.

Causes of physiological disorders can be identified by examining:

- Where symptoms first appear on a plant—on new leaves, old leaves or all over?

- The pattern of any discolouration or yellowing—is it all over, between the veins or around the edges? If only the veins are yellow deficiency is probably not involved.
- Note general patterns rather than looking at individual plants—are the symptoms distributed throughout a group of plants of the same type growing together. In the case of a deficiency all of the plants should be similarly effected, although distribution will depend on past treatments applied to the soil.
- Soil analysis, such as determining pH, can help to confirm the presence of physiological disorders. Recent conditions, such as heavy rains, dry spells, frosts, etc., may also help to determine the cause of plant disorders.

Weather Damage

Figue: *Frosted seedhead of* Echinacea purpurea

Frost and cold are major causes of crop damage to tender plants, although hardy plants can also suffer if new growth is exposed to a hard frost following a period of warm weather. Symptoms will often appear overnight, affecting many types of plants. Leaves and stems may turn black, and buds and flowers may be discoloured, and frosted blooms may not produce fruit. Many annual plants, or plants grown in frost free areas, can suffer from damage when the air temperature drops below 40 degrees Fahrenheit (4 degrees Celsius). Tropical plants may begin to experience cold damage when the temperature is 42 to 48 °F (5 to 9 °C), symptoms include wilting of the top of the stems and/or leaves, and blackening or softening of the plant tissue.

Frost or cold damage can be avoided by ensuring that tender plants are properly hardened before planting, and that they are not planted too early in the season, before the risk of frost has passed. Avoid planting susceptible plants in frost pockets, or where they will receive early morning sun. Protect young buds and bloom with horticultural fleece if frost is forecast. Cold, drying easterly winds can also severely inhibit spring growth even without an actual frost, thus adequate shelter or the use of windbreaks is important.

Drought can cause plants to suffer from water stress and wilt. Adequate irrigation is required during prolonged hot, dry periods. Rather than shallow daily watering, during a drought water should be directed towards the roots, ensuring that the soil is thoroughly soaked two or three times a week. Mulches also help preserve soil moisture and keep roots cool. Heavy rains, particularly after prolonged dry periods, can also cause roots to split, onion saddleback (splitting at the base), tomatoes split and potatoes to become deformed or hollow. Using mulches or adding organic matter such as leaf mold, compost or well rotted manure to the soil will help to act as a 'buffer' between sudden changes in conditions. Water-logging can occur on poorly drained soils, particularly following heavy rains. Plants can become yellow and stunted, and will tend to be more prone to drought and diseases. Improving drainage will help to alleviate this problem.

Hail can cause damage to soft skinned fruits, and may also allow brown rot or other fungi to penetrate the plant. Brown spot markings or lines on one side of a mature apple are indicative of a spring hailstorm.

Plants affected by salt stress are unable to take water from soil, due to an osmotic imbalance between soil and plant.

Nutrient Deficiencies

Poor growth and a variety of complaints such as leaf discolouration (chlorosis) can be caused by a lack of plant foods. This may be due to shortages of necessary nutrients, or because the nutrients are present but not available to the plant. The latter can be caused by incorrect pH, shortages of water or an excess of another nutrient. Generally, the key to avoiding nutrient deficiencies is to ensure that the soil is healthy and contains plenty of well rotted organic matter rather than by feeding or treating individual plants.

Antagonism (Phytopathology)

In phytopathology, antagonism refers to the action of any organism that suppress or interfere the normal growth and activity of a plant pathogen, such as the main parts of bacteria or fungi.

These organisms can be used for pest control and are referred to as "Biological Control Agents''. They may be predators, parasites, parasitoides, or pathogens that attack harmful insect, weed or plant disease or any other organism in its vicinity . the inhibitory substance is highly specific in its action affecting only a specific species. many soil microorganisms are antagonistic .they secrete a potent enzyme which destroys other cells by digesting their cell walls and degrade the cellular material as well as d released protoplasmic material serves as a nutrient for the inhibitor organism for example Aspergillus has an antagonistic effect on Penicillium and Cladosporium . Trichoderma has an effect on actinomycetes. Pseudomanas shoe antagonism on Cladosporium.

Antagonism – Mechanism

- Antibiosis example- enzymes, toxins, antibiotics.
- Direct parasitism example- biotrophic or necrotrophic.
- competition example- for nutrients.
- Induced resistance (indirect).

Blackheart (Plant Disease)

Figure: *A tomato with blackheart rot is shown on the right.*

Blackheart is a non-parasitic disease of the fruit of cultivated plants, such as tomatoes, that causes them to rot from the inside.

External signs of the disease may or may not be present. The internal blackening develops in plants exposed to environmental pathology such as drought conditions or poor soil.

Blossom-end Rot

Blossom-end rot is another sign of plant disease that may accompany blackheart in the same plant. A black rot originates around the stem area of the plant fruit (the "blossom-end") and spreads over the fruit as a dark, hardened area. The causes of blossom-end rot and blackheart are the same.

Causes and Treatment

The most common cause of blossom-end rot and/or blackheart is exposure to a prolonged hot, or drought period. These plant pathologies may also be caused by an extremely low calcium mineral content in the soil.

Plant fruit that is afflicted by either blossom-end rot or blackheart should be picked and discarded, leaving the healthy fruit on the plant. While the condition is not contagious and it will not spread from fruit to fruit, removing affected fruit will enable the plant to concentrate strength towards the healthier parts of the plant.

Additional water should be added to a plant suffering from the effects of a hot, dry spell. If low calcium in the soil is the cause, this can be treated with calcium-rich fertilizers, such as gypsum, limestone, or ground eggshells.

- Boron deficiency

Boron (B) deficiency is the most widespread micronutrient deficiency around the world and causes large losses in crop production both quantitatively and qualitatively. Boron deficiency affects vegetative and reproductive growth of plants resulting in inhibition of cell expansion, death of meristem and reduced fertility.

Plants contain Boron both in a water-soluble and insoluble form. In intact plants, the amount of water-soluble boron fluctuates with the amount of boron supplied, while insoluble boron does not. The appearance of a boron deficiency coincides with the decrease of water-insoluble boron. It appears that the insoluble boron is the functional form while the soluble boron represents the surplus.

Boron is essential for the growth of higher plants. The primary function of the element is to provide structural integrity to the cell wall in plants. Other functions likely include the maintenance of the plasma membrane and other metabolic pathways.

Symptoms

Symptoms include dying growing tips and bushy stunted growth, extreme cases may prevent fruit set. Crop-specific symptoms include;

- *Apple*- interacting with calcium, may display as "water core", internal areas appearing frozen
- *Beetroot*- rough, cankered patches on roots, internal brown rot.
- *Cabbage*- distorted leaves, hollow areas in stems.
- *Cauliflower*- poor development of curds, and brown patches. Stems, leafstalks and midribs roughened.
- *Celery*- leaf stalks develop cracks on the upper surface, inner tissue is reddish brown.
- *Celeriac*- causes brown heart rot
- *Pears*- new shoots die back in spring, fruits develop hard brown flecks in the skin.
- *Strawberries*- Stunted growth, foliage small, yellow and puckered at tips. Fruits are small and pale.
- *Swede (rutabaga)* and *turnip*- brown or gray concentric rings develop inside the roots.
- *Arecaceae* (*Palm Tree*) - brown spots on fronds & lower productivity.references/

Soil Conditions

Boron is present in the soil in many form, the most common being Boric Acid (H_3BO_3). An adequate amount of boron in the soil is 12 mg/kg. If the boron content of the soil drops below 0.14 mg/kg then boron deficiency is likely to be observed. Boron deficiency is also observed in basic soils with a high pH because in basic conditions boric acid exists in an undissociated form which the plant is unable to absorb. Soils with low organic matter content (<1.5%) are also susceptible to boron deficiency. Highly leached sandy soils are also characteristic of boron deficiency because the boron will not be retained in the soil. Boron toxicity is also possible if the boron content of the soil is high enough that the plant cannot cope with the excess boron. The levels at which boron is toxic to plants varies with different species of plants.

Boron Requirements

Boron is an essential micronutrient which means it is essential for plant growth and development, but is required in very small quantities. Although Boron requirements vary among crops, the optimum boron content of the leaves for most crops is 20-100 ppm. Excess boron can result in boron toxicity and the toxicity level varies between plants.

Treatment

Boric acid (16.5%boron), borax (11.3% boron) or SoluBor (20.5% boron) can be applied to soils to correct boron deficiency. Typical applications of actual boron are about 1.1 kg/hectare or 1.0 lb/acre but optimum levels of boron vary with plant type.

Borax, Boric Acid or Solubor can be dissolved in water and sprayed or applied to soil as a dust. Excess boron is toxic to plants so care must be taken to ensure correct application rate and even coverage. Leaves of many plants are damaged by boron; therefore, when in doubt, only apply to soil.

Application of boron may not correct boron deficiency in alkaline soils because even with the addition of boron, it may remain unavailable for plant absorption. Continued application of boron may be necessary in soils that are susceptible to leaching such as sandy soils. Flushing soils containing toxic levels of boron with water can remove the boron through leaching.

Functions

Once boron has been absorbed by the plant and incorporated into the various structures that require boron, it is unable to disassemble these structures and re-transport boron through the plant resulting in boron being a non-mobile nutrient. Due to translocation difficulties the youngest leaves often show deficiency symptoms first.

Cell Wall

Boron is part of the dRG-II-B complex which is involved in the cross linking for pectin located in the primary cell wall and the middle lamella of plant cells. This cross linking is thought to stabilize the matrix of plant cell walls.

Carbohydrate Metabolism

Protein Synthesis

Germination and Pollination: The B requirement is much higher for reproductive growth than for vegetative growth in most plant species. Boron increases flower production and retention, pollen tube elongation and germination, and seed and fruit development.

A deficiency of B can cause incomplete pollination of corn or prevent maximum pod-set in soybeans.

Sugar Translocation

Photosynthesis transforms sunlight energy into plant energy compounds such as sugars. For this process to continue in plants, the sugars must be moved away from the site of their development, and stored or used to make other compounds.

Boron increases the rate of transport of sugars (which are produced by photosynthesis in mature plant leaves) to actively growing regions and also in developing fruits. Boron is essential for providing sugars which are needed for root growth in all plants and also for normal development of root nodules in legumes such as alfalfa, soybeans and peanuts.

- *Calcium deficiency :* Calcium (Ca) deficiency is a plant disorder that can be caused by insufficient calcium in the growing medium, but is more frequently a product of low transpiration of the whole plant or more commonly the affected tissue. Plants are susceptible to such localized calcium deficiencies in low or nontranspiring tissues because calcium is not transported in the phloem. This may be due to water shortages, which slow the transportation of calcium to the plant, poor uptake of calcium through the stem, or can be caused by excessive usage of potassium or nitrogen fertilizers.

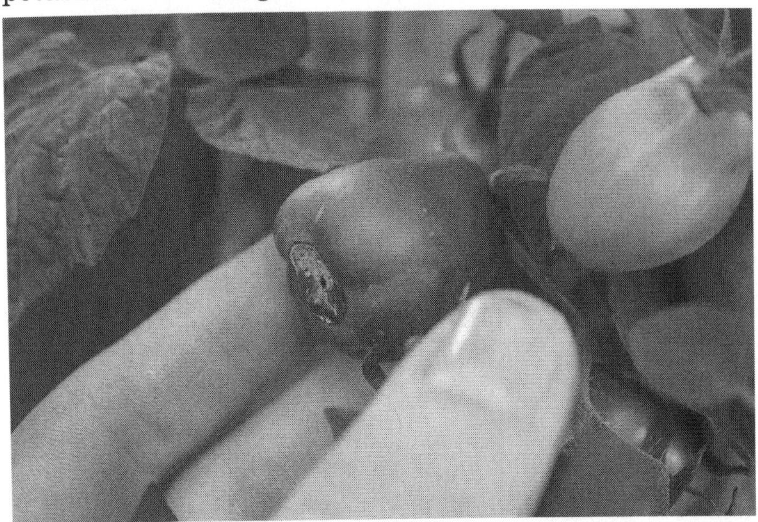

***Figure:** Calcium deficiency (blossom end rot) on a tomato*

Causes

Acidic, sandy, or coarse soils often contain less calcium. Uneven soil moisture and overuse of fertilizers can also cause calcium deficiency. At times, even with sufficient calcium in the soil, it can be in an insoluble form and is then unusable by the plant or it could be attributed to a "transport protein". Soils containing high phosphorus are particularly susceptible to creating insoluble forms of calcium.

Symptoms

***Figure:** Blossom end rot on a grape tomato*

Calcium deficiency symptoms appear initially as localized tissue necrosis leading to stunted plant growth, necrotic leaf margins on young leaves or curling of the leaves, and eventual death of terminal buds and root tips. Generally, the new growth and rapidly growing tissues of the plant are affected first. The mature leaves are rarely if ever affected because calcium accumulates to high concentrations in older leaves.

Crop-specific symptoms include:

Apple : 'Bitter pit' – fruit skins develop pits, brown spots appear on skin and/or in flesh and taste of those areas is bitter. This usually occurs when fruit is in storage, and Bramley apples are particularly susceptible. Related to boron deficiency, "water cored" apples seldom display bitter pit effects.

Cabbage and Brussels sprouts

Internal browning and "tip burn"

Carrot: 'Cavity spot' – oval spots develop into craters which may be invaded by other disease-causing organisms.

Celery: Stunted growth, central leaves stunted.

Tomatoes and Peppers: 'Blossom end rot' – Symptoms start as sunken, dry decaying areas at the blossom end of the fruit, furthest away from the stem, not all fruit on a truss is necessarily affected. Sometimes rapid growth from high-nitrogen fertilizers may exacerbate blossom end rot.

Treatment

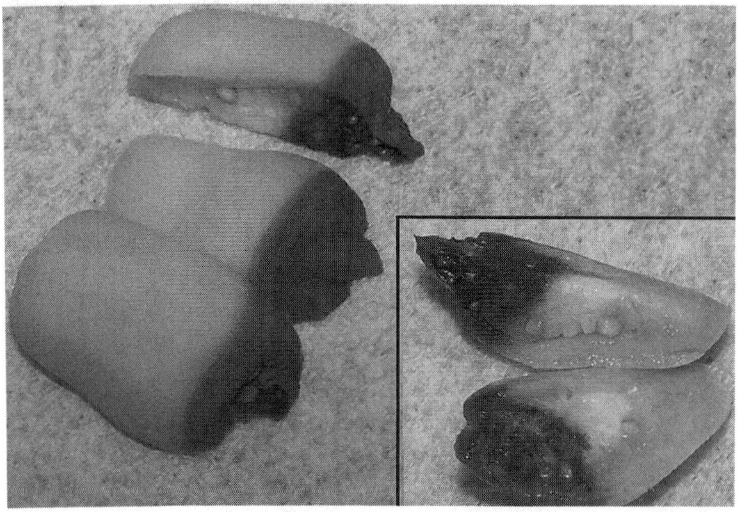

Figure: *Dissection of Grape Tomato With Blossom end Rot*

Calcium deficiency can sometimes be rectified by adding agricultural lime to acid soils, aiming at a pH of 6.5, unless the subject plants specifically prefer acidic soil. Organic matter should be added to the soil to improve its moisture-retaining capacity. However, because of the nature of the disorder (i.e. poor transport of calcium to low transpiring tissues), the problem cannot generally be cured by the addition of calcium to the roots. In some species, the problem can be reduced by prophylactic spraying with calcium chloride of tissues at risk.

Plant damage is difficult to reverse, so corrective action should be taken immediately, supplemental applications of calcium nitrate at 200 ppm nitrogen, for example. Soil pH should be tested, and corrected if needed, because calcium deficiency is often associated with low pH.

Iron deficiency

Iron (Fe) deficiency is a plant disorder also known as "lime-induced chlorosis". It can be confused with manganese deficiency. A deficiency in the soil is rare but iron can be unavailable for absorption if soil pH is not between about 5 and 6.5. A common problem is when the soil is too alkaline (the pH is above 6.5). Also, iron deficiency can develop if the soil is too waterlogged or has been overfertilised. Elements like calcium, zinc, manganese, phosphorus, or copper can tie up iron if they are present in high amounts.

***Figure:** Iron deficiency showing chlorotic leaves in a lemon tree. Compare yellow clorotic leaves with the green non chlorotic leaves at left of this image*

Iron is needed to produce chlorophyll, hence its deficiency causes chlorosis. For example, iron is used in the active site of glutamyl-tRNA reductase, an enzyme needed for the formation of 5-Aminolevulinic acid which is a precursor of heme and chlorophyll.

Symptoms

Symptoms include leaves turning yellow or brown in the margins between the veins which may remain green, while young leaves may appear to be bleached. Fruit would be of poor quality and quantity. Any plant may be affected, but raspberries and pears are particularly susceptible, as well as most acid-loving plants such as azaleas and camellias.

Treatment

Iron deficiency can be avoided by choosing appropriate soil for the growing conditions (e.g., avoid growing acid loving plants on lime soils), or by adding well-rotted manure or compost. If iron deficit chlorosis is suspected then check the pH of the soil with an appropriate test kit or instrument. Take a soil sample at surface and at depth. If the pH is over seven then consider soil remediation that will lower the pH towards the 6.5 - 7 range. Remediation includes: i) adding compost, manure, peat or similar organic matter (warning. Some retail blends of manure and compost have pH in the range 7 - 8 because of added lime. Read the MSDS if available. Beware of herbicide residues in manure. Source manure from a certified organic source.) ii) applying Ammonium Sulphate as a Nitrogen fertilizer (acidifying fertilizer due to decomposition of ammonium ion to nitrate in the soil and root zone) iii) applying elemental Sulphur to the soil (oxidizes over the course of months to produce sulphate/sulphite and lower pH).

Note: adding acid directly e.g. sulphuric/hydrochloric/citric acid is dangerous as you may mobilize metal ions in the soil that are toxic and otherwise bound. Iron can be made available immediately to the plant by the use of iron sulphate or iron chelate compounds. Two common iron chelates are Fe EDTA and Fe EDDHA. Iron sulphate (Iron(II)_sulfate) and iron EDTA are only useful in soil up to PH 7.1 but they can be used as a foliar spray (Foliar_feeding). Iron EDDHA is useful up to PH 9 (highly alkaline) but must be applied to the soil and in the evening to avoid photodegradation. EDTA in the soil may mobilize Lead, EDDHA does not appear to.

- *Magnesium deficiency:* Magnesium (Mg) deficiency is a detrimental plant disorder that occurs most often in strongly acidic, light, sandy soils, where magnesium can be easily leached away. Magnesium is an essential macronutrient found from 0.2-0.4% dry matter and is necessary for normal plant growth.

Figure: *A plant with Magnesium Deficiency*

Excess potassium, generally due to fertilizers, further aggravates the stress from the magnesium deficiency.

Magnesium has an important role in photosynthesis because it forms the central atom of chlorophyll. Therefore, without sufficient amounts of magnesium, plants begin to degrade the chlorophyll in the old leaves. This causes the main symptom of magnesium deficiency, chlorosis, or yellowing between leaf veins, which stay green, giving the leaves a marbled appearance.

Due to magnesium's mobile nature, the plant will first break down chlorophyll in older leaves and transport the Mg to younger leaves which have greater photosynthetic needs. Therefore, the first sign of magnesium deficiency is the chlorosis of old leaves which progresses to the young leaves as the deficiency continues. Magnesium also is a necessary activator for many critical enzymes, including ribulosbiphosphate carboxylase (RuBisCO) and phosphoenolpyruvate carboxylase (PEPC), both essential enzymes in carbon fixation.

Thus low amounts of Mg lead to a decrease in photosynthetic and enzymatic activity within the plants. Magnesium is also crucial in stabilizing ribosome structures, hence, a lack of magnesium causes depolymerization of ribosomes leading to pre-mature aging of the plant. After prolonged magnesium deficiency, necrosis and dropping of older leaves occurs. Plants deficient in magnesium also produce smaller, woodier fruits.

Magnesium deficiency may be confused with zinc or chlorine deficiencies, viruses, or natural aging since all have similar symptoms. Adding Epsom salts (diluted to 8.5 oz. per 2.2 gal. of water) or crushed dolomitic limestone to the soil can rectify magnesium deficiencies. For a more organic solution, applying home-made compost mulch can prevent leaching during excessive rainfall and provide plants with sufficient amounts of nutrients, including magnesium.

- *Manganese deficiency :* Manganese (Mn) deficiency is a plant disorder that is often confused with, and occurs with, iron deficiency. Most common in poorly drained soils, also where organic matter levels are high. Manganese may be unavailable to plants where pH is high.

Affected plants include onion, apple, peas, French beans, cherry and raspberry, and symptoms include yellowing of leaves with smallest leaf veins remaining green to produce a 'chequered' effect. The plant may seem to grow away from the problem so that younger leaves may appear to be unaffected. Brown spots may appear on leaf surfaces, and severely affected leaves turn brown and wither.

Prevention can be achieved by improving soil structure. Do not over-lime.

- *Nitrogen deficiency:* All plants require sufficient supplies of macronutrients for healthy growth, and nitrogen (N) is a nutrient that is commonly in limited supply. Nitrogen deficiency in plants can occur when organic matter with high carbon content, such as sawdust, is added to soil. Soil organisms use any nitrogen to break down carbon sources, making N unavailable to plants. This is known as "robbing" the soil of nitrogen. All vegetables apart from nitrogen fixing legumes are prone to this disorder.

Nitrogen deficiency can be prevented in the short term by using grass mowings as a mulch, or foliar feeding with manure, and in the longer term by building up levels of organic matter in the soil. Sowing green manure crops such as grazing rye to cover soil over the winter will help to prevent nitrogen leaching, while leguminous green manures such as winter tares will fix additional nitrogen from the atmosphere.

Detection

The visual symptoms of nitrogen deficiency mean that it can be relatively easy to detect in some plant species. Symptoms include poor plant growth, and leaves that are pale green or yellow because they are unable to make sufficient chlorophyll. Leaves in this state are said to be chlorotic. Lower leaves (older leaves) show symptoms first, since the plant will move nitrogen from older tissues to more important younger ones. Nevertheless, plants are reported to show nitrogen deficiency symptoms at different parts. For example Nitrogen deficiency of tea is identified by retarded shoot growth and yellowing of younger leaves.

However, these physical symptoms can also be caused by numerous other stresses, such as deficiencies in other nutrients, toxicity, herbicide injury, disease, insect damage or environmental conditions. Therefore, nitrogen deficiency is most reliably detected by conducting quantitative tests in addition to assessing the plants visual symptoms. These tests include soil tests and plant tissue test.

Plant tissue tests destructively sample the plant of interest. However, nitrogen deficiency can also be detected non-destructively by measuring chlorophyll content.

Chlorophyll content tests work because leaf nitrogen content and chlorophyll concentration are closely linked, which would be expected since the majority of leaf nitrogen is contained in chlorophyll molecules. Chlorophyll content can be detected with a Chlorophyll content metre;

a portable instrument that measures the greenness of leaves to estimate their relative chlorophyll concentration.

Chlorophyll content can also be assessed with a chlorophyll fluorometre, which measures a chlorophyll fluorescence ratio to identify phenolic compounds that are produced in higher quantities when nitrogen is limited. These instruments can therefore be used to non-destructively test for nitrogen deficiency.

- Phosphorus deficiency

Phosphorus deficiency is a plant disorder that is most common in areas of high rainfall, especially on acid, clay or poor chalk soils. Cold weather can cause a temporary deficiency.

All plants may be affected, although this is an uncommon disorder. Particularly susceptible are carrots, lettuce, spinach, apples, currants and gooseberries. Symptoms include poor growth, and leaves that turn blue/green but not yellow—oldest leaves are affected first. Fruits are small and acid tasting. Phosphorus deficiency may be confused with nitrogen deficiency. Undersides of tomato plant leaves, and the veins and stems, may turn purple. stiff, stunted plants with purlish tinge.

It can be controlled by applying organic sources of phosphorus such as rock phosphate or a high-phosphorus fish fertilizer applied as a soil drench or foliar feed.

Plants that are naturally adapted to low levels of available soil phosphorus, however, are more likely to suffer from phosphate poisoning: the key is to provide the right level for any particular plant type, neither too high nor too low.

Potassium Deficiency

Soil Type and Potassium Deficiency

Potassium deficiency, also known as potash deficiency, is a plant disorder that is most common on light, sandy soils, because potassium ions (K^+) are highly soluble and will easily leach from soils without colloids. Potassium deficiency is also common in chalky or peaty soils with a low clay content. It is also found on heavy clays with a poor structure.

Role of Potassium in Plants

The main role of potassium is to provide the appropriate ionic environment for metabolic processes in the cytosol, and as such functions as a regulator of various processes including growth regulation. Plants require potassium ions (K^+) for protein synthesis and for the opening and closing of stomata, which is regulated by proton pumps to make surrounding guard cells either turgid or flaccid. A deficiency of potassium ions can impair a plant's ability to maintain these processes. Potassium also functions in other physiological processes such as photosynthesis, protein synthesis, activation of some enzymes, phloem solute transport of photoassimilates into source organs, and maintenance of cation:anion balance in the cytosol and vacuole.

Symptoms of Potassium Deficiency

Typical symptoms of potassium deficiency in plants include brown scorching and curling of leaf tips as well as chlorosis (yellowing) between leaf veins. Purple spots may also appear on the leaf undersides. Plant growth, root development, and seed and fruit development are usually reduced in potassium-deficient plants. Often, potassium deficiency symptoms first appear on older (lower) leaves because potassium is a mobile nutrient, meaning that a plant can allocate potassium to younger leaves when it is K deficient. Deficient plants may be more prone to frost damage and disease, and their symptoms can often be confused with wind scorch or drought. The deficiency is most common in several important fruit and vegetable crops; notably potatoes, brassicas, tomatoes, apples, currants, gooseberries, and raspberries. Sugar beets, cereals, and clover are also commonly affected. Specific symptoms for each of these plants are as follows:

In potatoes, tuber size is much reduced and crop yield is low. The leaves of the plant appear dull and are often blue-green in colour with intervenal chlorosis. Leaves will also develop small, dark brown spots on the undersides and a bronzed appearance on the upper surfaces.

In brassicas, leaves are blue-green in colour and may have a low degree of intervenal chlorosis. Scorching along the outside edges of leaves is common, and leaves are often tough in texture due to slow growth.

In tomatoes, the stems are woody and growth is slow. Leaves are blue-green in colour, and the intervenal area often fades to a pale gray colour. Leaves may also have a bronzed appearance and yellow and orange patches may develop on some of the leaflets. Fruits often ripen unevenly and sometimes have green patches near the stalks.

In apples, leaves are scorched around the edges, and intervenal chlorosis is common. Apple fruits often have a slightly acidic or woody taste.

In gooseberries, currants, and raspberries, dieback of shoots and branches is common and although the plant may produce many blossom buds in the early stages of deficiency, fruit yields turn out low and the fruits are of poor quality.

Potassium Deficiency and Plant Disease

For many species, potassium-deficient plants are more susceptible to frost damage and certain diseases than plants with adequate potassium levels. Increased disease resistance associated with adequate potassium levels indicates that potassium has roles in providing disease resistance, and increasing the potassium levels of deficient plants have been shown to decrease the intensity of many diseases. However, increasing potassium concentration above the optimal level does not provide greater disease resistance. In agriculture, some cultivars are more efficient at K uptake due to genetic variations, and often these plants have increased disease resistance. The mechanisms involved with increased host resistance and potassium include a decreased cell permeability and decreased susceptibility to tissue penetration. Silica, which is accumulated in greater quantities when adequate potassium is present, is incorporated into cell walls, strengthening the epidermal layer which functions as a physical barrier to pathogens. Potassium has also been implicated to have a role in the proper thickening of cell walls.

Prevention and Cure

Prevention and cure can be achieved in the shorter term by adding potassium-specific fertilizer, often called potash, which consists of K_2CO_3. Rock potash may be a good solution because it has a high potassium content but is released slowly to reduce overdose. Common forms of inorganic fertilizers include potassium nitrate, potassium sulfate, and monopotassium phosphate. Other potassium-rich treatments include feeding with home-made comfrey liquid, adding seaweed meal, composted bracken, compost rich in decayed banana peels, or other organic potassium-rich fertilizers. Wood ash has high potassium content, but should be composted first as it is in a highly soluble form. Adequate moisture is necessary for effective potassium uptake; low soil water reduces K uptake by plant roots. Liming acidic soils will also increase K retention in the soil by reducing leaching. In the longer term the soil structure should be improved by adding plenty of well rotted compost or manure. Additionally, improving soil structure will help reduce leaching.

- Shortage of trace elements such as molybdenum can also cause disorders such as whiptail in cauliflower.

A micronutrient deficiency (or trace mineral deficiency) is a physiological plant disorder which occurs when a micronutrient is deficient in the soil in which a plant grows. Micronutrients are distinguished from macronutrients (such as nitrogen, phosphorus, and potassium) by the relatively low quantities needed by the plant. It also represents a human physiological disorder which occurs when a person's diet does not contain required nutrients and/or when illnesses (such as diarrhoea or malaria) cause rapid loss of nutrients through feces or vomit. Untreated, sub-clinical deficiencies can manifest as serious physical disabilities or life-threatening disorders, such as Beriberi (thiamine deficiency), scurvy (vitamin C) or xerophthalmia (vitamin A).

A number of elements are known to be needed in these small amounts for proper plant growth and development. Nutrient deficiencies in these areas can adversely affect plant growth and development. Some of the best known trace mineral deficiencies include: boron deficiency, calcium deficiency, iron deficiency, magnesium deficiency, and manganese deficiency.

Micronutrient Deficiency can also be found in people. Micronutrient deficiencies in children are associated with 10% of all children's deaths. Micronutrient Deficiencies are when a child is lacking essential vitamins or minerals such as Vitamin A, iron, and zinc. These deficiencies are caused by a long-term lack of nutritious food or they can be caused by infections such as worms.

Molybdenum

Molybdenum is a Group 6 chemical element with the symbol Mo and atomic number 42. The name is from Neo-Latin *Molybdaenum*, from Ancient Greek Ìüëõâäïò *molybdos*, meaning lead, since its ores were confused with lead ores. Molybdenum minerals have been known into prehistory, but the element was discovered (in the sense of differentiating it as a new entity from the mineral salts of other metals) in 1778 by Carl Wilhelm Scheele. The metal was first isolated in 1781 by Peter Jacob Hjelm.

Molybdenum does not occur naturally as a free metal on Earth, but rather in various oxidation states in minerals. The free element, which is a silvery metal with a gray cast, has the sixth-highest melting point of any element. It readily forms hard, stable carbides in alloys, and for this reason most of world production of the element (about 80%) is in making many types of steel alloys, including high strength alloys and superalloys.

Most molybdenum compounds have low solubility in water, but the molybdate ion MoO2–4 is soluble and forms when molybdenum-containing minerals are in contact with oxygen and water. Industrially, molybdenum compounds (about 14% of world production of the element) are used in high-pressure and high-temperature applications, as pigments, and as catalysts.

Molybdenum-containing enzymes are by far the most common catalysts used by some bacteria to break the chemical bond in atmospheric molecular nitrogen, allowing biological nitrogen fixation. At least 50 molybdenum-containing enzymes are now known in bacteria and animals, although only bacterial and cyanobacterial enzymes are involved in nitrogen fixation, and these nitrogenases contain molybdenum in a different form from the rest. Owing to the diverse functions of the various other types of molybdenum enzymes, molybdenum is a required element for life in all higher organisms (eukaryotes), though not in all bacteria.

Characteristics

Physical Properties: In its pure form, molybdenum is a silvery-grey metal with a Mohs hardness of 5.5. It has a melting point of 2,623 °C (4,753 °F); of the naturally occurring elements, only tantalum, osmium, rhenium, tungsten, and carbon have higher melting points. Weak oxidation of molybdenum starts at 300 °C (572 °F). It has one of the lowest coefficients of thermal expansion among commercially used metals. The tensile strength of molybdenum wires increases about 3 times, from about 10 to 30 GPa, when their diametre decreases from ~50–100 nm to 10 nm.

Isotopes

There are 35 known isotopes of molybdenum, ranging in atomic mass from 83 to 117, as well as four metastable nuclear isomers. Seven isotopes occur naturally, with atomic masses of 92, 94, 95, 96, 97, 98, and 100. Of these naturally occurring isotopes, only molybdenum-100 is unstable. Molybdenum-98 is the most abundant isotope, comprising 24.14% of all molybdenum. Molybdenum-100 has a half-life of about 10^{19} y and undergoes double beta decay into ruthenium-100. Molybdenum isotopes with mass numbers from 111 to 117 all have half-lives of approximately 150 ns. All unstable isotopes of molybdenum decay into isotopes of niobium, technetium, and ruthenium.

As also noted below, the most common isotopic molybdenum application involves molybdenum-99, which is a fission product. It is a parent radioisotope to the short-lived gamma-emitting daughter radioisotope technetium-99m, a nuclear isomer used in various imaging applications in medicine. In 2008, the Delft University of Technology applied for a patent on the molybdenum-98-based production of molybdenum-99.

Compounds and Chemistry

Oxidationstate	*Example*
–2	$Na_2[Mo_2(CO)_{10}]$
0	$Mo(CO)_6$
+1	$Na[C_6H_6Mo]$
+2	$MoCl_2$
+3	$Na_3[Mo(CN)]_6$
+4	MoS_2
+5	$MoCl_5$
+6	MoF_6

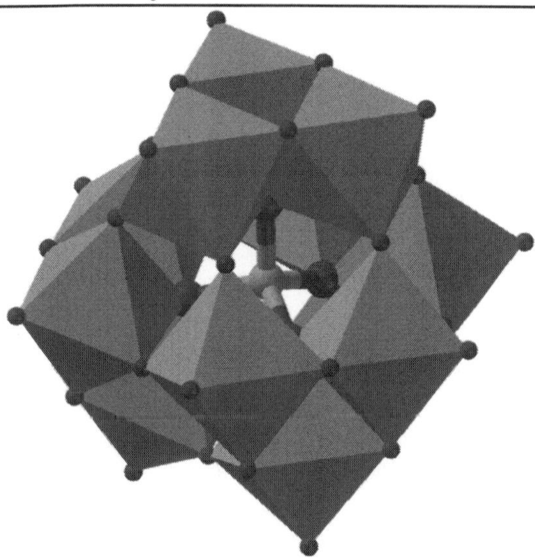

Figure: *Keggin structure of the phosphomolybdate anion ($P[Mo_{12}O_{40}]^{3"}$), an example of a polyoxometalate*

Molybdenum is a transition metal with an electronegativity of 2.16 on the Pauling scale and a standard atomic weight of 95.95 g/mol. It does not visibly react with oxygen or water at room temperature, and the bulk oxidation occurs at temperatures above 600 °C, resulting in molybdenum trioxide:

$2\ Mo + 3\ O_2 \rightarrow 2\ MoO3$

The trioxide is volatile and sublimates at high temperatures. This prevents formation of a continuous protective oxide layer, which would stop the bulk oxidation of metal. Molybdenum has several oxidation states, the most stable being +4 and +6. The chemistry and the compounds show more similarity to those of tungsten than that of chromium. An example is the instability of molybdenum(III) and tungsten(III) compounds as compared with the stability of the chromium(III) compounds. The highest oxidation state is common in the molybdenum(VI) oxide (MoO_3), whereas the normal sulfur compound is molybdenum disulfide MoS_2.

Molybdenum(VI) oxide is soluble in strong alkaline water, forming molybdates (MoO_4^{2-}). Molybdates are weaker oxidants than chromates, but they show a similar tendency to form complex oxyanions by condensation at lower pH values, such as $[Mo_7O_{24}]^{6-}$ and $[Mo_8O_{26}]^{4-}$. Polymolybdates can incorporate other ions into their structure, forming polyoxometalates. The dark-blue phosphorus-containing heteropolymolybdate $P[Mo_{12}O_{40}]^{3-}$ is used for the spectroscopic detection of phosphorus. The broad range of oxidation states of molybdenum is reflected in various molybdenum chlorides:

- Molybdenum(II) chloride $MoCl_2$ (yellow solid)
- Molybdenum(III) chloride $MoCl_3$ (dark red solid)
- Molybdenum(IV) chloride $MoCl_4$ (black solid)
- Molybdenum(V) chloride $MoCl_5$ (dark green solid)
- Molybdenum(VI) chloride $MoCl_6$ (brown solid)

The structure of the $MoCl_2$ is composed of $Mo_6Cl_8^{4+}$ clusters with four chloride ions to compensate the charge.

Like chromium and some other transition metals, molybdenum is able to form quadruple bonds, such as in $Mo_2(CH_3COO)_4$. This compound can be transformed into $Mo_2Cl_8^{4-}$, which also has a quadruple bond.

The oxidation state 0 is possible with carbon monoxide as ligand, such as in molybdenum hexacarbonyl, $Mo(CO)_6$.

History

Molybdenite—the principal ore from which molybdenum is now extracted—was previously known as molybdena. Molybdena was confused with and often utilized as though it were graphite. Like graphite, molybdenite can be used to blacken a surface or as a solid lubricant. Even when molybdena was distinguishable from graphite, it was still confused with the common lead ore PbS (now called galena); the name comes from Ancient Greek Ìüëõâäïò *molybdos*, meaning *lead.* (The Greek word itself has been proposed as a loanword from Anatolian Luvian and Lydian languages).

Although apparent deliberate alloying of molybdenum with steel in one 14th-century Japanese sword (mfd. ca. 1330) has been reported, that art was never employed widely and was later lost. In the West in 1754, Bengt Andersson Qvist examined molybdenite and determined that it did *not* contain lead, and thus was not the same as galena.

By 1778 Swedish chemist Carl Wilhelm Scheele stated firmly that molybdena was (indeed) not galena nor graphite. Instead, Scheele went further and correctly proposed that molybdena was an ore of a distinct new element, named *molybdenum* for the mineral in which it resided, and from which it might be isolated. Peter Jacob Hjelm successfully isolated molybdenum by using carbon and linseed oil in 1781.

For about a century after its isolation, molybdenum had no industrial use, owing to its relative scarcity, difficulty extracting the pure metal, and the immaturity of appropriate metallurgical techniques. Early molybdenum steel alloys showed great promise in their increased hardness, but efforts to manufacture them on a large scale were hampered by inconsistent results and a tendency towards brittleness and recrystallization. In 1906, William D. Coolidge filed a patent for rendering molybdenum ductile, leading to its use as a heating element for high-temperature furnaces and as a support for tungsten-filament light bulbs; oxide formation and degradation require that molybdenum be physically sealed or held in an inert gas. In 1913, Frank E. Elmore developed a flotation process to recover molybdenite from ores; flotation remains the primary isolation process

During the first World War, demand for molybdenum spiked; it was used both in armor plating and as a substitute for tungsten in high speed steels. Somc British tanks were protected by 75 mm (3 in) manganese steel plating, but this proved to be ineffective. The manganese steel plates were replaced with 25 mm (1 in) molybdenum steel plating allowing for higher speed, greater maneuverability, and better protection. The Germans also used molybdenum-doped steel for heavy artillery. This was because traditional steel melted at the heat produced by enough gunpowder to launch a one ton shell. After the war, demand plummeted until metallurgical advances allowed extensive development of peacetime applications. In World War II, molybdenum again saw strategic importance as a substitute for tungsten in steel alloys.

Occurrence and Production

Molybdenum is the 54th most abundant element in the Earth's crust and the 25th most abundant element in its oceans, with an average of 10 parts per billion; it is the 42nd most abundant element in the Universe. The Russian Luna 24 mission discovered a molybdenum-bearing grain (1 × 0.6 μm) in a pyroxene fragment taken from Mare Crisium on the Moon. The comparative rarity of molybdenum in the Earth's crust is offset by its concentration in a number of water-insoluble ores, often combined with sulfur, in the same way as copper, with which it is often found. Though molybdenum is found in such minerals as wulfenite ($PbMoO_4$) and powellite ($CaMoO_4$), the main commercial source of molybdenum is molybdenite (MoS_2). Molybdenum is mined as a principal ore, and is also recovered as a byproduct of copper and tungsten mining.

Historically, the Knaben mine in southern Norway, opened in 1885, was the first dedicated molybdenum mine. It closed from 1973–2007, but is now reopened. Large mines in Colorado (such as the Henderson mine and the Climax mine) and in British Columbia yield molybdenite as their primary product, while many porphyry copper deposits such as the Bingham Canyon Mine in Utah and the Chuquicamáta mine in northern Chile produce molybdenum as a byproduct of copper mining.

The world's production of molybdenum was 250,000 tonnes in 2011, the largest producers being China (94,000 t), United States (64,000 t), Chile (38,000 t), Peru (18,000 t) and Mexico (12,000 t). The total reserves are estimated at 10 million tonnes, and are mostly concentrated in China (4.3 mt), US (2.7 mt) and Chile (1.2 mt). By continent, 93% of world molybdenum production is about evenly split between North America, South America (mainly in Chile), and China. Europe and the rest of Asia (mostly Armenia, Russia, Iran and Mongolia) produce the remainder.

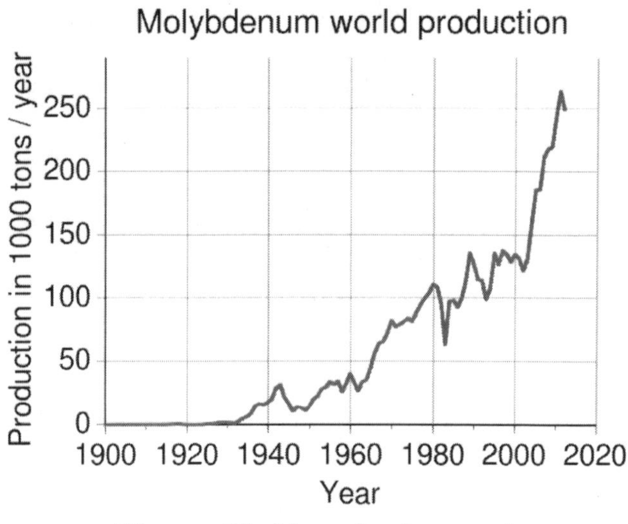

Figure: *World production trend*

In molybdenite processing, the molybdenite is first heated to a temperature of 700 °C (1,292 °F) and the sulfide is oxidized into molybdenum(VI) oxide by air:

$$2\ MoS_2 + 7\ O_2 \rightarrow 2\ MoO_3 + 4\ SO_2$$

The oxidized ore is then either heated to 1,100 °C (2,010 °F) to sublimate the oxide, or leached with ammonia, which reacts with the molybdenum(VI) oxide to form water-soluble molybdates:

$$MoO_3 + 2\ NH_4OH \rightarrow (NH_4)_2(MoO_4) + H_2O$$

Copper, an impurity in molybdenite, is less soluble in ammonia. To completely remove it from the solution, it is precipitated with hydrogen sulfide.

Pure molybdenum is produced by reduction of the oxide with hydrogen, while the molybdenum for steel production is reduced by the aluminothermic reaction with addition of iron to produce ferromolybdenum. A common form of ferromolybdenum contains 60% molybdenum.

Molybdenum has a value of approximately $30,000 per tonne as of August 2009. It maintained a price at or near $10,000 per tonne from 1997 through 2003, and reached, due to increased demand, a peak of $103,000 per tonne in June 2005. In 2008 the London Metal Exchange announced that molybdenum would be traded as a commodity on the exchange.

Applications

Alloys: About 86% of molybdenum produced is used in metallurgical applications such as alloys, with the rest of molybdenum used as compounds in chemical applications. Estimated fractional global industrial use of molybdenum is structural steel 35%, stainless steel 25%, chemicals 14%, tool & high-speed steels 9%, cast iron 6%, molybdenum elemental metal 6%, and superalloys, 5%.

The ability of molybdenum to withstand extreme temperatures without significantly expanding or softening makes it useful in applications that involve intense heat, including the manufacture of armor, aircraft parts, electrical contacts, industrial motors and filaments.

Most high-strength steel alloys (example 41xx steels) contain 0.25% to 8% molybdenum. Despite such small portions, more than 43,000 tonnes of molybdenum are used as an alloying agent each year in stainless steels, tool steels, cast irons and high-temperature superalloys.

Molybdenum is also used in steel alloys for its high corrosion resistance and weldability. Molybdenum contributes further corrosion resistance to "chrome-moly" type-300 stainless steels (high-chromium steels that are corrosion-resistant already due to their chromium content, but in reality "chrome-moly" is not a type-300 stainless, but instead a 41xx series that is not stainless) and especially so in the so-called superaustenitic stainless steels (such as alloy AL-6XN). Molybdenum acts by increasing lattice strain, thus increasing the energy required to dissolve out iron atoms from the surface. Because of its lower density and more stable price, molybdenum is sometimes used instead of tungsten. An example is the 'M' series of high-speed steels such as M2, M4 and M42 as substitution for the 'T' steel series, which contain tungsten. Molybdenum can be implemented both as an alloying agent and as a flame-resistant coating for other metals. Although its melting point is 2,623 °C (4,753 °F), molybdenum rapidly oxidizes at temperatures above 760 °C (1,400 °F) making it better-suited for use in vacuum environments.

TZM (Mo (~99%), Ti (~0.5%), Zr (~0.08%) and some C) is a corrosion-resisting molybdenum superalloy that resists molten fluoride salts at temperatures above 1300C. It has about twice the strength of pure Mo, and is more ductile and more weldable, yet in tests it resisted corrosion of a standard eutectic salt (FLiBe) and salt vapors used in molten salt reactors for 1100 hours with so little corrosion that it was difficult to measure.

Other molybdenum-based alloys that do not contain iron have only limited applications. For example, because of the corrosion resistance against molten zinc, both pure molybdenum and the molybdenum/tungsten alloy (70%/30%) are used for piping, stirrers and pump impellers which come into contact with molten zinc.

Other Applications as the Pure Element

- Molybdenum powder is used as a fertilizer for some plants, such as cauliflower.
- Elemental molybdenum is also used in NO, NO_2, NO_x analyzers in power plants for pollution controls. At 350 °C (662 °F) the element acts as a catalyst for NO_2/NO_x to form only NO molecules for consistent readings by infrared light.
- Molybdenum anodes replace tungsten in certain low voltage X-ray sources, for specialized uses such as mammography.
- The radioactive isotope molybdenum-99 is used to generate technetium-99m, which is used for medical imaging.

Compounds (14% of Global Use)

- Molybdenum disulfide (MoS_2) is used as a solid lubricant and a high-pressure high-temperature (HPHT) antiwear agent. It forms strong films on metallic surfaces and is a common additive to HPHT greases — in the event of a catastrophic grease failure, a thin layer of molybdenum prevents contact of the lubricated parts. It also has semiconducting properties with distinct advantages over traditional silicon or graphene in electronics applications. MoS_2 is also used as a catalyst in hydrocracking of petroleum fractions containing nitrogen, sulfur and oxygen.

- Molybdenum disilicide ($MoSi_2$) is an electrically conducting ceramic with primary use in heating elements operating at temperatures above 1500 °C in air.
- Molybdenum trioxide (MoO_3) is used as an adhesive between enamels and metals. Lead molybdate (wulfenite) co-precipitated with lead chromate and lead sulfate is a bright-orange pigment used with ceramics and plastics.
- Ammonium heptamolybdate is used in biological staining procedures.
- Molybdenum coated soda lime glass is used for CIGS solar cell fabrication.
- Phosphomolybdic acid is a stain used in thin layer chromatography.
- Molybdenum-99 is a parent radioisotope to the daughter radioisotope technetium-99m, which is used in many medical procedures. The isotope is handled and stored as the molybdate.

Biological Role

Biochemistry

The most important role of the molybdenum in living organisms is as a metal heteroatom at the active site in certain enzymes. In nitrogen fixation in certain bacteria, the nitrogenase enzyme, which is involved in the terminal step of reducing molecular nitrogen, usually contains molybdenum in the active site (though replacement of Mo with iron or vanadium is also known). The structure of the catalytic centre of the enzyme is similar to that in iron-sulfur proteins: it incorporates a Fe_4S_3 and multiple $MoFe_3S_3$ clusters.

In 2008, evidence was reported that a scarcity of molybdenum in the Earth's early oceans was a limiting factor for nearly two billion years in the further evolution of eukaryotic life (which includes all plants and animals) as eukaryotes cannot fix nitrogen, and must therefore acquire most of their oxidized nitrogen suitable for making organic nitrogen compounds, or the organics themselves (like proteins) from prokaryotic bacteria. The scarcity of molybdenum resulted from the relative lack of oxygen in the early ocean. Most molybdenum compounds have low solubility in water, but the molybdate ion MoO_4^{2-} is soluble and forms when molybdenum-containing minerals are in contact with oxygen and water. Once oxygen made by early life appeared in seawater, it helped dissolve molybdenum into soluble molybdate from minerals on the sea bottom, making it available for the first time to nitrogen-fixing bacteria, and allowing them to provide more fixed usable nitrogen compounds for higher forms of life. In 2013, Steven Benner suggested it was possible that boron and molybdenum catalyzed the production of RNA on Mars with life being transported to Earth via a meteorite around 3 billion years ago.

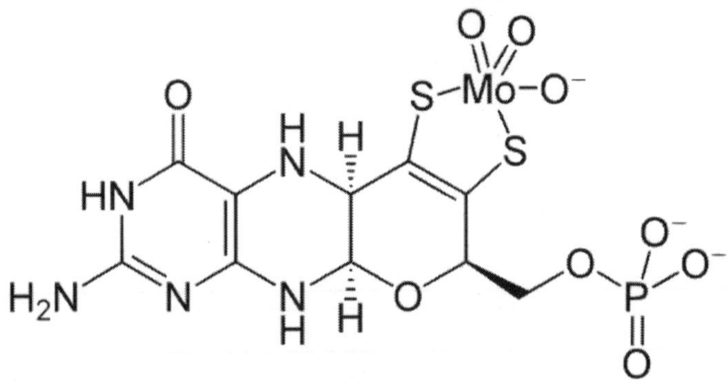

Figure: *The molybdenum cofactor (pictured) is composed of a molybdenum-free organic complex called molybdopterin, which has bound an oxidized molybdenum atom through adjacent sulfur (or occasionally selenium) atoms.*

Although oxygen once promoted nitrogen fixation via making molybdenum available in water, it also directly poisons nitrogenase enzymes. Thus, in Earth's ancient history, after oxygen arrived in large quantities in Earth's air and water, organisms that continued to fix nitrogen in aerobic conditions were required to isolate and protect their nitrogen-fixing enzymes in heterocysts, or similar structures protecting them from too much oxygen. This structural isolation of nitrogen fixation reactions from oxygen in aerobic organisms continues to the present.

Though molybdenum forms compounds with various organic molecules, including carbohydrates and amino acids, it is transported throughout the human body as MoO_4^{2-}. At least 50 molybdenum-containing enzymes were known by 2002, mostly in bacteria, and their number is increasing with every year; those enzymes include aldehyde oxidase, sulfite oxidase and xanthine oxidase. In some animals, and in humans, the oxidation of xanthine to uric acid, a process of purine catabolism, is catalyzed by xanthine oxidase, a molybdenum-containing enzyme. The activity of xanthine oxidase is directly proportional to the amount of molybdenum in the body. However, an extremely high concentration of molybdenum reverses the trend and can act as an inhibitor in both purine catabolism and other processes. Molybdenum concentrations also affect protein synthesis, metabolism and growth.

In animals and plants a tricyclic compound called molybdopterin (which, despite the name, contains no molybdenum) is reacted with molybdate to form a complete molybdenum-containing cofactor called molybdenum cofactor. Save for the phylogenetically-ancient molybdenum nitrogenases discussed above, which fix nitrogen in some bacteria and cyanobacteria, all molybdenum-using enzymes so far identified in nature use the molybdenum cofactor. Molybdenum enzymes in plants and animals catalyze the oxidation and sometimes reduction of certain small molecules, as part of the regulation of nitrogen, sulfur and carbon cycles.

Human Dietary Intake and Deficiency

The human body contains about 0.07 mg of molybdenum per kilogram of weight. It occurs in higher concentrations in the liver and kidneys and in lower concentrations in the vertebrae. Molybdenum is also present within human tooth enamel and may help prevent its decay.

The average daily intake of molybdenum varies between 0.12 and 0.24 mg, depending on the molybdenum content of the food. Pork, lamb, and beef liver each have approximately 1.5 parts per million of molybdenum. Other significant dietary sources include green beans, eggs, sunflower seeds, wheat flour, lentils, cucumbers and cereal grain. Acute toxicity has not been seen in humans, and the toxicity depends strongly on the chemical state. Studies on rats show a median lethal dose (LD_{50}) as low as 180 mg/kg for some Mo compounds. Although human toxicity data is unavailable, animal studies have shown that chronic ingestion of more than 10 mg/day of molybdenum can cause diarrhea, growth retardation, infertility, low birth weight and gout; it can also affect the lungs, kidneys and liver. Sodium tungstate is a competitive inhibitor of molybdenum. Dietary tungsten reduces the concentration of molybdenum in tissues.

Dietary molybdenum deficiency from low soil concentration of molybdenum has been associated with increased rates of esophageal cancer in a geographical band from northern China to Iran. Compared to the United States, which has a greater supply of molybdenum in the soil, people living in these areas have about 16 times greater risk for esophageal squamous cell carcinoma.

Molybdenum deficiency has also been reported as a consequence of non-molybdenum supplemented total parenteral nutrition (complete intravenous feeding) for long periods of time. It results in high blood levels of sulfite and urate, in much the same way as molybdenum cofactor deficiency. However, presumably since pure molybdenum deficiency from this mechanism is seen primarily in adults, the neurological consequences have not been as marked as for the congenital cofactor deficiency.

In humans, four enzymes depend on molybdenum: sulfite oxidase, xanthine oxidoreductase, aldehyde oxidase, and mitochondrial amidoxime reductase. People severely deficient in molybdenum have poorly functioning sulfite oxidase and are prone to toxic reactions to sulfites in foods.

Related Diseases

A congenital molybdenum cofactor deficiency disease, seen in infants, results in interference with the ability of the body to use molybdenum in enzymes. It causes high levels of sulfite and urate, and neurological damage. The cause is the inability of the body to synthesize molybdenum cofactor, a heterocyclic molecule that binds molybdenum at the active site in all known human enzymes that use molybdenum.

Copper-molybdenum Antagonism

High levels of molybdenum can interfere with the body's uptake of copper, producing copper deficiency. Molybdenum prevents plasma proteins from binding to copper, and it also increases the amount of copper that is excreted in urine. Ruminants that consume high amounts of molybdenum develop symptoms including diarrhea, stunted growth, anemia and achromotrichia (loss of hair pigment). These symptoms can be alleviated by the administration of more copper into the system, both in dietary form and by injection. The condition, as an effective copper deficiency, can be aggravated by excess sulfur.

Copper reduction or deficiency can also be deliberately induced for therapeutic purposes by the compound ammonium tetrathiomolybdate, in which the bright red anion tetrathiomolybdate is the copper-chelating agent. Tetrathiomolybdate was first used therapeutically in the treatment of copper toxicosis in animals. It was then introduced as a treatment in Wilson's disease, a hereditary copper metabolism disorder in humans; it acts both by competing with copper absorption in the bowel and by increasing excretion. It has also been found to have an inhibitory effect on angiogenesis, potentially via the inhibition of copper ion dependent membrane translocation process involving a non-classical secretion pathway. This makes it an interesting investigatory treatment for cancer, age-related macular degeneration, and other diseases featuring excessive blood vessel deposition.

Precautions

Molybdenum dusts and fumes, which can be generated by mining or metalworking, can be toxic, especially if ingested (including dust trapped in the sinuses and later swallowed). Low levels of prolonged exposure can cause irritation to the eyes and skin. Direct inhalation or ingestion of molybdenum and its oxides should be avoided. OSHA regulations specify the maximum permissible molybdenum exposure in an 8-hour day as 5 mg/m^3. Chronic exposure to 60 to 600 mg/m^3 can cause symptoms including fatigue, headaches and joint pains.

Calcium Deficiency (Plant Disorder)

Calcium (Ca) deficiency is a plant disorder that can be caused by insufficient calcium in the growing medium, but is more frequently a product of low transpiration of the whole plant or more commonly the affected tissue. Plants are susceptible to such localized calcium deficiencies in low or nontranspiring tissues because calcium is not transported in the phloem. This may be due to water shortages, which slow the transportation of calcium to the plant, poor uptake of calcium through the stem, or can be caused by excessive usage of potassium or nitrogen fertilizers.

Causes

Acidic, sandy, or coarse soils often contain less calcium. Uneven soil moisture and overuse of fertilizers can also cause calcium deficiency. At times, even with sufficient calcium in the soil, it can be in an insoluble form and is then unusable by the plant or it could be attributed to a "transport protein". Soils containing high phosphorus are particularly susceptible to creating insoluble forms of calcium.

Carpellody

Carpellody is a plant disorder that results in misshapen fruits caused by abnormal development of the ovule-bearing part of the flower in angiosperms. It is commonly called "cat face" and is specific to papayas. Some fruits resemble female fruits, but the disorder can cause severely deformed fruits with deep longitudinal ridges. Fruits are generally more rounded and are unmarketable because of their small size and poor eye appeal.

Causes

Carpellodic fruits may be caused by low night temperatures combined with high moisture and high nitrogen levels. These factors may be what cause the stamens to develop abnormally and develop into carpel-like fleshy structures and cause the resulting deformed fruit.

Chlorosis

In botany, chlorosis is a condition in which leaves produce insufficient chlorophyll. As chlorophyll is responsible for the green colour of leaves, chlorotic leaves are pale, yellow, or yellow-white. The affected plant has little or no ability to manufacture carbohydrates through photosynthesis and may die unless the cause of its chlorophyll insufficiency is treated, although some chlorotic plants, such as the albino *Arabidopsis thaliana* mutant *ppi2*, are viable if supplied with exogenous sucrose.

Chlorosis is derived from the word Chloris or from the Greek *khloros* meaning 'greenish-yellow', 'pale green', 'pale', 'pallid', or 'fresh'.

In viticulture, the most common symptom of poor nutrition in grapevines is the yellowing of grape leaves caused by chlorosis and the subsequent loss of chlorophyll. This is often seen in vineyard soils that are high in limestone such as the Italian wine region of Barolo in the Piedmont, the Spanish wine region of Rioja and the French wine regions of Champagne and Burgundy. In these soils the grapevine often struggles to pull sufficient levels of iron which is a needed component in the production of chlorophyll.

Causes

Chlorosis is typically caused when leaves do not have enough nutrients to synthesise all the chlorophyll they need. It can be brought about by a combination of factors including:

- a specific mineral deficiency in the soil, such as iron or magnesium
- deficient nitrogen and/or proteins
- a soil pH at which minerals become unavailable for absorption by the roots
- poor drainage (waterlogged roots)
- damaged and/or compacted roots
- pesticides and particularly herbicides may cause chlorosis, both to target weeds and occasionally to the crop being treated.
- exposure to sulphur dioxide
- ozone injury to sensitive plants [2]

However, the exact conditions vary from plant type to plant type. For example, Azaleas grow best in acidic soil and rice is unharmed by waterlogged soil.

Chlorosis in Grape Vines

Like many other plants, grape vines are susceptible to chlorosis, and symptoms of iron deficiency tend to be common on soils rich in limestone. In the wake of The Great French Wine Blight, when European *Vitis vinifera* were affected by *Phylloxera*, chlorosis became a greater problem in viticulture. To deal with the Phylloxera blight, *V. vinifera* was grafted onto rootstock based on American species of the *Vitis* genus, such as *Vitis riparia, Vitis rupestris, Vitis berlandieri*. However, many of these were less adapted to the lime-rich soils that were common in France's vineyards, in particular many of those that produced wines of top quality.

Many grafted vines in lime-rich vineyards therefore showed signs of iron deficiency, and in France this specific form of chlorosis was termed *chlorose calcaire*. The problem was largely overcome by the selection of lime-resistant American vines as basis for hybrid vines used for rootstock material. However, since such rootstocks may be less than optimal in other respects, it is necessary for the viticulturalist to balance the need for chlorosis resistance against other viticultural needs. This is illustrated by one of the most common lime-resistant rootstocks, 41 B, which is a hybrid between *V. vinifera* cultivar Chasselas and *V. berlandieri*, which generally has a sufficient, but not extremely high, Phylloxera resistance.

Treatments

Specific nutrient deficiencies (often aggravated by high soil pH) may be corrected by supplemental feedings of iron, in the form of a chelate or sulphate, magnesium or nitrogen compounds in various combinations.

False Vivipary

False vivipary is an abnormal condition found in many types of plants in which a plantlet is produced where the flower should appear. It is not a completely understood topic, but some say it could be caused by a hormonal mistake. The plantlet which appears can be rooted and grown like normal plants. This abnormal behaviour can occasionally be seen in many types of carnivorous plants.

Fertilizer Burn

Fertiliser burn is defined as leaf scorch resulting from over-fertilisation, usually referring to excess nitrogen salts

Fertiliser burn is the result of desiccation of plant tissues due to osmotic stress, creating a state of hypertonicity.

Fertilizers vary in their tendency to burn roughly in accordance with their salt index.

Treatment

Fertiliser burn can be remedied by soaking the soil with water to flush out excess salts. It can be prevented by applying only dilute or controlled-release fertilizer products

Ice Pruning

Ice pruning is the natural process of selective vegetative pruning on the windward side of a plant, executed by the impact of ice and snow particles driven by wind. The process is sometimes termed snow pruning. The time scale required for this phenomenon is typically over several growing seasons. The characteristic asymmetry of an ice-pruned plant is achieved only if the prevailing winds during the snow season have a definite directional bias, as shown on a wind rose.

Ice pruning is seen in high latitudes and altitudes, such as high mountain slopes and locations more than 50 degrees of latitude from the equator. In parts of northern Canada, forests dominated by Black Spruce have been noted for containing individual trees that are distinctively ice-pruned.

Salt Pruning

Salt pruning is the process by which saline mists generated by seawater are driven ashore by winds and thus over time alter the shape of trees or shrubs. The process degrades foliage and branches on the windward side of the plant that faces the body of saline water, more than it does the foliage on the landward side. The resultant growth form is asymmetrical, appearing "swept back" away from the ocean.

Examples: There are numerous examples worldwide of this phenomenon. In the eastern United States on Long Island occurrences of salt-pruned *Quercus stellata* are observable in Flax Marsh. In San Diego County, California, a colony of *Pinus torreyana* has been salt-pruned by spray from the Pacific Ocean.

The logo of the Torrey Pines Golf Course in La Jolla, California, features a salt-pruned Torrey pine.

Leaf Scorch

Leaf scorch (also called leaf burn, leaf wilt, and sun scorch) is defined as a browning of plant tissues, including leaf margins and tips, and yellowing or darkening of veins which may lead to eventual wilting and abscission of the leaf.Causes

Leaf scorch can be caused by soil compaction, transplant shock, nutrient deficiency, drought, salt toxicity, and herbicide injury.

Treatment

Affected plants may sometimes recover through watering and fertilization (if the cause is not over-fertilization). Light pruning may also help to reduce the water-pumping load on the roots and xylem. In the case of leaf scorch through over-fertilization, recovery may take time, requiring a treatment of a slow leaching process through drip irrigation over 24–48 hours.

Prevention

Reversal of symptoms and damage can be enacted through the following cultural practices:

- Pruning sprouts and affected areas
- Avoiding frequent, light waterings which promote unhealthy root systems
- Watering heavily to promote deep root systems
- Avoiding over-fertilization

Yellow Vein Mosaic Disease of Okra and its Management Strategies

The most common disease afflicting the okra plant is verticillium wilt, often causing a yellowing and wilting of the leaves. Other diseases include powdery mildew in dry tropical regions, leaf spots, and root-knot nematodes.

Vein-clearing / yellow vein mosaic of Okra and Management: Bhendi yellow vein mosaic virus

Symptoms:

- Yellowing of the entire network of veins in the leaf blade is the characteristic symptom.
- In severe infections the younger leaves turn yellow, become reduced in size and the

plant is highly stunted.

- In a field, most of the plants may be diseased and the infection may start at any stage of plant growth. Infection restricts flowering and fruits, if formed, may be smaller and harder.

The virus is transmitted by the whitefly, *Bemisia tabaci.*

Disease Management of Horticultural Crops Though Organics

16. Saffron corn rot disease and its biological management
17. Ecofriendly management of fusarial wilts of vegetable and fruit crops

19. Cultural practices: an ecological and economical approach for plant disease management
20. Management of chilli fruit rot and Anthracnose in Kashmir Valley
21. Plant disease resistance genes: concepts and connections
22. Root rot: A threat to apples in Kashmir Valley
24. Nonpathogenic disorders in tomato and their management
25. Nonpathogenic diseases of potato tubers
27. Bubble diseases of white button mushroom and their management
28. Cloning sequencing and transformation of coat protein gene of papaya Ringspot virus
29. Characterization and management of papaya ring spot virus

Role of Nutrients in Plant Disease Management

Plant diseases are a major limiting factor in agricultural production. Most growers use high amounts of chemicals to control plant diseases, unaware to the fact that mineral nutrition has an important role in disease control. In addition to the economical consequences, the use of pesticides raises environmental and food safety concerns.

All essential plant nutrients influence the health of plants and their susceptibility to disease. Plants suffering a nutrient stress will be more susceptible to diseases, while adequate crop nutrition makes plants more tolerant of or resistant to disease.

The resistance of plants to diseases is mainly related to genetics. However, the ability of the plant to express its genetic resistance to a particular disease is affected by mineral nutrition.

Some nutrients have a greater impact on plant diseases than others. However, it should be noted that a particular nutrient may have opposite impacts on different diseases and in different environments, i.e. the same nutrient may decrease the incidence of one disease, but increase the incidence of others.

Soil pH, nitrogen forms and the availability of nutrients play a major role in disease management. Nutrient manipulation can be achieved directly by applying adequate fertilizers, or indirectly, through the use of different cultural practices, such as liming for pH adjustment, irrigation, organic amendments, tillage etc.

Integrating variety selection, cultural practices, chemical treatments and mineral nutrition is found to be the best approach to plant disease control.

How Can Mineral Nutrition Prevent Plant Disease?

Mineral nutrition can affect two primary resistance mechanisms:

1. Formation of mechanical barriers (e.g. thickness of cell walls).
2. Synthesis of natural defence compounds (e.g. antioxidants, phytoalexins and flavenoids).

Basically, all diseases have a cycle of 3-5 parts, and any disease can be prevented or suppressed if that cycle is interrupted. Different pathogens have different infection mechanisms.

Fungi penetrate the surface cells (epidermis), by passing between the cells or through them. The cell walls present a physical resistance to the fungus and stronger cell walls can prevent the infection. Certain nutrients, like Calcium, play a major role in the ability of the plant to develop stronger cell walls and tissues.

Plants exude various compounds. When some nutrients are below a certain level, the released compounds contain higher amounts of sugars and amino acids that stimulate the establishment of the fungus.

Mineral nutrition also affects the production of antifungal compounds in plants.

Bacteria invade the plant tissue through wounds, sucking insects and through the stomata. Then they spread within the intercellular spaces. The bacteria release enzymes that dissolve the plant tissue. Calcium is known in its ability to inhibit such enzymes.

The ability of bacteria to spread within the plant tissues depends on the strength of the internal cells, which is greatly affected by mineral nutrition.

Another mechanism in which bacteria spread within the plant is in the xylem (the vessels that transport water in the plant). The bacteria forms slime within the vessels and block them. As a result, stems and leaves wilt and die. Certain plant nutrients suppress the ability of bacteria to form this slime.

Viruses are transmitted to plants by sucking insects and fungi. It was found that silicon, although it is not a plant nutrient, inhibits the feeding ability of some sucking pests like aphids. As a result, virus infection is reduced.

Cultural Practices and Soil Condition

Many plant diseases are categorized as high pH diseases or low pH diseases, high ammonium or nitrate types, high or low moisture etc. Cultural conditions that influence the form of nitrogen also affect soil pH, and therefore disease development. For example, diseases such as Verticillium wilt of vegetables, Phymatotrichum root rot of cotton, Thielaviopsis root rot of tobacco, are associated with alkaline soils.

Scab is reduced in lower soil pH. Sulfur and NH_4^+ are being used to decrease soil pH and reduce scab, while Ca, K, and NO_3^- may increase scab.

Same Nutrient, Different Effect

Different oxidation forms of the same nutrient often have opposing effects on plant disease. This is true mainly for nitrogen, sulfur, manganese and iron. For example, Nitrate (NO_3^-) and Ammonium (NH_4^-) nitrogen forms have different metabolic pathways and therefore different effects on plant disease.

Additional Examples

Excess Nitrogen: Adequate nitrogen levels are necessary for disease resistance. However, excess nitrogen may promote favourable conditions to plant disease. Some of the reasons for this are:

- Excess nitrogen promotes thinner and weaker cell walls
- Increased plants density and therefore high humidity and low light conditions.
- Excess nitrogen delays maturity of plant tissues and therefore increases the risk of disease infection and development.
- Unbalanced N :K ratio affects both yield and disease resistance.

Potassium and K:Ca Ratio

Potassium is a very important nutrient in plant disease prevention, as it is involved in many cellular processes that influence disease severity. It was also found effective in the prevention of bacteria, fungi and nematodes. However, excess potassium and high K:Ca ratio might result in calcium deficiency and reduced resistance to diseases.

Biological Control of Foliar Pathogens of Crop Plants

Generally, the culprit is one of several fungal leaf diseases that begin to express symptoms as the growing season progresses Fungal leaf diseases of ornamental trees often are frustrating problems. This may seem surprising because foliar diseases are not usually life-threatening. Most otherwise-healthy trees can withstand several seasons of partial defoliation without suffering significant injury. Trees produce an abundance of leaves, so some loss of foliage, in itself, is not a major cause for concern. The loss of many leaves early in the growing season, however, may result in depletion of food reserves as the tree attempts to refoliate. This depletion, particularly if repeated for several years, may leave the tree vulnerable, lacking the energy to defend itself and the reserves to survive additional stresses. Fortunately, heavy and repeated early season defoliation is more the exception than the rule for fungal foliage diseases.

While defoliation may not be a threat to the tree's life, it can be a threat to your job. Most ornamental trees can survive 20 to 40 percent defoliation with little or no significant injury, but the tree's owner may notice as little as 5 percent defoliation. Leaves are one of the most visible features of a tree, and people quickly notice any change in appearance. Ten-percent defoliation or discolouration may be cause for alarm for the owner and a call to action. Thus, you are managing foliar diseases primarily as aesthetic problems rather than life-threatening ones.

Another difficulty with managing foliar disease is timing. Once the symptoms are visible, it is usually too late for effective treatment in the current season. This is a never-ending source of frustration to tree owners.

One additional source of difficulty with managing foliar diseases is that they typically present you with symptoms rather than signs. Symptoms are changes in the normal appearance of the plant (abnormally small or yellowing leaves, for example) while a sign is physical evidence of the disease agent itself (in the case of fungi, mushrooms, conks or some vegetative portion-hyphae-of the fungus). Signs are the "smoking gun" for disease identification. Once you have identified the sign, you know the disease. Unfortunately, signs are not always visible or available. Bacteria, viruses and other organisms are too small to see even with the aid of a hand lens. Fungi, while larger, still produce few visible signs other than fruiting structures, and these may not be present at the time of identification.

Fungi are among the most common foliar-disease agents, but they are not the only disease organism or disorder to cause leaf symptoms. So how do you determine that the symptoms are due to a fungus? Again, this is not always an easy task, but don't make the mistake of "picture-booking" your way through the diagnostic process and then try to spray your way out of the problem. There is a saying in the medical profession that applies to the tree-care profession as well: "Prognosis without diagnosis is malpractice." Always make sure you have identified the problem before you have identified the treatment!

How do you decide if the symptoms are due to a fungus? You have several means of separating symptoms of fungal leaf diseases from those of bacteria. Fungal leaf diseases tend to develop circular or halo patterns on the leaves with the dead interior area a tan colour and the surrounding dying tissue a yellow or red colour. The affected tissue often has a dry, papery feel. Bacterial leaf diseases often have a more slimy texture, and the disease appears as small, angular spots. Viruses, another possible disease agent, often show up as vein clearing with surrounding tissue remaining green or developing a mosaic-a patchwork of green and yellow. They may also appear as ring spots or flecks. These are only the general patterns of symptoms for each of the major disease organisms and numerous exceptions occur.

Foliar symptoms may also originate from disorders, which are non-living or abiotic factors. Many disorders produce foliar symptoms that mimic those associated with several fungal diseases. Thus, people often blame leaf fungal diseases for problems that have their origin elsewhere. Just because the symptoms appear on the leaves does not mean that this is where the basic problem resides. Browning leaves may be due to a leaf fungus or perhaps a fungus in another part of the tree. The symptoms might even be due to drought, construction within the rooting area, de-icing salts or herbicides, among many other possible agents. With disorders, you'll never see any signs because there is no pathogen. You must examine the pattern of the symptoms. Uniform injury, either on the leaf or within the canopy, is usually an indication of a disorder. Air pollution and drought injury on conifers, for example, will often show symptoms of needle tip burn with a distinctive boundary between living and dead tissue. Fungi and other living organisms produce a more random pattern of symptoms. Only scattered needles will be affected and to various lengths. The boundary between living and dead tissue will also be less distinct.

Sometimes, leaf curl or discolouration is not due to either a disease or disorder. These are called false symptoms-particular species or cultivar traits that may appear unusual. For example, it's easy to confuse the deep-yellow variegation of a 'Rainbow' dogwood (Cornus florida 'Rainbow') leaf with a potential problem if you are not familiar with the cultivar's ornamental characteristics. Finally, always remember that problems are usually multidimensional rather than caused by a single factor. A tree may have discoloured leaves due to a scab and air pollution (and you only can manage one of these).

Know the life cycles Managing fungal leaf diseases not only requires correct identification but also some knowledge about the life cycle. Too often, applicators correctly identify the disease but apply the fungicide treatments at the wrong time. Many fungicides are protectants that provide a protective shield to prevent the fungus from penetrating the plant tissue. Protectants have little or no effect after the fungus enters the plant and becomes established. Therefore, to know when to start fungicide applications and how many to apply, you need to understand the leaf-disease cycle.

The typical life cycle begins in the spring. Spores produced from fruiting structures on dead, fallen leaves or dead twigs still in the canopy become air-borne and infect young, expanding leaves-the most susceptible foliage. The spores germinate and the fungus infects the new leaf and sometimes the young, succulent twigs. Later in the spring and early summer, fruiting structures form on the infected leaves and new infections occur, particularly if cool, moist conditions persist. In late summer or fall, the infected leaves fall and the cycle repeats the following year. If fungicide treatments begin in the summer when the symptoms appear, it's too late. In addition, if treatments are limited to only one or two sprays in the spring and cool, moist conditions persist into the summer, the disease may still develop. Treating fungus diseases requires good timing and persistence.

While fungi are responsible for many of our leaf problems, the problems are different throughout the country. However, several common types of fungal leaf diseases occur widely and understanding the symptoms and management of these types will be helpful when you are working on a specific disease.

Common fungal leaf diseases of deciduous trees Anthracnose. Anthracnose diseases probably are the best-known foliar fungal diseases of deciduous trees. They affect many ornamental trees including major shade-tree genera such as sycamore (Platanus spp.), oak (Quercus spp.), maple (Acer spp.), elm (Ulmus spp.) and ash (Fraxinus spp.). Anthracnose actually is a general term describing symptoms such as dead irregular areas that form along and between the main vein of the leaf. The leaves may also become curled and distorted and twigs may die back. The fungus overwinters in infected twigs and the petioles of fallen leaves, and the spores disseminate in the spring by wind and splashing rain. The disease, while unsightly, rarely results in the tree's death. Sycamores and other trees often withstand many years of partial defoliation. However, one anthracnose disease is more serious.

Dogwood anthracnose (Discula destructiva) is a relatively recent landscaping problem. Pathologists first identified it in the late 1970s, and it has since become a major disease problem on flowering dogwood (C. florida) and Pacific dogwood (C. nuttallii). While Chinese dogwood (C. kousa) is supposedly resistant, some pathologists have reported infections of this species as well. Dogwood anthracnose differs from other anthracnose diseases in that it can kill trees, rather than merely disfigure them. The disease begins with the leaves developing brown spots surrounded by reddish-brown halos. The flower bracts may develop similar symptoms. The disease also spreads to the twigs with the tips of infected shoots turning gray-tan and then dying. These declining branches often support an abundance of water sprouts. Eventually the tree may die from the disease.

Leaf blisters result in the blistering, curling and puckering of leaf tissue. Oak leaf blister (Taphrina caerulescens) is a common blister disease of oaks, particularly the red-oak subgenus, which includes red oak (Quercus rubra) and pin oak (Q. palustris) among others. The symptoms begin as a slight yellowing of the infected leaf followed by round, raised blisters. These turn brown, and the infected leaves fall prematurely. This fungus overwinters as spores on the buds.

Leaf spot is another common fungal leaf disease. The general symptoms include dead spots with a defined boundary between living and dead tissue. The dead tissue often separates from the surrounding living tissue creating a "shot-hole" appearance on the infected leaves.

Tar spot (Rhytisma spp.) is a leaf disease with initial symptoms similar to leaf spot. The disease is most common on red (Acer rubrum) and silver maple (A. saccharinum), but it can occur on a wide range of maple species from sugar (A. saccharum) and Norway (A. platanoides) to bigleaf maple (A. macrophyllum). The symptoms begin in the spring as small greenish-yellow spots on the upper leaf surface that, by mid-summer, progress to black tar-like spots about 0.5 inch in size. The disease is not fatal to the tree, but the appearance of the tar spots alarms some tree owners. A major outbreak in New York about 10 years ago left many maples completely defoliated by mid-August.

Powdery mildew fungi infect most species of deciduous woody plants. The typical symptoms of this disease are small dusty-white or gray patches that develop by mid-summer. These patches continue to enlarge during the summer, and the entire leaf may eventually appear white. By late summer, tiny brown to black fruiting structures usually develop in these patches. The disease overwinters on the fallen leaves or as mycelium in infected buds. In spring, the fruiting structures release wind- and rain-dispersed spores from the leaves. This fungus grows best in warm, moist conditions. The mycelium, unlike most other foliar diseases, grows on the surface of the leaves rather than within the leaf. Infected leaves may have less photosynthetic capability, but the problem is generally aesthetic. Although this disease rarely results in sufficient injury to warrant treatments, its high visibility is a frequent cause of concern to tree owners.

Rusts are unusual fungi in that they may require two unrelated hosts to complete their life cycle. Rust fungi produce fruiting bodies on host species "A," which releases spores that only infect host species "B." The fungi in "B" produces spores that only infect "A," and the cycle repeats. The fungi do cause some injury in both hosts, but often only one of the two species has any ornamental value. The rust disease that affects an ornamental tree may have an alternate host that is just an incidental plant in the landscape, or even a crop or a weed. Obviously, a simple solution would be the elimination of the alternate host. While some have attempted this strategy, for example the attempted eradication of Ribes spp. for control of white pine blister rust (Cronartium ribicola), it usually is an impractical undertaking.

Removal of the alternate host also is common recommendation for a rust that affects one of our most popular flowering trees, the crabapple (Malus spp.). However, here again, it generally proves to be impractical. Cedar-apple rust (Gymnosporangium juniperi-virginianae) is a problem familiar to most people that have a crabapple in their yard. Although many tree owners are aware that the disease must spend part of its life on cedar, to break the cycle you would not only need to remove the redcedars in your yard but also all those within a radius of up to a mile-obviously an impractical task.

Sooty mold is a common concern to tree owners but it is not an infectious disease. The black soot that appears on leaves is a fungus that lives on the honeydew secreted by aphid and scale insects. The fungus does not live in the leaf, it lives on it. Any other symptoms such as yellowing or curling leaves are due to the insects that produce the honeydew, not the mold. While heavy levels of sooty mold can block enough light to reduce photosynthesis by up to 70 percent, this primarily is an aesthetic problem

Common fungal needle diseases of conifers Foliage problems on conifers are a little more serious than on deciduous trees. Conifers can not refoliate during the same season so the impact can be great. And unlike deciduous trees, the needles are important sites of food reserves for conifers. Fortunately most foliar diseases infect either the older needles or the new ones, so some foliage remains, and the trees rarely die from these diseases. However, an infected tree can look so bad that you may hope that it dies! Needle diseases are of three types: needle cast, needle blight and needle rust.

Needle casts are a common problem with many conifers throughout the country. These diseases affect only the needles on newly formed shoots, but the symptoms are not evident until the following spring. The infected needles develop spots that turn tan to reddish-brown. The fungal fruiting structures emerge on these needles and are usually large enough to be visible to the eye. Rhizosphaera needle cast (Rhizosphaera kalkhoffii) is probably one of the most common cast diseases because it infects one of the most ubiquitous trees in the landscape, Colorado blue spruce (Picea pungens). A season after infection, the needles turn reddish brown to purple, with the fruiting structures appearing as rows of small dots running lengthwise along the needles. These infected needles are cast (dropped) in the fall. Trees infected with this disease for many years may only have the current year's needles remaining rather than the 5- to 8-year complement of needles a healthy spruce maintains.

Sphaeropsis blight (Sphaeropsis sapinea), formerly called diplodia twig blight, is one of the most common needle blights, thought it is more a shoot blight. It infects the young, succulent shoots and needles on two-and three-needled pines with Austrian pine (Pinus nigra), ponderosa pine (P. ponderosa) and Scots pine (P. sylvestris) being the most susceptible. This disease typically does not show up until the trees reach maturity. Infected trees have stunted, twisted candle growth with the expanding needles becoming straw-coloured and then brown. Sphaeropsis twig blight produces small black fruiting structures on the cone scales and at the base of the infected needles.

Several rust diseases affect conifers. Spruce needle rust (Chrysomixa spp.) and pine needle rust (Coleosporum asterum) to name two. Fortunately, with the exception of white pine blister rust-really a canker rust rather than a needle rust-these diseases are not a major problem in ornamental landscapes.

Cultural control measures for common fungus problems Foliage must remain wet for a period of several hours or more for spores to germinate, so persistently cool, moist weather is ideal for fungal leaf-disease development. While you can do nothing about the weather, irrigating in the early evening and maintaining dense plantings can achieve the same environmental conditions that favour disease. Therefore, always maintain good air circulation and never irrigate leaves in the evening-they may not dry completely until morning. Pruning can thin the plant to allow better air circulation. (Pruning also is important for removing dead, infected twigs that, for some diseases, are a source of spores in the spring. However, in most instances, pruning is more cosmetic than disease control.)

Raking fallen leaves is a common recommendation for removing the primary source of next year's infections. While this may be effective in certain instances, it is nearly impossible to eliminate all the infected leaves from a property, let alone the fallen leaves from adjacent properties. In addition, many fungal leaf diseases overwinter in the buds, twigs and cones remaining in the tree, so raking may have a minimal effect on the level of spore production.

Deciding if chemical treatments are needed Fungal leaf diseases generally do not require chemical treatments because they usually are not life-threatening. Here are some of the questions you should consider before deciding to apply a treatment.

Have you correctly identified the disease agent? Has the tree withstood several years of partial or complete defoliation due to this disease? Has the tree recently experienced heavy stress from other pests or disorders? Is the disease life-threatening to the tree? Is the host tree so valuable or visible that the loss of some foliage will concern the client? Is the client aware that several treatments may be necessary this growing season and that the disease may come back again next year? If you answered "no" to any of these questions, it may not be appropriate for you to treat-at least without additional information.

Chemical treatments for common fungus problems Treatment of many fungal leaf diseases is with protectants. This requires maintaining a chemical barrier between the leaf and the fungus, which means repeated applications at fixed intervals-usually 7 to 10 days-beginning with bud break for deciduous trees or the formation of the candle for conifers. You typically need to make three applications, though more may be necessary if cool, moist conditions persist into the summer.

Several systemic fungicides also are available for treatment of fungal leaf diseases. You can trunk-inject these products into the plant, which then translocates them to the leaves and shoots. Because systemics are not subject to weathering, they can provide a longer period of treatment than other fungicides. Adjacent homeowners appreciate these products because they produce no drift. However, be sure to use the proper techniques to minimize wounding to the tree. And remember, you are wounding the tree when you inject, so the benefits should always outweigh the risks. Also, timing for systemic injections may be different than for the typical foliar application. In some instances, the most effective time to treat is fall, as with thiabendazole to manage sycamore anthracnose.

Fungal leaf diseases are a challenge to identify and manage. Avoid the temptation to quickly reach for the sprayer at the first appearance of leaf symptoms. First, identify the agent or agents responsible for the symptoms. Then, if it is a fungal leaf disease, determine the potential damage to the tree's health and appearance. And finally, remember that it's too late to do much right now, but it's just about right to begin planning for spring treatments.

Ecofriendly management of diseases in Mulberry

Morus, a genus of flowering plants in the family Moraceae, comprises 10–16 species of deciduous trees commonly known as mulberries growing wild and under cultivation in many temperate world regions.

The closely related genus *Broussonetia* is also commonly known as mulberry, notably the Paper Mulberry, *Broussonetia papyrifera*. Mulberries are swift-growing when young, but soon become slow-growing and rarely exceed 10–15 m (33–49 ft) tall. The leaves are alternately arranged, simple, often lobed, more often lobed on juvenile shoots than on mature trees, and serrated on the margin.

Depending on the species, they can be monoecious or dioecious.

The mulberry fruit is a multiple fruit, 2–3 cm (0.79–1.18 in) long. Immature fruits are white, green, or pale yellow. In most species, the fruits turn pink and then red while ripening, then dark purple or black, and have a sweet flavor when fully ripe. The fruits of the white-fruited cultivar are white when ripe; the fruit in this cultivar is also sweet but has a very mild flavor compared with the darker variety.

Species

The taxonomy of *Morus* is complex and disputed. Over 150 species names have been published, and although differing sources may cite different selections of accepted names, only 10–16 are generally cited as being accepted by the vast majority of botanical authorities. *Morus* classification is even further complicated by widespread hybridisation, wherein the hybrids are fertile.

The following species are generally accepted:

- *Morus alba* L. – White Mulberry (E Asia)
- *Morus australis* Poir. – Chinese Mulberry (SE Asia)
- *Morus celtidifolia* Kunth (Mexico)
- *Morus insignis* (S America)
- *Morus mesozygia* Stapf – African Mulberry (S and C Africa)
- *Morus microphylla* – Texas Mulberry (Mexico, Texas (USA))
- *Morus nigra* L. – Black Mulberry (SW Asia)
- *Morus rubra* L. – Red Mulberry (E N America)

Uses and Cultivation

The ripe fruit is edible and is widely used in pies, tarts, wines, cordials and tea. The fruit of the black mulberry, native to southwest Asia, and the red mulberry, native to eastern North America, have the strongest flavor, which has been likened to 'fireworks in the mouth'.

The fruit of the white mulberry, an east Asian species which is extensively naturalized in urban regions of eastern North America, has a different flavor, sometimes characterized as refreshing and a little tart, with a bit of gumminess to it and a hint of vanilla. In North America the white mulberry is considered an invasive exotic and has taken over extensive tracts from native plant species, including the red mulberry. The mature plant contains significant amounts of resveratrol, particularly in stem bark. The fruit and leaves are sold in various forms as nutritional supplements. Unripe fruit and green parts of the plant have a white sap that may be toxic, stimulating, or mildly hallucinogenic.

Black, red, and white mulberry are widespread in southern Europe, the Middle East, northern Africa and Indian Subcontinent where the tree and the fruit have names under regional dialects. Jams and sherbets are often made from the fruit in this region. Black mulberry was imported to Britain in the 17th century in the hope that it would be useful in the cultivation of silkworms. It was much used in folk medicine, especially in the treatment of ringworm. Mulberries are also widespread in Greece, particularly in the Peloponnese, which in the Middle Ages was known as Morea (Greek: ÌùñéÜò, *Morias*), deriving from the Greek word for the tree (Greek: ÌïõñéÜ, *Ìouria*).

Mulberries can be grown from seed, and this is often advised as seedling-grown trees are generally of better shape and health, but they are most often planted from large cuttings which root readily. The mulberry plants which are allowed to grow tall with a crown height of 5–6 feet from ground level and a stem girth of 4–5 inches or more is called tree mulberry. They are specially raised with the help of well-grown saplings 8–10 months old of any of the varieties recommended for rain-fed areas like S-13 (for red loamy soil) or S-34 (black cotton soil) which are tolerant to drought or soil-moisture stress conditions. Usually, the plantation is raised and in block formation with a spacing of 6 feet x 6 feet, or 8 feet x 8 feet, as plant to plant and row to row distance. The plants are usually pruned once a year during the monsoon season (July – August) to a height of 5–6 feet and allowed to grow with a maximum of 8–10 shoots at the crown. The leaves are harvested 3–4 times a year by a leaf picking method under rain-fed or semi-arid conditions, depending on the monsoon.

The tree branches pruned during the fall season (after the leaves have fallen) are cut and used to make durable baskets supporting agriculture and animal husbandry.

Some North American cities have banned the planting of mulberries because of the large amounts of pollen they produce, posing a potential health hazard for some pollen allergy sufferers.

Silk Industry

***Figure:** A silkworm,* Bombyx mori, *feeding on a mulberry tree*

Mulberry leaves, particularly those of the white mulberry, are ecologically important as the sole food source of the silkworm (*Bombyx mori*, named after the mulberry genus *Morus*), the pupa/cocoon of which is used to make silk. Other Lepidoptera larvae—which include the common emerald, the lime hawk-moth, and the sycamore moth—also sometimes eat the plant.

Pigments

Mulberry fruit contains anthocyanins, which may prove useful as dietary modulators of mechanisms for various diseases Anthocyanins are responsible for the attractive colours of fresh plant foods, including orange, red, purple, black, and blue. These colours are water-soluble and easily extractable, yielding natural food colourants. Due to a growing demand for natural food colourants, their significance in the food industry is increasing.

A cheap and industrially feasible method has been developed to extract anthocyanins from mulberry fruit which could be used as a fabric tanning agent or food colourant of high colour value (above 100).

Scientists found that, of thirty-one Chinese mulberry cultivars tested, the total anthocyanin yield varied from 148 mg to 2725 mg per litre of fruit juice. It was also found that all the sugars, acids, and vitamins of the fruit remained intact in the residual juice after removal of the anthocyanins, so the juice could be used to produce products such as juice, wine, and sauce.

Anthocyanin content depends on climate and area of cultivation and is particularly high in sunny climates. This finding holds promise for tropical countries that grow mulberry trees as part of the practice of sericulture to profit from industrial anthocyanin production through the recovery of anthocyanins from the mulberry fruit.

This offers a challenging task to the mulberry germplasm resources for

- exploration and collection of fruit yielding mulberry species;
- their characterization, catalogueing, and evaluation for anthocyanin content by using traditional as well as modern means and biotechnology tools;
- developing an information system about these cultivars or varieties;
- training and global coordination of genetic stocks;
- evolving suitable breeding strategies to improve the anthocyanin content in potential breeds by collaboration with various research stations in the field of sericulture, plant genetics, and breeding, biotechnology and pharmacology.

Mulberry Pest and Diseases

1. Powdery mildew :

Causal organism: Phyllactinia corylea

Symptom:

- White powdery patches appear on the lower surface of leaf which is gradually increased
- and cover whole leaf surface.

Affected leaves turn yellowish and defoliate prematurely. Peak season: October-November

Control measure: Foliar spray of 0.2% Sulfex 80 WP 2 g/l. Lower surface of the leaves should be thoroughly drenched

Safe period: 15 days

2. Leaf rust :

Causal organism: Peridiospora mori

Symptom:

- Several small pin head shaped brown postules appear on the lower surface of mature leaves
- Reddish brown spot appear on the upper surface of the infected leaves.
- Severely infected leaves turn yellowish and margin of the leaves become dry.

Peak season: February-March

Control measure: Foliar spray of 0.2% Blitox 50 WP or 0.2% Bavistin 50 WP.

Safe period: 15 days.

2. Leaf spot :

Causal organism: Cercospora moricola

Symptom:

- circular light brown spots appear on both sides of the leaves.
- The adjacent spots unite together to form a larger spot
- The necrotic tissues of such spots drop out and form the characteristics ýÿshot holesýÿ.
- Highly infected leaves defoliate prematurely

Peak season: Rainy and winter season.

Control measure:

- Avoid dense planting.
- Collect and burn unused infected leaves after pruning.
- Spray 0.1% Bavistin when disease symptom appear 2-3 times at ten days interval. Safe period: 7 days.

3. Sooty mould:

Causal organism : A group of ascomycetes and deuteromycetes fungi.

Symptom : Thick black coating developed on the upper surface of the leaves.

Peak Season : August-December.

Control measure :

4. Bacterial blight :

- Foliar spray of 0.02% Monocrotophos on 15th and 30th day of pruning to control.

- Spray 0.2% Indofil M-45 75 WP to check the growth of saprophytic fungi.

Safe Period : 15 days

Foliar diseases of mulberry reduce the yield and quality of leaf thereby affecting silkworm rearing especially during rainy and winter seasons. The cumulative loss due to major foliar diseases is upto 15-18%, besides deteriorating the leaf quality. The following technologies were developed for the control of major foliar diseases to avoid the leaf yield loss.

Leaf spot : The disease caused by a fungus, Cercospora moricola is more prevalent during rainy and winter seasons. The symptoms are brownish necrotic, irregular spots on the leaf surface which enlarge, coalesce and leave the characteristic shot hole. Foliar spray of 0.2% Bavistin (Carbendazim 50% WP) solution (2 g Bavistin in 1 litre water) has been found effective.

1. Root knot Disease

It is caused by a bacteria, Pseudomonas syringae pv. mori/ Xanthomonas campestrispv. mori and is common during rainy season when there is high humidity and temperature. It shows numerous blackish brown irregular water soaked patches on the leaves resulting in curling and rotting of leaves. Step-up pruning (30 cm above the ground) during rainy season in high rainfall areas and spraying 0.2% Streptomycin or Dithane M45 (Mancozeb 75% WP) with safe period of 2-3 days are recommended.

Methodology:

- 150-180 litres of fungicide solution is required for one-acre garden. The quantity is obtained by dissolving 300-320 g/ml of chemicals in 150-180 litres of water.
- Ist spray is to be given 30-35 days after pruning/leaf harvesting. 2nd spray has to be given 10-15 days after first spray, if the disease is not controlled. Safe period is 5 days.

Root knot is one of the major diseases limiting crop production throughout the world. It can occur any time of the year mainly in sandy soils low in organic matter. The severity of the disease increases with increased age of the garden. The estimated yield loss due to the disease is 15-30%. Infected plants become weak and predisposed to other diseases while severely infected plants ultimately die.

Causative organism: The organism causing root knot disease is a nematode Meloidogyne incognita, an endoparasite inhabiting mulberry roots.

Symptoms:

- Stunted growth
- Poor and delayed sprouting
- Reduced leaf size and yield
- Chlorosis and marginal necrosis of leaves, yellowing and wilting of leaves in spite of adequate soil moisture availability
- Death of plants in severe cases

Symptoms on the underground parts :

- Formation of gall/knots on roots
- Reduced and stubby root system.
- Retarded root growth.
- Necrotic lesions on the root surfaces and death of active rootlets.

Chemical control : Chemical methods of nematode management become necessary in sick soils and heavily infested gardens as they give quick results. Furadon 3 G (40 kg/ha) can be applied either in furrows or broadcast to the soil after light harrowing followed by irrigation. The leaves from treated plots can be fed to silkworms after 45 days.

2. Botanicals to control Mulberry Diseases

A water soluble compound, isolated from a plant is dissolved in water and the solution is applied by foliar spray on mulberry plants twice at an interval of 7 days. The treatment reduces root-knot and tukra disease of mulberry. No residual toxicity exists in mulberry leaves of treated plants after a couple of weeks. The leaves could then be fed to silkworm larvae safely. The technology is cost-effective. Patent is pending. The compound was first tested in pots then in fields and finally in farmers fields.

Ecofriendly Methods for the Control of Certain Diseases of Tea (Camellia Sinensis)

Camellia sinensis is the species of plant whose leaves and leaf buds are used to produce the popular beverage tea. It is of the genus *Camellia*, a genus of flowering plants in the family Theaceae. White tea, yellow tea, green tea, oolong, pu-erh tea and black tea are all harvested from this species, but are processed differently to attain different levels of oxidation. Kukicha (twig tea) is also harvested from *Camellia sinensis*, but uses twigs and stems rather than leaves. Common names include tea plant, tea shrub, and tea tree (not to be confused with *Melaleuca alternifolia*, the source of tea tree oil).

There are two major varieties used for tea: Chinese tea, *Camellia sinensis* var. *sinensis*, and Assam tea, *Camellia sinensis* var. *assamica.*

Nomenclature and Taxonomy

The name *Camellia* is taken from the Latinized name of Rev. Georg Kamel, SJ (1661–1706), a Czech-born Jesuit lay brother, pharmacist, and missionary to the Philippines. Carl Linnaeus chose his name in 1753 for the genus to honour Kamel's contributions to botany (although Kamel did not discover or name this plant, or any *Camellia*, and Linnaeus did not consider this plant a *Camellia* but a *Thea*). Robert Sweet shifted all formerly *Thea* species to the *Camellia* genus in 1818. The name *sinensis* means *from China* in Latin.

Four varieties of *Camellia sinensis* are recognized. Of these, *C. sinensis* var. *sinensis* and *C. sinensis* var. *assamica* (JW Masters) Kitamura are most commonly used for tea, and *C. sinensis* var. *pubilimba* Hung T. Chang and *C. sinensis* var. *dehungensis* (Hung T. Chang & BH Chen) TL Ming are sometimes used locally.

Cultivars

Cultivars of *C. sinensis* include:

- Benifuuki
- Fushun
- Kanayamidori
- Meiryoku
- Saemidori
- Okumidori
- Yabukita

Description

Camellia sinensis is native to East, South and Southeast Asia, but it is today cultivated across the world in tropical and subtropical regions. It is an evergreen shrub or small tree that is usually trimmed to below 2 m (6.6 ft) when cultivated for its leaves. It has a strong taproot. The flowers are yellow-white, 2.5–4 cm (0.98–1.57 in) in diametre, with 7 to 8 petals.

The seeds of *Camellia sinensis* and *Camellia oleifera* can be pressed to yield tea oil, a sweetish seasoning and cooking oil that should not be confused with tea tree oil, an essential oil that is used for medical and cosmetic purposes, and originates from the leaves of a different plant.

The leaves are 4–15 cm (1.6–5.9 in) long and 2–5 cm (0.79–1.97 in) broad. Fresh leaves contain about 4% caffeine. The young, light green leaves are preferably harvested for tea production; they have short white hairs on the underside. Older leaves are deeper green. Different leaf ages produce differing tea qualities, since their chemical compositions are different. Usually, the tip (bud) and the first two to three leaves are harvested for processing. This hand picking is repeated every one to two weeks.

Cultivation

Camellia sinensis is mainly cultivated in tropical and subtropical climates, in areas with at least 127 cm. (50 inches) of rainfall a year. Tea plants prefer a rich and moist growing location in full to part sun, and can be grown in USDA climate zones 7 - 9. However, the clonal one is commercially cultivated from the equator to as far north as Cornwall on the UK mainland. Many high quality teas are grown at high elevations, up to 1,500 metres (4,900 feet), as the plants grow more slowly and acquire more flavour.

Tea plants will grow into a tree if left undisturbed, but cultivated plants are pruned to waist height for ease of plucking. Two principal varieties are used, the small-leaved Chinese variety plant (*C. sinensis sinensis*) and the large-leaved Assamese plant (*C. sinensis assamica*), used mainly for black tea.

Chinese Teas

The Chinese plant (sometimes called *C. sinensis* var. *sinensis*) is a small-leafed bush with multiple stems that reaches a height of some 3 metres. It is native to southeast China. The first tea plant to be discovered, recorded and used to produce tea three thousand years ago, it yields some of the most popular teas.

C. sinensis var. *waldenae* was considered a different species, *Camellia waldenae* by SY Hu, but it was later identified as a variety of *C. sinensis*. This variety is commonly called Waldenae Camellia. It is seen on Sunset Peak and Tai Mo Shan in Hong Kong. It is also distributed in Guangxi Province, China.

Indian Teas

There are Three Main Kinds of Tea Produced in India: Assam tea comes from the northeastern section of the country. This heavily forested region is home to much wildlife, including the rhinoceros. Tea from here is rich and full-bodied. It was in Assam that the first tea estate was established, in 1837.

Darjeeling – the Darjeeling region is cool and wet, and tucked in the foothills of the Himalayas. Tea plantations reach 2200 metres. The tea is delicately flavoured, and considered to be one of the finest teas in the world. The Darjeeling plantations have 3 distinct harvests, termed 'flushes', and the tea produced from each flush has a unique flavour. First (spring) flush teas are light and aromatic, while the second (summer) flush produces tea with a bit more bite. The third, or autumn flush gives a tea that is lesser in quality.

Nilgiris tea comes from a part of India that is almost as high as Darjeeling. This southern Indian region has elevations between 1,000 and 2,500 metres. The flavours of Nilgiris teas are subtle and rather gentle. They are frequently blended with other, more robust teas.

Figure: *Seed bearing fruit of* Camelia sinensis

Pests and Diseases

The tea plant is subject to attack from at least 150 insect species and 380 fungus diseases. In northeast India, where 125 pests and 190 fungi have been detected, losses from pests and diseases have been estimated at 67 million pounds (30 million kg) of tea per annum. More than 100 pests and 40 diseases occur in the tea fields of Japan. Sri Lanka, where estates are close together or contiguous, has recorded many blights and suffered serious losses. Africa has little trouble with blights; the tea mosquito (*Helopeltis theivora*) is the only serious pest. The Caucasus, with a climate similar to that of Japan, grows the China variety of plant and has no serious pests or blights.

Blight control has become highly developed. Northeast Indian scientists have issued a list of 40 approved proprietary pesticides. Some of these pesticides cannot be applied during the plucking season; others require that the two subsequent rounds of weekly pluckings be discarded.

Tea leaves are eaten by some herbivores, like the caterpillars of the willow beauty (*Peribatodes rhomboidaria*), a geometre moth.

Blister Blight

Blister blight is a seasonal disease and can spread very rapidly by means of air borne spores. Mature tea in general and teas recovering from pruning in particular suffer from blister blight. The fungus attacks young succulent growth on all teas especially when the environment is foggy, moist and cool. A single spore can produce a blister when deposited on tender shoots in about 10-21 days from the time of infection.

Tea growing areas	*Periods of outbreak*
Darjeeling	July to September
Assam	March to May and November to December
Dooars	September to November

Control Measures

This disease can be controlled by adopting chemical as well as cultural control strategies simultaneously. Some of the predisposing factors are dull, humid, cloudy and cool weather prevailing for several days, heavy shade etc. In such conditions prevails, monitoring of blister prone sections should be thoroughly undertaken. The shades should be thinned.

Normally 4-6 rounds Copper are necessary in Darjeeling. In Assam and Dooars, the spraying can be discontinued once the damp weather condition is over and it becomes hot and sunny for a few consecutive days. The order of priority for chemical treatment is:

nurseries > pruned young tea > early medium pruned tea > light pruned tea > unprune tea.

Black Rot

The disease is caused by fungi Corticium theae and Corticium invisum. Both the fungi produce similar effect on the bush and some times occur together. They attack the maintenance leaves, causing gradual deterioration in the health of the bush and consequent loss of crop. The disease is first noticed in May/June/July although germination from the sclerotial stage starts in March/April. The fungus remains active right up to September and then over winters in the sclerotial form in the cracks and crevices of the stem. Transformation in to sclerotial stage is a means for survival during the unfavourable weather condition. When favourable condition prevails, they germinate and re-infect.

Cultural Control :

1. Pruning or skiffing of the severely affected sections.
2. Improvement of aeration by lopping side branches and 'matidals'.
3. Thinning out of the dense shade.
4. Improvement of drainage.
5. Alkaline wash after pruning.
6. Adoption of shorter pruning cycle in chronic and severely affected sections.

Chemical Control: Thorough drenching of all the stems and decomposing pruning litters below the bush with copper oxychloride (COC) one week after the first spell of rainfall in February/March. Spray two blanket rounds of COC or Copper hydroxide or Hexaconazole at 15 days interval during May-June. Spot spraying of these two fungicides should continue till September if the disease incidence is severe. Application of two monthly rounds of COC or Carboxin during winter will inhibit sclerotia formation.

Control by Bioagents: Two fortnightly rounds of 10% Bacillus subtilis broth can be applied in lieu of chemical fungicide during both active and dormant phase of the disease. Spraying should be directed towards small stems and under surfaces of the leaves, which bear fungal pathogens.

Branch Canker and Thorny Stem Blight: Branch canker and Thorny stem blight are important stem diseases of tea prevailing in plains of Assam and Darjeeling hills, respectively. Branch canker is caused by Poria hypobrunnea and the causal organism of thorny stem blight is Tunstallia aculeata. These pathogens are typical wound parasites and enter the tea plants normally through heavy pruning cuts. Hence it is necessary to protect the fresh wounds with protective paints to stop the entry of these pathogens.

Control Measures: For an effective control of these diseases careful removal of all dead wood at each pruning and subsequent protection of the pruning cuts with protective paints/sprays is necessary. These protective paints may be bituminous paint, COC or Trichoderma bioagents. During recent times Trichoderma treatment has been found to be the best method of protecting the pruning cuts from the damage of both Poria and Tunstallia. For painting of large pruning cuts soon after rejuvenation and medium pruning 80 litres of 20% Trichoderma spore suspension using 16L or kg of the bioagent formulation/ha is required (considering 8000 plants/ha). For spraying of entire bush frame after light pruning/cut across pruning, 400 litres of 5% Trichoderma spore suspension using 20L or kg of the bioagent formulation/ha is required using a NMD 60/450 nozzle.

Red Rust

It is a disease of young stems mostly, the ultimate symptoms of which normally manifest on the leaves. The causal organism of this disease is Cephaleuros parasiticus, an alga, causes severe damage especially to the young teas. Tissues of the stem are killed in patches and cause dieback. The leaves of the infected branches variegate with yellow patches. The alga produces brick red or orange coloured fructifications in patches on the infected stems of the tea bush.

Cultural Control:

1. The predisposing factors to be identified and corrected are i) poor drainage, ii) low soil fertility, particularly potash, iii) improper soil acidity, iv) inadequate shade and v) continuous use of green crops like Tephrosia candida, T. vogelli etc.
2. Pruning of severely affected sections.

Chemical Control: In susceptible areas young teas should be sprayed with four rounds of COC/Copper hydroxide during mid April to mid July. First two rounds at 15 days interval and the subsequent rounds at monthly interval. Young shade trees/ green crops in nurseries should be painted with COC to avoid fresh infection.

2% MOP and Urea can be sprayed as a rehabilitary measure in the severely affected sections. Spraying of fungicides should be directed towards the young stems and laterals bearing rusty fructifications.

Dieback of Primaries and Seed Decay: During the last couple of years, dieback of primaries in certain succulent TV clones and seed decay in some biclonal seed stocks have been causing serious concern. The causal organism of these maladies is identified as Fusarium solani, a parasitic fungus. Generally this pathogenic fungus is predominant in the humid temperate climate of North East India. The susceptible clones and seed stocks to Fusarium solani are recorded as TV 19, TV 23, TV25, TV 26, TV 29, S3A1 parent of seed stock 491 and both the parents of seed stock 520. Symptoms of these maladies are as follows:

On primaries : Blackening of the petioles of leaves, which gradually extends to the nodes and inter-nodes, followed by wilting of the primaries. White cottony fungal growth is seen on the dying tissues, which turns brown at maturity accompanying with development of small pink perithecia of Necteria

On seeds : Blackening of fruit carp followed by immature cracking and dropping of seeds. Seed becomes light pinkish with powdery coverage due to the fungus sporulation in the later stage of infestation.

Control Measures: Several chemicals have been evaluated for controlling Fusarium solani both in laboratory as well as in field. In these trials Carbendazim, Hexaconazole and Tridemorph and their combinations were found effective in controlling the pathogen.

Primary Root Diseases

The most common primary root diseases of tea plants in the plains of North East India are "Charcoal stump rot" and "Brown root rot"."Black root rot" is occasionally found in Darjeeling gardens while Red root rot is of rare occurrence. The root diseases can spread by means of direct contact or from left over diseased woody material. However, the spread of Charcoal stump rot and Black root rots may take place by air borne spores as well.

The diseased materials may also be carried by rain to healthy teas. Unfortunately, there is no known method of detecting root infection by examining the above ground portion of the plant. Depending on the age and size of the bush it may require six months to four years to kill a infected plant. The tea roots remain in close proximity and hence by the time a dead plant is noticed, it is possible that its immediate neighbours have also become infected. Adoption of proper cultural practices is a vital point for control of primary root diseases as no effective chemicals are available at the moment to control these pathogens.

Cultural Control:

1. When the disease occurs on individual bush or in patch, the diseased plant together with a ring of apparently healthy bushes surrounding it should be uprooted removing all bits of roots.
2. When a large area is infected the diseased patches should be isolated by 90 -100 cm deep and 30 cm wide trench surrounding the area until uprooting can be undertaken. This trench should be connected to the nearest drain.
3. In case of old tea, due for uprooting, only the diseased plant is to be uprooted and the whole area should be kept isolated by an encircling trench as described above. The area should be uprooted two years before replanting.
4. All vacant areas thus created should be kept under a green crop for two yearsbefore infilling or replanting.

Biological Control: Trichoderma bio agents have been found effective in controlling primary root diseases namely, Charcoal stump rot and Brown root rots. Planting pit treatment for infilling and replanting, 10% spore suspension should be used for treating the excavated soil of planting pits @ 100ml/pit. Alternatively, compost or well rotten cattle manure should be mixed thoroughly with the spore suspension and the treated manure should be mixing thoroughly with the excavated soil of the planting pit @ 200g/pit. The treatment will require 30L or kg of Trichoderma biocide/ha.

For top dressing around infills already planted in mature tea (1 year or below) should be sprayed around the collar region on the round @ 100 ml/plant using 5% spore suspension during Apr/May or Sep/ Oct before the soil gets dry. Forking of soil is a must before spraying.

Secondary Root Diseases

Violet Root Rot and Diplodia: Violet root rot caused by Spherostilbe repens and attacks of Diplodia sp. on weak plants are the common secondary root diseases of tea. Violet root rot disease is very common on stiff clayey waterlogged soils. It can easily be controlled by improvement of drainage. Diplodia fungus can live on dead stem and root tissues, which are deficient in reserved starch material. Healthy plants are not affected by this fungal pathogen. No chemical treatment is necessary for controlling both these diseases.

1. Use of biocontrol agents in integrated management of Leaf Rot in Root (Wilt) affected coconut

The coconut palm is affected by a number of diseases, some of which are lethal while others gradually reduce the vigour of the palm causing severe loss in yield.

Bud Rot

Symptoms:

1. The earlier symptom is the yellowing of one or two younger leaves. Black spots appear on spindle leaves.
2. In the later stages the spindle withers and drops down.
3. The tender leaf base and soft tissues of the crown rot into a slimy mass of decayed material emitting foul smell.
4. Ultimately the entire crown falls down and the palm dies.

Control Measures:

1. Remove all affected tissue of the crown and apply Bordeaux paste on cut end and provide a protective covering till normal

shoot emerges (Dissolve 100 gm of copper sulphate and 100 gm of quick lime each in 500ml. water separately and mix to form 1 litre of Bordeaux paste).

2. Spray 1% Bordeaux mixture on the crown of the neighbouring palms as a prophylatic measure.
3. Spray with 1% Bordeaux mixture during May and September if the disease occur frequently.
4. Cut and burn severely affected palms which cannot be saved.

Leaf Rot

Symptoms:

1. Blackening and shriveling up of distal ends of the leaflets in the central spindle and younger leaves which later break off in bits.
2. Gradual weakening of the tree resulting in decline yield.

Control Measures :

1. Improve general condition of palms through proper manuring and management.
2. Remove the decayed portions of spindle leaf and upmost two leaves only.
3. Pour fungicide solution of Hexaconazol (Contaf 5E) - 2ml or Mancozeb (Dithane M45/Indifil M45) - 3g in 300ml water per palm to the base of spindle leaf.
4. Apply 20g Phorate 10G mixed with 200g fine sand around the base of spindle leaf.
5. 2-3 rounds of spraying is sufficient in case of mild infection.

Stem Bleeding

Symptoms:

1. Exudation of reddish brown liquid through cracks developing on the stem.
2. Decaying of tissues at bleeding point
3. Vigour and yield declining.
4. Development of big holes inside the trunk

Control Measures:

1. Chisel affected tissue and dress the wound with 5% Calixin (5ml in 100ml water). Apply coal tar after 2 days. Burn off chiselled pieces.
2. Avoid any mechanical injury to the stem.
3. To avoid spread of disease on to upper portion of trunk, root feeding with 5% calixin may be adopted 3 times a year - April-May, Spetember-October and January-February.
4. Along with 50kg organic manure, apply 5kg neem cake containing the antagonistic fungi, Trichoderma culture to the basin during September.
5. Provide adequate irrigation during summer and drainage during rainy season.
6. Apply recommended doses of organic manures and chemical fertilizers.
7. Coconut stem boring insects like Xyleborus, Diocalandra should be controlled by applying Carbaryl 50% WP on the trunk @ 3g per litre water.

Root (wilt) Disease:

Symptoms :

1. Abnormal bending or ribbing of the leaflets, flaccidity of the leaves, general yellowing followed by marginal necrosis of the leaflets.
2. Abnormal shedding of buttons, reduced leaves and crown, gradual reduction in yield.

Control Measures: Being a non-lethal, debilitating disease an integrated approach is to be followed for management of root wilt.

1. Recommended control measures for leafrot disease as this disease gets superimposed on most of the root wilt affected palms.
2. Follow integrated nutrient management apply organic manure @ 50kg / palm / year. apply balanced dose of chemical fertilizers i.e. 500g Nitrogen (1.1kg urea), 300g Phosphorus (1.7kg Mussouriphos), 1000g Potassium (1.7kg Muriate of potash) in two splits - 1/3rd during April-May and 2/3rd during September-October under rainfed condition and in four splits during January, April, July and October under irrigated condition. In addition to this 500g MgO (1kg Magnesium sulphate) also has to be applied along with second dose of fertilizer application.
3. Cut and remove disease advanced, uneconomical palms yielding less than 10 nuts per palm per year.
4. Grow green manure crops - cowpea, sunhemp (Crotalaria juncea), Mimosa invisa, Calapagonium mucanoides, Pueraria phaseoloides

etc. may be sown in coconut basins during April-May and incorporated during September-October.

5. Irrigate coconut palms with at least 250 litre water in a week.
6. Adopt suitable inter/mixed cropping in coconut gardens.
7. Provide adequate drainage facilities.

Tanjavur wilt

Symptoms:

1. Decay of finer nuts, withering, yellowing, drooping and drying of outer whorl of leaves and exudation of reddish brown fluid at the base of the trunk.

Control Measures:

1. Practice growing banana as intercrop in coconut.
2. Root feeding with Calixin (2ml in 100ml water) once in 3 months.
3. Drench the basin with 25 litre of 0.1% Calixin.
4. Apply neem cake containing Trichoderma @ 5kg /palm/year.
5. Apply recommended dose of organic manures.
6. Avoid flood irrigation and follow drip irrigation.
7. Practice clean cultivation and burn off diseased plant pests.
8. Isolate infected palms by taking trenches of 1m depth and 0.5m width around the palm at 1.5m away from it.

Management of Fungal Disease by Application of Arbuscular Mycorrhizal Fungi

An arbuscular mycorrhizal fungus (plural *mycorrhizae* or *mycorrhizas*, aka *AM fungi*) is a type of mycorrhiza in which the fungus penetrates the cortical cells of the roots of a vascular plant. Arbuscular mycorrhizas (AMs) are characterized by the formation of unique structures, arbuscules and vesicles by fungi of the phylum Glomeromycota (AM fungi). AM fungi (AMF) help plants to capture nutrients such as phosphorus, sulfur, nitrogen and micronutrients from the soil. It is believed that the development of the arbuscular mycorrhizal symbiosis played a crucial role in the initial colonisation of land by plants and in the evolution of the vascular plants.

It has been said that it is quicker to list the plants that do not form mycorrhizae than those that do. This symbiosis is a highly evolved mutualistic relationship found between fungi and plants, the most prevalent plant symbiosis known, and AM is found in 80% of vascular plant families in existence today.

The tremendous advances in research on mycorrhizal physiology and ecology over the past 40 years have led to a greater understanding of the multiple roles of AMF in the ecosystem. This knowledge is applicable to human endeavours of ecosystem management, ecosystem restoration, and agriculture.

Evolution of Mycorrhizal Symbiosis

Paleobiology: Both paleobiological and molecular evidence indicate that AM is an ancient symbiosis that originated at least 460 million years ago. AM symbiosis is ubiquitous among land plants, which suggests that mycorrhizas were present in the early ancestors of extant land plants. This positive association with plants may have facilitated the development of land plants.

The Rhynie chert of the lower Devonian has yielded fossils of the earliest land plants in which AM fungi have been observed. The fossilized plants containing mycorrhizal fungi were preserved in silica.

The Early Devonian saw the development of terrestrial flora. Plants of the Rhynie chert from the Lower Devonian (400 m.yrs ago) were found to contain structures resembling vesicles and spores of present *Glomus* species. Colonized fossil roots have been observed in *Aglaophyton major* and *Rhynia*, which are ancient plants possessing characteristics of vascular plants and bryophytes with primitive protostelic rhizomes.

Intraradical mycelium was observed in root intracellular spaces, and arbuscules were observed in the layer thin wall cells similar to palisade parenchyma. The fossil arbuscules appear very similar to those of existing AMF. The cells containing arbuscules have thickened walls, which are also observed in extant colonized cells.

Mycorrhizas from the Miocene exhibit a vesicular morphology closely resembling that of present Glomerales. The need for further evolution may have been lost due to the readily available food source provided by the plant host. However, it can be argued that the efficacy of signalling process is likely to have evolved, which could not be easily detected in the fossil record. A finetuning of the signalling processes would improve coordination and nutrient exchange between symbionts while increasing the fitness of both the fungi and the plant symbionts.

The nature of the relationship between plants and the ancestors of arbuscular mycorrhizal fungi is contentious. Two hypotheses are:

- Mycorrhizal symbiosis evolved from a parasitic interaction that developed into a mutually beneficial relationship.
- Mycorrhizal fungi developed from saprobic fungi that became endosymbiotic.

Both saprotrophs and biotrophs were found in the Rhynie Chert, but there is little evidence to support either hypothesis. There is some fossil evidence that suggests that the parasitic fungi did not kill the host cells immediately upon invasion, although a response to the invasion was observed in the host cells. This response may have evolved into the chemical signalling processes required for symbiosis. In both cases, the symbiotic plant-fungi interaction is thought to have evolved from a relationship in which the fungi was taking nutrients from the plant into a symbiotic relationship where the plant and fungi exchange nutrients.

Molecular Evidence

Increased interest in mycorrhizal symbiosis and the development of sophisticated molecular techniques has led to the rapid development of genetic evidence. Wang *et al.* (2010) performed an intensive investigation of three widely occurring plant genes that encode for a signal transduction cascade vital for communication with order Glomales fungal partners (DMI1, DMI3, IPD3). Sequences of these three genes were obtained from all major clades of modern land plants (including liverworts, the most basal group), and the maximum probability phylogeny of the three genes was in complete agreement with the current land plant phylogenies. These findings imply that the mycorrhizal genes must have been present in the common ancestor of land plants, and that these genes must have been vertically inherited since the colonization of land by plants.

Physiology

Presymbiosis: The development of AM fungi prior to root colonization, known as presymbiosis, consists of three stages: spore germination, hyphal growth, host recognition and appressorium formation.

Spores of the AM fungi are thick-walled multi-nucleate resting structures. The germination of the spore does not depend on the plant, as spores have been germinated under experimental conditions in the absence of plants both *in vitro* and in soil. However, the rate of germination can be increased by host root exudates. AM fungal spores germinate given suitable conditions of the soil matrix, temperature, carbon dioxide concentration, pH, and phosphorus concentration.

The growth of AM hyphae through the soil is controlled by host root exudates known as strigolactones, and the soil phosphorus concentration.

Low-phosphorus concentrations in the soil increase hyphal growth and branching as well as induce plant exudation of compounds that control hyphal branching intensity.

The branching of AM fungal hyphae grown in phosphorus media of 1 mM is significantly reduced, but the length of the germ tube and total hyphal growth were not affected. A concentration of 10 mM phosphorus inhibited both hyphal growth and branching. This phosphorus concentration occurs in natural soil conditions and could thus contribute to reduced mycorrhizal colonization.

Root exudates from AMF host plants grown in a liquid medium with and without phosphorus have been shown to affect hyphal growth. Pre-germinated surface-sterilized spores of *Gigaspora magarita* were grown in host plant exudates. The fungi grow in the exudates from roots starved of phosphorus had increased hyphal growth and produced tertiary branches compared to those grown in exudates from plants given adequate phosphorus. When the growth-promoting root exudates were added in low concentration, the AM fungi produced scattered long branches. As the concentration of exudates was increased, the fungi produced more tightly clustered branches. At the highest-concentration arbuscules, the AMF structures of phosphorus exchange were formed.

This chemotaxic fungal response to the host plants exudates is thought to increase the efficacy of host root colonization in low-phosphorus soils. It is an adaptation for fungi to efficiently explore the soil in search of a suitable plant host.

Further evidence that AM fungi exhibit host-specific chemotaxis: Spores of *Glomus mosseae* were separated from the roots of a host plant, nonhost plants, and dead host plant by a membrane permeable only to hyphae. In the treatment with the host plant, the fungi crossed the membrane and always emerged within 800 μm of the root. Whereas in the treatments with nonhost plants and dead plants, the hyphae did not cross the membrane to reach the roots. This demonstrates that arbuscular mycorrhizal fungi have chemotaxic abilities that enable hyphal growth towards the roots of a potential host plant.

Molecular techniques have been used to further understand the signalling pathways that occur between arbuscular mycorrhizae and the plant roots. In the presence of exudates from potential host plant roots, the AM undergoes physiological changes that allow it to colonize its host. AM fungal genes required for the respiration of spore carbon compounds are triggered and turned on by host plant root exudates. In experiments, there was an increase in the transcription rate of 10 genes half-hour after exposure and an even greater rate after 1 hour. A morphological growth response was observed 4 hours after exposure. The genes were isolated and found to be involved in mitochondrial activity and enzyme production. The fungal respiration rate was measured by O_2 consumption rate and increased by 30% 3 hours after exposure to root exudates. This indicates that AMF spore mitochondrial activity is positively stimulated by host plant root exudates. This may be part of a fungal regulatory mechanism that conserves spore energy for efficient growth and the hyphal branching upon receiving signals from a potential host plant.

When arbuscular mycorrhizal fungal hyphae encounter the root of a host plant, an appressorium (an infection structure) is formed on the root epidermis. The appressorium is the structure from which the hyphae can penetrate into the host's parenchyma cortex. The formation of appressoria does not require chemical signals from the plant. AM fungi could form appressoria on the cell walls of "ghost" cells in which the protoplast had been removed to eliminate signalling between the fungi and the plant host. However, the hyphae did not further penetrate the cells and grow in towards the root cortex, which indicates that signalling between symbionts is required for further growth once appressoria are formed.

Symbiosis

Once inside the parenchyma, the fungus forms highly branched structures for nutrient exchange with the plant called "arbuscules". These are the distinguishing structures of arbuscular mycorrhizal fungus. Arbuscules are the sites of exchange for phosphorus, carbon, water, and other nutrients. There are two forms: *Paris* type is characterized by the growth of hyphae from one cell to the next; and *Arum* type is characterized by the growth of hyphae in the space between plant cells. The choice between *Paris* type and *Arum* type is primarily determined by the host plant family, although some families or species are capable of either type.

The host plant exerts a control over the intercellular hyphal proliferation and arbuscule formation. There is a decondensation of the plant's chromatin, which indicates increased transcription of the plant's DNA in arbuscule-containing cells. Major modifications are required in the plant host cell to accommodate the arbuscules. The vacuoles shrink and other cellular organelles proliferate. The plant cell cytoskeleton is reorganized around the arbuscules.

There are two other types of hyphae that originate from the colonized host plant root. Once colonization has occurred, short-lived runner hyphae grow from the plant root into the soil. These are the hyphae that take up phosphorus and micronutrients, which are conferred to the plant. AM fungal hyphae have a high surface-to-volume ratio, making their absorptive ability greater than that of plant roots. AMF hyphae are also finer than roots and can enter into pores of the soil that are inaccessible to roots. The third type of AMF hyphae grows from the roots and colonizes other host plant roots. The three types of hyphae are morphologically distinct.

Nutrient Uptake and Exchange

AM fungi are obligate symbionts. They have limited saprobic ability and are dependent on the plant for their carbon nutrition. AM fungi take up the products of the plant host's photosynthesis as hexoses.

The transfer of carbon from the plant to the fungi may occur through the arbuscules or intraradical hyphae. Secondary synthesis from the hexoses by AM occurs in the intraradical mycelium. Inside the mycelium, hexose is converted to trehalose and glycogen. Trehalose and glycogen are carbon storage forms that can be rapidly synthesized and degraded and may buffer the intracellular sugar concentrations. The intraradical hexose enters the oxidative pentose phosphate pathway, which produces pentose for nucleic acids.

Lipid biosynthesis also occurs in the intraradical mycelium. Lipids are then stored or exported to extraradical hyphae where they may be stored or metabolized. The breakdown of lipids into hexoses, known as gluconeogenesis, occurs in the extraradical mycelium. Approximately 25% of the carbon translocated from the plant to the fungi is stored in the extraradical hyphae. Up to 20% of the host plant's photosynthate carbon may be transferred to the AM fungi. This represents a considerable carbon investment in mycorrhizal network by the host plant and contribution to the below-ground organic carbon pool.

An increase in the carbon supplied by the plant to the AM fungi increases the uptake of phosphorus and the transfer of phosphorus from fungi to plant Phosphorus uptake and transfer is also lowered when the photosynthate supplied to the fungi is decreased. Species of AMF differ in their abilities to supply the plant with phosphorus. In some cases, arbuscular mycorrhizae are poor symbionts, providing little phosphorus while taking relatively high amounts of carbon.

The benefit of mycorrhizas to plants is mainly attributed to increased uptake of nutrients, especially phosphorus. This increase in uptake may be due to increase surface area of soil contact, increased movement of nutrients into mycorrhizae, a modification of the root environment, and increased storage. Mycorrhizas can be much more efficient than plant roots at taking up phosphorus. Phosphorus travels to the root or via diffusion and hyphae reduce the distance required for diffusion, thus increasing uptake. The rate of inflow of phosphorus into mycorrhizae can be up to six times that of the root hairs. In some cases, the role of phosphorus uptake can be completely taken over by the mycorrhizal network, and all of the plant's phosphorus may be of hyphal origin. Less is known about the role of nitrogen nutrition in the arbuscular mycorrhizal system and its impact on the symbiosis and community. While significant advances have been made in elucidating the mechanisms of this complex interaction, much investigation remains to be done.

The available phosphorus concentration in the root zone can be increased by mycorrhizal activity. Mycorrhizae lower the rhizosphere pH due to selective uptake of NH_4^+ (ammonium-ions) and release of H^+ ions. Decreased soil pH increases the solubility of phosphorus precipitates. The hyphal uptake of NH_4^+ also increases the flow of nitrogen to the plant as NH_4^+ is absorbed to the soil's inner surfaces and must be taken up by diffusion.

Ecology

Biogeography: Arbuscular mycorrhizal fungi are most frequent in plants growing on mineral soils, and are of extreme importance for plants growing in nutrient-deficient substrates such as in volcanic and sand dune environments. The populations of AM fungi is greatest in plant communities with high diversity such as tropical rainforests and temperate grasslands where they have many potential host plants and can take advantage of their ability to colonize a broad host range. There is a lower incidence of mycorrhizal colonization in very arid or nutrient-rich soils. Mycorrhizas have been observed in aquatic habitats; however, waterlogged soils have been shown to decrease colonization in some species. Arbuscular mycorrhizal fungi are found in 80% of plant species and have been surveyed on all continents except Antarctica. The biogeography of Glomeromycota is influenced by dispersal limitation, environmental factors such as climate, soil series and soil pH and plant community,. While previous evidence suggests that AM fungi are not specialists on their host species, current studies have indicated that at least some fungi taxa are host specialists.

Response to Plant Communities

The specificity, host range, and degree of colonization of mycorrhizal fungi are difficult to analyze in the field due to the complexity of interactions between the fungi within a root and within the system. There is no clear evidence to suggest that arbuscular mycorrhizal fungi exhibit specificity for colonization of potential AM host plant species as do fungal pathogens for their host plants. This may be due to the opposite selective pressure involved. In parasitic relations, the host plant benefits from mutations that prevent colonization, whereas, in a symbiotic relationship, the plant benefits from mutation that allow for colonization by AMF. However, plant species differ in the extent and dependence on colonization by certain AM fungi, and some plants may be facultative mycotrophs, while others may be obligate mycotrophs. Recently, mycorrhizal status has been linked to plant distributions, with obligate mycorrhizal plants occupying warmer, drier habitats while facultative mycorrhizal plants occupy larger ranges of habitats.

The ability of the same AM fungi to colonize many species of plants has ecological implications. Plants of different species can be linked underground to a common mycelial network. One plant may provide the photosynthate carbon for the establishment of the mycelial network that another plant of a different species can utilize for mineral uptake. This implies that arbuscular mycorrhizae are able to balance below-ground intra–and interspecific plant interactions.

Since Glomeromycota fungi live inside plant roots, they can be influenced substantially by their plant host and in return affect plant communities as well. Plants can allocate up to 30% of their photosynthate carbon to AM fungi and in return AM fungi can acquire up to 80% of plant phosphorus and nitrogen. The diversity of AM fungal communities has been positively linked to plant diversity, plant productivity and herbivory. Arbuscular mycorrhizal fungi can be influenced by small scale interactions with the local plant community. For example, the plant neighbourhood around a focal plant can alter AM fungal communities as can the order of plant establishment within sites.

AM Fungi and Plant Invasion

During invasions of plant species, the AM fungal community and biomass can be drastically altered. In the majority of cases AM fungal biomass and diversity decrease with invasions. However, some mycotrophic plant species may actually increase AM fungal diversity during invasion.

The mycorrhizal status of invasive plant species often varies between regions. For example, in the United Kingdom and central Europe recently invasive plants are more frequently obligately mycorrhizal than expected, while invasive plants in California were found to be less frequently mycorrhizal than expected.

Interactions Between AM fungi and other Plant Symbionts

All symbionts within a plant host interact, often in unpredictable ways. A recent meta-analysis indicated that plants colonized by both

AM fungi and vertically transmitted endophytes often are larger than plants independently colonized by these symbionts. However, this relationship is context-dependent as AM fungi can interact synergistically with fungal endophytes inhabiting the leaves of their host plant, or antagonistically,. Similar ranges of interactions can occur between AM fungi and ectomycorrhizal fungi and dark septate endophytes.

Response to Environmental Gradients

Arbuscular mycorrhizal fungi vary across many environmental gradients. The tolerance of AM fungi to freezing and drying is known to shift between AM fungal taxa. AM fungi become less prevalent and diverse at higher soil nutrient and moisture concentrations, presumably because both plants allocate less carbon to AM fungi and AM fungi reallocate their resources to intradical hyphae in these environmental conditions. Over the long term, these environmental conditions can even create local adaptation between plant hosts, AM fungi and the local soil nutrient concentrations. Along elevational gradients AM composition often becomes less diverse on mountain tops than at lower elevations, but this effect is driven by the composition of plant species.

Rhizosphere Ecology

The rhizosphere is the soil zone in the immediate vicinity of a root system.

Arbuscular mycorrhizal symbiosis affects the community and diversity of other organisms in the soil. This can be directly seen by the release of exudates, or indirectly by a change in the plant species and plant exudates type and amount.

Mycorrhizae diversity has been shown to increase plant species diversity as the potential number of associations increases. Dominant arbuscular mycorrhizal fungi can prevent the invasion of non-mycorrhizal plants on land where they have established symbiosis and promote their mycorrhizal host.

Recent research has shown that AM fungi release an unidentified diffusional factor, known as the myc factor, which activates the nodulation factor's inducible gene MtEnod11. This is the same gene involved in establishing symbiosis with the nitrogen fixing, rhizobial bacteria (Kosuta *et al.* 2003). When rhizobium bacteria are present in the soil, mycorrhizal colonization is increased due to an increase in the concentration of chemical signals involved in the establishment of symbiosis (Xie *et al.* 2003). Molecules similar to Nod factors were isolated from AM fungi and were shown to induce MtEnod11, lateral root formation and enhance mycorrhization. Effective mycorrhizal colonization can also increase the nodulations and symbiotic nitrogen fixation in mycorrhizal legumes.

The extent of arbuscular mycorrhizal colonization and species affects the bacterial population in the rhizosphere. Bacterial species differ in their abilities to compete for carbon compound root exudates. A change in the amount or composition of root exudates and fungal exudates due to the existing AM mycorrhizal colonization determines the diversity and abundance of the bacterial community in the rhizosphere.

The influence of AM fungi on plant root and shoot growth may also have indirect effect on the rhizosphere bacteria. AMF contributes a substantial amount of carbon to the rhizosphere through the growth and degeneration of the hyphal network. There is also evidence to suggest that AM fungi may play an important role on mediating the plant species' specific effect on the bacterial composition of the rhizosphere.

Glomeromycota and Global Change

Global change is affecting AM fungal communities and interactions between AM fungi and their plant hosts. While it is generally accepted that interactions between organisms will affect their response to global change, we still lack the ability to predict the outcome of these interactions in future climates. In recent meta-analyses, AM fungi were found to increase plant biomass under drought conditions and decrease plant biomass under simulated nitrogen deposition studies,. Arbuscular mycorrhizal fungi themselves have been shown to increase their biomass in response to elevated atmospheric CO2

Plants That Lack Arbuscular Mycorrhiza

Members of the mustard family (Brassicaceae), like cabbage, cauliflower, canola, and crambe, don't have arbuscular mycorrihizal fungi on their roots.

Molecular Genetic Analyses of Arbuscular Mycorrhizal Fungi

In the past ten years there have been spectacular advances in molecular genetic technologies and tools. These advances allow microbial and mycorrhizal ecologists to ask new and exciting questions about the ecological and evolutionary roles of arbuscular mycorrhizal (AM) fungi as individuals, in communities and ecosystems. Genetic analyses of AM fungi have been used to explore the genetic structure of single spores using multilocus genotyping, AM fungal diversity and adaptation across multiple grassland communities, all the way up to a global investigation of AM fungal diversity, which greatly increased the described molecular diversity within the phylum Glomeromycota.

All the recent advances in molecular genetics clearly permit the analysis of microbial communities at much finer and functional scales and potentially with more confidence than previous methods. The classical AM fungal identification method of spore extraction from soil and further spore morphological analysis is fraught with complicating issues due to the various strategies and forms of AM fungi, e.g., lack of sporulation in certain species, seasonality, high unculturability, possible misidentification (human error), and new evidence of multinucleate spores and high genetic variation within clonal AM species. Because of these various problems, in the past researchers likely misrepresented the true composition of AM fungal communities present at any one point in time or place.

Additionally, by following the traditional extraction, culture and microscopic identification methods, there is no way to determine active/functioning AM fungal populations, which are likely the most important when attempting to relate plant-AM symbiotic interactions and mechanisms to ecological or ecosystem function. This is especially true in the case of root colonization analyses, which can determine percentage of roots colonized by AM fungi.

The major problem with this analysis is in field soils, which contain multiple species of AM fungi in association with a target plant at the same time. The identification of the associated fungal symbionts is impossible without the use of molecular methods. Though genetic analysis of AM fungal communities has advanced a great deal in the past decade, the methodology is not yet completely refined. Below is an overview of the methods used in molecular genetic analyses of AM fungi, along with applications to research, future directions and some of their problems.

Overview of Methods

DNA/RNA: Genetic analyses of AM fungi from soil and root samples range in their applicability to answer ecological or phylogenetic questions. DNA analyses utilize various nuclear markers to describe AM fungi and represent different regions of the nuclear ribosomal operon (18S rRNA) found in all eukaryotic organisms. The DNA analysis of AM fungi using these markers began in the early 1990s and are continuing to be developed today. The small subunit (SSU) rRNA gene, the internal transcribed spacer (ITS) gene, and the large subunit (LSU) rRNA gene are currently the most common DNA markers used. The SSU region has been used most frequently in ecological studies, while the ITS and LSU regions have been predominantly used in taxonomic constructions of the phylum Glomeromycota.

General Procedure

The first step of all molecular genetic analyses is the preparation and/or preservation of a sample. In the case of AM fungi, samples typically come in the form of soil or roots that will contain AM spores, hyphae and/or various AM colonization structures. Sample preservation will vary depending on the desired analysis (DNA or RNA). For analysis of DNA, samples should either be processed immediately or kept frozen prior to nucleic acid extraction. For analysis of RNA, samples should be cryogenically frozen ("196 °C) almost immediately upon collection, or stored in an RNA stabilization and preservation reagent (e.g. RNAlater). The next step is to extract the desired nucleic acids from the sample, which can be performed manually using various published extraction methods or by using one of the many commercially available DNA/RNA extraction kits. Due to the labile nature of RNA, synthesis of complementary DNA (cDNA) using extracted RNA as a template is performed for further analysis. For most molecular genetic sequencing methods of AM fungi a PCR step is required to increase the total amount of target DNA/RNA/cDNA. There are many PCR conditions proposed for analysis of AM fungi and some of the most accessible are briefly summarized below.

PCR Methods

From Öpik et al.:

- Reaction mixture:
  - 20 ìl Qiagen HotStarTaq Master mix
  - 0.23 ìM of each primer (NS31 and AM1, more on AM fungal specific primers below)
  - 2 ìl template DNA
- PCR:
  - Run on an MWG AG Biotech Primus 96 Plus thermocycler
  - 15 minutes at 99 °C
  - 5 cycles of 30 seconds at 42 °C
  - 60 seconds at 72 °C
  - 45 seconds at 92 °C
  - 35 cycles of 30 seconds at 65 °C
  - 60 seconds at 72 °C
  - 45 seconds at 92 °C
  - 30 seconds at 65 °C

  - o 10 minutes at 72 °C
  - o PCR products then separated by gel electrophoresis on 1.5% agarose gel in 0.5 x TBE
  - o Separated PCR products were then purified using the Qiagen QIAquick Gel Extraction kit

From Krüger et al.:

- Reaction mixture:
  - o 0.02 U ìl^{-1} Phusion polymerase
  - o 1X Phusion buffer with 1.5 mM $MgCl_2$
  - o 200 ìM of each dNTP
  - o 0.5 ìM of each primer : SSUmAf-LSUmAr and SSUmCf-LSUmBr
- PCR:
  - o Thermal cycling was performed in an Eppendorf Mastercycler Gradient
  - o 5 minute initial denaturation at 99 °C
  - o 40 cycles of 10 second denaturation at 99 °C
  - o 30 seconds annealing at 60 °C
  - o 1 minute elongation at 72 °C
  - o 10 minute final elongation
  - o To visualize PCR product, load onto 1%agarose gel with 1x sodium borate buffer at 220 V, and stain with ethidium bromide (1 ìg ml^{-1})

Primer Selection for Arbuscular Mycorrhizal Fungi

One difficulty with the genetic analysis of arbuscular mycorrhizal fungi has been the selection of ideal, comprehensive, and repeatable primers or primer sets. Currently there are four common AM fungal specific markers/primers used in genetic sequencing to describe AM fungal communities in a sample, ideally to species level identification. These sequence markers are designed for the nuclear ribosomal RNA (rRNA) in the 18S region and are either used individually or in some combination.

The partial small subunit (SSU), the partial large subunit (LSU), and the internal transcribed spacer (ITS1, 5.8S, ITS2) are the regions used for genetic sequencing of AMF. Additionally, there are 'primer sets' that incorporate a combination of these different regions into one target primer for AMF, these include the "Krüger " and the "Redecker" primers. The "Krüger" primer utilizes the partial SSU, the ITS, and the partial LSU regions, while the "Redecker" primer utilizes the partial SSU and the ITS.

Currently, there is no consensus as to which primers or primer sets, being used with varying degrees of success, repeatability and species-level resolution, are best for molecular genetic analysis of AMF. Additionally, the current advances and coming changes in genetic sequencing technology, e.g. Sanger, to 454 pyrosequencing, to Illumina HiSeq/MiSeq, can force researchers to only use certain primers.

The large size of the "Krüger" (~1500bp) and "Redecker" (~900bp) primer sets prohibit use with newer sequencing technology (e.g. Illumina MiSeq) as opposed to 454 pyrosequencing that is capable of these long read lengths.

Though Roche Diagnostics has announced the discontinuation of the 454 platform for 2016, it is still commonly used in genetic analyses.

Perhaps new 'all-inclusive' AM specific primers should be created to support the new technologies for as descriptive a molecular analysis from the "Kruger" primer set using 454 pyrosqeuncing, as shown below. The reverse may also be true, where molecular technologies should be developed with both long read lengths (which would allow for large primer sets) as well as sequencing depth.

Kohout et al. present a study using all of the aforementioned primers/primer sets on identical plant samples using 454 sequencing analysis. Results of their experiment are summarized below.

- "Kruger" primers yielded relatively higher diversity parametres than other comparable primers (LSU, ITS2)
- "Kruger" primers showed significantly higher Shannon diversity measures than did SSU primer
- "Redecker" primers yielded the most different, but maybe most descriptive community composition of all primers tested. This may be explained by the ability of the "Redecker" primers to find less abundant AMF lineages such as the Claroideoglomeraceae or the Paraglomeraceae
- LSU primers had a strong bias towards Glomeraceae, excluding other families
- SSU primers had a bias towards Glomeraceae and underestimated the presence of different families within the Glomeromycota, including the Claroideoglomeraceae, Diversisporaceae and Paraglomeraceae

Family	*MOTUs from SSU*	*MOTUs from LSU*	*MOTUs from ITS2*	*MOTUs from "Kruger"*	*MOTUs from "Redecker"*
Glomeraceae	14	19	17	21	8
Claroideoglomeraceae	1	2	2	3	5
Archaeosporales	1	1	1	1	2
Diversisporaceae	1	0	2	3	2
Paraglomeraceae	0	0	0	0	2
Acaulosproraceae	0	0	1	1	1

MOTU = Molecular operational taxonomic unit, synonymous with OTU or phylotype.

qPCR and qRT-PCR

Real-time PCR or quantitative PCR (qPCR), is becoming a well-established method to quickly amplify and simultaneously quantify targeted AM fungal DNA from biological samples (plant roots or soils). Fairly recent developments in qPCR markers allow researchers to explore the relative abundance of AM fungal species within roots in greenhouse experiments as well as in the field to identify local AM fungal communities.

qPCR markers for arbuscular mycorrhizal fungi will consist of AM specific primers and fluorescently labellled hydrolysis probes. These AM specific primers (discussed above) can be chosen by the researcher and this decision is typically guided by the question at hand, resources available, and willingness to troubleshoot in the lab.

Microarray

DNA microarray analysis is currently being used in AM fungal research to simultaneously measure the expression of many genes from target species or experimental samples. The most common tool or method is to use functional gene array (FGA) technology, a specialized microarray that contains probes for genes that are functionally important in microbial processes such as carbon, nitrogen or phosphorus cycling. FGAs have the ability to simultaneously examine many functional genes. This technique is typically used for general analysis of functional microbial genes, but when complemented with genetic sequencing, inferences can be made about the connection between fungal community composition and microbial functionality.

PLFA/NLFA

Specific organismal chemical signatures can be used to detect biomass of more cryptic organisms, such as AM fungi or soil bacteria.

Lipids, more specifically phospholipids and neutral lipids, contain fatty acids connected to a glycerol backbone. The fatty acid composition of organisms varies, and the proportions of specific fatty acids can be organism specific. For example, in AM fungi the proportion of the fatty acids, 16:1ù5 and 18:1ù7, in the phospholipid portion account for approximately 58% of total fatty acid composition. The fatty acid, 16:1ù5 is the most commonly used acid to characterize AM fungi in soils and can be used as a strong indicator of mycelial biomass in soil sample.

Neutral lipid fatty acid analysis of AM fungi is typically looked upon as a method to indicate energy storage, but most importantly, the ratio of NLFA (16:1ù5) to PLFA (16:1ù5) can potentially be used to indicate nutritional status of AM fungi in soils. Energy is mainly stored in AM fungi as neutral lipids in storage structures like spores and vesicles. Because of this NLFA correlates quite well with the number of spores in a given volume of soil. The ratio of NLFA concentration to PLFA concentration (active mycelia) can then give the proportion of carbon allocated to storage structures (spores, measured as NLFA).

Problems with lipid fatty acid analyses include the incomplete specificity of fatty acids to AM fungi, the species- or genera-specific variation in fatty acid composition can complicate analysis in systems with multiple AM fungal species (e.g. field soil), the high background levels of certain fatty acid concentration in soils, and that phospholipids are correlated to an organism's membrane area, and the surface to volume ratio can vary widely between organisms such as bacteria and fungi. More work must be done to identify the efficacy of this method in field soils with many genera and species of AM fungi to discern the methods ability to discriminate between many varying fatty acid compositions.

Future Research Directions with AM Fungi

An exciting prospect for future analysis of AM fungi is the use of stable isotope probes. Stable isotope probing (SIP) is a technique that can be used to determine the active metabolic function of individual taxa within a complex system of microbes. This level of specificity, linking microbial function and phylogenetics, has not been achieved previously in microbial ecology. This method can also be used independently of classical culture methods in microbial ecology, allowing for *in situ* analysis of functional microbes.

SIP Method

SIP, more explicitly DNA/RNA-based SIP, uses stable-isotope enriched substrates, such as ^{13}C, ^{15}N, or $H_2{}^{18}O$, and then analyzes the 'labelled' markers using species specific DNA or RNA markers. The analysis of labelled DNA is performed by separating unlabelled and labelled DNA on a cesium chloride gradient formed in an ultra centrifuge. Because all microbial organisms are capable of importing water into their cells, the use of $H_2{}^{18}O$ stable isotope probing is a very exciting new method that can shed light on questions microbial ecologists and biologists have struggled with answering for years, in particular, what are the active microbial organisms in my system? The $H_2{}^{18}O$, or heavy water method will target all organisms that are actively growing, and induce little influence on growth itself. This would be especially true with most greenhouse experiments with arbuscular mycorrhizas because plants must be watered anyway, and water does not directly select for organisms with specific metabolic pathways, as would happen when using ^{13}C and^{15}N.

Little has been done with this method in arbuscular mycorrhizal experiments, but if proven to work in a controlled experiment, and with further refinement of DNA/RNA fungal community analyses techniques, this may be a viable option to very specifically determine the actively growing portion of AM fungal species across growing seasons, with different plant hosts or treatments, and in the face of climate change.

Phytoremediation

The use of arbuscular mycorrhizal fungi in ecological restoration projects (phytoremediation) has been shown to enable host plant establishment on degraded soil and improve soil quality and health. Disturbance of native plant communities in desertification-threatened areas is often followed by degradation of physical and biological soil properties, soil structure, nutrient availability, and organic matter. When restoring disturbed land, it is essential to replace not only the above ground vegetation but also the biological and physical soil properties.

A relatively new approach to restoring land and protecting against desertification is to inoculate the soil with arbuscular mycorrhizal fungi with the reintroduction of vegetation. A long-term study demonstrated that a significantly greater long-term improvement in soils' quality parametres was attained when the soil was inoculated with a mixture of indigenous arbuscular mycorrhizal fungi species compared to the noninoculated soil and soil inoculated with a single exotic species of AM fungi. The benefits observed were an increased plant growth and soil nitrogen content, higher soil organic matter content, and soil aggregation. The improvements were attributed to the higher legume nodulation in the presence of AMF, better water infiltration, and soil aeration due to soil aggregation.

Inoculation with native AM fungi increased plant uptake of phosphorus, improving plant growth and health. The results support the use of AM fungi as a biological tool in the restoration of biotopes to self-sustaining ecosystems.

Agriculture

Many modern agronomic practices are disruptive to mycorrhizal symbiosis. There is great potential for low-input agriculture to manage the system in a way that promotes mycorrhizal symbiosis.

Conventional agriculture practices, such as tillage, heavy fertilizers and fungicides, poor crop rotations, and selection for plants that survive these conditions, hinder the ability of plants to form symbiosis with arbuscular mycorrhizal fungi.

Most agricultural crops can perform better and are more productive when well-colonized by AM fungi. AM symbiosis increases the phosphorus and micronutrient uptake and growth of their plant host (George *et al.* 1992). Management of AM fungi is especially important for organic and low-input agriculture systems where soil phosphorus is, in general, low, although all agroecosystems can benefit by promoting arbuscular mycorrhizae establishment.

Some crops that are poor at seeking out nutrients in the soil are very dependent on AM fungi for phosphorus uptake. For example flax, which has poor chemotaxic ability, is highly dependent on AM-mediated phosphorus uptake at low and intermediate soil phosphorus concentrations (Thingstrup *et al.* 1998).

Proper management of AMF in the agroecosystems can improve the quality of the soil and the productivity of the land. Agricultural practices such as reduced tillage, low phosphorus fertilizer usage, and perennialized cropping systems promote functional mycorrhizal symbiosis.

Tillage

Tillage reduces the inoculation potential of the soil and the efficacy of mycorrhizaes by disrupting the extraradical hyphal network (Miller *et al.* 1995, McGonigle & Miller 1999, Mozafar *et al.* 2000).

By breaking apart the soil macro structure, the hyphal network is rendered non-infective (Miller *et al.* 1995, McGonigle & Miller 1999). The disruption of the hyphal network decreases the absorptive abilities of the mycorrhizae because the surface area spanned by the hyphae is greatly reduced. This, in turn, lowers the phosphorus input to the plants that are connected to the hyphal network.

In reduced-tillage system, heavy phosphorus fertilizer input may not be required as compared to heavy-tillage systems. This is due to the increase in mycorrhizal network, which allows mycorrhizae to provide the plant with sufficient phosphorus (Miller *et al.* 1995).

Phosphorus Fertilizer

The benefits of AMF are greatest in systems where inputs are low. Heavy usage of phosphorus fertilizer can inhibit mycorrhizal colonization and growth. As the soil's phosphorus levels available to the plants increases, the amount of phosphorus also increases in the plant's tissues, and carbon drain on the plant by the AM fungi symbiosis become non-beneficial to the plant (Grant 2005).

A decrease in mycorrhizal colonization due to high soil-phosphorus levels can lead to plant deficiencies in other micronutrients that have mycorrhizal-mediated uptake such as copper (Timmer & Leyden 1980).

Perennialized Cropping Systems

Cover crops are grown in the fall, winter, and spring, covering the soil during periods when it would commonly be left without a cover of growing plants.

Mycorrhizal cover crops can be used to improve the mycorrhizal inoculum potential and hyphal network (Kabir and Koide 2000, Boswell *et al.*1998, Sorensen et al. 2005).

Since AM fungi are biotrophic, they are dependent on plants for the growth of their hyphal networks. Growing a cover crop extends the time for AM growth into the autumn, winter, and spring. Promotion of hyphal growth creates a more extensive hyphal network.

The mycorrhizal colonization increase found in cover crops systems may be largely attributed to an increase in the extraradical hyphal network that can colonize the roots of the new crop (Boswell *et al.* 1998). The extraradical mycelia are able to survive the winter, providing rapid spring colonization and early season symbiosis (McGonigle and Miller 1999). This early symbiosis allows plants to tap into the well-established hyphal network and be supplied with adequate phosphorus nutrition during early growth, which greatly improves the crop yield.

Soil Quality

Restoration of native AM fungi increases the success of ecological restoration project and the rapidity of soil recovery. AM fungi enhance soil aggregate stability is due to the production of extraradical hyphae and a soil protein known as glomalin.

Glomalin-related soil proteins (GRSP) have been identified using a monoclonal antibody (Mab32B11) raised against crushed AMF spores. It is defined by its extraction conditions and reaction with the antibody Mab32B11.

There is other circumstantial evidence to show that glomalin is of AM fungal origin. When AM fungi are eliminated from soil through incubation of soil without host plants, the concentration of GRSP declines. A similar decline in GRSP has also been observed in incubated soils from forested, afforested, and agricultural land and grasslands treated with fungicide.

Glomalin is hypothesized to improve soil aggregate water stability and decrease soil erosion. A strong correlation has been found between GRSP and soil aggregate water stability in a wide variety of soils where organic material is the main binding agent, although the mechanism is not known. The protein glomalin has not yet been isolated and described, and the link between glomalin, GRSP, and arbuscular mycorrhizal fungi is not yet clear.

Ecofriendly Management of Wilt of Guava by VAM Technology

One of the most gregarious of fruit trees, the guava, *Psidium guajava* L., of the myrtle family (Myrtaceae), is almost universally known by its common English name or its equivalent in other languages. In Spanish, the tree is *guayabo*, or *guayavo*, the fruit *guayaba* or *guyava*. The French call it *goyave* or *goyavier*; the Dutch, *guyaba*, *goeajaaba*; the Surinamese, *guave* or *goejaba;* and the Portuguese, *goiaba* or *goaibeira*. Hawaiians call it guava or *kuawa*. In Guam it is *abas*. In Malaya, it is generally known either as guava or *jambu batu*, but has also numerous dialectal names as it does in India, tropical Africa and the Philippines where the corruption, *bayabas,* is often applied. Various tribal names–*pichi, posh, enandi, etc.*–are employed among the Indians of Mexico and Central and South America.

Description

A small tree to 33 ft (10 in) high, with spreading branches, the guava is easy to recognize because of its smooth, thin, copper-coloured bark that flakes off, showing the greenish layer beneath; and also because of the attractive, "bony" aspect of its trunk which may in time attain a diametre of 10 in (25 cm). Young twigs are quadrangular and downy. The leaves, aromatic when crushed, are evergreen, opposite, short-petioled, oval or oblong-elliptic, somewhat irregular in outline; 2 3/4 to 6 in (7-15 cm) long, I 'A to 2 in (3-5 cm) wide, leathery, with conspicuous parallel veins, and more or less downy on the underside.

Faintly fragrant, the white flowers, borne singly or in small clusters in the leaf axils, are 1 in (2.5 cm) wide, with 4 or 5 white petals which are quickly shed, and a prominent tuft of perhaps 250 white stamens tipped with pale-yellow anthers.

The fruit, exuding a strong, sweet, musky odour when ripe, may be round, ovoid, or pear-shaped, 2 to 4 in (5-10 cm) long, with 4 or 5 protruding floral remnants (sepals) at the apex; and thin, light-yellow skin, frequently blushed with pink. Next to the skin is a layer of somewhat granular flesh, 1/8 to 1/2 in (3-12.5 mm) thick, white, yellowish, light- or dark-pink, or near-red, juicy, acid, subacid, or sweet and flavorful. The central pulp, concolourous or slightly darker in tone, is juicy and normally filled with very hard, yellowish seeds, 1/8 in (3 min) long, though some rare types have soft, chewable seeds. Actual seed counts have ranged from 112 to 535 but some guavas are seedless or nearly so.

When immature and until a very short time before ripening, the fruit is green, hard, gummy within and very astringent.

Figure: *Plate L: GUAVA,* Psidium guajava

The fruit, exuding a strong, sweet, musky odour when ripe, may be round, ovoid, or pear-shaped, 2 to 4 in (5-10 cm) long, with 4 or 5 protruding floral remnants (sepals) at the apex; and thin, light-yellow skin, frequently blushed with pink. Next to the skin is a layer of somewhat granular flesh, 1/8 to 1/2 in (3-12.5 mm) thick, white, yellowish, light- or dark-pink, or near-red, juicy, acid, subacid, or sweet and flavorful. The central pulp, concolourous or slightly darker in tone, is juicy and normally filled with very hard, yellowish seeds, 1/8 in (3 min) long, though some rare types have soft, chewable seeds. Actual seed counts have ranged from 112 to 535 but some guavas are seedless or nearly so. When immature and until a very short time before ripening, the fruit is green, hard, gummy within and very astringent.

Origin and Distribution

The guava has been cultivated and distributed by man, by birds, and sundry 4-footed animals for so long that its place of origin is uncertain, but it is believed to be an area extending from southern Mexico into or through Central America. It is common throughout all warm areas of tropical America and in the West Indies (since 1526), the Bahamas, Bermuda and southern Florida where it was reportedly introduced in 1847 and was common over more than half the State by 1886. Early Spanish and Portuguese colonizers were quick to carry it from the New World to the East Indies and Guam. It was soon adopted as a crop in Asia and in warm parts of Africa. Egyptians have grown it for a long time and it may have travelled from Egypt to Palestine. It is occasionally seen in Algeria and on the Mediterranean coast of France. In India, guava cultivation has been estimated at 125,327 acres (50,720 ha) yielding 27,319 tons annually.

Apparently it did not arrive in Hawaii until the early 1800's. Now it occurs throughout the Pacific islands. Generally, it is a home fruit tree or planted in small groves, except in India where it is a major commercial resource. A guava research and improvement programme was launched by the government of Colombia in 1961. In 1968, it was estimated that there were about 10 million wild trees (around Santander, Boyacá, Antioquia, Palmira, Buga, Cali and Cartago) bearing, 88 lbs (40 kg) each per year and that only 10% of the fruit was being utilized in processing. Bogotà absorbs 40% of the production and preserved products are exported to markets in Venezuela and Panama.

Brazil's modern guava industry is based on seeds of an Australian selection grown in the botanical garden of the Sao Paulo Railway Company at Tatu. Plantations were developed by Japanese farmers at Itaquera and this has become the leading guava-producing area in Brazil. The guava is one of the leading fruits of Mexico where the annual crop from 36,447 acres (14,750 ha) of seedling trees totals 192,850 tons (175,500 MT). Only in recent years has there been a research programme designed to evaluate and select superior types for vegetative propagation and large-scale cultivation.

In Florida, the first commercial guava planting was established around 1912 in Palma Sola. Others appeared at Punta Gorda and Opalocka. A 40-acre (16 ha) guava grove was planted by Miami Fruit

Industries at Indian-town in 1946. There have been more than two dozen guava jelly manufacturers throughout the state. A Sarasota concern was processing 250 bushels of guavas per day and a Pinellas County processor was operating a 150-bushel capacity plant in 1946. There has always been a steady market for guava products in Florida and the demand has increased in recent years with the influx of Caribbean and Latin American people.

The guava succumbs to frost in California except in a few favourable locations. Even if summers are too cool–a mean of 60° F (15.56° C)– in the coastal southern part of the state, the tree will die back and it cannot stand the intense daytime heat of interior valleys.

In many parts of the world, the guava runs wild and forms extensive thickets–called "guayabales" in Spanish–and it overruns pastures, fields and roadsides so vigorously in Hawaii, Malaysia, New Caledonia, Fiji, the U.S. Virgin Islands, Puerto Rico, Cuba and southern Florida that it is classed as a noxious weed subject to eradication.

Nevertheless, wild guavas have constituted the bulk of the commercial supply. In 1972, Hawaii processed, for domestic use and export, more than 2,500 tons (2,274 MT) of guavas, over 90% from wild trees. During the period of high demand in World War II, the wild guava crop in Cuba was said to be 10,000 tons (9,000 MT), and over 6,500 tons (6,000 MT) of guava products were exported.

Cultivars

Formerly, round and pear-shaped guavas were considered separate species–*P. pomiferum* L. and *P. pyriferum* L.–but they are now recognized as mere variations. Small, sour guavas predominate in the wild and are valued for processing.

'Redland', the first named cultivar in Florida, was developed at the University of Florida Agricultural Research and Education Centre, Homestead, and described in 1941. Very large, with little odour, white-fleshed and with relatively few seeds, it was at first considered promising but because of its excessively mild flavor, low ascorbic acid content, and susceptibility to algal spotting, it was abandoned in favour of better selections.

'Supreme' came next, of faint odour, thick, white flesh, relatively few, small seeds, high ascorbic acid content and ability to produce heavy crops over a period of 8 months from late fall to early spring.

'Red Indian', of strong odour, medium to large size, round but slightly flattened at the base and apex, yellow skin often with pink blush; with medium thick, red flesh of sweet flavor; numerous but small seeds; agreeable for eating fresh; fairly productive in fall and early winter.

'Ruby', with pungent odour, medium to large size; ovate; with thick, red flesh, sweet flavor, relatively few seeds. An excellent guava for eating fresh and for canning; fairly productive, mainly in fall and early winter.

'Blitch' (a seedling which originated in West Palm Beach and was planted at Homestead)–of strong odour, medium size, oval, with light-pink flesh, numerous, small seeds; tart, pleasant flavor; good for jelly.

'Patillo' (a seedling selection at DeLand propagated by a root sucker and from that by air-layer and planted at Homestead)–of very mild odour, medium size, ovate to obovate, with pink flesh, moderate number of small seeds; subacid, agreeable flavor; good for general cooking. (As grown in Hawaii it is highly acid and best used for processing).

'Miami Red' and 'Miami White', large, nearly odourless and thick-fleshed, were released by the University of Miami's Experimental Farm in 1954.

In early 1952, Dr. J.J. Ochse imported into Florida air-layers of a seedless guava from Java. All died. In September 1953, the writer received air-layers from Saharanpur, India. One survived and was turned over to the Agricultural Research and Education Centre, Homestead. Four more were ordered from Coimbatore but arrived dead. Willim Whitman brought in a grafted plant from Java in 1954 which grew well, fruited and was the source of propagating material. In 1955, Whitman obtained a plant of a seedless guava from Cuba and it bore its first fruit in 1957. Seedless guavas are the result of low fertility of pollen grains and self-incompatibility. The fruits tend to be malformed and the trees are scant bearers. Applications of gibberellic acid increase fruit size, weight and ascorbic acid content but induce prominent ridges on the surface.

Among early California cultivars were:

'Webber' (formerly 'Riverside'), of medium-large size, pale-yellowish flesh, good flavor and 9.5% sugar.

'Rolfs', of medium size with pink flesh; of good quality and containing 9% sugar.

'Hart', fairly large, with pale-yellow flesh, and 8% sugar content.

Currently, some rare fruit fanciers grow the Florida-developed 'Red Indian' and 'White Indian'; also 'Detwiler' and 'Turnbull'.

In 1975, a guava trial project was undertaken at the Maroochy Horticultural Research Station in southeastern Queensland, beginning with 5 strains from Hawaii. By 1981, 4 selections (GA9-39R1T2', 'GA11-56T7', GA11-56R5T2' and 'GA11-564T1') seemed to hold promise for processing and 2 selections ('GA11-56T3' and 'GA11-56R1T1') for marketing fresh. They were all vegetatively propagated and tested as to performance.

The green-skinned, acid, 'GA11-56' and another Hawaiian selection, '1050', yellow-skinned and mild in flavor and odour, are being grown commercially for processing in New South Wales.

In India much attention is given the characteristics of local and introduced guava cultivars and their suitability for various purposes. Among common white-fleshed cultivars are:

'Apple Colour'–of medium size, slightly oblate; deep-pink skin, creamy-white flesh, moderate amount of seeds, very sweet flavor (0.34-2.12% acid, 9 to 11.36% sugar); heavy bearer; good keeping quality; good for canning.

'Behat Coconut'–large, with thick white flesh, few seeds; poor for canning.

'Chittidar'–medium to large, round-ovate, white-fleshed, mild acid-sweet flavor; bears moderately well; keeps well; good for canning.

'Habshi'–of medium size with thick, white flesh, few seeds; halves good for canning.

'Lucknow 42'–of medium size, roundish, with creamy-white, soft flesh; sweet, pleasant flavor; very few seeds; good quality; bears heavily; keeps fairly well; not suitable for canning.

'Lucknow 49'–medium-large with cream-white, thick flesh, few seeds; acid-sweet; good quality; heavy bearer; high in pectin and good for jelly; halves good for canning.

'Safeda'–of medium size, with very thin skin, thick, white flesh, few seeds. Outstanding quality for canning. A famous guava, widely planted, but susceptible to wilt and branches are brittle and break readily.

'Smooth Green'–of medium size, with thick white flesh, few, small, hard seeds. Halves are firm, good for canning.

'Allahabad'–large, white-fleshed, with few, medium-sized, fairly hard seeds.

'Karela'–medium-large, pear-shaped, furrowed, rough-skinned, with soft, granular, white flesh; sweet, rich, pleasant flavor. Poor bearer. Not popular.

'Nagpur Seedless'–small to medium, often irregular in shape; white-fleshed.

'Seedless' (from Allahabad)–medium to large, pear-shaped to ovoid; with thick white flesh, firm to soft, sweet. Light bearer; poor keeper.

A seedless type at Poona, India, was found to be a triploid with 33 chromosomes in place of the usual 22.

Other white-fleshed guavas with poor canning qualities are: 'Dharwar', 'Mirzapuri', 'Nasik', 'Sindh', and 'White Supreme X Ruby'.

Among red-fleshed cultivars in India there are:

'Anakapalle'–small, with thin, red flesh, many seeds; not suitable for canning.

'Florida Seedling'–small, with thin, red, acid flesh; many seeds; not suitable for canning.

'Hapi'–medium to large, with red flesh.

'Hybrid Red Supreme'–large, with thin, red, acid flesh; moderate amount of seeds; not suitable for canning.

'Kothrud'–of medium size with medium thick, red flesh; moderate amount of seeds; not suitable for canning.

'Red-fleshed'–of medium size with many (about 567) fairly soft seeds; high in pectin and good for jelly; not suitable for canning.

Among other Indian cultivars are: 'Banaras', 'Dholka', 'Hasijka', 'Kaffree', and 'Wickramasekara'. The latter is a small fruit and poor bearer.

Indian breeders have crossed the guava with its dwarf, small-fruited relative, *P. guineense* Sw., with a view to reducing tree size and enhancing hardiness and yield.

In Egypt, a cultivar named 'Bassateen El Sabahia' has long been the standard commercial guava. Efforts have been made to improve quality and yield and to this end selections were made from 300 seedlings. The most promising selection was tested and introduced into cultivation in 1975 under the name 'Bassateen Edfina'. It is pear-shaped, of medium size, sometimes pink-blushed, with thick, white flesh, few seeds, good flavor and higher ascorbic acid content than the parent. It bears well over a long season.

In Puerto Rico, over 100 promising selections were under observation in 1963.

Numerous cultivated clones identified only by number have been evaluated for processing characters. Others have been tested and rated for resistance to *Glomerella* disease. Among the few named cultivars are 'Corozal Mixta', 'Corriente', and 'Seedling 57-6-79'.

In Trinidad, a large, white-fleshed type is known as 'Cayenne'.

In 1967, French horticulturists made a detailed evaluation of 11 guava cultivars grown at the Neufchateau Station in Guadeloupe:

'Elisabeth'–large, round, pink-fleshed, very acid; good for processing.

'Red' X 'Supreme' X 'Ruby'–large, ovoid, with deep-pink flesh; agreeable for eating fresh.

'Large White'–large, round, white-fleshed; low sugar content, astringent; can be useful as filler in preserves.

'Acid Speer'–large, round, with pale-yellow flesh; acid; recommended only as source of pectin.

'Red' X 'Supreme' X 'Ruby' X 'White'–large to very large, pear-shaped, with creamy-white flesh; good for eating fresh and for juice and nectar.

'Pink Indian'–of medium size, red-fleshed; agreeably acid; good for eating fresh and for processing.

'Red Hybrid'–medium, sub-ovoid, red-fleshed; medium quality.

'Supreme' X 'Ruby'–medium, sub-ovoid, white-fleshed; unremarkable except for high productivity.

'Stone'–small, ovoid, with deep-pink flesh; attractive and of agreeable flavor for eating fresh.

'Supreme'–small, ovoid, with pale-yellow, pink-tinged flesh; sweet; good for sherbet and paste; very productive.

'Patricia'–very small, ovoid, salmon-fleshed; attractive; good to eat fresh but quickly loses its distinct strawberry flavor; good for sirup; very productive.

Between 1948 and 1969, 21 guava cultivars from 7 countries were introduced into Hawaii. Some have been test planted and evaluated at the Waimanalo Experimental Farm. Four sweet, white-fleshed, thick-walled cultivars were rated as commercially desirable: 'Indonesian White', 'Indonesian Seedless', 'Lucknow 49', and 'No. 6363' (a 'Ruby' X 'Supreme' hybrid from Florida). Lower ratings were given four others of this group: 'Apple' (too musky and seedy); 'Allahabad Safeda' (too bumpy of surface); 'Burma' (too seedy) and 'Hong Kong White' (too seedy). Of the sweet, pink-fleshed, thick-walled cultivars examined, 'Hong Kong Pink' was preferred. Second choice was 'No. 6362' (a seedling of a 'Ruby' X' Supreme' cross in Florida). 'No. 7199', a seedling of a

'Stone Acid' X 'Ruby' cross in Florida, was considered too musky. Among acid, non-musky, thick-walled guavas, 'Beaumont', a Hawaiian selection, is large and pink-fleshed. 'Pink Acid' (#7198), from a Florida cross of 'Speer' and 'Stone Acid', has dark-pink flesh and few seeds. These cultivars are employed in breeding programmes in Hawaii. In 1978, a new cultivar, 'Ka Hua Kula', selected from 1,200 seedlings of 'Beaumont', was released and recommended for commercial guava puree. The fruit is large, with thick, deep-pink flesh, and fewer seeds than 'Beaumont', and is less acid. It is also a heavier bearer.

In Colombia, the cultivars 'Puerto Rico', 'Rojo Africano', and 'Agrio', all yield over 2,200 fruits annually. Other high-yielding cultivars being evaluated are 'White', 'Red', 'D-13', 'D-14', and 'Trujillo 2'.

Collecting guava cultivars is a hobby of Mr. Arthur Stockdale, Finca Catalina, Zitacuaro, Mexico. He is said to have some very superior selections in his grove.

Pollination

The chief pollinator of guavas is the honeybee (*Apis mellifera*). The amount of cross-pollination ranges from 25.7 to 41.3%.

Climate

The guava thrives in both humid and dry climates. In India, it flourishes up to an altitude of 3,280 ft (1,000 m); in Jamaica, up to 3,906 ft (1,200 m); in Costa Rica, to 4,590 ft (1,400 m); in Ecuador, to 7,540 ft (2,300 m). It can survive only a few degrees of frost. Young trees have been damaged or killed in cold spells at Allahabad, India, in California and in Florida. Older trees, killed to the ground, have sent up new shoots which fruited 2 years later. The guava requires an annual rainfall between 40 and 80 in (1,000-2,000 mm); is said to bear more heavily in areas with a distinct winter season than in the deep Tropics.

Soil

The guava seems indiscriminate as to soil, doing equally well on heavy clay, marl, light sand, gravel bars near streams, or on limestone; and tolerating a pH range from 4.5 to 9.4. It is somewhat salt-resistant. Good drainage is recommended but guavas are seen growing spontaneously on land with a high water table–too wet for most other fruit trees.

Propagation

Guava seeds remain viable for many months. They often germinate in 2 to 3 weeks but may take as long as 8 weeks. Pretreatment with sulfuric acid, or boiling for 5 minutes, or soaking for 2 weeks, will hasten germination. Seedlings are transplanted when 2 to 30 in (5-75 cm) high and set out in the field when 1 or 2 years old. Inasmuch as guava trees cannot be depended upon to come true from seed, vegetative propagation is widely practiced.

In Hawaii, India and elsewhere, the tree has been grown from root cuttings. Pieces of any roots except the smallest and the very large, cut into 5 to 10 in (12.5-20 cm) lengths, are placed flat in a prepared bed and covered with 2 to 4 in (5-10 cm) of soil which must be kept moist. Or one can merely cut through roots in the ground 2 to 3 ft (0.6-0.9 m) away from the tree trunk; the cut ends will sprout and can be dug up and transplanted.

By another method, air-layers of selected clones are allowed to grow 3 to 5 years and are then sawn off close to the ground. Then a ring of bark is removed from each new shoot; root-inducing chemical is applied. Ten days later, the shoots are banked with soil to a height 4 to 5 in (10-12.5 cm) above the ring. After 2 months, the shoots are separated and planted out. Pruned branches may serve as propagating material. Cuttings of half-ripened wood, 1/4 to 1/2 in (6-12.5 mm) thick will root with bottom heat or rooting-hormone treatment. Using both, 87% success has been achieved. Treated softwood cuttings will also root well in intermittent mist.

In Trinidad, softwood, treated cuttings have been rooted in 18 days in coconut fibre dust or sand in shaded bins sprayed 2 or 3 times daily to keep humidity above 90%. Over 100,000 plants were produced by this method over a 2-year period. Under tropical conditions (high heat and high humidity), mature wood 3/4 to 1 in (2-2.5 cm) thick and 1 1/2 to 2 ft (45-60 cm) long, stuck into 1-ft (30-cm) high black plastic bags filled with soil, readily roots without chemical treatment.

In India, air-layering and inarching have been practiced for many years. However, trees grown from cuttings or air-layers have no taproot and are apt to be blown down in the first 2 or 3 years. For this reason, budding and grafting are preferred.

Approach grafting yields 85 to 95% success. Trials have been made of the shield, patch and Forkert methods of budding. The latter always gives the best results (88 to 100%). Vigorous seedlings 1/2 to 1 in (1.25-2.5 cm) thick are used as rootstocks. The bark should slip easily to facilitate insertion of the bud, which is then tightly bound in place with a plastic strip and the rootstock is beheaded, leaving only 6 to 8 leaves above the bud. About a month later, an incision is made halfway through 2 or 3 in (5-7.5 cm) above the bud and the plant is bent over to force the bud to grow. When the bud has put up several inches of growth, the top of the rootstock is cut off immediately above the bud. Sprouting of the bud is expedited in the rainy season.

At the Horticultural Experiment and Training Centre, Basti, India, a system of patch budding has been demonstrated as commercially feasible. A swollen but unsprouted, dormant bud is taken as a 3/4 x 3/8 in (2 x 1 cm) patch from a leaf axil of previous season's growth and taped onto a space of the same size cut 6 to 8 in (15-20 cm) above the ground on a 1-year-old, pencil-thick seedling during the period April-June. After the bud has "taken", 1/3 is cut from the top of the seedling; 2-3 weeks later, the rest of the top is cut off leaving only 3/4 to 1 1/4 in (2-3.2 cm) of stem above the bud. This method gives 80 to 90% success. If done in July, only 70%.

In Hawaii, old seedling orchards have been topworked to superior selections by patch budding on stump shoots.

Culture

Guava trees are frequently planted too close. Optimum distance between the trees should be at least 33 ft (10 m). Planting 16 1/2 ft (5 m) apart is possible if the trees are "hedged". The yield per tree will be less but the total yield per land area will be higher than at the wider spacing. Some recommend setting the trees 8 ft (2.4 m) apart in rows 24 ft (7.3 m) apart and removing every other tree as soon as there is overcrowding.

Where mass production is not desired and space is limited, guava trees can be grown as cordons on a wire fence. Rows should always run north and south so that each tree receives the maximum sunlight. Exudates from the roots of guava trees tend to inhibit the growth of weeds over the root system.

Light pruning is always recommended to develop a strong framework, and suckers should also be eliminated around the base. Experimental heading-back has increased yield in some cultivars in Puerto Rico. In Palestine, the trees are cut back to 6 1/2 ft (2 m) every other spring to facilitate harvesting without ladders. Fruits are borne by new shoots from mature wood. If trees bear too heavily, the branches may break. Therefore, thinning is recommended and results in larger fruits.

Guava trees grow rapidly and fruit in 2 to 4 years from seed. They live 30 to 40 years but productivity declines after the 15th year. Orchards may be rejuvenated by drastic pruning.

The tree is drought-tolerant but in dry regions lack of irrigation during the period of fruit development will cause the fruits to be deficient in size. In areas receiving only 15 to 20 in (38-50 cm) rainfall annually, the guava will benefit from an additional 2,460 cm (2 acre feet) applied by means of 8 to 10 irrigations, one every 15-20 days in summer and one each month in winter.

Guava trees respond to a complete fertilizer mix applied once a month during the first year and every other month the second year (except from mid-November to mid-January) at the rate of 8 oz (227 g) per tree initially with a gradual increase to 24 oz (680 g) by the end of the second year.

Nutritional sprays providing copper and zinc are recommended thrice annually for the first 2 years and once a year thereafter. In India, flavor and quality of guavas has been somewhat improved by spraying the foliage with an aqueous solution of potassium sulfate weekly for 7 weeks after fruit set.

Control of Wild Trees

Large trees that have overrun pastures are killed in Fiji with 2,4-D dicamba or 2,4,5-T in diesel fuel or old engine oil. Extensive wild stands of young trees are best burned. Cutting results in regrowth with multiple stems.

Cropping and Yield

The fruit matures 90 to 150 days after flowering. Generally, there are 2 crops per year in southern Puerto Rico; the heaviest, with small fruits, in late summer and early fall; another, with larger fruits, in late winter and early spring. In northern India, the main crop ripens in mid-winter and the fruits are of the best quality.

A second crop is home in the rainy season but the fruits are less abundant and watery. Growers usually withhold irrigation after December or January or root-prune the trees in order to avoid a second crop. The trees will shed many leaves and any fruits set will drop. An average winter crop in northern

India is about 450 fruits per tree. Trees may bear only 100-300 fruits in the rainy season but the price is higher because of relative scarcity despite the lower quality. Of course, yields vary with the cultivar and cultural treatment.

Experiments have shown that spraying young guava trees with 25% urea plus a wetting agent will bring them into production early and shorten the harvest period from the usual 15 weeks to 4 weeks.

Handling and Keeping Quality

Ripe guavas bruise easily and are highly perishable. Fruits for processing may be harvested by mechanical tree-shakers and plastic nets. For fresh marketing and shipping, the fruits must be clipped when full grown but underripe, and handled with great care.

After grading for size, the fruits should be wrapped individually in tissue and packed in 1 to 4 padded layers with extra padding on top before the cover is put on. They have been successully shipped from Miami to wholesalers in major northern cities in refrigerated trucks at temperatures of 45° to 55° F (7.22°-12.78° C). It is commonly said that guavas must be tree-ripened to attain prime quality, but the cost of protecting the crop from birds makes early picking necessary. It has been demonstrated that fruits picked when yellow-green and artificially ripened for 6 days in straw at room temperature developed superior colour and sugar content.

Guavas kept at room temperature in India are normally overripe and mealy by the 6th day, but if wrapped in pliofilm will keep in good condition for 9 days. In cold storage, pliofilm-wrapped fruits remain unchanged for more than 12 days. Wrapping checks weight loss and preserves glossiness. Unwrapped 'Safeda' guavas, just turned yellow, have kept well for 4 weeks in cold storage at 47° to 50° F (8.33°-10° C) and relative humidity of 85-95%, and were in good condition for 3 days thereafter at room temperature of 76° to 87° F (24°-44° C).

Fruits coated with a 3% wax emulsion will keep well for 8 days at 72° to 86° F (22.2°-30° C) and 40 to 60% relative humidity, and for 21 days at 47° to 50° F (8.3°-10° C) and relative humidity of 85-90%. Storage life of mature green guavas is prolonged at 68° F (20° C), relative humidity of 85%, less than 10% carbon dioxide, and complete removal of ethylene.

Researchers at Kurukshetra University, India, have shown that treatment of harvested guavas with 100 ppm morphactin (chlorflurenol methyl ester 74050) increases the storage life of guavas by controlling fungal decay, and reducing loss of colour, weight, sugars, ascorbic acid and non-volatile organic acids. Combined fungicidal and double-wax coating has increased marketability by 30 days.

Australian workers report prolonged life and reduced rotting in storage after a hot water dip, but better results were achieved by dipping in an aqueous benomyl suspension at 122° F (50° C). Higher temperatures cause some skin injury, as does a guazatine dip which is also a less effective fungicide.

Fruits sprayed on the tree with gibberellic acid 20-35 days before normal ripening, were retarded nearly a week as compared with the untreated fruits. Also, mature guavas soaked in gibberellic acid off the tree showed a prolonged storage life.

Trials at Haryana Agricultural University, Hissar, India, showed that weekly spraying with 1.0% potassium sulfate–1.6 gals (6 litres) per tree–beginning 7 days after fruit set and ending just before harvesting at the pale-green stage, delays yellowing, retains firmness and flavor beyond normal storage life.

Food technologists in India found that bottled guava juice (strained from sliced guavas boiled 35 minutes), preserved with 700 ppm SO_2, lost much ascorbic acid but little pectin when stored for 3 months without refrigeration, and it made perfectly set jelly.

Pests and Diseases

Guava trees are seriously damaged by the citrus flat mite, *Brevipalpus californicus* in Egypt. In India, the tree is attacked by 80 insect species, including 3 bark-eating caterpillars *(Indarbella* spp.) and the guava scale, but this and other scale insects are generally kept under control by their natural enemies. The green shield scale, *Pulvinaria psidii,* requires chemical measures in Florida, as does the guava white fly, *Trialeurodes floridensis,* and a weevil, *Anthonomus irroratus,* which bores holes in the newly forming fruits.

The red-banded thrips feed on leaves and the fruit surface. In India, cockchafer beetles feed on the leaves at the end of the rainy season and their grubs, hatched in the soil, attack the roots. The larvae of the guava shoot borer penetrates the tender twigs, killing the shoots. Sometimes aphids are prevalent, sucking the sap from the underside of the leaves of new shoots and excreting honeydew on which sooty mold develops.

The guava fruit worm, *Argyresthia eugeniella,* invisibly infiltrates hard green fruits, and the citron plant bug, *Theognis gonagia,* the yellow beetle, *Costalimaita ferruginea,* and the fruit-sucking bug, *Helopeltis antonii,* feed on ripe fruits. A false spider mite, *Brevipalpus phoenicis,* causes surface russeting beginning when the fruits are half-grown.

Fruit russeting and defoliation result also from infestations of red-banded thrips, *Selenothrips rubrocinctus.* The coconut mealybug, *Pseudococcus nipae,* has been a serious problem in Puerto Rico but has been effectively combatted by the introduction of its parasitic enemy, *Pseudaphycus utilis.*

Soil-inhabiting white grubs require plowing-in of an approved and effective pesticide during field preparation in Puerto Rico. There are other minor pests, but the great problems wherever the guava is grown are fruit flies.

The guava is a prime host of the Mediterranean, Oriental, Mexican, and Caribbean fruit flies, and the melon fly–*Ceratitis capitata, Dacus dorsalis, Anastrepha ludens, A. suspensa,* and *Dacus cucurbitae.* Ripe fruits will be found infested with the larvae and totally unusable except as feed for cattle and swine.

To avoid fruit fly damage, fruits must be picked before full maturity and this requires harvesting at least 3 times a week. In Brazil, choice, undamaged guavas are produced by covering the fruits with paper sacks when young (the size of an olive). Infested fruits should be burned or otherwise destroyed. In recent years, the Cooperative Extension Service in Dade County, Florida, has distributed wasps that attack the larvae and pupae of the Caribbean fruit fly and have somewhat reduced the menace.

In Puerto Rico, up to 50% of the guava crop (mainly from wild trees) may be ruined by the uncontrollable fungus, *Glomerella cingulata,* which mummifies and blackens immature fruits and rots mature fruits. *Diplodia natalensis* may similarly affect 40% of the crop on some trees in South India.

Fruits punctured by insects are subject to mucor rot (caused by the fungus, *Mucor hiemalis)* in Hawaii. On some trees, 80% of the mature green fruits may be ruined.

Algal spotting of leaves and fruits (caused by *Cephaleuros virescens)* occurs in some cultivars in humid southern Florida but can be controlled with copper fungicides. During the rainy season in India, and the Province of Sancti Spiritus, Cuba, the fungus, *Phytophthora parasitica, is* responsible for much infectious fruit rot. *Botryodiplodia* sp. and *Dothiorella* sp. cause stem-end rot in fruits damaged during harvesting. *Macrophomina* sp. has been linked to fruit rot in Venezuela and *Gliocladium roseum* has been identified on rotting fruits on the market in India.

In Bahia, Brazil, severe deficiency symptoms of guava trees was attributed to nematodes and nematicide treatment of the soil in a circle 3 ft (0.9 in) out from the base restored the trees to normal in 5 months. Zinc deficiency may be conspicuous when the guava is grown on light soils. It is corrected by two summer sprayings 60 days apart with zinc sulphate.

Wilt, associated with the fungi *Fusarium solani* and *Macrophomina phaseoli*, brings about gradual decline and death of undernourished 1-to 5-year-old guava trees in West Bengal. A wilt disease brought about by the wound parasite, *Myxosporium psidii*, causes the death of many guava trees, especially in summer, throughout Taiwan.

Wilt is also caused by *Fusarium oxysporum* f. *psidii* which invades the trunk and roots through tunnels bored by the larvae of *Coelosterna* beetles. Anthracnose (*Colletotrichum gloeosporioides*) may attack the fruits in the rainy season. *Pestalotia psidii* sometimes causes canker on green guavas in India and rots fruits in storage.

Severe losses are occasioned in India by birds and bats and some efforts are made to protect the crop by nets or noisemakers.

Bibliography

Alfred Steferud: *Diseases of Fruits and Nuts*, Biotech Books, Delhi, 2005.

Anderson, E. F.: *The Cactus Family*, Timber Press, Portland, Oregon, 2001.

Andrews L.: *Citrus Production - Orange*, St. Augustine, Trinidad and Tobago, 1990

Arora, Chandralekha : *Advances in Horticulture Science Research*, Oxford Book Company, Delhi, 2009.

Benson, L. H.: *The Cacti of the United States and Canada*. Stanford University Press, Stanford, California, 2001.

Benu Singh: *A Modern Book on Forestry and Horticulture*, Vista International Pub, Delhi, 2010.

Bhandari, M.M.: *The Flora of the Indian Desert*. Scientific Publisher, Jodhpur, 1990.

Bhattacharjee, Supriya Kumar: *Advances in Ornamental Horticulture*, Pointer, Delhi, 2006.

Bhutani, R.C.: *Fruit and Vegetable Preservation*, Biotech Books, Delhi, 2003.

Chadha, K.L. : *Advances in Horticulture*, Malhotra Publishing House, Delhi, 2006.

Collymore L.: *Fruit Production in Barbados*, Port of Spain, Trinidad and Tobago, 1996.

Dabholkar, A.R. : *General Plant Breeding*, Concept, Delhi, 2006.

Doijode S. D.: *Seed Germination in Fruits*, New Delhi, Malhotra Publishers, 1993.

Goswami, C L : *Plant Physiology : Research Methods*, Scientific, Delhi, 2007.

Julia F.: *Fruits of Warm Climates*, Miami, Julia F. Morton Publisher, 1987.

Kumar Supriya : *Amenity Horticulture, Biotechnology and Post-Harvest Technology*, Pointer, Delhi, 2006.

Little, E.L.: *Common Fuelwood Crops*, Communi-Tech Assoc., Morgantown, West Virginia, 1983.

Malins A.: *Postharvest Handling of Pineapple and Mango,* Port of Spain, Trinidad and Tobago, 1992.

Mitra S.: *Postharvest Physiology and Storage of Tropical and Subtropical Fruits,* Oxon, CABI, 1997.

Morton, Julia F.: *Fruits of Warm Climates.* Wipf and Stock Publishers, 1987.

Pareek O. P.: *Advances in Horticulture: Fruit Crops,* New Delhi, Malhotra Publishing House, 1993.

Paul B. : *Food Biotechnology in Ethical Perspective,* Aspen, CO: Aspen Publishers, 1997.

Peter, K.V. : *Basics of Horticulture,* New India, Delhi, 2009.

Popenoe, W.: *Manual of Tropical and Subtropical Fruits.* Hafner Press, 1974.

Proctor F. J.: *Onions in Tropical Regions,* Kent, Natural Resources Institute, 1990.

Purohit, S.S. : *Organic Farming : Biocontrol and Biopesticide Technology,* Agrobios, Delhi, 2008.

Saini, M.L. : *Plant Breeding and Crop Improvement,* CBS, Delhi, 1997.

Saroj, P.L. : *Advances in Arid Horticulture,* International Book Dist, Delhi, 2004.

Sharma, V.K. : *Advances in Horticulture : Strategies, Production, Plant Protection and Value Addition,* Deep and Deep, Delhi, 2004.

Singh Z.: *Propagation and Rootstocks of Citrus,* New Delhi, Malhotra Publishers, 1993.

Thomas E.: *Fruit Production in St. Kitts and Nevis,* Port of Spain, IICA, 1996.

Tyagi, I.D. : *Plant Breeding and Genetics at a Glance,* South Asian, Delhi, 2005.

Index

H

I

L

M

N

O

P

R

S

T

W

Y

Handbook of Biological Control and Horticultural Crops

Handbook of
Biological Control and Horticultural Crops

Vol. 3

Dr. J.S. Bohra

AGROTECH PRESS

JAIPUR ■ NEW DELHI

AGROTECH PRESS

Regd. Off. "Bony Residency", Gate No. 2
Opp. Tilak Public School, Vishweswariya Nagar,
Gopalpura Road, Jaipur - 302018, India
Ph: 0141 - 2761280, Telefax: 2761381
E-mail : agrotechpress@gmail.com

Delhi Office: 102, 1st Floor, Satyam House,
4327/3, Ansari Road, Darya Ganj,
New Delhi - 110002, India
Ph: 011-45652440
visit us: www.agrotechpress.com

Distributed By
KEYA INTERNATIONAL PRESS LLP
E-mail: keyainternationalpressllp@gmail.com

Published in India, 2025

ISBN : 9789384568559 (3 Vols. Set)

Typeset by Thomson Press

Printed and bound in India by Thomson Press Ltd.,
New Delhi

Contents

Preface

With the advent of synthetic insecticides, easy control of insect pests appeared at hand. However, it soon became obvious that there were problems associated with the use of insecticides. Some insect pests became resistant, and some non target organisms were adversely affected, and pest resurgence occurred. Additionally, environmental and health concerns arose. Today, the protection of food and fibre crops from insect, mite, disease, and weed pests in conventional agricultural systems still relies primarily on the use of chemical pesticides. However, continued reliance solely on conventional pesticides has drawbacks. The need to develop alternatives to conventional pesticides may be more acute in some cases than in others, for e.g. that of nematodes. Biological pest control provides an important route to environmentally harmless plant protection. This route reduces risks, e.g. of becoming aware too late of unfortunate side effects of plant protection agents. To realize the practical potential of biocontrol for agricultural production, it will be imperative to unite knowledge of mechanistic interactions with an appreciation of the complexity of the agroecosystem. Understanding the recognition, signaling, and cooperative and antagonistic interplay between the biological partners will lead to strategies to direct the outcome of the interaction more precisely and consistently. Knowledge of the events leading to disease control may suggest modifications in the timing, placement, and formulation of the biocontrol agent to achieve maximum disease control. Similarly, that knowledge may indicate situations in which certain biocontrol agents will not be successful and may thus lead to wiser choices to tailor the agent to the physical, chemical, and biological context into which it will be introduced.

Nematode species can be difficult to distinguish, and although over 25,000 have been described, of which more than half are parasitic, the total number of nematode species has been estimated to be about 1 million. Depending on the species, a nematode may be beneficial or detrimental to plant health. From agricultural and horticulture perspectives, the two categories of nematodes are the predatory ones, which will kill garden pests like cutworms and corn earworm moths, and the pest nematodes, like the root-knot nematode, which attack plants, and those that act as vectors spreading plant viruses between crop plants. Predatory nematodes can be bred by soaking a specific recipe of leaves and other detritus in water, in a dark, cool place, and can even be purchased as an organic form of pest control. Nematodes are usually considered pests because of the diseases they cause in humans and animals and the economic impact they have on many agricultural products. There are, however, a small but significant number of beneficial entomogenous nematodes, i.e., nematodes associated with insects. In addition to insects, nematodes can parasitize spiders, leeches, annelids, crustaceans and mollusks. Some of these entomopathogenic nematodes are of considerable interest because of their potential as biological control agents of pest insects.

Considering the limitations of chemical control of plant nematodes, there has been a growing awareness to develop biological control measures which brings about a reasonably good degree of reduction in nematode population and at the same time ensure sustainability of production, cost effectiveness and healthy environment. Commercial scale production of biopesticides for the biological control of plant nematodes in the recent past has been quite intense. Several package of practices have been developed nationally and internationally for nematode management using commercially available bio-inputs especially microorganisms. Widespread concern about the consequences of living nematodes and the growth of pigeonpea crop in conventional pesticide use has resulted in increased field soil. Natural products derived from neem including leaf, kernel and seed powders, seed extracts and oil have been reported to control several agricultural pests including plant-parasitic nematodes.

While it is unlikely that biological control will completely replace chemical pesticides in the foreseeable future, we can expect that there will be some decline in the use of chemicals, particularly in developed countries. Thus far, most approaches have involved the single antagonist concept, although a biological systems approach, where disease is suppressed from several angles, might provide a better alternative. Similarly, the use of biological control agents could be used as one component of an integrated management program to achieve the best possible results. The future of biological control is promising, but this tactic will constitute just one of many pest management options. Many obstacles will need to be overcome before biological control can reach its full potential. Traditionally, educational and extension efforts have not provided growers with the necessary information to help identify natural enemies or determine their presence in fields. Therefore this book is an earnest effort to collect relevant information related to biomanagement of nematodes that should be accessible to bio-practitioners, agriculture scientists, botanists etc.

Dr. J.S. Bohra

1

Introduction

The nematodes or roundworms constitute the phylum Nematoda. They are a diverse animal phylum inhabiting a very broad range of environments. Nematode species can be difficult to distinguish, and although over 25,000 have been described, of which more than half are parasitic, the total number of nematode species has been estimated to be about 1 million. Unlike cnidarians and flatworms, nematodes have tubular digestive systems with openings at both ends.

The richness and variety of life in soil and debris is remarkable and a few grams of soil may contain microscopic life forms numbered in the millions. These include fungi and bacteria as major components and also large numbers of protozoans, tardigrades, rotifers, nematodes and other tiny things. Competition for nutrients and the close physical proximity amongst different groups of organisms in this environment has been conducive to the development of many predatory and parasitic associations.

The relationship between fungi and nematodes is not only one of the most fascinating but it has led us unexpectedly in a biological direction that is beginning to explain more clearly the complexities of wood decay and the CARBON CYCLE. You should understand right away that most introductory biology texts have a pretty poor version of the various biological components involved in the carbon cycle and the roles they play. Fungi can capture nematodes in a variety of ways but the most sophisticated and perhaps the most dramatic is called the constricting ring. An erect branch from a hypha curves round and fuses with itself to form a three-celled ring about 20-30 microns in diametre (30/1000th mm). When a nematode "swims" into a ring it triggers a response in the fungus and the three cells expand rapidly inwards with such power that they constrict the body of the nematode victim and hold it securely with no chance to escape. It takes only 1/10th of a second for the ring cells to inflate to their maximum size.

Remarkably, the constricting ring cells go to three times their volume with a two times increase in surface area of wall and membrane in the blink of an eye. At the same time the osmotic pressure (OP) of the cell is apparently maintained to the point where it can crush the nematode.

Transmission Electron Microscopy by others has shown that there is preformed reserve wall material and preformed reserve membrane material located at the inside wall of each ring cell. Others have also shown that you can "tickle" the inside walls of the ring with a needle and this mechanical stimulation can trigger the ring. It has also been also shown that the rings could be artificially triggered by heat stimulation (e.g. hot scalpel in the vicinity of the ring).

Our scanning electron microscopy studies, resulting in the picture shown above, showed that there is a rip in the outer wall of the ring cells, and that a new wall has expanded into the centre of the ring.

The above results suggested that there is a "hair trigger" line of weakness running along the inside wall of the constricting ring cells. When this is touched by a nematode or any other mechanical force the cell is triggered with the following sequence of events.

1. The line of weakness breaks.
2. The osmotic pressure of the cell results in a rapid intake of water over the surface of the cell.
3. The cell expands in the direction of the break with a resulting 3X increase in cell volume.
4. The reserve wall and membrane materials are located at the point of the break and reorganize to form new wall and membrane.
5. The three times increase in volume of the ring cells must result in a three times decrease in osmotic pressure of the ring cells at the time of expansion.
6. The ring cells "cushion" around the body of the nematode and hold it fast but do not constrict it at that point in time.

7. There is a rapid conversion of reserve materials in the rings cells to allow the OP of the cell to recover. When the OP of the rings cells exceeds that of the nematode cells then the nematode wall will collapse and the body will be constricted.
8. In its struggles to escape the nematode will often put its tail into a second ring and be held imobile.
9. Once captured the rings cells will germinate and invasive hyphae penetrate into the living body of the nematode, grow throughout the body and digest the contents. This will be accomplished within 12-24 hours. The food gained from this will be translocated elsewhere for further growth and production of more traps or for producing the reproductive bodies of the fungus which in this case is called *Arthrobotrys anchonia.*

Both of these nematode groups carry within their bodies insect-pathogenic bacteria — Xenorhabdus in the case of steinernematids and Photorhabdus in the case of heterorhabditids. Steinernematid and heterorhabditid nematodes are termed entomopathogenic because of their association with these bacteria.

The relationship between the nematodes and bacteria is a true obligate mutualism because the bacterium needs the nematode to carry it into the insect body cavity. The nematode needs the bacterium to create conditions in the insect suitable for its reproduction and growth, and as food. The bacteria are safe to vertebrates and only occur in association with these nematodes and infected insects. They have never been detected living freely in soil. The bacteria produce pigments, so that insects infected with heterorhabditid nematodes turn a brick-red or maroon colour, and those infected with steinernematids turn ochre, tan, or brown. Nematode-infected insect cadavers do not smell putrid, and the insect cuticle stays intact until very late in the infection process.

Life History

The infective third-stage juvenile (sometimes referred to as dauer or IJ) is the only life stage of the nematode that exists outside of the host insect. The IJ is the stage that is purchased in commercial products. The IJ is nonfeeding and is more resistant than other stages to environmental conditions. It carries the bacteria in its intestinal tract. Upon locating a suitable insect host, the infective juvenile enters through natural openings (anus, spiracles, mouth) of the insect and penetrates into the insect body cavity. There it releases its bacteria, which multiply and kill the host insect.

The nematodes develop into adults on bacterial cells in the insect cadaver and reproduce. As resources of the insect are depleted and crowding occurs, IJ are produced. The IJ emerge from the cadaver to search for new insect hosts in the soil. The reproductive potential of entomopathogenic nematodes is very high. Thousands of nematodes can be produced from a single infected insect host. The time from infection of the insect until infective juveniles emerge takes about two weeks in the laboratory. Under natural conditions the recycling time may vary depending on environmental conditions and the susceptibility of the host insect.

Life Cycle of Insect-Parasitic Nematodes

1. Infective juvenile nematodes in soil enter insect body through natural openings.
2. Nematodes enter insect body cavity.
3. Nematodes develop into adults.
4. Nematodes reproduce and produce offspring.
5. Infective juvenile nematodes leave the dead insect and seek a new insect host.

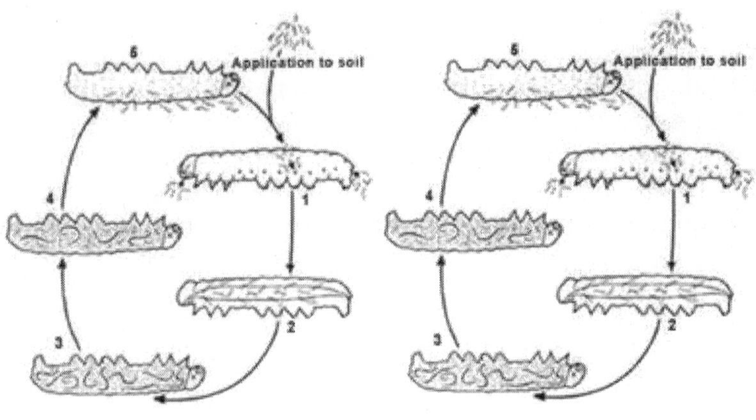

Figure: *Life Cycle of Insect-Parasitic Nematodes*

Application of Insect-Parasitic Nematodes

Some nematodes that are commercially available are *Steinernema carpocapsae*, *S. feltiae*, *S. riobrave* , *Heterorhabditis bacteriophora*, *H. marelatus*, and *H. megidis* . These nematodes are most commonly used for management of soil insect pests in high value crops-for example, in home lawns and gardens, turf, nurseries, citrus, cranberries, and mushrooms. Because nematodes are living organisms, their successful use is influenced by environmental conditions. Nematodes need adequate moisture, temperatures within the tolerance levels for the specific nematode, and protection from UV radiation (direct sunlight) during application.

These nematodes, like most soil nematodes, are actually semi-aquatic. Their natural home is in the water film that surrounds soil particles. The IJ are sensitive to destruction by bright sunlight and desiccation if they are sprayed on plant foliage unless they are especially formulated for that use. The most common usage is in soil, although in some instances they have been successfully applied above-ground to insect tunnels or mines in plant tissue. Nematodes are formulated as suspensions in liquid, on sponge, in gels, or as semidry granules. The main application approach for use is as an aqueous suspension at a typical rate of 2.5 billion/hectare (1 billion/acre), but this rate varies depending on the crop. They can be applied with conventional chemical application equipment, but nozzle filters or screens smaller than 50 mesh will clog and it is best to remove screens in nozzles when applying nematodes with a back-pack sprayer or spray rig. Care should be taken when using hydraulic pumps that have high internal pressure and high shear force as these will shred the nematodes.

Nematodes tend to settle in the tank, so agitation must be provided for uniform application. Nematodes can be killed by excessive tank agitation through sparging (recirculation of a portion of spray mix) or excessive mechanical stirring that is used to keep the nematodes in suspension. Pump pressure and temperature above 300 pounds per square inch and 85°F, respectively, will kill nematodes.

It is best to apply entomopathogenic nematodes to moist soil in the early morning or late evening when air temperatures are between 60 and 85°F. A pre-application irrigation can be applied to moisten the soil and a post-application irrigation can be applied to wash any nematodes on plant surfaces to the soil surface. The post-application irrigation should be applied before spray droplets dry and must provide sufficient (0.1-0.25 inches) water to allow the nematodes to move into the upper soil layers, out of sun or drying air exposure. Applications can be made before or even during a rainfall to wash nematodes to the soil surface. Successful application of nematodes is influenced by spray volume. Most nematode labels suggest volumes of two to six gallons of spray per 1000 square feet (87-260 gallons per acre). This is achievable for many boom sprayers and lawn shower nozzle sprayers that are equipped with sufficiently large nozzles. Some turf applicators use shower nozzles that deliver 1-1.5 gallons of spray per 1000 square feet.

When lower spray volumes are used, pre- and post-application irrigation can be adjusted to counteract the problem of low volume sprays and to assist in moving the nematodes to the soil and off exposed surfaces.

Nematodes can also be applied with irrigation. However, some irrigation systems, especially low volume trickle systems, may not move water fast enough to keep nematodes suspended. When in doubt, check periodically by taking a sample at the emitters to determine if live nematodes are being moved through the system.

Conservation of Entomopathogenic Nematodes

Entomopathogenic nematodes occur worldwide, and they have been found throughout the U.S. from many different soil types and habitats, both natural and managed. Native nematodes may play an important role in the regulation of insect populations in some systems, but the level of disturbance in agricultural systems may require the use of non-native nematodes that are tolerant to practices being used. Some studies suggest that entomopathogenic nematodes are more abundant in minimum- or no-till plots than in conventionally-tilled plots, and in less disturbed systems, such as orchards or woodlots. Different nematode species are differentially tolerant to soil disturbance. Survival of entomopathogenic nematodes is higher in mulched soil than in bare soil, probably related to soil environmental factors — for example, soil structure, temperature and moisture.

Research has been conducted to determine the effects of pesticides on entomopathogenic nematodes. Some pesticides and fertilizers are less harmful to entomopathogenic nematodes than others. Compatibility with chemicals is usually provided on product labels or packaging information that accompanies purchased nematodes.

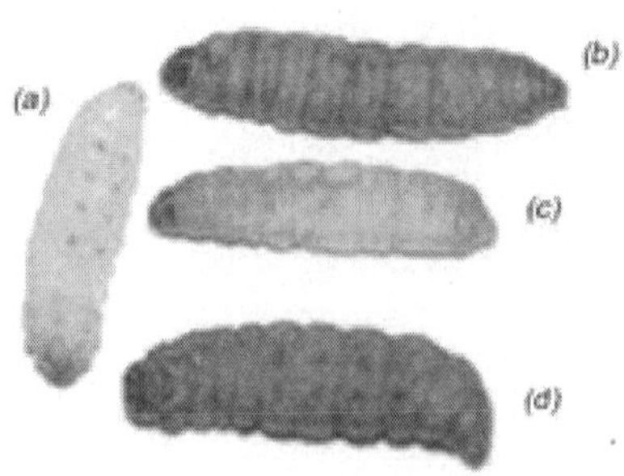

Figure: *(a) Healthy (cream); (b)* Steinernema feltiae-*infected (brown); (c)* Steinernema carpocapsae - *infected (tan) and (d)* Heterorhabditis bacteriophora - *infected (red) larvae of* Galleria mellonella.

Efficacy of Entomopathogenic Nematodes

To date, levels of control achieved by the application of nematodes has been mixed, with some great successes - mostly in controlled systems — e.g. nursery containers, mushroom houses — and some failures. The efficacy of nematodes is affected by environmental conditions — they need adequate but not excessive moisture, temperatures within the tolerance levels for the particular nematode, and protection from UV radiation during application (apply early in morning or in the evening). Most failures in efficacy of field applications are related to a poor match between the nematode species and target insect pest. Species of nematodes vary in their host range and in their host-finding behaviour. Some nematodes, for example, Steinernema glaseri and Heterorhabditis bacteriophora, are very active in the soil and search a relatively large area for a host insect, whereas the widely available nematode, Steinernema carpocapsae, is relatively sedentary and tend to sit and wait for a host insect to pass by in close proximity. Steinernema carpocapsae is classified as an ambusher and is most suitable for mobile pests, for example armyworms and cutworms. Steinernema riobrave moves well through the soil. It was originally found in the Rio Grande Valley of Texas, is adapted to warm soils and is more tolerant to tilled soils than are some other nematode species. Heterorhabditis bacteriophora is highlyparasitic on some lepidopteran andcoleopteran larvae. It is classified as a cruiser and is effective against sedentary insect hosts, for example, white grubs. Heterorhabditis megidis is a parasite of black vine weevil and various other soil insects.

A nematode that is an active searcher (cruiser) will be more effective at finding a sedentary insect host, for example white grubs, than will a sedentary nematode (ambusher). The relatively sedentary nematodes are effective at infecting active insect hosts, such as cutworms or mole crickets. Some known appropriate pathogen-host targets are S. glaseri against the Japanese beetle; S. scapterisci against mole crickets; S. riobrave against cutworms and other noctuid larvae, pupae and citrus weevils; and S. feltiae against sawfly larvae and fungus gnat larvae.

As with any purchased natural enemy, quality of the product can affect efficacy. Quality of the product can be affected by batch, and shipping, storage, and application conditions. Nematodes are living organisms and are subject to destruction by excessive cold or heat, and lack of moisture or oxygen. A small sample of the mixed product should be checked with a hand lens or magnifying glass to observe living, moving nematodes. Nematodes that are very straight and motionless may be dead, and therefore, ineffective.

Nematodes Are Highly Diverse, Complex and Specialized Metazoans

Nematodes are the haiku among multicellular animals, combining endless variation with a deceptively simple underlying anatomical pattern. In the search for maximally informative models, this reputation for underlying simplicity has sometimes caused C. elegans *to be erroneously described as a typical representative of all nematodes, or as a typical example of a simple metazoan. However, simplicity lies very much in the eye of the beholder: nematodes are highly diverse in almost every respect, including for example morphology. A small sampler of some of the many other kinds of remarkable nematode adaptations will further illustrate this point.*

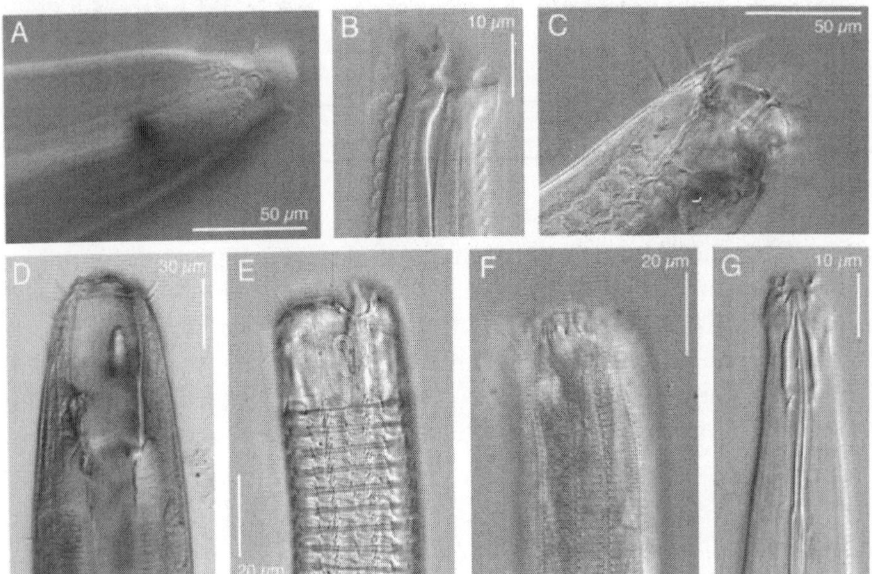

Figure: *Examples of divergence in anterior morphology of some freeliving nematodes. A.* Thoracostoma *sp (Enoplina). B.* Acromoldavicus mojavicus *(Tylenchina: Cephalobomorpha). C.* Enoploides *sp. (Enoplina). D.* Pontonema *cf.* parpapilliferum *(Oncholaimina). E.* Ceramonema *sp. (Plectida). F.* Latronema *sp. (Chromadorida). G.* Actinca irmae *(Dorylaimida). Click on a picture to open a small video clip (200–600 Kb), or on a letter to open a large clip (2–5 Mb) of the depicted nematode. Use the left and right arrow keys on your keyboard to focus up and down. These clips were produced with Video Capture and Editing microscopy as described in De Ley & Bert.*

Two *Cryonema* species permanently live at freezing point inside lacunae in arctic ice; one of them preys on other lacunary nematodes. Nematodes of the subfamily Stilbonematinae are covered with a dense "fur" of species-specific ectosymbiotic sulfur-oxidizing bacteria, allowing them to thrive at the redox boundary layer in sulfur-rich marine sediments. *Oncholaimus* mates by traumatic insemination: males inject sperm through the female cuticle, females then develop specialized internal structures for sperm transfer to the reproductive system - or for evacuation of excess sperm into the intestine. The entomopathogen *Steinernema tami* produces dimorphic sperm, with 50–100 ìm wide megaspermatozoa functioning as self-propelled spermatophores carrying the 2 ìm wide microspermatozoa on their surface. *Mehdinema alii* uses male crickets as vectors for transmission between female cricket hosts, female nematodes give birth to fully formed dauers while males have a motile copulatory claw extruding through a separate postcloacal opening (Luong et al., 2000). The millipede gut inhabitant *Zalophora* is an intra-intestinal predator and cannibal of other gut nematodes.

Some recent discoveries in nematodes extend the range of biological processes and properties known for all Metazoa. For example, mitochondrial mRNA transcription in *Teratocephalus lirellus* involves insertional editing of polyA motifs, a process otherwise only known in protists. Mitochondrial DNA of the potato cyst nematode *Globodera pallida* is arranged in at least six different mini-circles, each of which carries only a subset of the mitochondrial genome, resembling conditions occurring in some plants and fungi but not other metzoans. Conversely, closer scrutiny of increasingly distant relatives of *C. elegans* has revealed a number of aspects in which some nematodes are much less different from other Metazoa than *C. elegans* is. A very recent case in point is the discovery that early development in *Tobrilus diversipapillatus* passes through a classical metazoan coeloblastula stage, unlike any other nematode species studied so far. Clearly, our appreciation of nematode diversity will continue to change as we explore the biology of more species that have so far received little attention, and as we encounter more of the potentially vast number of unknown species.

Changing perspectives in nematode phylogeny and classification

In the face of overwhelming diversity, a phylogenetic framework is needed to underpin meaningful comparisons across taxa and to generate hypotheses on the evolutionary origins of interesting properties and processes. Our understanding of nematode relationships has a varied and at times turbulent history, reflecting not only wider developments in phylogenetics, but also the expertise and perspectives of those few systematists who produce comprehensive classifications. Until recently, the data for most nematode phylogenies consisted of relatively few morphological characters derived primarily from light microscopy and often by individual effort of the lone taxonomist. Molecular phylogenetics, bioinformatics and digital communication technologies have substantially altered the dynamics of nematode systematics, creating conditions where collaborative strategies are much more productive than individual effort. This approach was exemplified by the analyses of Blaxter et al. (1998) based on small subunit (SSU) rDNA sequences of 53 nematode species.

Just seven years later, SSU rDNA sequences are available in public databases for more than 600 nematode species. The basic topology obtained by Blaxter et al. appears to remain quite robust, although a number of important groups remain to be included. Also, it is increasingly clear that some important aspects of nematode phylogeny cannot be resolved by SSU data alone. Two of the major features of SSU analyses have independently confirmed suspicions held by several proponents of earlier morphological systems, i.e., that many important anatomical features have arisen repeatedly during evolution, and that one of the two traditional classes (Secernentea) is deeply phylogenetically embedded within the other (Adenophorea). In an effort to translate the implications of SSU rDNA sequences into classification, De Ley & Blaxter (2002, 2004) proposed a system based primarily on the molecular backbone of SSU phylogenies, but also incorporating other characters. The result combines elements from many previous systems, and introduces some new features with respect to ranking (e.g., use of infraorders). For the sake of convenience, we will follow the nomenclature of this system here, in order to outline the major features of SSU-based nematode phylogenies.

The Backbone of the Nematode Tree

The molecular data published to date confirm the presence of three early nematode lineages, corresponding to the previously recognized subclasses Chromadoria, Dorylaimia and Enoplia. The exact order of appearance of these three lineages is not yet resolved. It seems likely that Enoplia appeared first, and it is even possible that Dorylaimia and/ or Chromadoria could have originated within Enoplia. On the other hand, SSU data also allow for the possibility that Dorylaimia diverged first, which is an intriguing possibility because all known Dorylaimia are absent from marine habitats. A "Dorylaimia first" topology would therefore imply that the ancestor of all nematodes was perhaps a freshwater organism, and not a marine animal as more commonly assumed. Within Chromadoria, a number of clades have arisen in a series of successive bifurcations and radiations. These clades are classified as separate orders, including for example the predominantly marine Chromadorida and Desmodorida, as well as orders that have also diversified extensively in freshwater sediments (e.g., Monhysterida) and moist soils (e.g., Plectida). In addition, the chromadorian monophylum also includes the clade formerly ranked as class Secernentea, a hugely successful radiation of predominantly terrestrial nematodes. SSU phylogenies place this taxon at the crown of Chromadoria and as sister group to the order Plectida. For this reason, De Ley & Blaxter, classified it instead as order Rhabditida, thereby greatly expanding the contents of this taxon compared to all previous systems.

Major Features of Enoplea

The phylum Nematoda occurs in an incredibly wide spectrum of ecological habitats and natural histories, ranging from e.g., deep sea sediments to arid deserts, or from interstitial bacterivores to obligate parasites with multiple intermediate hosts. Several of its constituent clades cover a large subset of this ecological spectrum, but interestingly none of them appears by itself capable of covering the full ecological range of the phylum, especially so within the class Enoplea.

This suggests that the evolution of ecological adaptations within each nematode taxon was constrained by limitations on the rates of change in genes and ecophysiology, or by competitive exclusion from habitats previously colonized by other taxa, or both.

The non-marine occurrence and the present diversity of Dorylaimia both suggest that these could have been the first nematodes to conquer freshwater and terrestrial habitats. Such an early origin could explain the great diversity within this subclass, which includes not only the mostly freeliving Doryaimida and Mononchida, but also remarkable animal parasites such as Mermithida and Trichinellida. Mermithids are highly unusual among metazoan parasites in that they actually leave the host before reaching adulthood - a property otherwise only found in the phylum Nematomorpha.

The present existence of such exceptional biologies hints at much greater past diversity. The most successful surviving clade within Dorylaimia, however, is the order Dorylaimida. This includes many species of large predators/omnivores, as well as the plant-parasitic family Longidoridae, of which some species transmit plant viruses. The evolutionary radiation of dorylaims appears to have resulted in large part from functional diversification of the odontostyle, a protrusible, hollow and often needle-like tooth used for puncturing and emptying food items. Although predation or plant feeding are well documented for larger dorylaims, the food sources of most smaller species (with much smaller odontostyles) remain unknown.

Enoplia are especially diverse in marine habitats, but multiple lineages are also found in freshwater sediments and/or moist soils. One of these lineages includes marine, freshwater and terrestrial taxa, suggesting that early Enoplia were characterized by much greater osmotic tolerance than early Dorylaimia. Most enoplian clades include large predators with big hooks or teeth in more or less complex arrangements, as well as interesting sensory structures such as eyespots and a unique type of stretch receptors (= metanemes; Lorenzen,. Enoplia are especially interesting phylogenetically because of the occurrence of features that are presumably ancestral within nematodes, such as a highly indeterminate mode of development and retention of the nuclear envelope in mature spermatozoa. No Enoplia are known to have adapted to terrestrial environments subject to extreme temperatures, nor are there any surviving lineages that parasitize animals (with the possible exception of a few enigmatic taxa of uncertain position). The one enoplian order that has clearly undergone extensive evolution in soils is the order Triplonchida, which includes plant parasites such as *Trichodorus*. These are convergent with dorylaims in a number of respects, e.g., they also have a protrusible tooth for feeding (called an onchiostyle) and several species are known to act as virus vectors. Molecular data have shown that the triplonchid clade includes freeliving nematodes such as *Tobrilus* and *Prismatolaimus*, even though these are morphologically quite divergent from trichodorids.

Features and Diversity of Chromadorea

The Chromadoria include at least four lineages that have attained greater habitat breadth than all Enoplia and most Dorylaimia. Chromadoria are common in marine sediments, but they have also flourished in terrestrial habitats e.g., those subject to frequent episodes of rapid de- and rehydration, such as mosses and lichens or extremely xeric and/or cryogenic soils.

Throughout Chromadoria, cuticular structure has undergone a wide range of evolutionary modifications, sometimes resulting in strikingly decorative ornamentations. A key adaptation within the order Rhabditida was the development of a chemically impermeable cuticle, which clearly contributed to their success as parasites, colonizers and extremophiles. Another subject of striking modification in Chromadoria is the pharynx, which is structurally much more diverse in this subclass than in Enoplia or Dorylaimia. This diversity revolves mostly around the evolution of one or more rounded muscular bulbs, which has apparently allowed for more compact body designs. Freeliving Chromadoria are on average noticeably smaller than Enoplia and Dorylaimia, correlated with a greater preponderance of rapidly reproducing bacterial feeders. As a possible corollary, many species (such as *C. elegans*) have also evolved compressed generation times and other adaptations for rapid dispersal and efficient colonization of eutrophic environments. Several chromadorian lineages have independently evolved curved, swiveling teeth in can-opener-like arrangements. These are used to e.g., pry open the silicate frustule of diatomaceous algae, or to slice the cuticle of other nematodes. Some of these species are among the smallest known predatory nematodes. The most distant relatives of *C. elegans* that can presently be efficiently cultured with *C. elegans*-like methods are certain bacterivorous species of the orders Monhysterida and Plectida.

Within the order Rhabditida, a major factor has been the development of a modified juvenile stage specifically adapted to long-term survival. A true dauer stage is rarely reported outside of the suborder Rhabditina (which includes *C. elegans*) but a "proto-dauer" actually occurs in several other lineages, including some panagrolaims as well as the morphologically bizarre genus *Myolaimus*. Appearance of a proto-dauer in the early evolution of Rhabditida probably set the stage for subsequent specialization into the non-feeding, highly modified and highly dispersive dauer stage of Rhabditina. True dauers are often capable of seeking out and hitching rides phoretically on larger animals, which has in turn allowed multiple invasions of the internal organs of other animals. At least three Rhabditida lineages have independently evolved major zooparasitic radiations. A fourth lineage has not only given rise to zooparasitic species, but also radiated into the most diverse group of plant parasites and fungal feeders among nematodes. These are the tylenchs, equipped with a protrusible stomatosylet that is convergent with, but clearly different from, the odontostyle of dorylaims and the onchiostyle of trichodorids. SSU sequences have confirmed the previously unpopular hypothesis that their closest relatives are the morphologically very dissimilar cephalobs. Both groups are therefore now united in the suborder Tylenchina - another example of a more drastic change induced by the new phylogenies.

Analogous to zooparasitic Rhabditida, the life cycle of some parasitic tylenchs includes an infective juvenile stage. However, most tylenchs and cephalobs do not have one single dispersive and enduring stage, but are instead capable of surviving harsh conditions throughout most of their life cycle. This has enabled them to compete very successfully with dorylaims in even the driest and coldest terrestrial environments. One secret to this success could be constitutive postembryonic expression of traits that first evolved and appeared in the proto-dauer stage only: tylenchs and cephalobs are in some respects quite dauer-like throughout their life cycles. Interestingly, some cephalobs have also developed strikingly complex labial structures, presumably for collecting and ingesting scarce bacteria in poor soils.

The third major lineage within Tylenchina are the panagrolaims, a less clearly circumscribed amalgam of freeliving opportunists, fermentation specialists, insect pathogens and animal parasites. Although there are numerous morphological intermediates between panagrolaims and cephalobs, robust molecular resolution is still missing. One interesting character that supports monophyly of Tylenchina is axis determination in the early zygote, which does not depend on sperm entry in these nematodes. Parthenogenesis appears to be much more common in Tylenchina than hermaphroditism, and Goldstein et al. speculated that this could in fact be linked to the mechanism of axis determination.

Within the suborder Rhabditina, diversity segregates into four major groups: strongylids, diplogasterids, bunonematids and rhabditids (*sensu stricto*). The latter group includes *C. elegans* as well as many other species that differ morphologically in details of the male and female reproductive system. Strongylids were traditionally placed in their own order, on the basis of their importance as animal parasites and their morphological complexity as adults. However, they actually arose from within rhabditids, as is clear from SSU sequences, juvenile morphology and male genital characters. Compared to rhabditids, diplogasterids are characterized by a shift of pharyngeal pumping function to the median bulb, with concomitant muscle and valve reduction in the basal bulb. This arrangement is superficially very similar to the pharynx of Tylenchina, but ultrastructural and molecular data strongly indicate that the resemblance is purely convergent. The group includes the "satellite model" *Pristionchus pacificus*, in a colourful array of bacterivores, fungivores, animal parasites and "can-opener" type predators. The closest relatives of diplogasterids appear to be certain species that exhibit the morphology typical of rhabditids, but also the truly perplexing bunonematids. The latter include some of the most unusual anatomies among nematodes, with complex cuticular ornamentations arranged in dorsoventral symmetry, *i.e.* at right angles to the bilateral symmetry of the internal organs. They appear to be specifically adapted for life along surfaces within decomposing material, and provide a most appropriate conclusion to this chapter's quick tour through nematode diversity.

Important Nematode Pests

Nematodes are microscopic roundworms that live in many habitats. At least 2 500 species of plant-parasitic nematodes have been described, characterized by the presence of a stylet, which is used for penetration of host plant tissue. Most attack roots and underground parts of plants, but some are able to feed on leaves and flowers.

Plant-parasitic nematodes are of great economic importance. However, because most of them live in the soil, they represent one of the most difficult pest problems to identify, demonstrate and control. Their effects are commonly underestimated by farmers, agronomists and pest management consultants, but it has been estimated that some 10 percent of world crop production is lost as a result of plant nematode damage.

Although many nematodes have been found associated with small-grained cereals, only a few of them are considered economically important. Those of importance include: (i) cereal cyst nematodes, *Heterodera* spp.; (ii) root lesion nematodes, *Pratylenchus* spp.; (iii) root knot nematodes, *Meloidogyne* spp.; (iv) seed gall nematode, *Anguina tritici*; and (v) stem nematode, *Ditylenchus dipsaci*. Each of these is described and discussed below.

Management of nematodes may be approached by using a combination of methods in an integrated pest management system or may involve only one of these methods. Some of the most commonly practised methods will be discussed, including crop rotation, the use of resistant and tolerant cultivars, cultural practices and chemicals. It is important to stress that the most appropriate control method will be determined by the nematode involved and the economic feasibility of implementing the possible control(s). These will be discussed briefly for each nematode.

Cereal Cyst Nematodes

Distribution

The cereal cyst nematodes, *Heterodera* spp., are a group of several closely related species and are considered to be one of the most important groups of plant-parasitic nematodes on a worldwide basis. The most commonly recorded species of economic importance on cereals is *H.*

avenae, which has been detected in many countries, including Australia, Canada, Israel, South Africa, Japan and most European countries (Kort, 1972), as well as India and countries within North Africa and West Asia, including Morocco, Tunisia, Pakistan and Libya (Sikora, 1988), and recently Algeria and Saudi Arabia. Although its distribution is global, much of the research has been confined to Europe, Canada, Australia and India.

Heterodera avenae is the principal species on temperate cereals, while another important cereal species, *H. latipons*, is essentially only Mediterranean in distribution, being found in Syria, Israel, Cyprus, Italy and Libya (Kort, 1972). However, it is also known to occur in northern Europe. Other *Heterodera* species known to be of importance to cereals include: *H. hordecalis* in Sweden, Germany and the United Kingdom; *H. zeae*, which is found in India, Pakistan and Iraq; *H. filipjevi* in Russia and Turkey (Nicol *et al.*, unpublished data); and various others, including *H. mani*, *H. bifenestra* and *H. pakistanensis*, and an unrelated species of cyst nematode, *Punctodera punctata* (Sikora, 1988). Other cyst nematode species have been found on cereals, but they have not been shown to be economically important.

Most of these species are difficult to differentiate easily and require a strong taxonomic under-standing of morphological traits of cysts or juveniles. Recent molecular techniques, such as random fragment length polymorphism (RFLP) of the ribosomal DNA, have enabled solid taxonomic differentiation among several entities of the cereal cyst nematode complex.

Biology

The host range of *H. avenae* is restricted to graminaceous plants. There is sexual dimorphism with the male remaining worm-like, whereas the female becomes lemon-shaped and spends its life inside or attached to the root. The adult white female is clearly visible on roots with the swollen body, about 1 mm across, protruding from the root surface. Eggs are retained within the female's body, and after the female has died, the body wall hardens to a resistant brown cyst, which protects the eggs and juveniles. The eggs within the cyst remain viable for several years (Kort, 1972). *Heterodera avenae* has only one generation per year, with the hatch of eggs determined largely by temperature.

The symptoms produced on the roots are different dependent on the host. Wheat attacked by *H. avenae* shows increased root production such that the roots have a 'bushy knotted' appearance usually with several females visible at each knot as illustrated in Plate 55. Oat roots are shortened and thickened, while barley roots appear less affected. Other species of *Heterodera* also appear to produce host-specific symptoms on the roots of cereals. For example, in Israel *H. latipons* did not produce knotted roots as *H. avenae*. Above-ground symptoms of *H. avenae* appear early in the season as pale green patches of plants with fewer tillers. Patches may vary in size from 1 m^2 to 100 m^2 or more. In France, successful detection of *H. avenae* in wheat fields was achieved with the use of radiothermometry. It is possible that this technique could be extended to thermography, which could improve the detection of cereal cyst nematode attacks in large areas.

Heterodera avenae is the best known species, but is polymorphous with many pathotypes. The induction or suppression of dormancy (diapause) by different temperatures regulates the hatching of *H. avenae* juveniles. In Mediterranean climates, the diapause is obligate and durable, acting when the climate is hot and dry and being suppressed when the soil temperature falls and moisture rises. The diapause requirements in other climates with *Heterodera* species are less well understood but they are essential to understanding the biology and control of those species.

To date, the pathotypes of *H. avenae* have been recognized with the test developed by Andersen and Andersen designated The International Cereal Test Assortment for Defining Cereal Cyst Nematode Pathotypes, which has been modified by Rivoal and Cook . In these tests, it is quite difficult to make clear-cut distinctions between resistance and susceptibility based on the number of cysts alone. Further, pathotypes may also occur in mixtures, which complicates delineation of the pathotype in a particular sample.

Economic Importance

Heterodera avenae has been associated with economic levels of damage exclusively in light soils. However, it can cause economic damage irrespective of soil type when the intensity of cereal cropping exceeds a certain limit (Kort, 1972). Yield losses due to this nematode are: 15 to 20 percent on wheat in Pakistan (Maqbool, 1988); 40 to 92 percent on wheat and 17 to 77 percent on barley in Saudi Arabia; and 20 percent on barley and 23 to 50 percent on wheat in Australia.

Little is known about the economic importance of the species *H. latipons*, even though it was first described in 1969. Field studies in Cyprus indicated a 50 percent yield loss on barley. Because the cysts are similar in size and shape, it is possible that previous findings of this recently described nematode species have erroneously been attributed to the economically important *H. avenae*. In West Asia and North Africa, *H. latipons* has been found on wheat and barley in four countries. It has also recently been confirmed in Turkey (Nicol *et al.*, unpublished data). It has also been reported from several Mediterranean countries associated with the poor growth of wheat. Unfortunately, this nematode has not been studied in detail, and information on its host range, biology and pathogenicity is scarce; none-theless, it is suspected to be an important constraint on barley and durum wheat production in temperate semi-arid regions. Other cyst nematodes, such as *P. punctata* and *H. hordecalis,* have been described from roots of cereals in several countries, but their distribution and economic importance is unknown.

Control

One of the most efficient methods of controlling *H. avenae* is with grass-free rotations using non-host crops. In long-term experiments, non-host or resistant cereal frequencies of 50 percent (80 percent in lighter soils) keep populations below damaging thresholds. Clean fallow and/or deep summer ploughing reduce the population density of the nematode but are not always environmentally sound.

Cultivar resistance is considered one of the best methods for nematode control and has been found to be successful in several countries such as Australia, Sweden and France on a farm scale. However, it has also been observed that the use of resistance, especially derived from single dominant genes, may cause a disequilibrium in the biological communities and possibly ecological replacement with other nematodes, such as *Pratylenchus*. Another potential concern is the breakdown of resistance sources with repeated use. This has occurred in France with the resistant oat cultivar Panema and the appearance of a new *H. avenae* pathotype.

In order for cultivar resistance to be effective and durable, a sufficient understanding of the number of species and pathotypes within species is essential. The International Cereal Test Assortment for Defining Cereal Cyst Nematode Pathotypes offers classification of pathotype variation; pathotypes from Australia and India are often distinct from those in Europe. Although useful, a pathotype scheme for a species complex based on interaction with three cereal genera will not easily describe extensive variation in virulence. Furthermore, to date there are few molecular or other diagnostic methods that can provide consistent and reliable pathotype and pathogenicity differentiation.

Molecular technology has also been applied to identify markers for various cereal cyst nematode resistance genes using techniques such as RFLP and polymerase chain reaction (PCR) in both barley and wheat. Furthermore, many of the wild grass relatives have been introgressed into a hexaploid wheat background for breeding purposes. Many of these have had molecular work applied to identify the location and the possibility to produce markers to the known gene(s). More details about introgressions, substitutions and molecular characterization of these materials can be found in McIntosh *et al.* (2001). Some of these markers are actively being implemented in marker-assisted selection and pyramiding of gene resistance in Australian cereal breeding programmes against *H. avenae*, pathotype Ha13. This is an example where there is sufficient under-standing of the biology of the pathogen and genetic control of the resistance so that both conventional breeding and the modern tools of molecular biology can be combined to the advancement of controlling this disease. Such potential exists for other nematodes, but will require a similar understanding and combining of related skill base.

The utilization of these identified sources, and possibly of other as yet unidentified sources of resistance, is country-specific and dependent on the number and types of *Heterodera* species and pathotypes that need to be controlled.

Many developing countries unfortunately have limited resources and/or expertise to establish this information, and current control methods are based on understanding the response of local cultivars to the pathogen(s). For example, in Israel all locally grown wheat and barley cultivars tested against *H. avenae* and *H. latipons* are excellent hosts. However, the oat cultivars tested were extremely poor hosts to *H. avenae* but good hosts to *H. latipons*. In Mediterranean countries, such as Algeria, Spain, Israel and southern France, oats appeared generally to be a poor host for *H. avenae*, in comparison to northern Europe where they are considered to be a good host, suggesting the possibility that the nematode has developed host race types. In order to make the best use of existing research findings, greater collaboration between research institutions and countries where the nematode is considered important is essential. An excellent example of this is the most recent report by Rivoal *et al.* (2001), which offers a great start to unravelling the complexity of *Heterodera* populations and the existing knowledge of resistant sources and their possible uses in controlling the cereal cyst nematode in different regions of the world.

Nematode Genus / Species Comparisons		
Helicotylenchus	Cephalobidae	Mononchidae

Nematophagous fungi are microfungi that can capture, kill and digest nematodes. They use special mycelial structures, the so called traps, or spores to trap vermiform nematodes or hyphal tips to attack nematode eggs and cysts before penetration of the nematode cuticle, invasion and digestion. The more than 200 species known differ in saprophytic/parasitic ability. Although many of the trap forming and egg parasitic fungi can survive in soil saprophytically, the endoparasites are mostly more dependent on nematodes as nutrients. Molecular data have greatly improved our understanding of the function of the fungi in respect of their taxonomy, their physiological/biochemical activity and their ecology, including their function as biocontrol agents.

Key Concepts:

- There are three major groups of nematophagous fungi: the nematode trapping fungi that capture free living nematodes using specialised morphological structures (i.e. traps), the endoparasitic fungi that infect nematodes using adhesive spores, and the egg and cyst parasitic fungi that infect these stages with their hyphal tips.
- The nematophagous growth habit has evolved several times and independently since nematophagous fungi are found in all major groups of fungi.
- The nematophagous fungi display a large diversity and extensive plasticity of infection structures. In many species, the formation of trapping structures is initiated in response to signals from the environment including compounds secreted by the host nematode.
- Isolation of more virulent strains, and the development of better formulations and fermentation techniques have significantly improved the efficiency of nematophagous fungi as biological control agents against plant and animal parasitic nematodes.
- Nematophagous fungi infect nematodes through a sequence of events including adhesion of the trapping structures to the host nematode surface, followed by penetration, killing and digestion of the nematode tissues.
- The secretion of extracellular enzymes such as serine proteases and chitinases are required for successful parasitism of free living nematodes and nematode eggs by nematophagous fungi.

Tylenchina

The Tylenchina is a group of nematodes supported as monophyletic by molecular analyses. The clade lacks clear morphological synapomorphies, but potential ancestral characters include a stegostom without epithelial interradial cells or glottoid part, the absence of a median bulb, mono-prodelphy, and the absence of a bursa.

Superorder Tylenchina. The female esophagus is composed of a procorpus, metacorpus and a glandular postcorpus with an intervening isthmus between the metacorpus and the glandular regions. A dorsal esophageal gland oriface opens in the anterior procorpus. The metacorpus generally has a valve. The stylet in males and females are generally with three basal knobs. The male feeding apparatus is atrophied in some taxa. The lip region is distinct or undifferentiated from the general body contour. Transverse body annualation is generally interrupted by longitudinal incisures. Except in aberrant males, the stoma are armed with a knobbed axial stylet. The glandular postcorpus shows varying degrees of development; it may join the intestine directly or overlap the anterior intestine for some distance. Females have one or two genital branches. The oviduct has two rows of seven cells. Males generally have caudal alae and paired spicules. The deirids are present or absent.

```
Tylenchina
    |  i. s.: Drilonematomorpha [Drilonematida, Drilonematoidea]
LB02
    |           |--Drilonematidae LB02
    |           |--Ungellidae LB02
    |           |--Homungellidae LB02
    |           |--Pharyngonematidae LB02
    |           `--Creagrocercidae LB02
    |--Panagrolaimomorpha LB02
    |    |--Strongyloidoidea LB02
    |    `--Panagrolaimidae [Panagrolaimoidea] LB02
    |         |--Turbatrix aceti [=Anguillula aceti] H02
    |         |--Panagrellus H02
    |         |    |--P. redivivus W02
    |         |    `--P. silusiae W02
    |         `--Panagrolaimus Fuchs 1930 E86
    |              |--P. davidi W02
    |              |--P. rigidus L02
    |              `--P. ruffoi Andrassy 1962 E86
    `--+--Tylenchomorpha [Tylenchida] LB02
       |    |  i. s.: Parasitaphelenchus MM02
       |    |         Prothallonema MM02
       |    |--Sphaerularioidea LB02
       |    |--Aphelenchoidea LB02
       |    |--Criconematoidea LB02
       |    |--Tylenchoidea LB02
       |    `--Myenchoidea LB02
       `--Cephalobomorpha [Cephalobina, Cephaloboidea] LB02
            |--Elaphonematidae LB02
            |--Osstellidae LB02
            |--Alirhabditidae LB02
            |--Bicirronematidae LB02
            `--Cephalobidae LB02
                 |--Acrobeloides nanus H02
                 `--Cephalobus Bastian 1865 E86
                      |--C. dubius G02
                      |--C. lentus G02
                      |--C. persegnis MM02
                      |--C. quinilineatus J02
                      `--C. troglophilus Andrassy 1967 E86
```

Rhabditina

Subclass Rhabditia is mostly composed of parasitic nematodes (particularly in the Strongylida), though there are some free-living species as well (particularly in the Rhabditida). Phasmids (posterior sensory structures) are well-developed, while amphids (anterior sensory structures) are poorly developed or absent in this group.

In an alternate classification system, they are treated as suborder Rhabditina, with the orders listed here being ranked as infraorders. Also, the Diplogasterida, which are sometimes considered a monotypic subclass, are probably better placed in the Rhabditia. On the other hand, the old placement of the Ascaridida in Rhabditia instead of Spiruria seems unwarranted. The Rhabditida contain a number of families which are probably better placed in the Tylenchia; alternatively, the latter group may entirely be merged with the Rhabditia.

Phylum Nematoda

Nematodes are found in almost all habitats, but are often overlooked because most of them are microscopic in size. For instance, a square yard of woodland or agricultural habitat may contain several million nematodes. Many species are highly specialized parasites of vertebrates, including humans, or of insects and other invertebrates. Other kinds are plant parasites, some of which can cause economic damage to cultivated plants. Nematodes are particularly abundant in marine, freshwater, and soil habitats. One study in Colorado estimated that nematodes consumed about as much grass as a prairie dog colony.

Soil Nematodes

Soil is an excellent habitat for nematodes, and 100 cc of soil may contain several thousand of them. Because of their importance to agriculture, much more is known about plant-parasitic nematodes than about the other kinds of nematodes which are present in soil. Most kinds of soil nematodes do not parasitize plants, but are beneficial in the decomposition of organic matter. These nematodes are often referred to as free-living nematodes. Juvenile or other stages of animal and insect parasites may also be found in soil. Although some plant parasites may live within plant roots, most nematodes inhabit the thin film of moisture around soil particles. The rhizosphere soil around small plant roots and root hairs is a particularly rich habitat for many kinds of nematodes.

Classification

Nematodes are roundworms in the Phylum Nematoda. Various authorities distinguish among 16 to 20 different orders within this phylum. Only about 10 of these orders regularly occur in soil, and four orders (Rhabditida, Tylenchida, Aphelenchida, and Dorylaimida) are particularly common in soil.

More than 15,000 species and 2,200 genera of nematodes had been described by the mid-1980s. Although the plant-parasitic nematodes are relatively well-known, most of the free-living nematodes have not been studied very much. Therefore there is a high probability that most soil habitats will contain undescribed species of free-living nematodes. Identification of these groups is extremely difficult, and there are only a few nematode taxonomists in the world who can formally describe new species of free-living nematodes to science. Therefore most nematode ecologists identify soil nematodes only to family or genus.

Feeding Habits

Soil-inhabiting nematodes can also be classified according to their feeding habits. This classification is particularly useful to ecologists in understanding the positions of nematodes in soil food webs. Several important feeding groups of nematodes commonly occur in most soils. In addition, algivores (feed on algae) and various stages of insect and animal parasites occasionally are found in soil. The nematode feeding groups are called trophic groups by some authors.

Herbivores. These are the plant parasites, which are relatively well known. This group includes many members of the order Tylenchida, as well as a few genera in the orders Aphelenchida and Dorylaimida. The mouthpart is a needlelike stylet which is used to puncture cells during feeding. Ectoparasites remain in the soil and feed at the root surface. Endoparasites enter roots and can live and feed within the root.

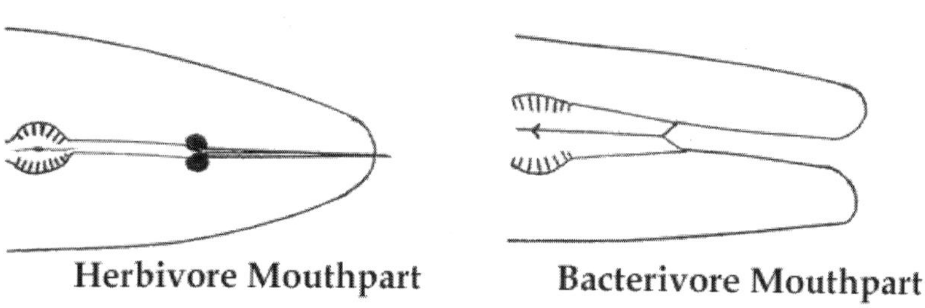

Figure: *Diagram of head regions of a herbivore (left) and nematode bacterivore (right). In the herbivore, the mouthpart is modified into a stylet for puncturing plant cells. In the bacterivore, the mouth or stoma is a hollow tube.*

Bacterivores. Many kinds of free-living nematodes feed only on bacteria, which are always extremely abundant in soil. In these nematodes, the “mouth”, or stoma, is a hollow tube for ingestion of bacteria. This group includes many members of the order Rhabditida as well as several other orders which are encountered less often. These nematodes are beneficial in the decomposition of organic matter.

Fungivores. This group of nematodes feeds on fungi and uses a stylet to puncture fungal hyphae. Many members of the order Aphelenchida are in this group. Like the bacterivores, fungivores are very important in decomposition.

Predators. These nematodes feed on other soil nematodes and on other animals of comparable size. They feed indiscriminately on both plant parasitic and free-living nematodes. One order of nematodes, the Mononchida, is exclusively predacious, although a few predators are also found in the Dorylaimida and some other orders. Compared to the other groups of nematodes, predators are not common, but some of them can be found in most soils.

Omnivores. The food habits of most nematodes in soil are relatively specific. For example, bacterivores feed only on bacteria and never on plant roots, and the opposite is true for plant parasites. A few kinds of nematodes may feed on more than one type of food material, and therefore are considered omnivores. For example, some nematodes may ingest fungal spores as well as bacteria. Some members of the order Dorylaimida may feed on fungi, algae, and other animals.

Unknown. Since free-living nematodes have not been studied very much, the food habits of some of them are unknown. The microscopic size of these animals presents additional difficulties. For example, it can be very difficult to distinguish whether a nematode is feeding on dead cells from a plant root or on fungi growing on the cell surface. Sometimes a nematode showing this feeding behaviour may be classified simply as a root or plant associate.

Community Composition (Back to Top)

In one pasture in south-central Florida, herbivores comprised almost half of the soil nematode community, but bacterivores and fungivores were also well-represented. This habitat provided many fibrous roots as a food source for herbivores, but other nematode groups, particularly bacterivores, may predominate in other habitats.

The composition of the soil nematode community depends on the vegetation present, as well as on soil type, season, soil moisture level, amount of soil organic matter, and many other factors. Because they are responsive to so many different factors, it is believed that nematodes may be useful bioindicators of the condition of the soil environment.

Decomposition

Free-living nematodes are very important and beneficial in the decomposition of organic material and the recycling of nutrients in soil. Nematode bacterivores and fungivores do not feed directly on soil organic matter, but on the bacteria and fungi which decompose organic matter. The presence and feeding of these nematodes accelerate the decomposition process. Their feeding recycles minerals and other nutrients from bacteria, fungi, and other substrates and returns them to the soil where they are accessible to plant roots.

Plant Classification

In order to study the billions of different organisms living on earth, biologists have sorted and classified them based on their similarities and differences. This system of classification is also called a taxonomy and usually features both English and Latin names for the different divisions.

All plants are included in one so-called kingdom (Kingdom *Plantae*), which is then broken down into smaller and smaller divisions based on several characteristics, including:

- Whether they can circulate fluids (like rainwater) through their bodies or need to absorb them from the moisture that surrounds them;
- How they reproduce (e.g., by spores or different kinds of seeds); and
- Their size or stature.

The majority of the 260,000 plant species are flowering herbs. To describe all plant species, the following divisions (or phyla) are most commonly used to sort them. The first grouping is made up of plants that are non-vascular; they cannot circulate rainwater through their stems and leaves but must absorb it from the environment that surrounds them. The remaining plant species are all vascular (they have a system for circulating fluids). This larger group is then split into two groups: one that reproduces from spores rather than seeds, and the other that reproduces from seeds.

Nematodes are the most speciose phylum after the arthropods, they occur in nearly every habitat including as parasites in all sorts of plants and animals, (they don't like dry places however). One species is known that can live in old vinegar (Turbatrix aceti)and another that as only been found in German beer mats. Though only about 80 000 species have been described some scientists estimate there may be as many as a million species all told. They can occur in very dense numbers in the soil and rotting vegetation, as many as 90 000 have been found in a single rotting apple, while millions occur in the top 3cm (1 inch) of a square metre of good quality soil. While there are a huge number of free living Nematodes there are also a large number of parasitic species, many of which cause diseases to man and other animals as well as to plants, nearly every living organism has been found to be parasitised by one species of nematode or another. Most nematodes are reasonably small, they range in size from 100 micrometres in length (1/10th of a mm or 1/250th of an in) to the female Giant Nematode *Dioctophyme renale* which may be up to 1 metre, or 3 ft long.

Free living nematodes are long thin worms with transparent and typically curled bodies, parasitic species have a variety of less streamline shapes relating to their degenerate parasitic life styles, one unifying characteristic that makes the phylum unique is the lack of cilia or flagella, even the sperm of nematodes are amoeboid. Nematodes as parasites have been known for a long time and the earliest recorded literary mention of them is an Egyptian papyrus from 1500 BC, they are also mentioned by the ancient Greeks Aristotle and Hippocratis the father of scientific medicine.

Characteristics of Nematoda:-

1. Bilaterally symmetrical, and vermiform.
2. Body has more than two cell layers, tissues and organs.
3. Body cavity is a pseudocoel, body fluid under high pressure.
4. Body possesses a through gut with a subterminal anus.
5. Body covered in a complex cuticle.
6. Has a nervous system with pharyngeal nerve ring.
7. Has no circulatory system (no blood system)
8. Reproduction normally sexual and gonochoristic.
9. Feed on just about everything.
10. Live just about everywhere, many species are endoparasites.

Ecology

Nematodes live in a vast variety of habitats, ecologically they can be divided into free living forms and parasitic forms. Free living forms have a simple life cycle involving 4 juvenile instars on the path from egg to adult. Parasitic species have developed a wide range of variations on this basic theme. The variations involve whether there is a secondary host and the amount of time spent in one or either hosts. There is also considerable variability in the way that they move from one host species to another. thus while many species lay eggs that pass out of the primary host with the faeces where they are eaten by the secondary host which then gets eaten in turn by the primary host after the Nematodes have developed. Because it is not always totally reliable that the secondary host will be eaten just as the Nematode larvae have developed into the infective stage many species have the ability to encyst themselves in the muscle or cuticle of their secondary hosts.

Some species use another animal to transport them from one host to another thus *Wuchereria bancrofti* releases minute live young called 'microfilaria' into the primary hosts blood stream rather than eggs into the digestive tract. These microfilaria get ingested by mosquitoes when they feed on an infected person. Inside the mosquito they live in the mosquitoes gut where they develop until the Larva 3 stage wait for the mosquito to bite another host whereupon they enter the host via the mosquitoes proboscis sheath and the wound it makes in the hosts skin.

Nematodes in Mankind

Human beings, along with all other living things are host to numerous Nematode parasites. The most common of these is *Ascaris lumbricoides* with an estimated 700 million people effected globally, this Nematode is not normally fatal and in low numbers may have very little effect on adults, however in heavy doses it can be quite debilitating, especially for children. The Nematodes infecting mankind include several species of filarial worms, the most important of these are *Wuchereria bancrofti* and *Brugia malayi* which are very similar and cause lymphatic filariasis, *Onchocerca volvulus* which causes River Blindness and Loa loa which causes Loiasis. Other species are *Dranunculus medinensis* known as Guinea Worm, *Trichinella spiralis* causing Trichinosis, *Necator americanus* and *Ancylostoma duodenale* causing Hookworm, *Enterobius vermicularis* causing Pinworms and *Trichuris trichuria* causing Whipworm or Trichuriasis.

Anatomy

Basically a Nematode is a long hollow tube within which is another tube, the alimentary canal and the reproductive organs. Nematodes are round in cross section, this is because unlike the other worms that below them in the phyla table they maintain their body fluids under great pressure (on average internal pressure in a nematode equals 70mm of mercury or 1.49 PSI, with a maximum recorded value of 125mm of mercury or 2.41 PSI). To contain this high pressure nematodes have an extremely tough, yet elastic and flexible cuticle. This cuticle consists of up to 9 layers of proteinaceous fibres, with 3 layers being easily discerned, these are called, from the outside in, the cortex, the matrix layer and the fibre layer. Despite its complexity the Nematode cuticle is permeable to both water and gases, so respiration occurs through it. Beneath the cuticle is a hypodermis and a layer of longitudinal muscle. The combination of the flexure of these muscles with the high pressure of the system produces a characteristic whip-like wriggle that Nematodes use to swim. Scientifically this is called undulatory propulsion with sinusoidal waves passing back along the body.

At the anterior (head) end there is a mouth which has 3 lips behind which predatory species possess a few teeth, this leads to a pharynx which is triangular in cross section. Because of the high pressure within the body unsupported organs such as the intestines tend to collapse in much the same way that an uninflated bicycle tube tends to become oval or flat in cross section when laid flat on the table. The pharynx of Nematodes is an efficient pump and forces food into the intestines, there is a one way valve between the intestines and the pharynx. The pharynx can, when this valve is closed, be used to suck liquid food into the mouth. Digestion is rapid and faeces are expelled under pressure. This pressure is so great that the parasitic nematode Ascaris lumbricoides *which is about 12cm to 18cm long (5 to 7 inches) may shoot its faeces 60cm or 2 feet into the air.*

Nematodes, especially free living forms generally have a reasonably well developed nervous system. This is comprised of a circum-pharyngeal nerve ring made up from 4 nerve ganglia from which 6 longitudinal nerves extend down through the body to the various parts of the gut and the reproductive organs. There are also 6 shorter nerves which extend forwards from the circum-pharyngeal ganglia towards the mouth. Nematodes have no circulatory or respiratory organs and the excretion of metabolic waste is via two simple ducts or tubules which have no nephridia or flame cells.

Nematodes are copiously reproductive and most of their body cavity, which is a pseudocoelom is filled with paired sets of reproductive organs, either ovaries or testes. Males and females copulate and the male introduces sperm to the females vagina with the help of 2 stiff horny spicules that are a part of his cloaca. Fertilisation is internal and females lay eggs over a prolonged time period, thus a female Ascaris lumbricoides *may lay her eggs at the rate of 200,000 per day and have had a total 27 million eggs within her at the start of her reproductive career. Young nematodes hatch from these eggs and go through 4 moults before they become adults.*

***Table:** Classification of the Nematoda*

Phylum	***Class***	***Order***
Nematoda	***Adenophorea***	
		Enoplida
		Isolaimida
		Mononchida
		Dorylaimida
		Trichocephalida
		Mermithida
		Muspiceida
		Araeolaimida
		Chromadorida
		Desmoscolecida
		Desmodorida
		Monhysterida
	Secernentea	
		Rhabdita
		Strongylida
		Ascaridida
		Spirurida
		Camallanida
		Diplogasterida
		Tylenchida
		Aphelenchida

Basal Order Monhysterida

The Monhysterida are an order in the phylum Nematoda. Usually the stoma in Monhysterida is funnel shaped and lightly cuticularized, but it is sometimes spacious and heavily cuticularized and they always have protrusible teeth. They are either simply spiral or circular.

Class Dorylaimea

Class Enoplea

Enoplea is a class of nematodes, one of two classes that make up the phylum Nematoda in current taxonomy. It is considered to be a more ancestral group than the other class, Chromadorea, and researchers have referred to its members as the "ancestrally diverged nematodes", compared to the "more recently diverged nematodes" of Chromadorea.

Figure: *Enoplea are distinguished from Chromadorea by a number of characteristics. The enoplean esophagus is cylindrical or "bottle-shaped", compared to the bulbous chromadorean esophagus. Enopleans have pocket-like amphids, while chromadoreans have amphids shaped like slits, pores, coils, or spirals. An enoplean is smooth or marked with fine lines, while a chromadorean may have rings, projections, or setae. The enoplean excretory system is simple, sometimes made up of a single cell, while chromadoreans have more complex, tubular systems, sometimes with glands.*

Ecology

Several orders of enopleans are mainly freshwater animals, and several include marine species. Many enopleans are parasites of plants and animals, including humans. The orders Triplonchida and Dorylaimida include plant parasitic nematodes that are vectors of plant pathogens. The orders Mermithida and Marimermithida include parasites of invertebrates. The orders Dioctophymatida, Trichinellida, and Muspiceida include parasites of vertebrates such as birds and mammals. Examples are *Trichinella spiralis*, a nematode known for causing trichinosis in humans who consume it in undercooked pork, and whipworms (genus *Trichuris*), which are parasites of mammals, including cats, dogs, and humans.

Class Secernentea

Secernentea are the main class of nematodes, characterised by numerous caudal papillae and an excretory system possessing lateral canals. Like all nematodes, they have no circulatory or respiratory system. The Secernentea contain several significant parasitic species among the Rhabditia and Spiruria. *Caenorhabditis elegans*, the famous laboratory animal, is one of the many harmless members of the Rhabditia.

Systematics

Subclasses and orders of Secernentea are:

- Subclass Rhabditia (paraphyletic?)
  - Rhabditida
  - Strongylida
- Subclass Spiruria
  - Ascaridida
  - Camallanida (sometimes included in Spirurida)
  - Drilonematida (sometimes included in Spirurida)
  - Oxyurida (= Rhabdiasida)
  - Rhigonematida (formerly in Tylenchia)
  - Spirurida
- Subclass Diplogasteria (may belong in Rhabditia)
  - Diplogasterida
- Subclass Tylenchia (may belong in Rhabditia)
  - Aphelenchida
  - Tylenchida

Some families traditionally considered to be Rhabditida seem to be closer to the Tylenchida. If the Tylenchia are to be maintained as separate, they probably will be included therein. These are also called as Phasmidians as they have the phasmids which are the posterior glandulo sensory and excretory structure.

"Chromadorea" Assemblage

The Chromadorea are a class of the roundworm phylum, Nematoda. They contain a single subclass (Chromadoria) and several orders. With such a redundant arrangement, the Chromadoria are liable to be split up if the orders are found to form several clades, or abandoned if they are found to constitute a single radiation.

Formerly, they were treated as a subclass in the paraphyletic "Adenophorea" assemblage, which has been by and large abandoned by modern authors. It is also suspected that the Chromadorea may not be monophyletic as delimited here; at least the Monhysterida seem to be a distinct and far more ancient lineage than the rest. Members of this class' bodies usually have annules, their amphids elaborate and spiral, and they all have three esophageal glands. They usually live in marine sediments, although they can live elsewhere. They have more sophisticated pharynxs than most roundworms.

Biological Control Attributes

Preliminary field observations in the native and introduced range of GWSS resulted in the decision to prioritize egg parasitoids for evaluation as biological control agents. Dr Serguei Triapitsyn of the University of California, Riverside, and Larry Bezark of the CDFA made the first trips to Mexico in search of GWSS egg parasitoids in 1999. In 2000 and following years, with the help of the USDA-APHIS (Dr Simmons), USDA-ARS (Dr Walker Jones), and the Universdad Autonomata de Tamaulipas (Dr. Svetlana Myartseva), Collections of proconiine egg masses were made in NE Mexico. Permits were issued to transport the egg masses to quarantine facilities in California and Texas where they were incubated and examined daily for the emergence of egg parasitoids. Any parasitoids emerging were identified then exposed to fresh GWSS egg masses to ensure that they could successfully develop on egg masses obtained from Californian populations of GWSS.

Once stable populations of egg parasitoid species had been established in quarantine, they were evaluated for their suitability as GWSS biological control agents. Attributes such as longevity, fecundity, and host specificity were examined before submitting requests to the USDA-APHIS for permission to release the selected biological control agents into California.

There are other biological control agents undergoing evaluation for use as biological control agents. These include parasitoids from North and South America.

Gonatocerus Triguttatus

When first explorations were made in Mexico and SE USA, this insect was the most commonly collected natural enemy of GWSS in its native range. It produces one wasp per GWSS egg and offspring are female sex-biased.

This parasitoid can successfully parasitize the native smoke tree sharpshooter but does not fare well in desert conditions where the native sharpshooter is commonly found. This takes 12-14 days to develop from egg to adult. Adults mate on emergence then females disperse to search for GWSS eggs to parasitize. Adults feed on sugar sources to increase longevity. This insect was selected because of its importance as a natural enemy in the GWSS native range.

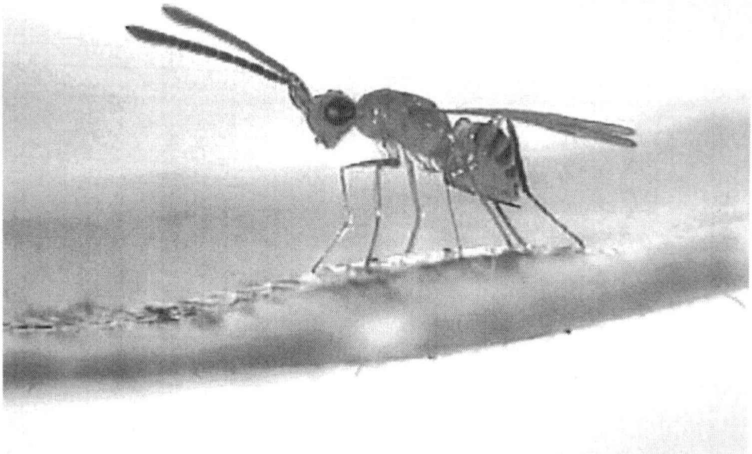

Figure: *Gonatocerus triguttatus— This wasp can successfully parasitize the native smoke tree sharpshooter.*

Gonatocerus Ashmeadi

Populations of *G. ashmeadi* already exist in California. Research carried out by Drs. Stouthamer and Hoddle suggest that it is not native to California but arrived in California prior to 1980, long before the arrival of GWSS, and became established on the smoke tree sharpshooter. When GWSS arrived in California, the parasitoid switched back to its supposed evolutionary host where is now the most important natural enemy of GWSS in California.

Figure: *Gonatocerus ashmeadi —A wasp used to help in the fight against GWSS.*

Populations of *G. ashmeadi* were recovered in SE USA and NE Mexico and, after appropriate screening, were released into California. The rationale for the release of exotic populations of *G. ashmeadi* was that this population might have attributes that made is a more efficient GWSS parasitoid than the California population that may have evolved to being a more effective natural enemy of the smoke tree sharpshooter. It is hoped that the best attributes of each population would eventually combine, resulting in higher levels of parasitism of GWSS. Unfortunately, it is almost impossible to discriminate between native and non-native populations of *G.ashmeadi* so testing such a hypothesis is currently not possible.

Gonatocerus ashmeadi and *G. triguttatus* have similar life histories. Although the former appears to be competitively superior under laboratory conditions, there may be biotic (e.g. host plant factors) and abiotic (e.g. climatic conditions) where *G. triguttatus* may be a more effective biological control agent. As both species probably coevolved, it is unlikely that they occupy exactly the same niche.

Gonatocerus Walkerjonesi

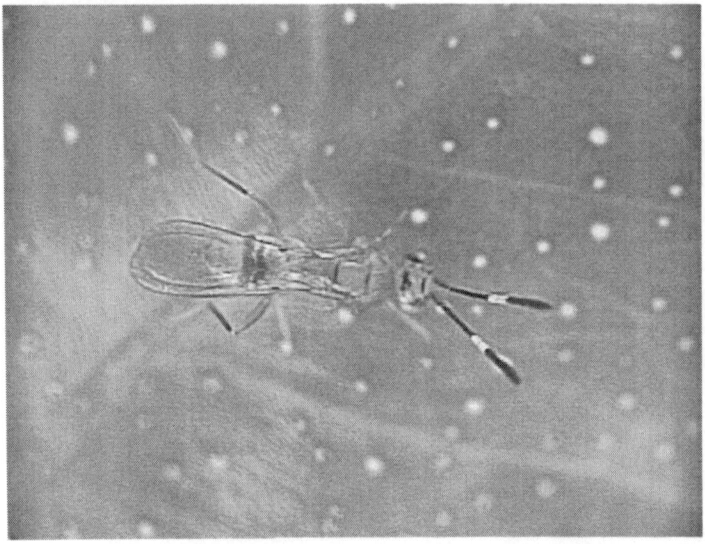

Figure: *Gonatocerus walkerjonesi—A wasp used to help in the fight against GWSS.*

A species of *Gonatocerus* was recovered from Mexico in 2000 and evaluated as previously described. This insect was identified as being identical to a California species of *Gonatocerus* and so when permits were issued for its release, populations of field insects from California were added to augment production rate. It was some years later that the two populations mixed were, in fact, two different species and that the Californian species, *G. walkerjonesi* was previously undescribed. It is believed that the Mexico species, *G. morrilli*, soon became extinct in the wasp production system due to being outnumbered by additions of *G. walkerjonesi* to the colonies. Consequently, *G. walkerjonesi* was released into California for four years before the colony was reinitiated with *G. morrilli* from Mexico.

Gonatocerus Morrilli

Releases of the true *G. morrilli* were initiated in 2005. This insect has a considerably faster development rate than any of the previously described congeners, emerging about two days earlier. This faster lifecycle raised the possibility that the insect could build up populations faster in the spring prior to the build up of other egg parasitoids. Although it has not been long since releases were initiated, we are already getting positive signs that this insect has become established in California.

Gonatocerus Fasciatus

When this insect was first collected in Louisiana, there were high hopes that it would prove an efficient early season biological control agent. In its native range, the insect becomes the predominant natural enemy of GWSS eggs for over a month before it is replaced by *G. ashmeadi*.

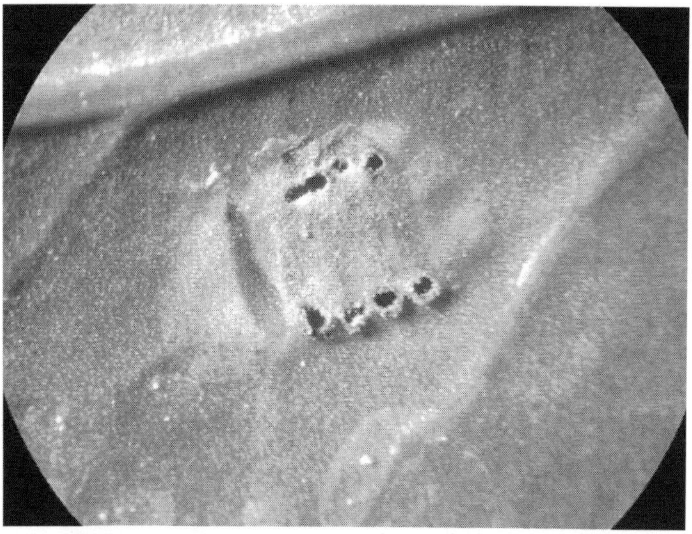

***Figure:** A GWSS egg mass that—was ps possibly have emerged from.*

The insect has one major attribute that benefit it over the other species of *Gonatocerus* described; it is gregarious, in other words, more than one wasp emerges from each GWSS egg parasitized. Initially it was thought that only two wasps emerged from each GWSS egg but it was soon found that up to seven wasps could emerge from each egg. The other individuals emerging from one of the two emergence holes cut at each end of the GWSS egg. The ability to produce more than one offspring from each GWSS egg favours the rapid build up of populations of this insect, especially when host density is low (i.e. early in the year) so search time for egg masses is a major investment. *Gonatocerus fasciatus* is, not surprisingly, smaller than its congerers and consequently has a shorter adult longevity of about a week compared to two weeks.

This insect was permitted for release into California and the CDFA spent three years attempting to get it established. During this time several discoveries were made about this insect. Under laboratory conditions, they successfully parasitized the smaller native blue-green sharpshooter but this was never observed in the field. Also, it was discovered that this insect was, after all, native to California; individual specimens have been repeatedly collected in Northern California where it is supposed to attack sharpshooter species confined to that area.

Anagrus Epos

The search for a biological control agent that could overwinter lead to the discovery of *A. epos* in Minnesota on the border of North Dakota. A literature search revealed that this species had previously been permitted for release as a biological control agent of the beet leafhopper but later publications corrected this as a misidentification of a new species. These insects are extremely small so there are few characteristics that can be used to discriminate between species.

Figure: Anagrus epos—*can produce up to 14 individuals from one GWSS egg.*

Although *A. epos* is in the same family as *Gonatocerus*, the Mymaridae has a considerably different life history. It is believed to be univoltine in its native range; its evolutionary host is a smaller sharpshooter than GWSS in the genus *Cuerna*. The optimal conditions for development of this species are considerably lower than for *Gonatocerus* spp and development time can take over a month. Nonetheless, the ability to overwinter and produce up to 14 individuals from one GWSS egg makes it an interesting candidate for GWSS biological control. Because *A. epos* can parasitize other sharpshooter species, studies are underway to develop production techniques that employ hosts that are easier to rear than GWSS, producing up to 14 individuals from one GWSS egg.

North America

Recent field exploration trips were made by researchers at Texas A&M University into Mexico. Several species of egg parasitoids were recovered, one of which is a new species and is currently undergoing evaluation in quarantine at UC Riverside.

Closer to home, during the process of monitoring field release sites, a new species of *Gonatocerus* was discovered at one location in Orange County in 2002. In a publication describing the insect, it was suggested that it is exotic to California as it has a number of characteristics that suggest its origin to be Central America. Since then, *G. morgani* has spread southward along the coast to San Diego. Attempts are underway to establish a colony of the parasitoid so that releases can be made into areas that are geographically isolated from its current range, for instance Ventura which is isolated by the city of Los Angeles and inland areas that are isolated by mountain and hill ranges.

Similar work is being carried out on the trichogrammatids *Ufens* spp that are desert specialists and currently confined to the desert regions east of Los Angeles. These insects have proved to be well nigh impossible to rear and so wasps emerging from field collected parasitized egg masses are collected and relocated to distant release sites. Currently releases have been confined to native vegetation in the Central Valley as these insects appear to be most effective at parasitizing eggs laid in native host plants and in desert environments. Most areas in the Central Valley that are not under irrigation are considered to be desert regions and there are strict restrictions on the use of conventional insecticidal treatments in these areas.

South America

One of the more controversial aspects of biological control being evaluated by the PDCP is the introduction of exotic natural enemies for control of a sharpshooter with which they have had no evolutionary association. This technique is called neoclassical biological control and has had its successes but also has critics. Several South American species of egg parasitoid have been collected from proconiine sharpshooter eggs and sent to California and Texas quarantine facilities for evaluation. Two concerns have been raised regarding the release of biological control agents from South America. The first is that the agents would not add to the level of biological control being attained in California from agents and natural enemies already present. In fact, some have suggested that the introduction of a new species might even depress the level of control due to competition effects between parasitoids. The second is that there is, however small, the risk that the introduced agent may be inadvertently introduced into the native range of GWSS and might disrupt the ecological balance of native proconiine sharpshooters and their natural enemies. These issues need to be addressed and such risks balanced against the potential benefits to be gained by introducing a neoclassical biological control agent.

Biological Control for the Public

There are natural control factors that limit the increase in numbers of an organism in an ecosystem. These naturally occurring limiting factors prevent most organisms from becoming pests. This process is known as Natural Control. Natural control consists of two major components, environmental factors (abiotic) and natural enemies (biotic). When natural enemies are killed by man's actions in any habitat or when the pests are introduced to new habitats without their natural enemies, natural control often fails and results in pest outbreaks. To remedy the situation caused by his own actions, man can play an active role in reintroducing the natural enemies to control pests. The use of natural enemies by man to control pests is Biological Control. Thus the difference between Biological Control and Natural control is that the former is brought about through man's action while the latter is not.

Despite the successful use of biological control in the control of insects and weeds for over a hundred years, this concept is still not very well known among the general public. Even in major universities, the concepts and methodology of biological control are not taught in the more general introductory courses, but only in specialized courses that usually have small enrollment. Consequently, less than 10 percent of the students attend classes that include the concept of biological control. This lack of knowledge of biological control among students, growers, and the general public throughout the State of Virginia is readily perceived in my dealings with them during the past 27 years. Very few are aware of the valuable potentials of biological control as a pest problem solving technique, or of the availability of research information on the topic. This unfortunate situation can be readily rectified through a more effective method of dissemination of information.

Pest Outbreaks and Population Dynamics

Pest outbreaks are the result of natural or man-induced factors. Natural factors include migration and environmental changes that allow the pest to increase in numbers. Man-induced factors include the introduction of a new species, a change in crop management, an alteration of the habitat, side effects of pesticide use, monoculture or continuous cropping that provides a concentrated food source for species that can survive and reproduce on them. One of the most common factors is the introduction of insect species by man, directly or indirectly, to new areas without their natural enemies. In the absence of their natural enemies, the pests survive and reproduce rapidly in their new environment. This is often referred to as "expression of their Biotic potentials".

Biotic Potential is the inherent characteristics of an organism to survive and reproduce. Without restraining factors, they will soon colonize the whole world. However, there are various factors that limit the growth of any organism. These natural control factors act as restraining forces that keep a species from unlimited growth. In any habitat, natural control factors suppress pest outbreaks. When the natural control factors are removed, then pest outbreaks occur. These natural control factors are often referred to as environmental resistance.

Environmental Resistance is the sum total of all mortality factors (abiotic physical factors such as temperature, and biotic factors such as natural enemies) in the organism's environment which prevents it from expressing its full capacity to reproduce. Mortality factors can be broadly divided into two groups, biotic and abiotic.

Biotic factors are living factors such as the effects of other organisms

Abiotic factors are environmental factors that include climate, soil, air, space, and light.

Concept of Biological Control

Biological control is the study and use of natural enemies by man as control agents for reducing pest populations and number. In 1919, Smith defined it as the use of natural enemies to control insect pests. Harley in 1985 broadened the concept to the study and utilization of parasites, predators, and pathogens to regulate the populations of pests. Biological control agents must be living organisms so that they can seek out the target pest. They can be natural or genetically improved organisms.

Ecological Principles of Biological Control

The basic principles of biological control are that:

- One living organism (biological control agent) can be used to control another living organism.
- Some biological control agents have a limited host range (host specificity) while others do not.

To be effective in biological control, an agent must be host specific so that they do not affect nontarget organisms.

Effects of Biological Control Agents

Biological control agents exert either a direct or indirect effect through their actions. In the formal case, biological control agents attack and kill the pest directly. In the indirect action, the agents weaken the hosts so that they are unable to reproduce at their normal rate.

Natural Enemies

The term natural enemy refers to living organisms that are normally found in the same habitat as the target pest. They keep the pest numbers down. Natural enemies, also called beneficial organisms, can be grouped into four types:

1. Parasite — an organism that lives and feeds in or on a host.
2. Parasitoid — an insect parasite that destroys its host.
3. Predator — a living organism that feeds upon other organisms that are smaller and weaker than itself.
4. Pathogen — a microorganism that causes disease in its host.

Parasites and Predators

There are some major differences between how parasites and predators act. It is useful to know some of their characteristics. In general, parasites are more host specific, that is, they have a narrow host range. They are also better adapted with the host in the same environment, and in many cases the parasite and host have evolved together. Thus parasites are synchronized with the host as well as its environment. A parasite requires fewer hosts for its development and survival. It feeds less than a predator and the young do not need to search for food. Parasite eggs are usually laid into the target host and as the egg hatches, the immature stage will feed inside the host and can complete its development in a single host. Predators need to have a wide range of prey. They destroy large numbers of prey quickly and thus their spectacular successes are often noticed. They are more likely to show an immediate impact. However, because they are mobile and move away after they have fed, the effects of many predators are often overlooked.

Natural enemies work quietly, often unnoticed and so their importance is often overlooked. Frequently their effectiveness as biological control agents becomes noticeable only after they have been eliminated through the wide use of pesticides that kill not only the target pests, but these natural enemies as well. The lack of appreciation of natural enemies is not a recent occurrence. As early as 1880, Forbes indicated that "...the annihilation of all the established 'enemies' of a species would, as a rule, have no effect to increase its final numbers". This obviously has been proven incorrect. The importance of natural enemies is now widely recognized among pest management specialists, and current pest control practices emphasize the preservation of natural enemies.

Attributes of Natural Enemies

Some of the more desirable characteristics of natural enemies are:

- Good searching ability — ability to locate the target host
- Host specificity — attacks only the target pest
- High reproduction rate— high rate of increase
- Short life cycle — shorter life cycle than host
- Adaptability — adapts well to host habitats
- Host synchrony — well adapted to different stages of life cycle of target host
- Sustainability — able to maintain itself after reducing host population.

Examples of the importance of natural enemies in Virginia

Some examples of natural enemies that keep major pests of Cole crops under control in Virginia are shown below:

Natural enemy	*Pest*
Cotesia orobenae (wasp)	Cross-striped cabbageworm
Virus	Cabbage looper
Diadegma insulare (wasp)	Diamondback moth

Types of Biological Control

There are three major types of applied biological control where man plays a role: Augmentative, Classical, and Neoclassical.

1. Augmentation is the increase of native agents for control of native or exotic pests. Usually there is a lack of synchrony where the pest occurs early in the season and the natural enemies are absent. Thus, releasing the natural enemy early in the season will help to ensure that there is no scarcity of natural enemies when the pests first appear. This process is called augmentation.
2. Classical biological control is the use of exotic biological control agents imported from its native home into the target area against exotic pests that have arrived without their natural enemies. In the absence of their natural enemies, the exotic pests multiply rapidly and spread to become a major pest. Thus, classical biological control attempts to introduce the natural enemies also into the new area so that they will reestablish equilibrium with the pest and keep it under control. This is often referred to as an old association because the natural enemy and pest were in the same place or ecosystem.
3. Neoclassical biological control is the use of exotic biological control agents against a native pest. Since they were not in the same area before, this is referred to as a new association.

Pests most suitable for Biological Control

Not all pests can be controlled by this technique. The pests most suitable are exotic pests that have been introduced into this country without their natural enemies. The goal in biological control is to find out what the effective natural enemies are, where they can be found, and to try to import them from their native area into the affected areas to control the pests.

Biological Control is most suited for single, dominant, exotic pests that are not closely related to indigenous beneficial species. It is not appropriate for a complex of pests, or pests with many related members, which are of economic importance because it may attack them as well.

How to import Biological Control Agents

Before any biological control agent or natural enemy can be released, permission needs to be obtained from the Federal and State Governments. The USDA, APHIS is the federal agency responsible for granting permission. For individual states, the State Entomologist reviews requests for release of biological control agents. Host specificity studies of potential biological control agents carried out by research scientists are submitted to the federal and state agencies for review. Only when approval is obtained can any exotic biological control agent be released in the USA.

History of Biological Control

Parasites and predators have long been recognized in agriculture as biological control agents. Predation (use of predators) was first noted in ancient China where the citrus growers used predaceous ants (*Oecephylla smaragdina*) to control pests of citrus crops. Parasitism was first reported in 1602 when a parasite (*Apanteles glomeratus*) was used to control the common cabbage butterfly (*Pieris rapae*).

Early use of biological control in the USA began in mid nineteenth century. In 1856, Asa Fitch used parasites against the wheat midge, and in 1867 Benjamin Walsh continued using parasites against the same pest. C. V. Riley introduced parasites to control the plum curculio in 1870 and the cabbage butterfly in 1883. Crawford in 1886 and Koebele in 1887 introduced parasites to control the cottony cushion scale.

Procedures of Biological Control

The major steps involved in biological control are briefly described as follows:

1. First we need to determine whether the pest is suitable for biological control since not all pests are suitable for this method of control.
2. Next we select suitable and effective natural enemies (biological control agents) for pests that are considered appropriate for biological control.
3. We conduct safety tests to ensure that the potential biological control agents will only attack the target pests and not turn out to be pests themselves. This means that the host range has to be carefully studied.
4. We develop techniques to rear the biological control agents so that there will be sufficient for distribution if they are approved for release.
5. Upon approval for release, we conduct field establishment studies of the biological control agents.
6. Lastly we conduct evaluation surveys to determine how effective the biological control agents are in controlling the target pest and also to determine whether nontarget speceis are affected.

Major Landmarks in Biological Control as a Discipline

The first important success in the biological control of insects occurred in 1888 when the ladybeetle *Rodolia cardinalis* was imported from Australia and New Zealand for the control of the cottony cushion scale in the USA. The first major success in the biological control of a weed was in 1940 where the cactus moth *Cactoblastis cactorum* introduced from Argentina controlled the prickly pear cactus in Australia.

There are many examples of successes in the use of biological control. Perhaps the best documented example is the use of the vedalia ladybeetle to control the cottony cushion scale in California in 1888. Since that time, many more successes have been reported. Some examples are shown in the following table.

Examples of Successful Biological Control Programmes

Pest	*Place/Country*
Cottony cushion scale	USA
Oriental fruitfly	Hawaii
White grub	Hawaii
Rhinoceros beetle	Mauritius
Screwworm fly	USA
Alfalfa weevil	USA
Musk thistle	USA

Successes in Biological Control of Insects

Biological control Agent	*Host (Pest)*
Pediobius foveolatus (Parasitic wasp -Eulophidae)	Mexican bean beetle
Microctonus aethiopoides (Parasitic wasp - Braconidae)	Alfalfa weevil
Rodolia purchasi (Predator,ladybeetle - Coccinellidae)	Cottony cushion scale

Successes in Biological Control of Weeds

Biological control Agent	*Host (Pest*
Cactoblastis cactorum (Cactus moth - Phycitidae)	Prickly pear cactus
Chrysolina quadrigemina (Leaf beetle - Chrysomelidae)	Klamath weed
Rhinocyllus conicus (Thistle weevil -Curculionidae)	Musk thistle

Success of Pathogens in Biological Control

Pathogen	Host (Pest)
Nuclear polyhedrosis virus (Virus)	Cabbage looper
Bacillus thuringiensis (Bacterium)	Cabbageworm
Colletotrichum gloeosporioides (Fungus)	Northern jointvetch

General Rules for Releasing Biological Control Agents

There are some general rules governing the release of biological control agents. These include number and size of the releases, and where the releases should be made. The releases should be in each climatic zone that is occupied by the host so that the natural enemy will have a chance to develop in all areas where the host occurs. The releases should be large enough to ensure rapid establishment. Often more than one release in an area is needed to have successful establishment. Releases should be synchronized with the occurrence of an abundance of the preferred host growth stage, i.e. when the preferred host stage is present in large numbers.

It is important to make sure that the biological control agents are not adversely affected by ongoing activities in the area of release. Thus, temperature during the time of day that encourages the biological control agent to stay and not disperse should be taken into consideration for release. Also, pesticide applications that are routinely carried out in the area before release should be avoided so that they will not kill the biological control agents. To find out whether establishment has occurred, adequate recovery collections should be carried out. These should be made several times in the season and in as many sites as possible.

Time span in Biological Control

One of the commonly asked questions is how long it will take for biological control to be easily observed. In general, for effective control, it will take at least three years or three host generations. Thus, biological control is a long-term project and not a short-term process.

Single versus Multiple Biological Control Agents

The optimum number of species of biological control agents to be released for control of a given pest has generated much discussion. There are two schools of thought. One group supports the release of the "best" single biological control agent to decrease competition between species. The other supports the idea of using a number of species of biological control agents, as their combined effects would be greater than that of a single biological control agent. Decision on whether more than one biological control agent should be used should be based on two primary criteria:

- How complete is the geographic coverage of the host environment by the selected natural enemy?
- Is the natural enemy superior to all others over the whole range?

Multiple natural enemies should be used if the net impact is greater than that of a single natural enemy.

Research Emphasis in Classical Biological Control

The most important aspect of classical biological control is the selection of effective biological control agents that can be safely used in the release area without fear of non-target hosts being attacked. How successfully this can be done depends on appropriate host specificity tests to determine the potential host range. Major concerns are for host specificity, effectiveness, conflict of interest, and threat to endangered species.

Guidelines for Host Specificity Assessment

There are well-established guidelines for the host specificity testing of biological control agents of weeds, but guidelines for predators, parasites and pathogens of insect pests are not as well developed.

Host Specificity Studies

These studies are carried out to determine which organisms are likely to be attacked in the release area. Two tests that are involved are:

- o Host acceptance — to find out which hosts are accepted for egg laying
- o Host suitability — to ascertain whether feeding on a specific host can support the complete development of the biological control agent.

Host specificity tests can be conducted in the laboratory under quarantine in the target area or in its native habitat in the field.

The quarantine lab tests are to determine the physiological host range or maximum potential impact of the natural enemy being tested. These involve:

- o Feeding (starvation) tests that determine whether the natural enemy will feed or starve on the test host.
- o Developmental tests to determine whether the natural enemy can successfully complete development on the host and produce fertile progeny by the formation and maturation of eggs.

Field-tests in the native habitat are to determine the actual or ecological range. Only a fraction of the potential host range that is revealed in the lab test is likely to be shown in the field.

Advantages of Biological Control

- Free of side effects
- Safe to handle or use
- Occurs naturally
- High degree of host specificity
- Cost effective
- Self perpetuation
- Searching ability
- Survive at low host density

Disadvantages of Biological Control

- Slow to achieve results
- Impact often not dramatic
- Partial success
- Can be complex

Why is Biological Control not widely practiced?

- Need for quick fix
- Lack of patience
- Potent pesticides
- Industry advocates
- Easy application of pesticides

Problems associated with dependence on use of Pesticides

We cannot depend on pesticides to solve pest problems permanently due to:

- Reduced availability
- Pesticide resistance
- Environmental concerns

Reasons for using the Biological Control Method

Biological control should be implemented whenever possible because it does not pollute the environment. It uses natural enemies of the pests for their control. Most of the two major categories of pests in the USA, arthropods (insects and mites) and weeds, are exotic and since the use of their natural enemies has not been fully exploited, biological control should be the preferred method. Some of the reasons for use of biological control are:

- o Environmentally sound
- o Economically feasible
- o Time tested technique
- o Not fully exploited
- o Potential for use because natural enemies exist for all organisms
- o Role in sustainable agriculture
- o Federal mandate

Biological Control Organizations

The current important biological control organizations are:

1. International Organization of Biological Control of Noxious Animals and Plants (IOBC) formed in 1971. Its headquarters is in France.
2. International Institute of Biological Control (IIBC) formed in 1990. It is based in the United Kingdom.
3. National Biological Control Institute (NBCI), USA formed in 1990 and is based in Maryland.

Glossary of Biological Control Terms

The following are some of the more common terms used in biological control:

Abiotic refers to factors that include climate, soil, air, space, and light.

APHIS = Animal and plant Inspection Service.

Augmentation is the increase of native agents for control of native or exotic pests.

Biological Control is the use of natural enemies by man to control pest Biotic refers to effects of other organisms Biotic Potential is the inherent characteristics of an organism to survive and reproduce Environmental Resistance is the sum total of all mortality factors (physical such as temperature and biotic such as natural enemies) in the environment of an organism that prevents it from expressing its full capacity to reproduce.

Classical biological control is the use of exotic biological control agents imported from its native home into the target area against exotic pests that have arrived without their natural enemies.

Density dependent relationship refers to mortality factors that destroy a larger percentage of the target population as the target population increases. Density-Independent relationship refers to mortality factors that destroy a fixed percentage of the target population regardless of changes in the target density.

Host Selection is the process in which an organism selects its host. It has four phases. First is locating the host habitat, followed by finding the host in the habitat, selecting a suitable host for laying the eggs and finally, the successful development of the parasite in the host that results in producing fertile offspring.

Hyperparasite — a parasite that attacks another parasite.

IIBC = International Institute of Biological Control.

IOBC = International Organization of Biological Control of Noxious Animals and Plants.

Natural Control is the sum effect of natural control factors that limit the increase of an organism in an ecosystem.

NBCI = National Biological Control Institute.

Neoclassical biological control is the use of exotic biological control agents against a native pest

New Association refers to natural enemies that have not evolved with the target pest.

Old association refers to natural enemies that have evolved with the target pest in its place of origin.

Ovisorption is the process in which ripe eggs are not deposited, but are absorbed back into the body. This occurs when the female does not get enough proteins or when the female fails to locate a suitable host for laying the egg.

Parasite — an organism that lives and feeds in or on a host.

Parasitoid — an insect parasite that destroys its host.

Pathogen — a microorganism that causes disease in its host.

Predator — a living organism that feeds upon other organisms that are smaller and weaker than itself.

Preoviposition period is the interval between emergence of adult female and when it lays the first egg.

Provigenic species are those that lay their eggs in a short period of its life, but produces no more eggs during the rest of its life span.

Races or strains — members of the same race or strain look similar and have the same characteristics, but differ from other races in function and activity. Synovigenic species are those that lay their eggs throughout the adult stage. The female needs to feed and requires a protein source from honeydew, plant nectaries, or from feeding on the host.

Root-knot Nematode

Onychophora

The velvet worms (Onychophora — literally "nail bearers", also known as Protracheata) are a minor ecdysozoan phylum with ~180 species. These obscurely segmented organisms have tiny eyes, antennae, multiple pairs of legs and slime glands. They have variously been compared to worms with legs, caterpillars and slugs. Most common in tropical regions of the Southern Hemisphere, they prey on smaller animals such as insects, which they catch by squirting an adhesive mucus. In modern zoology, they are particularly renowned for their curious mating behaviour and for bearing live young.

The two extant families of velvet worms are Peripatidae and Peripatopsidae. They show a peculiar distribution, with the peripatids being predominantly equatorial and tropical, while the peripatopsids are all found in what used to be Gondwana. Formerly considered part of Tracheata, velvet worms are now considered close relatives of the Arthropoda and Tardigrada, with which they form the taxon Panarthropoda. This makes them of palaeontological interest, as they can help reconstruct the ancestral arthropod.

Anatomy

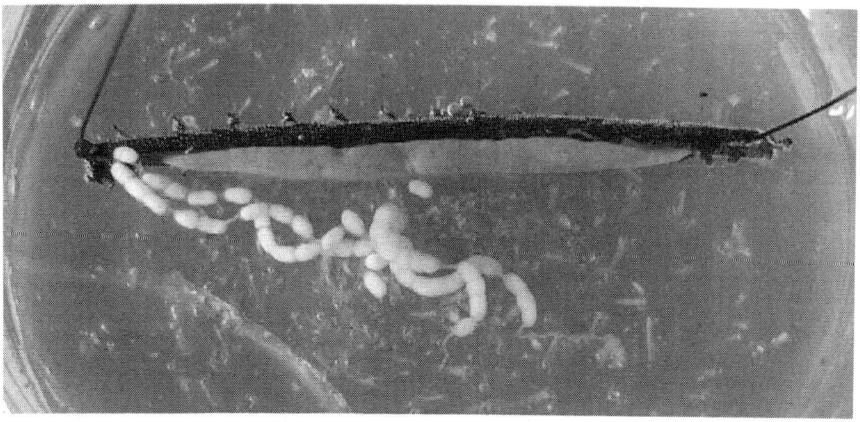

***Figure:** A dissection of* Euperipatoides kanangrensis. *The width of the dish is approx. 4 cm. The two ovaries, full of stage II embryos, are floating to the bottom of the image. The brown mass of the viscera is partially protruding from the body cavity.*

Velvet worms are segmented creatures with a flattened cylindrical body cross-section and rows of unstructured body appendages known as lobopods (informally: stub feet). The animals grow to between 0.5 and 20 cm (.2 to 8 in), with the average being about 5 cm (2 in), and have between 13 and 43 pairs of legs.

Their skin consists of numerous, fine transverse rings and is often inconspicuously coloured orange, red or brown, but sometimes also bright green, blue, gold or white, and occasionally patterned with other colours.

Segmenting—outwardly inconspicuous and identifiable only in the regular spacing of the pairs of legs—is visible in the regular arrangement of skin pores, excretion organs and concentrations of nerve cells. The individual body sections are largely unspecialised; even the head develops only a little differently from the abdominal segments.

Segmentation is apparently specified by the same gene as in other groups of animals and is activated in each case, during embryonic development, at the rear border of each segment and in the growth zone of the stub feet.

Body Appendages

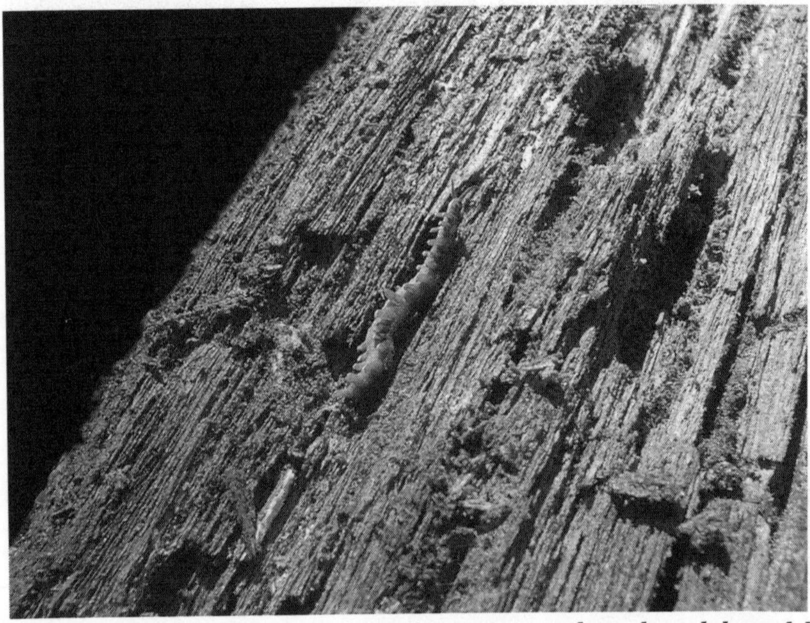

Figure: Euperipatoides kanangrensis, *overturned to show lobopod legs*

The stub feet that characterise the velvet worms are conical, baggy appendages of the body, which are internally hollow and have no joints. Although the number of feet can vary considerably between species, their structure is basically very similar. Rigidity is provided by the hydrostatic pressure of their fluid contents, and movement is usually obtained passively by stretching and contraction of the animal's entire body. However, each leg can also be shortened and bent by internal muscles. Due to the lack of joints, this bending can take place at any point along the sides of the leg.

In some species, two different organs are found within the feet:

- Crural glands are situated at the shoulder of the legs, extending into the body cavity. They open outwards at the crural papillae—small wart-like bumps on the belly side of the leg—and secrete chemical messenger materials called pheromones. Their name comes from the Latin *cruralis* meaning "of the legs".
- Coxal vesicles are pouches located on the belly side of the leg, which can be everted and probably serve in water absorption. They are only found within the family Peripatidae and are named from *coxa*, the Latin word for "hip".

On each foot is a pair of retractable, hardened (sclerotised) chitin claws, which give the taxon its scientific name: Onychophora is derived from the Greek *onyches*, "claws"; and *pherein*, "to carry". At the base of the claws are three to six spiny "cushions" on which the leg sits in its resting position and on which the animal walks over smooth substrates. The claws are used mainly to gain a firm foothold on uneven terrain.

Apart from the pairs of legs, there are three further body appendages, which are at the head and comprise three segments:

Antennae

- On the first head segment is a pair of slender antennae, which serve in sensory perception. They probably do not correspond directly to the antennae of the Arthropoda, but perhaps rather with their "lips" or labrum. At their base is found a pair of simple eyes, except in a few blind species. In front of these, in many Australian species, are various dimples, the function of which is not yet clear. It appears that in at least some species, these serve in the transfer of sperm-cell packages (spermatophores).

Mouthparts

- On the belly side of the second head segment is the labrum, a mouth opening surrounded by sensitive "lips". In the velvet worms, this structure is a muscular outgrowth of the throat, so, despite its name, it is probably not homologous to the labrum of the Arthropoda and is used for feeding. Deep within the oral cavity lie the sharp, crescent-shaped "jaws", or mandibles, which are strongly hardened and resemble the claws of the feet, with which they are probably homologous; early in development, the jaw appendages have a similar position and shape to the subsequent legs. The jaws are divided into internal and external mandibles and their concave surface bears fine denticles. They move backward and forward in a longitudinal direction, tearing apart the prey.

The surface of the mandibles is smooth, with no ornamentation. The cuticle in the mandibles (and claws) is distinct from the rest of the body. It has an inner and outer component; the outer component has just two layers (whereas body cuticle has four), and these outer layers (in particular the inner epicuticle) are dehydrated and strongly tanned, affording toughness.

Slime glands

- On the third head segment, to the left and right of the mouth, are two openings designated "oral papillae". Within these are a pair of large, heavily internally branched slime glands. These lie roughly in the centre of the body and secrete a sort of milky-white slime, which is used to ensnare prey and for defensive purposes. Sometimes the connecting "slime conductor" is broadened into a reservoir, which can buffer pre-produced slime. The slime glands themselves are probably modified crural glands.

All three structures correspond to an evolutionary origin in the leg pairs of the other segments.

Skin and Musculature

Figure: *A member of the Peripatidae family.*

Unlike the arthropods, velvet worms do not possess a rigid exoskeleton. Instead, their fluid-filled body cavity acts as a hydrostatic skeleton, similarly to many unrelated soft-bodied animals that are cylindrically shaped, for example sea anemones and various worms. Pressure of their incompressible internal bodily fluid on the body wall provides rigidity, and muscles are able to act against it.

The body wall consists of a non-cellular outer skin, the cuticula; a single layer of epidermis cells forming an internal skin; and beneath this, usually three layers of muscle, which are embedded in connective tissues.

The cuticula is about a micrometre thick and covered with fine villi. In composition and structure, it resembles the cuticula of the arthropods, consisting of á-chitin and various proteins, although not containing collagen. It can be divided into an external epicuticula and an internal procuticula, which themselves consist of exo- and endo-cuticula. This multi-level structure is responsible for the high flexibility of the outer skin, which enables the velvet worm to squeeze itself into the narrowest crevices. Although outwardly water-repellent, the cuticula is not able to prevent water loss by respiration, and, as a result, velvet worms can only live in microclimates with high humidity to avoid desiccation.

The surface of the cuticula is scattered with numerous fine papillae, the larger of which carry visible villi-like sensitive bristles. The papillae themselves are covered with tiny scales, lending the skin a velvety appearance (from which the common name is likely derived). It also feels like dry velvet to the touch, for which its water-repellent nature is responsible. Moulting of the skin (ecdysis) takes place regularly, around every 14 days, induced by the hormone ecdysone. The inner surface of the skin bears a hexagonal pattern.

At each moult, the shed skin is replaced by the epidermis, which lies immediately beneath it; unlike the cuticula, this consists of living cells. Beneath this lies a thick layer of connective tissue, which is composed primarily of collagen fibres aligned either parallel or perpendicular to the body's longitudinal axis. Within the connective tissue lie three continuous layers of unspecialised smooth muscular tissue. The relatively thick outer layer is composed of annular muscles, and the similarly voluminous inner layer of longitudinal muscles. Between them lie thin diagonal muscles that wind backward and forward along the body axis in a spiral. Between the annular and diagonal muscles exist fine blood vessels, which lie below the superficially recognisable transverse rings of the skin and are responsible for the pseudo-segmented markings.

Beneath the internal muscle layer lies the body cavity. In cross-section, this is divided into three regions by so-called dorso-ventral muscles, which run from the middle of the underbelly through to the edges of the upper side: a central midsection and on the left and right, two side regions that also include the legs.

The colouration of Onychophora is generated by a range of pigments. The solubility of these pigments is a useful diagnostic character: in all arthropods and tardigrades, the body pigment is soluble in ethanol. This is also true for the Peripatidae, but in the case of the Peripatopsidae, the body pigment is insoluble in ethanol.

Haemocoel and Circulation

The body cavity is known as a "pseudocoel", or haemocoel. Unlike a true coelom, a pseudocoel is not fully enclosed by a cell layer derived from the embryonic mesoderm. A coelom is, however, formed around the gonads and the waste-eliminating nephridia.

As the name *haemocoel* suggests, the body cavity is filled with a blood-like liquid in which all the organs are embedded; in this way, they can be easily supplied with nutrients circulating in the blood. This liquid is colourless as it does not contain pigments; for this reason, it only serves a limited role in oxygen transport. Two different types of blood cells (or haemocytes) circulate in the fluid: amoebocytes and nephrocytes. The amoebocytes probably function in protection from bacteria and other foreign bodies; in some species, they also play a role in reproduction. Nephrocytes absorb toxins or convert them into a form suitable for elimination by the nephridia.

The haemocoel is divided by a horizontal partition, the diaphragm, into two parts: the pericardial sinus along the back and the perivisceral sinus along the belly. The former encloses the tube-like heart, and the latter, the other organs. The diaphragm is perforated in many places, enabling the exchange of fluids between the two cavities.

The heart itself is a tube of annular muscles consisting of epithelial tissues, with two lateral openings (ostia) per segment. While it is not known whether the rear end is open or closed, from the front, it opens directly into the body cavity. Since there are no blood vessels, apart from the fine vessels running between the muscle layers of the body wall and a pair of arteries that supply the antennae, this is referred to as an open circulation.

The timing of the pumping procedure can be divided into two parts: diastole and systole. During diastole, blood flows through the ostia from the pericardial sinus (the cavity containing the heart) into the heart. When the systole begins, the ostia close and the heart muscles contract inwards, reducing the volume of the heart. This pumps the blood from the front end of the heart into the perivisceral sinus containing the organs. In this way, the various organs are supplied with nutrients before the blood finally returns to the pericardial sinus via the perforations in the diaphragm. In addition to the pumping action of the heart, body movements also have an influence on circulation.

Respiration

Oxygen uptake occurs to an extent via simple diffusion through the entire body surface, with the coxal vesicles on the legs possibly being involved in some species. However, of most importance is gas exchange via fine unbranched tubes, the tracheae, which draw oxygen from the surface deep into the various organs, particularly the heart. The walls of these structures, which are less than three micrometres thick in their entirety, consist only of an extremely thin membrane through which oxygen can easily diffuse. The tracheae originate at tiny openings, the spiracles, which themselves are clustered together in dent-like recesses of the outer skin, the atria. The number of "tracheae bundles" thus formed is on average around 75 per body segment; they accumulate most densely on the back of the organism.

Unlike the arthropods, the velvet worms are unable to control the openings of their tracheae; the tracheae are always open, entailing considerable water loss in arid conditions. Water is lost twice as fast as in earthworms and forty times faster than in caterpillars. For this reason, velvet worms are dependent upon habitats with high air humidity.

Digestive System

The digestive tract begins slightly behind the head, the mouth lying on the underside a little way from the frontmost point of the body. Here, prey can be mechanically dismembered by the mandibles with their covering of fine toothlets. Two salivary glands discharge via a common conductor into the subsequent "throat", which makes up the first part of the front intestine. The saliva that they produce contains mucus and hydrolytic enzymes, which initiate digestion both within and outside the mouth. Historically, the salivary glands probably evolved from the waste-elimination organs known as nephridia, which are found homologously in the other body segments.

The throat itself is very muscular, serving to absorb the partially liquified food and to pump it, via the oesophagus, which forms the rear part of the front intestine, into the central intestine. Unlike the front intestine, this is not lined with a cuticula but instead consists only of a single layer of epithelial tissue, which does not exhibit conspicuous indentation as is found in other animals. On entering the central intestine, food particles are coated with a mucus-based peritrophic membrane, which serves to protect the lining of the intestine from damage by sharp-edged particles. The intestinal epithelium secretes further digestive enzymes and absorbs the released nutrients, although the majority of digestion has already taken place externally or in the mouth. Indigestible remnants arrive in the rear intestine, or rectum, which is once again lined with a cuticula and which opens at the anus, located on the underside near to the rear end.

Excretory Organs

In almost every segment is a pair of excretory organs called nephridia, which are derived from coelom tissue. Each consists of a small pouch that is connected, via a flagellated conductor called a nephridioduct, to an opening at the base of the nearest leg known as a nephridiopore. The pouch is occupied by special cells called podocytes, which facilitate ultrafiltration of the blood through the partition between haemocoelom and nephridium.

The composition of the urinary solution is modified in the nephridioduct by selective recovery of nutrients and water and by isolation of poison and waste materials, before it is excreted to the outside world via the nephridiopore. The most important nitrogenous excretion product is the water-insoluble uric acid; this can be excreted in solid state, with very little water. This so-called uricotelic excretory mode represents an adjustment to life on land and the associated necessity of dealing economically with water.

A pair of former nephridia in the head were converted secondarily into the salivary glands, while another pair in the final segment of male specimens now serve as glands that apparently play a role in reproduction.

Sensory Organs

The entire body—including the stub feet—is littered with numerous papillae: warty protrusions that carry a mechanoreceptive bristle (responsive to mechanical stimuli) at the tip, each of which is also connected to further sensory nerve cells lying beneath. The mouth papillae, the exits of the slime glands, probably also have a function in sensory perception.

Sensory cells known as "sensills" on the "lips" or labrum respond to chemical stimuli and are known as chemoreceptors. These are also found on the two antennae, which can be regarded as the velvet worm's most important sensory organs. Except in a few (typically subterranean) species, one simply constructed eye (ocellus) lies laterally, just underneath the head, behind each antenna. This consists of a chitinous ball lens, a cornea and a retina and is connected to the centre of the brain via an optic nerve. The retina comprises numerous pigment cells and photoreceptors; the latter are easily modified flagellated cells, whose flagellum membranes carry a photosensitive pigment on their surface.

The rhabdomeric eyes of the Onychophora are thought to be homologous with the median ocelli of arthropods; this would imply that the last common ancestor of arthropods bore only median ocelli. However, the innervation shows that the homology is limited: the eyes of Onychophora form behind the antenna, whereas the opposite is true in arthropods.

Reproductive Organs

Both sexes possess pairs of gonads, opening via a channel called a gonoduct into a common genital opening, the gonopore, which is located on the rear ventral side. Both the gonads and the gonoduct are derived from true coelom tissue.

In females, the two ovaries are joined in the middle and to the horizontal diaphragm. The gonoduct appears differently depending on whether the species is live-bearing or egg-laying. In the former, each exit channel divides into a slender oviduct and a roomy "womb", the uterus, in which the embryos develop.

The single vagina, to which both uteri are connected, runs outward to the gonopore. In egg-laying species, whose gonoduct is uniformly constructed, the genital opening lies at the tip of a long egg-laying apparatus, the ovipositor. The females of many species also possess a sperm repository called the receptacle seminis, in which sperm cells from males can be stored temporarily or for longer periods.

Males possess two separate testes, along with the corresponding sperm vesicle (the vesicula seminalis) and exit channel (the vasa efferentia). The two vasa efferentia unite to a common sperm duct, the vas deferens, which in turn widens through the ejaculatory channel to open at the gonopore. Directly beside or behind this lie two pairs of special glands, which probably serve an auxiliary reproductive function; the rearmost glands are also known as anal glands.

A penis-like structure has so far only been found in males of the genus *Paraperipatus* but has not yet been observed in action. As previously mentioned, males of many Australian species exhibit special structures on the head, which apparently take over certain tasks in transferring sperm to the females. In the species *Euperipatoides rowelli*, sperm is collected by these structures, and, when a female is encountered, the worm inserts its head in the vagina.

Distribution and Habitat

Velvet worms live in all tropical habitats and in the temperate zone of the Southern Hemisphere, showing a circumtropical and circumaustral distribution. Individual species are found in Central and South America; the Caribbean islands; equatorial West Africa and South Africa; northeastern India; Indonesia and parts of Malaysia; New Guinea; Australia; and New Zealand. Fossils have been found in Baltic amber, indicating that they were formerly more widespread in the Northern Hemisphere when conditions were more suitable.

Figure: Euperipatoides kanangrensis, *enrolled due to sudden exposure (i.e. being plucked from its log into the desiccating daylight)*

All extant velvet worms are terrestrial (land-living) and prefer dark environments with high air humidity. They are found particularly in the rainforests of the tropics and temperate zones, where they live among moss cushions and leaf litter, under tree trunks and stones, in rotting wood or in termite tunnels. They also occur in unforested grassland, if there exist sufficient crevices in the soil into which they can withdraw during the day.

Two species live in caves, a habitat to which their ability to squeeze themselves into the smallest cracks makes them exceptionally well-adapted and in which constant living conditions are guaranteed. Since the essential requirements for cave life were probably already present prior to the settlement of these habitats, this may be described as exaptation. Agriculture has apparently made available new habitats for velvet worms; in any case, they are found in man-made cocoa and banana plantations in South America and the Caribbean.

Because the danger of desiccation is greatest during the day and in dry weather, it is not surprising that velvet worms are usually most active at night and during rainy weather. Under cold or dry conditions, they actively seek out crevices in which they shift their body into a resting state. Velvet worms are negatively phototactical: they are repelled by bright light sources. The largest measured population density is very low, at approximately ten individuals per square metre; velvet worms are often difficult to find in their natural habitat.

Slime

The slime of the Onychophora is forcefully squirted from a pair of slime glands (oral papillae) in defence against predators and to capture prey. The slime glands, positioned on the sides of the head below the antennae, are a pair of highly modified limbs and typically propel the slime around a centimetre. The slime can be propelled up to four centimetres, although accuracy drops with range, which is usually much shorter than this. One squirt usually suffices to snare a prey item, although larger prey may be further immobilised by smaller squirts targeted at the limbs; additionally, the fangs of spiders are sometimes targeted.

The slime, which can account for up to 11% of the organism's dry weight, is 90% water; its dry residue consists mainly of proteins—primarily a collagen-type protein. 1.3% of the slime's dry weight consists of sugars, mainly glactosamine. The slime also contains lipids and the surfactant nonylphenol. Onychophora are the only organisms known to produce this latter substance.

The proteinaceous composition accounts for the slime's high tensile strength and stretchiness. Upon ejection, it forms a net of threads about 20 μm in diametre, with evenly spaced droplets of viscous adhesive fluid along their length. It subsequently dries, shrinking, losing its stickiness, and becoming brittle. Onychophora will eat and "reuse" any dried slime.

The lipid and nonylphenol constituents may serve one of two purposes. They may line the ejection channel, stopping the slime from sticking to the organism when it is secreted; or they may slow the drying process long enough for the slime to reach its target.

It takes an onychophoran around 24 days to replenish an exhausted slime repository.

Behaviour

Locomotion

Velvet worms move in a slow and gradual motion that makes them difficult for prey to notice. Their trunk is raised relatively high above the ground, and they walk with non-overlapping steps. To move from place to place, the velvet worm crawls forward using its legs; unlike in arthropods, both legs of a pair are moved simultaneously. The claws of the feet are only used on hard, rough terrain where a firm grip is needed; on soft substrates, such as moss, the velvet worm walks on the foot cushions at the base of the claws.

Figure: Peripatoides *sp., clearly showing the stub feet*

The actual locomotion is achieved less by the exertion of the leg muscles than by local changes of body length. This can be controlled using the annular and longitudinal muscles. If the annular muscles are contracted, the body cross-section is reduced, and the corresponding segment stretches, since its volume must remain constant due to the incompressible behaviour of the coelom's liquid contents; this is the usual mode of operation of the hydrostatic skeleton as also employed by other worms.

Due to the stretching, the legs of the segment concerned are lifted and swung forward. Local contraction of the longitudinal muscles then shortens the appropriate segment, and the legs, which are now in contact with the ground, are moved to the rear. This part of the locomotive cycle is the actual leg stroke that is responsible for forward movement. The individual stretches and contractions of the segments are coordinated by the nervous system such that contraction waves run the length of the body, each pair of legs swinging forward and then down and rearward in succession. The organisms can reach speeds of up to four centimetres per second.

Social Interaction

The brains of Onychophora, though small, are very complex; consequently, the organisms are capable of rather sophisticated social interactions. Behaviour may vary from genus to genus, so this article reflects the most studied genus, *Euperipatoides*.

They form social groups of up to fifteen individuals, usually closely related, which will typically live and hunt together. Groups usually live together, an example in drier regions would be in a region of the moist interior of a rotting log. Group members are extremely aggressive towards individuals from other logs. Dominance is achieved through aggression and maintained through submissive behaviour. After a kill, the dominant female always feeds first, followed in turn by the other females, then males, then the young. Social hierarchy is established by a number of interactions: higher-ranking individuals will chase, bite and crawl on top of their subordinates. Juveniles never engage in aggressive behaviour, but climb on top of adults, which tolerate their presence on their backs. When assessing other individuals, individuals often measure one another up by running their antennae down the length of the other individual. Once hierarchy has been established, paired individuals will often cluster together to form an aggregate; this is fastest in male-female pairings, followed by pairs of females, then pairs of males. Hierarchy is quickly established between individuals from a single group, but not between organisms from different groups; these are substantially more aggressive and very rarely climb one another or form aggregates.

Distribution

Individuals within an individual log are usually closely related; especially so with males. This may be related to the intense aggression between unrelated females.

Feeding

Velvet worms are ambush predators, hunting only by night, and are able to capture animals at least their own size, although it may take almost all of their slime-secreting capacity to capture a large prey item. They feed on almost any small invertebrates, including woodlice (Isopoda), termites (Isoptera), crickets (Gryllidae), book/bark lice (Psocoptera), cockroaches (Blattidae), millipedes and centipedes (Myriapoda), spiders (Araneae), various worms and even large snails (Gastropoda). Depending on their size, they eat on average every one to four weeks. They are considered to be ecologically equivalent to centipedes (Chilopoda).

Potential victims can be detected from up to four centimetres away and are investigated by the gentle application of the antennae. If they are judged to be a suitable size, slime is ejected to immobilise the prey item. The most energetically favourable prey are two-fifths the size of the hunting onychophoran. The onychophoran bites into the prey and injects saliva, which further reduces motion and may initiate digestion of the prey item's innards. Ninety percent of the time involved in eating a specimen is spent ingesting it; re-ingestion of the slime used to trap the insect is performed while the onychophoran locates a suitable place to puncture the prey, and this phase accounts for around 8% of the feeding time, with the remaining time evenly split between examining, squirting and injecting the prey.

Onychophora probably do not primarily use vision to detect their prey; although their tiny eyes do have a good image-forming capacity, their forward vision is obscured by their antennae; their nocturnal habit also limits the utility of eyesight. Air currents, formed by prey motion, are thought to be the primary mode of locating prey; the role of scent, if any, is unclear. The animals literally creep up on their prey, with their smooth, gradual and fluid movement escaping detection by predators. Once they reach their prey, they touch it very softly with their antennae to assess its size and nutritional value. After each poke, the antenna is hastily retracted to avoid detection by the prey individual. This investigation may last anywhere upwards of ten seconds, until the velvet worm makes a decision as to whether to attack it—or until it disturbs the prey and flees. Hungry Onychophora spend less time investigating their prey and are quicker to apply their slime. Once slime has been squirted, Onychophora will stop at nothing to hound down and devour their prey, in order to recoup the energetic investment. They have been observed to spend up to ten minutes searching for removed prey, after which they return to their slime to eat it. In the case of smaller prey, they may opt not to slime it at all.

Subsequently, a soft part of the prey item (usually a joint membrane in arthropod prey) is identified, punctured with a bite from the jaws, and injected with saliva. This kills the prey very quickly and begins a slower process of digestion. While the onychophoran waits for the prey to digest, it salivates on its slime and begins to eat it (and anything attached to it). It subsequently tugs and slices at the earlier perforation to allow access to the now-liquefied innards of its prey. In social groups, the dominant female is the first to feed, not permitting competitors access to the prey item for the first hour of feeding. Subsequently, subordinate individuals begin to feed. The number of males reaches a peak after females start to leave the prey item. After feeding, individuals clean their antennae and mouth parts before re-joining the rest of their group. Because it takes so long to ingest a prey item, hunting mainly happens around dusk; the onychophorans will abandon their prey at sunrise.

This predatory way of life is probably a consequence of the velvet worm's need to remain moist. Due to the continual risk of desiccation, often only a few hours per day are available for finding food. This leads to a strong selection for a low cost-benefit ratio, which can barely be achieved with a herbivorous diet.

Reproduction and Life-cycle

Almost all species of velvet worm reproduce sexually. The sole exception is *Epiperipatus imthurni*, of which no males have been observed; reproduction instead occurs by parthenogenesis.

All species are in principle sexually distinct and bear, in many cases, a marked sexual dimorphism: the females are usually larger than the males and have, in species where the number of legs is variable, more legs. The females of many species are fertilized only once during their lives, which leads to copulation sometimes taking place before the reproductive organ of the females are fully developed. In such cases, for example at the age of three months in *Macroperipatus torquatus*, the transferred sperm cells are kept in a special reservoir, where they can remain viable for longer periods.

Fertilization takes place internally, although the mode of sperm transmission varies widely. In most species, for example in the genus *Peripatus*, a package of sperm cells called the spermatophore is placed into the genital opening of the female. The detailed process by which this is achieved is in most cases still unknown, a true penis having only been observed in species of the genus *Paraperipatus*. In many Australian species, there exist dimples or special dagger- or axe-shaped structures on the head; the male of *Florelliceps stutchburyae* presses a long spine against the female's genital opening and probably positions its spermatophore there in this way. During the process, the female supports the male by keeping him clasped with the claws of her last pair of legs. The mating behaviour of two species of the genus *Peripatopsis* is particularly curious. Here, the male places two-millimetre spermatophores on the back or sides of the female. Amoebocytes from the female's blood collect on the inside of the deposition site, and both the spermatophore's casing and the body wall on which it rests are decomposed via the secretion of enzymes. This releases the sperm cells, which then move freely through the haemocoel, penetrate the external wall of the ovaries and finally fertilize the ova. Why this self-inflicted skin injury does not lead to bacterial infections is not yet understood (though likely related to the enzymes used to deteriorate the skin or facilitate the transfer of viable genetic material from male to female).

Velvet worms are found in egg-laying (oviparous), egg-live-bearing (ovoviviparous) and live-bearing (viviparous) forms.

- Ovipary occur solely in the Peripatopsidae, often in regions with erratic food supply or unsettled climate. In these cases, the yolk-rich eggs measure 1.3 to 2.0 mm and are coated in a protective chitinous shell. Maternal care is unknown.

- The majority of species are ovoviviparous: the medium-sized eggs, encased only by a double membrane, remain in the uterus. The embryos do not receive food directly from the mother, but are supplied instead by the moderate quantity of yolk contained in the eggs—they are therefore described as lecithotrophic. The young emerge from the eggs only a short time before birth. This probably represents the velvet worm's original mode of reproduction, i.e., both oviparous and viviparous species developed from ovoviviparous species.
- True live-bearing species are found in both families, particularly in tropical regions with a stable climate and regular food supply throughout the year. The embryos develop from eggs only micrometres in size and are nourished in the uterus by their mother, hence the description "matrotrophic". The supply of food takes place either via a secretion from the mother directly into the uterus or via a genuine tissue connection between the epithelium of the uterus and the developing embryo, known as a placenta. The former is found only outside the American continent, while the latter occurs primarily in America and the Caribbean and more rarely in the Old World. The gestation period can amount to up to 15 months, at the end of which the offspring emerge in an advanced stage of development. The embryos found in the uterus of a single female do not necessarily have to be of the same age; it is quite possible for there to be offspring at different stages of development and descended from different males. In some species, young tend to be released only at certain points in the year.

A female can have between 1 and 23 offspring per year; development from fertilized ovum to adult takes between 6 and 17 months and does not have a larval stage. This is probably also the original mode of development. Velvet worms have been known to live for up to six years.

Ecology

Predators and Parasites

The velvet worm's important predators are primarily various spiders and centipedes, along with rodents and birds, such as, in Central America, the Clay-coloured Thrush (*Turdus grayi*). Hemprichi's Coral Snake (*Micrurus hemprichii*) feeds almost exclusively on velvet worms. For defence, some species roll themselves reflexively into a spiral, while they can also fight off smaller opponents by ejecting slime.

Various mites (Acari) are known as ectoparasites, which infest the skin of the velvet worm. Skin injuries are usually accompanied by bacterial infections, which are almost always fatal.

Conservation Status

Figure: *Recently discovered* Eoperipatus totoro *is known from only one location in southern Vietnam*

The global conservation status of velvet worm species is difficult to estimate; many species are only known to exist at their type locality (the location at which they were first observed and described). The collection of reliable data is also hindered by low population densities, their typically nocturnal behaviour and possibly also as-yet undocumented seasonal influences and sexual dimorphism.

To date, only eleven species have been studied in sufficient detail to enable population estimates, of which three—*Opistopatus roseus, Speleoperipatus spelaeus* and *Peripatopsis leonina*—are considered critically endangered, the last being probably already extinct. Two species—*Macroperipatus insularis* and *Tasmanipatus anophthalmus*—are assessed by the IUCN as endangered, while four further species are listed as threatened.

The primary threat comes from destruction and fragmentation of velvet worm habitat due to industrialisation, draining of wetlands and "slash and burn" for agriculture. Many species also have naturally low population densities and closely restricted geographic ranges; as a result, relatively small localised disturbances of important ecosystems can lead to the extinction of entire populations or species. Collection of specimens for universities or research institutes also plays a role on a local scale.

There is a very pronounced difference in the protection afforded to velvet worms between regions: in some countries, such as South Africa, there are restrictions on both collecting and exporting, while in others, such as Australia, only export restrictions exist. Many countries offer no specific safeguards at all. Tasmania has a protection programme that is unique worldwide: one region of forest has its own velvet worm conservation plan, which is tailored to a particular velvet worm species.

Phylogeny

In their present forms, the velvet worms are probably very closely related to the arthropods, a very extensive taxon that incorporates, for instance, the crustaceans, insects and arachnids. They share, among other things, an exoskeleton consisting of á-chitin and non-collagenous proteins; gonads and waste-elimination organs enclosed in true coelom tissue; an open blood system with a tubular heart situated at the rear; an abdominal cavity divided into pericardial and perivisceral cavities; respiration via tracheae; and similar embryonic development. Segmentation, with two body appendages per segment, is also shared. However, antennae, mandibles and oral papillae are probably not homologous to the corresponding features in arthropods, i.e., they probably developed independently.

Another closely related group are the comparatively obscure water bears (Tardigrada); however, due to their very small size, these lack some characteristics of the velvet worms and arthropods, such as blood circulation and separate respiratory structures. Together, the velvet worms, arthropods and water bears form a monophyletic taxon, the Panarthropoda, i.e., the three groups collectively cover all descendants of their last common ancestor.

Due to certain similarities of form, the velvet worms were usually grouped with the water bears to form the taxon Protoarthropoda. This designation would imply that both velvet worms and water bears are not yet as highly developed as the arthropods. Modern systematic theories reject such conceptions of "primitive" and "highly developed" organisms and instead consider exclusively the historical relationships among the taxa.

For a long time, velvet worms were also considered related to the annelids. They share, among other things, a worm-like body; a thin and flexible outer skin; a layered musculature; paired waste-elimination organs; as well as a simply constructed brain and simple eyes. Decisive, however, was the existence of segmentation in both groups, with the segments showing only minor specialisation. The parapodia appendages found in annelids therefore correspond to the stump feet of the velvet worms.

Within the Articulata concept developed by Georges Cuvier, the velvet worms therefore formed an evolutionary link between the annelids and the arthropods: worm-like precursors first developed parapodia, which then developed further into stub feet as an intermediate link in the ultimate development of the arthropods' appendages. Due to their structural conservatism, the velvet worms were thus considered "living fossils". This perspective was expressed paradigmatically in the statement by the French zoologist A. Vandel:

Onychophorans can be considered highly evolved annelids, adapted to terrestrial life, which announced prophetically the Arthropoda. They are a lateral branch which has endured from ancient times until today, without important modifications.

Modern taxonomy does not study criteria such as "higher" and "lower" states of development or distinctions between "main" and "side" branches—only family relationships indicated by cladistic methods are considered relevant. From this point of view, several common characteristics still support the Articulata concept—segmented body; paired appendages on each segment; pairwise arrangement of waste-elimination organs in each segment; and above all, a rope-ladder-like nervous system based on a double nerve strand lying along the belly.

An alternative concept, most widely accepted today, is the so-called Ecdysozoa hypothesis. This places the annelids and Panarthropoda in two very different groups: the former in the Lophotrochozoa and the latter in the Ecdysozoa. Mitochondrial gene sequences also provide support for this hypothesis.

Proponents of this hypothesis assume that the aforementioned similarities between annelids and velvet worms either developed convergently or were primitive characteristics passed unchanged from a common ancestor to both the Lophotrochozoa and Ecdysozoa. For example, in the first case, the rope-ladder nervous system would have developed in the two groups independently, while in the second case, it is a very old characteristic, which does not imply a particularly close relationship between the annelids and Panarthropoda.

The Ecdysozoa concept divides the taxon into two, the Panarthropoda into which the velvet worms are placed, and the sister group

Cycloneuralia, containing the threadworms (Nematoda), horsehair worms (Nematomorpha) and three rather obscure groups: the mud dragons (Kinorhyncha); penis worms (Priapulida); and brush-heads (Loricifera).

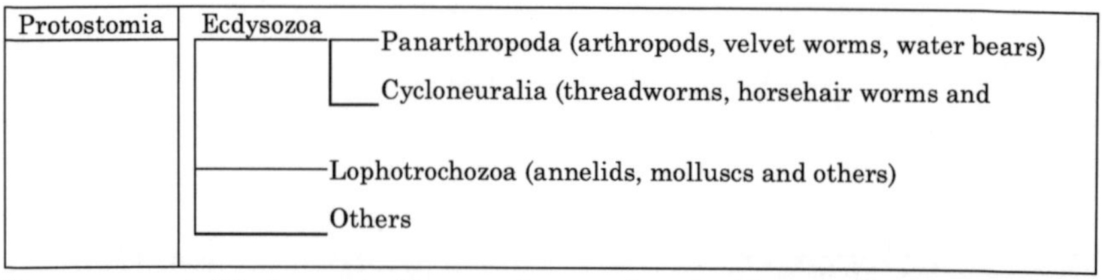

ProtostomiaEcdysozoaPanarthropoda (arthropods, velvet worms, water bears)Cycloneuralia (threadworms, horsehair worms and others)Lophotrochozoa (annelids, molluscs and others)Others

Particularly characteristic of the Cycloneuralia is a ring of "circumoral" nerves around the mouth opening, which the proponents of the Ecdysozoa hypothesis also recognise in modified form in the details of the nerve patterns of the Panarthropoda. Both groups also share a common skin-shedding mechanism (ecdysis) and molecular biological similarities. One problem of the Ecdysozoa hypothesis is the velvet worms' subterminal mouth position: unlike in the Cycloneuralia, the mouth is not at the front end of the body, but lies further back under the belly. However, investigations into their developmental biology, particularly regarding the development of the head nerves, suggest that this was not always the case and that the mouth was originally terminal (situated at the tip of the body). This is supported by the fossil record.

The "stem-group arthropod" hypothesis is very widely accepted, but some trees suggest that the onychophorans may occupy a different position; their brain anatomy is more closely related to that of the chelicerates than to any other arthropod.

Evolutionary History

Fossils from the early Cambrian bear a striking resemblance to the velvet worms. These fossils, known collectively as the lobopodians, were marine and may represent a stem group to the onychophorans or a grade from which arthropods and onychophora arose. They are found in the Cambrian, Ordovician (possibly), Silurian and Pennsylvanian periods.

Historically, all fossil Onychophora and lobopods were lumped into the taxon Xenusia, further subdivided by some authors to the Paleozoic Udeonychophora and the Mesozoic/Tertiary Ontonychophora; living Onychophora were termed Euonychophora.

Importantly, none of the Cambrian fossils bear features that distinctively unite them with the Onychophora; none can be confidently assigned to the onychophoran crown or even stem group.

Terrestrialization

It is not clear when the transition to a terrestrial existence was made, but it is considered plausible that it took place between the Ordovician and late Silurian—approximately 490 to 430 million years ago—via the intertidal zone.

The low preservation potential of the non-mineralised Onychophora means that they have a sparse fossil record. Stem-group members include *Ilyodes* (=*Helenodora*, Carboniferous), *Tertiapatus dominicanus* and *Succinipatopsis balticus* (Tertiary); crown group representatives are known only from amber—there is a single, partial specimen from the Cretaceous, and a more comprehensive record in Eocene deposits from 40 million years ago.

But, in fact, some of these amber-borne specimens lack slime papillae and separate feet, thus may belong in the stem group.

Taphonomy

The vagaries of the preservation process can make fossils difficult to interpret.

Experiments on the decay and compaction of onychophora demonstrate difficulties in interpreting fossils; certain parts of living onychophora are only visible in certain conditions.

The mouth may or may not be preserved; claws may be re-oriented or lost; leg width may increase or decrease; and mud may be mistaken for organs.

More significantly, features seen in fossils may be artefacts of the preservation process: for instance, "shoulder pads" may simply be the second row of legs compressed coaxially onto the body; branching "antennae" may in fact be produced through decay.

Taxonomy

The modern velvet worms form a monophyletic group, incorporating all the descendants of their common ancestor.

Important common derivative characteristics (synapomorphies) include, for example, the mandibles of the second body segment and the oral papillae and associated slime glands of the third; nerve strands extending along the underside with numerous cross-linkages per segment; and the special form of the tracheae.

By 2011, some 180 modern species, comprising 49 genera, had been described; the actual number of species is probably about twice this. According to more recent study, 82 species of Peripatidae and 115 species of Peripatopsidae have been described thus far. However, among these 197 species, 20 are nomina dubia due to major taxonomic inconsistencies. The best-known is the type genus *Peripatus*, which was described as early as 1825 and which, in English-speaking countries, stands representative for all velvet worms.

All genera are assigned to one of two families, the distribution ranges of which do not overlap but are separated by arid areas or oceans:

- The Peripatopsidae exhibit relatively many characteristics that are perceived as original or "primitive". They have between 13 and 25 pairs of legs, behind or between the last of which is the genital opening (gonopore). Both oviparous and ovoviviparous, as well as genuinely viviparous, species exist, although the Peripatopsidae essentially lack a placenta. Their distribution is circumaustral, encompassing Australia, Tasmania, New Zealand, New Guinea, South Africa and Chile.
- The Peripatidae exhibit a range of derivative features. They are longer, on average, than the Peripatopsidae and also have more leg pairs, numbering between 22 and 43—the gonopore is always between the penultimate pair. There are no oviparous species—the overwhelming majority are viviparous. The females of many viviparous species develop a placenta with which to provide the growing embryo with nutrients. Distribution of the Peripatidae is restricted to the tropical and subtropical zones; in particular, they inhabit Central and northern South America, several Caribbean islands, West Africa, northern India, Malaysia and various Indonesian islands.

2

Methods to Reduce Nematode Populations

There are numerous general management practices to take into consideration when determining how to manage nematodes. These practices include crop rotation with a nonhost or poor-host crop, resistant varieties, fallowing and chemical controls.

Not all of these methods will yield the same results, but they will each provide varying levels of nematode management. Most of these methods, with the exception of chemical controls, will gradually reduce the amount of nematodes in a field, whereas chemical control is more apt to quickly reduce nematodes. When possible, each method should be combined with other methods to produce the best results.

Four Practices Commonly Used to Effectively Combat Nematodes

Crop Rotation	The most important aspect of an effective crop rotation is to find crops that are poor hosts for the nematodes commonly associated with tomatoes. Because numerous species of nematodes are commonly found in one field, it's important to select a crop that is a nonhost for root-knot nematodes, the most damaging nematode in tomato production. Hairy indigo, American jointvetch, sorghum, rye, barley, wheat and pangola digitgrass have shown to be effective in limiting populations of root-knot nematodes.

Contd...

Resistant Varieties	Resistant varieties are useful for control of some nematodes and diseases in tomatoes. Nematodes fail to develop and reproduce in the roots of resistant varieties, giving tomatoes a better chance to grow as they would normally. Growers should not rely on the same variety for disease control year after year and should change varieties to slow down the development of new problems. Additionally, not all nematode-resistant varieties of tomatoes have proven to be consistently resistant, so other practices like crop rotation and chemical controls should be used in concert with the resistant varieties.
Fallowing	Having clean land is often a very effective method of reducing nematodes. Populations of nematodes hinge on the availability of a food source; without a crop to feed on, nematode levels will dwindle due to starvation. However, weeds and volunteer crops need to be controlled in a plowed, fallow field as nematodes have a wide host range. If properly managed, a fallow field can decrease the level of nematodes.
Chemical Controls	Nematicides, such as Telone® II soil fumigant, are proven to reduce nematode populations and, in turn, improve fruit production. Telone II is injected into the soil as a liquid and immediately converts to a gas, which creates a zone of protection around developing roots. As a fumigant, Telone II moves through the soil on its own, rather than requiring water or incorporation for movement.

Plant-parasitic nematodes cause major economic losses to horticultural and field crops in the U.S. Current control methods rely largely on nematicides (both fumigant and non-fumigant types) for most crops and pre-plant soil fumigation with the broad-purpose fumigant methyl bromide for high value fruit and vegetable crops. However, there are serious concerns about the use of nematicides in terms of food safety and environmental quality issues.

In addition, methyl bromide is proposed to be withdrawn from use in the U.S. in 2005. Alternative environmentally compatible control strategies that use bio-based options (including resistant, non-host, or antagonistic crop cultivars and biological control agents) are urgently needed to replace high-risk chemical nematicides. Importance of work and consequences if not done. Identification, characterization, and enhancement of resistant crop germplasm, antagonistic crops cover crops, and biocontrol agents are essential for the development of bio-based management options against plant-parasitic nematodes. Assessment of the effects of such management techniques on diversity of both plant-parasitic and free-living nematodes will increase our understanding of soil health and aid in development of sustainable crop management. If this work is not done, crop production will continue to move offshore where environmental standards for pesticide use are less stringent and our supply of safe, inexpensive grains, fruits, and vegetables will dwindle. Moreover, local farmers will lose market share and have reduced opportunities for economic development.

Nematode Management in Organic Agriculture

Nematodes are usually microscopic in size and are classified as unsegmented worms, belonging to the Phylum Nematoda. Plant-parasitic nematodes are a concern for growers of agricultural or garden crops. These plant-parasitic nematodes will mainly feed on the roots of plants. A few kinds will feed on foliage but this not common. Many other kinds of nematodes are present in the soil as well.

These include decomposers, predators, insect parasites, and animal parasites. Some nematodes are aquatic and do not affect terrestrial plants. Other nematodes act as decomposers, predators, and insect parasites. In farming systems, nematode predators and parasites of insects are beneficial, while nematode parasites of animals and plants are considered pests in agriculture. Beneficial nematodes that feed on either bacteria or fungi help in nutrient cycling by accelerating the decomposition of organic matter. Predatory nematodes may keep harmful nematodes at lower levels by feeding on them. Entomopathogenic nematodes (nematodes harmful to insects) may help to reduce numbers of some insect pests by infecting them with bacteria. The objectives of this publication are to provide information on plant parasitic nematodes causing damage in organic agriculture and to introduce methods for their management.

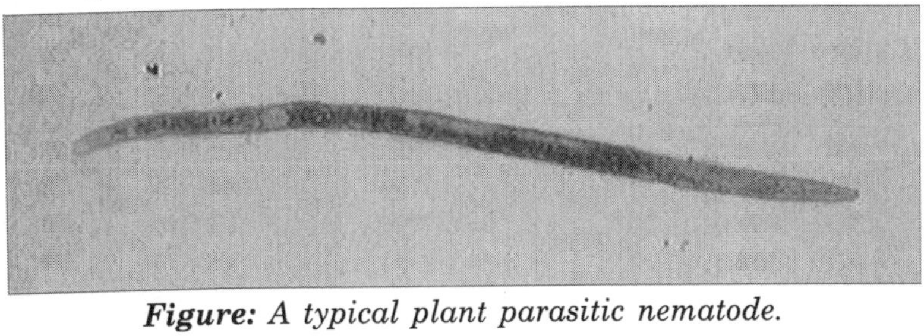

Figure: *A typical plant parasitic nematode.*

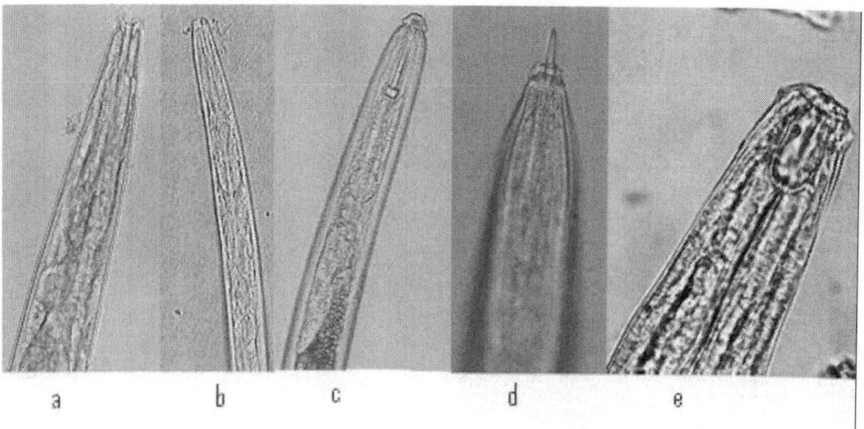

Figure: *Different nematodes: a) Bacterivore, b) Fungivore, c) Herbivore, d) Omnivore, e) Predator*

Nematode Damage and Detection

Since most plant-parasitic-nematodes feed on plant roots, symptoms are comparable to nutrient or water deficiency. These can be yield loss, stunting, yellowing, wilting, symptoms of nutrient deficiency, and malformations of the root (including tubers, peanuts) caused by direct feeding damage. In addition, invasion by plant-parasitic nematodes often provides an infection route for other organisms such as bacteria or fungi, since nematode activity creates an entryway into the root that would otherwise not be available.

Nematode reproduction is very fast. On average, a typical life cycle may be only 30 days at summer temperatures. This means that even if nematode numbers are low at the beginning of the growing season, nematode populations can rapidly increase and can become harmful to the crop in a relatively short period of time.

Often, the most damaging nematodes in the southeastern United States and the tropics are root-knot nematodes (*Meloidogyne* spp.). These nematodes are pests of nearly all major crops and are therefore widespread. Damage can be directly observed by examining the roots, because root-knot nematodes produce galls or knot-like swellings along the plant roots.

These galls cannot not be easily removed because they are part of the plant root tissue. In contrast, nodules caused by beneficial nitrogen fixing bacteria can be easily removed. Another method to distinguish nematode galls from nodules is to slice them in half and expose to air, which will cause nitrogen-fixing nodules to appear pink or red if the bacterial colony is active, or green or brown if the colony is inactive.

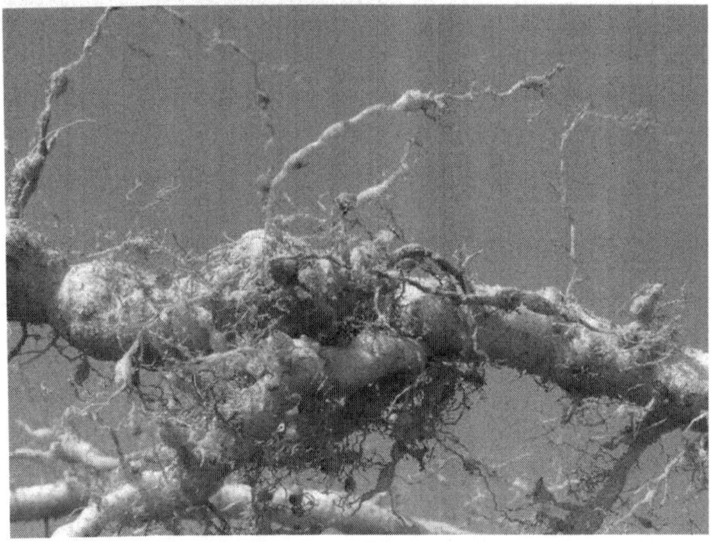

Figure: *Root galls caused by root-knot nematodes.*

Evidence of direct feeding, apparent in the form of galls and other root malformations leads to the damage symptoms described above. The invasive stage of the root-knot nematode life cycle is the juvenile root-knot nematode, which can freely move through the soil and will enter the root of a suitable host plant. Only the female is capable of establishing a feeding site. The female will become immobile, causing the plant to form giant cells for feeding, which essentially appears as a gall on the root. The presence of root-knot nematodes can be detected by the presence of these characteristic galls.

Soil Samples. Other plant-parasitic nematodes do not cause these gall symptoms, and so soil samples should be taken to determine which nematodes are present in a field. Even when it is clear that root-knot nematodes are to blame for the damage caused, it may be beneficial to find out if other nematodes are present to determine which treatments should be employed and will work best.

Nematode Management

In nematode management it is important to remember that nematodes can move only very short distances on their own. Therefore nematodes are mainly spread through lack of sanitation and movement of infected soil and planting material. In order to limit a build-up of nematodes, planting equipment and tools should be properly cleaned, and in extreme cases could only be used for the same field. Furthermore only soil and planting material free of nematodes should be used, because once nematodes are introduced into a field they cannot be eradicated. After harvest infected plants should be destroyed to prevent the build-up of nematodes on these crop residues and therefore in the soil. Once they become established in a site, the nematodes will persist there, and management will be required on a regular basis.

Resistant Plants and Rotation Crops. Nematode management is primarily a pre-planting activity. In order to protect the crop most activities must be started two or three months before the scheduled planting date. Several pre-plant treatments are available for the organic farmer. The choice of a suitable crop cultivar can be a critical decision. Host plant resistance achieved by traditional breeding programmes can be a valuable protection against some nematodes.

Two terms that are often used when talking about host plant resistance to nematodes are "tolerance" and "resistance." Tolerance means that the plant can withstand some damage caused by nematodes without experiencing significant yield reduction. In contrast, resistance means that nematode reproduction is very low or non-existant on the plant. Both provide protection for the crop plant, but the next crop following a tolerant plant could be damaged by the nematodes that survived on the tolerant plant. Different plant species or even cultivars of the same plant species can exhibit varying degrees of resistance or tolerance. According to the University of Missouri-Columbia extension service, the following vegetable plants are recommended as reasonably resistant to root-knot nematodes: broccoli, brussel sprouts, mustard, chives, cress, garlic, leek, groundcherry, and rutabaga. In contrast, Globe artichokes, Jerusalem artichoke, asparagus, sweet corn, horseradish, some lima bean varieties, onion, and rhubarb are considered to be tolerant. However, the Alabama Cooperative Extension System provides additional information on resistant crops. Cowpea or black-eyed peas, which can be used as a rotation crop to reduce nematode numbers, has several cultivars that are quite different in their ability to support reproduction of the root-knot nematode Meloidogyne incognita. *Planted into the sandy soils of Florida, the cultivars Tennessee Brown, Mississippi Silver, and California Blackeye #5 inhibited reproduction more effectively than the cultivar Purple Knuckle (intermediate). Population densities were high after planting the cultivars Whippoorwill, Pinkeye Purplehull, and Texas Purplehull. Iron Clay is another cowpea cultivar that is known to reduce population levels of* M. incognita.

Crop rotation utilizes crops that are a poor or non-host to the nematodes found in an agricultural field. These crops can either be plants that provide a secondary cash crop grown in between cycles of the primary cash crop, or they could be cover crops that are not primary but provide benefits to the farming system such as nitrogen enrichment, nematode reduction, or possible additional income. In either case, nematode numbers are reduced simply because nematodes are deprived of a suitable host crop.

This does not mean that nematode densities are reduced indefinitely, but a successful crop rotation should reduce nematode levels enough so that a following susceptible crop will produce sufficient yields and survive until the end of its regular growing season. Popular cover crops are sorghum, sorghum- sudangrass, different grains such as oat and rye, many grasses, marigold, cowpea, and some tropical legumes such as sunn hemp (*Crotalaria juncea*) and velvetbean (*Mucuna* spp.). These cover crops are useful to reduce root-knot nematode population densities. Even though these cover crops are effective against root-knot nematodes, they might increase other harmful plant-parasitic nematodes that are also present in the soil.

For example, densities of the sting nematode (*Belonolaimus longicaudatus*) and stubby-root nematode (*Paratrichodorus minor*) increased on SX-17 sorghum-sudangrass. Mississippi Silver cowpea was able to reduce *P. minor* and *M. incognita* but increased *B. longicaudatus*. The most effective cover crop that simultaneously reduced root-knot, stubby-root, and sting nematodes was determined to be velvetbean. Results vary in different locations as well. In Kenya, a plant related to sunn hemp (*C. ochroleuca*), which successfully reduced populations of root-knot nematodes appeared to increase the occurrence of lesion nematodes (*Pratylenchus zeae*), whereas jack bean (*Canavalia ensiformis*), hyacinth bean (*Lablab purpureus*), and velvet bean appeared to reduce both nematodes.

Response of nematodes is greatly influenced by the cultivar of a crop plant species. Therefore even if the same plant species is used, results may vary if the cultivar is different. Sorghum is often recommended as a cover crop to decrease population levels of root-knot nematodes, and is widely used for this purpose.

However, use of an uncommonly used cultivar (Asgrow Chaparral) actually increased population levels of root-knot nematodes (*M. incognita*) in a field experiment in North Florida. Furthermore, reactions of plants seem also to be site dependent. Cover crops that decrease nematode populations in one field may not be effective in another, depending on local nematode species and their local populations and variants present. Since weeds can harbor nematodes, it is important to have a dense cover crop stand that blocks out weeds.

Table: *Cover crops, which demonstrated low or intermediate susceptibility to* M. arenaria *race 1,* M. incognita *race 1, and* M. javanica .

Common Name	***Scientific Name***	***Cultivar***
cowpea	*V. unguiculata*	'Mississippi Silver'
		'Iron Clay'
sunnhemp	*Crotalaria juncea*	
American jointvetch	*Aeschynomene americana*	
marigold	*Tagetes minuta*	
	T. patula	'Dwarf Primrose'
velvetbean	*Mucuna deeringina*	
sorghum-sudangrass	*Sorghum bicolour x S. sudanese*	'SX-17'
jack bean	*Canavalia ensiformis*	
partridge pea	*Cassia fasciculata*	
sesame	*Sesamum indicum*	'Sesaco 16'
		'Paloma'

Tillage

Tillage and the practice of fallowing fields may appear as alternatives to cover crops for nematode management. Tillage inverts and mixes soil and exposes deeper soil layers to the sun. This practice is meant to kill nematodes by desiccation, since nematodes depend on moisture for survival. This practice may kill some of the nematodes that are in the upper soil layers, however it will not reach nematodes that have retreated into moderate or deeper soil layers. Nematodes can retreat to depths greater than 12 inches (30 cm) in Florida soils, and can migrate upward once a susceptible host is planted. Once a field has been fallowed, nematodes will move into deeper soil layers to avoid drying and may enter an inactive stage that enables them to survive periods without food and in addition protects them from desiccation. McSorley and Gallaher (1993) compared the effects of tillage versus crop rotation on nematode densities in tropical corn (*Zea mays*) cultivar Pioneer X304C in North Florida. Tillage did not have significant effects on nematode densities, while a rotation crop of sorghum cultivar DeKalb BR64 reduced population levels of root-knot nematodes (*M. incognita*). These results demonstrate that tillage is not a reliable method for nematode management in Florida. Fallowing can deprive nematodes of food, if the field is kept 100 % weed free, and therefore can reduce population levels. But clean fallowing is a poor practice for soil conservation and nutrient runoff. In the absence of plants that keep soil in place, wind will be able to carry some of the topsoil away. Additional soil will be lost with rainwater runoff. Rain will also carry soil nutrients into lakes and rivers, which will degrade water quality and interfere with aquatic ecosystems.

Solarization. A promising technique is the use of heat to decrease not only nematode densities, but also other harmful organisms and weed seeds. This can involve pasteurization, steaming, or solarization of the soil before planting. Of these, solarization is probably the most practical. It involves the covering of the soil with clear plastic. Transparent plastic sheets allow short-wave radiation from the sun to penetrate the plastic. Once the light passes through the plastic and is reflected from the soil, the wavelength becomes longer and cannot escape through the plastic.

The trapped light facilitates heating of the soil to temperatures detrimental to most living organisms. There are different types of plastic sheets available, mainly differing in their thickness (insulation) and ability to let light through (transparency). Black, opaque, or translucent plastics are not suitable for solarization. Thin, transparent plastic sheets appear to achieve the best results. Katan (1981) recommended thicknesses between 25 to 30 μm. Temperatures can rise anywhere between 35 and 60 C (95-140 F) during the summer months when air temperatures are close to 32 C (89.6 F) or higher. However, cloud cover and rain limit solar radiation and may diminish the success of solarization. Soil temperatures only rise to detrimental levels in the first 10 to 30 cm of soil (Katan 1987), and even in this range temperatures drop off as depth increases. The plastic has to be sealed to prevent air movement underneath the plastic, which would prevent temperatures from rising sufficiently.

Solarization

The soil should remain covered for a minimum of four weeks, but increasing solarization time improves effectiveness. This helps to heat the soil at a greater depth, which means that more nematodes will be affected by it. Furthermore it also ensures that an adequate accumulation of solarized hours is achieved, which is important in regions like Florida where the sky is often overcast during the summer months.

A combination of a suitable cover crop and solarization seems to achieve best results. Wang et al. (2006) showed that following a cowpea cover crop with solarization accomplished better results than solarization alone. In fact, the effectiveness of this combined treatment was comparable to methyl bromide fumigation, which in conventional agriculture was the most effective treatment used against soil-borne diseases and pests. The disadvantage of solarization is its negative impact on beneficial soil organisms, since they will meet the same fate as their harmful counterparts. But recovery is usually attained quickly through rapid recolonization. Furthermore, other beneficials such as *Bacillus, Pseudomonas,* and *Trichoderma* are able to survive the high temperatures generated by solarization. Solarization generates considerable plastic waste, which, when farming with a sustainable approach in mind, is a drawback. However, plastic sheets could be used multiple times.

Biological Control. Biological control is the management of plant-parasitic nematodes by living organisms such as bacteria, fungi, predatory nematodes, or other invertebrates. Biological control is mainly accomplished by attempting to build-up beneficial organisms through the use of various soil amendments. The introduction of beneficial soil organisms to the soil has only been attempted successfully in a few instances. *Pasteuria* spp., which are bacterial parasites of various plant-parasitic nematodes occurring naturally in Florida soils, may be promising biological control agents. In a 7-year experiment with tobacco, Weibelzahl-Fulton et al. (1996) showed that *M. incognita* and *M. javanica* were suppressed exclusively by *P. penetrans*.

Tobacco plants treated with *P. penetrans* had fewer galls, egg masses, and eggs than plants that either did not receive *P. penetrans* or lacked the biological control agent due to autoclaving the soil in which the plants were grown. Nematode-trapping fungi are also potential candidates for biological control. Their adhesive knobs, rings, or net structures trap nematodes and kill them. Other types of fungi may parasitize nematode eggs. Neither fungi nor *Pasteuria* spp. are available for widespread commercial use at this moment. But they can be found in a healthy soil environment, as can predators such as mites and predatory nematodes. Recent research has shown that a wide range of nematode natural enemies occur in Florida soils. Most predators are generalists, meaning they feed on a variety of prey species during their lifetime. For biological control purposes, generalist predators are a disadvantage, because they will not only feed on the key pest, but on other suitable organisms as well including beneficials. Furthermore if there is not enough food available they will disperse, possibly causing the pest to rebound. On the other hand, generalist predators can keep many different kinds of pests at low population densities. Currently, no predators are commercially available for augmentive releases for nematode control in vegetable production systems.

Organic Amendments in Relation to Biological Control

Biological control is difficult in soil, because it is a complex environment. Many of the possible organisms that could provide biological control lack specificity and therefore will not focus on a particular organism and may even interfere with beneficials. Therefore biological control of nematodes is achieved mainly by conservation of existing biological control; meaning that the soil environment is modified to aid the survival and reproduction of nematode natural enemies that are already present. Primarily this is accomplished through the addition of organic amendments. Organic amendments can improve the soil environment to aid biological control, benefit general plant health by helping with water retention and providing additional nutrients, and affect nematodes directly and negatively through detrimental decomposition products. The impact of organic amendments on nematodes is often inconsistent and unpredictable. In most cases when organic amendments are applied, they are helpful mainly as a plant nutrient source and do not directly aid in nematode management. However, even if nematodes are unaffected by the added amendments, plant health may improve due to other favourable properties of amendments. McSorley and Gallaher (1995) observed that amendments consisting of yard waste compost (four to six-month-old wood fragments, leaves, grass clippings) used as mulch or incorporated into the soil improved yield of squash and okra even though plants experienced heavy galling. Although nematode densities in composted treatments were similar to those in non-amended soil plants benefited greatly from the improved retention of soil moisture. Kimpinski et al. (2003) showed similar results. Amendments (compost and manure) increased yield in barley and potatoes in Canada, but this was not necessarily due to reduced nematode numbers.

The addition of organic amendments may stimulate the entire soil food web, including beneficial biological control organisms. When soil is amended with nutrient-rich organic matter, microorganisms may immobilize the released nutrients. Free-living nematodes that feed on these bacteria and fungi may play an important role in the release of nutrients tied up in such a way. Soil-inhabitating microorganisms and nematodes therefore may act together to influence the rate of decomposition and release of nitrogen from crop residues, including nitrogen-rich leguminous crops like sunn hemp. Wang et al. (2004) investigated effects of sunn hemp decomposition on the soil nematode community. Litterbags that allowed nematodes to pass in and out were filled with sunn hemp hay and buried in soil. Periodic analysis of the bags showed that sunn hemp decomposition was fairly rapid and most was completed within two weeks. Population densities of nematodes feeding on bacteria increased greatly during decomposition, and this was followed by an increase in omnivorous nematodes that fed on them. However, plant-parasitic nematodes were neither affected by the sunn hemp decomposition nor by the omnivorous nematodes. In fact, plant-parasitic nematode numbers increased over time with increasing crop biomass, which may indicate that sunn hemp residue does not have a direct detrimental effect on plant-parasitic nematodes. In another recent study, Chellemi (2006) used urban plant debris (green waste from public landfill deposited by homeowners and landscape companies) to amend pepper fields in southeast Florida. As a result, the combined density of plant-parasitic nematodes was reduced, and other disease-causing plant pathogens (*Pythium* spp. and *Phytophthora* spp.) were reduced as well. These examples illustrate the potential use of organic amendments for direct or indirect management of nematodes, but they also indicate that outcomes can be complex and unpredictable.

Management of Infected Plants

Once plants are infected with nematodes, there is little that can be done to remove or reduce nematodes. Therefore, prevention and sanitation are critical to controlling nematodes. The improvement of plant health is an important cultural technique to lessen detrimental effects on plants caused by plant-parasitic nematodes. Proper irrigation, fertilization and organic amendments as surface mulches or soil incorporated are important. Furthermore, removal of weed hosts and old crop plants immediately after harvest can reduce nematode densities for the future. Nematodes are a long-term pest, which cannot be eradicated once they become established in a site. They can only be kept at low levels with carefully selected management tactics that are often specific to the managed site.

There are a variety of additional methods that may have some effect on nematodes. These include methods such as use of rhizobacteria, chitin, sesame residues, flooding, or microwave energy. Some of these (flooding, microwaving) may be restricted to specialized situations. Performance of these methods may be inconsistent or ineffective in some situations. Amendments such as chitin or sesame residue may provide nutrients that are beneficial to overall plant health regardless of any effects on nematodes. In these cases only small test areas should be subjected to new materials to determine their effectiveness under local conditions. Many of the methods described above such as host plant resistance and rotation are more dependable under a variety of conditions.

Nematode Management in Tomatoes, Peppers, and Eggplant

Plant parasitic nematodes, are small microscopic roundworms that live in the soil and attack the roots of plants. Crop production problems induced by nematodes, therefore, generally occur as a result of root dysfunction, reducing rooting volume and foraging and utilization efficiency of water and nutrients. Many different genera and species of nematodes can be important to crop production in Florida. In many cases a mixed community of plant parasitic nematodes is present in a field, rather than having a single species occurring alone. In general, the most widespread and economically important nematode species include the root-knot nematode, *Meloidogyne* spp., and sting nematode, *Belonolaimus longicaudatus*. The host range of these nematodes, as with others, includes most if not all of the commercially grown vegetables within the state. Yield reductions can be extensive but vary significantly between plant and nematode species. In addition to the direct crop damage caused by nematodes, many of these species have also been shown to predispose plants to infection by fungal or bacterial pathogens or to transmit virus diseases, which contribute to additional yield reductions.

***Figure:** The simultaneous occurrence of both root-knot nematode (*Meloidogyne *spp.) and Fusarium wilt disease (*Fusarium oxysporum*) causing enhanced disease development and tomato yield loss.*

Biology and Life History

Most species of plant parasitic nematodes have a relatively simple life cycle consisting of the egg, four larval stages, and the adult male and female. Development of the first stage larvae occurs within the egg where the first molt occurs. Second stage larvae hatch from eggs to find and infect plant roots or in some cases foliar tissues. Host finding or movement in soil occurs within surface films of water surrounding soil particles and root surfaces. Depending on species, feeding will occur along the root surface or in other species like root-knot, young larval stages will invade root tissue, establishing permanent feeding sites within the root. Second stage larvae will then molt 3 times to become adult male or female. For most species of nematodes, as many as 50-100 eggs are produced per female, while in others such as root-knot, upwards of 2000 may be produced. Under suitable environmental conditions, the eggs hatch and new larvae emerge to complete the life cycle within 4 to 8 weeks depending on temperature. Nematode development is generally most rapid within an optimal soil temperature range of 70 to 80°F.

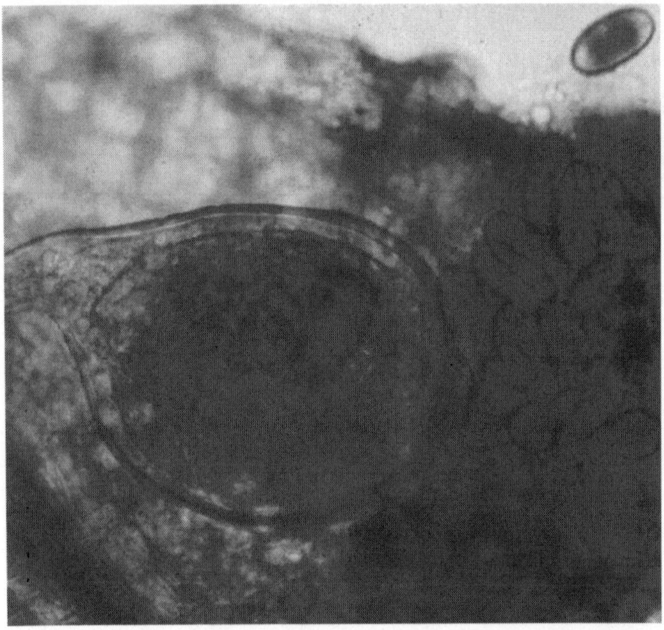

Figure: *Adult female of the root-knot nematode (*Meloidogyne *sp.) with attached egg mass.*

Symptoms

Typical symptoms of nematode injury can involve both aboveground and below ground plant parts. Foliar symptoms of nematode infestation of roots generally involve stunting and general unthriftiness, premature wilting and slow recovery to improved soil moisture conditions, leaf chlorosis (yellowing), and other symptoms characteristic of nutrient deficiency. An increased rate of ethylene production, thought to be largely responsible for symptom expression in tomato, has been shown to be closely associated with root-knot nematode root infection and gall formation. Plants exhibiting stunted or decline symptoms usually occur in patches of nonuniform growth rather than as a overall decline of plants within an entire field.

Figure: *Wilting, yellowing, and stunting of pepper due to the root-knot nematode (*Meloidogyne *sp.). Note the formation of galls on roots.*

Figure: *Wilting of pepper due to massive root infection by the root-knot nematode (*Meloidogyne *sp.).*

The time in which symptoms of plant injury occur is related to nematode population density, crop susceptibility, and prevailing environmental conditions. For example, under heavy nematode infestation, crop seedlings or transplants may fail to develop, maintaining a stunted condition, or die, causing poor or patchy stand development.

Under less severe infestation levels, symptom expression may be delayed until later in the crop season after a number of nematode reproductive cycles have been completed on the crop. In this case aboveground symptoms will not always be readily apparent early within crop development, but with time and reduction in root system size and function, symptoms become more pronounced and diagnostic.

Root symptoms induced by sting or root-knot nematodes can oftentimes be as specific as aboveground symptoms. Sting nematode can be very injurious, causing infected plants to form a tight mat of short roots, oftentimes assuming a swollen appearance.

New root initials generally are killed by heavy infestations of the sting nematode, a symptom reminiscent of fertilizer salt burn. Root symptoms induced by root-knot cause swollen areas (galls) on the roots of infected plants.

Gall size may range from a few spherical swellings to extensive areas of elongated, convoluted, tumorous swellings which result from exposure to multiple and repeated infections. Symptoms of root galling can in most cases provide positive diagnostic confirmation of nematode presence, infection severity, and potential for crop damage.

Damage

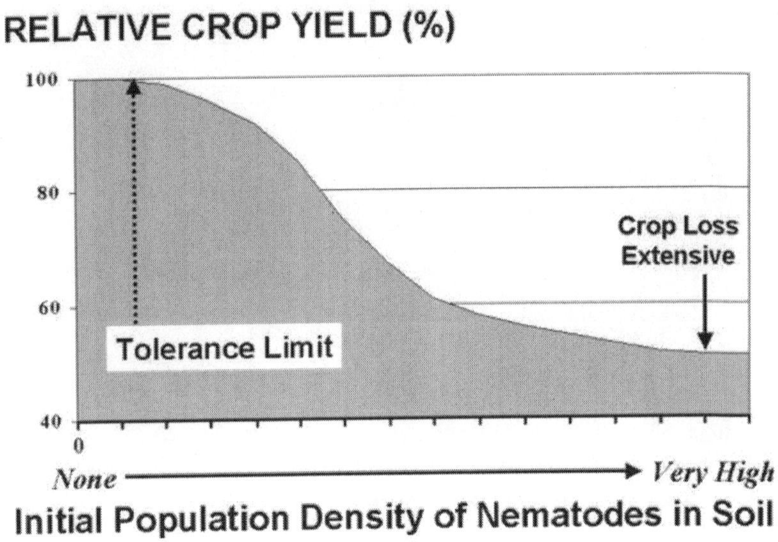

Figure: *Typical nematode induced crop damage relationship in which crop yields, expressed as a percentage of yields that would be obtained in the absence of nematodes, decline with increased population density of nematodes in soil. The tolerance level is identified as the initial or minimal soil population density at which crop damage is first observed.*

For most crop and nematode combinations the damage caused by nematodes has not been accurately determined. Most vegetable crops produced in Florida are susceptible to nematode injury, particular by root and sting nematodes. Plant symptoms and yield reductions are often directly related to preplant infestation levels in soil and to other environmental stresses imposed upon the plant during crop growth. As infestation levels increase so then do the amount of damage and yield loss. In general, the mere presence of root-knot or sting nematodes suggests a potentially serious problem, particularly on sandy ground during the fall when soil temperatures favour high levels of nematode activity. At very high levels, typical of those that might occur under doubling cropping, plants may be killed. Older transplants, unlike direct seed, may tolerate higher initial population levels without incurring as significant a yield loss.

Field Diagnosis and Sampling

Because of their microscopic size and irregular field distribution, soil and root tissue samples are usually required to determine whether nematodes are causing poor crop growth or to determine the need for nematode management. For nematodes, sampling and management are a preplant or postharvest consideration because if a problem develops in a newly planted crop, currently no postplant corrective measures are available to rectify the problem completely once the nematode becomes established. Nematode density and distribution within a field must therefore be accurately determined before planting, guaranteeing that a representative sample is collected from the field. Nematode species identification is currently only of practical value when rotation schemes or resistant varieties are available for nematode management. This information must then be coupled with some estimate of the expected damage to formulate an appropriate nematode control strategy.

Advisory or Predictive Sample (Prior to Planting)

Samples taken to predict the risk of nematode injury to a newly planted crop must be taken well in advance of planting to allow for sample analysis and treatment periods if so required. For best results, sample for nematodes at the end of the growing season, before crop destruction, when nematodes are most numerous and easiest to detect. Collect soil and root samples from 10 to 20 field locations using a cylindrical sampling tube or, if unavailable, a trowel or shovel. Since most species of nematodes are concentrated in the crop rooting zone, samples should be collected to a soil depth of 6 to 10 inches. Sample in a regular pattern over the area, emphasizing removal of samples across rows rather than along rows. One sample should represent no more than 10 acres for relatively low-value crops and no more than 5 acres for high value crops. Fields which have different crops (or varieties) during the past season or which have obvious differences either in soil type or previous history of cropping problems should be sampled separately. Sample only when soil moisture is appropriate for working the field, avoiding extremely dry or wet soil conditions.

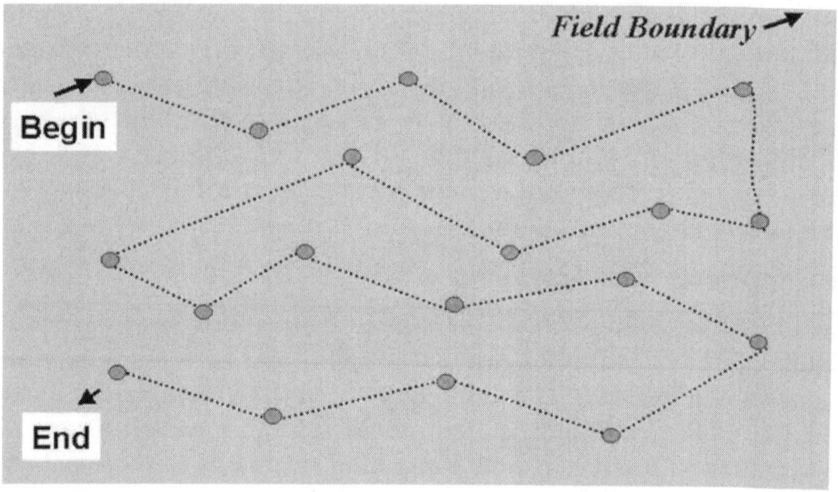

Figure: *Figure: Suggested pattern for collecting preplant soil samples for nematode analysis based on compositing soil from 10-20 field locations.*

Diagnostics on Established Plants

Roots and soil cores should be removed to a depth of 6 to 10 inches from 10 to 20 suspect plants. Avoid dead or dying plants, since dead or decomposing roots will often harbor few nematodes. For seedlings or young transplants, excavation of individual plants may be required to insure sufficient quantities of infested roots and soil. Submission of additional samples from adjacent areas of good growth should also be considered for comparative purposes.

For either type of sample, once all soil cores or samples are collected, the entire sample should then be mixed thoroughly but carefully, and a 1 to 2 pint subsample removed to an appropriately labelled plastic bag. Remember to include sufficient feeder roots. The plastic bag will prevent drying of the sample and guarantee an intact sample upon arrival at the laboratory. Never subject the sample(s) to overheating, freezing, drying, or to prolonged periods of direct sunlight. Samples should always be submitted immediately to a commercial laboratory. If sample submission is delayed, then temporary refrigerated storage at temperatures of 40 to 60°F is recommended.

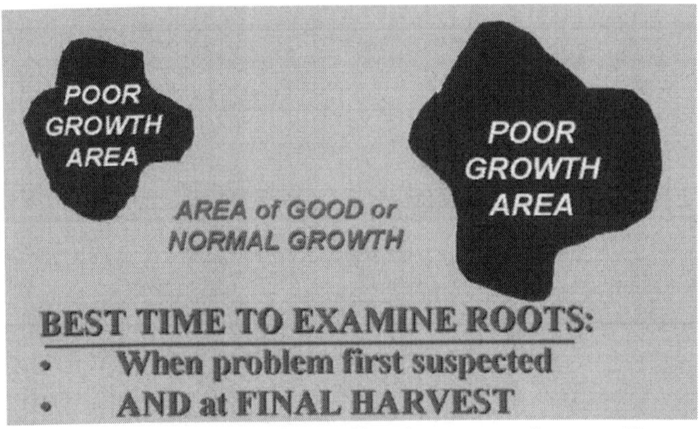

Figure: *Suggested strategy for collecting post-plant soil samples for nematode analysis comparing sampling results from areas of good and poor plant growth.*

Recognizing that the root-knot nematode causes the formation of large swollen areas or galls on the root systems of susceptible crops, relative population levels and field distribution of this nematode can be largely determined by simple examination of the crop root system for root gall severity. Root gall severity is a simple measure of the proportion of the root system which is galled. Immediately after final harvest, a sufficient number of plants should be carefully removed from soil and examined to characterize the nature and extent of the problem within the field. In general, soil population levels increase with root gall severity. This form of sampling can in many cases provide immediate confirmation of a nematode problem and allows mapping of current field infestation. As inferred previously, the detection of any level of root galling usually suggests a nematode problem for subsequent planting of a susceptible crop, particularly within the immediate areas from which the galled plant(s) were recovered.

Figure: *Field diagnosis for the presence and severity of root-knot nematodes via visual examination of uprooted plant root systems for root galls.*

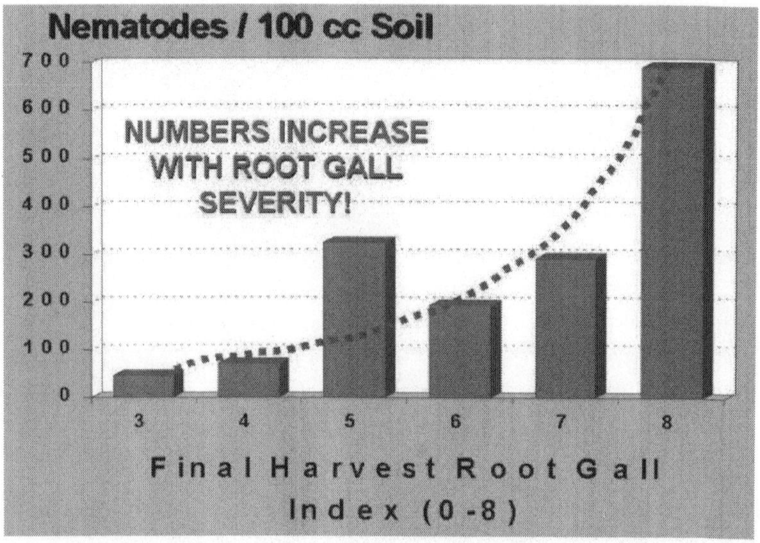

Figure: *Generalized relationship between the number of root-knot nematode juveniles in soil based on the level and severity of plant root galling.*

General Management Considerations

Currently nematode management considerations include crop rotation of less susceptible crops or resistant varieties, cultural and tillage practices, use of transplants, and preplant nematicide treatments. Where practical, these practices are generally integrated into the summer or winter 'off-season' cropping sequence. It should be recognized that not all land management and cultural control practices are equally effective in controlling plant parasitic nematodes and varying degrees of nematode control should be expected.

These methods, unlike other chemical methods, tend to reduce nematode populations gradually through time. Farm specific conditions, such as soil type, temperature, and moisture, can be very important in determining whether different cultural practices can be effectively utilized for nematode management.

Cultural Practices

Crop Rotation

For crop rotation to be effective, crops unsuitable for nematode infection, growth, or reproduction must be introduced into the rotation sequence. In most of Florida it is not uncommon to observe a multispecies community of nematodes all occurring within the same field. Under these circumstances it may not be possible to find a rotation or cover crop that will effectively reduce populations of all nematode pests, particularly if root-knot and sting nematodes occur in combination In this case, crop rotations detrimental to root-knot, which is generally the most difficult to control, should be selected. In some cases, resistant crop varieties are available, which can be used within the rotation sequence to minimize problems to some species of root-knot but not sting nematodes.

Use of poor or nonhost cover crops within the rotation sequence, may in some cases offer an effective approach to nematode control. Four leguminous cover crops adaptable for managing soil populations of sting or root-knot nematode include Iron Clay Cowpea (*Vigna unguiculata* . cv. 'Iron Clay'), sunn hemp (*Crotalaria juncea),* hairy indigo (*Indigofena hirsuta*) and American jointvetch (*Aeschynomene americana*). Sorghum is also a popular cover crop restoring large amounts of soil organic matter, but is a good host for sting nematode but not root-knot. Most of the small grains commonly used as winter cover crops in central and north Florida, such as rye, barley, wheat, or oats, can support limited reproduction of root-knot nematode. To avoid an increase in root-knot populations, these crops should only be planted when soil temperatures are below 65°F, a threshold temperature for nematode activity. Cover crop rotations with some pastureland grasses (particularly pangola digitgrass, and to some extent bahiagrass, and Bermuda grass) have significantly reduced, but not eliminated, root-knot nematodes. In north Florida, long term (6- to 9-year) pastureland rotations have allowed economic melon production within root-knot infested fields. It should be recognized that as the crop rotation period is shortened or eliminated, nematode problems will intensify accordingly. Other perennial legumes currently under evaluation may play an important role in future nematode management programmes.

For cover crops to be most effective stands must be established quickly and undesirable weeds that can serve as alternative hosts must be controlled. Given that many different weeds serve as alternative plant hosts to nematodes (i.e., nutsedges), it may not be possible to manage root-knot nematode with crop rotation unless an integrated programme to manage weeds is also considered and implemented within the field. With many cover crops, rapid stand establishment has been a significant problem. Similarly, economic crop rotation sequences are often further complicated by lack of crop management skills, specialized equipment to grow and harvest the crop, or by the lack of closely located processing facilities or markets. In some cases other measures should be considered such as fallowing, which is usually as efficient as crop rotation for reducing field infestations of nematodes.

Fallowing

Clean fallow during the off-season is probably the single most important and effective cultural control measure available for nematodes. When food sources are no longer readily available, soil population densities of nematodes gradually decline with death occurring as a result of starvation. Due to the wide host range of many nematode species, weeds and crop volunteers must be controlled during the fallow period to prevent nematode reproduction and further population increase. At least two discing operations are generally required to maintain clean fallow soil conditions during the interim period between crops. Fallowing by use of herbicides to deplete nematode populations is a much slower process because the soil is not disturbed, thereby subjecting nematodes from deeper soil layers to the drying action of sun and wind. The unfavourable effects of fallowing on soil organic matter and soil structure is usually more than compensated for by the level of nematode control achieved and the resulting increase in crop productivity. When soil erosion is a potentially serious problem other measures should be considered.

Biological Control

At present there are no effective, commercially available, biological control agents which can be successfully used to control nematodes.

Plant Resistance

Use of nematode-resistant crop varieties has not been extensively evaluated in Florida, but is often viewed as the foundation of a successful integrated nematode management programme on all high value crops in which methyl bromide is currently used. Commercially available nematode-resistant varieties are currently available only for tomato, pepper, southernpea, and sweet potato. In a resistant variety, nematodes fail to develop and reproduce normally within root tissues, allowing plants to grow and produce fruit even though nematode infection of roots occurs. Some crop yield loss can still occur, however, even though the plants are damaged less and are significantly more tolerant of root-knot infection than that of a susceptible variety.

In tomato, a single dominant gene (subsequently referred to as the Mi gene) has been widely used in plant breeding efforts and varietal development, which confers resistance to all of the economically importance species of root-knot nematode found in Florida, including *Meloidogyne incognita, arenaria,* and *javanica.* Commercially resistant fresh market varieties, climatically and horticulturally adapted for Florida, are available in a number of tomato varieties (e.g, 'Sanibel').

Unfortunately, in previous research with resistance tomato varieties, the resistance has often failed as a result of the heat instability or apparent temperature sensitivity of the resistant Mi gene. For example, previous research has demonstrated threshold soil temperatures and incremental reductions in nematode resistance with each degree above 78°F, such that at 91°F tomato plants are fully susceptible. This would suggest that in Florida use of these varieties may have to be restricted to spring plantings when cooler soil temperatures prevail.

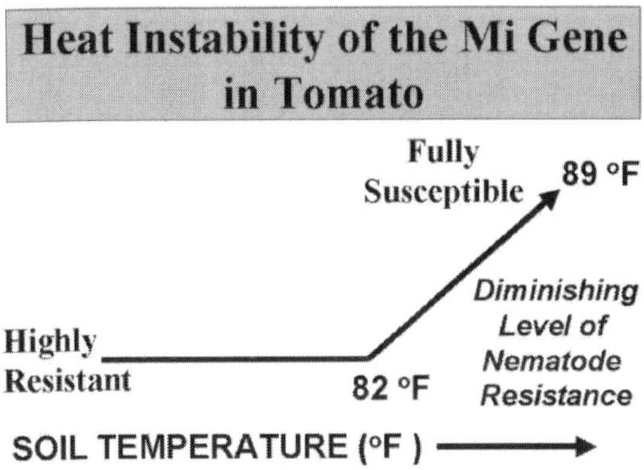

Figure: *Diagrammatic representation showing the complete loss of root-knot nematode resistance conferred by the Mi gene in tomato with increasing soil temperature.*

In pepper, two newly developed root-knot nematode resistant varieties (Carolina Belle and Carolina Wonder) were released from the USDA Vegetable Research Laboratory for commercial seed increase in April 1997. Both varieties are open pollinated, and homozygous for the N root-knot nematode resistant gene. Preliminary research has demonstrated that these varieties confer a high degree of resistance to the root-knot nematode; however, expression of resistance is heat sensitive. Further research is necessary to characterize the usefulness of these varieties under the high soil temperature conditions of Florida. Like tomato, use of these varieties may have to be restricted to spring plantings when cooler soil temperatures prevail.

In addition to problems of heat instability, the continuous or repeated planting of resistant plant varieties will almost certainly select for virulent races of *Meloidogyne* capable of overcoming the resistance. Therefore, the duration and/or utility of the resistance may be time-limited. In previous studies with resistant tomatoes, resistance breaking nematode races have been shown to develop within 1 to 3 years. Since new races of the nematode can develop so rapidly, a system of integrated control usually mandates the rotation of resistant and non-resistant varieties to slow the selection process for new virulent races. Previous trials in Florida have demonstrated the capacity of some species or races of root-knot to reproduce and inflict damage upon a resistant tomato variety. The results of these experiments have demonstrated that even with a resistant variety, which were damaged less than that of a susceptible variety, some consideration of initial soil population levels of the root-knot nematode must be observed to minimize tomato yield losses. Given that significant yield losses can still occur, combined efforts to manage soil populations to low levels prior to planting must still be considered, particularly if tomatoes are planted as a fall crop. If this situation develops, the combination of a nematicide and resistant variety may also comprise an option to reduce nematode populations to nondamaging levels.

Soil Amendments

Many different types of amendments and composted materials have been applied to soil to suppress populations of plant parasitic nematode and improve crop yield and plant health. Animal manures, poultry litter, and disk-incorporated cover crop residues are typical examples of soil amendments used in agriculture to improve soil quality and as a means for enhancing biocontrol potential of soil. Some amendments that contain chitin and inorganic fertilizers that release ammoniacal nitrogen into soil suppress nematode populations directly and enhance the selective growth of microbial antagonists of nematodes. More recently, composted municipal wastes and sludges have been used to amend soil to improve soil fertility, organic matter content, water-holding capacity, nutrient retention, and cation exchange capacity.

Suppression of soilborne pathogens via the incorporation or simple mulching of composted amendments is reputedly based on enhanced microbial activity and increased numbers of antagonists generated by decomposition of the amendment in soil. Soils with a diversity of beneficial microorganisms are more suppressive to pathogens than soils with little or no biological diversity. Other possible mechanisms for pathogen suppression by composts include direct inhibition of the pathogen or reduced infectivity of the organisms into the plant host. Population increases of beneficial organisms in soil appear to be the direct result of environmental changes brought about by the amendments after addition to soil. This suggests that to sustain soil suppressiveness, amendments must be periodically reapplied to maintain the soil environment conducive to antagonists.

The level to which soilborne pest and disease control can be achieved is not only related to the type of material but to the age of the compost. Nematode and disease suppression has been repeatedly demonstrated with composted municipal yard wastes containing significant quantities of tree bark. If the compost is immature, the product may not only be difficult to handle and have an offensive odor, but may contain salts and metabolites toxic to plants. For example, weed suppression has been demonstrated with some types of immature composted materials that contain and or produce organic acids with phytotoxic properties. Other studies have shown that soils amended with different sources of composted municipal wastes were disease suppressive as the long as they were relatively fresh (< 6 months), but as the composted municipal waste was aged, disease suppressiveness was lost. In other Florida studies, application of composted municipal wastes at rates up to 120 tons per acre have not been shown to be pesticidal in activity, but actually dramatically increased populations of nematodes and other disease organisms such as *Fusarium* and *Phytopthora* spp. Nematode population increases were directly related to increases in plant growth and root system size with amendment application rate.

Previous studies in Florida were conducted to determine the extent to which increasing application rates of a municipal solid composted waste affect the ability of tomato plants to tolerate root infection by species of root-knot nematode (*Meloidogyne* spp.) These studies showed that in a sandy soil poor in organic matter content (less than 2%), tomato yields could be increased significantly with soil amendments in both nematode free or nematode infested soil. The impact of the root-knot nematode on tomato yield was effectively constant, however, suggesting that application of the soil amendment did not enhance the ability of tomato plants to tolerate infection by the root-knot nematode. Much of the previous and ongoing research in Florida also seems to indicate that the major effects of soil amendments to crop yields appear to be less related to nematode or soil pathogen control than to enhanced plant nutrition and nutrient and water availability.

It is not clear at this time and preliminary stage of university field research whether benefits to crop growth after the initial crop following soil amendment application can be expected. Previous studies showed no response in second crop tomato yields (double crop) following amendment application rates from 15 to 120 tons per acre. Disappearance of nutrients and soil organic matter content appears to be very rapid in the hot, moist soils of Florida. Preliminary research suggests that reapplication of the amendments may have to be made on 1-2 year basis to sustain crop growth and yield benefits. In summary, the high rates of application (tons/acre) and attendant costs required for crop response and nematode control for many different types of organic amendments, and the apparently rapid losses of the materials in soil appear to preclude use of these materials primarily to homeowner or small farm operations at this time. However, with additional research and advances in application technology and use efficiency, use of soil amendments may become an integral component of Florida crop production systems.

Flooding

Flooding has been shown to suppress nematode populations. Alternating 2 to 3 week cycles of flooding and drying have proven to be more effective than long, continuous flooding cycles. At present, only limited areas within the state are situated to take advantage of flooding as a viable means of nematode control. Given the growing concern about aquifer depletion, salt water intrusion, and water use inefficiencies, it seems unlikely that Florida water management officials will continue to permit flooding within these areas in the future.

Soil Solarization

Soil solarization is a nonchemical technique in which transparent polyethylene tarps are laid over moist soil for a 6 to 12 week period to heat noncropped soils to temperatures lethal to nematodes and other soil-borne pathogens. Soil temperatures are magnified by the trapping of incoming solar radiation under the clear, polyethylene panels. To be effective, soils must be wetted and maintained at high soil moisture content to increase the susceptibility (thermal sensitivity) of soil borne pests and thermal conductivity of soil. Wet mulched soils increase soil temperatures due primarily to the elimination of heat loss by evaporation and upward heat convection, in addition to a greenhouse effect by prohibiting dissipation of radiation from the soil. At the end of the solarization period the clear plastic is painted with a white latex paint to allow continued use of the plastic as a mulch cover for the production of vegetables on raised beds.

Soil solarization, the use of clear plastic panels as bed coverings to heat non-cropped soils to temperatures lethal to nematodes and other soilborne pathogens.

Soil Decontamination

A 2008 study used a solar cell to generate an electric field for electrokinetic (EK) remediation of cadmium-contaminated soil. The solar cell could drive the electromigration of cadmium in contaminated soil, and the removal efficiency that was achieved by the solar cell was comparable with that achieved by conventional power supply. In Korea, various remediation methods of soil slurry and groundwater contaminated with benzene at a polluted gas station site were evaluated, including a solar-driven, photocatalyzed reactor system along with various advanced oxidation processes (AOP). The most synergistic remediation method incorporated a solar light process with TiO2 slurry and H2O2 system, achieving 98% benzene degradation, a substantial increase in the removal of benzene.

Attempts were made to use solar energy for controlling disease agents in soil and in plant material already in the ancient civilization of India. In 1939, Groashevoy, who used the term "solar energy for sand disinfection," controlled Thielaviopsis basicola upon heating the sand by exposure to direct sunlight.

Soil solarization is the third approach for soil disinfestation; the two other main approaches, soil steaming and fumigation; were developed at the end of the 19th century. The idea of solarization was based on observations by extension workers and farmers in the hot Jordan Valley, who noticed the intensive heating of the polyethylene-mulched soil. The involvement of biological control mechanisms in pathogen control and the possible implications were indicated in the first publication, noticing the very long effect of the treatment. In 1977, American scientists from the University of California at Davis reported the control of Verticillium in a cotton field, based on studies started in 1976, thus denoting, for the first time, the possible wide applicability of this method.

The use of polyethylene for soil solarization differs in principle from its traditional agricultural use. With solarization, soil is mulched during the hottest months (rather than the coldest, as in conventional plasticulture which is aimed at protecting the crop) in order to increase the maximal temperatures in an attempt to achieve lethal heat levels.

In the first 10 years following the influential 1976 publication, soil solarization was investigated in at least 24 countries and has been now been applied in more than 50, mostly in the hot regions, although there were some important exceptions. Studies have demonstrated effectiveness of solarization with various crops, including vegetables, field crops, ornamentals and fruit trees, against many pathogens, weeds and a soil arthropod. Those pathogens and weeds which are not controlled by solarization were also detected. The biological, chemical and physical changes that take in solarized soil during and after the solarization have been investigated, as well as the interaction of solarization with other methods of control. Long-term effects including biological control and increased growth response were verified in various climatic regions and soils, demonstrating the general applicability of solarization. Computerized simulation models have been developed to guide researchers and growers whether the ambient conditions of their locality are suitable for solarization.

Studies of the improvement of solarization by integrating it with other methods or by solarizing in closed glasshouses, or studies concerning commercial application by developing mulching machines were also carried out.

The use of solarization in existing orchards (e.g. controlling Verticillium in pistachio plantations) is an important deviation from the standard preplanting method and was reported as early as 1979.

How to Solarize Soil

The area to be solarized should be level and free of debris and large clods. The soil should be watered if the soil is dry. The soil should not be saturated but have the crumbly damp look that the soil would have at planting. Place clear plastic over the soil surface and bury the edges in a foot deep trench. The plastic can be clear construction grade plastic and vary in thickness from 1 to 6 mils. The thinner plastic (1-2 mils) will allow better soil heating since it will reflect less solar energy. The plastic should be left in place for 4-6 weeks. Solarization is most effective when done in June and July. However, depending on your geographic location, solarization may also be effective in May, August, and September.

The most successful use of soil solarization appears to occur in heavier (loamy to clay soils) rather than sandy soils. Soils with poor water holding capacity and rapid drainage can significantly inhibit heat transfer to deeper soil horizons. Loss of pest control is directly correlated with soil depth. The depth to which lethal temperature can be achieved (6 to 8 inches) is also dependent on the intensity and duration of sunlight and ambient temperature. At present, the only time to consider soil solarization for pest control is during our hot, summer and early fall months, which fortunately are 'off-season' in most peninsular Florida vegetable row crops. Unfortunately our summers are also our wettest period of the year with frequent afternoon rain showers that have a cooling effect on the soil.

Many different pests have been suppressed and or controlled by soil solarization, particularly within arid environments with intense sunshine, and limited cloud cover and rainfall. Previous studies in

Florida have demonstrated that soil solarization can also be effective in a subtropical environment. Plant parasitic nematodes have generally proved to be more difficult to control with soil solarization, as have some weed pests such as crabgrass in a central Florida study. The results of preliminary experiments also suggest the potential for selection pressures towards a buildup of heat tolerant individuals that may serve to reduce soil solarization efficacy after repeated use as a nematode control tactic. In some studies, effective use of solarization for nematode control has required an integrated systems approach, coupling solarization with other chemical or nonchemical approaches. For example, the combined use of soil solarization with a nematicide has improved nematode control and crop yield. In addition, use of virtually impermeable, photo-selective plastic mulches may also complement low dose fumigant treatments to reduce weed germination and growth in the event of extended periods of cloud cover occurring during the solarization regime. At this time, further research is needed demonstrating soil solarization pest control activity and consistency in the various geographical regions of Florida where vegetable crops are grown.

Other Cultural Practices

Other cultural measures that reduce nematode problems include rapid destruction of the infested crop root system following harvest. Fields that are disced as soon as possible after the crop is harvested will not only prevent further nematode population growth but subject existing populations to dissipation by sun and wind. Use of nematode free tomato transplants is also recommended since direct seeded plants are particularly susceptible since they are vulnerable to injury for a longer duration, during an early, but critical period of crop development. Since nematodes can be carried in irrigation water that has drained from an infested field, growers should avoid use of ditch or pond waters for irrigation or spray mixtures. In most cases, a combination of these management practices will substantially reduce nematode population levels but will rarely bring them below economically damaging levels. This is especially true of lands that are continuously planted to susceptible crop varieties. In these cases some form of pesticide assistance will still usually be necessary to improve crop production.

Antagonistic Crops

A large proportion of the food production in many tropical and subtropical countries comes from smallholders farming relatively small parcels of land. Crops may be grown together in mixed cultivations often following a system that has been long established and generally successful. In the era of "progress" signalled by the development of monoculture with new varieties, crop protection aids and inorganic fertilizers, it is regrettable that in hindsight the knowledge of yesterday contained in the practices and methods of these smallholders has not been given more attention.

The trend towards urbanization has generated the need for regular supplies of food to be produced from farms and smallholdings close to cities, or with reasonable access to a road network. Intensification of food-crop production to meet this consumer demand has caused various problems for agronomists, soil scientists and crop protectionists, particularly if the land is in continuous cultivation.

The traditional system of agriculture relied on the use of long periods of fallow between crops. During this time the natural vegetation regenerated, soil fertility was restored and populations of pests and diseases declined. Traditional farming systems are usually of mixed crops (even including trees or woody perennials) that provide an ecosystem in which total crop loss due to pests and diseases is rare. Nevertheless, subsistence farmers will usually grow enough to cover the possibility of losing up to 40 percent of production before or after harvest. The problems of a rapidly increasing human population for land use has led to the abandonment of traditional systems, and these are now practised only in sparsely populated countries.

Crop losses due to pests and diseases appear to be more frequent as improved crop varieties are grown in monoculture on the same land with the aid of pesticides and inorganic fertilizers. In many instances, crop damage has increased as a result of the inadvertent introduction of alien pests on crop plants. The distribution of several nematode species has been increased in this way, e.g. *Radopholus similis* and *Pratylenchus* spp. on bananas, *Tylenchulus semipenetrans* on citrus and *Globodera* spp. on potatoes.

In considering potential strategies for controlling nematodes there are certain practices that can be implemented once the life history and the behaviour of the pest is known. Plant-parasitic nematodes occur in greatest abundance in the rooting zones of crops, mostly in the top 30 cm of soil; nematodes may also penetrate to deeper soil layers particularly in unfavourable conditions (Prot, 1980) only to return to the surface layers when crop growth resumes. The free-living stages of some species survive the unfavourable season through physiological or structural adaptation, or survive for several seasons as eggs, as in the case of the potato-cyst nematode.

Nematodes are usually at their most vulnerable at the time when they are actively searching for host roots and when surviving unfavourable growing seasons. The survival of free-living stages in the absence of hosts can be longer than originally thought, and fallow periods of less than 12 months may be insufficient to control those species that reproduce rapidly. The main abiotic factors that influence free-living nematodes and eggs are heat, desiccation and anaerobis. Plant-parasitic nematodes will not survive long periods at temperatures above 40°C. These temperatures may be reached on bare soil in some countries during the hot season, but heat penetration may not extend below 10 cm.

The technique of increasing soil temperatures by solarization can be of value in controlling nematodes, but this is only possible where there are long periods of uninterrupted sunshine. Solarization is also practised for controlling soil-borne fungal pathogens. Nematodes are aquatic animals that inhabit the films of water that surround soil particles. As soils dry, the volume of soil water decreases and the ability of nematodes to move is impaired, eventually free-living nematodes may die as the soil moisture decreases. The extent to which species have adaptations to survive desiccation is still a matter for further study. By regulating the metabolism in times of stress some species are capable of anhydrobiotic survival. The contrivance of bare fallows that deprive nematodes of a host may not always be a worthwhile control strategy especially as fallowing in hot, dry areas can deplete soil organic matter and the levels of beneficial soil biota. In such situations for specific nematode pests, a better alternative might be cover crops with non-host leguminous trees or shrubs such as *Crotalaria* spp.

The control of nematodes by flooding has been advocated in certain locations. When flooded all the soil pore spaces are filled and the oxygen supply for the soil microflora and fauna becomes limiting. Many plant-parasitic nematodes are intolerant of oxygen starvation and soon die. Similar effects are observed when nematodes are stored in deep water in the laboratory, but often they can be revived by aeration. To be effective in the field therefore, the anaerobiosis has to be of sufficient duration to kill the nematodes. In Côte d'Ivoire banana fields are flooded for at least five weeks. Flooding is not a technique that can be widely used for nematode control and whether it is equally efficacious against all plant-parasitic species is not known.

The development of strategies for nematode control requires a thorough understanding of the growth of the host plant and the biology of the nematode. In some climates it is possible to avoid (or contain) damage by planting the crop at a time when nematode invasion will be slow, or when plant growth will be faster than that of the parasite. Damage to transplanted short-cycle crops can be limited if seedlings are produced in nematode-free nurseries or in soil-less potting media. In general, protection from invasion during the early growth of a crop will enable the plant to produce a yield acceptable T most farmers even if it is planted in nematode-infested soil, particularly if the crop receives adequate supplies of water and nutrients. This will not prevent the development of large nematode populations on the root systems and in the soil that will influence the subsequent crop. Brown showed that treating wheat with low nematicide doses at seeding gave a substantial yield improvement, resulting from protection from the invasion of cereal-cyst nematodes during the establishment of the crop.

The concept of growing companion plants with nematicidal root exudates alongside susceptible crops is attractive, but the actual benefits may be questionable or at least difficult to quantify. Species of *Tagetes, Crotalaria* and castor bean (*Ricinus communis*) have been widely used. Where *Tagetes* was grown as a companion to tomato in a rotation trial in the United Republic of Tanzania, yields were significantly less than with other treatments, indicating that such a cropping practice was not likely to be adopted by farmers.

The use of organic mulches for managing nematodes has been widely studied and there are various examples of plant debris having beneficial effects on plant growth. In small garden plots that receive heavy mulches of organic waste, nematodes are unlikely to be a major problem. Subsistence gardens are therefore well suited to the needs of householders, who, as Hussey (1990) reported, will usually grow sufficient quantities to meet immediate needs and allow for crop losses due to pests and diseases. Regular mulching is realistic for subsistence gardeners but becomes less feasible if the area of cultivation is large. To be effective, the recommended mulches may have to be used at application rates of 5 to 10 tonnes per hectare. The acquisition and distribution of such volumes is not likely to be practised by farmers growing crops on a large scale.

Future Developments

If chemical treatments are unavailable or inappropriate, the management of plant-parasitic nematodes will have to depend on alternative strategies. Those farmers in the Near East region who routinely use methyl bromide or non-fumigant nematicides may in future have to seek alternative treatments. To help them do so, government-supported research will be necessary, particularly for control strategies not requiring the regular purchase of proprietary products. This type of research is unlikely to be funded by manufacturing companies.

New research will require multidisciplinary collaboration to enable the integration of traditional practices with new ideas created by the scientists. Through more research on nematode biology and the environment, nematologists must consider several possibilities:

- How nematode life cycles may be interrupted.
- How microbial activity at the root-soil interface, which could lessen nematode invasion, can be promoted.
- Identification of compounds that interfere with reception of stimuli or repel nematodes.
- The selection or breeding of resistant varieties, or varieties that tolerate nematode infestation, and that can produce acceptable yields.

In conducting field experiments that evaluate the effect of cropping systems on nematode populations, scientists should plan to suit the needs of the farmer. Many research station trials are designed to produce results for ready analysis and presentation in journals; in consequence, treatments may be selected to suit the needs of the scientist. Most farmers grow short-cycle, annual and perennial crops in close proximity. Such mixed cropping practices are rarely undertaken in field experimentation by nematologists.

Any new cultural, microbial or management approaches that are developed must be within the capabilities of the farmer and meet the necessary environmental and economic requirements. New strategies will be considered as successful if they lower nematode damage thresholds. The farmer will be satisfied if these treatments are reliable, practicable and economically justified; his customer, the consumer, will be content if the product has the desired quality, contains no toxic residues, and is fairly priced.

Biofumigation

Bio-fumigation is based on incorporating soil amendment (fresh plant mass, manure) into the soil, which will release chemical substances, known as isothiocyanates (ITC's), able to suppress soil-borne pests and diseases, plus a soil heater to enhance biological activities. Plants from Cruciferae family (cabbage, radish, cauliflower etc.) release large amount of these toxic to soil-borne pests and diseases substances - in the soil and are considered the best material for bio-fumigation.

How to do it:

- Incorporate the fresh mass into the soil. This can be done directly if the mass is coming from grown crop or plant mass taken from elsewhere and brought into the plot or field. If the mass is transported to the field, the soil should be well prepared before the incorporation. During transportation and storage of these organic materials in the field, care must be taken not to lose the gases produced from biodegradation, by covering the piles of the bio-fumigant with plastic until the time of application.

A dose of 50 t / ha is recommended, although when problems with nematodes or fungi are very serious, 100 t / ha should be applied, a dose that can be reduced by choosing a cultivation techniques such as application in furrows.

The bio-fumigant should be distributed uniformly, so that no concentration of pathogens will appear that could create problems for the crop. Once the biofumigant is distributed, it should be incorporated immediately into the soil.

- Water the field, if possible by sprinkling, until the soil is saturated, although watering can be done by flooding, or drip irrigation can be installed.
- Cover the soil surface tightly with a transparent plastic film for at least 2 weeks to retain the gases produced from the biodegradation of the organic matter. This could be the same plastic as the one used for soil solarisation.
- The film is removed 3-4 weeks after and the soil slightly removed in order to permit the gases to escape from soil.
- Planting of the interested crop can be done 24 hours later.

Using Bio-fumigation against Nematodes

Different mustards (e.g. Brassica juncea var integrifolia or Brassica juncea var juncea) should be used as intercrop on infested fields. As soon as mustards are flowering they are mulched and incorporated into the soil. While incorporated plant parts are decomposing in a moist soil, nematicidal compounds of this decomposing process do kill nematodes. Two weeks after incorporating plant material into the soil a new crop can be planted or sown (it takes about 2 weeks for the plant material to decompose and stop releasing phytotoxic substances = chemicals poisonous to plants).

Mexican marigold, also known as Tagetes, has been successfully used in the control of root-knot nematode in roses by a Kenyan Farmer.

It is recommended to alternate the use of agricultural residues with green manure, especially from brassicae, using 5-8 kg square-metres of green matter, although combinations of legumes and grass can be applied. In the case of the use of green manure cultivated in the same field, fast growing plants should be used to be incorporated at least 30 days after having been planted, to avoid the increase of pathogen populations.

Planting mustard after bio-fumigation can serve as bio-indicators of possible phytotoxicity, because the germination of their seeds is sensitive to phytotoxic substances. At the same time they are very sensitive to nematodes and permit the detection of areas in the crop where biofumigation is not effective. They act like trap plants, and like bio-fumigants when incorporated into the soil.

In Spain, successful application of bio-fumigation was achieved in strawberries, peppers, cucurbits, tomato, brassicae, cut flowers, citrus and banana. Bio-fumigation has also been recently applied to Swiss chard and carrot crops. The most utilized biofumigants have been goat, sheep and cow manure, and residues from rice, mushroom, brassicae and gardens.

The effectiveness of bio-fumigation in controlling nematodes, fungi, insects, bacteria, and weeds is nearly the same as with the use of conventional pesticides. Bio-fumigation may also regulate viral problems by controlling vector organisms.

Biofumigation Against Bacterial Wilt

From 1999 research on using bio-fumigation for control of bacterial wilt has been carried out in Australia and the Philippines by the Autralian Centre for International Agricultural Research (ACIAR). The Centre with collaborators have been working to identify suitable brassicas for tropical environments and to evaluate them in the field, initially on experimental farms and more recently on commercial farms and smallholder farmers fields. There are many varieties and they differ enourmously in the level of disease-suppressing chemicals they produce.

In recent years Australian banana growers have started growing brassica green manures for nematode control, and in the US farmers are using mustard green manure crops to replace synthetic fumigation for potatoes with huge savings on costs.

Growing of radish, mustard and broccoli have reduced bacterial wilt significantly (50-60%) in most of the experiments, though researchers believe the treatment is more effective on sandy soil than on heavy clay soil.

Methods to Increase Microbial Abundance Andor Activity

Nematophagous fungi parasitize nematodes. The possibility of using these fascinating fungi for biological control of plant- and animal-parasitic nematodes has resulted in increased interest in their biology. The following images can be used to introduce this concept to students.

Endoparasitic fungi use their spores to infect nematodes. The spores adhere to the nematode cuticle or, in some species, are ingested together with food. The endoparasitic fungi are obligate parasites of nematodes and spend their entire vegetative lives inside infected nematodes. Examples of these fungi are *Catenaria anguillulae* (zoospores), *Drechmeria coniospora* (adhesive conidia), and *Nematoctonus concurrens* (adhesive spores).

Nematode-trapping fungi use hyphal trapping devices to capture nematodes. These traps can have either adhesive or mechanical function. In contrast to the endoparasites, the nematode-trapping fungi can live as saprophytes in soil. Examples of nematode-trapping fungi with different trapping devices are *Arthrobotrys oligospora* (adhesive networks), *Dactylaria candida* (adhesive knobs), *Monacrosporium cionopagum* (adhesive branches), *Arthrobotrys dactyloides* (constricting rings), and *Nematoctonus leiosporus* (hourglass traps).

3

Nematode Trapping Fungi and Nematophagous Fungi

Nematode trapping fungi, or "nematophageous fungi," are carnivorous fungi that have developed methods and structures that enable them to successfully trap and consume nematodes. Nematode trapping fungi are responsible for keeping the nematode population in check and are an important part of the subsoil ecosystem. These fungi prey on nematodes and are in turn consumed by organisms on the next trophic level. Nematophageous fungi use several methods to hunt their prey. These methods include living within the nematode and slowly consuming them as well as spreading diseases through nematode populations. The fungi also live in the soil and set traps for the nematodes to squirm into.

Biological Interaction

Nematophagous fungi and nematodes share a special predator-prey relationship. The nematodes, like many other soil inhabitants, secrete chemicals as they travel through the soil. Fungi that prey upon these nematodes have found a way to detect and respond to the existence of these chemicals.The fungi have also developed many sophisticated ways to trap the nematodes by attacking from both outside and within the nematode.

Detecting the Nematodes

In order to effectively hunt its prey, many species of fungi have developed the ability to detect some of the chemicals that nematodes use to communicate and grow. These chemicals are detected when the nematodes wander too close to the fungus. And if the fungus is in need of nutrients that are not provided in its diet, it will create traps in the area in which it detected the chemicals, thus effectively hunting its prey.

Secretion by Nematodes

Nematodes produce a highly conserved family of small molecules called ascarosides.These small molecules are unique to the nematodes and the fungi are able to detect them. Different species of nematode produce different varieties of ascarosides.

Detection by Fungi

Traps to ensnare the nematode are created when these ascarosides are detected.(4) Rather than being proactive and using valuable energy constructing trap structures that may never be used, the fungus in question acts in a reactive manner. Traps are constructed in the area in which the ascarosides were detected. This act however, is not the norm. This chemical eavesdropping phenomenon is done by a reserved few species of fungi such as *Arthrobotrys oligospora*. And also, only when the fungus is nutrient deprived. Such structures are energy intensive, so the fungus must be hungry and know where its prey is located before it will set its traps.

Trapping the Nematodes

Fungi use many methods in order to ensnare their prey. These fungi can strike from the outside as well as from the inside.

Hyphal Rings

One method used by fungi to trap nematodes is the fungal ring. The fungus produces hyphae that end in an open constricting loop. When a nematode swims through this loop, the loop suddenly fills with water. This sudden change in the physiology of the loop causes the diametre of the inside of the loop to narrow and in turn constricts around the nematode. Within 24 hours, hyphae form from within the loop and penetrate the nematode as it begins to digest it.

Adhesive Hyphae

Another method commonly used by nematophagous fungi is the use of adhesive hyphae. The fungi who utilize this method create adhesive knobs or nets out of adhesive hyphae. The fungus creates such structures in order to trap the nematode within the structure, and by utilizing the adhesive hyphae the nematode becomes easily ensnared thus providing the fungus with a nitrogen rich meal.

Nematotoxic Compounds

Nematode trapping fungi can use either or both of the methods listed above in order to trap their prey. But some fungi such as the Pleurotaceae family of mushrooms also take it a step further. This family of fungi utilize adhesive knobs to catch nematodes, but once the nematode is caught the fungus secretes a nematacide that kills the nematode and prevents its escape. This extra step is not used by all nematode trapping fungi, but it is utilized by some. It is an extra step of insurance that guarantees that those nematodes that are caught are unable to escape.

Attacking from Within

While many kinds of fungi use traps to catch nematodes, there are also others that live within the nematode and cause damage from there.

Endoparasitic Fungi

Fungi that live within the nematode can infect the nematode at any stage depending on the species of fungus. The infection can be done perorally as in through the mouth or percutanously which means through the skin. From within the nematode, the fungus can consume nutrients meant for the nematode's growth and spread to other nematodes when able. These endoparasitic fungi can spread throughout a community in a measurably short time and weaken a whole population of nematodes throughout the process.

Niche

Due to the overwhelming abundance of nematodes throughout the world it is easier to say where nematodes cannot be found than to list the many places that they can be found. The interaction between fungus and nematode is limited to environments where they can both thrive.

Nematode Habitats

Nematodes survive and thrive in almost every habitat on earth. They are present on both land and in aquatic environments. The only stipulation that can limit the spread of nematodes is availability of moisture and prey.

Aquatic Habitats

Nematodes can be found in great abundance in both fresh and sea water. Nematodes can infect aquatic plants and animals. There is no discernible biological difference between nematodes that live in an aqueous environment and soil nematodes. Soil nematodes also swim but in the water filled pores between soil particles.

Terrestrial Habitats

Nematodes can be found in nearly soil on earth. As long as a minimal amount of soil moisture and some sort of organism is present in the soil, it is certain that nematodes can be found there.

Fungal Habitats

Fungi can be found in almost every soil and water source in the world. Although not as enormously widespread and survivable as nematodes, fungi are nearly ever present. Fungi are important to the breakdown of organic compounds and require organic compounds, moisture, and proper pH to survive.

Aquatic Habitats

Fungi can be easily found in both fresh and saltwater sources. Fungi can live and grow in rivers, ponds, and oceans. Purely terrestrial fungi also often rely on such water sources to aid in spore dispersal as well as dispersal of its mycelial elements.

Terrestrial Habitats

Soil is the most common environment for fungi. Fungi are present in various soil types all over the world. They are vital to innumerable food webs and work to break down organic matter.

Microbial Processes

This microbial interaction is important due to its effect on subsoil trophic levels. By consuming nematodes, the fungi creates biomass that can be consumed by other organisms. And in consuming the nematodes it also helps to control the present population keeping the nematodes in check.

Benefits

Having nematophagous fungi in the soil confers many benefits both environmentally and economically. Nematodes can affect both plants and animals. They cause several diseases through the use of their feeding stylet as well as some speciesof the worm living within the host.

Benefits to Environment

Nematophagous fungi are predators and can be limited by the amount of prey that is available. This means that it's population is constantly in check by its prey. It also keeps the nematode population under control. This allows a wide variety of plants to grow, even those that are especially susceptible to nematodes. The mass that is gained by the fungus by consuming the nematodes is also beneficial. This mass provides a food source to other larger organisms that are higher up the food chain. Having this microbial interaction promotes the cycling of nutrients and diversifies the food web. These nematodes are also innumerable worldwide which results in a stable supply for soil ecosystems. And constant replenishment through the means of soil movement.

Economic Benefits

Nematodes are present in nearly every square foot of soil worldwide. Plant pathogenic nematodes can also be detrimental to crops. The nematodes are ever present in the fields stealing nutrients from the crops. Nematophagous fungi however, can intervene and help to destroy many of these nematodes before they can cause even more damage. If nematophagous fungi can be weaponized even further against this threat, then we will be able to see an even more noticeable difference. Plant parasitic nematodes feed through a needle-like mouthpiece called a stylet. This mouthpiece is injected into the host and the nematode ingests the nutrients of its host. This feeding action not only deprives the host of nutrients but also creates wounds that can be entryways for invading microbes to colonize the plant and spread disease. The feeding action of nematodes can also spread disease by itself. By feeding from an infected host then moving to an uninfected host, the nematode can directly inject the disease into the new host. The damage done by nematode feeding is oftentimes unnoticed. Seeing that the plant hosts will usually still grow, even though this growth can be stunted or slow. And a drop in yield can often be seen versus uninfested crops. The number of nemataphageous fungi in the soil does not by any means limit the population of these nematodes. But they are a step in the right direction while trying to create an IPM plan to counter the nematodes.

Key Microorganisms

There are many different varieties of nematophagous fungi. And many have yet to be studied extensively. They are present nearly everywhere in the world and have evolved to hunt nematodes in their own way. These nematophagous fungi have evolved alongside the nematodes that they consume and have over time developed methods that are most useful to catch the nematode species that they hunt the most.

Varieties of Nematode Trapping Fungi

Fungal Ring Users

Notable fungi that use a fungal ring include *Drechslerella anchonia*, *Drechslerella dactyloides*, and *Drechslerella brochopaga*.

Adhesive Hyphae Users

Fungi that utilize adhesive hyphae include *Dactylellina candida, Drechmeria coniospora,* and *Arthrobotrys oligospora.*

Nematotoxin Users

Notable nematotoxin users include *Coprinus comatus*, *Stropharia rugosoannulata*, *Arthrobotrys dactyloides*, and the Pleurotaceae family.

Endoparasitic Fungi

Fungi that survive as spores in the soil until the time is right include *Drechmeria coniospora*, *Pythium caudatum*, and *Harposporium leptospira*.(7)

Current Research

Due to the over abundance of nematodes and the huge economic loss that is caused by them each year, many experiments are under way in order to find a way to alleviate this damage and promote plant health.

The purpose of this experiment was to determine the effectiveness of the fungus *Verticillium chlamydosporium* against the nematode against *eloidogyne arenaria* under greenhouse conditions using tomato plants as a host. It was decided to use three isolates of the fungus. This fungus was rinsed in a shaken liquid culture of Czapek Dox broth for a week at 18 degrees Celsius. The fungal liquid was then added to a peat/sandy soil and allowed to grow in the greenhouse at 23 degrees Celsius for two weeks.

The first experiment involved adding second stage juveniles to the soil of the tomato plants. After 50 days, nematode populations were estimated. For the second experiment, sand and sand/bran mixtures were set up and aliquots of .5 and 5 mL of the nematodes were added to the soil after they had been mixed with fungal particles. This experiment was to determine how well fungal particles of a specific fungal isolate affected the nematodes.

The last experiment involved infecting the soil of tomato plants with nematodes but also adding a substantial amount of a fungal isolate. This experiment was to determine if the implementation of either added material affected the plant's growth. Throughout these experiments it was plain to see through the data that was collected that the isolate that was used in the experiments did provide an 80 percent reduction of nematode populations. This isolate also did not cause harm to the plants that it had been planted near.

This experiment was conducted to determine which kind of culture of *Paecilomyces lilacinu*(a nematophagous fungus) was most effective against which varieties of nematodes. Seventeen varieties of nematodes were used throughout this experiment and all of these nematodes were exposed to the filtered filtrate of the fungus. The test was to conclude whether or not the filtrate paralyzed the nematode.

This was conducted by exposing all of the nematodes(50 at a time in 20 microlitres of water) to the filtrate. After 24 hrs, the nematodes were probed with a sharp needle in order to test for paralysis. Estimates were taken after this 24 hr period in order to determine the number of paralyzed individuals. Several types of medium were also tested to see which method produced the most toxin. The study found that toxin production was not influenced by light or pH, and that the toxin only killed efficiently against the Heteroderidae family.

This experiment investigated how well the bacteria-feeding nematode *P. redivivus* was attracted to living mycelia of different nematophagous fungi. Methods were developed that allowed the detection of both attraction and repulsion as well as the magnitude of each interaction. The fungi that were involved in this experiment were allowed to grow to nearly the edge of their plates and were grown on various kinds of agar. The nematophagous fungi were also tested with traps present as well as with out traps. The nematode used in the experiment was the bacteria-feeding nematode *P. redivivus* and it was reared and provided in a way that made it constant throughout the experiment. The attraction assay consisted of adding a drop of the nematode suspension which contained about 100 specimens onto a small round section of the plate after the plate had been dried for 2-3 days. After 24 hours, the remaining nematodes were counted and the attraction/repulsion towards the hyphae was measured. The attraction intensity assay was conducted in the same way as the other experiment, but the nematodes were counted hourly in order to determine how quickly the nematodes trevalled towards or away form the hyphae. The experiment ended with a bank of valuable data. It was found through all of the fungi present whether or not traps were formed spontaneously from the introduction of nematodes, attraction vs repulsion, and also attraction vs repulsion after killing the fungi using U.V. radiation.

Nematophagous Fungi

So far we know more than 160 fungus species that live on Nematodes partially or entirely. Nematophagous fungi are found in almost every natural soil and aditionally in a number of other substrates. They play an important role as antagonists of nematodes. Therefore there is great interest in research of possibilities to use them in biological plant protection.

Nematode-trapping Fungi (Predators, Predaceous Fungi)

Predators we call fungi that develop their mycelium predominantly outside their host, which means that they are also able to live as saprophytes and trap their prey on vegetative hyphae or in special traps. This means that only mobile nematode stages are caught.

The predators are devided into several groups concerning the trapping mechanisms. Except for some species that are able to develop adhesive substance on any part of their mycelium, predacious fungi develop special traps: adhesive three-dimensional networks, adhesive columnar branches, adhesive knobs, constricting rings and non-constricting rings.

At the place of adhesion, a mortiferous excrescence perforates the cuticle of the nematode (in the case of most predators - but not of Nematoctonus - forming a globular structure called infection bulb), then giving rise to assimilative hypha growing through the body of the prey, which is digested by enzymes secreted by the fungus and after that resorbed (

Endoparasitic Nematophagous Fungi

Except for the development and spreading of the spores, the entire life cycle of the endoparasites takes place inside the host. Depending on the species, any nematode stages can be attacked. The way of infection can be either peroral (e. g. Harposporium anguillulae LOHDE) or - in most species - percutanous. Only the second group is of significance for biological control of phytopathogenous nematodes, because eelworms with a mouth prick cannot swallow whole spores, instead, some of them are able to suck out the conidia of the fungus.

Nematophagous Fungi as Root Endophytes

Nematophagous and entomopathogenic fungi (NEF) comprise an important group of fungal parasites of invertebrates (FPI). NEF belong to a wide range of fungal taxa, but most of them are anamorphic fungi and facultative parasites. These fungi can infect, kill and digest nematodes and insects, respectively, which we will call their canonical, or normal, hosts. These hosts have barriers to the environment (eggshells and cuticles) that have common structural features. Therefore, the infection cycles share common strategies (e.g. adhesion to the host) or metabolites (e.g. proteases and chitinases for host penetration). Some species (e.g. *Lecanicillium lecanii*) can even be isolated from both infected nematodes and insects. The NEF may also infect other organisms (other fungi and plants) apart from their canonical hosts in a similar or different mode.

Endophyte

An endophyte is an endosymbiont, often a bacterium or fungus, that lives within a plant for at least part of its life without causing apparent disease. Endophytes are ubiquitous and have been found in all the species of plants studied to date; however, most of these endophyte/ plant relationships are not well understood. Many economically important grasses (e.g., *Festuca* spp., *Lolium* spp., *Zea*) carry fungal endophytes which enhance their growth, may improve the plant's to tolerate abiotic stresses such as drought, as well as improve their resistance to insect and mammalian herbivores.Transmission

Endophytes may be transmitted either vertically (directly from parent to offspring) or horizontally (among individuals). Vertically transmitted fungal endophytes are typically considered sterile and transmit via fungal hyphae penetrating the host's seeds (e.g., *Neotyphodium*). Since their reproductive fitness is intimately tied to that of their host plant, these fungi are often mutualistic. Conversely, horizontally transmitted fungal endophytes are fertile, and reproduce through asexual or sexual spores that can be spread by wind and/or insect vectors. Because of the complex mating systems of fungi, some endophytes normally considered to be vertically transmitted may occasionally become sexually reproductive, producing spores that can be transmitted horizontally (e.g., *Epichloë*). Some endophytic fungi are actually latent pathogens or saprotophs that only become active and reproduce when their host plants are stressed or begin to senesce, respectively

Endophyte-host Interactions

Endophytes may benefit host plants by preventing pathogenic organisms from colonizing them. Extensive colonization of the plant tissue by endophytes creates a "barrier effect", where the local endophytes outcompete and prevent pathogenic organisms from taking hold. Endophytes may also produce chemicals which inhibit the growth of competitors, including pathogenic organisms. Some bacterial endophytes have proven to increase plant growth. The presence of fungal endophytes can cause higher rates of water loss in leaves. However, certain fungal endophytes help plants survive drought and heat. Fungal endophyte-related host benefits are common phenomena, and have been the focus of much research, particularly among the grass endophytes.

Applications

The wide range of compounds produced by endophytes have been shown to combat pathogens and even cancers in animals including humans. One notable endophyte with medicinal benefits to humans was discovered by Gary Strobel: *Pestalotiopsis microspora*, an endophytic fungus of *Taxus wallachiana* (Himalayan Yew) was found to produce taxol. Endophytes are also being investigated for roles in agriculture and biofuels production.

Inoculating crop plants with certain endophytes may provide increased disease or parasite resistance while others may possess metabolic processes that convert cellulose and other carbon sources into "myco-diesel" hydrocarbons and hydrocarbon derivatives. *Piriformospora indica* is an interesting endophytic fungus of the order Sebacinales, the fungus is capable of colonising roots and forming symbiotic relationship with every possible plant on earth. *P. indica* has also been shown to increase both crop yield and plant defence of a variety of crops(barley, tomato, maize etc.) against root-pathogens.

It is speculated that there may be many thousands of endophytes useful to mankind but since there are few scientists working in this field, and since forests and areas of biodiversity are rapidly being destroyed, many useful endophytes for curing disease might be permanently lost for medicinal use before they are discovered. The effects of climate change on endophytes is being investigated. Studies of plants grown at different climates or at increased carbon dioxide levels have different distributions of endophytic species.

Identification

While many endophytes are known to colonize multiple species of plants, some are host specific. Endophytic species are very diverse; it is thought that only a small minority of all existing endophytes have been characterized. A single leaf of a plant can harbor many different species of endophytes, both bacterial and fungal.

Endophytes can be identified in several ways, usually through amplifying and sequencing a small piece of DNA. Some endophytes can be cultured from a small piece of their host plant in an appropriate growth medium. Not all endophytes present can be cultured in this way because amplification of ground up plant tissue using fungal or bacterial specific primers has revealed the existence cryptic species. Some grass endophytes can be seen as coiled tubes of hyphae under the microscope at 400X following clearing of the leaf sheaths in ethanol and staining with aniline blue.

Diversity of Fungal Endophytes

Fungal endophytes are generally from the phylum Ascomycota. Some specific examples of which are found in orders hypocreales and xylareales of the Sordariomycete(Pyrenomycete) class. Additionally the class of Loculoascomycetes have endophoytes.

Microbial Inoculant

Microbial inoculants also known as soil inoculants are agricultural amendments that use beneficial endophytes (microbes) to promote plant health. Many of the microbes involved form symbiotic relationships with the target crops where both parties benefit (mutualism). While microbial inoculants are applied to improve plant nutrition, they can also be used to promote plant growth by stimulating plant hormone production.

Research into the benefits of inoculants in agriculture extends beyond their capacity as biofertilizers. Microbial inoculants can induce systemic acquired resistance (SAR) of crop species to several common crop diseases (provides resistance against pathogens). So far SAR has been demonstrated for powdery mildew (*Blumeria graminis* f. sp. *hordei*), take-all (*Gaeumannomyces graminis* var. *tritici*), leaf spot (*Pseudomonas syringae*) and root rot (*Fusarium culmorum*).

Bacterial

Rhizobacterial Inoculants

The rhizobacteria commonly applied as inoculants include nitrogen-fixers and phosphate-solubilisers which enhance the availability of the macronutrients nitrogen and phosphorus to the host plant. Such bacteria are commonly referred to as plant growth promoting rhizobacteria (PGPR).

Nitrogen-fixing Bacteria

The most commonly applied rhizobacteria are *Rhizobium* and closely related genera. *Rhizobium* are nitrogen-fixing bacteria that form symbiotic associations within nodules on the roots of legumes. This increases host nitrogen nutrition and is important to the cultivation of soybeans, chickpeas and many other leguminous crops. For non-leguminous crops, *Azospirillum* has been demonstrated to be beneficial in some cases for nitrogen fixation and plant nutrition.

For cereal crops, diazotrophic rhizobacteria have increased plant growth, grain yield, nitrogen and phosphorus uptake (Galal *et al.*, 2003), and nitrogen, phosphorus and potassium content. Rhizobacteria live in root nodes, and are associated with legumes.

Phosphate-solubilising Bacteria

To improve phosphorus nutrition, the use of phosphate-solubilising bacteria (PSB) such as *Agrobacterium radiobacter* has also received attention. As the name suggests, PSB are free-living bacteria that break down inorganic soil phosphates to simpler forms that enable uptake by plants.

Fungal Inoculants

Several different fungal inoculants have been explored for their benefits to plant nutrition. The most commonly investigated fungi for this purpose are the arbuscular mycorrhizae (AM). Other endophytic fungi, such as *Piriformis indica* can also be beneficial.

Composite Inoculants

The combination of strains of Plant Growth Promoting Rhizobacteria has been shown to benefit rice and barley (*Hordeum*, Belimov *et al.* (1995a)). The main benefit from dual inoculants is increased plant nutrient uptake, from both soil and fertiliser (Bashan *et al.*, 2004; Belimov *et al.* 1995a). Interestingly, multiple strain inoculants have also been demonstrated to increase total nitrogenase activity compared to single strain inoculants, even when only one strain is diazotrophic.

PGPR and arbuscular mycorrhizae in combination can be useful in increasing wheat growth in nutrient poor soil and improving nitrogen-extraction from fertilised soils. In salinised soils, Rabie (2005) found that inoculating AM-infected *Vicia faba* plants with *Azospirillum brasilense* amplified the beneficial effects of AM inoculation.

Fungi and Vertebrates

At least 30 species of Australian mammals have been found to eat fungal fruiting bodies. The animals form quite a varied group: Mountain Pygmy-possum, various native rodents, various macropods (kangaroos, wallabies, pademelons, Quokka), Yellow-bellied Glider, Mountain Brushtail Possum, Common Brushtail Possum, Bettongs, Potoroos, Bilby, Bandicoots and the Common Wombat.

These animals are distributed over a wide variety of habitats, from rainforest to desert. Given the lack of comprehensive dietary studies, the current list of fungal-eating native mammals is undoubtedly incomplete. In some cases there have been sightings of fungal consumption, but most of the evidence comes from fungal spores found in animal droppings. The spores of many fungal species are fairly robust and able to pass, undamaged, through an animal's digestive system and out with the faeces.

Figure: *Long-nosed Potoroo (*Potorous tridactylus*)*

While we know that at least 30 species of native mammals eat fungi, the level of consumption is largely unknown. A number of these mammals are undoubtedly only occasional eaters of fungi. However, fungal consumption by some of the smaller mammals has been well-studied in south-eastern Australia and it has been found that the truffle-like fungi feature heavily in the diets of these mammals. The eating of fungi is well-developed in the rat-kangaroos (a marsupial group which includes the bettongs and potoroos). The Long-nosed Potoroo (*Potorous tridactylus*) of western Victoria eats over 40 different species of fungi, the majority of which are truffle-like fungi. The rare Long-footed Potoroo (*Potorous longipes*) of eastern Victoria and southeast NSW eats at least 33 species and the Tasmanian Bettong (*Bettongia gaimardi*) eats at least 49 species. For Potoroos and Bettongs in general, fungi (mostly truffle-like) constitute at least 30-40% of the animals' diets throughout the year - but there is considerable variation between the species. For example, truffle-like fruiting bodies constitute 80-90% of the diet of the Long-footed Potoroo, regardless of season.

In the three-way (tree-fungus-marsupial) association it is clear that there is a benefit to all parties. Tree and fungus each benefit from the other through their mycorrhizal association. The animal uses the fungus for food, but then helps disperse the fungal spores.

Elsewhere in the world many animals are known to eat fungi and the truffle-like fungi are especially attractive to various animals - including various small rodents, deer, elk and bears. Squirrels will dig up truffle-like fungi and leave them to dry on tree limbs. In South America opossums and pigeons eat *Cyttaria* fruiting bodies. Siberian Jays have been found to feed heavily on fungal fruiting bodies early in winter - with the fungi apparently pilfered from squirrel caches.

Other birds have also been seen pecking at various types of fungal fruiting bodies, but in some cases the birds may be after any larvae inside. However, in the Kuwaiti desert there is one truffle-like species (*Phaeangium lefebvrei*) which is deliberately scratched out and eaten by at least eleven species of migratory birds - and the Bedouin Arabs use the fruiting bodies for bait when they want to catch live birds.

Figure: *Thaxterogaster porphyreum*

New Zealand lacks the small mammals, that are so widespread in Australia, but has a number of ground-dwelling birds occupying similar ecological niches. While there is still much to be learnt about the New Zealand truffle-like fungi, a number of species produce very strikingly coloured fruiting bodies, for example *Paurocotylis pila, Thaxterogaster porphyreum* and *Weraroa erythrocephala*. When these appear in the leaf litter of the forest floor, they look much like the seeds of liliaceous plants or the fruits of conifers like *Podocarpus* and its relatives. It is possible that the various ground-feeding birds are attracted to these colourful fruiting bodies, mistaking them for plant seeds or berries, and then eat them. Later the fungal spores would pass out in the bird droppings, so the birds would act as dispersal agents, filling the role played by the various small mammals in Australia.

Do some Australian birds actively seek out truffle-like fungi? There certainly are colourful truffle-like fruiting bodies in Australia, a number of which do protrude above the ground and resemble berries in the leaf litter. Furthermore, a number of berry-eating birds come to mind – for example, Cassowary, Brush Turkey and Emu. On one occasion an Eastern Yellow Robin has been happy to take truffle-like fungi flicked to it.

Fungal Pathogen Genomics and Proteomics

Fungi and oomycetes are the causal agents of many of the world's most serious plant diseases and are unique among the microbial pathogens in being able to breach the intact surfaces of host plants, rapidly establishing infections that can have disastrous consequences for large-scale agricultural production. The ability to cause plant disease is not a common trait among the many saprotrophic and mutualistic fungal species, but it is a very widespread one, occurring throughout the fungal kingdom. Recently, there have been a number of studies published describing the genome sequences of a diverse set of fungi and oomycetes, including one published in this issue of *The Plant Cell*, and this provides an opportunity to review what we have learned so far from sequencing the genomes of pathogenic and free-living fungi and also to look forward to the mass of genome sequence information that is likely to be generated in the next few years. The deployment of low-cost, high-throughput DNA sequencing technologies and large-scale functional genomics to eukaryotic plant pathogens will provide new insight into their biology and into the evolution of pathogenicity.

The Rapid Pace of Fungal Genome Sequencing

There are currently genome sequences from 12 plant pathogenic fungi and two oomycetes that are publicly available, out of a total of >40 fungal genomes that have now been sequenced. A significantly higher number of projects is in progress, and the dramatic fall in the price of DNA sequencing, coupled with the new high-throughput sequencing technologies, strongly suggest that there will soon be many more sequences, both of new species and of isolates of the same species, supplying information regarding gene inventory conservation, pathogenic adaptation, evolutionary history, and allelic variability of virulence-associated genes. Here, we have brought together some observations based on analysis of the fungal and oomycete genomes that have been reported so far.

Evolutionary Genomics of the Eukaryotic Phytopathogens

With increased genome sampling from a range of fungal species, it is now becoming possible to resolve a fungal phylogeny using multiple concatenated gene alignments. Strikingly, this approach consistently has shown that nutritional modes of life and cellular characteristics are a mosaic across the fungal phylogeny. For example, fungi that form spores with flagella (chytrids) are not monophyletic and do not form a single unified branch, suggesting that loss of the cilium/flagellum occurred at least four times throughout the kingdom fungi. Furthermore, the yeast and filamentous mode of growth and the ability to parasitize plant tissue are both interdispersed traits across the fungal kingdom. Comparison of 42 fungal taxa focusing mainly on ascomycetes and extended to represent the diversity of the eukaryotes suggests that plant pathogenic mechanisms rose separately on at least five occasions within the fungi and a total of seven times across the eukaryotes. Of course, increased genome and taxon sampling may alter the model presented, but clearly the rise of phytopathogenesis has involved multiple phenotype acquisitions and a complex evolutionary history, suggesting that even within relatively closely related ascomycete groups, pathogenic and saprotrophic organisms are likely to have many key differences. The Pezizomycota, which include many phytopathogenic fungal species, was only represented by four genomes in the study, two of which were phytopathogens, yet this analysis suggested that *Magnaporthe grisea* has gained 3240 novel genes and has undergone 193 unique duplications and 1860 unique gene losses.

The other phytopathogen sampled, *Fusarium graminearum*, had a similar strikingly novel gene complement compared with its sister ascomycetes, including 2513 novel genes, 305 unique duplications, and 711 unique losses. Comparative genome analysis of the *Stagonospora* genome, when considered with the genomic comparisons discussed above, and the differential complements of genes across the ascomycete phylogeny, suggest that the rise of ascomycete phytopathogenesis has involved radically different evolutionary genomic changes. Increased genome sampling is therefore required to fully understand the nature and evolutionary background of phytopathogenic genomic innovations.

Phytopathogenesis is not a lifestyle limited to the fungi, however, as there are two other major groups of eukaryotic phytopathogenic microbes branching in very different places in the eukaryotic phylogeny. The oomycetes (such as *Phytophthora* sp and *Hyaloperonospora*) group along with the brown algae and the diatoms within the heterokonts (stramenopiles), while the plasmodiophorids (e.g., *Plasmodiophora brassicae*, which cause a range of plant diseases, including clubroot of brassicas) group together with the heterotrophic cercozoan protozoa within the Rhizaria. It has recently been demonstrated that lateral gene transfers have occurred between the fungi and the oomycetes. Based upon the predicted function of these lateral gene transfers, it seems likely that they were important for the acquisition of osmotrophy and perhaps pathogenesis. For example, one strong candidate for gene transfer from the fungal lineage to the oomycetes encodes a putative protocatechuate, 3,4-dioxygenase â-subunit (*PcaH*), a key functional component of the â-ketoadipate pathway, which would provide a means to use aromatic compounds from the environment. Such acquisitions may have been important for the origin of the oomycete plant pathogens that descended from a phototrophic ancestor bearing a plastid acquired from the endosymbiosis of a red algae. The study by Richards et al. (2006) demonstrates that distantly related plant pathogens may have striking genetic similarities within an unrelated genomic background. Horizontal gene transfer, however, continues to impact the host range and virulence of pathogens. Ability to synthesize the ToxA host-specific toxin, for example, was transferred laterally from *Stagonospora nodorum* to the wheat tan spot fungus *Pyrenophora tritici-repentis* at some point in the recent past, prior to 1941, allowing tan spot to emerge as a significant disease on wheat cultivars containing the *Tsn1* gene .

Synteny

The availability of an increasing number of fungal genome sequences has allowed comparisons of gene order or synteny between different species. Generally the large degree of evolutionary distance between the species where genomes have been sequenced has meant that no significant synteny, or gene order conservation, has been detected between species of a different genus. Comparisons between three species of *Aspergillus* have shown that a majority of each genome could be mapped to conserved syntenic blocks in at least one other genome, but there have been extensive rearrangements. There are also large regions lacking synteny, particularly subtelomeric regions, which in the aspergilli are enriched for secondary metabolite biosynthetic genes, thought to play a role in niche adaptation and opportunistic pathogenicity. It can be hypothesized that rapid rearrangement of such subtelomeric regions might facilitate species-specific evolution of these genes. There are some examples of small-scale synteny between fungal species. For example, the quinate cluster consists of seven coregulated genes that control the catabolism of quinate. This cluster is widespread amongst filamentous ascomcyetes. Although the clustering of these genes is conserved amongst filamentous ascomycetes, there is no conservation of gene order or orientation.

It is clear that with increased genomic data we will be able to pinpoint true lineage-specific pathogenic innovations and broader pathogenic mechanisms conserved among many species, to track the rise and spread of key pathogenic factors (such as effector proteins), and to identify both specific and general targets for disease intervention. Some examples of characteristics of the gene inventories of phytopathogenic fungi and oomycetes are described in the following sections.

Cell Surface Receptors

To successfully infect a host plant, plant pathogenic fungi and oomycetes must make appropriate responses to a variety of environmental cues, including the chemical and physical characteristics of the host plant surface. This is achieved via cell surface receptors that respond to external stimuli and relay that information into the cell. The genomes of animals contain large numbers of G-protein–coupled receptors (GPCRs), which are characterized by having seven transmembrane helices. They bind exogenous ligands and relay signals into the cell by activating intracellular signaling pathways via conformational changes in G-proteins.

The genomes of the model ascomycete yeast species *Saccharomyces cerevisiae* and *Schizosaccharomyce pombe* contain only three and four GPCRs, respectively. Analysis of the genomes of filamentous phytopathogenic ascomycete fungi, however, reveals a significant number of additional putative GPCRs, as many as 61 in the rice blast fungus *M. grisea* and 84 in the head blight pathogen *F. graminearum*. Furthermore, a novel class of putative GPCRs has only been found so far in the genomes of filamentous ascomycetes. This group of putative receptors includes Pth11, a plasma membrane protein that activates appressorium differentiation in response to inductive surface cues in *M. grisea*. The Pth11 protein contains an extracellular epidermal growth factor (EGF)–like domain at the N terminus followed by seven transmembrane domains.

There are 12 putative GPCRs that contain the extracellular EGF-like domain (known as the CFEM domain) in the *M. grisea* genome. Comparative genome analysis reveals that there has been an expansion in CFEM domain–containing receptor-encoding genes in the genome of *M. grisea* when compared with the nonpathogenic ascomycete fungi *Neurospora crassa* and *Aspergillus nidulans*. Analysis of the genome of the Dothideomycete wheat pathogen *S. nodurum* reveals only three Pth11-like proteins in contrast with other ascomycete pathogens. Two of these putative GPCR genes were studied further and did not have a virulence-associated phenotype when corresponding deletion mutant strains were created, suggesting diversification and redundancy in

GPCR function among the fungi. Interestingly, the genome of *M. grisea* also contains six genes that code for receptors similar to a cAMP receptor from the amoebozoan *Dictyostelium discoideum*, suggesting the capacity to respond to an exogenous cAMP signal. Three similar genes are found in the saprotrophic fungus *N. crassa*, indicating that this may be a conserved character among sordariomycetes. Interestingly, the genomes of the oomycete phytopathogens *Phytophthora sojae* and *Phytophthora ramorum*, which contain 24 putative GPCR genes, also contain four homologs of the *Dictyostelium* cAMP receptor in addition to 12 putative GPCRs, each with a C-terminal intracellular phosphatidylinositol-4-phosphate 5-kinase domain similar to RpkA from *Dictyostelium*, which might allow GPCR-mediated signal transduction through phosphoinositide second-messenger synthesis rather than via heterotrimeric G-proteins.

In *M. grisea*, Pth11 is thought to act upstream of a cAMP-dependent signaling pathway that is required for infection-related development. The cAMP response pathway operates across the fungal kingdom, including within *S. cerevisiae*. However, in *S. cerevisiae*, protein kinase A acts downstream of the glucose-sensing Gpr1 GPCR and the genome of this yeast contains no Pth11 homologs. Thus, pathogenic fungi, such as *M. grisea*, have acquired the ability to respond to a variety of environmental cues using novel classes of receptors that activate conserved intracellular signaling pathways, giving new triggers to old weapon systems.

Secreted Proteins

Fungi and oomycetes, although phylogenetically very distantly related, are both osmotrophic microorganisms. This means that they live by secreting into the external environment enzymes that degrade polymers encountered there, such as cellulose, lipid, protein, and lignin, and transport the resulting simple sugars, amino acids, and fatty acids into the growing cell for use.

Therefore, to a large degree the ecological niche that a fungus or oomycete occupies is defined by the products that it secretes, be they small molecules such as toxins, proteinaceous effectors to perturb host signaling or metabolism, or the array of hydrolytic enzymes that digest extracellular macromolecules into smaller subunits that the fungus can use. The secretome is defined as the set of soluble secreted proteins encoded by a genome. The secretome for a fungal species is usually predicted by analyzing the set of translated gene sequences using bioinformatics tools, such as SignalP 3.0, which detects the presence of a signal peptide in the N terminus of a protein, or WoLF PSORT, which uses a variety of known signal motifs as models to predict the subcellular localization of a protein. However, these tools rely on correct prediction of gene structure (especially the 52 -end) and have generally been trained on secreted proteins from the model yeast *S. cerevisiae*, such that the resulting predicted secretome may not be very accurate in the absence of direct experimental proof of secretion. Nevertheless, secretome sets based on genome sequence have been predicted for a number of fungal species, although the different methodologies applied for each species means that comparisons must be treated with caution.

It is clear that a diverse array of proteolytic and carbohydrate-degrading enzymes is secreted by numerous fungi, reflecting the variety of nutritional sources they can use. The various ecological niches occupied by different fungal species are reflected in the gene families present in the secretome.

For example, the white-rot basidiomycete fungus *Phanerochaete chrysosporium* secretes enzymes involved in lignin degradation, such as lignin and manganese-dependent peroxidases, copper radical oxidases, extracellular FAD oxidases, and multicopper oxidases. The secretome of *Aspergillus oryzae*, a fungus used in the production of traditional Japanese fermentation foods such as soy sauce, contains large numbers of á-amylases, pectinolytic enzymes, and proteases. These enzymes are essential for the ability of the fungus to ferment rice and soybean products.

Many species of plant-pathogenic fungi secrete a battery of glycoside hydrolases, polysaccharide lyases, and esterases, which the fungi use to degrade the host-plant cell wall for nutrition and in some cases to aid entry into plant tissue. Many saprophytic fungi gain nutrition from decaying plant material and therefore also secrete cell wall degrading enzymes. From this relatively limited sample and crude comparison, there seems no obvious correlation between the ability to infect plants and the number of cell wall–degrading enzymes present within the genome.

Saprophytic species, such as *A. nidulans*, produce more glycosyl hydrolases and polysaccharide lyases than any of the phytopathogenic species. Notably, the biotrophic phytopathogen *Ustilago maydis* produces considerably fewer cell wall–degrading enzymes than either *M. grisea* or *F. graminearum*. This is consistent with the necessity of minimizing plant cell wall damage (and the release of cell wall fragments that may trigger plant defence responses) for the biotrophic lifestyle of *U. maydis*. Although *M. grisea* has a clear and extended biotrophic stage of development, it is likely that the formation of necrotic lesions and conidiogenesis is essentially a necrotrophic stage of development that require the action of cell wall hydrolases.

Effectors

Infection of the host plant by biotrophic and hemibiotrophic fungi and oomycetes also requires the secretion of protein effectors that suppress plant defences and alter cellular metabolism to suit the needs of the invading pathogen. The delivery of such proteins to the plant cell is not trivial and requires traversal of the plant plasma membrane. Bacterial phytopathogens deliver effector proteins into host cells via a specialized type III secretion system. A large group of potential effectors have been found in the genomes of the oomycete pathogens *P. sojae* and *P. ramorum* (<“350 in each genome). These proteins share a putative host cell targeting motif with effector proteins encoded by four known avirulence genes cloned from *P. sojae*, *Phytophthora infestans*, and the oomycete obligate biotroph *Hyaloperonospora parasitica* that are recognized in the host cytoplasm by intracellular resistance gene products (and therefore must be delivered into plant cells).

The protein sequences of these effectors all share an RXLR EER motif at their N terminus. This is similar to the RXLXE/Q host cell targeting signal that is required for the translocation of proteins from the malaria parasite *Plasmodium falciparum* into host red blood cells. The RXLR EER motif appears to function as a signal that mediates trafficking of effector proteins into host cells, perhaps acting through a translocation protein in the plant plasma membrane or by stimulating endocytosis in a host plant cell at the haustorial interface. With the availability of numerous genomes, it is now possible to study the evolutionary history of host cell targeting mechanisms by comparing the phylogeny of host effector proteins with the presence and absence of N-terminal motifs shown to mediate trafficking of effector proteins. This approach will define whether functional motifs, such as RXLR-EER motifs, are ancient and conserved eukaryotic protein properties and present in other heterokonts, alveolates, and Rhizaria or instead are the product of recent convergent evolution.

So far, RXLR EER motifs have not been reported in proteins secreted by fungal phytopathogens, and identifying a host cell targeting sequence in fungal effectors is a major goal for future analysis of fungal phytopathogen genomes. Nevertheless, groups of potential secreted effectors, without identifiable host cell targeting motifs, have been identified in the secretome of the rice blast fungus *M. grisea*. These include 21 proteins containing a novel variant of a chitin binding motif. Chitin binding proteins, such as Avr4 from *Cladosporium fulvum*, have been shown to protect the fungal cell wall from plant chitinases. Interestingly, this particular chitin binding domain-encoding gene only seems to have evolutionarily related genes present in animals; there seems to be no Avr4 homolog in any sequenced fungal genome so far sampled. There is also a family of novel proteins that may interact with rice GTP binding proteins and four proteins that are similar to the necrosis-inducing peptide (Npp1) of *P. infestans*. However, homologs of this peptide are also found in the secretomes of saprophytes, such as *A. nidulans*, but not in the basidiomycete biotroph *U. maydis*. The two sequenced oomycete genomes both contain a large family of Npp1-like proteins (40 members in *P. ramorum* and 29 in *P. sojae*). These oomycetes also secrete large numbers of protein toxins from other families, including Crn (crinkling and necrosis-inducing) and PcF proteins. By contrast, the genomes of *Phytophthora* species contain smaller numbers of cytochrome P450s and nonribosomal peptide synthases (NRPSs) than filamentous fungi and no polyketide synthases. This suggests that oomyctes may synthesize fewer toxic small molecules than fungi and instead rely more on protein toxins for necrotrophic growth.

The diversity of proteins found in the secretome is striking, and 70% of proteins secreted by *U. maydis*, for instance, cannot be assigned a function. Of these unknown proteins, two-thirds have no known homologs even among other fungi. It is also notable that 18.6% of secreted *U. maydis* proteins are arranged in 12 gene clusters, and most of these are highly expressed in a coordinated manner within infected plant tissue. Deletion of five of these clusters resulted in changes in virulence, suggesting that many of these proteins are involved in pathogenicity. Sequence comparisons between two *Phytophthora* species suggest that secreted proteins are evolving more rapidly than the proteome overall, because 17 and 11% of *P. sojae* and *P. ramorum* secreted proteins, respectively, are unique compared with 9 and 4% of proteins in the total proteome of the two species. The rapid diversification of the secretomes is also evident from the large number of proteins (77%) that occur in multigene families and the diversifying selection evident among the secreted RXLR effectors of oomycetes.

A large number of secreted proteins from the glume blotch pathogen *S. nodorum* do not have homologs in any other sequenced fungi, providing more evidence for lineage-specific novel secretome diversity. This pattern is consistent with the large differences in gene loss, duplication, and acquisition across the fungi. As sequencing technology become more accessible, it is more likely that different strains of the same species (perhaps with different host ranges) will be sequenced. This is important because it will allow the identification of variations between strains, pinpointing genes that show a history of rapid variation, potentially indicative of host–pathogen coevolution and/or diversifying selection under this selective pressure). Such genes are therefore candidates for further research because they are likely to encode proteins involved in interactions with the host plant. When coupled with large-scale gene functional analysis studies that are now possible in fungal pathogens, this will allow rapid functional definition of effector-encoding genes.

Secondary Metabolic Pathways

Toxins are important virulence factors for a number of fungal diseases. Nonspecific toxins, such as the trichothecenes produced by *Fusarium* species, damage cells of a number of plant species, whereas host-specific toxins, such as the toxins produced by different pathotypes of *Alternaria alternata*, determine the host range of the fungus. Some fungal toxins are small peptides (for example, the Npp1-like proteins referred to in the previous section), but many are secondary metabolites that are low molecular weight molecules with a variety of structures.

Necrotrophic pathogens use toxins to elicit plant cell death and derive nutrition from dead plant tissue. Toxin biosynthetic pathways have been elucidated in a many fungal species, for example, the synthesis of sirodesmin in *Leptosphaeria maculans*, the blackleg pathogen of canola. A number of key enzyme families are involved in toxin biosynthesis (as well as other secondary metabolic pathways), including PKSs, NRPSs, and cytochrome P450 monooxygenases. Ascertaining the number of genes encoding enzymes from these families is useful in predicting the capacity of an organism to produce secondary metabolites (including toxins). It is notable that there are relatively low numbers of genes encoding these enzymes in the genomes of the oomycete phytopathogens compared with the true fungi. As mentioned in the previous section, the *Phytophthora* species secrete a large variety of small toxic proteins and may well produce a smaller number of secondary metabolite toxins than phytopathogenic fungi. The biotroph *U. maydis* also has very few PKS- and cytochrome-P450–encoding genes in its genome. Although some saprophytic fungi have lower numbers of PKS- and NRPS-encoding genes in their genomes than the phytopathogens, the genome of the saprophyte *A. nidulans* has more PKS and NRPS-encoding genes than the pathogen *M. grisea*. The genome of *A. oryzae* contains 30 PKS-encoding genes. Phylogenomic analysis of PKS-encoding genes has shown a wide divergence between different fungal genomes, even in the same genus, reflecting perhaps the variety of different secondary metabolites that these species produce. However, comparisons of the number of PKS and NRPS families between fungi reveals far more than the capacity of a fungus to produce toxins. This is because members of these enzyme families are involved in the synthesis of many other secondary metabolites with potential roles in fungal development or perhaps as effectors within plant cells, altering cell morphology and perturbing cell signaling.

For example, fungal metabolites are able to perturb calcium signaling (cyclosporin), sterol synthesis (lovastatin), microtubule polymerization (taxol), actin assembly and disassembly (cytochalasin and phalloidin), Golgi function (brefeldin A), and a host of other cellular processes. Of course, some fungi additionally produce homologs of plant growth regulators, such as gibberellins, which were first discovered in *Gibberella fujikuroi*, demonstrating the capacity of fungal metabolites to fulfill effector-type roles in host plants.

PKSs are also involved in the synthesis of a number of pigments, including melanin, and NRPSs are necessary for the synthesis of iron-chelating siderophores. The genome of the rice blast pathogen *M. grisea* also encodes a number of PKS-NRPS hybrid enzymes. One of these, Ace1, is involved in the synthesis of a secondary metabolite that is recognized by rice cultivars carrying the *Pi33* resistance gene and therefore acts, genetically, as an avirulence gene. *ACE1* is only expressed in appressoria during fungal penetration of rice leaves, consistent with an effector function for the secondary metabolic product during plant infection. The PKS-NRPS hybrid genes are composed of several unique protein domains. This gene and gene architecture are found in a number of filamentous ascomycetes but appear to be absent from the hemiascomycete yeasts (e.g., *Saccharomyces*) and the basidiomycetes (e.g., *Ustilago*), suggesting that the PKS-NRPS hybrid enzymes were the product of a unique gene fusion that occurred in a primitive branch of the filamentous ascomycete evolutionary tree.

Proteomics

Proteomics is the large-scale study of proteins, particularly their structures and functions. Proteins are vital parts of living organisms, as they are the main components of the physiological metabolic pathways of cells. The term *proteomics* was first coined in 1997 to make an analogy with genomics, the study of the genome. The word *proteome* is a blend of *protein* and *genome*, and was coined by Marc Wilkins in 1994 while working on the concept as a PhD student.

The proteome is the entire set of proteins, produced or modified by an organism or system. This varies with time and distinct requirements, or stresses, that a cell or organism undergoes. Proteomics is an interdisciplinary domain formed on the basis of the research and development of the Human Genome Project; it is also emerging scientific research and exploration of proteomes from the overall level of intracellular protein composition, structure, and its own unique activity patterns. It is an important component of functional genomics.

While *proteomics* generally refers to the large-scale experimental analysis of proteins, it is often specifically used for protein purification and mass spectrometry.

Complexity of the Problem

After genomics and transcriptomics, proteomics is the next step in the study of biological systems. It is more complicated than genomics because an organism's genome is more or less constant, whereas the proteome differs from cell to cell and from time to time. Distinct genes are expressed in different cell types, which means that even the basic set of proteins that are produced in a cell needs to be identified.

In the past this phenomenon was done by mRNA analysis, but it was found not to correlate with protein content. It is now known that mRNA is not always translated into protein, and the amount of protein produced for a given amount of mRNA depends on the gene it is transcribed from and on the current physiological state of the cell. Proteomics confirms the presence of the protein and provides a direct measure of the quantity present.

Post-translational Modifications

Not only does the translation from mRNA cause differences, but many proteins are also subjected to a wide variety of chemical modifications after translation. Many of these post-translational modifications are critical to the protein's function.

Phosphorylation

One such modification is phosphorylation, which happens to many enzymes and structural proteins in the process of cell signaling. The addition of a phosphate to particular amino acids—most commonly serine and threonine mediated by serine/threonine kinases, or more rarely tyrosine mediated by tyrosine kinases—causes a protein to become a target for binding or interacting with a distinct set of other proteins that recognize the phosphorylated domain.

Because protein phosphorylation is one of the most-studied protein modifications, many "proteomic" efforts are geared to determining the set of phosphorylated proteins in a particular cell or tissue-type under particular circumstances. This alerts the scientist to the signaling pathways that may be active in that instance.

Ubiquitination

Ubiquitin is a small protein that can be affixed to certain protein substrates by enzymes called E3 ubiquitin ligases. Determining which proteins are poly-ubiquitinated helps understand how protein pathways are regulated. This is, therefore, an additional legitimate "proteomic" study. Similarly, once a researcher determines which substrates are ubiquitinated by each ligase, determining the set of ligases expressed in a particular cell type is helpful.

Additional Modifications

Listing all the protein modifications that might be studied in a "proteomics" project would require a discussion of most of biochemistry. Therefore, a short list illustrates the complexity of the problem. In addition to phosphorylation and ubiquitination, proteins can be subjected to (among others) methylation, acetylation, glycosylation, oxidation and nitrosylation. Some proteins undergo all these modifications, often in time-dependent combinations. This illustrates the potential complexity of studying protein structure and function.

Distinct Proteins are Made Under Distinct Settings

Even studying a particular cell type, that cell may make different sets of proteins at different times, or under different conditions. Furthermore, as mentioned, any one protein can undergo a wide range of post-translational modifications.

Therefore a "proteomics" study can become complex, even if the topic of the study is restricted. In more ambitious settings, such as when a biomarker for a tumor is sought – when the proteomics scientist is obliged to study sera samples from multiple cancer patients – the amount of complexity that must be dealt with is as great as in any modern biological project.

Limitations of Genomics and Proteomics Studies

Proteomics gives a different level of understanding than genomics for many reasons:

- the level of transcription of a gene gives only a rough estimate of its *level of translation* into a protein. An mRNA produced in abundance may be degraded rapidly or translated inefficiently, resulting in a small amount of protein.

- as mentioned above many proteins experience *post-translational modifications* that profoundly affect their activities; for example some proteins are not active until they become phosphorylated. Methods such as phosphoproteomics and glycoproteomics are used to study post-translational modifications.
- many transcripts give rise to more than one protein, through alternative splicing or alternative post-translational modifications.
- many proteins form complexes with other proteins or RNA molecules, and only function in the presence of these other molecules.
- protein degradation rate plays an important role in protein content.

Reproducibility. Proteomics experiments conducted in one laboratory are not easily reproduced in another. For instance, Peng et al. have identified 1504 yeast proteins in a proteomics experiment of which only 858 were found in a similar previous study. Further, the previous study identified 607 proteins that were not found by Peng et al. This translates to a reproducibility of 57% (Peng vs. Washburn) to 59% (Washburn vs. Peng).

Methods of Studying Proteins

Protein Detection with Immunoassays

ELISA. The Enzyme-Linked Immunosorbent Assay has been used for decades to detect and quantitatively measure proteins in samples.

MSIA. The use of mass spectrometry is the gold standard for both discovery and quantitative proteomics. Randall Nelson pioneered the use of immunoassays with mass spectrometry in 1995.

SISCAPA. Stable Isotope Standard Capture with Anti-Peptide Antibodies, is a term introduced by Leigh Anderson

Identifying Proteins that are Post-translationally Modified

One way a particular protein can be studied is to develop an antibody specific to that modification. For example, there are antibodies that only recognize certain proteins when they are tyrosine-phosphorylated, known as phospho-specific antibodies. Also, there are antibodies specific to other modifications. These can be used to determine the set of proteins that have undergone the modification of interest.

For sugar modifications, such as glycosylation of proteins, certain lectins have been discovered that bind sugars. These too can be used.

A more common way to determine post-translational modification of interest is to subject a complex mixture of proteins to electrophoresis in *two-dimensions*, which simply means that the proteins are electrophoresed first in one direction, and then in another, which allows small differences in a protein to be visualized by separating a modified protein from its unmodified form. This methodology is known as "two-dimensional gel electrophoresis". Recently, another approach has been developed called PROTOMAP, which combines SDS-PAGE with shotgun proteomics to enable detection of changes in gel-migration, such as those caused by proteolysis or post translational modification.

Determining the Existence of Proteins in Complex Mixtures

Classically, antibodies to particular proteins or to their modified forms have been used in biochemistry and cell biology studies. These are among the most common tools used by practicing biologists today.

For proteomic study, more recent techniques such as matrix-assisted laser desorption/ionization (MALDI) have been employed for rapid determination of proteins in particular mixtures and increasingly electrospray ionization (ESI).

More recently, thermal protease resistance was exploited in a novel proteomic assay called Fast parallel proteolysis (FASTpp), which enabled detection of specific proteins in E. coli lysate and might be used in the future to detect proteomic perturbations in cancer or mechanistic effects of point mutations.

Establishing Protein–protein Interactions

Most proteins function in collaboration with other proteins, and one goal of proteomics is to identify which proteins interact. This is especially useful in determining potential partners in cell signaling cascades.

Several methods are available to probe protein–protein interactions. The traditional method is yeast two-hybrid analysis. New methods include protein microarrays, immunoaffinity chromatography followed by mass spectrometry, dual polarisation interferometry, Microscale Thermophoresis and experimental methods such as phage display and computational methods.

Practical Applications of Proteomics

One major development to come from the study of human genes and proteins has been the identification of potential new drugs for the treatment of disease. This relies on genome and proteome information to identify proteins associated with a disease, which computer software can then use as targets for new drugs. For example, if a certain protein is implicated in a disease, its 3D structure provides the information to design drugs to interfere with the action of the protein. A molecule that fits the active site of an enzyme, but cannot be released by the enzyme, inactivates the enzyme.

This is the basis of new drug-discovery tools, which aim to find new drugs to inactivate proteins involved in disease. As genetic differences among individuals are found, researchers expect to use these techniques to develop personalized drugs that are more effective for the individual.

Proteomics is also used to reveal complex plant-insect interactions that help identify candidate genes involved in the defensive response of plants to herbivory.

Biomarkers

Early researchers identified key concepts and developed tactics for multiple-option management of nematodes. Although the emphasis on integrated pest management over the past three decades has promoted strategies and tactics for nematode management, comprehensive studies on the related soil biology-ecology are relatively recent. Traditional management tactics include host resistance (where available), cultural tactics such as rotation with nonhosts, sanitation and avoidance, and destruction of residual crop roots, and the judicious use of nematicides.

There have been advances in biological control of nematodes, but field-scale exploitation of this tactic remains to be realized. New technologies and resources are currently becoming central to the development of sustainable systems for nematode-pest-crop management: molecular diagnostics for nematode identification, genetic engineering for host resistance, and the elucidation and application of soil biology for general integrated cropping systems.

The latter strategy includes the use of nematode-pest antagonistic cover crops, animal wastes, and limited tillage practices that favour growth-promoting rhizobacteria, earthworms, predatory mites, and other beneficial organisms while suppressing parasitic nematodes and other plant pathogens. Certain rhizobacteria may induce systemic host resistance to nematodes and, in some instances, to foliage pathogens. The systems focusing on soil biology hold great promise for sustainable crop-nematode management, but only a few research programmes are currently involved in this labour-intensive endeavour.

The National Institutes of Health has defined a biomarker as "a characteristic that is objectively measured and evaluated as an indicator of normal biological processes, pathogenic processes, or pharmacologic responses to a therapeutic intervention."

Understanding the proteome, the structure and function of each protein and the complexities of protein–protein interactions is critical for developing the most effective diagnostic techniques and disease treatments in the future. For example, proteomics is highly useful in identification of candidate biomarkers (proteins in body fluids that are of value for diagnosis), identification of the bacterial antigens that are targeted by the immune response, and identification of possible immunohistochemistry markers of infectious or neoplastic diseaases.

An interesting use of proteomics is using specific protein biomarkers to diagnose disease. A number of techniques allow to test for proteins produced during a particular disease, which helps to diagnose the disease quickly. Techniques include western blot, immunohistochemical staining, enzyme linked immunosorbent assay (ELISA) or mass spectrometry. Secretomics, a subfield of proteomics that studies secreted proteins and secretion pathways using proteomic approaches, has recently emerged as an important tool for the discovery of biomarkers of disease.

Proteogenomics

In what is now commonly referred to as proteogenomics, proteomic technologies such as mass spectrometry are used for improving gene annotations. Parallel analysis of the genome and the proteome facilitates discovery of post-translational modifications and proteolytic events, especially when comparing multiple species (comparative proteogenomics).

Current Research Methodologies

Fluorescence two-dimensional differential gel electrophoresis (2-D DIGE) can be used to quantify variation in the 2-D DIGE process and establish statistically valid thresholds for assigning quantitative changes between samples.

Comparative proteomic analysis can reveal the role of proteins in complex biological systems, including reproduction. For example, treatment with the insecticide triazophos causes an increase in the content of brown planthopper (*Nilaparvata lugens* (Stål)) male accessory gland proteins (Acps) that can be transferred to females via mating, causing an increase in fecundity (i.e. birth rate) of females. To identify changes in the types of accessory gland proteins (Acps) and reproductive proteins that mated female planthoppers received from male planthoppers, researchers conducted a comparative proteomic analysis of mated *N. lugens* females. The results indicated that these proteins participate in the reproductive process of *N. lugens* adult females and males. Proteome analysis of *Arabidopsis peroxisomes* has been established as the major unbiased approach for identifying new peroxisomal proteins on a large scale.

There are many approaches to characterizing the human proteome, which is estimated to contain between 20,000 and 25,000 non-redundant proteins. The number of unique protein species will likely increase by between 50,000 and 500,000 due to RNA splicing and proteolysis events, and when post-translational modification are also considered, the total number of unique human proteins is estimated to range in the low millions.

In addition, first promising attempts to decipher the proteome of animal tumors have recently been reported.

Structural Proteomics

Structural proteomics includes the analysis of protein structures at large-scale. It compares protein structures and helps identify functions of newly discovered genes. The structural analysis also helps to understand that where drugs bind to proteins and also show where proteins interact with each other. This understanding is achieved using different technologies such as X-ray crystallography and NMR spectroscopy.

Expression Proteomics

Expression proteomics includes the analysis of protein expression at larger scale. It helps identify main proteins in a particular sample, and those proteins differentially expressed in related samples—such as diseased vs. healthy tissue. If a protein is found only in a diseased sample then it can be a useful drug target or diagnostic marker. Proteins with same or similar expression profiles may also be functionally related. There are technologies such as 2D-PAGE and mass spectrometry that are used in expression proteomics.

Interaction Proteomics

It is the analysis of protein interactions at larger scale. The characterization of protein-protein interactions are useful to determine the protein functions and it also explains the way proteins assemble in bigger complexes. Technologies such as affinity purification, mass spectrometry, and the yeast two-hybrid system are particularly useful in interaction proteomics. Proteome analysis techniques are not simple and straightforward as those used in transcriptomics. The benefit of proteomics, however, is that it deals with the real functional molecules of the cells. It is known that Strong gene expression results in an abundant mRNA but it does not necessarily mean that the corresponding protein is also abundant. In proteomics things are not so simple as one gene does not always produce the same protein. The genes usually consist of a series of sub structures, which are called exons. These sub structures can be joined in a variety of ways, which helps to give momentum to a whole series of very similar but different proteins. Further increasing complications, once proteins are made, they are ornamented with different other chemicals.

These chemicals can be phosphate, sugars or fats. The affect of the decorations is so severe on the function of protein, for example phosphate normally behaves as an on-off switch and sugars usually tells the proteins where to go and attach in the cell. Therefore, it was comparatively very simple and easy to sequence human genome as there are only 46 molecules and they are made up of 4 building blocks or letters (A,C,G,T) where as proteins have 20 building blocks and each of which can be customized or ornamented after the protein is built. Hence, proteome have to deal with ca. 30,000 genes that can be arranged to give some 800,000 proteins that can be modified and decorated with over 300 different chemicals. Additionally, proteomics also describe the nature of proteins, where they are being produced in a particular cell type and at a specific time. The way they are modified in the cell and the location where they are modified and also with whom they are in contact. Finally, the most difficult thing, is to determine the function of the protein.

Proteomics and System Biology

Proteomics has recently come into the act as a promising force to transform biology and medicine. It is becoming increasingly apparent that changes in mRNA expression correlate poorly with protein expression changes. Proteins changes enormously in patterns of expressions across developmental and physiological responses. Proteins also face changes on the act of environmental perturbations. Proteins are the actual effectors driving cell behaviour. The field of proteomics strives to characterize protein structure and function, protein-protein,protein-nucleic acid, protein-lipid, and enzyme-substrate interactions, protein processing and folding, protein activation, cellular and sub-cellular localization, protein turnover and synthesis rates, and even promoter usage. Integrating proteomic data with information such as gene, mRNA and metabolic profiles helps in better understanding of how the system works.

Current Proteomic Technologies

Proteomics has steadily gained momentum over the past decade with the evolution of several approaches. Few of these are new and others build on traditional methods. Mass spectrometry-based methods and micro arrays are the most common technologies for large-scale study of proteins.

Mass Spectrometry and Protein Profiling

There are 2 mass spectrometry based methods currently used for protein profiling. The more established and widespread method uses high resolution, two-dimensional electrophoresis to separate proteins from different samples in parallel, followed by selection and staining of differentially expressed proteins to be identified by mass spectrometry. Despite the advances in 2DE and its maturity, it has its limits as well. The central concern is the inability to resolve all the proteins within a sample, given their dramatic range in expression level and differing properties.

The second quantitative approach uses stable isotope tags to differentially label proteins from two different complex mixtures. Here, the proteins within a complex mixture are labelled first isotopically, and then digested to yield labelled peptides. The labelled mixtures are then combined, peptides separated by multidimensional liquid chromatography and analyzed by tandem mass spectrometry. Isotope coded affinity tag (ICAT) reagents are the widely used isotope tags. In this method, the cysteine residues of proteins get covalently attached to the ICAT reagent, thereby reducing the complexity of the mixtures omitting the non-cysteine residues. Quantitative proteomics using stable isotopic tagging is an increasingly useful tool in modern development. Firstly, chemical reactions have been used to introduce tags into specific sites or proteins, for the purpose of probing specific protein functionalities. The isolation of phosphorylated peptides have been achieved using isotopic labelling and selective chemistries to capture the fraction of protein among complex mixture. Secondly, the ICAT technology was used to differentiate between partially purified or purified macromolecular complexes such as large RNA polymerase II pre initiation complex and the proteins complexed with yeast transcription factor. Thirdly, ICAT labelling was recently combined with chromatin isolation to identify and quantify chromatin-associated proteins. Finally

ICAT reagents are useful for proteomic profiling of cellular organelles and specific cellular fractions. Another quantitative approach is the Accurate Mass and Time (AMT) tag approach developed by Richard D. Smith and coworkers at Pacific Northwest National Laboratory. In this approach increased throughput and sensitivity is achieved by avoiding the needed for tandem mass spectrometry, and making use of precisely determined separation time information and highly accurate mass determinations for peptide and protein identifications.

Protein Chips

Balancing the use of mass spectrometres in proteomics and in medicine is the use of protein micro arrays. The aim behind protein micro arrays is to print thousands of protein detecting features, for the interrogation of biological samples. Antibody arrays are an example in which a host of different antibodies are arrayed to detect their respective antigens from a sample of human blood. Another approach is the arraying of multiple protein types for the study of properties like protein-DNA, protein-protein and protein-ligand interactions. Ideally, the functional proteomic arrays would contain the entire complement of the proteins of a given organism. The first version of such arrays consisted of 5000 purified proteins from yeast deposited onto glass microscopic slides. Despite the success of first chip, it was a greater challenge for protein arrays to be implemented. Proteins are inherently much more difficult to work with than DNA. They have a broad dynamic range, less stable than DNA and their structure is difficult to preserve on glass slides, though they are essential for most assays. The global ICAT technology has striking advantages over protein chip technologies.

Reverse Phased Protein Microarrays

This is a promising and newer microarray application for the diagnosis, study and treatment of complex diseases such as cancer. The technology merges laser capture microdissection (LCM) with micro array technology, to produce reverse phase protein microarrays. In this type of microarrays, the whole collection of protein themselves are immobilized with the intent of capturing various stages of disease within an individual patient. When used with LCM, reverse phase arrays can monitor the fluctuating state of proteome among different cell population within a small area of human tissue. This is useful for profiling the status of cellular signaling molecules, among a cross section of tissue that includes both normal and cancerous cells. This approach is useful in monitoring the status of key factors in normal prostate epithelium and invasive prostate cancer tissues. LCM then dissects these tissue and protein lysates were arrayed onto nitrocellulose slides, which were probed with specific antibodies. This method can track all kinds of molecular events and can compare diseased and healthy tissues within the same patient enabling the development of treatment strategies and diagnosis. The ability to acquire proteomics snapshots of neighbouring cell populations, using reverse phase microarrays in conjunction with LCM has a number of applications beyond the study of tumors. The approach can provide insights into normal physiology and pathology of all the tissues and is invaluable for characterizing developmental processes and anomalies.

Bioinformatics for Proteomics

There is a large amount of proteomics data being collected with the help of high throughput technologies such as mass spectrometry and microarray. It would often take weeks or months to analyze the data and perform comparisons by hand. For this reason, biologists and chemists are collaborating with computer scientists and mathematicians to create programmes and pipeline to computationally analyze the protein data. Using bioinformatics techniques, researchers are capable of faster analysis and data storage. The applications of bioinformatics based proteomics includes medicine, disease diagnosis, biomarker identification, and many more.

Protein Identification

Mass spectrometry and microarray produce peptide fragmentation information but do not give identification of specific proteins present in the original sample. due to the lack of specific protein identification, past researchers were forced to decipher the peptide fragments themselves. However, there are currently programmes available for protein identification. These programmes take the peptide sequences output from mass spectrometry and microarray and return information about matching or similar proteins. This is done through algorithms implemented by the programme which perform alignments with proteins from known databases such as UniProt and PROSITE to predict what proteins are in the sample with a degree of certainty. Three such programmes are MASCOT, PeptideMass, and SPIRE.

Protein Structure

The biomolecular structure forms the 3D configuration of the protein. Understanding the proteins structure aids in identification of the proteins interactions and function. It used to be that the 3D structure of proteins could only be determined using X-ray crystallography and NMR spectroscopy. Now, through bioinformatics, there are computer programmes that can predict and model the structure of proteins. These programmes use the chemical properties of amino acids and structural properties of known proteins to predict the 3D model of sample proteins. This also allows scientists to take a look at protein interactions on a larger scale. In addition, biomedical engineers are developing methods to factor in the flexibility of protein structures to make comparisons and predictions.

Post-Translational Modifications

Unfortunately, most programmes available for protein analysis are not written for proteins that have undergone post-translational modifications. Some programmes will accept post-translational modifications to aid in protein identification, but then ignore the modification during further protein analysis. It is important to account for these modifications since they can effect the proteins structure. In turn, computational analysis of post-translational modifications as gained the attention of the scientific community. The current post-translational modification programmes are only predictive. Chemists, biologists and computer scientists are working together to create and introduce new pipelines that allow for analysis of post-translational modifications that have been experimentally identified for their effect on the proteins structure and function.

Computational Methods in Studying Protein Biomarkers

One example of the use of bioinformatics and the use of computational methods is the study of protein biomarkers. Computational predictive models have shown that extensive and diverse feto-maternal protein trafficking occurs during pregnancy and can be readily detected non-invasively in maternal whole blood. This computational approach circumvented a major limitation, the abundance of maternal proteins interfering with the detection of fetal proteins, to fetal proteomic analysis of maternal blood. Computational models can use fetal gene transcripts previously identified in maternal whole blood to create a comprehensive proteomic network of the term neonate. Such work shows that the fetal proteins detected in pregnant woman's blood originate from a diverse group of tissues and organs from the developing fetus. The proteomic networks contain many biomarkers that are proxies for development and illustrate the potential clinical application of this technology as a way to monitor normal and abnormal fetal development.

An information theoretic framework has also been introduced for biomarker discovery, integrating biofluid and tissue information. This new approach takes advantage of functional synergy between certain biofluids and tissues with the potential for clinically significant findings not possible if tissues and biofluids were considered individually. By conceptualizing tissue-biofluid as information channels, significant biofluid proxies can be identified and then used for guided development of clinical diagnostics. Candidate biomarkers are then predicted based on information transfer criteria across the tissue-biofluid channels. Significant biofluid-tissue relationships can be used to prioritize clinical validation of biomarkers.

Emerging Trends in Proteomics

A number of emerging concepts have the potential to improve current features of proteomics. Obtaining absolute quantification of proteins and monitoring post translational modifications are the two tasks that impacts the understanding of protein function in healthy and diseased cells. Advances in quantitative proteomics would clearly enable more in-depth analysis of cellular systems. For many cellular events the protein concentrations do not change, rather their function is modulated by post transitional modifications (PTM). Methods of monitoring PTM are an underdeveloped area in proteomics. Selecting a particular subset of protein for analysis substantially reduces protein complexity, making it advantageous for diagnostic purposes where blood is the starting material. Another important aspect of proteomics, yet not addressed, is that, proteomics methods should focus on studying proteins in the context of the environment. The increasing use of chemical cross linkers, introduced into living cell to fix protein-protein, protein-DNA and other interactions may ameliorate this problem partially. The challenge is to identify suitable methods of preserving relevant interactions. Another goal for studying protein is to develop more sophisticated methods to image proteins and other molecules in living cells and real time.

Human Plasma Proteome

Characterizing human plasma proteome has become a major goal in proteomics arena. The plasma proteome is without doubt the most complex proteome in the human body. It contains immunoglobulin, cytokines, protein hormones, secreted proteins and indicative of infection on top of resident, hemostatic proteins. It also contains tissue leakage proteins due to the blood circulation through different tissues in the body. The blood thus contains information on physiological of all tissues and combined with its accessibility makes the blood proteome invaluable for medical purposes. With the recent advancement in proteomics, characterizing the proteome of blood plasma is a daunting challenge.

Temporal and spatial dynamics further complicated the study of human plasma proteome. The turnover of some proteins is quite faster than others and protein content of artery may substantially vary from that of vein. All these differences make even the simplest proteomic task of catalogueing the proteome seem out of reach. To tackle this problem, priorities needs to be established. Capturing the most meaningful subset of proteins among the entire proteome to generate diagnostic tool is one such priority. Secondly, since cancer is associated with enhanced glycosylation of proteins, methods that focus on this part of proteins will also be useful. Again: multiparametre analysis best reveals a pathological state. As these technologies improve, the disease profiles should be continually related to respective gene expression changes.

Soil and Rhizosphere Environment

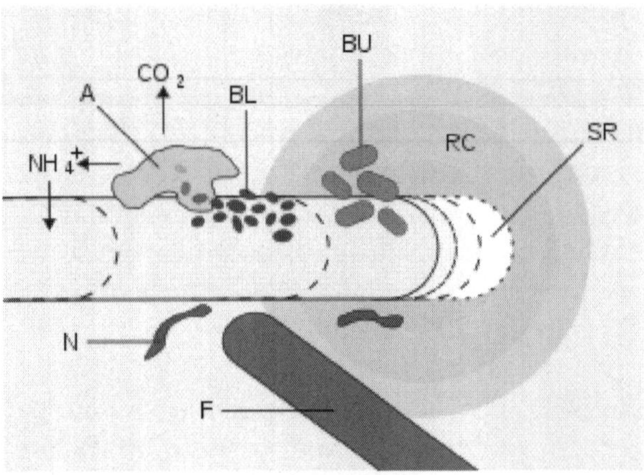

***Figure:** An illustration of the rhizosphere. A=Amoeba consuming bacteria; BL=Energy limited bacteria; BU=Non-energy limited bacteria; RC=Root derived carbon; SR=Sloughed root hair cells; F=Fungal hyphae; N=Nematode worm*

The rhizosphere is the narrow region of soil that is directly influenced by root secretions and associated soil microorganisms. Soil which is not part of the rhizosphere is known as bulk soil. The rhizosphere contains many bacteria that feed on sloughed-off plant cells, termed *rhizodeposition*, and the proteins and sugars released by roots. Protozoa and nematodes that graze on bacteria are also more abundant in the rhizosphere. Thus, much of the nutrient cycling and disease suppression needed by plants occurs immediately adjacent to roots.

Secretions

Plants secrete many compounds into the rhizosphere which serve different functions. Strigolactones, secreted and detected by mycorhizal fungi, stimulate the germination of spores and initiate changes in the mycorhiza that allow it to colonize the root. The parasitic plant, *Striga* also detects the presence of strigolactones and will germinate when it detects them; they will then move into the root, feeding off the nutrients present. Symbiotic Nitrogen-fixing bacteria, such as *Rhizobium* species, detect compounds like flavonoids secreted by the roots of leguminous plants and then produce nod factors which signal to the plant that they are present and will lead to the formation of root nodules. In these nodules bacteria, sustained by nutrients from the plant, convert nitrogen gas to a form that can be used by the plant. Non-symbiotic (or "free-living") nitrogen-fixing bacteria may reside in the rhizosphere just outside the roots of certain plants (including many grasses), and similarly "fix" nitrogen gas in the nutrient-rich plant rhizosphere. Even though these organisms are thought to be only loosely associated with plants they inhabit, they may respond very strongly to the status of the plants. For example, nitrogen-fixing bacteria in the rhizosphere of the rice plant exhibit diurnal cycles that mimic plant behaviour, and tend to supply more fixed nitrogen during growth stages when the plant exhibits a high demand for nitrogen. Some plants secrete allelochemicals from their roots which inhibit the growth of other organisms. For example garlic mustard produces a chemical which is believed to prevent mutualisms forming between the trees and mycorhiza in mesic North American temperate forests.

The Rhizosphere - Roots, Soil and Everything in Between

Soil is one of the last great scientific frontiers and the rhizosphere is the most active portion of that frontier in which biogeochemical processes influence a host of landscape and global scale processes. A better understanding of these processes is critical for maintaining the health of the planet and feeding the organisms that live on it. There is a small but concerted effort under way to harness the root system of plants in an attempt to increase yield potentials of staple food crops in order to meet the projected doubling in global food demand in the next 50 years. These efforts are being done in the face of a changing global climate and increasing global population which will inevitably require more productively grown food, feed and fibre on less optimal (and often infertile) lands; a condition already encountered in many developing countries. Meeting the global challenges of climate change and population growth with a better understanding and control of rhizosphere processes will be one of the most important science frontiers of the next decade for which a diverse, interdisciplinary trained workforce will be required.

4

Classification and Function of Root Derived Products

Newman examined a variety of plant species and estimated that roots can release anywhere from 10 to 250 mg C /g root produced or about 10-40% of their total photosynthetically fixed carbon. The C released is in both organic (e.g., low molecular weight organic acids) and inorganic (e.g., HCO_3) forms, however, the organic forms are the most varied and can have the most influence on the chemical, physical and biological processes in the rhizosphere. The composition and amount of the released compounds is influenced by many factors including plant type, climactic conditions, insect herbivory, nutrient deficiency or toxicity, and the chemical, physical and biological properties of the surrounding soil. The root products imparted to the surrounding soil are generally called rhizodeposits. Rhizodeposits have been classified based on their chemical composition, mode of release, or function but are classically defined to include sloughed-off root cap and border cells, mucilage, and exudates.

Roots exert a tremendous amount of pressure (>7kg/cm^2 or ~100 psi) at the growing root tips in order to push their way through the soil. Helping to lubricate and protect the root during growth, root cap and epidermal cells secrete mucilage, a viscous, high molecular weight, insoluble, polysaccharide-rich material. Beyond lubrication, the mucilage also provides protection from desiccation, assists in nutrient acquisition, and most notably binds soil particles together forming aggregates which improve soil quality by increasing water infiltration and aeration. Also serving to protect the root tip, cells lining and capping the root meristem are programmed to release (slough-off) which helps to reduce frictional forces that would otherwise damage the root. Interestingly, the cells that are sloughed off continue to function and secrete mucilage for several days and have been shown to attract beneficial microorganisms, serve as "bait" for root pathogens, and sequester toxic metals (e.g. Al^{3+}).

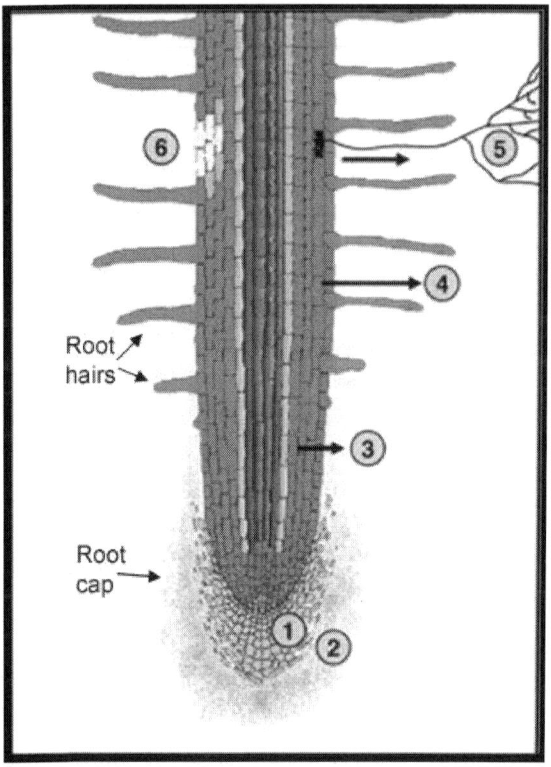

Figure: *Schematic of a root showing the 6 major regions of rhizodeposits. 1 loss of cap and border cells, 2 loss of insoluble mucilage, 3 loss of soluble root exudates, 4 loss of volatile organic carbon, 5 loss of C to symbionts, 6 loss of C due to death and lysis of root epidermal and cortical cells.*

Root exudates include both secretions (including mucilage) which are actively released from the root and diffusates which are passively released due to osmotic differences between soil solution and the cell, or lysates from autolysis of epidermal and cortical cells. The organic compounds released through these processes can be further divided into high and low molecular weight (HMW and LMW, respectively). By weight, the HMW compounds which are those complex molecules that are not easily used by microorganisms (e.g. mucilage, cellulose) make up the majority of C released from the root; however, the LMW compounds are more diverse and thus have a wider array of known or potential functions. The list of specific LMW compounds released from roots is very long, but can generally be categorized into organic acids, amino acids, proteins, sugar, phenolics and other secondary metabolites which are generally more easily used by microorganisms. The cocktail of chemicals released is influenced by plant species, edaphic and climactic conditions which together shape and are shaped by the microbial community within the rhizosphere.

There is still very little known about the role that a majority of the LMW compounds play in influencing rhizosphere processes. A growing body of literature is beginning to lift the veil on the many functions of root exudates as a means of acquiring nutrients (e.g. acquisition of Fe and P), agents of invasiveness (i.e. allelopathy) or as chemical signals to attract symbiotic partners (chemotaxis) (e.g. rhizobia and legumes) or the promotion of beneficial microbial colonization on root surfaces (e.g. *Bacillus subtilis, Pseudomonas florescence*).

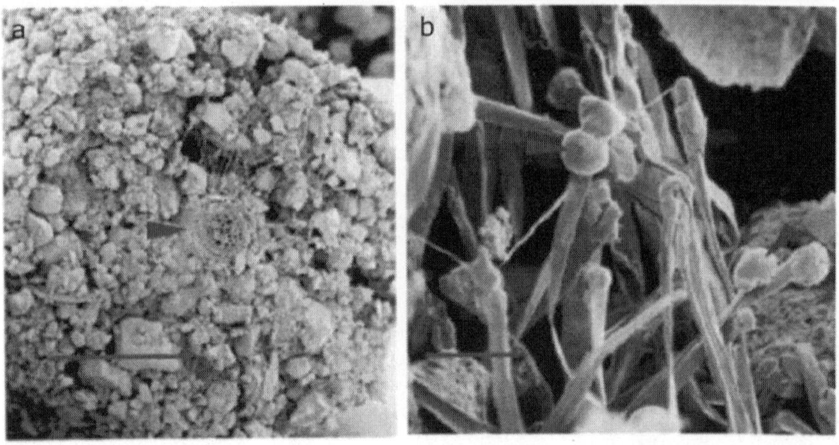

Figure: *A branch root (red arrow) of buckwheat (Fagopyron esculentum) frozen together with its surrounding soil and then imaged using scanning electron microscopy to reveal the intimate contact between the root and soil. Note the root hairs extending out and exploring the surrounding soil. Bar = 600 ìm. b) Red arrows point to droplets of root exudates released from the tips of root hairs on the surface of broom corn (Sorghum sp.). Bar = 50 ìm.*

Root Exudates and Mineral Nutrition

Plants respond to nutrient deficiency by altering root morphology, recruiting the help of microorganisms and changing the chemical environment of the rhizosphere. Components in root exudates assist plants in accessing nutrients by acidifying or changing the redox conditions within the rhizosphere or directly chelating with the nutrient. Exudates can liberate nutrients via dissolution of insoluble mineral phases or desorption from clay minerals or organic matter where they are released into soil solution and can then be taken up by the plant. The nutrients most limiting to plant growth are nitrogen and phosphorus. Even thought 78% of the Earth's atmosphere is composed of nitrogen (N_2 gas), it is in a form that is only utilizable by nitrogen-fixing organisms. As such, inorganic forms of N (NO_3^-,NH_4^+) that can be used by plants are added to soils. The availability of nitrogen in most soils is low because of the leaching losses of soluble nitrate (NO_3^-) with infiltrating rainwater, fixation of ammonium (NH_4^+) in clays and soil organic matter and bacterial denitrification. Plants respond differently depending on the form of nitrogen in the soil. Ammonium has a positive charge, and thus the plant expels one proton (H^+) for every NH_4^+ taken up resulting in a reduction in rhizosphere pH. When supplied with NO_3^-, the opposite can occur where the plant releases bicarbonate (HCO_3^-) which increases rhizosphere pH. These changes in pH can influence the availability of other plant essential micronutrients (e.g., Zn, Ca, Mg).

Phosphate (PO_4^{3-}), the form of P used by plants, is highly insoluble in soils, binding strongly to Ca, Al and Fe oxide, and soil organic matter rendering much of the P unavailable to plants. Under P deficiency, plants have evolved special mechanisms to obtain PO_4^{3-} which depend on plant type (dicot vs. monocot), species and genotype. Plant roots can exude organic acids such as malic and citric acids into the rhizosphere which effectively reduced rhizosphere pH and solubilize P bound in soil minerals. Pigeon pea uses another mechanism by releasing piscidic acid in response to P deficiency which chelates Fe in $FePO_4$, releasing the PO_4^{3-} (Raghothama, 1999). Plants also liberate PO_4^{3-} from organic sources by releasing enzymes such as acid phosphatase.

Iron deficiency elicits a response from plants which generally differs depending on whether the plant is a dicot or monocot. Dicots respond to Fe deficiency by releasing protons into the soil environment and increasing the reducing capacity of the rhizodermal cells. In monocots, Fe deficiency triggers the release of phytosiderophores such as mugienic acid which is a non-proteinogenic amino acid with extremely high affinity for Fe. The phytosiderophore chelates strongly with Fe and is then brought back to the root via diffusion where plasma membrane transporters specific to the chelated Fe shuttle it into the cells. Plants using the former method of Fe acquisition are called Strategy I and those using the latter Strategy II.

Key Microbe-Plant Relationships

Rhizodeposits make the rhizosphere a desirable niche for microbial communities to proliferate. One teaspoon of bare or tilled soil contains more microorganisms than there are people on Earth, however, the rhizosphere can have 1000-2000 times that number (10^{10}-10^{12} cells per gram rhizosphere soil) making it a pretty crowded place. Nutrient availability for microbial growth is higher in the rhizosphere compared to the bulk soil; however, many microbes are competing for these nutrients, some more successfully than others.

The plant rhizosphere -microbe relationships that have received the most attention include those of Rhizobia bacteria and their symbiotic plant partners, mychorrhizal fungi associations, and beneficial plant growth promoting rhizobacteria (PGPR) each of which will be discussed in more detail below.

Legume-Rhizobia Symbiosis

The observation that including leguminous plants in crop rotations results in better yields of non-legumes goes back many centuries to the Romans and Greeks. It was not until 1888, however, that Hellriegel and Wilfarth (1888) definitively proved that the cause of the improved yield was from the conversion of atmospheric dinitrogen ($N_{2(g)}$) into ammonia, a form usable by the plant, by rhizobium bacteria (*Rhizobium leguminosarum*) housed within nodules on the roots of leguminous plants. This discovery inspired Lorenz Hiltner who spent his career researching and developing rhizobial inoculants to improve agricultural productivity eventually leading to his development of the Rhizosphere concept. The interaction between rhizobia and legumes has been exhaustively studied since that time, elucidating several facets of the interaction.

What we have learned is that the dialogue between the rhizobium and the plant begins first with a chemical signal (flavonoid) released when the plant is under N starved conditions. The flavonoid signal induces nodulation genes (nod genes) in the rhizobia which encode enzymes necessary to produce a chemical response (lipochitooligosaccharide) to the plant known as a nod factor. The nod factors initiate a cascade of developmental processes in the plant root which allow for the invasion of the bacteria and formation of the nodule in which the bacteria are eventually housed. The structure of the nod factors are largely species specific and are one of the reasons for the host specificity observed between rhizobia and their plant partners. This specificity is quite astonishing considering that one gram of soil (about the size of a raisin) contains >10^9 bacteria from 10,000 different species, from which a miniscule amount have developed the ability to symbiotically cohabitate with plants.

Mycorrhizal Fungi and Nutrient Acquisition

Mycorrhizae (from the Greek words for fungus and root) is a general term describing a symbiotic relationship between a soil fungus and plant root. Unlike rhizobia and their legume partners, mycorrhizal associations are ubiquitous and relatively nonselective, occurring in ~80% of angiosperms and in all gymnosperms (Wilcox, 1991). The ability for plants to form symbiotic associations with mycorrhizae occurred early in the evolution of plants (~450 million years ago) compared to legumes (~60 million years ago), which is likely why mycorrhizae are now ubiquitous throughout the plant kingdom. Although parasitic and neutral relations exist, a majority of these associations are beneficial both to the host plant and the colonizing fungi. Mycorrhiza assist plants in obtaining water, phosphorus and other micronutrients (e.g., Zn and Cu) from the soil and in return receive sustenance (carbon) from the plant.

There are two broad categories of mycorrhizal associations with plant roots, ectomycorrhiza and endomycorrhiza, which are differentiated by how they physically interface with the plant. The ectomycorrhizae (EM) occur mainly in the roots of woody plants (i.e. forest trees) and form a dense hyphal covering (fungal sheath or mantel) over the root tip from which hyphae grow into the intercellular spaces forming a net (Hartig net) of hyphae around the root cortex cells, but do not penetrate the cell walls. In contrast, the endomycorrhizae fungal hyphae grow into the root cortex and enter the cells forming fan-like, highly branched structure known as an arbuscule that remain separated from the cytoplasm by the plant plasma membrane.

The endomycorrhiza can be further divided into the more widespread arbuscular mycorrhiza (AM) and the specialized orchid and ericoid mycorrhizas which, as the name implies, are colonizers of orchids and ericoid (e.g., cranberry) plant species. The AM fungal associations are the most abundant of all mycorrhizal associations. In both cases, the Hartig net and the arbuscules increase the contact area between the fungus and the plant through which the transfer of nutrients to the plant and carbon to the fungus occurs. Unlike the ectomycorrhiza, the endomycorrhiza are wholly dependent on the plant for their carbon and when associations occur, both endomycorrhiza and ectomycorrhiza can demand up to 20-40% of the total photosynthetically fixed carbon the plant produces.

The chemical dialogue that initiates mycorrhizal associations with plants is intricate and not as well understood as that of the Rhizobia. In part, the lack of understanding rests in the fact that endomycorrhiza are obligate symbionts, meaning they cannot be grown independent of their plant hosts, which makes it difficult for scientists to study them in the lab under controlled conditions. For this reason, more is known about the development of the ectomycorrhizal symbiosis because the two partners can be grown independently. In both cases, there is evidence that the fungus can sense the presence of volatiles (CO_2), or chemicals in plant root exudates which then initiates hyphal growth and branching. Conversely, chemical signals from mycorrhizea which are perceived by the plant have long been suspected and only recently has the structure of one of these chemicals in the ecotmycorrhizal association been fully characterized.

The chemical signals are aptly called *Myc*-factors which, interestingly, are similar in structure to the nod factors produced by rhizobia in the rhizobia-legume symbiosis. It is still unclear whether the fungus produces the signals all the time or only when it senses the plant is near. Regardless, fungal spores in the soil remain in a resting state until one of these stimuli initiates germination and hyphal growth. If close to a root, hyphal growth and branching will increase and *Myc*-factors will initiate root branching presumably to improve the odds of root-fungus interception. After making contact with the plant root, infection begins by either the formation of a hyphopodium (in arbuscular mycorrhizae) or growth between dermal cells (in ectomycorrhizea). If a root is not in the vicinity, the hyphae stops growing and the spore returns to a vegetative state; subsisting off of its triacylglyceride and glycogen reserves. Initiation of hyphal growth can occur multiple times until root interception is achieved.

To capture nutrients for the plant, both the ecto- and endomycorrhizae extend hyphae centimetres into the soil resulting in a 10 fold increase in the effective root surface area and a 2-3 fold increase in the uptake of phosphorus (and other nutrients) per unit root length compared to non-mycorrhizal plants. It is not only the amount of hyphae that helps in nutrient acquisition, but also their small size (<200 mm) that enables them to access small soil pores and cracks that the plant root would otherwise not be able to access. The network of fungal hyphae emanating from the plant roots also has a tremendous impact on soil quality. The mycorrhizal hyphae promote soil aggregate formation and stability via biological, physical and biochemical mechanisms which reduces soil erosion and increases soil aeration and water infiltration, which together improves plant productivity. Interestingly, the dense, intertwined network of fungal hyphae forms a "common mycorrhizal network" (CMN), in which hyphae from mycorrhizae infecting two or more plants are interconnected. Through the CMN, plants have been shown to share nutrients and mediate the interactions between plants which are not immediately sharing the same space. The diversity of these interactions and the mycorrhizal fungus taking part in them have been shown to have a great deal of influences on plant biodiversity, ecosystem function and stability.

Plant Growth Promoting Rhizobacteria (PGPR)

Plant growth promoting rhizobacteria were first defined by Kloepper and Schroth (1978) as organisms that, after being inoculated on seeds, could successfully colonize plant roots and positively enhance plant growth. To date, there have been over two dozen genera of nonpathogenic rhizobacteria identified. Plant growth promotion can be shown to work directly on the plant in the absence of root pathogens by the release of plant growth stimulating compounds (e.g. phytohormones such as auxins or cytokinins) and improvement in mineral uptake (e.g. siderophore release increasing Fe availability). Plant growth promotion can also occur indirectly by control of pathogens (biocontrol) via synthesis of antibiotics or secondary metabolite-mediated induced systemic resistance (ISR).

Microbial colonization of the plant root surface is not uniform, but instead occurs in patches along the root, ultimately covering ~15- 40% of the total plant root surface.

The density and structure of the microbes on the root surface are dictated by nutrient availability and physicochemical variations throughout the root surface. Root exudates can serve as a food source and chemoattractant for microbes which then attach to the root surface and form microcolonies. Common sites for bacterial attachment and colonization are at epidermal cell junctions, root hairs, axial groves, cap cells, and sites of emerging lateral roots. Microcolonies can eventually grow into larger biofilms consisting of multiple layers of bacteria which are encased in an exopolymeric matrix.

In many cases the effectiveness of rhizobacteria at promoting growth occurs in a density dependant manner. Once a critical microbial density is reached, the biofilms then begin to act in unison, in a process known as quorum sensing, coordinating the release of compounds that aid in the promotion of plant growth via the direct and indirect mechanism discussed previously.

Root System Architecture — Sensing and Foraging

The spatial extent and influence of the rhizosphere are dictated by root system architecture (RSA). Root system architecture is very malleable, determined by plant species and occurring in response to changing climactic, biological and edaphic conditions. The distribution of nutrients in soils is heterogeneous or patchy and there is evidence plants can "sense" the presence of nutrients and allocate more resources to the root system and direct root growth towards or within these patches.

Examples include increased root depth in drought tolerant beans, wheat, and maize and an increase of dense shallow root systems leading to topsoil *foraging* for P in plants under low P conditions. The nutrients themselves and the activity of specific molecular pathways involved in their acquisition may act to induce hormonal signals triggering increases in root density and length.

Because of the underground nature of plant roots, uncovering the mechanisms responsible for root foraging has been difficult, but full annotation of the *Arabidopsis* genome and a complete map of root cell types have dramatically increased our understanding of root biology and these processes.

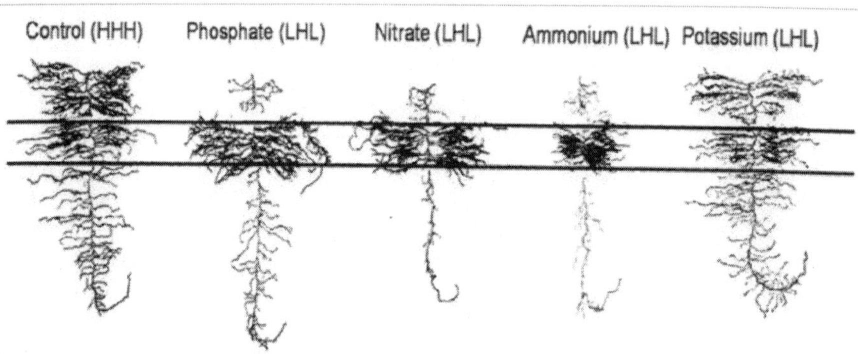

***Figure:** Changes in roots system architecture (RSA) of barely (*Hordeum vulgare*) in response to zones of high phosphate, nitrate, ammonium and potassium availability.*

Root Exudates

Plant exudates include saps, gums, latex, resin, and sometimes nectar is considered an exudate. Plant roots exude a variety of molecules into the rhizosphere, including acids, sugars, polysaccharides and ectoenzymes; this can account for 40% of root carbon. Exudation of these compounds has various benefits to the plant and to the microorganisms of the rhizosphere.

Exudates vs. Transudates

	Transudate	***Exudate***
Main causes	Increased hydrostatic pressure, Decreased colloid osmotic pressure	Inflammation
Appearance	Clear	Cloudy
Specific gravity	< 1.012	> 1.020
Protein content	< 25 g/L	> 29 g/L
$\frac{\text{fluid protein}}{\text{serum protein}}$	< 0.5	> 0.5
Difference of albumin content with blood albumin	> 1.2 g/dL	< 1.2 g/dL
$\frac{\text{fluid LDH}}{\text{upper limit for serum}}$	< 0.6 or < $\frac{2}{3}$	> 0.6 or > $\frac{2}{3}$
Cholesterol content	< 45 mg/dL	> 45 mg/dL

There is an important distinction between transudates and exudates. Transudates are caused by disturbances of hydrostatic or colloid osmotic pressure, not by inflammation. They have a low protein content in comparison to exudates. Medical distinction between transudates and exudates is through the measurement of the specific gravity of extracted fluid. Specific gravity is used to measure the protein content of the fluid. The higher the specific gravity, the greater the likelihood of capillary permeability changes in relation to body cavities. For example, the specific gravity of the transudate is usually less than 1.012 and a protein content of less than 2 g/100 mL (2 g%). Rivalta test may be used to differentiate an exudate from a transudate. It is not clear if there is a distinction in the difference of transudates and exudates in plants.

Plant Nematode Interactions

Plant parasitic nematodes have developed the capacity to sense and respond to chemical signals of host origin and the ability to orientate towards plant roots enhances the nematode's chance of survival. Root exudates contain a range of compounds which mediate belowground interactions with pathogenic and beneficial soil organisms. Chemical components of root exudates may deter one organism while attracting another and these compounds alter nematode behaviour and can either attract nematodes to the roots or result in repellence, motility inhibition or even death. In vitro, plant signals present in root exudates, trigger a rapid alteration of the surface cuticle of Meloidogyne incognita and the same changes were also induced by indole-acetic acid (IAA). IAA binds to the chemosensory organs of M. incognito and it is possible that IAA acts as a signal that orientates the nematode on the root surface in the rhizosphere and/or inside the root tissue and thereby promotes nematode infection.

Nematodes feed on all parts of the plant, including roots, stems, leaves, flowers and seeds. Nematodes feed from plants in a variety of ways, but all use a specialized spear called a stylet. Note the differences in stylet length and shape. The size and shape of the stylet is used to classify nematodes and also can be used to infer their mode of feeding. All three nematodes are ectoparasites, but *Belonolaimus* and *Longidorus* feed deep within the roots using their long stylets, while *Helicotylenchus* feeds on the exterior of the root or partially burrows into the root to feed using its short stout stylet.

Often nematodes withdraw the contents of plant cells, killing them. When this type of feeding occurs, large lesions are formed in the plant tissue. Some nematodes do not kill the plant cells they feed upon but "trick" the plant cells to enlarge and grow, thus producing one or more nutrient-rich feeding cells for the nematode. These feeding cells enable long term feeding associations, and form by repeated nuclear division in the absence of cell division (giant cells) or by the incorporation of adjacent cells into a syncytium formed by the breakdown of neighbouring cell walls. Collectively, nematodes can feed on almost any plant cell type, and form a variety of feeding cell types. The number of feeding cells can vary from one to a half dozen depending on the nematode species.

Many plant-parasitic nematodes feed on the roots of plants. The feeding process damages the plant's root system and reduces the plant's ability to absorb water and nutrients. Typical nematode damage symptoms are a reduction of root mass, a distortion of root structure and/or enlargement of the roots. Nematode damage of the plant's root system also provides an opportunity for other plant pathogens to invade the root and thus further weakens the plant. Direct damage to plant tissues by shoot-feeding nematodes includes reduced vigor, distortion of plant parts, and death of infected tissues depending upon the nematode species.

The aboveground symptoms of nematode damage to roots are relatively nondescript, including nutrient deficiency, incipient wilt, stunting, poor yield and sometimes plant death. Few diagnostic signs and symptoms of plant damage by nematodes exist except root galls, cysts, "nematode wool," and seed galls. Thus, damage to crops by root-infesting nematodes often goes unnoticed by growers. Field patterns of nematode damage to roots begin in a small area and spreads radially from the initial infection site, often assisted by farm equipment. The only way to accurately diagnose nematode disease is to sample soil and plant material from suspected sites and extract nematodes for analysis.

Nematodes are extracted from the soil by floating them in water to remove heavy soil particles and then catching the floating nematodes on sieves with fine pore sizes. Plant tissues infected with motile nematodes can be incubated in a Baermann funnel or moist chambers to collect nematodes that will exit the tissues.

Most plant parasitic nematodes are soilborne root pathogens, but a few species feed primarily upon shoot tissues. The majority of plant parasitic nematode species are in the class Chromodorea, order Rhabditida (formerly placed in the order Tylenchida). There are seven major types of nematode feeding strategies used by plant parasitic nematodes.

Ectoparasites: The first feeding type is the ectoparasitic mode, in which the nematode remains outside of the plant and uses its stylet to feed from the cells of the plant roots. Nematodes that use this strategy can graze on numerous plants, making it easier for them to switch hosts, but their added mobility makes them very susceptible to environmental fluctuations and predators. Ectoparasitic nematodes can have extremely long stylets, which assist them in feeding deep within the plant root on nutrient rich plant cells. Some of these nematodes induce the plant to form an enlarged cell(s) that the nematode feeds from for an extended period of time. Note, in all life cycle diagrams in this article the abbreviation J=juvenile and the number refers to the stage of the nematode and M=molt and refers to how many molts the nematode has completed.

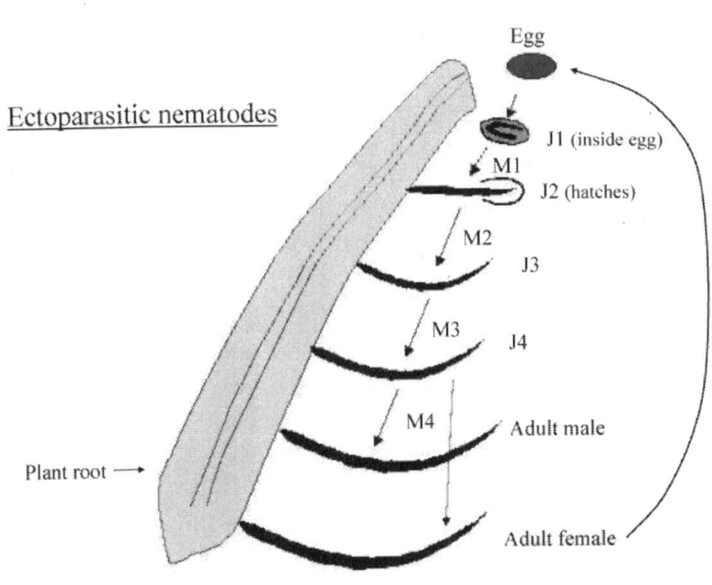

The above life cycle is typical for a nematode in the class Enoplea, but most nematodes in the class Chromadorea undergo their first molt in the egg and hatch as J2. All motile nematode stages are capable of feeding from the plant. The nematodes feed, undergo four molts into adults, mate and lay eggs. Some nematodes that use this feeding strategy can form terminal galls in the roots and cause severe stunting of the root system. This enoplean nematode is particularly problematic because it can harbor and transmit plant viruses. Only ectoparasitic nematodes in the class Enoplea transmit viruses, but because of the virus transmission, this type of nematode in low numbers can still be very damaging to plants

Migratory endoparasites: Migratory endoparasitic nematodes spend much of their time migrating root tissues destructively feeding on plant cells. These nematodes cause massive plant tissue necrosis because of their migration and feeding. When they feed from the plant, they simply suck out the plant cell cytoplasm using their stylet, killing the plant cell and moving ahead of the lesion. They make no permanent feeding cells. In a typical life cycle, the nematode hatches from the egg as a second-stage juvenile and starts feeding on the plant. The nematodes feed, molt and reproduce primarily within the plant tissue. All motile stages are capable of feeding from the plant and they are able to move into the soil in search of new roots to invade. Because these nematodes create extensive wounds in the plant root, secondary infection by bacteria and fungi can often occur, further damaging the root system. Below is a typical life cycle for a migratory endoparasite. Examples of migratory endoparasitic nematodes are *Pratylenchus* (lesion nematode), *Radopholus* (burrowing nematodes) and *Hirschmanniella* (rice root nematode).

Sedentary endoparasites: The most damaging nematodes in the world have a sedentary endoparasitic life style. The two main nematodes in this group are the cyst nematodes (*Heterodera* and *Globodera*) and the root-knot nematodes (*Meloidogyne*). In these nematodes, the J2 invades the plant near the tip of a root and migrates through the tissue to the developing vascular cells. These nematodes are completely embedded in the root during their initial stages of development, but later the cyst nematodes protrude from the root. The J2 nematodes inject secretions into and around the plant cells to stimulate the formation of large feeder cell(s), which they non-destructively feed on throughout their life cycle. Feeding cells of root-knot nematodes (giant cells) form by repeated nuclear division in the absence of cell division. Feeding cells of cyst nematodes form by the incorporation of neighbouring cells into a syncytium formed by the breakdown of neighbouring cell walls. Once the feeding cells are formed, the juvenile nematode rapidly becomes sedentary because their somatic muscles atrophy. The juveniles feed, enlarge and molt three times to the adult stage. The large feeding cells formed by these nematodes plug the vascular tissue of the plant making it susceptible to water stress. Female sedentary endoparasites enlarge considerably into a saclike shape and are capable of laying large numbers of eggs. Eggs are typically laid outside the nematode in a gelatinous egg mass, but in cyst nematodes most eggs are retained inside the female body. Both types of nematode have the same basic feeding strategy, but many cyst nematodes have an obligate sexual cycle, whereas the most common species of root-knot nematodes reproduce largely by parthenogenesis.

Biological Control of Pest Insects With Nematodes.

Beneficial Nematodes naturally occur in soil and are used to control soil pest insects and whenever larvae or grubs are present. Like all of our products, it will not expose humans or animals to any health or environmental risks. Beneficial nematodes only attack soil dwelling insects and leave plants and earthworms alone. The beneficial nematodes enters the larva via mouth, anus or respiratory openings and starts to feed. This causes specific bacteria to emerge from the intestinal tract of the nematode. These spread inside the insect and multiply very rapidly. The bacteria convert host tissue into products which can easily be taken up by the nematodes. The soil dwelling insect dies within a few days. Beneficial nematodes are a totally safe biological control in pest insects. The Beneficial nematodes are so safe the EPA has waived the registration requirements for application.

Nature's Best Way of Killing Grubs and Japanese Beetles.

Though they are harmless to humans, animals, plants, and healthy earthworms, beneficial nematodes aggressively pursue insects. The beneficial nematodes can be used to control a broad range of soil inhabiting insects and above ground insects in their soil inhabiting stage of life. More than 200 species of pest insects from 100 insect families are susceptible to these nematodes. When they sense the temperature and carbon dioxide emissions of soil-borne insects, beneficial nematodes move towards their prey and enter the pest through its body openings. The nematodes carry an associated bacterium (Xenorhabdus species) that kills insects fast within 48 hours. The bacteria is harmless to humans and other organisms and cannot live freely in nature. Several generations of nematodes may live and breed within the dead insect, feeding on it as a food source. When the food source is gone, they migrate into the soil in search of a new host. When the pest population is eliminated, the beneficial nematodes die off and biodegrade. Beneficial nematodes are so effective, they can work in the soil to kill the immature stages of garden pests before they become adults.

Beneficial Nematodes Infest Grubs and Other Pest Insects That are Known to Destroy Lawns and Plants.

The Nematodes are effective against grubs and the larval or grub stage of Japanese Beetles, Northern Masked Chafer, European Chafer, Rose Chafer, Fly larvae, Oriental Beetles, June Beetles, Flea beetles, Bill-bugs, Cut-worms, Army worms, Black Vine Weevils, Strawberry Root Weevils, Fungus Gnats, Sciarid larvae, Sod Web-worms, Girdler, Citrus Weevils, Maggots and other Dip-tera, Mole Crickets, Iris Borer, Root Maggot, Cabbage Root Maggot and Carrot Weevils.

Beneficial nematodes belong to one of two genera: Steinernema and Heterorhabditis are commercially available in the U.S. Steinernema is the most widely studied beneficial nematode because it is easy to produce. Heterorhabditis is more difficult to produce but can be more effective against certain insects, such as th white grubs, and Japanese beetles.

How Beneficial Nematodes Work:

The life cycle of beneficial nematodes consists of six distinct stages: an egg stage, four juvenile stages and the adult stage. The adult spends its life inside the host insect. The third juvenile stage, called a dauer, enters the bodies of insects (usually the soil dwelling larval form. Some nematodes seek out their hosts, while others wait for the insect to come to them. Host seeking nematodes travel through the soil the thin film of water that coats soil particles. They search for insect larvae using built-in homing mechanisms that respond to changes in carbon dioxide levels and temperature. They also follow trails of insect excrement. After a single nematode finds and enters an insect through its skin or natural openings, the nematode release a toxic bacteria that kills its host, usually within a day or two. In less than two weeks the nematodes pass through several generations of adults, which literally fill the insect cadaver. Steinernema reproduction requires at least two dauer nematodes to enter an insect, but a single Heterorhabditis can generate offspring on its own. The nematodes actively searches for insect larvae. Once inside the larva the nematodes excretes specific bacteria from its digestive trac before it starts to feed. The bacteria multiply very rapid and convert the host tissue into products that the nematodes take up and use for food. The larva dies within a few days and the colour changes from white-beige to orange-red or red-brown. The nematodes multiply and develop within the dead insect. As soon as the nematodes are in the infectious third stage, they leave the old host and start searching for new larvae. Infected grubs turn colour from white-beige to red brown 2-4 days after application and becomes slimy. After a few weeks, dead larvae disintegrate completely and are difficult to find.

Beneficial nematodes are also very effective against termites, German cockroaches, flies, ant, and fleas.

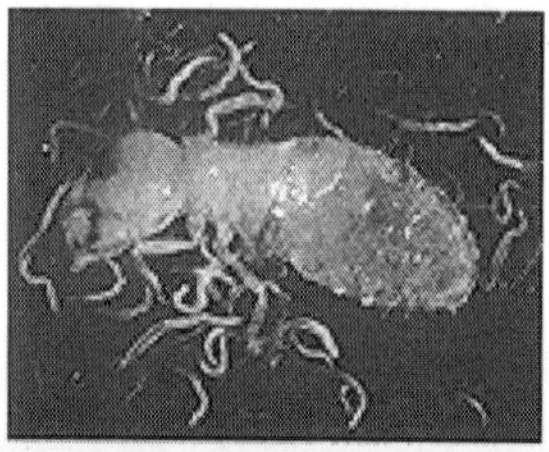

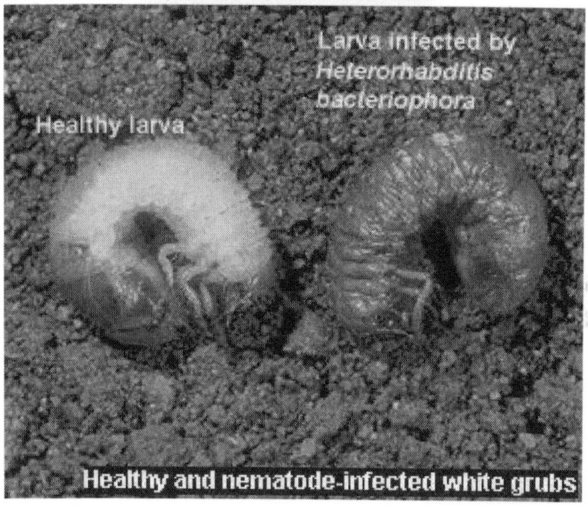

Healthy and nematode-infected white grubs

Application

Beneficial Nematodes are very easy to use. Mix with water and spray or sprinkle on the soil along garden plants or lawn. Put the contents of the Beneficial nematodes in a bucket of water and stir to break up any lumps, and let the entire solution soak for a few minutes. Application can be made using a water-can, irrigation system, knapsack or sprayer. On sprayer use a maximum pressure to avoid blockage, all sieves should be removed. The sprayer nozzle opening should be at least 1/2 mm. Evenly spread the spraying solutions over the ground area to be treated. Continuous mixing should take place to prevent the nematodes from sinking to the bottom. After application keep the soil moist during the first two weeks for the nematodes to get establish.

For a small garden the best method is using a simple sprinkling or water can to apply the Beneficial nematodes to the soil. Apply nematodes before setting out transplants; for other pest insects, Japanese Beetles and grubs, apply whenever symptomatic damage from insects is detected. Best to apply water first if soil is dry.Application and amount for 50 and 100 Mil. Nematodes. The 50 Mil. + nematodes are packed in an inert carrying material that will dissolve in water when mixed. You can use a watering can, pump sprayer; hose end sprayer and irrigation system, backpack sprayers, or motorized sprayer. The 50 and 100 Mil. Nematodes mix ½ teaspoon per gallon of water.

The Large yard size: 1/2 Acre Size (50 Million) you can use up to 50 Gallons of water The Acre size 100 Mil. Nematodes you can use up to 100 Gallons of water. o For covering up to 800 square feet, place approximately 1½ teaspoons of dry nematodes in the hose end sprayer container, and then fill to the 3 gallon mark. (for Garden size) o Evenly spread the solution over the ground areas to be treated. o Continuous mixing should take place to prevent the nematodes from sinking to the bottom of the container.

To avoid blockages, remove all filters from the sprayer. o You can sprinkle the soil with water again after application to move the nematodes into the soil. o Apply nematodes as soon as possible for best product performance. o Keep the soil most for the first week after application. Application for the 10 Mil. garden size Nematodes. The 10 Mil. nematodes are placed in a sponge. Place the entire sponge in a gallon of water, squeeze for a few minutes to get the nematodes out of the sponge and into the water. Discard the sponge and pour the gallon of water into the sprayer or water can and apply to the soil.

Proper Storage and Handling is Essential to Nematode Health.

Always follow the package instructions for the best method of mixing nematodes. Formulations vary depending on the species and target insect. Nematodes can be stored in the refrigerator up to a month (not the freezer) before they are mixed with water, but once the nematodes are diluted in water, they cannot be stored. Also, nematodes shouldn't be stored in hot vehicles, or left in spray tanks for long periods of time.

Nematodes need moisture in the soil for movement (if the soil is too dry or compact, they may not able to search out hosts) and high humidity if they are used against foliage pests. Watering the insect-infested area before and after applying nematodes keeps the soil moist and helps move them deeper into the soil. Care should be taken not to soak the area because nematodes in too much water cannot infect.

Exposure to UV Light or very High Temperatures can Kill Nematodes

Apply nematodes in the early evening or late afternoon when soil temps are lower and UV incidence is lower as well (cloudy or rainy days are good too). Nematodes function best with soil temperatures between 48F° and 93F° day time temperatures.

Application is Usually Easy

In most cases, there is no need for special application equipment. Most nematodes species are compatible with pressurized, mist, electrostatic, fan and aerial sprayers! Hose-end sprayers, pump sprayers, and watering cans are effective applicators as well. Nematodes are even applied through irrigation systems on some crops. Check the label of the nematode species to use the best application method. Repeat applications if the insect is in the soil for a longer period of time. There is no need for masks or specialized safety equipment. Insect parasitic nematodes are safe for plants and animals (worms, birds, pets, children). Because they leave no residues, application can be made anytime before a harvest and there is no re-entry time after application.

How to use Beneficial Nematodes

For the home gardener, localized spraying is probably the quickest and easiest way to get the nematodes into the soil. Producers ship beneficial nematodes in the form of dry granules, powder type clay, and sponges. All of these dissolve in water and release the millions of nematodes. Each nematode ready to start searching for an insect in your lawn or garden. Nematodes should be sprayed on infested areas at the time when pests is in the soil. Timing is important, or else you will have to repeat the application. Northern gardeners should apply the nematodes in the spring, summer and fall, when the soil contains insect larvae. Most of the beneficial nematodes are adaptive to cold weather. In fact , the very best time to control white grubs is in the spring and fall. If your in a warmer climate, beneficial nematodes are most effective in the summer.

Fertilizers should be avoided roughly 2 weeks prior to and after nematode application, because they may be adversely affected by high nitrogen content.

Some pesticides work well with nematodes when their mutual exposure is limited while other pesticides may kill nematodes. Check labels or specific fact sheets to find out. Some chemicals to avoid are bendiocarb, chlorpyrifos, ethoprop, and isazophos. Fungicides to avoid are anilazine, dimethyl benzyl, ammonium chloride, fenarimol, and mercurous chloride. The herbicides, 2,4-D and trichlopyr and nematicide, fenamiphos, should be avoided as well. During hot weather release nematodes in the evening or afternoon when temperature is cooler. Release once or twice a year or until infestation subsides. Nematodes are shipped in the infectious larvae stage of their life cycle and can be stored in the refrigerator for up to 4 weeks. Always release very early in the morning or late in the late afternoon.

Beneficial nematodes seek out and kill all stages of harmful soil-dwelling insects. They can be used to control a broad range of soil-inhabiting insects and above-ground insects in their soil-inhabiting stage of life.

Parasitic nematodes are beneficial for eliminating pest insects. First, they have such a wide host range that they can be used successfully on numerous insect pests. The nematodes' nonspecific development, which does not rely on specific host nutrients, allows them to infect a large number of insect species.

Nematodes enter pest bugs while they are still alive, then they multiply inside the bugs (which eventually die) and finally burst out of the dead bodies.

The number of nematodes inside a single bug (depending on the species) ranges from 5,000 to 10,000. Although you can barely see one young nematode with your naked eye, large groups of these tiny wigglers pouring out of the dead insects are easy to see. Then the nematodes wriggle off to find other insects to "invade," starting the whole cycle all over again.

Second, nematodes kill their insect hosts within 48 hours. As mentioned earlier, this is due to enzymes produced by the Xenorhabdus bacteria.

Also, the infective juveniles can live for some time without nourishment as they search for a host.

Finally, there is no evidence that parasitic nematodes or their symbiotic bacteria can develop in vertebrates. This makes nematode use for insect pest control safe and environmentally friendly.

The United States Environmental Protection Agency (EPA) has ruled that nematodes are exempt from registration because they occur naturally and require no genetic modification by man. Beneficial nematodes can be an excellent tool in the lawn and garden to control certain pest insects. They can be used with organic gardening and are safe for kids and pets.

Endo-β-1,4-Glucanase Expression in Compatible Plant–Nematode Interactions

Feeding cells induced by cyst nematodes (*Globodera* and *Heterodera* spp) and root-knot nematodes (RKN; *Meloidogyne* spp), termed syncytia and giant cells, respectively, are formed from host root cells during parasitism to sustain the growth and reproduction of the nematode. Motile second-stage juveniles (J2) of the nematode penetrate plant roots and migrate to the vascular cylinder, where the feeding cells serve as the sole nutritive source for the subsequent sedentary parasitic life stages of these nematodes. Evidence suggests that stylet (hollow feeding spear) secretions originating from three large unicellular esophageal gland cells of RKN and cyst nematodes play essential roles during parasitism of plant roots, including the induction of feeding cells. Nematode secretions may alter the development of affected host plant cells directly or indirectly. This modification of normal plant cell development causes plant cells to redifferentiate into unique cell types strictly for the benefit of the nematode and is accompanied by multiple changes in plant gene expression.

Giant cells and syncytia have several characteristics in common. In both cell types, there is an increase in metabolic activity and cytoplasmic density, the large central vacuole is reduced to several smaller vacuoles, organelles proliferate, individual cells hypertrophy, cell walls thicken, and finger-like protuberances (ingrowths) form along walls adjacent to the xylem vessels to increase membrane surface area for solute uptake. The nuclei within these cells enlarge, develop an amoeboid appearance, have a very prominent nucleolus, and are polyploid. In giant cells of RKN, the nuclei are stimulated to divide in the absence of cell division, resulting in enlarged plant root cells containing hundreds of nuclei.

The syncytium of cyst nematodes also is multinucleate, but it arises by a different mechanism than that of giant cells. Within the initial syncytial cell, the plasmodesmatal openings gradually widen, and wall degradation is initiated at pit fields. As the initial syncytial cell enlarges, the cell wall gaps expand and neighbouring protoplasms fuse. Progressive cell wall dissolution allows the syncytium to expand longitudinally along the length of the vascular cylinder (extending as far as 2 to 3 mm) and can incorporate up to 200 plant cells.

The extensive cell wall architectural modifications (i.e., thickening, ingrowth, disassembly, and dissolution) that occur in cyst nematode and RKN feeding cells likely are mediated by the activity of both cell wall biosynthetic and cell wall–degrading enzymes. The origin of these enzymes, especially the cell wall–degrading enzymes, has been in question since the earliest microscopic examinations of nematode-induced feeding cells. Early biochemical investigations suggested that plant-parasitic nematodes produced cell wall–degrading enzymes, and several investigators hypothesized that the secretion of such enzymes might be responsible for the cell wall dissolution observed in feeding sites. Other researchers speculated that the anatomical features of feeding cells were more supportive of a role for cell wall–degrading enzymes of plant origin, based on the observation that cell wall dissolution appears to be a controlled process that does not occur adjacent to the nematode stylet but instead takes place distal to the nematode as the syncytium extends in both directions along the length of the root. Although cell wall–degrading enzymes of plant origin have been implicated in this process, there is no biochemical or molecular evidence to support this hypothesis.

Recently, this issue was revisited by the confirmation of two types of endo-â-1,4-glucanase (EGase) genes, *eng-1* and *eng-2*, that are expressed endogenously within the subventral esophageal gland cells of both cyst nematodes and RKN. Developmental expression of the HG-ENG-1 and HG-ENG-2 EGases of the soybean cyst nematode and the GT-ENG-1 and GT-ENG-2 EGases of the tobacco cyst nematode (TCN) supports a role for these enzymes in cell wall degradation during the penetration and intracellular migration of J2 through root tissue.

The EGases were shown to be regulated developmentally and expressed primarily in J2 within eggs before hatching, through the late J2 parasitic stage, and again only in adult males that regain their vermiform shape and migrate out of the root. Immunolocalization of soybean cyst nematode EGases in cross-sections of soybean roots at 24 hr after inoculation with infective J2 showed that the HG-ENG-2 protein is secreted from the nematode stylet into plant tissue and is detectable along the migratory path of the nematode. In addition to a role in migration, secreted cyst nematode EGases may play a role in the initial stages of syncytia formation, because the sedentary parasitic J2 contain EGase in their subventral esophageal glands. A role for cell wall–degrading enzymes of plant origin in developing syncytia is suggested by the fact that nematode EGases are not expressed in later parasitic stages as the syncytium continues to expand along the vasculature. Secretion of RKN EGases has not been evaluated in planta, but they are expressed in J2 and adult female life stages. The function of RKN EGases is unclear, but it may be related to the cell wall architecture of giant cells.

Upregulation of Tobacco EGases in Nematode-Infected Roots

To compare the relative expression of individual tobacco EGase genes in noninfected and nematode-infected (RKN or TCN) tobacco root tissue, we analyzed the presence of EGase mRNA by RT-PCR. Gene-specific primer sets designed for each of the five tobacco EGase cDNA sequences amplified fragments of the following sizes: *Ntcel2*, 181 bp; *Ntcel4*, 350 bp; *Ntcel5*, 245 bp; *Ntcel7*, 209 bp; and *Ntcel8*, 300 bp. These same size fragments were not amplified from tobacco genomic DNA or from control reactions that did not contain reverse transcriptase, excluding the possibility of contaminating genomic DNA in the results. Equivalent amounts of either RNA or root tissue were used for the synthesis of first-strand cDNAs, and the relative amounts of amplification products were consistently higher in nematode-infected roots than in uninfected roots for each tobacco EGase gene in three separate experiments. When the entire uninfected root PCR reaction was loaded on a gel, only a faint band could be detected (data not shown). DNA gel blots of amplified fragments hybridized with probes for each tobacco EGase cDNA confirmed the identity of the products. The three genes showing the highest expression levels (*Ntcel2*, *Ntcel7*, and *Ntcel8*) were selected for further analysis.

To determine the spatial expression pattern of tobacco *Ntcel2*, *Ntcel7*, and *Ntcel8* EGases in nematode-infected root tissue, we conducted mRNA in situ hybridizations on sections cut at weekly intervals from developing tobacco galls after inoculation with RKN. RT-PCR experiments indicated that tobacco EGase expression levels were relatively low; therefore, serial sections were cut to a thickness of 20 ìm to increase the transcript volume for detecting hybridization signals. Strong labelling was detected within the giant cells and lateral root primordia in sections probed with the antisense *Ntcel7* riboprobe but not in sections probed with a sense *Ntcel7* riboprobe. Sections probed with tobacco 18S rRNA sense and antisense riboprobes were used as controls. Strong hybridization signal was detected in all gall tissues with the antisense, but not the sense, 18S rRNA probe.

One of the most distinguishing characteristics of feeding cells in plants induced by RKN and cyst nematode parasitism is the extensive remodelling of their cell walls. Cell expansion, cell elongation, cell wall biosynthesis, and cell wall dissolution are all physiological processes that have been observed indirectly within nematode-induced feeding cells. The incorporation of new plant cells via cell wall dissolution is a hallmark of cyst nematode–induced syncytia. This report describes the identification of several EGase genes expressed in roots of tobacco, and combined with the temporal and spatial expression of plant EGases and nematode EGase secretion in planta, it provides molecular confirmation of the origin of cell wall–degrading enzymes involved in feeding cell formation. Antiserum specific for TCN EGases confirmed the secretion of nematode EGases into plant tissue at 24 hr after infection during the migration of infective J2 through root tissue. TCN EGases were detected within the secretory gland cells as well as within plant tissue along the migratory path of the J2 through the root cortex. EGase was detected outside the tip of the nematode head on the cell wall in contact with the migrating nematode, similar to what was observed for the soybean cyst nematode. Once TCN J2 became sedentary adjacent to root vascular tissue (48 hr after infection), no EGase secreted from the nematode was detected outside or inside the initial syncytial cells. By 96 hr after infection, no TCN EGase was detectable within either the nematodes or the expanding syncytia. Although the limits of detection and localization of secreted nematode EGase may warrant further evaluation, the lack of detection in both the nematode and the plant at this stage was consistent with the developmental cessation of EGase expression in TCN observed previously. Because nematode-induced syncytia continue to expand for several weeks after this time, it is likely that the cell wall–modifying enzymes involved in this process were of plant origin.

There is considerable evidence of a role for plant EGases during plant developmental processes involving changes in cell wall architecture, ranging from cell wall expansion to dissolution. Five tobacco cDNA clones (*Ntcel2, Ntcel4, Ntcel5, Ntcel7,* and *Ntcel8*), each encoding different EGases, were obtained from nematode-infected roots by RT-PCR using degenerate primers to conserved amino acid domains of plant EGases. Only *Ntcel2, Ntcel7,* and *Ntcel8* were expressed at high enough levels for analyses. Interestingly, all three tobacco EGases appear to be upregulated in response to parasitism by both RKN and TCN. Tobacco *Ntcel7* and *Ntcel8* were expressed exclusively within RKN-induced giant cells and TCN-induced syncytia and root tips. Initially, this result was somewhat surprising, because extensive cell wall dissolution does not contribute to the formation of RKN-induced giant cells, as is the case with syncytia. Although giant cells are not formed by the cell wall dissolution of adjacent cells, they must modify their cell wall architecture to expand and elongate.

Nematode feeding cell development is a complex, highly regulated process. The recruitment of putative plant EGases by both cyst nematodes and RKN as components of cell wall modifications in feeding cells is consistent with observations that these nematodes induce numerous changes in plant gene expression during parasitism. It is hypothesized that nematodes modify the temporal and spatial expression of plant genes whose wild-type functions have been diverted to serve the parasitic needs of the nematode. These gross modifications in plant gene expression highlight the potential value of using plant–nematode interactions as a tool to investigate plant cellular biology.

Prey-predator Interactions

A predator is an organism that eats another organism. The prey is the organism which the predator eats. Some examples of predator and prey are lion and zebra, bear and fish, and fox and rabbit. The words "predator" and "prey" are almost always used to mean only animals that eat animals, but the same concept also applies to plants: Bear and berry, rabbit and lettuce, grasshopper and leaf.

Predator and prey evolve together. The prey is part of the predator's environment, and the predator dies if it does not get food, so it evolves whatever is necessary in order to eat the prey: speed, stealth, camouflage (to hide while approaching the prey), a good sense of smell, sight, or hearing (to find the prey), immunity to the prey's poison, poison (to kill the prey) the right kind of mouth parts or digestive system, etc.

Likewise, the predator is part of the prey's environment, and the prey dies if it is eaten by the predator, so it evolves whatever is necessary to avoid being eaten: speed, camouflage (to hide from the predator), a good sense of smell, sight, or hearing (to detect the predator), thorns, poison (to spray when approached or bitten), etc.

Obligate Multitrophic Relationships

Arthropod herbivores are considered as pests in many agricultural systems. They damage, by feeding or as vectors of plant diseases, plants that are meant for human use or consumption. In the war against pests, humans have developed various methods to prevent herbivores from doing extensive damage.

The simplest method was killing all life that threatened our crops with pesticides. This worked ..., for a while. We rapidly discovered that nature is well equipped to counteract the most destructive chemicals. A few resistant individuals survived that could reproduce and build up new populations of damaging insects that have become resistant to pesticides. An alternative method of pest control is to exploit what already exists in nature: Almost all arthropod herbivores have natural enemies that can be used in what is known as 'biological control'. For this we need to understand the relationship between the plant, the herbivores and their natural enemies, as well as the factors that can influence these interactions.

Multitrophic Interactions

Information about multitrophic level interactions can provide an essential foundation for designing effective biological control, and for improving the efficacy and understanding the suppression of herbivore populations in biological control. In the last two decades, the study of trophic interactions between organisms has evolved from simple plant-herbivore or prey-predator interactions to a more complex approach involving three or more trophic levels. This complex approach addresses the complexity of food webs much more realistically. Research on multitrophic interactions aims to identify the forces that regulate populations. In general, interactions between trophic levels can constitute bottom-up forces (controlled by resources) or top-down forces (controlled by predators and antagonists).

However, many multitrophic interactions are more complex than the linear bottom-up or top-down interactions. For instance, food webs can be characterised in two dimensions: vertically, in which they form a food chain of rarely more than four or five trophic levels, or horizontally, i.e. within one trophic level. Furthermore, ecological interactions between two species are often (indirectly) mediated by a third species of the same or another trophic level. This is found for example in intra-guild interactions, where two species that share a host or prey also engage in a trophic interaction with each other, or in apparent competition, where two species, that do not come into direct contact, interact because they share a natural enemy, adding complexity to the study of these food webs.

In insects, tritrophic interactions between plants, herbivores and their natural enemies are among the most studied multitrophic interactions. In this food chain, the first trophic level, represented by the plant, influences the herbivore (the second trophic level) by its quality and quantity as a food-source. On the third trophic level, natural enemies limit herbivore populations by mortality. However, interactions between herbivores and their host plants and between herbivores and their natural enemies can only be understood when considered all together within a tritrophic context. For instance, the first trophic level (plant) can also influence the efficiency of the third trophic level (natural enemy) by providing shelter, mediating host/prey accessibility and availability, providing host/prey finding cues, influencing host/prey suitability, and providing supplemental food sources for natural enemies. Vice-versa, natural enemies may 'help' the plants, using the benefits named above, thus limiting the herbivore population more than would be otherwise possible.

To further understand fluctuations in predator or parasitoid populations and the level of herbivore suppression, not only tritrophic interactions have to be examined, but also the impact of higher-level natural enemies. Predatory and parasitic insects are attacked by their own suite of predators, parasitoids and pathogens, which constitute the fourth trophic level. The impact of these higher trophic levels on natural enemies of herbivores has received relatively little attention. They may exert a significant negative effect on plant-fitness by removing parasitoids or predators of the herbivores (top-down regulation). But not only top-down effects are to be expected. It has been shown recently that bottom-up forces may also play a role in mediating interactions involving plants, herbivores, parasitoids and hyperparasitoids. Harvey *et al* . have demonstrated that qualitative differences in herbivore diet can differently affect the performance of interacting organisms across four trophic levels.

Hyperparasitoids

Hyperparasitoids are also called secondary insect parasitoids as they develop at the expense of insect primary parasitoids. In other words, a hyperparasitoid attacks another insect that is itself parasitic on a host insect, which is often a herbivore, and is therefore part of the fourth trophic level. The great majority of hyperparasitoids are members of the order Hymenoptera, a few species belonging to the Diptera and the Coleoptera. There exist several types of hyperparasitism. Obligate hyperparasitoids can develop only in or on a primary parasitoid. On the other hand, facultative hyperparasitoids can develop as either primary or secondary parasitoids.

The immature hyperparasitoid can in turn be attacked by a conspecific or another species of hyperparasitoid. This is called tertiary parasitism, or if this tertiary parasitoid is itself the host, quaternary parasitism. These types are rare, and Gordh hypothesised that tertiary and quaternary hyperparasitism is too precarious to evolve as an obligate trophic strategy, as depletion of host resources causes a significant decrease in the size of tertiary and quaternary parasitoids, which makes these types of hyperparasitism less profitable. The last type of hyperparasitism is heteronomy. Heteronomous species (or adelphoparasites) produce females as primary parasitoids and males as hyperparasitoids, often on the females of the same species or other primary parasitoids. These species are also called autoparasitoids Like parasitoids, larvae of endophagous hyperparasitoids feed inside the host, whereas ectophagous species feed externally. Koinobiont hyperparasitoid species allow their host to continue development after oviposition, and idiobionts attack non-growing or non-feeding host stages and/orarrest the development of the host by paralysis of killing during oviposition.

Hyperparasitism has a wide taxonomic distribution among insects. However, none of the parasitoid families consists exclusively of hyperparasitoids, although within families hyperparasitism may follow phylogenetic lines. This suggests that hyperparasitism has evolved independently several times in different taxa. Obligate hyperparasitism could have evolved in at least two ways: a) via facultative hyperparasitism as an opportunistic trade-off to use herbivore or parasitoid hosts, and/ or if the hyperparasitic species frequently encounters already parasitised hosts; or b) by a host shift where a primary parasitoid of one host becomes a secondary parasitoid of another species. This host transfer is facilitated if the usual primary and new secondary hosts share physiological and/or ecological attributes. One of the reasons why hyperparasitism may have evolved in a multitrophic context might be in order to avoid the sequestration of plant toxins in the host. Compared to the herbivore, the primary parasitoid may be a less toxic resource, especially after voiding the meconium (Vet, pers. comm).

Influence of Hyperparasitism on Primary Parasitoid Populations

Interactions between hyperparasitoids and primary parasitoids have been primarily studied in biological control situations because these systems are often less complex than natural ecosystems and the economical value of the crops justifies research done on pest control. Traditionally, hyperparasitoids have been thought to have a negative effect on primary parasitoid populations. There are several ways in which hyperparasitoids can influence primary parasitoid populations: directly through mortality, or indirectly by changing the behaviour of parasitoids or herbivores.

Theoretically, if a large fraction of a parasitoid population is attacked by hyperparasitoids, an increase in the herbivore's equilibrium density should be expected. If that fraction becomes large, the herbivore population may escape control by the primary parasitoid entirely. Mathematical models have given variable results. The majority of models predict an increase of the herbivore density. On the other hand,

Beddington and Hammond (1977) predicted that in a stable host - primary parasitoid - hyperparasitoid system, hyperparasitism weakens biological control, but when the system is unstable, the presence of a hyperparasitoid may dampen the oscillations and may enable a stable three-species equilibrium to be attained.

This may benefit biological control, altering the system from one in which the pest exhibits periodic outbreaks to one of continuous sub-economic densities. It is not known how realistic these models are because of a lack of information on the biology and behaviour of hyperparasitoids. It is often assumed that hyperparasitoids and primary parasitoids have similar life histories and information about primary parasitoids is extrapolated to the next trophic level.

In the literature, high levels of hyperparasitism have often been reported. In an agro-ecosystem, the mortality of parasitoids due to hyperparasitism can even reach 100%. It is often assumed that as hyperparasitism increases, the greater the negative impact on herbivore control by primary parasitoids. Indeed, in several studies the low level of biological control by parasitoids has been repeatedly attributed to the high level of hyperparasitism.

However, in other cases, hyperparasitoids had little or no influence on biological control, even when the level of hyperparasitism was high. These differences have been explained based on the timing of hyperparasitoid attack during the season and synchronisation between primary parasitoid and hyperparasitoid.

In conclusion, these studies have produced little definitive evidence regarding the impact of hyperparasitism on the regulating capacity of parasitoids in biological control. Luck *et al* . emphasised that percent mortality is not necessarily a good measure of a mortality's importance without knowing the levels of other sources of mortality, and the interactions between different sources of mortality.

Regardless of the impact of a hyperparasitoid on a parasitoid population, a hyperparasitoid cannot affect biological control if biological control does not exist in the first place. Therefore, the impact of primary parasitoids on herbivores in the absence of hyperparasitoids should be studied first. For example, Mackauer and Völkl suggested that the degree of aphid colony exploitation primarily results from the wasp's foraging efficiency and oviposition decisions, instead of hyperparasitism. Furthermore, aestivation, a high mortality due to other factors than hyperparasitism (e.g intra-guild predation) and dispersal can also result in low levels of primary parasitoid abundance.

Host Plant and Aphid

The potato, *Solanum tuberosum* L. var. Norland was used as the host plant to rear the potato aphid, *Macrosiphum euphorbiae* (Thomas). This polyphagous aphid is a serious pest on potato crops and damages the plant directly by diverting photosynthate and indirectly by acting as plant virus disease vector. Also, the honeydew that is excreted as a waste product of aphid feeding causes mold to grow on the leaves.

Aphids live in colonies in shaded areas on leaves, stems and blossoms of plants. In nature, the potato aphid reproduces asexually during the summer, giving birth to female nymphs. At the end of summer, sexual forms are produced and aphids overwinter in the egg stage. In the laboratory potato aphid colonies are easily maintained asexually on potato plants. Adults are usually without wings under controlled conditions. Winged adults develop in response to high population densities, decline of host plant quality and changes in environmental conditions.

Primary Aphid Parasitoid

The primary parasitoid *Aphidius nigripes* Ashmead belongs to the subfamily Aphidiinae (Hymenoptera: Braconidae). It is the dominant parasitoid of the potato aphid in North America. It is a solitary species and the females attack the aphid in the nymphal or adult stage. As it has a koinobiont development, the parasitised aphid continues to live and grow or reproduce, although reproduction is often diminished or repressed by aphidiine wasps. The larval instars feed on the aphid, destroying the remaining tissues and ultimately killing the host. Before completing its development, the parasitoid larva spins a cocoon inside or under the empty aphid skin. At this stage, the aphid skin becomes indurate and the typical "mummy" is formed. The pupal stage of the parasitoid develops within the mummy.

Aphid Hyperparasitoids

The hyperparasitoids that are used in this study were chosen as representatives of the four principal aphid hyperparasitoid families: *Asaphes suspensus* Walker (Pteromalidae), *Dendrocerus carpenteri* (Curtis) (Megaspilidae), *Alloxysta victrix* (Westwood) (Alloxystidae), and *Syrphophagus aphidivorus* (Mayr) (Encyrtidae). These species were chosen because while they all naturally exploit *Aphidius* spp., they possess different biological attributes and host ranges, which are described in detail in the sections below. All are found attacking hosts in the chosen model system of potato, *M. euphorbiae* , *A. Nigripes*.

Development Mode

Aphid hyperparasitoids can be divided according to their development mode: *A. victrix* and *S. aphidivorus* are koinobionts, which means that after oviposition their living host continues its development. In contrast, *D. carpenteri* and *A. suspensus* are idiobionts. During oviposition they kill or paralyse their host which stops its development. Koinobiosis is associated with endoparasitism (the egg is placed inside the host) and idiobiosis with ectoparasitism (the egg is placed on the surface of the host).

Host Stages

Aphid hyperparasitoids can attack the immature parasitoid within the aphid at different stages of its development. Mummy hyperparasitoids, such as *D. carpenteri* and *A. suspensus* , attack the parasitoid prepupa or pupa after it has killed the aphid and the mummy is formed (hereafter called aphid mummy). Hyperparasitoids, such as *A. victrix* , attack parasitoid larva in live aphids before mummification (hereafter called parasitised aphid). *Alloxysta victrix* does not use a venom to paralyse its host, so after parasitisation the primary parasitoid larva continues to feed and grow.

It is only after the aphid is mummified by the primary parasitoid larva, that the egg hatches and the hyperparasitoid larva starts to feed endophagously until it kills and completely consumes the host. Although *Syrphophagus aphidivorus* is an endophagous koinobiont hyperparasitoid like *A. victrix* , it has a dual oviposition behaviour that is atypical for aphid hyperparasitoids. It has the capacity to attack both parasitoid larva in a live aphid or parasitoid prepupa or pupa in a mummified aphid. Furthermore, it prefers to oviposit in mummified aphids, which is not found in other endophagous koinobiont hyperparasitoids of aphids.

Host Range

There are differences in host range between the four selected aphid hyperparasitoid species. *A. victrix* is considered to have the most restricted host range. Its potential host range includes several aphidiine parasitoid species, and some authors include aphelinid species, but *A. victrix* seems to prefer Aphidiinae. *Dendrocerus carpenteri* and *A. suspensus* are generalist hyperparasitoids of four to five genera of aphidiine and aphelinid primary parasitoids. They can also be tertiary parasitoids of their own species or of other aphid hyperparasitoids. *Syrphophagus aphidivorus* attacks at least four genera of aphidiine and aphelinid primary parasitoids.

New Paradigms for Nematode Management

Andean Tuber Crops

The Andean region of South America is one of the eight centres of diversity of cultivated plants described by Vavilov (1951) and has recently become recognized as an important area for minor-crop development and germplasm conservation. There is also strong evidence that the southern Peruvian Andes is one of the four areas of the world where the independent invention of agriculture took place.

The potato has been the subject of international crop development and is now commonly grown throughout the world. There are, however, many other important food crops that were domesticated in the Andes but that are poorly known scientifically. The subject of this paper is three of these crops: *Ullucus tuberosus* Caldas (Basellaceae; *Oxalis tuberosa* Mol. (Oxalidaceae); and *Tropaeolum tuberosum* R. & P. (Tropaeolaceae). Each of these crops is a potential new crop for other areas of the world.

The Andean Crop Complex

Numerous other root and tuber crops have been domesticated in the Andes; *Arracacia xanthorriza* Bancr. (Apiaceae), *Canna edulis* Ker-Gawl. (Cannaceae), *Lepidium meyenii* Walp. (Brassicaceae), *Mirabilis expanse* R & P (Nyctaginaceae) and *Polymnia sonchifolia* Poepp. & Endl. (Asteraceae). No discussion of Andean crop resources would be complete without mentioning the global potential of these and other crops that are part of the agricultural heritage of the Andean region. Other important crops include the high protein pseudograins, *Chenopodium quinoa* Willd. (Chenopodiaceae), *C. pallidicaule* Heller, and a high protein legume, *Lupinus mutabilis* Sweet (Fabaceae). As a group, these tuber, grain legume and other crops have been among the primary food sources in the highland Andean region for centuries. A National Academy of Sciences report on Andean crops has presented general information on these and other crops (NRC 1989).

Description

These herbaceous crops are all annually propagated for their starchy tubers. Tubers are produced below ground on axillary stolons which enlarge to form terminal tubers. *Ullucus tuberosus,* ("Ullucu" or "lisas"), produces smooth spherical tubers 2-10 cm across or curved and elongate to 25 cm long. *Oxalis tuberosus* ("Oca") and *Tropaeolum tuberosum* ("Mashua" or "Añu") despite being in different botanical families bear many morphological similarities. Both produce elongate tubers slightly roughened from the enlarged scale leaves.

Pre-history

Sources for dating the presence and importance of Andean tubers include illustrations on wooden vessels (keros), ceramic urns, and sculptures. Images of *Oxalis tuberosa* and *Ullucus tuberosus* have been documented on keros from the early post-conquest era in Southern Peru. A “pacheco” urn dated to 950 (before present) BP from the central Peruvian highlands is decorated with paintings of all three of these tubers. Botanical material from several coastal Peruvian archaeological sites has also been identified as containing starch grains, vessels and xylem elements of all three of the crops discussed in this paper. Evidence of *Oxalis tuberosa* and *Ullucus tuberosus* dating to ±4,250-4,050 BP has been recorded in the dry coastal dessert of Peru (Martins 1976). One highland archaeological site, Tres Ventanas at 3,925 m has yielded material of *Ullucus tuberosus* reputed to be from a time period 10,000 years BP.

Current Distribution

These three crops are cultivated in the Andean region from Venezuela to Argentina. Outside of the Andean region *Oxalis tuberosa* is cultivated commercially in Mexico and New Zealand. Compared to potatoes, which are now cultivated in 130 countries around the world, these tubers are essentially still unknown outside of the Andean region. The crop characteristics that have, up to now, caused this limited distribution will be examined below.

Agricultural Requirements

As with most tuber crops, all three Andean tuber crops discussed here are propagated vegetatively. Whole tubers are selected from storage for annual planting. No effort is made to select larger or more vigorous appearing tubers. In fact, smaller tubers are preferred for planting in Peru as they are less valuable for food.

As these tuber crops have been vegetatively propagated from thousands of years there has been little selection pressure for sexual seed production, a common pattern in most vegetative crops. Seed production is uncommon in two of these tubers. *Ullucus tuberosus* and *Oxalis tuberosa.*

Ullucus tuberosus was long thought to be have been a completely sterile plant. However, fruit production has recently been independently documented and germination has been observed for some true seed. True viable seed has only recently been documented in *Oxalis tuberosa* while *Tropaeolum tuberosum* is known to occasionally produce fruits with viable seed (J. Rea pers. commun.).

Environment and Habitat

All three tuber crops are adapted to high Andean altitudes of 2,400-4,000 m (4,200) where introduced Old World crops are not well adapted. The few other crops found at the upper attitudinal limits of agriculture in the Andes are bitter potatoes, (*Solanum* spp.) "maca" (*Lepidium meyenii*) and barley. The high altitude Andean ecosystem is typified by steep terrain, strong winds, shallow soil and bare rock surfaces with high water run-off. Low minimum temperatures and large ranges in diurnal temperatures impose severe stress on plants.

Cultural Practices

Cultural practices for all three tubers are similar to potatoes. Planting is done in rows or hills 80-100 cm apart with plants spaced 40-60 cm apart in the rows (Rea 1975). Most planting and harvest is by hand, with mechanization rare or unknown. Monoculture predominates, but interplanting of several tuber species in one field is common. In many instances Andean farmers will allow a mixture containing two or three tuber crops to be planted in one field. Often this mixture consists of several different clones of each species. Such mixed fields may later be sorted into tuber types during harvest or before cooking. Mixed planting with legumes or grain crops is occasionally practiced.

Cropping Cycle

The length of cropping cycle is quite variable from the northern to southern Andes. In Colombia, planting takes place from March to August followed by harvest 180-360 days later. In Equador, planting starts in December in some areas and may continue until July depending on local climate. In Peru, tubers are most commonly planted in November and harvested in June. Irrigation is practiced in some areas. The variation in cropping cycle reflects not only the different climatic and day length regimes of the Andes, but also the variation between different clones of each species.

All three crops are short day length plants. *Ullucus tuberosus* requires 11-13.5 hours of day length for stolon formation and tuberization. Tubers fail to form under longer days. In each species tubers are borne on axillary stolons produced below, or rarely above ground which are positively geotropic and enlarge into terminal tubers underground.

Diseases

Andean farmers report that few pests or diseases bother these crops. However, viruses can be a problem as in many clonally propagated crops. *Ullucus tuberosus* may contain complexes of three or four viruses. These viruses are identified as a potevirus (PMV/U, a tobamovirus (TMV/U) a polyvirus, (*Ullucus* mosaic virus, UMV), and a newly recognized comovirus (*Ullucus* virus C). All four viruses have been experimentally eliminated from *Ullucus tuberosus* using meristem-tip culture and chemotherapy. Eliminating viruses dramatically increases plant vigor and yield. Further research on this topic is needed. Material exchanged internationally will have to be disease free.

Yields

Yields vary with the cultural method but are similar for all three species with *Ullucus tuberosus* and *Oxalis tuberosa* producing about 7-10 tons/hectare. In Peru, *Ullucus tuberosus* and *Oxalis tuberosa* are each planted on 15,000 hectares annually (King 1987). Annual production figures are difficult to determine as much of the crop is cultivated in small fields in remote areas.

Nearly all production is for fresh market consumption. Traditional dishes, especially soups and stews are prepared from all three tubers. *Ullucus tuberosus* has a smooth, somewhat mucilaginous taste, similar to okra and makes excellent soups. *Oxalis tuberosa*, has a sweet acid flavor. A very small portion of the *Ullucus tuberosus* crop is canned and exported to major cities of the United States. One of the important factors for people within and outside the Andean region is the nutritional value of the crops.

Nutritional Value

Nutritional analysis of these tubers has shown that there is great variation in both the quantity and quality of protein within and between cultivars of the three species. Clearly much variability exists in the protein content of these species. Recently conducted amino acid profiles of these three species has revealed that they contain a good balance of essential amino acids, with the limiting amino acids being valine and tryptophan.

Comparative Nutritional Value

The mean proximate nutritional value of these three crops and several other root and tuber crops utilized around the world are compared. These data show that *Ullucus tuberosus* compares favourably to *Ipomea batatas* (sweet potato), and *Manihot esculenta* (cassava) in protein content *Oxalis tuberosa* is nearly equal to *Solanum tuberosum* (potato) and is superior in caloric value to *Dioscorea* spp. (true yams). The nutritional value of these Andean tuber crops then is good when compared with staple root and tuber crops eaten around the world.

Secondary Compounds

Each of the three species has been found to contain low levels of secondary compounds. *Ullucus tuberosus,* contains saponin (Hegnauer 1964), *Oxalis tuberosa* contains oxalic acid (King 1988) and *Tropaeolum tuberosum* contains a mustard oil, p-methoxybenzyl isothiocyanate, which has been used in Andean ethnomedicine (Johns and Towers 1981). The compound of concern to human nutrition is oxalic acid. The level of oxalic acid in Andean, Mexican and New Zealand cultivars of *Oxalis tuberosa* has been shown to vary between 1.2 to 51.3 mg/100 gram fresh material (King 1988). Even at the higher levels the oxalic acid content does not pose a threat to human health unless a persons diet consisted of 100% *Oxalis tuberosa.* Potatoes contain 2-7 mg/100 g fresh weight of oxalic acid and spinach (*Spinacia oleracea*) contains 356 to 780 mg/100 g fresh material.

Germplasm

International Collections

In recent years there have been several germplasm collecting expeditions in Ecuador, Peru and Bolivia. Germplasm of all three of these species is now being stored and evaluated in germplasm banks in Ecuador and Peru.

The Ecuadorian germplasm collection is maintained and evaluated at the Instituto Nacional de Investigaciones Agropecuarias (INIAP), Santa Catalina research station.

The largest number of germplasm accessions is in Peru, with a total of five ex-situ germplasm banks located in Cajamarca, Huancayo, Ayacucho, Cuzco and Puno. These five germplasm banks are coordinated through the Proyecto Investigacion de sistemas Agropecuarias Andinos (PISA). In Peru, there is also a collection of all three species at the University of San Marcos laboratory of genetic resources and biotechnology This duplicate collection from Peruvian germplasm banks is maintained in-vitro. The primary aim of this group is to eliminate viruses from the three species.

Finnish researchers at the University of Turku have also collected germplasm of *Ullucus tuberosus* in Ecuador, Peru and Bolivia which is being used for research on variation patterns in this species. There are no other known germplasm banks for the two to three cultivars of *Oxalis tuberosa* cultivated in Mexico and New Zealand. Mexican farmers, with only limited genetic diversity available, have expressed interest in testing some of the numerous cultivars from the Andean zone.

Oxalis tuberosa is the most well represented of the crops, 1205 accessions being maintained in six germplasm banks in Ecuador and Peru. The next largest in number of accessions is *Ullucus tuberosus,* followed by *Tropaeolum tuberosum.*

Diversity

The large number of germplasm accessions of these three species is likely to contain much duplication. Research on the biochemical differentiation of these accessions is being developed for application to the large number of accessions of *Oxalis tuberosa.*

There is, however, a high degree of intraspecific diversity within all three of these species and agronomic characterization of these cultivars is being conducted at germplasm banks in Ecuador and Peru. In addition, The International Board for Plant Genetic Resources (IBPGR) is currently supporting a scientist to assist in the efforts to distinguish genotypes and characterize the collections.

Potential

Traditionally all three crops have been confined to their naive Andean home. Attempts were made to introduce them into Europe during the early part of the 19th century but these were largely unsuccessful the plants remaining little more than curiosities. Recently, there has been an effort to select adapted clones for trial in other regions of the World. These efforts are aimed at increasing the crop options available to farmers, increasing diversity of food sources and developing crops for marginal growing regions. Suitable areas include high altitude regions in low latitudes and oceanic or insular climates with long cool growing seasons.

Possibilities also exist for treating these tubers as winter crops in mild climates of the southern and northern hemispheres. The success of introduction efforts will depend partly on careful selection of germplasm adapted to specific environments. The requirement of short days for tuber formation may limit potential production areas. The existing variation of different clones in response to day length gives promise that new cultivars adapted to long days or day neutral clones might be selected. Currently, these tubers are under trial in Nepal for growing in the Himalayas. With limited protection *Ullucus tuberosus* can produce a crop of tubers at Turku, Finland.

Production of true seed will be necessary for breeding. Given the occurrence of occasional fruit production in each of these species the probability of identifying mechanisms for seed failure is likely Cytological studies of *Ullucus tuberosus* indicate that abnormal meiosis resulting in incomplete microsporocyte formation may account for the rarity of seed formation.

Additional exploration and germplasm collection should be undertaken, particularly at the southernmost portion of cultivation where clones adapted to longer day are more likely to be found. Exploration should occur in Argentina and Chile as the Andean tuber complex is reported from the island of Chilöe. In order to investigate seed biology, accessions should be obtained for each wild progenitor, especially for *Ullucus tuberosus* Caldas subsp. *aborigineus* (Brücher) Sperling.

The Andean potato was originally adapted to the same environment and had the same day length requirements as these three tubers. Given the success of the potato, there is a strong possibility of successfully introducing these three tubers in new growing regions.

Oca Oxalis Tuberosa Molina

Oxalis tuberosa (Oxalidaceae) is a perennial herbaceous plant that overwinters as underground stem tubers. These tubers are known as uqa in Quechua, hispanicized oca, as New Zealand yam and a number of other alternative names. The plant was brought into cultivation in the central and southern Andes for its tubers, which are used as a root vegetable. The plant is not known in the wild, but populations of wild *Oxalis* species that bear smaller tubers are known from four areas of the central Andean region. Oca was introduced to Europe in 1830 as a competitor to the potato and to New Zealand as early as 1860. In New Zealand, oca has become a popular table vegetable and is simply called yam.

***Figure:** Yellow and purple* Oxalis tuberosa *(oca) tubers*

Cultural Significance

Grown primarily by Quechua and Aymara farmers, oca has been a staple of rural Andean diets for centuries. Of all Andean root and tuber crops, oca is presently second only to potato in area planted within the Central Andean region. Oca is important to local food security because of its role in crop rotations and its high nutritional content.

Diversity

Andean farmers cultivate numerous varieties of oca. Oca diversity may be described with respect to morphological characters, local cultivar names, or molecular markers.

Morphological Characters

Oca morphotypes are distinguished by foliar, floral, fruit, stem, and tuber characteristics, as described in the International Plant Genetic Resources Institute's document on oca descriptors. The morphological diversity of oca tubers, in particular, is astounding.

Tubers range from 25 to 150 mm in length by 25 mm in width; skin and flesh colour may be white, cream, yellow, orange, pink, red, and/or purple and distributed in range of patterns.

Local Cultivar Names

Oca-growing communities often name varieties based primarily on tuber morphology and secondarily on flavor. For example, common names may include ushpa negra (black ash) or puka panti (red *Cosmos peucedanifolius*). Great inconsistency of nomenclature has been reported within and among communities.

Molecular Markers

Numerous studies have additionally described oca diversity through molecular approaches to study protein and genetic variation. Molecular markers, such as allozymes (e.g., del Río, 1999) and inter-simple sequence repeats (e.g., Pissard et al., 2006), show oca diversity to be low relative to other crops, probably because of its vegetative mode of propagation.

While genetic differentiation corresponds well with folk classification, cluster analyses indicate that folk cultivars are not perfect clones, but rather genetically heterogeneous groupings.

Edibility

Oca is cultivated primarily for its edible stem tuber, but the leaves and young shoots can be eaten as a green vegetable also. Andean communities have various methods to process and prepare tubers, and in Mexico oca is eaten raw with salt, lemon, and hot pepper.

The flavour is slightly tangy, and texture ranges from crunchy (like a carrot) when raw or undercooked, to starchy or mealy when fully cooked.

Use Categories

Oca is fairly high in oxalates, concentrated in the skin, and the bioavailability of oxalate appears to be similar to spinach. Significant variation in oxalate concentration exists among varieties, and this variation distinguishes two oca use-categories recognized by Andean farmers.

One use-category, sour oca, contains cultivars with high levels of oxalic acid. Farmers process these tubers to form a usable storage product, called khaya in Quechua. To prepare khaya, tubers are first soaked in water for approximately one month. Then they are left outside during hot, sunny days and cold, freezing nights until they become completely dehydrated. This process is similar to the preparation of chuñu from bitter potatoes. Cultivars in this use category are referred to in Quechua as *khaya* (name of the dried, processed product) or p'usqu (sour/fermented), and in Aymara as luk'i.

The other use-category, sweet oca, contains cultivars with lower levels of oxalic acid. The traditional Andean preparation methods for this use-category are also geared towards reducing the oxalate level of the harvested vegetable, but without dehydration. This is done by exposure to sunlight, which decreases the organic acid content and thereby increases the sweet taste of the oca. Once exposed to sunlight, oca can be boiled, baked or fried. In the Andes it is used in stews and soups, served like potatoes, or can be served as a sweet. Cultivars in this category are referred to in Quechua as wayk'u (boiling), misk'i (sweet/delicious) and in Aymara as q'ini.

Sour oca and sweet oca form distinct genetic clusters based on AFLP data. This suggests the possibility of distinct evolutionary histories for each use-category.

Nutrition

The table to the right displays the nutritional content for fresh and dried oca. Oca is a valuable source of vitamin C, potassium (included in value for ash), and iron. It also provides some protein, with valine and tryptophan its limiting amino acids. Cultivars vary greatly in nutritional content, so these measures should be taken only as approximates.

Cultivation

Oca is one of the important staple crops of the Andean highlands, due to its easy propagation, and tolerance for poor soil, high altitude and harsh climates.

Distribution

Oca is planted in the Andean region from Venezuela to Argentina, from 2800 to 4100 metres above sea level. Its highest abundance and greatest diversity are in central Peru and northern Bolivia, the probable area of its domestication.

Climate Requirements

Oca needs a long growing season, and is day length dependent, forming tubers when the day length shortens in autumn (around March in the Andes). In addition, oca requires climates with average temperatures of approximately 10 to 12 °C (ranging between 4 and 17 °C) and average precipitation of 700 to 885 millimetres per year.

In areas with harsh winter climates, early frosts may cut back the foliage before the tubers have a chance to form. In tropical areas where the days are unchanging in length, oca will not set a crop successfully.

Soil Requirements

Oca grows with very low production inputs, generally on plots of marginal soil quality, and tolerates acidities between about pH 5.3 and 7.8. In traditional Andean cropping systems, it is often planted after potato and therefore benefits from persisting nutrients applied to, or left over from, the potato crop.

Propagation

Oca is usually propagated vegetatively by planting whole tubers.

Propagation by seed is possible but is rarely used in practice. Sexual propagation is complicated by several factors. First, like many other species in the genus *Oxalis*, oca flowers exhibit tristylous heterostyly and are consequently subject to auto-incompatibility.

Furthermore, on the rare occasion that oca plants do produce fruit, their loculicidal capsules dehisce spontaneously, making it difficult to harvest seed. Oca flowers are pollinated by insects (e.g., genera *Apis*, *Megachile*, and *Bombus*). Data regarding the frequency of volunteer hybrids and farmers' subsequent incorporation of them has not yet been published.

Cropping Factors

Oca tuber-seeds are planted in the Andes in August or September and harvested from April to June. The first flowers bloom around three to four months after planting, and the tubers also begin to form then. Between planting and harvesting, the oca crop requires little tending, except for a couple of weedings and hillings.

Oca is a component of traditional crop rotations and is usually planted in a field directly after the potato harvest. A common sequence in this rotation system may be one year of potato, one year of oca, one year of oats or faba beans, and two to four years fallow. Within this system, q'allpa is a Quechua term that signifies soil previously cultivated and prepared for planting of a new crop.

The cultural practice is similar to potatoes. Planting is done in rows or hills 80–100 cm apart with plants, spaced 40–60 cm apart in the rows. Monoculture predominates, but interplanting with several tuber species, mashua and olluco, in one field is common in Andean production. Often this intercoppng consists of several different varieties of each species. Such mixed fields may later be sorted into tuber types during harvest or before cooking.

Harmine found in root secretions of *Oxalis tuberosa* has been found to have insecticidal properties.

Yields

Yields vary with the cultural method. Annals from Andean countries report about 7-10 tonnes per hectare for *Oxalis tuberosa* production. But with adequate inputs and virus free propagation material, oca production can range from 35 to 55 tonnes per hectare.

Limitations

Pests and diseases limit the production of oca. Crops in the Andes are often infected with viruses, causing chronic yield depression. Adequate techniques to remove viruses have to be applied before the varieties can be used outside the Andean region. Cultivation is also constrained the Andean potato weevil (*Premnotrypes spp*), ulluco weevil (*Cylydrorhinus spp*), and oca weevil, the identification of which remains uncertain (possibly *Adioristidius*, *Mycrotrypes*, or *Premnotrypes*). These weevils often destroy entire crops. Further notable pests are nematodes.

As already mentioned day length restrictions and the presence of oxalates can also be considered limiting factors. Scientists work with specific breeding, selection, and virus cleaning programmes on these purposes.

Agricultural Potential

Potential distribution to other suitable ecogeographical zones of, for example, Asia and Africa may be possible. The cultivation and use of a fleshy pink variety of *Oxalis tuberosa* in New Zealand already indicates a wider utilization and agricultural interest than has been previously recognized.

Conservation Efforts

A number of ongoing ex situ and in situ conservation projects currently focus on the preservation of *Oxalis tuberosa* diversity.

Quinoa Chenopodium Quinoa

Quinoa, a species of goosefoot *(Chenopodium),* is a grain crop grown primarily for its edible seeds. It is a pseudocereal rather than a true cereal, as it is not a member of the true grass family. As a chenopod, quinoa is closely related to species such as beetroots, spinach and tumbleweeds.

It is high in protein, lacks gluten, and is tolerant of dry soil.

Quinoa (the name is derived from the Spanish spelling of the Quechua name *kinwa*) originated in the Andean region of Ecuador, Bolivia, Colombia and Peru, where it was domesticated 3,000 to 4,000 years ago for human consumption, though archaeological evidence shows a non-domesticated association with pastoral herding some 5,200 to 7,000 years ago.

Similar *Chenopodium* species, such as pitseed goosefoot (*Chenopodium berlandieri*) and fat hen (*Chenopodium album*), were grown and domesticated in North America as part of the Eastern Agricultural Complex before maize agriculture became popular. Fat hen, which has a widespread distribution in the Northern Hemisphere, produces edible seeds and greens much like quinoa, but in smaller quantities.

The nutrient composition is favourable compared with common cereals. Quinoa seeds contain essential amino acids like lysine and acceptable quantities of calcium, phosphorus, and iron.

After harvest, the seeds must be processed to remove the coating containing the bitter-tasting saponins. The seeds are in general cooked the same way as rice and can be used in a wide range of dishes. The leaves are eaten as a leaf vegetable, much like amaranth, but commercial availability of quinoa greens is limited.

Biology

Chenopodium quinoa is a dicotyledonous annual plant usually about 1–2 metres (3.3–6.6 ft) high. It has broad, generally pubescent, powdery, smooth (rarely) to lobed leaves normally arranged alternately. The woody central stem is branched or unbranched depending on the variety and may be green, red or purple. The flowering panicles arise from the top of the plant or from leaf axils along the stem. Each panicle has a central axis from which a secondary axis emerges either with flowers (amaranthiform) or bearing a tertiary axis carrying the flowers (glomeruliform). The green hypogynous flowers have a simple perianth and are generally bisexual and self-fertilizing. The fruits are about 2 millimetres (0.079 in) in diametre and of various colours—from white to red or black, depending on the cultivar.

Natural Distribution

Chenopodium quinoa is believed to have been domesticated in the Peruvian Andes from wild or weed populations of the same species. There are non-cultivated quinoa plants (*Chenopodium quinoa* var. *melanospermum*) that grow in the area it is cultivated; these may either be related to wild predecessors, or they could be descendants of cultivated plants.

Saponin Content

Figure: *Red quinoa, cooked.*

In their natural state, the seeds have a coating of bitter-tasting saponins, making them unpalatable. Most of the grain sold commercially has been processed to remove this coating. This bitterness has beneficial effects during cultivation, as it is unpopular with birds and therefore requires minimal protection. The genetic control of bitterness involves quantitative inheritance; lowering the saponin content through selective breeding to produce sweeter, more palatable varieties are complicated by about 10% cross-pollination.

The toxicity category rating of quinoa saponins treats them as mild eye and respiratory irritants and as a low gastrointestinal irritant. The saponin is a toxic glycoside, a main contributor to its hemolytic effects when combined directly with blood cells. In South America, quinoa saponin has many uses, including as a detergent for clothing and washing and as an antiseptic for skin injuries. High levels of oxalic acid are in the leaves and stems of all species of the *Chenopodium* genus, and are also in the related genera of the *Amaranthaceae* family. The risks associated with quinoa are minimal, provided it is properly prepared and the leaves are not eaten to excess.

Nutritional Value

Quinoa was important to the diet of pre-Columbian Andean civilizations. Quinoa grain has been called a superfood. Protein content is very high for a cereal/pseudo-cereal (14% by mass), but not as high as most beans and legumes. The protein content per 100 calories is higher than brown rice, potatoes, barley and millet, but is less than wild rice and oats. Nutritional evaluations indicate that quinoa is a source of complete protein. Other sources claim its protein is not complete but relatively high in essential amino acids. The grain is a good source of dietary fibre and phosphorus and is high in magnesium and iron. It is a source of calcium, and thus is useful for vegans and those who are lactose intolerant. It is gluten-free and considered easy to digest. Because of these characteristics, it is being considered a possible crop in NASA's Controlled Ecological Life Support System for long-duration human occupied space flights.

The grain may be germinated in its raw form to boost its nutritional value, provided that the grains are rinsed thoroughly to remove any saponin. It has a notably short germination period: only 2–4 hours in a glass of clean water is enough to make it sprout and release gases, as opposed to 12 hours with wheat. This process, besides its nutritional enhancements, softens the seeds, making them suitable to be added to salads and other cold foods.

Cultivation

Climate Requirements

The plant's growth is highly variable due to a high complexity of different subspecies, varieties and landraces (domesticated plants or animals adapted to the environment in which they originated). However, in general it is undemanding and altitude-hardy. It is grown from coastal regions (Chile) to over 4,000 m (13,000 ft) in the Andes near the equator, with most of the cultivars being grown between 2,500 metres (8,200 ft) and 4,000 metres (13,000 ft). Depending on the variety, optimal growing conditions are in cool climates with temperatures that vary between 25 °F ("4 °C) during the night to near 95 °F (35 °C) during the day. Some cultivars can withstand lower temperatures without damage. Light frosts normally do not affect the plants at any stage of development, except during flowering. Mid-summer frosts during flowering, often occurring in the Andes, lead to sterilization of the pollen. Rainfall conditions are highly variable between the different cultivars, ranging from 300 to 1,000 millimetres (12 to 39 in) during growing season. Growth is optimal with well-distributed rainfall during early growth and development and dry conditions during seed maturation and harvesting.

Quinoa has been cultivated in the United States, primarily in the high elevation San Luis Valley (SLV) of Colorado where it was introduced in 1982. In this high-altitude desert valley, maximum summer temperatures rarely exceed 85 °F (29 °C) and night temperatures are about 45 °F (7 °C). Due to the short growing season, North American cultivation requires short-maturity varieties, typically of Bolivian or Chilean origin.

Sowing

Quinoa plants do best in sandy, well-drained soils with a low nutrient content, moderate salinity, and a soil pH of 6 to 8.5.

The seedbed must be well prepared and drained to avoid waterlogging. In the Andes, the seeds are normally broadcast over the land and raked into the soil. Sometimes it is sown in containers of soil and transplanted later.

Cultivation Management

Yields are maximised when 150 to 180 lb (68 to 82 kg) N/acre is available. The addition of phosphorus does not improve yield. In eastern North America, it is susceptible to a leaf miner that may reduce crop success and which also affects the common weed and close relative *Chenopodium album*, but *C. album* is much more resistant.

Harvesting and Handling

Quinoa grain is usually harvested by hand and rarely by machine, because the extreme variability of the maturity period of most Quinoa cultivars complicates mechanization. Harvest needs to be precisely timed to avoid high seed losses from shattering, and different panicles on the same plant mature at different times. The seed yield (often around 3 t/ha up to 5 t/ha) is comparable to wheat yields in the Andean areas. In the United States, varieties have been selected for uniformity of maturity and are mechanically harvested using conventional small grain combines. The plants are allowed to stand until they are dry and the grain has reached a moisture content below 10%. Handling involves threshing the seedheads and winnowing the seed to remove the husk. Before storage, the seeds need to be dried in order to avoid germination. Dry seeds can be stored raw until washed or mechanically processed to remove the pericarp to eliminate the bitter layer containing saponins.

History and Culture

Early History

Quinoa was first domesticated by Andean peoples around 3,000 to 4,000 years ago. It has been an important staple in the Andean cultures where the plant is indigenous but relatively obscure in the rest of the world. The Incas, who held the crop to be sacred, referred to it as *chisaya mama* or "mother of all grains", and it was the Inca emperor who would traditionally sow the first seeds of the season using "golden implements". During the Spanish conquest of South America, the colonists scorned it as "food for Indians", and suppressed its cultivation, due to its status within indigenous religious ceremonies. The conquistadors forbade quinoa cultivation for a time and the Incas were forced to grow wheat instead.

Rising Popularity and Crop Value

The grain has become increasingly popular in the United States, Canada, Europe, China and Japan where it is not typically grown, increasing crop value. Between 2006 and early 2013 quinoa crop prices tripled. In 2011, the average price was US$3,115 per ton with some varieties selling as high as $8,000 per ton. This compares with wheat prices of $9 per bushel (about $340 per ton). Since the 1970s, producers' associations and cooperatives have worked towards greater producer control of the market. The higher prices make it harder for people to purchase, but also brings a livable income for farmers and enables many urban refugees to return to working the land.

The popularity of quinoa grain in non-indigenous regions has raised concerns over food security. Due to continued widespread poverty in regions where it is produced and because few other crops are compatible with the soil and climate in these regions, it has been suggested that the inflated price disrupts local access to food supplies. In 2013, *The Guardian* compared it to asparagus cultivated in Peru, a cash crop criticized for excessive water use, as "feeding our apparently insatiable 365-day-a-year hunger for this luxury vegetable[...]" It has been suggested that, as people rise above subsistence-level income, they choose higher-status Western processed foods. However, anthropologist Pablo Laguna states that farmers are still saving a portion of the quinoa crop for their own use, and that the high prices affect nearby city dwellers more, but consumption in cities has traditionally been lower. According to Laguna, the net benefit of increased revenue for farmers outweighs the costs, saying that it is "very good news for small, indigenous farmers". The transformation from a healthy staple food for farming families and communities into a product that is held to be worth too much to keep for oneself and one's family is an ongoing process. It is seen as a valuable resource that can bring in far greater amounts of cheap, low nutrient foods such as pasta and rice.

It used to be seen as a peasant food that provided farming families with a very important source of nutrition, but now occupies a spectrum from an everyday food of urban Bolivia's middle class to a luxury food in the Peruvian capital of Lima where "it sells at a higher per pound price than chicken, and four times as much as rice". Efforts are being made in some areas to distribute it more widely and ensure that farming and poorer populations have access to it and have an understanding of its nutritional importance. These include incorporating it into free school breakfasts and in government provisions distributed to pregnant and nursing women in need.

Kosher Controversy

Quinoa has become popular in the Jewish community as a substitute for the leavened grains that are forbidden during the Passover holiday. Several kosher certification organizations refuse to certify it as being kosher for Passover, citing reasons including its resemblance to prohibited grains or fear of cross-contamination of the product from nearby fields of prohibited grain or during packaging.

In December 2013, the Orthodox Union, the world's largest kosher certification agency, announced it would begin certifying quinoa as kosher for Passover.

International Year of Quinoa

The United Nations General Assembly declared 2013 as the "International Year of Quinoa" in recognition of ancestral practices of the Andean people, who have preserved it as food for present and future generations, through knowledge and practices of living in harmony with nature. The objective is to draw the world's attention to the role that quinoa could play in providing food security, nutrition and poverty eradication, in support of achieving Millennium Development Goals.

The Food and Agriculture Organization of the United Nations is as the Secretariat of the international year. Bolivia has the presidency of the Coordination Committee and Ecuador, Peru and Chile share the vice presidency, with the rapporteurship in the hands of Argentina and France.

Most Important Nemotodes in Tuber and Grain Crops

Root and tuber crops are still living organisms after they have been harvested and losses that occur during storage arise mainly from their physical and physiological condition. The main causes of loss are associated with mechanical damage, physiological condition (maturity, respiration, water loss, sprouting), diseases and pests.

To ensure effective storage of root and tuber crops, these major causative factors need to be properly understood and, where appropriate, be properly controlled, taking into account the socio-economic factors which prevail in the areas of production and marketing.

Root and tuber crops need to be handled gently to minimize bruising and breaking of the skin because of its relatively soft texture compared, for example, to cereal grains. The effect of mechanical injury resulting in external and/or internal bruising and tissue discolouration is often underestimated. Severely damaged tubers should not be stored for three reasons;

- because of lower quality
- because of the increased risk of subsequent pathogenic losses.
- because of the risk of introducing disease organisms into sound produce.

Most mechanical damage occurs as a result of careless handling at harvest and during transport to and within a store since, generally in the tropics, food handling procedures are poorly developed and fresh produce is all too frequently treated as an inert object.

Careful harvesting and proper handling of roots and tubers is, therefore, an important step towards successful storage. Crops are most likely to be injured at harvest by the digging tools, which may be wooden sticks, machetes, hoes or forks.

Roots and tubers have the ability to heal their skin wounds when held at relatively high temperatures and humidities for few days after harvest whilst at the same time there is a general strengthening of the skin. The term "curing" refers to the operation of self-healing of wounds, cuts and bruises. There are two steps involved in the curing process:

- Cells suberization. The production of suberin and its deposition in cell walls;
- The formation of cork cambium. The production of cork tissue in the bruised area. The new cork tissue seals the cut or bruised areas and helps prevent the entrance of decaying organism while also reducing water loss.

Although wound healing is important in preventing the invasion by pathogens, it also important in limiting the rate of respiration and water loss. Since water loss in badly suberised and damaged tubers may be very high, it is important that the curing operation is carried out as soon as possible after harvest. This will encourage as complete healing as possible although the extent of healing will depend on the on the type of damage; the older the tuber, the less well a wound will heal. Four factors affect the healing of wounds;

- temperature of the commodity
- oxygen and carbon dioxide concentration within the commodity
- humidity within the commodity
- the use of sprout inhibitors

It is perhaps unfortunate that the curing operation when carried out properly involve conditions of high temperature and high humidity within the commodity, which also favour the rapid growth of microorganisms and commodity deterioration. Therefore it needs to be carried out very carefully and quickly and it follows that after curing further handling of the produce should be reduced to a minimum so as to avoid further injury that may result in further losses.

The length of time for proper curing cannot be precisely defined as it depends on many factors, such as condition of the crop at harvest, type of wound, season, storage temperature and relative humidity.

Proper Packaging and Handling

The ideal in packaging is to protect the produce from damage during handling, transport and storage and to provide containers of uniform size that are conveniently stacked and handled, easily accounted for in quantity and, where appropriate, in weight.

In many developing countries traditional baskets, and various types of trays or buckets are used for transporting produce to the house or to village markets. These are usually of low cost, made from readily available material and serve the purpose for transport over short distances. But, they have many disadvantages in large loads carried over long distances:

- they are difficult to clean when contaminated with decay organisms;
- they often lack rigidity and distort when stacked thus applying severe local pressure to their contents
- they are frequently very variable in shape and therefore are difficult to load, especially for long journeys.
- being of local manufacture they are often rather crude and may have sharp edges or splinters causing cuts and punctures to the commodity.

Many authorities have observed that produce being transported and marketed in commercial quantities needs better packaging in appropriately sized units if losses are to be minimised and to achieve economical use of transport.

The shape of packages is significant because of need to load for maximum capacity and stability. Round baskets, whether cylindrical or tapered, hold considerably less produce than boxes occupying the same cubic space; a cylindrical basket contains only 78.5% by volume compared with a rectangular box occupying the same space on a vehicle.

However, packaging can be a major item of expense in produce marketing, especially in developing countries where packaging industries are not well developed.

The selection of suitable containers for commercial scale marketing requires very careful consideration. Among the various types of packaging material that are available, natural and synthetic fibre sacks and bags as well as moulded plastic boxes seem to be more suitable and have greater promise for packaging roots and tubers and for their transport to distant markets.

Natural and Synthetic Fibre Sacks and Bags

Bags up to 5 kg capacity and sacks from 25 to over 50 kg can be made from natural fibres, jute or sisal, or from synthetic fibres of polypropylene or polyethylene. These sacks are ubiquitous and have been in continuous use for packaging and transporting root and tuber crops to markets for a considerable time.

Their main problem in many developing countries is that often they are too large and when filled (and rarely are they not filled absolutely full) they are too heavy for careful handling and are often dropped or thrown down resulting in severe bruising and damage to the contents. Finely woven sacks, particularly of synthetic material impair ventilation when stacked.

Moulded Plastic Boxes

Moulded high-density polythene boxes, allegedly reusable, are widely used for transporting perishable produce in many countries. They can be made to almost any specifications, are strong, rigid, smooth, easily cleaned and can be made to stack when full of produce and to nest when empty in order to conserve space. Their disadvantages are:

- they can be produced economically only in large numbers but are still costly;
- they have to be imported into most developing countries, adding to the cost and pressures on foreign exchange;
- there has to be a good organisation and sound reliable cooperation between the market traders, transport services and producers if there is to be a effective container return service.
- the containers, particularly the cheaper varieties, deteriorate rapidly when exposed to sunlight, unless treated with an ultraviolet inhibitor.

Before deciding on what packaging to use, the producer or trader has to ensure that the cost of packaging does not exceed its benefits. A decision should involve consultation between the transport contractors and market traders. The factors to be considered include;

- the type of produce;
- the level of losses occurring during marketing;
- the comparative cost of the present and improved packaging;
- the regularity of supply of the packaging material;
- the acceptance of the packaging method to the market.

Control of Temperature

Temperature has a great influence on many factors that cause loss during storage. It is the single most important factor affecting the rate of respiration, it also influences the rate of sprout growth, the development of rotting micro-organisms and insect infestation. The effect of different temperature regimes on the storage of potatoes. At 10°C, the rates of sprout development, rotting and respiration are shown to be moderate but at 4°C, sprouting is stopped, while rotting and respiration continue but at very low levels.

With the exception of highland areas, low temperature storage in the tropics within the range 10° to 15°C can only be envisaged by using a refrigeration process of some kind. At subsistence or small farmer level this is generally not practical because of cost implications and the technical support needed to sustain conventional refrigeration technology. Only in very dry areas is simple evaporative cooling at all successful but even this simple technology needs a prime mover to be operating almost continuously. Therefore, successful storage of roots and tubers in any sort of structure depends very much on natural ventilation to remove respiration heat, to remove carbon dioxide, which in concentration can lead to the breakdown of dormancy, and to keep the temperature of the crop as low as possible. Ventilation should be with the coolest possible air, night time ventilation is not only the coolest but has the highest relative humidity, so that water loss through transpiration is also held to a minimum.

Control of Sprouting

How long roots and tubers, with the exception of cassava, can be made to remain dormant with limited endogenous metabolic activity is, generally, the determining factor in how long the commodity can be stored. The end of dormancy leads to the initiation of sprouting which, in turn, means increased respiration and dry matter loss. Therefore, if the duration of storage is to be longer than the natural dormancy period an alternative method to prevent or delay sprouting is needed. One or more of three methods can be used:

- storage at low temperatures,
- the use gamma irradiation,
- the use of chemical sprout inhibitors.

Storage at Low Temperatures

For potatoes stored at 4°C sprout growth is negligible; for yams storage at 15° to 16°C prolonged the dormancy period for six months. But, socio-economic constraints in most developing countries will limit the use of refrigeration in the storage of roots and tubers at farmer level.

Gamma Irradiation

Gamma irradiation is known to induce lesions on nucleic acids and cellular proteins thus preventing them from multiplying. When applied at doses of about 7.5 head, gamma irradiation has been reported to inhibit sprouting of yams, potatoes and sweet potatoes However, this technique has not yet been applied on a commercial scale in the tropics and is unlikely to be of practical value to farmers because of the cost of the high technology involved. Furthermore it is a treatment that is not widely acceptable to consumers or a permitted treatment for food commodities in some developed countries.

Chemical Sprout Inhibitors

In stores using natural air ventilation, with relatively high ambient temperatures (20°C to 30°C such as are normally experienced in tropical and subtropical lowlands) and for any period of storage beyond the normal or natural end of the dormancy period, the use of sprout inhibiting chemicals is the only practical means of controlling sprouting. This treatment has proved effective on potatoes and sweet potatoes.

CIPC (isopropyl-N-chlorophenylcarbamate) is the chemical most commonly used as a sprout suppressant. It is mixed with an inert filler to give a concentration of about 2%, and applied as a dust on potatoes at a rate of 1-1.5kg per ton of produce and it has been very effective in delaying and reducing sprout growth. As sprout suppressants also inhibit the development of wound cork, they must be applied only after the curing operation has been completed or a few weeks after harvest when wounds have healed and any lesions would have naturally sealed off.

Sprout suppressants have almost all proved to be quite ineffective on yams. This is probably because yam tubers are unique amongst plants propagating vegetatively in not having pre-formed buds. Most sprout suppressants, such as CIPC affect the meristematic cells of the sprouting loci. On potato tuber these loci are well formed at the time of harvest and are located at the tuber surface. In yam sprout initials are formed only just before the end of dormancy and then rise from beneath the periderm. When a sprout inhibitor is applied on the yam tuber just after harvest there are no sprout loci on which the chemical can act. Once sprouting has been started, the application of sprout suppressant may then inhibit further growth of the sprout initials.

Control of the Spread of Diseases

Efforts to control the spread of diseases should aim to be preventive rather than curative. Simple and low cost preventive measures which to help control the incidence of post-harvest diseases include:

- gentle handling to minimise the risk of injury to tubers during harvesting, transport and storage;
- adequate cultural practices and especially using disease free planting material;
- good phytosanitary practices, including regular inspection of fields and premises, proper disposal of diseased tubers and plant debris, the cleaning and sterilising of implements, boxes, buildings, etc.

- pre-harvest crop application of chemicals to control the diseases in the growing crop;
- curing of the crop before storage;
- only storing produce that has been dried before putting into store, avoiding produce in store getting wet and storing at the optimum temperature.

Some diseases can be prevented or controlled by the direct application of chemicals to the produce by dipping the produce, applying sprays or dusts. To be effective chemical treatment requires the application of the appropriate compound, at the recommended dosage, by the most appropriate method and at the most suitable time. It is important to have a thorough knowledge of the pathogens being treated in order that the correct treatment can be selected and properly applied.

Thiabendazole and Benomyl are presently the most commonly used fungicides for post-harvest treatment of pathogenic diseases of roots and tubers. The best results are obtained when the chemicals are applied not later than three days after harvest and not after the pathogen is well established.

One drawback to these particular chemicals is that they are not always readily available to farmers or they are too expensive. There is a very real risk, which is frequently borne out in practice, that farmers who, for any reason, cannot obtain these chemicals will use what ever is at hand without regard to the danger of applying chemicals which are suitable for growing plants but which are dangerously toxic when applied directly to stored food produce, whether or not the produce will be treated, as a minimum being washed in potable water, before being consumed.

Control of Nematodes

Nematodes are well known to cause serious losses of roots and tubers, of both quantity and quality. There are three possible methods of control:

- Application of chemicals to soil and plants. This treatment has not proved to be economical;
- Treatment of propagative material prior to planting by immersion in nematicides or hot water (50°C for 15-60 minutes is reported to give good results;
- Cultural practices. The most common and reliable method. It involves growing alternative crops for several years which are not suitable hosts for the specific nematode pest. It is important

that the alternative crops really are unsuitable hosts for the nematode and that there are no ground keepers and weeds left growing that will permit a colony to survive.

Of course, the ideal solution to the problem of nematode infestation is to grow nematode-free seed material in soils from which nematodes have been eliminated.

Potato Rosary Nematode

The sugar beet arose in Northern Europe as a product of breeding research. It was derived from the domesticated beet (*Beta vulgaris* L.), which previously had been developed in the second and third centuries A.D. by selection from wild beets native to the Mediterranean sea coasts and adjacent Northern Europe coastal areas. The relatively recently developed sugar beet crop plant has become well-known for acquiring new pathogens via native plants in any areas into which it is introduced. This phenomenon began in the mid 1800s and included the most important disease problems encountered early in European sugar beet production such as the cyst nematode and Rhizoctonia root rot.

The same relationship was noted after sugar beet cultivation was introduced into the Western Hemisphere. A new nematode disease was first discovered from Nebraska in the late 1940s that had never been documented prior to the arrival of the sugar beet in North America. In fact it has yet to be identified from anywhere else in the world where sugar beets are produced.

False Root-Knot Nematode

The false root-knot nematode (*Nacobbus aberrans*) has been labelled with several different names in the literature, including the Nebraska root-galling nematode, Cobb's root-galling nematode, and the potato rosary nematode. Nematodes in the genus *Nacobbus* produce galls on the roots of their hosts that are similar in appearance to those caused by root-knot nematodes (*Meloidogyne* spp), providing the origin of this pathogen's common name. Basing diagnosis on symptoms alone, the disease could easily be mistakenly attributed to *Meloidogyne* and therefore members of this species are familiarly known as false root-knot nematodes. The pathogen was named in honour of Nathan Augustus Cobb = *Nacobbus*, the pioneering USDA pathologist who is also considered the Father of American Nematology.

New Nematode Pathogen in North America

Nathan Cobb was the first worker to observe specimens of what is now considered to be *Nacobbus* from Colorado in 1918, although he mistakenly documented them as *Heterodera schactii* (sugar beet cyst nematode). It is now known that the first species of *Nacobbus* was discovered and formally described by the prominent USDA nematologist Gerald Thorne. He found this nematode in 1935 infecting shadscale (*Atriplex confertifolia*), a native desert shrub on non-agricultural land in the foot hills of the Great Basin near Utah Lake, Utah, and named it *Anguillulina aberrans.*

The genus *Nacobbus* was eventually created in 1944, based on a new species found in southern California (*N. dorsalis*) and designated as the type species. Thorne's *Anguillulina aberrans* was eventually moved into the genus *Nacobbus*, thereby becoming *N. aberrans*, and was distinguished from *N. dorsalis* by several key morphological features. Although *N. dorsalis* is the type species of *Nacobbus*, it is rare and considered of minor importance today. There have been no confirmed reports from any economic plant species, and has been found infecting only the redstem filaree (*Erodium cicutarium*) in California.

New Sugar Beet Disease in North America

***Figure:** Young sugar beet roots infected by* N. aberrans. *(Photo: R. M. Harveson)*

The false root-knot nematode was first isolated from sugar beets in 1949, near Mitchell, Nebraska by University of Nebraska plant pathologist Max Schuster, although the disease symptoms had been recognized and mistaken for infection by the root-knot nematode for many years prior to that. Infection can occur on young plants, often resulting in early-season loss of stand, as well as stunted and severely reduced foliar and root growth throughout the season. In mid to late season, severely infected plants wilt and exhibit pronounced yellowing during hot periods of the day. The most conspicuous symptoms of *N. aberrans* on sugar beets are galls or swellings with a proliferation of lateral branches formed on mature taproots. These lateral roots may be studded with chains of irregular-shaped galls, giving them a beaded appearance, with numerous small rootlets emerging from galls.

Figure: *Wilting symptoms of a sugar beet plant (right) infected by* N. aberrans. *(Photo: R. M. Harveson)*

Figure: *Yellowing of sugar beet plants in a field infested with* N. aberrans *near harvest. Note also the large gaps in the foreground due to previous plant mortality. (Photo: R. M. Harveson)*

***Figure:** Mature sugar beet roots at harvest exhibiting the proliferation of lateral roots off the taproot, characteristic of the false root-knot nematode (*N. aberrans*). (Photo: R. M. Harveson)*

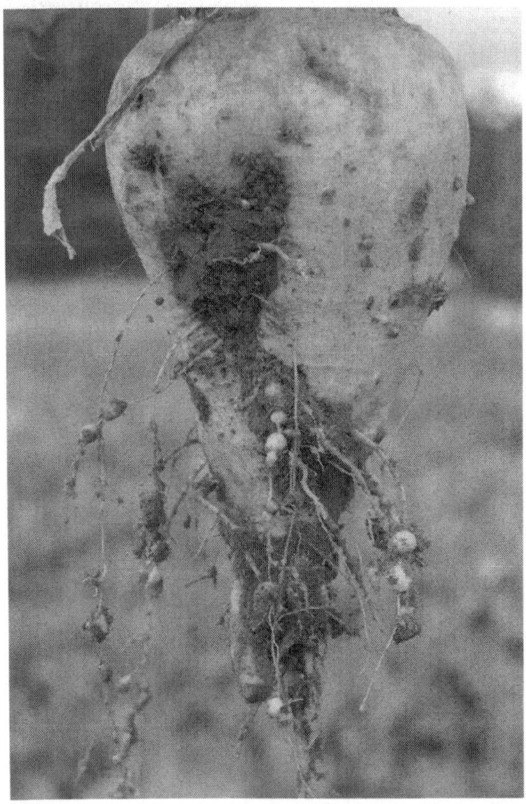

***Figure:** Mature sugar beet root exhibiting galls on lateral root branches, characteristic of the false root-knot nematode (*N. aberrans*).*

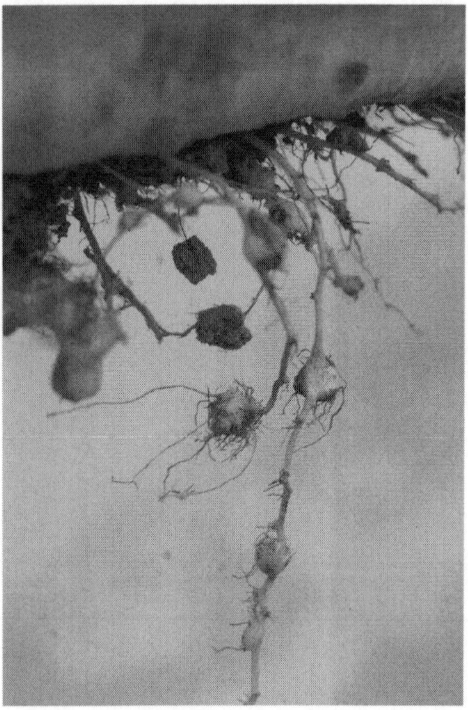

Figure: *Close-up image showing small rootlets emerging from galls of* N. aberrans-*infected sugar beet taproot.*

This disease was first described, characterized, and published from western Nebraska in 1956 by Schuster and Thorne. After their comprehensive investigations involving the pathogen's distribution, host range, life cycle, and pathology were completed, it was concluded that the sugar beet pathogen from Nebraska was distinct from the two species previously described as *N. dorsalis* and *N. aberrans*, and newly named *N. batatiformis,* based on the shape of the mature female. The female's body at maturity was tapered at both ends, resembling a sweet potato (*Ipomoea batatas*). This morphological feature further distinguished it from other gall-producing nematodes like *Meloidogyne* spp. that are characterized by pear-shaped bodies with a rounded posterior end. However, *N. batatiformis* was later merged with *N. aberrans*, retaining the latter name.

Distribution

A regional survey of sugar beet production in the late 1950s determined that although disease incidence was not as widespread as that of the sugar beet cyst nematode, the false root-knot pathogen was still identified from 16 counties in 6 states throughout the Great Plains, including Colorado, Kansas, Montana, Nebraska, South Dakota, and

Wyoming. Infestation of sugar beet fields by this nematode today is still limited to the central high plains east of the Rocky Mountains. Another earlier survey conducted by Schuster in western Nebraska exclusively over two years found the nematode infesting one third (32%) of 125 sugar beet fields monitored.

It is still considered a potentially serious pest in this region due to its widespread presence and because of the potential for serious yield reductions when infection does occur. Presently, Nebraska and Wyoming are the states most commonly and severely affected by this pest, with losses in sugar beets crops ranging from 10 to 20% being documented.

Pathogen Host Range

In the United States, *N. aberrans* is primarily an economic problem on sugar beet, but the pathogen can also affect numerous other vegetable hosts such as carrot, pea, lettuce, tomato, and numerous species in the mustard (broccoli, cabbage, rutabaga, radish, and turnip) and cucurbit (pumpkin, cucumber) families.

Fortunately, with the exception sugar beet, field crops grown in the central high plains of the United States under irrigation, including potatoes, corn, dry beans, sunflower, alfalfa, wheat, and barley, are not highly susceptible. Because of the broad host range on horticultural crops, the disease can still be problematic for city residents with backyard gardens, particularly with root vegetables like beets, turnips, radishes, and carrots.

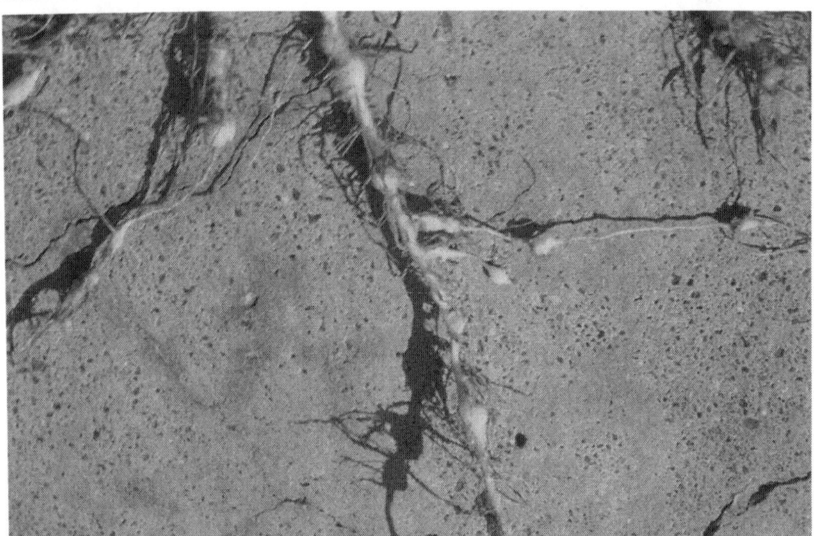

Figure: *Gall formation on* N. aberrans-*infected lambsquarters (*Chenopodium album*) root.*

Interestingly, the most commonly identified noneconomic hosts for the nematode in western Nebraska were determined to be three species of cacti, including two species of the prickly pear, *Opuntia* (*O. tortispina* and *O. fragilis*), and the spiny star cactus, *Escobaria vivapara.* All of these species are native to the central high plains region and other locations throughout western North America.

Nematode colonies isolated from both cacti and sugar beets each resulted in reciprocal infections with an identical colony reproduction and morphology. The false root-knot nematode has also been found commonly throughout the region infecting indigenous weed species, including various mustards (*Brassica* spp.), common lambsquarters (*Chenopodium album*), and Kochia (*Bassia scoparia),* as well as the introduced species common purslane (*Portulaca oleracea*), puncturevine (*Tribulus terrestris*), and Russian thistle (*Salsola tragis*).

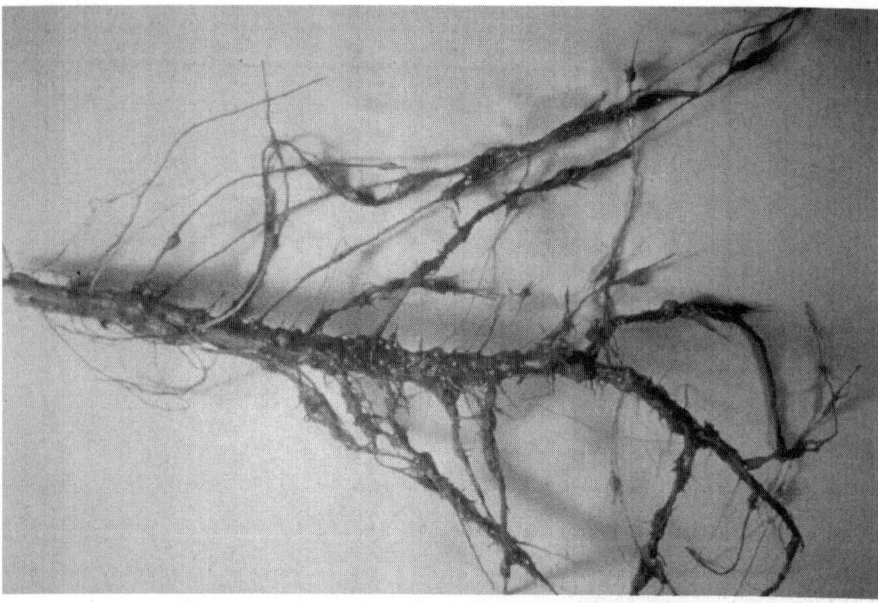

Figure: *Gall formation on* N. aberrans-*infected Kochia (*Bassia scoparia*) root.*

R. N. Inserra and colleagues determined from greenhouse studies that the Chenopodiacious weed, *Kochia* appeared to be more tolerant to the pathogen than sugar beet, suffering less damage in response to infection. Although these studies indicate that *Kochia* may be a poorer host for the nematode, they also provide evidence that this host can serve as a survival mechanism for *N. aberrans*, extending its longevity and maintenance in soils.

Taxonomic Confusion

The taxonomic history of *N. aberrans* has been controversial. Although most nematologists today recognize two primary species, *N. aberrans* and *N. dorsalis*, it is now becoming increasingly evident that a revision of the taxon is urgently needed as its species differences have not been fully resolved. This is further illustrated by reports in the literature over the years of three additional species from wide-ranging, contrasting locations, including the previously mentioned *N. batatiformis* from Nebraska, *N. serendipiticus* from greenhouse tomato production in Great Britain, and *N. serendipiticus bolivianus* from potatoes in Bolivia. In 1969, S. A. Sher transferred the latter three species into the heterogeneous *N. aberrans*, further complicating the taxonomy of this nematode.

However, at present, it is still generally accepted that the false root-knot nematode can be separated into at least three distinct groups or pathotypes, based on different host ranges and DNA characteristics. These pathotypes include a strain from the highland Andean regions of Argentina, Bolivia, Chile, and Peru that infects potatoes but not peppers. A bean strain from Mexico has also been found that damages beans, tomatoes, and peppers, but not sugar beet or potato. Lastly, the sugar beet strain previously discussed has been well characterized biologically and documented from the Great Plains of the United States. It infects sugar beets and many common vegetable crops, but not potato. Recent analyses of ITS sequences appear to confirm the validity of separating the pathogen into these specific subgroups, but no studies have compared populations from the Great Plains, Mexico, and Andes groups simultaneously.

Conclusions

The ability of the North American sugar beet pathotype of *N. aberrans* to parasitize nonindigenous host species makes it quite unique when compared to the two pathotypes from Mexico and South America that infect primarily the New World solanaceous (potato, peppers, and tomato) and leguminaceous (bean) crops.

It is clear that the genus *Nacobbus* (false root-knot nematode) is distinct from the various *Meloidogyne* species (root-knot nematodes) with which it has often been confused, and the species is also unquestionably endemic to the Americas. It has furthermore been demonstrated to occur in multiple, diverse areas of the Western Hemisphere, but is absent from any other locale in the world. The presence of the sugar beet strain of *N. aberrans* (initially named *N.*

batatiformis) occurring for more than a half century on native cacti from uncultivated rangeland in Scotts Bluff, Sioux, and Morrill counties of the western Nebraska Panhandle has provided convincing evidence that this nematode was indigenous to this region, but then adopted the sugar beet as its preferred host after the crop was introduced into North America.

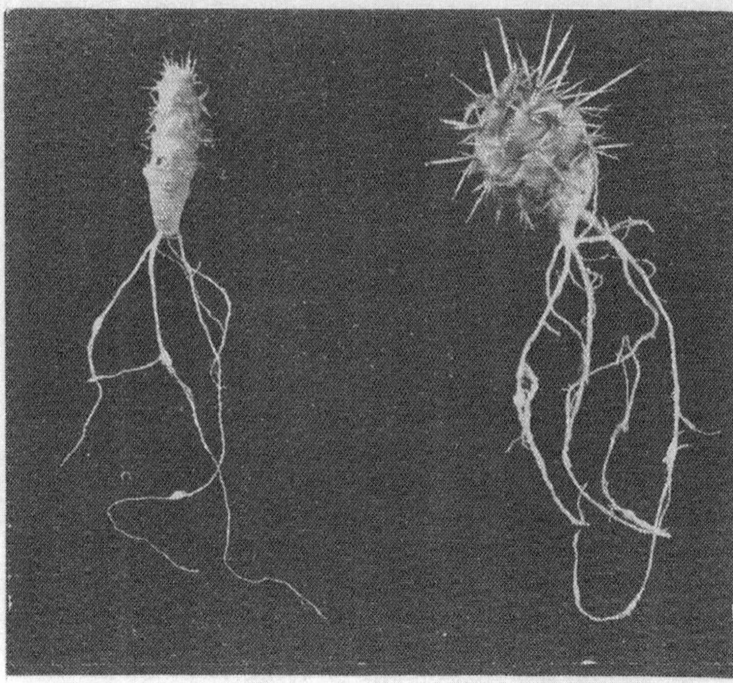

***Figure:** Gall formation on* N. aberrans-*infected prickly pear cactus (*Opuntia *spp.).*

5

Nematicides and Other Control Methods

A nematicide is a type of chemical pesticide used to kill plant-parasitic nematodes. Nematicides have tended to be broad-spectrum toxicants possessing high volatility or other properties promoting migration through the soil. Temik (Aldicarb), a carbamate insecticide marketed by Bayer CropScience, is an example of a commonly used commercial nematicide. It is important in potato production, where it has been used for control of soil-borne nematodes.

Aldicarb is a cholinesterase inhibitor, which prevents the breakdown of acetylcholine in the synapse. In case of severe poisoning, the victim dies of respiratory failure. It is no longer authorised for use in the EU and, in August, 2010, Bayer CropScience announced that it plans to discontinue Aldicarb by 2014. Human health safety and environmental concerns have resulted in the widespread deregistration of several other agronomically important nematicides. Prior to 1985, the persistent halocarbon DBCP was a widely used nematicide and soil fumigant. However, it was banned from use after being linked to sterility among male workers; the Dow Chemical company was subsequently found liable for more than $600 million in damages.

Natural Nematicides

Several natural nematicides are known. An environmentally benign garlic-derived polysulfide product is approved for use in the European

Union (under Annex 1 of 91/414) and the UK as a nematicide. Another common natural nematicide is obtained from neem cake, the residue obtained after cold-pressing the fruit and kernels of the neem tree. Known by several names in the world, the tree was first cultivated in India in ancient times and is now widely distributed throughout the world. The root exudate of marigold (*Tagetes*) is also found to have nematicidal action. Nematophagous fungi, a type of carnivorous fungi, can be useful in controlling nematodes, *Paecilomyces* being one example.

Besides chemicals, soil steaming can be used in order to kill nematodes. Superheated steam is induced into the soil, which causes almost all organic material to deteriorate.

Soil Steam Sterilization

Soil steam sterilization (soil steaming) is a farming technique that sterilizes soil with steam in open fields or greenhouses. Pests of plant cultures such as weeds, bacteria, fungi and viruses are killed through induced hot steam which causes their cell structure to physically degenerate. Biologically, the method is considered a partial disinfection. Important heat-resistant, spore-forming bacteria survive and revitalize the soil after cooling down. Soil fatigue can be cured through the release of nutritive substances blocked within the soil. Steaming leads to a better starting position, quicker growth and strengthened resistance against plant disease and pests. Today, the application of hot steam is considered the best and most effective way to disinfect sick soil, potting soil and compost. It is being used as an alternative to bromomethane, whose production and use was curtailed by the Montreal Protocol. "Steam effectively kills pathogens by heating the soil to levels that cause protein coagulation or enzyme inactivation."

Benefits of Soil Steaming

Soil sterilization provides secure and quick relief of soils from substances and organisms harmful to plants such as:

- Bacteria
- Viruses
- Fungi
- Nematodes and
- Other Pests

Further positive effects are:

- All weed and weed seeds are killed
- Significant increase of crop yields

- Relief from soil fatigue through activation of chemical – biological reactions
- Blocked nutritive substances in the soil are tapped and made available for plants
- Alternative to Methyl Bromide and other critical chemicals in agriculture

Steaming with Superheated Steam

Through modern steaming methods with superheated steam at 180–200 °C, an optimal soil disinfection can be achieved. Soil only absorbs a small amount of humidity. Micro organisms become active once the soil has cooled down. This creates an optimal environment for instant tillage with seedlings and seeds.

Additionally the method of integrated steaming can promote a target-oriented resettlement of steamed soil with beneficial organisms. In the process, the soil is first freed from all organisms and then revitalized and microbiologically buffered through the injection of a soil activator based on compost which contains a natural mixture of favourable microorganisms (e.g. Bacillus subtilis, etc.).

Different types of such steam application are also available in practice, including substrate steaming and surface steaming.

Surface Steaming

Several methods for surface steaming are in use amongst which are: area sheet steaming, the steaming hood, the steaming harrow, the steaming plough and vacuum steaming with drainage pipes or mobile pipe systems.

In order to pick the most suitable steaming method, certain factors have to be considered such as soil structure, plant culture and area performance. At present, more advanced methods are being developed, such as sandwich steaming or partially integrated sandwich steaming in order to minimize energy and cost as much as possible.

Sheet Steaming

Surface steaming with special sheets (sheet steaming) is a method which has been established for decades in order to steam large areas reaching from 15 to 400 m² in one step. If properly applied, sheet steaming is simple and highly economic. The usage of heat resistant, non-decomposing insulation fleece saves up to 50% energy, reduces the steaming time significantly and improves penetration. Single working step areas up to 400 m² can be steamed in 4–5 hours down to 25–30 cm depth / 90°C. The usage of heat resistant and non-decomposing synthetic insulation fleece, 5 mm thick, 500 gr / m^2, can reduce steaming time by about 30%. Through a steam injector or a perforated pipe, steam is injected underneath the sheet after it has been laid out and weighted with sand sacks.

The area performance in one working step depends on the capacity of the steam generator (e.g. steam boiler):

Steam capacity kg/h:	100	250	300	400	550	800	1000	1350	2000
Area m^2:	15-20	30-50	50-65	60-90	80-120	130-180	180-220	220-270	300-400

The steaming time depends on soil structure as well as outside temperature and amounts to 1-1.5 hours per 10 cm steaming depth. Hereby the soil reaches a temperature of about 85°C. Milling for soil loosening is not recommended since soil structure may become too fine which reduces its penetrability for steam. The usage of spading machines is ideal for soil loosening. The best results can be achieved if the soil is cloddy at greater depth and granulated at lesser depth. In practice, working with at least two sheets simultaneously has proven to be highly effective. While one sheet is used for steaming the other one is prepared for steam injection, therefore unnecessary steaming recesses are avoided.

Depth Steaming with Vacuum

Steaming with vacuum which is induced through a mobile or fixed installed pipe system in the depth of the area to be steamed, is the method that reaches the best penetration. Despite high capital cost, the fixed installation of drainage systems is reasonable for intensively used areas since steaming depths of up to 80 cm can be achieved.

In contrast to fixed installed drainage systems, pipes in mobile suction systems are on the surface. A central suction pipeline consisting of zinc-coated, fast-coupling pipes are connected in a regular spacing of 1.50 m and the ends of the hoses are pushed into the soil to the desired depth with a special tool.

The steaming area is covered with a special steaming sheet and weighted all around as with sheet steaming. The steam is injected underneath the sheet through an injector and protection tunnel. While with short areas up to 30 m length steam is frontally injected, with longer areas steam is induced in the middle of the beet using a T-connection branching out to both sides. As soon as the sheet is inflated to approximately 1m by the steam pressure, the suction turbine is switched on. First, the air in the soil is removed via the suction hoses. A vacuum is formed and the steam is pulled downward.

During the final phase, when the required steaming depth has been reached, the ventilator runs non-stop and surplus steam is blown out. To ensure that this surplus steam is not lost, it is fed back under the sheet. As with all other steaming systems, a post-steaming period of approximately 20–30 minutes is required. Steaming time is approximately 1 hour per 10 cm steaming depth. The steam requirement is approximately 7–8 kg/m^2.

The most important requirement, as with all steaming systems, is that the soil is well loosened before steaming, to ensure optimal penetration.

Negative Pressure Technique

"Negative Pressure technique generates appropriate soil temperature at a 60 cm depth and complete control of nematodes, fungi and weeds is achieved. In this technique, the steam is introduced under the steaming sheath and forced to enter the soil profile by a negative pressure. The negative pressure is created by a fan that sucks the air out of the soil through buried perforated polypropylene pipes. This system requires a permanent installation of perforated pipes into the soil, at a depth of at least 60 cm to be protected from plough."

Steaming With Hoods

A steaming hood is a mobile device consisting of corrosion-resistant materials such as aluminum, which is put down onto the area to be steamed. In contrast to sheet steaming, cost-intensive working steps such as laying out and weighting the sheets don't occur, however the area steamed per working step is smaller in accordance to the size of the hood.

Outdoors, a hood is positioned either manually or via tractor with a special pre-stressed 4 point suspension arm. Steaming time amounts to 30 min for a penetration down to 25 cm depth. Hereby a temperature of 90°C can be reached. In large stable glasshouses, the hoods are attached to tracks. They are lifted and moved by pneumatic cylinders. Small and medium-sized hoods up to 12m^2 are lifted manually using a tipping lever or moved electrically with special winches.

Combined Surface and Depth Injection of Steam (Sandwich Steaming)

Sandwich steaming, which was developed in a project among DEIAFA, University of Turin (Italy, www.deiafa.unito.it) and Ferrari Costruzioni Meccaniche, represents a combination of depth and surface steaming, offers an efficient method to induce hot steam into the soil.

The steam is simultaneously pushed into the soil from the surface and from the depth. For this purpose, the area, which must be equipped with a deep steaming injection system, is covered with a steaming hood. The steam enters the soil from the top and the bottom at the same time. Sheets are not suitable, since a high pressure up to 30 mm water column arises underneath the cover.

Sandwich steaming offers several advantages. On the one hand, application of energy can be increased to up to 120 kg steam per m^2/h. In comparison to other steaming methods up to 30% energy savings can be achieved and the usage of fuel (e.g. heating oil) accordingly decreases. The increased application of energy leads to a quick heating of the soil which reduces the loss of heat. On the other hand, only half of the regular steaming time is needed.

Comparison of sandwich steaming with other steam injection methods relating to steam output and energy demand(*):

Steaming method	*max. steam output*	*energy demand (*)*
Sheet steaming	6 kg/m^2h	about 100 kg steam/m^3
Depth steaming (Sheet + vacuum)	14 kg/m^2h	about 120 kg steam/m^3
Hood steaming (Alu)	30 kg/m^2h	about 80 kg steam/m^3
Hood steaming (Steel)	50 kg/m^2h	about 75 kg steam/m^3
Sandwich steaming	120 kg/m^2h	about 60 kg steam/m^3

(*) in soil max 30% moisture

Clearly, Sandwich steaming reaches the highest steam output at the lowest energy demand.

Partially Integrated Sandwich Steaming

The partial integrated Sandwich steaming is an advanced combined method for steaming merely the areas which shall be planted and purposely leaving out those areas which shall not be used. In order to avoid risk of re-infection of steamed areas with pest from unsteamed areas, beneficial organisms can directly be injected into the hygenized soil via a soil activator (e.g. special compost). The partial sandwich steaming unlocks further potential savings in the steaming process.

Container/Stack Steaming

Stack steaming is used when thermically treating compost and substrates such as turf. Depending on the amount, the material to be steamed is piled up to 70 cm height in steaming boxes or in small dump trailers. Steam is evenly injected via manifolds. For huge amounts, steaming containers and soil boxes are used which are equipped with suction systems to improve steaming results. Midget amounts can be steamed in special small steaming devices.

The amount of soil steamed should be tuned in a way that steaming time amounts to at most 1.5 h in order to avoid large quantities of condensed water in the bottom layers of the soil.

Steam Output kg/h:	100	250	300	400	550	800	1000	1350	2000
m^3/h about:	1.0-1.5	2.5-3.0	3.0-3.5	4.0-5.0	5.5-7.0	8.0-10.0	10.0-13.0	14.0-18.0	20.0-25.0

In light substrates, such as turf, the performance per hour is significantly higher.

History

Ancient civilizations in India and Egypt used steam, generated through the targeted usage of incident solar radiation on watered top soil, to sanitize and revive their arable land.

Modern soil steam sterilization was first discovered in 1888 (by Frank in Germany) and was first commercially used in the United States (by Rudd) in 1893 (Baker 1962). Since then, a wide variety of steam machines have been built to disinfest both commercial greenhouse and nursery field soils (Grossman and Liebman 1995).

In the 1950s, for example, steam sterilization technologies expanded from disinfestation of potting soil and greenhouse mixes to commercial production of steam rakes and tractor-drawn steam blades for fumigating small acres of cut flowers and other high-value field crops (Langedijk 1959). Today, even more effective steam technologies are being developed.

Application of hot steam

- In horticulture as well as nurseries for sterilization of substrates and top soil
- In agriculture for sterilization and treatment of food waste for pig fattening and heating of molasses
- In mushroom cultivation for pasteurization of growing rooms, sterilization of top soil and combined application as heating
- In wineries as combination boiler for sterilization and cleaning of storage tanks, tempering of mash and for warm water generation.
- In cities and communities for partial or area wide suppression of invasive Neophytes

R reniformis

Rotylenchulus reniformis, the reniform nematode, is a species of parasitic nematode of plants with a worldwide distribution in the tropical and subtropical regions.This nematode has a wide host range, infecting many species of plants around the world. It was first observed on the roots of cowpea in Hawaii, and was described as new species and new genus in 1940. Its specific epithet, *reniformis*, was inspired by the kidney shape of the adult female. There are now ten species classified in the genus, but *R. reniformis* is the only species of major economic importance to agriculture.

Distribution and Host Range

R. reniformis has been reported from thousands of localities in the Americas, Africa, Europe, Asia, and Australia. It has a wide host range that includes fruit trees, lentil, cotton, pigeon pea, tea, tobacco, soybean, pineapple, banana, okra, coconut, cabbage, sweet potato, alfalfa, corn, asparagus, palm, cucumber, tomato, squash, cassava, radish, eggplant, guava, melon, and ginger. Certain plants are considered to be non-hosts of the nematode, such as little barley, common barnyard grass, pangola grass, peppers, and some cultivars of black mustard, oat, spinach, and sugarcane.

Morphology

The reniform nematode has esophageal glands overlapping the intestine and a short stylet. The dorsal esophageal gland orifice is located posterior to the stylet knobs. The immature female is slender and may be spiral- or C-shaped in death. It is about 0.3 to 0.5 millimetres long. The mature female has a swollen, kidney-shaped body with a short tail, a short, thin stylet with rounded stylet knobs, a three-part esophagus, a long and narrow isthmus, and a well-developed metacarpus. The vulva is just behind the middle of the body. The male is vermiform: with a wormlike appearance. It has a weak stylet, curved spicules, and a pointed tail. The esophagus is reduced.

Life cycle

R. reniformis is sedentary semi-endoparasite on the roots of plants. The female penetrates the root and remains in one position at a permanent feeding site with its posterior end projecting from the root. The immature female is the infective agent, attacking the root and growing to maturity at its feeding site. Males and juveniles live in the soil; males are not parasites and do not feed. Under drought conditions the nematode can persist up to two years outside a host by entering an anhydrobiotic state.

The life cycle is 17 to 29 days long. The juvenile molts once while still inside the egg. The eggs hatch in 8 to 10 days. The juvenile molts three times to reach the immature stage. The immature female parasitizes the root for one to two weeks. During this time the male deposits sperm, which the female stores until her gonads mature. The nematode can also reproduce via parthenogenesis, without fertilization. Upon maturity the female exits the root and lays up to 200 eggs in a gelatinous matrix.

Host-parasite Relationship

When the immature female penetrates the root, a feeding tube forms from stylet secretions. Nematode infestation can cause symptoms in the plant that resemble those of moisture and nutrient deficiencies. It causes hypertrophy in the pericycle cells of seedling roots and in the periderm cells of the roots of older plants. Root growth slows and secondary root development is limited. Root necrosis has been observed in pineapple and banana. Shoot growth suppression and reduction of fruit quality has been observed in crops such as pineapple. Infested plants can become stunted and chlorotic. Wilt disease can follow when opportunistic fungi such as *Fusarium* and *Verticillium* infect the plants, a process observed in cotton. Secondary fungal infection can also cause root decay. Damage from *R. reniformis* is directly related to the number of nematodes present when the crop is planted. Variation among nematode populations, hosts, environmental conditions, and soil types may alter the threshold or economic injury level across the geographic distribution of the nematode.

Management

Some plants are resistant to this nematode. In susceptible taxa, chemical control is one of the most common management practices. Nematicides before and after planting can be effective. Intercropping and crop rotation, especially with resistant or non-host plants, is used as cultural control to improve soil, increase antagonistic microorganisms, and reduce nematode populations. French marigold (*Tagetes patula*) and sunn hemp (*Crotalaria juncea*) are recommended for these uses. The fungus *Purpureocillium lilacinum* has been used as an agent of biological pest control against the nematode in chickpea.

RNA Interference and Nematode Control

RNA interference (RNAi) is a biological process in which RNA molecules inhibit gene expression, typically by causing the destruction of specific mRNA molecules. Historically, it was known by other names, including *co-suppression, post transcriptional gene silencing* (PTGS), and *quelling.* Only after these apparently unrelated processes were fully understood did it become clear that they all described the RNAi phenomenon. Andrew Fire and Craig C. Mello shared the 2006 Nobel Prize in Physiology or Medicine for their work on RNA interference in the nematode worm *Caenorhabditis elegans*, which they published in 1998.

Two types of small ribonucleic acid (RNA) molecules – microRNA (miRNA) and small interfering RNA (siRNA) – are central to RNA interference. RNAs are the direct products of genes, and these small RNAs can bind to other specific messenger RNA (mRNA) molecules and either increase or decrease their activity, for example by preventing an mRNA from producing a protein. RNA interference has an important role in defending cells against parasitic nucleotide sequences – viruses and transposons. It also influences development.

The RNAi pathway is found in many eukaryotes including animals and is initiated by the enzyme Dicer, which cleaves long double-stranded RNA (dsRNA) molecules into short double stranded fragments of ~20 nucleotide siRNAs. Each siRNA is unwound into two single-stranded (ss) ssRNAs, respectively the passenger strand and the guide strand. The passenger strand is degraded and the guide strand is incorporated into the RNA-induced silencing complex (RISC). The most well-studied outcome is post-transcriptional gene silencing, which occurs when the guide strand pairs with a complementary sequence in a messenger RNA molecule and induces cleavage by Argonaute, the catalytic component of the RISC complex. In some organisms, this process spreads systemically, despite the initially limited molar concentrations of siRNA.

RNAi is a valuable research tool, both in cell culture and in living organisms, because synthetic dsRNA introduced into cells can selectively and robustly induce suppression of specific genes of interest. RNAi may be used for large-scale screens that systematically shut down each gene in the cell, which can help to identify the components necessary for a particular cellular process or an event such as cell division. The pathway is also used as a practical tool in biotechnology, medicine and insecticides.

Cellular Mechanism

RNAi is an RNA-dependent gene silencing process that is controlled by the RNA-induced silencing complex (RISC) and is initiated by short double-stranded RNA molecules in a cell's cytoplasm, where they interact with the catalytic RISC component argonaute. When the dsRNA is exogenous (coming from infection by a virus with an RNA genome or laboratory manipulations), the RNA is imported directly into the cytoplasm and cleaved to short fragments by Dicer. The initiating dsRNA can also be endogenous (originating in the cell), as in pre-microRNAs expressed from RNA-coding genes in the genome.

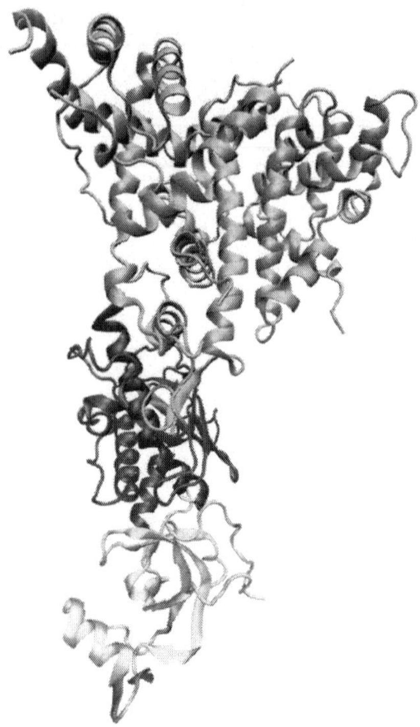

Figure: *The dicer protein from* Giardia intestinalis, *which catalyzes the cleavage of dsRNA to siRNAs. The RNase domains are coloured green, the PAZ domain yellow, the platform domain red, and the connector helix blue.*

The primary transcripts from such genes are first processed to form the characteristic stem-loop structure of pre-miRNA in the nucleus, then exported to the cytoplasm. Thus, the two dsRNA pathways, exogenous and endogenous, converge at the RISC complex.

dsRNA Cleavage

Endogenous dsRNA initiates RNAi by activating the ribonuclease protein Dicer, which binds and cleaves double-stranded RNAs (dsRNAs) to produce double-stranded fragments of 20–25 base pairs with a 2-nucleotide overhang at the 3' end. Bioinformatics studies on the genomes of multiple organisms suggest this length maximizes target-gene specificity and minimizes non-specific effects. These short double-stranded fragments are called small interfering RNAs (siRNAs). These siRNAs are then separated into single strands and integrated into an active RISC complex. After integration into the RISC, siRNAs base-pair to their target mRNA and cleave it, thereby preventing it from being used as a translation template.

Exogenous dsRNA is detected and bound by an effector protein, known as RDE-4 in *C. elegans* and R2D2 in *Drosophila*, that stimulates dicer activity. This protein only binds long dsRNAs, but the mechanism producing this length specificity is unknown. This RNA-binding protein then facilitates the transfer of cleaved siRNAs to the RISC complex.

In *C. elegans* this initiation response is amplified through the synthesis of a population of 'secondary' siRNAs during which the dicer-produced initiating or 'primary' siRNAs are used as templates. These 'secondary' siRNAs are structurally distinct from dicer-produced siRNAs and appear to be produced by an RNA-dependent RNA polymerase (RdRP).

MicroRNA

MicroRNAs (miRNAs) are genomically encoded non-coding RNAs that help regulate gene expression, particularly during development. The phenomenon of RNA interference, broadly defined, includes the endogenously induced gene silencing effects of miRNAs as well as silencing triggered by foreign dsRNA. Mature miRNAs are structurally similar to siRNAs produced from exogenous dsRNA, but before reaching maturity, miRNAs must first undergo extensive post-transcriptional modification. An miRNA is expressed from a much longer RNA-coding gene as a primary transcript known as a *pri-miRNA* which is processed, in the cell nucleus, to a 70-nucleotide stem-loop structure called a *pre-miRNA* by the microprocessor complex.

This complex consists of an RNase III enzyme called Drosha and a dsRNA-binding protein DGCR8. The dsRNA portion of this pre-miRNA is bound and cleaved by Dicer to produce the mature miRNA molecule that can be integrated into the RISC complex; thus, miRNA and siRNA share the same downstream cellular machinery downstream.

siRNAs derived from long dsRNA precursors differ from miRNAs in that miRNAs, especially those in animals, typically have incomplete base pairing to a target and inhibit the translation of many different mRNAs with similar sequences.

In contrast, siRNAs typically base-pair perfectly and induce mRNA cleavage only in a single, specific target. In *Drosophila* and *C. elegans*, miRNA and siRNA are processed by distinct argonaute proteins and dicer enzymes.

RISC Activation and Catalysis

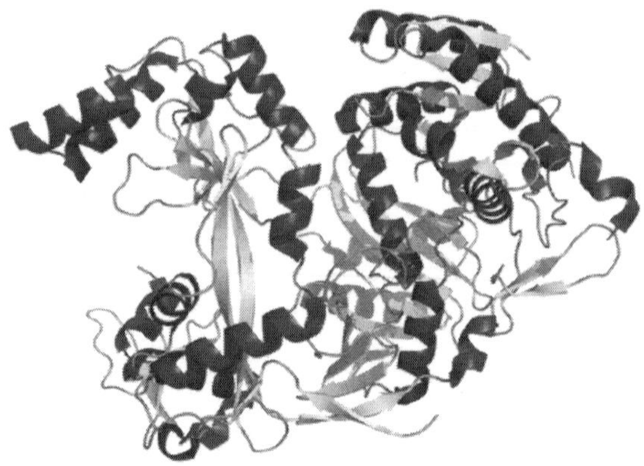

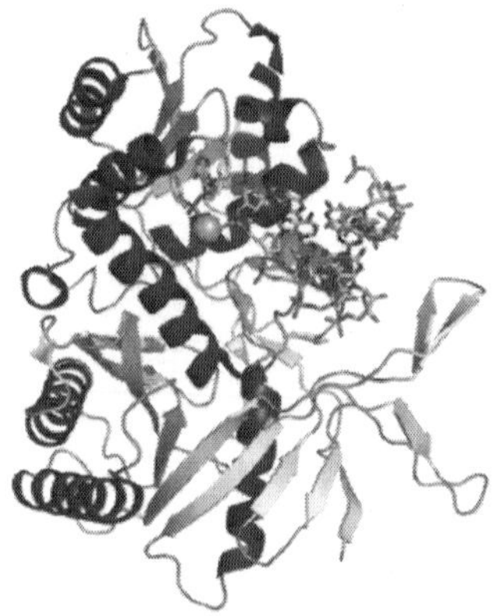

Figure: *Left: A full-length argonaute protein from the archaea species* Pyrococcus furiosus. Right: *The PIWI domain of an argonaute protein in complex with double-stranded RNA.*

The active components of an RNA-induced silencing complex (RISC) are endonucleases called argonaute proteins, which cleave the target mRNA strand complementary to their bound siRNA. As the fragments produced by dicer are double-stranded, they could each in theory produce a functional siRNA. However, only one of the two strands, which is known as the *guide strand*, binds the argonaute protein and directs gene silencing. The other *anti-guide strand* or *passenger strand* is degraded during RISC activation.

Although it was first believed that an ATP-dependent helicase separated these two strands, the process proved to be ATP-independent and performed directly by the protein components of RISC. However, an *in vitro* kinetic analysis of RNAi in the presence and absence of ATP showed that ATP may be required to unwind and remove the cleaved mRNA strand from the RISC complex after catalysis. The guide strand tends to be the one whose 5' end is less stably paired to its complement, but strand selection is unaffected by the direction in which dicer cleaves the dsRNA before RISC incorporation. Instead, the R2D2 protein may serve as the differentiating factor by binding the more-stable 5' end of the passenger strand.

The structural basis for binding of RNA to the argonaute protein was examined by X-ray crystallography of the binding domain of an RNA-bound argonaute protein. Here, the phosphorylated 5' end of the RNA strand enters a conserved basic surface pocket and makes contacts through a divalent cation (an atom with two positive charges) such as magnesium and by aromatic stacking (a process that allows more than one atom to share an electron by passing it back and forth) between the 5' nucleotide in the siRNA and a conserved tyrosine residue. This site is thought to form a nucleation site for the binding of the siRNA to its mRNA target. Analysis of the inhibitory effect of mismatches in either the 5' or 3' end of the guide strand has demonstrated that the 5' end of the guide strand is likely responsible for matching and binding the target mRNA, while the 3' end is responsible for physically arranging target mRNA into a cleavage-favourable RISC region.

It is not understood how the activated RISC complex locates complementary mRNAs within the cell. Although the cleavage process has been proposed to be linked to translation, translation of the mRNA target is not essential for RNAi-mediated degradation. Indeed, RNAi may be more effective against mRNA targets that are not translated. Argonaute proteins are localized to specific regions in the cytoplasm called P-bodies (also cytoplasmic bodies or GW bodies), which are regions with high rates of mRNA decay; miRNA activity is also clustered in P-bodies. Disruption of P-bodies decreases the efficiency of RNA interference, suggesting that they are a critical site in the RNAi process.

Transcriptional Silencing

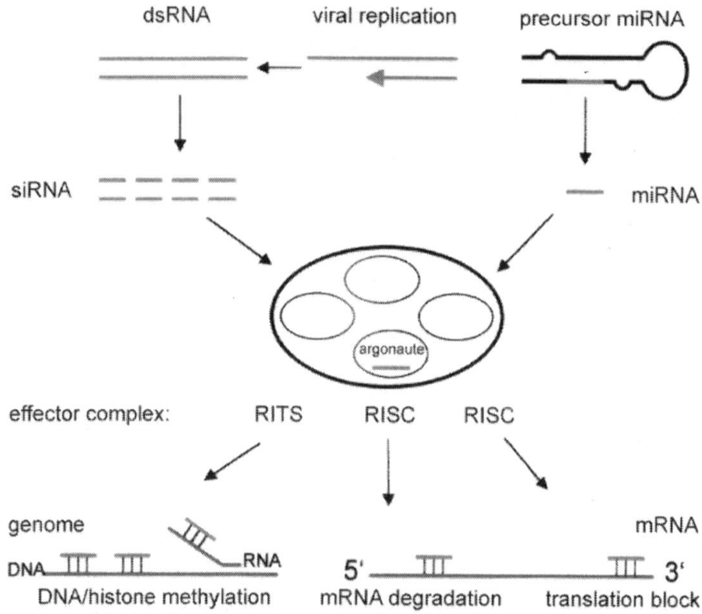

***Figure:** The enzyme dicer trims double stranded RNA, to form small interfering RNA or microRNA. These processed RNAs are incorporated into the RNA-induced silencing complex (RISC), which targets messenger RNA to prevent translation.*

Components of the RNAi pathway are used in many eukaryotes in the maintenance of the organization and structure of their genomes. Modification of histones and associated induction of heterochromatin formation serves to downregulate genes pre-transcriptionally; this process is referred to as RNA-induced transcriptional silencing (RITS), and is carried out by a complex of proteins called the RITS complex. In fission yeast this complex contains argonaute, a chromodomain protein Chp1, and a protein called Tas3 of unknown function. As a consequence, the induction and spread of heterochromatic regions requires the argonaute and RdRP proteins. Indeed, deletion of these genes in the fission yeast *S. pombe* disrupts histone methylation and centromere formation, causing slow or stalled anaphase during cell division. In some cases, similar processes associated with histone modification have been observed to transcriptionally upregulate genes.

The mechanism by which the RITS complex induces heterochromatin formation and organization is not well understood. Most studies have focused on the mating-type region in fission yeast, which may not be representative of activities in other genomic regions/organisms. In maintenance of existing heterochromatin regions, RITS forms a complex with siRNAs complementary to the local genes and stably binds local methylated histones, acting co-transcriptionally to degrade any nascent pre-mRNA transcripts that are initiated by RNA polymerase.

The formation of such a heterochromatin region, though not its maintenance, is dicer-dependent, presumably because dicer is required to generate the initial complement of siRNAs that target subsequent transcripts. Heterochromatin maintenance has been suggested to function as a self-reinforcing feedback loop, as new siRNAs are formed from the occasional nascent transcripts by RdRP for incorporation into local RITS complexes. The relevance of observations from fission yeast mating-type regions and centromeres to mammals is not clear, as heterochromatin maintenance in mammalian cells may be independent of the components of the RNAi pathway.

Crosstalk with RNA Editing

The type of RNA editing that is most prevalent in higher eukaryotes converts adenosine nucleotides into inosine in dsRNAs via the enzyme adenosine deaminase (ADAR). It was originally proposed in 2000 that the RNAi and A'!I RNA editing pathways might compete for a common dsRNA substrate. Some pre-miRNAs do undergo A'!I RNA editing and this mechanism may regulate the processing and expression of mature miRNAs. Furthermore, at least one mammalian ADAR can sequester siRNAs from RNAi pathway components. Further support for this model comes from studies on ADAR-null *C. elegans* strains indicating that A'!I RNA editing may counteract RNAi silencing of endogenous genes and transgenes.

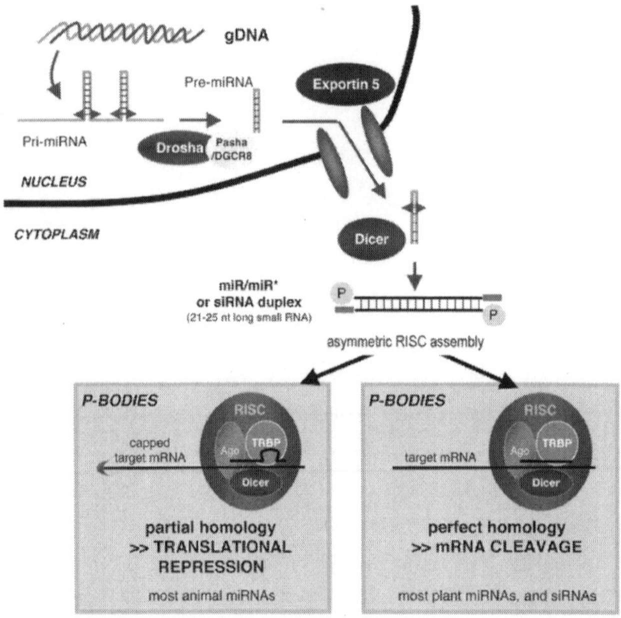

Figure: *Illustration of the major differences between plant and animal gene silencing. Natively expressed microRNA or exogenous small interfering RNA is processed by dicer and integrated into the RISC complex, which mediates gene silencing.*

Variation Among Organisms

Organisms vary in their ability to take up foreign dsRNA and use it in the RNAi pathway. The effects of RNA interference can be both systemic and heritable in plants and *C. elegans*, although not in *Drosophila* or mammals. In plants, RNAi is thought to propagate by the transfer of siRNAs between cells through plasmodesmata (channels in the cell walls that enable communication and transport).

Heritability comes from methylation of promoters targeted by RNAi; the new methylation pattern is copied in each new generation of the cell. A broad general distinction between plants and animals lies in the targeting of endogenously produced miRNAs; in plants, miRNAs are usually perfectly or nearly perfectly complementary to their target genes and induce direct mRNA cleavage by RISC, while animals' miRNAs tend to be more divergent in sequence and induce translational repression. This translational effect may be produced by inhibiting the interactions of translation initiation factors with the messenger RNA's polyadenine tail. Some eukaryotic protozoa such as *Leishmania major* and *Trypanosoma cruzi* lack the RNAi pathway entirely. Most or all of the components are also missing in some fungi, most notably the model organism *Saccharomyces cerevisiae.*

The presence of RNAi in other budding yeast species such as *Saccharomyces castellii* and *Candida albicans*, further demonstrates that inducing two RNAi-related proteins from *S. castellii* facilitates RNAi in *S. cerevisiae.*

That certain ascomycetes and basidiomycetes are missing RNA interference pathways indicates that proteins required for RNA silencing have been lost independently from many fungal lineages, possibly due to the evolution of a novel pathway with similar function, or to the lack of selective advantage in certain niches.

Related Prokaryotic Systems

Gene expression in prokaryotes is influenced by an RNA-based system similar in some respects to RNAi. Here, RNA-encoding genes control mRNA abundance or translation by producing a complementary RNA that anneals to an mRNA. However these regulatory RNAs are not generally considered to be analogous to miRNAs because the dicer enzyme is not involved.

It has been suggested that CRISPR interference systems in prokaryotes are analogous to eukaryotic RNA interference systems, although none of the protein components are orthologous.

Biological Functions

Immunity

RNA interference is a vital part of the immune response to viruses and other foreign genetic material, especially in plants where it may also prevent the self-propagation of transposons. Plants such as *Arabidopsis thaliana* express multiple dicer homologs that are specialized to react differently when the plant is exposed to different viruses.

Even before the RNAi pathway was fully understood, it was known that induced gene silencing in plants could spread throughout the plant in a systemic effect and could be transferred from stock to scion plants via grafting.

This phenomenon has since been recognized as a feature of the plant adaptive immune system and allows the entire plant to respond to a virus after an initial localized encounter. In response, many plant viruses have evolved elaborate mechanisms to suppress the RNAi response. These include viral proteins that bind short double-stranded RNA fragments with single-stranded overhang ends, such as those produced by dicer. Some plant genomes also express endogenous siRNAs in response to infection by specific types of bacteria. These effects may be part of a generalized response to pathogens that downregulates any metabolic process in the host that aids the infection process.

Although animals generally express fewer variants of the dicer enzyme than plants, RNAi in some animals produces an antiviral response. In both juvenile and adult *Drosophila*, RNA interference is important in antiviral innate immunity and is active against pathogens such as Drosophila X virus. A similar role in immunity may operate in *C. elegans*, as argonaute proteins are upregulated in response to viruses and worms that overexpress components of the RNAi pathway are resistant to viral infection.

The role of RNA interference in mammalian innate immunity is poorly understood, and relatively little data is available. However, the existence of viruses that encode genes able to suppress the RNAi response in mammalian cells may be evidence in favour of an RNAi-dependent mammalian immune response, although this hypothesis has been challenged as poorly substantiated. Maillard et al. and Li et al. provide evidence for the existence of a functional antiviral RNAi pathway in mammalian cells. Other functions for RNAi in mammalian viruses also exist, such as miRNAs expressed by the herpes virus that may act as heterochromatin organization triggers to mediate viral latency.

Downregulation of Genes

Endogenously expressed miRNAs, including both intronic and intergenic miRNAs, are most important in translational repression and in the regulation of development, especially on the timing of morphogenesis and the maintenance of undifferentiated or incompletely differentiated cell types such as stem cells.

The role of endogenously expressed miRNA in downregulating gene expression was first described in *C. elegans* in 1993. In plants this function was discovered when the "JAW microRNA" of *Arabidopsis* was shown to be involved in the regulation of several genes that control plant shape. In plants, the majority of genes regulated by miRNAs are transcription factors; thus miRNA activity is particularly wide-ranging and regulates entire gene networks during development by modulating the expression of key regulatory genes, including transcription factors as well as F-box proteins. In many organisms, including humans, miRNAs are linked to the formation of tumors and dysregulation of the cell cycle. Here, miRNAs can function as both oncogenes and tumor suppressors.

Upregulation of Genes

RNA sequences (siRNA and miRNA) that are complementary to parts of a promoter can increase gene transcription, a phenomenon dubbed RNA activation. Part of the mechanism for how these RNA upregulate genes is known: dicer and argonaute are involved, possibly via histone demethylation. miRNAs have been proposed to upregulate their target genes upon cell cycle arrest, via unknown mechanisms.

Evolution

Based on parsimony-based phylogenetic analysis, the most recent common ancestor of all eukaryotes most likely already possessed an early RNA interference pathway; the absence of the pathway in certain eukaryotes is thought to be a derived characteristic. This ancestral RNAi system probably contained at least one dicer-like protein, one argonaute, one PIWI protein, and an RNA-dependent RNA polymerase that may also have played other cellular roles.

A large-scale comparative genomics study likewise indicates that the eukaryotic crown group already possessed these components, which may then have had closer functional associations with generalized RNA degradation systems such as the exosome. This study also suggests that the RNA-binding argonaute protein family, which is shared among eukaryotes, most archaea, and at least some bacteria (such as *Aquifex aeolicus*), is homologous to and originally evolved from components of the translation initiation system.

The ancestral function of the RNAi system is generally agreed to have been immune defence against exogenous genetic elements such as transposons and viral genomes. Related functions such as histone modification may have already been present in the ancestor of modern eukaryotes, although other functions such as regulation of development by miRNA are thought to have evolved later.

RNA interference genes, as components of the antiviral innate immune system in many eukaryotes, are involved in an evolutionary arms race with viral genes. Some viruses have evolved mechanisms for suppressing the RNAi response in their host cells, particularly for plant viruses. Studies of evolutionary rates in *Drosophila* have shown that genes in the RNAi pathway are subject to strong directional selection and are among the fastest-evolving genes in the *Drosophila* genome.

Applications

Gene Knockdown

The RNA interference pathway is often exploited in experimental biology to study the function of genes in cell culture and *in vivo* in model organisms.

Double-stranded RNA is synthesized with a sequence complementary to a gene of interest and introduced into a cell or organism, where it is recognized as exogenous genetic material and activates the RNAi pathway.

Using this mechanism, researchers can cause a drastic decrease in the expression of a targeted gene. Studying the effects of this decrease can show the physiological role of the gene product. Since RNAi may not totally abolish expression of the gene, this technique is sometimes referred as a "knockdown", to distinguish it from "knockout" procedures in which expression of a gene is entirely eliminated.

Extensive efforts in computational biology have been directed towards the design of successful dsRNA reagents that maximize gene knockdown but minimize "off-target" effects. Off-target effects arise when an introduced RNA has a base sequence that can pair with and thus reduce the expression of multiple genes.

Such problems occur more frequently when the dsRNA contains repetitive sequences. It has been estimated from studying the genomes of humans, *C. elegans* and *S. pombe* that about 10% of possible siRNAs have substantial off-target effects.

A multitude of software tools have been developed implementing algorithms for the design of general mammal-specific, and virus-specific siRNAs that are automatically checked for possible cross-reactivity.

Depending on the organism and experimental system, the exogenous RNA may be a long strand designed to be cleaved by dicer, or short RNAs designed to serve as siRNA substrates. In most mammalian cells, shorter RNAs are used because long double-stranded RNA molecules induce the mammalian interferon response, a form of innate immunity that reacts nonspecifically to foreign genetic material.

Mouse oocytes and cells from early mouse embryos lack this reaction to exogenous dsRNA and are therefore a common model system for studying mammalian gene-knockdown effects. Specialized laboratory techniques have also been developed to improve the utility of RNAi in mammalian systems by avoiding the direct introduction of siRNA, for example, by stable transfection with a plasmid encoding the appropriate sequence from which siRNAs can be transcribed, or by more elaborate lentiviral vector systems allowing the inducible activation or deactivation of transcription, known as *conditional RNAi*.

Functional Genomics

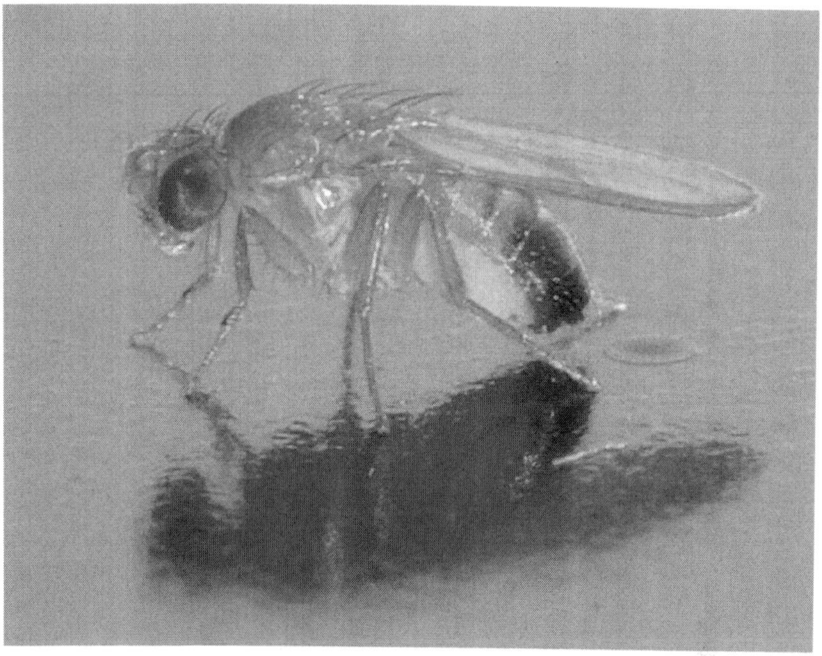

Figure: *A normal adult* Drosophila *fly, a common model organism used in RNAi experiments.*

Most functional genomics applications of RNAi in animals have used *C. elegans* and *Drosophila*, as these are the common model organisms in which RNAi is most effective. *C. elegans* is particularly useful for RNAi research for two reasons: firstly, the effects of gene silencing are generally heritable, and secondly because delivery of the dsRNA is extremely simple.

Through a mechanism whose details are poorly understood, bacteria such as *E. coli* that carry the desired dsRNA can be fed to the worms and will transfer their RNA payload to the worm via the intestinal tract.

This "delivery by feeding" is just as effective at inducing gene silencing as more costly and time-consuming delivery methods, such as soaking the worms in dsRNA solution and injecting dsRNA into the gonads. Although delivery is more difficult in most other organisms, efforts are also underway to undertake large-scale genomic screening applications in cell culture with mammalian cells.

Approaches to the design of genome-wide RNAi libraries can require more sophistication than the design of a single siRNA for a defined set of experimental conditions.

Artificial neural networks are frequently used to design siRNA libraries and to predict their likely efficiency at gene knockdown. Mass genomic screening is widely seen as a promising method for genome annotation and has triggered the development of high-throughput screening methods based on microarrays.

However, the utility of these screens and the ability of techniques developed on model organisms to generalize to even closely related species has been questioned, for example from *C. elegans* to related parasitic nematodes.

Functional genomics using RNAi is a particularly attractive technique for genomic mapping and annotation in plants because many plants are polyploid, which presents substantial challenges for more traditional genetic engineering methods.

For example, RNAi has been successfully used for functional genomics studies in bread wheat (which is hexaploid) as well as more common plant model systems *Arabidopsis* and maize.

Medicine

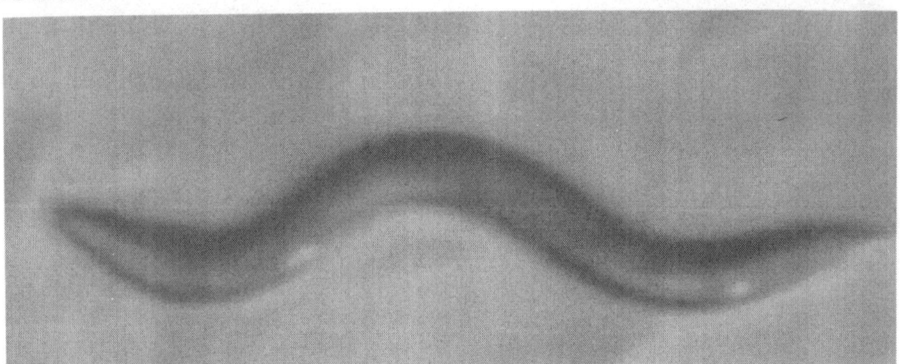

***Figure:** An adult* C. elegans *worm, grown under RNAi suppression of a nuclear hormone receptor involved in desaturase regulation. These worms have abnormal fatty acid metabolism but are viable and fertile.*

It may be possible to exploit RNA interference in therapy. Although it is difficult to introduce long dsRNA strands into mammalian cells due to the interferon response, the use of short interfering RNA has been more successful. Among the first applications to reach clinical trials were in the treatment of macular degeneration and respiratory syncytial virus. RNAi has also been shown to be effective in reversing induced liver failure in mouse models.

Antiviral

Potential antiviral therapies include topical microbicide treatments that use RNAi to treat infection (at Harvard Medical School; in mice, so far) by herpes simplex virus type 2 and the inhibition of viral gene expression in cancerous cells, knockdown of host receptors and coreceptors for HIV, the silencing of hepatitis A and hepatitis B genes, silencing of influenza gene expression, and inhibition of measles viral replication.

Potential treatments for neurodegenerative diseases have also been proposed, with particular attention to polyglutamine diseases such as Huntington's disease.

RNA interference-based applications are being developed to target persistent HIV-1 infection. Viruses like HIV-1 are particularly difficult targets for RNAi-attack because they are escape-prone, which requires combinatorial RNAi strategies to prevent viral escape.

Cancer

RNA interference is also a promising way to treat cancers by silencing genes differentially upregulated in tumor cells or genes involved in cell division.

A key area of research in the use of RNAi for clinical applications is the development of a safe delivery method, which to date has involved mainly viral vector systems similar to those suggested for gene therapy.

Due to safety concerns with viral vectors, nonviral delivery methods, typically employing lipid-based or polymeric vectors, are also promising candidates. Computational modelling of nonviral siRNA delivery paired with 'in vitro *and* in vivo *gene knockdown studies elucidated the temporal behaviour of RNAi in these systems.*

The model used an input bolus dose of siRNA and computationally and experimentally showed that knockdown duration was dependent mainly on the doubling time of the cells to which siRNA was delivered, while peak knockdown depended primarily on the delivered dose. Kinetic considerations of RNAi are imperative to safe and effective dosing schedules as nonviral methods of inducing RNAi continue to be developed.

Safety

Despite the proliferation of promising cell culture studies for RNAi-based drugs, some concern has been raised regarding the safety of RNA interference, especially the potential for "off-target" effects in which a gene with a coincidentally similar sequence to the targeted gene is also repressed.

A computational genomics study estimated that the error rate of off-target interactions is about 10%. One major study of liver disease in mice reported that 23 out of 49 distinct RNAi treatment protocols resulted in death.

Researchers hypothesized this alarmingly high rate to be the result of "oversaturation" of the dsRNA pathway, due to the use of shRNAs that have to be processed in the nucleus and exported to the cytoplasm using an active mechanism. Such considerations are under active investigation, to reduce their impact in the potential therapeutic applications.

RNAi in vivo delivery to tissues still eludes science—especially to tissues deep within the body. RNAi delivery is only easily accessible to surface tissues such as the eye and respiratory tract. In these instances, siRNA has been used in direct contact with the tissue for transport.

The resulting RNAi successfully focused on target genes. When delivering siRNA to deep tissues, the siRNA must be protected from nucleases, but targeting specific areas becomes the main difficulty. This difficulty has been combatted with high dosage levels of siRNA to ensure the tissues have been reached, however in these cases hepatotoxicity was reported.

Biotechnology

RNA interference has been used for applications in biotechnology and is nearing commercialization in others.

Foods

RNAi has been used to genetically engineer plants to produce lower levels of natural plant toxins. Such techniques take advantage of the stable and heritable RNAi phenotype in plant stocks. Cotton seeds are rich in dietary protein but naturally contain the toxic terpenoid product gossypol, making them unsuitable for human consumption. RNAi has been used to produce cotton stocks whose seeds contain reduced levels of delta-cadinene synthase, a key enzyme in gossypol production, without affecting the enzyme's production in other parts of the plant, where gossypol is itself important in preventing damage from plant pests. Similar efforts have been directed towards the reduction of the cyanogenic natural product linamarin in cassava plants.

No plant products that use RNAi-based genetic engineering have yet exited the experimental stage. Development efforts have successfully reduced the levels of allergens in tomato plants and fortification of plants such as tomatoes with dietary antioxidants.

Previous commercial products, including the Flavr Savr tomato and two cultivars of ringspot-resistant papaya, were originally developed using antisense technology but likely exploited the RNAi pathway.

Other Crops

Another effort decreased the precursors of likely carcinogens in tobacco plants. Other plant traits that have been engineered in the laboratory include the production of non-narcotic natural products by the opium poppy and resistance to common plant viruses.

Insecticide

RNAi is under development as an insecticide, employing multiple approaches, including genetic engineering and topical application. Cells in the midgut of many larvae take up the molecules and help spread the signal throughout the insect's body.

RNAi has varying effects in different species of Lepidoptera (butterflies and moths). Possibly because their saliva is better at breaking down RNA, the cotton bollworm, the beet armyworm and the Asiatic rice borer have so far not been proven susceptible to RNAi by feeding.

To develop resistance to RNAi, the western corn rootworm would have to change the genetic sequence of its Snf7 gene at multiple sites. Combining multiple strategies, such as engineering the protein Cry, derived from a bacterium called Bacillus thuringiensis (Bt), and RNAi in one plant delay the onset of resistance.

One unconfirmed 2012 paper detected small RNAs from food plants in the blood of mice and humans. The consequences of RNA insecticides in the human bloodstream have not been investigated.

Biological barriers—including saliva and blood enzymes and stomach acids may break down any ingested RNA. Critics charge that the human equivalent of the mouse diet in the study would be 33 kilograms of cooked rice a day. Two 2013 studies failed to detect RNAs in humans. Athletes consuming a diet of apples and bananas and monkeys consuming a fruit shake both appeared to be RNA-free.

Transgenic Plants

Transgenic crops have been made to express small bits of RNA, carefully chosen to silence crucial genes in target pests. RNAs exist that affect only insects that have specific genetic sequences. In 2009 a study showed RNAs that could kill any one of four fruit fly species while not harming the other three.

In 2012 Syngenta bought Belgian RNAi firm Devgen for $522 million and Monsanto paid $29.2 million for the exclusive rights to intellectual property from Alnylam Pharmaceuticals. The International Potato Centre in Lima, Peru is looking for genes to target in the sweet potato weevil, a beetle whose larvae ravage sweet potatos globally.

Other researchers are trying to silence genes in ants, caterpillars and pollen beetles. Monsanto will likely be first to market, with a transgenic corn seed that expresses dsRNA based on gene Snf7 from the western corn rootworm, a beetle whose larvae annually cause one billion dollars in damage in the United States alone. A 2012 paper showed that silencing Snf7 stunts larval growth, killing them within days. In 2013 the same team showed that the RNA affects very few other species.

Topical

Alternatively dsRNA can be supplied without genetic engineering. One approach is to add them to irrigation water. The molecules are absorbed into the plants' vascular system and poison insects feeding on them.

Another approach involves spraying RNA like a conventional pesticide. This would allow faster adaptation to resistance. Such approaches would require low cost sources of RNAs that do n ot currently exist.

Genome-scale Screening

Genome-scale RNAi research relies on high-throughput screening (HTS) technology. RNAi HTS technology allows genome-wide loss-of-function screening and is broadly used in the identification of genes associated with specific phenotypes. This technology has been hailed as the second genomics wave, following the first genomics wave of gene expression microarray and single nucleotide polymorphism discovery platforms. One major advantage of genome-scale RNAi screening is its ability to simultaneously interrogate thousands of genes. With the ability generate a large amount of data per experiment, genome-scale RNAi screening has led to an explosion data generation rates.

Exploiting such large data sets is a fundamental challenge, requiring suitable statistics/bioinformatics methods. The basic process of cell-based RNAi screening includes the choice of an RNAi library, robust and stable cell types, transfection with RNAi agents, treatment/incubation, signal detection, analysis and identification of important genes or therapeutical targets.

History

Example petunia plants in which genes for pigmentation are silenced by RNAi. The left plant is wild-type; the right plants contain transgenes that induce suppression of both transgene and endogenous gene expression, giving rise to the unpigmented white areas of the flower.

The discovery of RNAi was preceded first by observations of transcriptional inhibition by antisense RNA expressed in transgenic plants, and more directly by reports of unexpected outcomes in experiments performed by plant scientists in the United States and the Netherlands in the early 1990s.

In an attempt to alter flower colours in petunias, researchers introduced additional copies of a gene encoding chalcone synthase, a key enzyme for flower pigmentation into petunia plants of normally pink or violet flower colour.

The overexpressed gene was expected to result in darker flowers, but instead produced less pigmented, fully or partially white flowers, indicating that the activity of chalcone synthase had been substantially decreased; in fact, both the endogenous genes and the transgenes were downregulated in the white flowers. Soon after, a related event termed *quelling* was noted in the fungus *Neurospora crassa*, although it was not immediately recognized as related. Further investigation of the phenomenon in plants indicated that the downregulation was due to post-transcriptional inhibition of gene expression via an increased rate of mRNA degradation. This phenomenon was called *co-suppression of gene expression*, but the molecular mechanism remained unknown.

Not long after, plant virologists working on improving plant resistance to viral diseases observed a similar unexpected phenomenon. While it was known that plants expressing virus-specific proteins showed enhanced tolerance or resistance to viral infection, it was not expected that plants carrying only short, non-coding regions of viral RNA sequences would show similar levels of protection. Researchers believed that viral RNA produced by transgenes could also inhibit viral replication.

The reverse experiment, in which short sequences of plant genes were introduced into viruses, showed that the targeted gene was suppressed in an infected plant. This phenomenon was labelled "virus-induced gene silencing" (VIGS), and the set of such phenomena were collectively called *post transcriptional gene silencing.*

After these initial observations in plants, laboratories searched for this phenomenon in other organisms. Craig C. Mello and Andrew Fire's 1998 *Nature* paper reported a potent gene silencing effect after injecting double stranded RNA into *C. elegans*. In investigating the regulation of muscle protein production, they observed that neither mRNA nor antisense RNA injections had an effect on protein production, but double-stranded RNA successfully silenced the targeted gene. As a result of this work, they coined the term *RNAi*. This discovery represented the first identification of the causative agent for the phenomenon.

Novel Nematicides

Dimethyl disulfide (DMDS) is an organic chemical compound with the molecular formula CH_3SSCH_3 which is the simplest disulfide. It is a flammable liquid with an unpleasant, garlic-like odor.

Occurrence

Dimethyl disulfide along with dimethyl sulfide and dimethyl trisulfide have been confirmed as volatile compounds given off by the fly-attracting plant known as dead-horse arum (*Helicodiceros muscivorus*). These flies are attracted to the odor of fetid meat, they help pollinate this plant.

DMDS can be produced by the oxidation of methanethiol, e.g. with iodine:

$$2\ CH_3SH + I_2 \rightarrow CH_3SSCH_3 + 2\ HI$$

Chemical Reactions

Important reactions include chlorination giving methanesulfenyl chloride (CH_3SCl), methanesulfinyl chloride ($CH_3S(O)Cl$), and methanesulfonyl chloride (CH_3SO_2Cl) as well as oxidation with hydrogen peroxide or peracetic acid giving the thiosulfinate methyl methanethiosulfinate ($CH_3S(O)SCH_3$).

Uses

DMDS is used as a food additive in onion, garlic, cheese, meats, soups, savory flavors, and fruit flavors. Industrially, DMDS is used in oil refineries as a sulfiding agent.

DMDS is also an effective soil fumigant in agriculture, registered in many states in the U.S. as well as globally. In this capacity, DMDS is an important alternative in replacing methyl bromide, which is being phased out. This pesticide is marketed as “Paladin” by Arkema.

Food use

DMDS is used to alpha substitute 2-methylfuran-acrolein to produce a food stuff used in concentrations from 0.02 ppm up to 50 ppm.

Industrial Use

DMDS is a stable pale yellow liquid which works as an effective product in the sulfiding process because of its high sulfur content. It is the sulfiding industry’s reagent of choice because it offers more sulfur per pound of reagent when compared to its nearest competitor dimethyl sulfide.

DMDS also works as an effective product for operators in the petrochemicals industry who must protect their steam-cracking coils against the formation of coke and carbon monoxide.

DMDS is used in the preparation of 4-(methylthio)phenol which is used in the production of various pesticides. DMDS and chlorine are reacted with borontriflouride phenolate to produce 4-(methylthio) phenol.

Thiophene and DMDS are blended with combustible hydrocarbon fuel gas to impart a gassy odor to the fuel gas. DMDS is used as a sulfiding reagent to control catalyst activity.

523 Biological control

524 Cultural control

526 Sanitation weed management

Genetic resistance to nematodes in cotton

Resistance and tolerance mechanisms

H columbus and B longicaudatus

Technological Advances in Sustainable Management

In planta delivery of dsRNA to target genes of plant parasitic nematodes

The feeding strategy of sedentary endoparasitic nematodes

In planta RNAi as a Biotechnological Strategy

Future Prospects for RNAibased control of Plant Parasitic Nematodes

While trevalling by train or road, you may have noticed agricultural fields with patches of stunted crops, like the one in figure. Ever wondered what was wrong with these plants? Most probably, the stunted plants you observed were suffering infection by one or more of a variety of tiny round worms called nematodes.

Nematodes appeared about 700 million years ago and they are now the most numerous of all organisms, occupying all types of habitats. Several species of these round worms parasitize a wide variety of organisms including many important crop plants, cattle and human. To get an idea of their success as parasites, consider this: nearly half the human population suffers from nematode infection at some point during their life time, and world-wide agriculture looses over US$100 billion to these worms!

Despite the enormous impact nematodes have on our lives, we do not yet have an effective and safe method to control them. We have not made much progress towards understanding their biology either. One of the major hurdles is their parasitic style of living. Since they need a host to feed and reproduce, attempts to grow them in the laboratory have not been successful so far.

This has severely hampered investigations that rely on traditional biochemical and genetic tools – the two most common types of tools biologists employ to identify genes responsible for any biological process.

Nematodes: An Introduction

What are nematodes? As I mentioned earlier, these are round worms, but very different, biologically, from the earthworm that you may be familiar with. Unlike the earthworm, which has a segmented body, nematodes do not have segments.

Simply put, their body, which is usually transparent, consists of two tubes – intestine and the gonad – within a third tube, the outer cuticle or skin. Apart from these, they also possess nervous, excretory and chemosensory systems. Their size ranges from less than a millimetre to about a metre and contain a defined number of cells.

Not all nematodes are parasites. For example, the widely used experimental model *Caenorhabditis elegans* is a free-living soil nematode. Several features of this remarkably small (head-to-tail length – 1 mm) worm make it an ideal model organism for various experimental investigations. The adult body of every *C. elegans* worm contains exactly 959 cells.

These cells arise exactly in the same way in every one of the embryos following an invariant pattern of cell division. It takes about 2.5 days for a newly laid embryo to develop into an adult that lives for about 17 days. Primarily feeding on soil bacteria, each adult worm produces approximately 300 offsprings. Its genetic information content is 30 times smaller than the human. Since reviewing the immense body of literature available on the biology of *C. elegans* is beyond the scope of this article, I shall limit myself here to just one major recent development that is central to our work on parasitic nematodes.

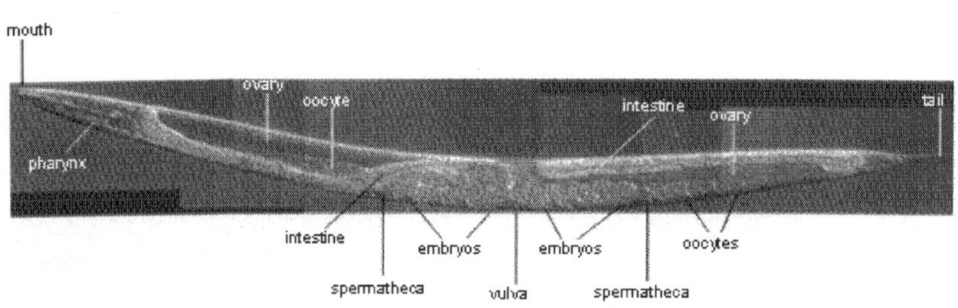

Figure 2. The free-living soil nematode *Caenorhabditis elegans*. Head-to-tail length is about 1 mm.

Before we continue the story on nematodes, let us take a short detour to molecular biology to familiarize ourselves with the central dogma of biology: genetic information – the information required for the formation and functioning of organisms – is encoded in molecules called DNA. This genetic code is ternary: DNA is a polymer consisting of four different monomers called shortly as A, T, G and C. The order, or sequence, of these four monomers in a given segment of DNA represents a unique genetic information.

For example, the genetic message contained in the sequence GAATTC is different from the one in AAGCTT. The genetic information is first transcribed from DNA into another type of molecules called mRNAs, which are then used for the synthesis of a third variety of biomolecules called proteins.

Most biological functions are carried out by proteins. A segment of DNA that contains a single genetic message is called as a gene. Most often, a gene represents the information to produce a single protein. In the cell, DNA exists in a double-stranded form, where the A in one strand interacts with the T in the other strand and G with C.

This A-T and G-C pairing allows us to predict the sequence of one strand from the other strand. The same principle is used by our cells to copy the genetic information as well as to transcribe genes to produce mRNAs, which are single-stranded.

RNA-mediated Interference

Now, back to worms: Fire and co-workers, working at the department of embryology at the Carnegie Institution of Washington in Baltimore, observed that the injection into *C. elegans* of double-stranded RNA (dsRNA) corresponding to any one of its genes specifically depleted the mRNA molecules transcribed from that gene.

In the absence of mRNA, the corresponding protein is not produced and, therefore, this ultimately results in the loss of function of the targeted gene. This phenomenon is termed as RNA-mediated interference or, shortly, RNAi. The cells of organisms such as nematodes, plants and human do not produce dsRNA. However, many viruses produce dsRNA.

Thus, the RNAi machinery seems to have evolved as a defence mechanism against viruses. Scientists have exploited RNAi as a powerful tool to disrupt the functions of specific genes to uncover their functions. Since its original discovery in *C. elegans*, RNAi has been observed in many other organisms as well. Now that the complete DNA sequence (genome) is available for many organisms, RNAi can be applied in all these organisms for revealing the biological functional information contained in their genomes. In fact, this has already revolutionized the field of functional genomics – an area of biology focused on the characterization of gene function in large scale.

RNA-mediated Interference: an Effective new Weapon Against Parasitic Nematodes

An amazing aspect of RNAi in *C. elegans* is that it can be triggered when the worm feeds on bacteria that produce the dsRNA of the worm's gene. Our group, at IIT-Kanpur, reasoned that a similar delivery of dsRNA of parasite's genes through host organisms is likely to induce RNAi in the parasites, and if we target genes that are essential for the survival of the parasite, then it may protect the host from infection.

In addition, this approach could be a powerful tool to the functional genomics of parasites. We engineered tobacco plants – a good plant model organism amenable for genetic transformation – to produce dsRNA of two essential genes of the parasitic nematode *Meloidogyne incognita*, which infects a wide range of agriculturally important plants. As we predicted, the transgenic tobacco plants very effectively resisted *M. incognita* infection.

A closer observation of the nematodes in these plants revealed that their development was severely impaired. Further, these nematodes were specifically deficient in the mRNA of targeted genes, indicating that the dsRNA produced in plants did indeed trigger RNAi response in the nematode.

Ongoing efforts in our laboratory aim to apply this technology to characterize the functions of plant-parasitic nematodes as well as to produce nematode-resistant transgenic crop plants.

Prey Resistance and Susceptibility to Predation

Biocontrol Potential of Dorylaim and Nygolaim Predators

Biocontrol potential of diplogasterid predators Integration of biological control with other methods of nematode management.

Root-knot nematodes are damaging to high value vegetable crops such as tomato. Yield losses can be >20% but total crop failure is not unknown.

In some cases cultivation of preferred crops is abandoned. Losses are worse if transplanted crops are grown in infested seedbeds. The incidence of bacterial wilt, Ralstonia (Pseudomonas) solanacearum, is greater in the presence of nematodes. The practical and economic problems to be resolved are:

(i) determining if farmers recognise root-knot nematodes and are willing to enter into participative activities that lead to the adoption of the techniques being developed, notwithstanding that such strategies may require more than one cropping cycle to become effective;

(ii) developing methods of mass producing and deploying bcas that are within the capability of the farmer;

(iii) determining the socio-economic conditions which favour adoption and the constraints to adoption.

Biological Control of Nematodes: Prospects and Opportunities

In the past 20 years three developments have occurred which have had significant effects on the prospects and opportunities for the biological control of plant-parasitic nematodes. First, several nematicides have been withdrawn from the market because of health and environmental problems associated with their production and use.

As a result of this, and increasing public concern over the use of pesticides in food production, there has been increased interest in the development of alternative methods of control, including the use of biological agents.

Second, it has been demonstrated in several soils that nematophagous fungi and bacteria increase under some perennial crops, and under those grown in monocultures, and so may control some nematode pests, including cyst and root-knot nematodes.

Such nematode-suppressive soils have been reported from around the world and include some of the best documented cases of effective biological control of nematode pests. Finally, a number of commercial products based on nematophagous fungi and bacteria have been developed, but all so far have had only limited success.

Their use has been based on empirical research, and it is instructive to consider what might be the key factors for a successful biological control agent for nematodes in order to identify the reasons for the general failure of the products that have been developed.

Biological control is more inconsistent, less effective and slower acting than control normally achieved with chemicals. Although improvements in performance might be expected from more research on individual agents, it seems likely that these limitations are inherent in most biological control agents and that their successful application will depend on integration with other control measures.

The most studied example of natural control of a plant-parasitic nematode concerns the decline of populations of the cereal-cyst nematode, *Heterodera avenae,* under monocultures of susceptible cereals in many soils throughout northern Europe (Kerry, 1982).

This is an example of an induced suppression in which the nematode increases to damaging population densities in the second and third cereal crops, but usually declines thereafter to infestations of <5 eggs/ g soil, which cause little loss of yield in northern European conditions; it is essential that the nematode is abundant in the early years of the monoculture to support the build-up of the microbial parasites.

The decline in nematode populations is mainly caused by two parasitic fungi, *Nematophthora gynophila* and *Verticillium chlamydosporium,* which attack the developing female on the root surface; in suppressive soils 95 to 97 percent of the females and eggs are destroyed.

Thus the natural control of cereal-cyst nematode in a range of soils is predictable and effective, but slow acting. Research on the manipulation of natural control has been limited and attention has concentrated on the introduction of specific agents to provide more rapid control that might be commercially exploitable.

It has proved difficult to develop a biological control agent that is effective worldwide for any soil-borne disease. Despite much research effort only two agents have had widespread success:

Phlebia gigantea for the control of *Heterobasidion annosum,* which spreads from tree stumps to the roots of adjacent trees, *and Agrobacterium radiobacter* that is applied as a root dip to transplants for control of crown gall caused by *A. tumefasciens.*

For both, the agents are applied in high concentrations as inundative treatments to a readily accessible site of action (the cut surface of a tree stump or the bare roots of a transplant) and protection from the diseases for a relatively short period provides long-term control (Deacon, 1991). In most situations where nematode control is required, there is a need to provide long-term protection to a relatively inaccessi- ble and growing root system without the use of inundative treatments that are likely to be impractical and uneconomic.

Hence the development of biological control agents for plant-parasitic nematodes is likely to be difficult and to require a detailed understanding of the biology and ecology of the agent and the nematode target.

The farming system in which biological control is applied has a marked effect on the way the agent is used (Davies, de Leij and Kerry, 1991). In general, growers in developed agriculture require an agent that can be applied to crops grown in monocultures over large areas using standard application machinery, so a formulated product with a good shelf-life that can be applied at low dosages is required; seed treatments are preferred for most arable crops. Little research has been done on the mass production and formulation of biological control agents for nematodes.

Some organisms, such as rhizosphere bacteria, can be applied as seed treatments, but such applications tend to provide short-term control and are only useful in reducing the invasion of roots by nematodes that have a single generation in the growing season. In subsistence farming systems, crops tend to be grown in mixed stands in relatively small areas (often less than one hectare) and labour inputs are often large.

As a consequence, relatively large application rates (up to one tonne per hectare) of an unformulated agent could be mixed into the soil by hand, as long as the organism could be produced cheaply and locally.

Thus, the initial exploitation of biological control agents for nematodes may be in developing countries (Hussey, 1990). However, if agents are only effective against specific nematode pests, and their efficacy is dependent on pest densities, then their effective use will require expert advice that may not be available in many developing countries.

Exploitation of Natural Enemies

The biology, ecology and potential of biological control agents for nematodes have been extensively reviewed in recent years. Nematologists have identified natural enemies with a range of modes of action similar to those currently studied by plant pathologists for the control of soil-borne diseases. Comments on the advantages and limitations of the major groups of organisms with potential as biological control agents are summarized.

It must be stressed that several organisms that are effective natural enemies of nematodes in the field, may have limited potential as biological control agents for application by growers. For example Nematophthora gynophila *is a causal agent of cereal-cyst nematode decline in many soils, but its limited host range, complex requirements for* in vitro *culture, and need for soil moisture levels to be at field capacity to ensure infection, mean that its potential use is too limited to warrant further development as a biological control agent.*

The stage in the nematode life cycle attacked by the control agent has a profound effect on the damage to the crop and the level of population control. Therefore, trapping fungi and rhizosphere bacteria that attack the second-stage juveniles of cyst and root-knot nematodes may significantly improve crop growth but is unlikely to prevent nematode populations increasing, especially those species that have more than one generation in a growing season.

In contrast, those parasites that attack developing females and eggs act like a partially resistant cultivar in that initial nematode invasion and plant damage is not prevented but multiplication of the nematode is significantly reduced. Root colonizing fungi such as the mycorrhizae and endophytic species such as Fusarium *spp. may reduce both nematode invasion and development.*

Sikora (1992) reviewed a range of control measures, including crop rotation, partial soil sterilization, soil amendments and nematicides, that could be combined to increase the activity of naturally occurring biological control agents.

Such measures could also be used to improve the performance of agents added to soil. Partial soil sterilization by methods such as solarization reduces nematode infestations and also reduces the competition from the residual soil microflora, enabling the biological control agent to establish more readily. Soil amendments may also reduce nematode infestations and, by providing an energy source, help to increase numbers of facultative parasites; pre-colonized substrates are most effective in establishing nematophagous fungi in soil.

Soils that are naturally suppressive to some nematodes may be used to shorten rotations of susceptible crops and improve the performance of nematicides and resistant cultivars.

The use of crop cultivars which are tolerant of nematode attack is likely to be very important for the successful deployment of biological control agents, that provide less effective nematode control than most nematicides.

Agents such as rhizosphere bacteria and rhizosphere competent fungi depend on root exudates for their proliferation on roots; exudation from roots differs markedly between plant species and cultivar and affects the efficacy of these agents.

Environmental Concerns

Soil conditions, nematode species, rate of development and density and host plant all have a considerable effect on the establishment and activity of biological control agents for nematodes. It is essential to understand these interactions for effective control.

In temperate zones soil temperatures range between 10° and 15°C for much of the growing season whereas in warmer climates temperatures of 20° to 25 °C are normal at 10 cm depth. Too often, research workers in temperate climates have tested the effects of biological control agents in pot experiments in greenhouses at temperatures considerably above those which occur in the field; data from such tests are of limited value. Temperature has a direct effect on the growth and sporulation of the agent and on the rate of development of the nematode target. Moisture in soil is rarely limiting for the growth of most fungi but affects spore dispersal, especially of zoospores.

Bacteria are more sensitive to soil moisture levels than fungi, but are unlikely to be affected by levels that enable nematodes to remain active. Soil texture and structure influence nematode activity and the growth and spread of micro-organisms. The great bulk of soil (2 500 tonnes per hectare) to plough depth (20 cm) makes the thorough incorporation of biological control agents difficult and broadcast treatments are unlikely to be economic.

The use of agents in arable crops is likely to depend on those organisms that proliferate in the rhizosphere from in-row applications or from seed treatments. The residual soil microflora competes with the introduced agent for scarce energy sources and can significantly affect the performance of the agent even when it is added to soil in a pre-colonized substrate.

Most plant-parasitic nematodes remain mobile throughout their life cycle and microbial parasites must produce traps or adhesive spores to infect them. Nematodes with sedentary stages such as cyst and root-knot nematodes may be parasitized from vegetative hyphae in the rhizosphere without the production of specialized infective structures apart from an appressorium.

Fungi such as Verticillium chlamydosporium *infect female cyst nematodes that are exposed on the root surface for several weeks during their maturation; depending on the time of infection, the fungus has a range of effects including a reduction in the number of mature females and their fecundity, and parasitism of the eggs. Usually only the eggs and not the females of root-knot nematodes are exposed in the rhizosphere, and these complete their embryonic development and hatch within about ten days at 25 °C. Hence, egg-parasitic fungi have less time to kill root-knot nematode eggs compared with those of cyst nematodes. Non-parasitic micro-organisms that degrade soil amendments and release nematicidal compounds, such as the bacterium which degrades chitin to produce ammonia, are likely to kill most nematodes in soil. Similarly, many endophytic fungi may compete with a wide range of nematodes in roots although those which are sedentary are likely to be more susceptible.*

The host plant influences the rate of nematode development and fecundity and the establishment of some natural enemies in the rhizosphere. Despite these effects the role of the host plant has rarely been considered in research on the biological control of nematodes.

Development of a Biological Control Agent for Nematodes - An Ecological Approach

Too often, biological control agents have failed because they have been used before a basic knowledge of their ecology and biology has been established. The importance of such knowledge can be seen from work at Rothamsted aimed at the development of V. chlamydosporium *as a biological control agent for root-knot nematodes. Stirling (1991) describes the design and methods used in the conduct of biological control experiments, which are more complex than those required to assess the efficacy of a nematicide. He identified five key aspects in setting up an experiment to evaluate a biological control agent and these are presented below in a slightly modified form.*

- The test organism and any organic amendment should be applied at practical application rates; 0.1 percent w/w soil is equivalent to 2.5 tonnes/ha and should represent a maximum dose. Tests should always be performed in a non-sterilized soil with a natural residual soil microflora.
- Appropriate treatments as well as an untreated control should be included if the organism is added with a substrate. These treatments should include the substrate alone, the organism alone, and the autoclaved colonized substrate. Too often, untreated controls are compared only with large applications of the organism and substrate and this does not allow separation of the effects of the agent from the effects of the substrate. In several tests reported in the literature, application of the substrate alone has decreased nematode populations to the same extent

as the substrate colonized by the agent, and there is no clear evidence of biological control.

- Population densities of the agent under test should be monitored to ensure that it has survived in soil throughout the period that activity against the nematode target is required. Such monitoring may require the development of selective media, which can be a difficult and time-consuming task.
- Nematode mortality caused by the organism under test should be measured to assess whether differences between nematode population densities in treated and untreated soil relate to the levels of kill caused by the agent. Infection levels are relatively straightforward to estimate for most parasites, but repeated sampling is required to determine total kills. The effects of agents which produce toxins or have indirect effects on nematodes through competition, the modification of root exudates or the colonization of feeding cells, can only be measured by assessing their impact on nematode development.
- The impact of the soil environment, host plant and nematode should be tested as these are likely to affect the efficacy of the biological control agent, and could account for the lack of activity of potential agents in specific test conditions.

In a review of the literature less than 15 percent of experiments purporting to demonstrate biological control caused by *Paecilomyces lilacinus* satisfied the above criteria (Kerry, 1990). Although this is an unsatisfactory situation that must be remedied, the difficulties in conducting carefully controlled and monitored experiments should not be underestimated.

Isolates of *V. chlamydosporium* differ markedly in their growth and sporulation *in vitro*, and in their virulence, saprophytic competitiveness and rhizosphere competence. Such differences between isolates of the same species of micro-organism are common and there is a need for simple laboratory-based screening methods to select the most promising isolates for further testing. *Verticillium chlamydosporium* isolates were collected from infested nematode females and eggs in suppressive soils around the world. Hence, they were isolated from the niche in which they would be required to be active as a biological control agent and from soils most likely to yield active isolates. Once isolated in pure culture using standard techniques, the *in vitro* growth requirements were determined.

Virulent isolates were selected by counting the number of nematode eggs parasitized after exposure to the fungus on agar in a standard test. Tests for proliferation in the rhizosphere and growth in soil were also used to select promising isolates. Although simple laboratory-based screens help eliminate many isolates that show insufficient activity to justify further testing, selected isolates will not necessarily be active in the field. As it may take 10 to 16 weeks to investigate adequately the performance of different isolates against cyst and root-knot nematodes in pot tests, relatively few can be screened.

The method of mass culturing of *V. chlamydosporium* for experiments can have a marked effect on the subsequent survival and proliferation of the fungus in soil. Inoculum produced in shaken liquid cultures consists mostly of hyphae and conidia, which require an energy source to ensure proliferation in soil (Kerry, 1987), whereas on solid media large numbers of chlamydospores are produced and these can be added to soil in aqueous suspension and rapidly establish the fungus.

The development of a semi-selective medium has enabled detailed studies to be made on changes in relative abundance of the fungus in soil and on roots. Some isolates of *V. chlamydosporium* may be extremely abundant in soil but unless they are capable of colonizing the rhizosphere they do not parasitize the eggs of root-knot nematodes. Growth in the rhizosphere differs markedly between plant species, e.g. tomato, cabbage and maize roots support much growth, whereas sorghum, pepper and cotton are poor hosts. Colonization by the fungus is confined to the rhizosphere and rhizoplane and there is no spread into root tissue; no lesions have been observed on roots grown in soil treated with *V. chlamydosporium* and there have been no detrimental effects on the growth of a range of crop species. A tenfold reduction (from 10^4 to 10^3 chlamydospores/g soil) in the amount of fungus applied to soil had no effect on the extent of colonization in the rhizosphere; the ability to proliferate on the root surface where the fungus is required to control nematodes is an important characteristic which may allow significant reductions in the amount of inoculum applied to soil.

The efficacy of *V. chlamydosporium* as a biological control agent for root-knot nematodes is affected by three key factors: the amount of fungus in the rhizosphere; the rate of development of eggs in the egg masses; and the size of the galls in which the female nematodes develop. In large galls female root-knot nematodes may produce egg masses which remain within the gall and are not exposed to parasitism by *V. chlamydosporium,* which is confined to the rhizosphere. Hence, *V.*

chlamydosporium is less effective in controlling root-knot nematodes in heavily infested soils and on highly susceptible crops because large galls are formed on the roots and many eggs escape parasitism. *Verticillium chlamydosporium is* unlikely to be useful in these situations where a grower would normally apply a nematicide. Also, at temperatures above 25°C eggs may complete their embryonic development and hatch before the fungus has completely colonized the egg mass; at 30°C about 30 percent of eggs of three root-knot species hatched and the second-stage juveniles escaped from the egg mass before the eggs were killed.

These studies in pot tests, if supported by field experiments, help to define the conditions in which *V. chlamydosporium* might be used successfully for control of root-knot nematodes. It is only from such detailed studies that the limitations and requirements of the fungus can be assessed.

Table: *The extent of colonization of the rhizosphere by* Verticillium chlamydosporium *on several plants grown on soil treated with 5 000 chlamydospores/g soil and the control of* Meloidogyne incognita

Host plant	*Rhizosphere colonization (cfu/cm² root)*[1]	*Control (%)*
Tomato	229	58
Maize	216	71
Cabbage	199	68
Potato	88	79
Tobacco	43	40
Pepper	29	36
Sorghum	25	0
Wheat	15	5

[1] cfu are the colony-forming units which develop on the semi-selective medium.

Verticillium chlamydosporium is not a replacement for nematicides but, despite its limitations, it may be a useful management tool. In most tests biological control agents have been applied to protect susceptible crops.

However, application of the fungus to a relatively poor host for the nematode on which small galls are produced so that most egg masses are exposed in the rhizosphere might provide more effective control;

soil population densities of the nematode would be reduced to non-damaging levels before a susceptible crop was planted. Bridge (1987) recommends the following rotation for the management of root-knot nematodes: susceptible host - poor host - poor host - resistant or non-host -susceptible crop. Application of the fungus before the first or second poor host may permit the shortening of rotations, or the replacement of the resistant or non-host crop, without increasing nematode damage to the susceptible crops. Currently, at Rothamsted Experimental Station, United Kingdom, poor hosts for root-knot nematodes are being screened for their ability to support *V. chlamydosporium* in their rhizospheres. It is hoped that by combining the host plant and the biological control agent, more effective and consistent control can be achieved.

The next 20 Years

The past 20 years have seen a significant increase in the number of scientists involved in research on the biological control of nematodes. Surveys and empirical tests are being replaced by quantitative experimentation and basic research on the modes of action, host specificity and epidemiology of selected organisms. Such basic information is essential for a realistic appraisal of the impact of molecular biology on the improvement of microbial agents and monitoring the spread and survival of released organisms, and for the development of rational strategies for control. Current experience suggests that biological control agents will not replace the use of nematicides but, integrated with other control measures including chemicals, they could play an important role in the development of integrated control strategies in both developed and developing agriculture. The urgent need to reduce the dependence on nematicides should provide the necessary impetus for the considerable amount of research and development still required to ensure the successful use of such agents.

Microbial Ecology and Nematode Control

Exploiting Genomics to Understand the Interactions

Plant Nematode Surfaces

Lectin Binding to the Banana-parasitic Nematode Radopholus Similis

In general, three basic strategies to create nematode resistance have been formulated: (a) the use or reinforcement of natural plant defence mechanisms, (b) the direct killing of nematodes, and (c) interference with the plant-nematode interaction. The first strategy includes cloning of natural resistance genes and reintroduction of these genes into susceptible plants.

This has been accomplished for the *Hs1$^{pro\text{-}1}$* gene of *Beta procumbens*, the wild relative of *B. vulgaris*, which is resistant to *Heterodera schachtii*. Transgenic plants had the same level of resistance as their wild-type counterparts.

In the second type of strategy, external and internal nematode structures are the target of lytic enzymes, such as chitinases and collagenases, and inhibitory or toxic proteins, such as protease-inhibitors or *Bacillus thuringiensis* endotoxins.

The aim of using chitinases is to increase the permeability of the nematode eggshell, which could be lethal to embryos. However, the chitin layer is not exposed directly to the environment, but is protected by a vitellin layer.

The same problem arises in the use of collagenases to degrade the cuticle of nematodes, as the cuticle is protected by the glycocalyx or surface layer.

Protease inhibitors are part of the natural defence response of plants and have already proved useful to control insect pests. Their use as anti-nematode proteins is based on the assumption that nematodes digest food with proteases. Results so far, for sedentary nematodes only, are encouraging. The application of *B. thuringiensis* toxins is hampered by the fact that they are highly species-specific.

Interference in the plant-nematode interaction can be accomplished by the expression of a repelling protein, an inhibition of the induction or maintenance of feeding structures by sedentary nematodes, or by the expression of a protein that hampers host recognition. No research has been done so far on the use of antisense RNA to inhibit the transcription products of nematode-induced genes, while negative effects on host plant roots have been observed after the local expression of the RNase barnase to kill feeding structures.

Antibodies, specifically targeted against nematode secretions responsible for feeding structure induction and maintenance, have not been successful in controlling *M. incognita* in transgenic tobacco.

An alternative is to direct antibodies towards the structures and secretions of nematodes, responsible for chemoreception of host signals. However, clear indications about such an effect are not available.

Alternative Strategies to Control Migratory Nematodes

Need for Novel Strategies

Research has generally been focused on resistance against sedentary endoparasitic nematodes, such as the root-knot nematode *Meloidogyne* spp., and the cyst nematodes *Globodera* spp. and *Heterodera* spp. The reasons are twofold. On the one hand, there is the agronomic impact: root-knot nematodes are an important threat to agricultural production in the tropics and sub-tropics, and cyst nematodes cause severe yield losses in the temperate regions.

The sedentary nematodes have, on the other hand, an intriguing biology: they are obligate parasites that spend the greater part of their life cycle inside the plant root, during which they engage in a long-term interaction with their host. They do not kill it, but induce genes and initiate a redifferentiation process. The reactions of the hosts are extensively characterised and often include a hypersensitive reaction. Moreover, *Arabidopsis thaliana* is used as a model host, as both *Meloidogyne* spp. and *Heterodera* spp. can complete their life cycle in its roots.

This is an advantage for the molecular genetical analysis of plant-nematode interactions, because the genome of *A. thaliana* is small, with a low number of repetitive DNA sequences, it has been sequenced, and genetic transformation has been proved to be efficient. The interaction between migratory endoparasitic nematodes and plants is far less studied, as appears from the existing strategies for the induction of resistance against nematodes.

Migratory nematodes do not interact in a sophisticated manner with their host tissue; they do not induce local, permanent feeding structures. Instead, they move themselves continuously through root tissue while feeding, which makes a localised, directed resistance mechanism, such as the inhibition of feeding structures, impossible. Moreover, all developmental stages of migratory endoparasitic nematodes, from secondary juveniles to adults, can move through root tissue as well as through the soil, in which they spend a great deal of their life cycle.

The most important nematode species that infect banana are migratory endoparasites, including *Radopholus similis*, *Pratylenchus* spp. and *Helicotylenchus multicinctus*. The development of valuable strategies against migratory nematodes is thus essential. Some of the aforementioned anti-nematode proteins are unsuccessful, while others are aimed at the feeding structure of sedentary nematodes. In this paper, lectins are presented as candidate anti-nematode proteins in an alternative strategy to control migratory endoparasitic nematodes.

Lectins as Candidate Anti-nematode Proteins

Plant lectins are proteins that possess at least one non-catalytic domain that binds reversibly to a specific mono- or oligosaccharide. In plants, seven different classes of lectins are defined according to their structure and evolution. They function mainly in the storage of nitrogen and in recognition processes.

Some plant lectins are involved in defence mechanisms against insects. In nematology, lectins have been used, in the first place, to characterize sugar moieties on nematode surfaces or in secretions via binding experiments. They have been shown to bind to several plant (mainly sedentary) parasitic nematodes at the excretion pore, the cuticle, the amphids or the head region in general, as well as at the spicules and the vulva.

A novel application in nematology is for testing the capacity of lectins to control nematodes. In contrast to the previously mentioned anti-nematode proteins, lectins are readily available and relatively easily applicable on a laboratory scale. They can be used in control strategies at two levels: expression inside the host root, or expression and secretion into the plant's rhizosphere.

It is assumed that lectins, by analogy with their effect in insects, have an intestine-associated activity in nematodes: the permeability of the digestive tract increases after ingestion of lectins.

This has not previously been tested *in vitro* because of the lack of artificial diets for phytophagous nematodes. However, Burrows and co-workers reported on the expression of *Galanthus nivalis* lectin (GNA) in transgenic potatoes at a level of 0.1-0.5% of total root proteins. One out of four of these transgenic lines reduced female *G. pallida* numbers by 50%. Recently, tests have been set up in the Laboratory of Tropical Crop Improvement of K.U.Leuven (Belgium), to screen transgenic *A. thaliana*, expressing different types of lectins, for resistance against migratory and sedentary endoparasitic nematodes of banana (K. Carlens, personal communication).

Marban-Mendoza and co-workers have done research on the control of root-knot nematodes of tomato by lectins. They found that an application of 12 μg of *Canavalia ensiformis* lectin (Con A) around the base of tomato plants, every week, during a period of 4 weeks, reduced root-knot formation by 75% compared with control plants. An explanation might be that lectins bind to the nematode surface and as a consequence have a toxic effect or inhibit the host-finding mechanism of nematodes. It has been substantiated that nematodes are attracted to host plants by chemotaxis.

The amphids, located in the head region, are considered to be the primary sensory organs for the perception of root signals. These chemical substances are captured by the amphidial secretions, which fill the amphidial channel, and are brought to the amphidial membranes by diffusion. Knowing this, Zuckerman formulated a strategy to create resistance against nematodes; blocking the host's signals or the signal perception should disturb the localisation of host plants. Not being able to detect a host plant, nematodes will wander about in the soil, with little chance of finding a food source.

A reduction of chemotaxis was accomplished in *Caenorhabditis elegans* by binding of lectins to the amphidial secretions. However, negative results were obtained by Forrest and co-workers: Con A had no effect on the capacity of *G. rostochiensis* juveniles to detect and reach pieces of agar soaked with potato root diffusates.

This strategy requires the presence of lectins in the surroundings of the rhizosphere of nematode host plants. Expression inside the plant cell is thus not sufficient and signal peptides are necessary for the transport of the proteins to the plant's apoplast and rhizosphere. Secretion of transgenic proteins has already been accomplished successfully in *N. tabacum* by Herbers and co-workers and Borisjuk and co-workers and in *A. thaliana* by Ziegler and co-workers. Preliminary experiments have been set up in the Laboratory of Tropical Crop Improvement of K.U.Leuven (Belgium) to test the efficiency of signal peptides in the transport of expressed transgenes in banana. So in the lectin secretion strategy, localised expression in root tissue is not necessary, and nematode infection is hampered before or at the moment of contact with the root. Moreover, this defence mechanism could also be effective for controlling the infective stage of sedentary nematodes, namely the juveniles, which are trying to locate suitable root cells to settle in and become sedentary.

Test of the Hypothesis

What needs to be tested is the hypothesis that the binding of lectins to the sensory organs of nematodes, the amphids, or their secretions can inhibit chemoreception of host signals, and in this way disturb the localisation of hosts by nematodes. Moreover, the binding of lectins to the surface of nematodes could limit their movement or could be toxic. In this paper, preliminary results of lectin binding to the banana-parasitic nematode *R. similis* are presented, to confirm that there is indeed an interaction between lectins and this nematode species. Additionally, results are shown of attempts to stain and visualise secretions of *R. similis*. The technique can then, in a later phase, be applied to test the effect of lectin binding on the production of secretions by the nematode.

Materials and Methods

Nematodes

A *Radopholus similis* population isolated from banana roots in Uganda was obtained from monoxenic alfalfa callus cultures for the lectin-binding experiments and the staining of nematode secretions. Under sterile conditions, nematodes were extracted from callus medium by a modified Baermann funnel and collected in siliconised Eppendorf tubes.

Lectin-binding Assay

The fluorescein isothiocyanate (FITC)-conjugated lectin of *Canavalia ensiformis* (Con A) was obtained from Sigma, as well as the colloidal gold-conjugated lectins of *Triticum aestivum* (WGA) (10 nm gold particles) and *Helix pomatia* (HPA) (5 nm gold particles). Nematodes were washed twice in Tris-buffer, pH 8 (15 mM Tris-HCl, 50 mM NaCl, 1 mM $CaCl_2$ and 1 mM $MnCl_2$), concentrated to 100 µl and incubated in 200 µl of Con A-FITC, dissolved in the same Tris-buffer at 500 µg/ml. After 2 h in the dark at 4°C, nematodes were held at 26°C for 0.5 h, washed extensively with Tris-buffer on a 25 µm-pore sieve and collected for analysis.

Treated nematodes were mounted on glass microscope slides and observed under a Nikon fluorescence microscope with an excitation wavelength of 450-490 nm and a barrier filter of 515 nm. For the binding experiment with gold-conjugated WGA and HPA, nematodes were washed twice in phosphate-buffered saline (PBS), pH 6.8, with 0.1 mM $CaCl_2$, 0.1 mM $MnCl_2$, 0.05% Tween-20 and 0.5% albumin (PBSAT). WGA and HPA were added to the nematode suspension at a dilution of 1:20 and incubated in the dark at room temperature for 4 h. Nematodes were washed with PBSAT on a 5 µm pore sieve, followed by an extensive rinse with distilled water. The surface of the sieve was dried and silver enhancement of the gold particles was applied (BB International). Nematodes were then collected with distilled water and mounted on glass microscope slides for observation under a Nikon light microscope. The control treatment for both experiments consisted of an incubation of *R. similis* in the appropriate buffer without lectins but with otherwise identical handling of the samples.

Staining of Nematode Secretions

Secretions of *R. similis* juveniles, females and males were stained according to the method of Premachandran and co-workers. Briefly, nematodes were concentrated to 100 µl in a siliconised Eppendorf tube and 200 µl of a staining solution containing Coomassie Brilliant Blue R (CBB-R) (ACROS) was added. A drop of the suspension was put on a glass microscope slide under a coverslip and then sealed with clear nail polish. Samples were incubated in the dark at room temperature for 24 h, followed by observation and analysis under a Nikon light microscope.

Results

Binding Experiment

Radopholus similis juveniles, females and males showed autofluorescence (FITC-filter) of the muscles in the head region, the muscles of the spicules, the oesophageal glands, and to some extent in the anterior part of the body. The pale yellow-green colour was, however, clearly distinguishable from the bright green colour of FITC-Con A. This lectin was found to bind to *R. similis* in the head region, but the principal lectin-binding sites were the pores or the secretions of the vulva and the spicules. Labelling of the excretion pore and the phasmids or their secretions was less frequent.

Table: *Binding of labelled lectins Con A, WGA and HPA to the head, excretion pore, vulva, spicules and phasmids of* R. similis

Binding site[a]	FITC-Con A	WGA-gold	HPA-gold
Excretion pore	+/-	+/-	+/-
Vulva	++	++	++
Spicules	++	-	+
Phasmids	+/-	+/-	+/-

++, strong binding; +, intermediate binding; +/-, less frequent binding; -, no binding [a] includes secretions

WGA and HPA, conjugated to colloidal gold particles, showed most extensive binding at the vulva of *R. similis* females. The spicules or their secretions were also stained when treated with HPA. Less frequent was the binding of WGA and HPA at the excretion pore, the phasmids and the pores of the amphids. In general, the resolution at 800 × magnification was not sufficient to make more detailed observations for Con A, WGA and HPA. Binding of these lectins to the pores or to the secretions of certain organs could not be distinguished.

Staining of Secretions

In preliminary tests different solutions of CBB-R were tested: (a) 1% CBB-R dissolved in water, (b) 0.2% CBB-R dissolved in 20% methanol, (c) 0.2% CBB-R dissolved in 40% methanol and 10% acetic acid, and (d) 0.1% CBB-R dissolved in 40% ethanol and 10% acetic acid. Solutions (c) and (d) gave clear and reproducible staining of the secretions of *R. similis*. It was decided to proceed with solution (c). The secretions in the head region of *R. similis* juveniles, females and males were stained as one big spot or as two distinct rays. These represent the secretions of the amphids, as was shown with the same staining solution for *H. schachtii*. The secretions of the excretory pore were also visible as one big spot or as a thin ray. Additionally, secretions were stained at the vulva, the spicules and the phasmids.

Bibliography

Alfred Steferud: *Diseases of Fruits and Nuts*, Biotech Books, Delhi, 2005.

Anderson, E. F.: *The Cactus Family*, Timber Press, Portland, Oregon, 2001.

Andrews L.: *Citrus Production - Orange*, St. Augustine, Trinidad and Tobago, 1990

Arora, Chandralekha : *Advances in Horticulture Science Research*, Oxford Book Company, Delhi, 2009.

Benson, L. H.: *The Cacti of the United States and Canada*. Stanford University Press, Stanford, California, 2001.

Benu Singh: *A Modern Book on Forestry and Horticulture*, Vista International Pub, Delhi, 2010.

Bhandari, M.M.: *The Flora of the Indian Desert*. Scientific Publisher, Jodhpur, 1990.

Bhattacharjee, Supriya Kumar: *Advances in Ornamental Horticulture*, Pointer, Delhi, 2006.

Bhutani, R.C.: *Fruit and Vegetable Preservation*, Biotech Books, Delhi, 2003.

Chadha, K.L. : *Advances in Horticulture*, Malhotra Publishing House, Delhi, 2006.

Collymore L.: *Fruit Production in Barbados*, Port of Spain, Trinidad and Tobago, 1996.

Dabholkar, A.R. : *General Plant Breeding*, Concept, Delhi, 2006.

Doijode S. D.: *Seed Germination in Fruits*, New Delhi, Malhotra Publishers, 1993.

Goswami, C L : *Plant Physiology : Research Methods*, Scientific, Delhi, 2007.

Julia F.: *Fruits of Warm Climates*, Miami, Julia F. Morton Publisher, 1987.

Kumar Supriya : *Amenity Horticulture, Biotechnology and Post-Harvest Technology*, Pointer, Delhi, 2006.

Little, E.L.: *Common Fuelwood Crops*, Communi-Tech Assoc., Morgantown, West Virginia, 1983.

Malins A.: *Postharvest Handling of Pineapple and Mango,* Port of Spain, Trinidad and Tobago, 1992.

Mitra S.: *Postharvest Physiology and Storage of Tropical and Subtropical Fruits,* Oxon, CABI, 1997.

Morton, Julia F.: *Fruits of Warm Climates.* Wipf and Stock Publishers, 1987.

Pareek O. P.: *Advances in Horticulture: Fruit Crops,* New Delhi, Malhotra Publishing House, 1993.

Paul B. : *Food Biotechnology in Ethical Perspective,* Aspen, CO: Aspen Publishers, 1997.

Peter, K.V. : *Basics of Horticulture,* New India, Delhi, 2009.

Popenoe, W.: *Manual of Tropical and Subtropical Fruits.* Hafner Press, 1974.

Proctor F. J.: *Onions in Tropical Regions,* Kent, Natural Resources Institute, 1990.

Purohit, S.S. : *Organic Farming : Biocontrol and Biopesticide Technology,* Agrobios, Delhi, 2008.

Saini, M.L. : *Plant Breeding and Crop Improvement,* CBS, Delhi, 1997.

Saroj, P.L. : *Advances in Arid Horticulture,* International Book Dist, Delhi, 2004.

Sharma, V.K. : *Advances in Horticulture : Strategies, Production, Plant Protection and Value Addition,* Deep and Deep, Delhi, 2004.

Singh Z.: *Propagation and Rootstocks of Citrus,* New Delhi, Malhotra Publishers, 1993.

Thomas E.: *Fruit Production in St. Kitts and Nevis,* Port of Spain, IICA, 1996.

Tyagi, I.D. : *Plant Breeding and Genetics at a Glance,* South Asian, Delhi, 2005.

Index

M

N

P

R

S

T

V

❑❑❑